Hydrophilic Interaction Liquid Chromatography (HILIC) and Advanced Applications

CHROMATOGRAPHIC SCIENCE SERIES

A Series of Textbooks and Reference Books

Editor:
NELU GRINBERG

Founding Editor:
JACK CAZES

1. Dynamics of Chromatography: Principles and Theory, *J. Calvin Giddings*
2. Gas Chromatographic Analysis of Drugs and Pesticides, *Benjamin J. Gudzinowicz*
3. Principles of Adsorption Chromatography: The Separation of Nonionic Organic Compounds, *Lloyd R. Snyder*
4. Multicomponent Chromatography: Theory of Interference, *Friedrich Helfferich and Gerhard Klein*
5. Quantitative Analysis by Gas Chromatography, Josef Novák
6. High-Speed Liquid Chromatography, *Peter M. Rajcsanyi and Elisabeth Rajcsanyi*
7. Fundamentals of Integrated GC-MS (in three parts), *Benjamin J. Gudzinowicz, Michael J. Gudzinowicz, and Horace F. Martin*
8. Liquid Chromatography of Polymers and Related Materials, *Jack Cazes*
9. GLC and HPLC Determination of Therapeutic Agents (in three parts), Part 1 *edited by Kiyoshi Tsuji and Walter Morozowich*, Parts 2 and 3 *edited by Kiyoshi Tsuji*
10. Biological/Biomedical Applications of Liquid Chromatography, *edited by Gerald L. Hawk*
11. Chromatography in Petroleum Analysis, *edited by Klaus H. Altgelt and T. H. Gouw*
12. Biological/Biomedical Applications of Liquid Chromatography II, *edited by Gerald L. Hawk*
13. Liquid Chromatography of Polymers and Related Materials II, *edited by Jack Cazes and Xavier Delamare*
14. Introduction to Analytical Gas Chromatography: History, Principles, and Practice, *John A. Perry*
15. Applications of Glass Capillary Gas Chromatography, *edited by Walter G. Jennings*
16. Steroid Analysis by HPLC: Recent Applications, *edited by Marie P. Kautsky*
17. Thin-Layer Chromatography: Techniques and Applications, *Bernard Fried and Joseph Sherma*
18. Biological/Biomedical Applications of Liquid Chromatography III, *edited by Gerald L. Hawk*
19. Liquid Chromatography of Polymers and Related Materials III, *edited by Jack Cazes*

20. Biological/Biomedical Applications of Liquid Chromatography, *edited by Gerald L. Hawk*
21. Chromatographic Separation and Extraction with Foamed Plastics and Rubbers, *G. J. Moody and J. D. R. Thomas*
22. Analytical Pyrolysis: A Comprehensive Guide, *William J. Irwin*
23. Liquid Chromatography Detectors, *edited by Thomas M. Vickrey*
24. High-Performance Liquid Chromatography in Forensic Chemistry, *edited by Ira S. Lurie and John D. Wittwer, Jr.*
25. Steric Exclusion Liquid Chromatography of Polymers, *edited by Josef Janca*
26. HPLC Analysis of Biological Compounds: A Laboratory Guide, *William S. Hancock and James T. Sparrow*
27. Affinity Chromatography: Template Chromatography of Nucleic Acids and Proteins, *Herbert Schott*
28. HPLC in Nucleic Acid Research: Methods and Applications, *edited by Phyllis R. Brown*
29. Pyrolysis and GC in Polymer Analysis, *edited by S. A. Liebman and E. J. Levy*
30. Modern Chromatographic Analysis of the Vitamins, *edited by André P. De Leenheer, Willy E. Lambert, and Marcel G. M. De Ruyter*
31. Ion-Pair Chromatography, *edited by Milton T. W. Hearn*
32. Therapeutic Drug Monitoring and Toxicology by Liquid Chromatography, *edited by Steven H. Y. Wong*
33. Affinity Chromatography: Practical and Theoretical Aspects, *Peter Mohr and Klaus Pommerening*
34. Reaction Detection in Liquid Chromatography, *edited by Ira S. Krull*
35. Thin-Layer Chromatography: Techniques and Applications, Second Edition, Revised and Expanded, *Bernard Fried and Joseph Sherma*
36. Quantitative Thin-Layer Chromatography and Its Industrial Applications, *edited by Laszlo R. Treiber*
37. Ion Chromatography, *edited by James G. Tarter*
38. Chromatographic Theory and Basic Principles, *edited by Jan Åke Jönsson*
39. Field-Flow Fractionation: Analysis of Macromolecules and Particles, *Josef Janca*
40. Chromatographic Chiral Separations, *edited by Morris Zief and Laura J. Crane*
41. Quantitative Analysis by Gas Chromatography, Second Edition, Revised and Expanded, *Josef Novák*
42. Flow Perturbation Gas Chromatography, *N. A. Katsanos*
43. Ion-Exchange Chromatography of Proteins, *Shuichi Yamamoto, Kazuhiro Naka-nishi, and Ryuichi Matsuno*
44. Countercurrent Chromatography: Theory and Practice, *edited by N. Bhushan Man-dava and Yoichiro Ito*
45. Microbore Column Chromatography: A Unified Approach to Chromatography, *edited by Frank J. Yang*
46. Preparative-Scale Chromatography, *edited by Eli Grushka*
47. Packings and Stationary Phases in Chromatographic Techniques, *edited by Klaus K. Unger*
48. Detection-Oriented Derivatization Techniques in Liquid Chromatography, *edited by Henk Lingeman and Willy J. M. Underberg*
49. Chromatographic Analysis of Pharmaceuticals, *edited by John A. Adamovics*

50. Multidimensional Chromatography: Techniques and Applications, *edited by Hernan Cortes*
51. HPLC of Biological Macromolecules: Methods and Applications, *edited by Karen M. Gooding and Fred E. Regnier*
52. Modern Thin-Layer Chromatography, *edited by Nelu Grinberg*
53. Chromatographic Analysis of Alkaloids, *Milan Popl, Jan Fähnrich, and Vlastimil Tatar*
54. HPLC in Clinical Chemistry, *I. N. Papadoyannis*
55. Handbook of Thin-Layer Chromatography, *edited by Joseph Sherma and Bernard Fried*
56. Gas–Liquid–Solid Chromatography, *V. G. Berezkin*
57. Complexation Chromatography, *edited by D. Cagniant*
58. Liquid Chromatography–Mass Spectrometry, *W. M. A. Niessen and Jan van der Greef*
59. Trace Analysis with Microcolumn Liquid Chromatography, *Milos Krejcl*
60. Modern Chromatographic Analysis of Vitamins: Second Edition, *edited by André P. De Leenheer, Willy E. Lambert, and Hans J. Nelis*
61. Preparative and Production Scale Chromatography, *edited by G. Ganetsos and P. E. Barker*
62. Diode Array Detection in HPLC, *edited by Ludwig Huber and Stephan A. George*
63. Handbook of Affinity Chromatography, *edited by Toni Kline*
64. Capillary Electrophoresis Technology, *edited by Norberto A. Guzman*
65. Lipid Chromatographic Analysis, *edited by Takayuki Shibamoto*
66. Thin-Layer Chromatography: Techniques and Applications: Third Edition, Revised and Expanded, *Bernard Fried and Joseph Sherma*
67. Liquid Chromatography for the Analyst, *Raymond P. W. Scott*
68. Centrifugal Partition Chromatography, *edited by Alain P. Foucault*
69. Handbook of Size Exclusion Chromatography, *edited by Chi-San Wu*
70. Techniques and Practice of Chromatography, *Raymond P. W. Scott*
71. Handbook of Thin-Layer Chromatography: Second Edition, Revised and Expanded, *edited by Joseph Sherma and Bernard Fried*
72. Liquid Chromatography of Oligomers, *Constantin V. Uglea*
73. Chromatographic Detectors: Design, Function, and Operation, *Raymond P. W. Scott*
74. Chromatographic Analysis of Pharmaceuticals: Second Edition, Revised and Expanded, *edited by John A. Adamovics*
75. Supercritical Fluid Chromatography with Packed Columns: Techniques and Applications, *edited by Klaus Anton and Claire Berger*
76. Introduction to Analytical Gas Chromatography: Second Edition, Revised and Expanded, *Raymond P. W. Scott*
77. Chromatographic Analysis of Environmental and Food Toxicants, *edited by Takayuki Shibamoto*
78. Handbook of HPLC, *edited by Elena Katz, Roy Eksteen, Peter Schoenmakers, and Neil Miller*
79. Liquid Chromatography–Mass Spectrometry: Second Edition, Revised and Expanded, *Wilfried Niessen*
80. Capillary Electrophoresis of Proteins, *Tim Wehr, Roberto Rodríguez-Díaz, and Mingde Zhu*
81. Thin-Layer Chromatography: Fourth Edition, Revised and Expanded, *Bernard Fried and Joseph Sherma*

82. Countercurrent Chromatography, *edited by Jean-Michel Menet and Didier Thiébaut*
83. Micellar Liquid Chromatography, *Alain Berthod and Celia García-Alvarez-Coque*
84. Modern Chromatographic Analysis of Vitamins: Third Edition, Revised and Expanded, edited by *André P. De Leenheer, Willy E. Lambert, and Jan F. Van Bocxlaer*
85. Quantitative Chromatographic Analysis, *Thomas E. Beesley, Benjamin Buglio, and Raymond P. W. Scott*
86. Current Practice of Gas Chromatography–Mass Spectrometry, *edited by W. M. A. Niessen*
87. HPLC of Biological Macromolecules: Second Edition, Revised and Expanded, *edited by Karen M. Gooding and Fred E. Regnier*
88. Scale-Up and Optimization in Preparative Chromatography: Principles and Bio-pharmaceutical Applications, *edited by Anurag S. Rathore and Ajoy Velayudhan*
89. Handbook of Thin-Layer Chromatography: Third Edition, Revised and Expanded, *edited by Joseph Sherma and Bernard Fried*
90. Chiral Separations by Liquid Chromatography and Related Technologies, *Hassan Y. Aboul-Enein and Imran Ali*
91. Handbook of Size Exclusion Chromatography and Related Techniques: Second Edition, *edited by Chi-San Wu*
92. Handbook of Affinity Chromatography: Second Edition, *edited by David S. Hage*
93. Chromatographic Analysis of the Environment: Third Edition, *edited by Leo M. L. Nollet*
94. Microfluidic Lab-on-a-Chip for Chemical and Biological Analysis and Discovery, *Paul C.H. Li*
95. Preparative Layer Chromatography, *edited by Teresa Kowalska and Joseph Sherma*
96. Instrumental Methods in Metal Ion Speciation, *Imran Ali and Hassan Y. Aboul-Enein*
97. Liquid Chromatography–Mass Spectrometry: Third Edition, *Wilfried M. A. Niessen*
98. Thin Layer Chromatography in Chiral Separations and Analysis, *edited by Teresa Kowalska and Joseph Sherma*
99. Thin Layer Chromatography in Phytochemistry, *edited by Monika Waksmundzka-Hajnos, Joseph Sherma, and Teresa Kowalska*
100. Chiral Separations by Capillary Electrophoresis, *edited by Ann Van Eeckhaut and Yvette Michotte*
101. Handbook of HPLC: Second Edition, *edited by Danilo Corradini and consulting editor Terry M. Phillips*
102. High Performance Liquid Chromatography in Phytochemical Analysis, *edited by Monika Waksmundzka-Hajnos and Joseph Sherma*
103. Hydrophilic Interaction Liquid Chromatography (HILIC) and Advanced Applications, *edited by Perry G. Wang and Weixuan He*

Hydrophilic Interaction Liquid Chromatography (HILIC) and Advanced Applications

Edited by

Perry G. Wang
Weixuan He

CRC Press
Taylor & Francis Group
Boca Raton London New York

CRC Press is an imprint of the
Taylor & Francis Group, an **informa** business

CRC Press
Taylor & Francis Group
6000 Broken Sound Parkway NW, Suite 300
Boca Raton, FL 33487-2742

First issued in paperback 2017

Version Date: 20110610

ISBN-13: 978-1-4398-0753-8 (hbk)
ISBN-13: 978-1-138-11339-8 (pbk)

Library of Congress Cataloging-in-Publication Data

Hydrophilic interaction liquid chromatography (HILIC) and advanced applications / editors, Perry G. Wang, Weixuan He.
p. cm. -- (Chromatographic science series ; v. 103)
Includes bibliographical references and index.
ISBN 978-1-4398-0753-8 (hardcover : alk. paper)
1. Hydrophilic interaction liquid chromatography. I. Wang, Perry G. II. He, Weixuan.

QD79.C454H93 2011
543'.84--dc23 2011017684

Visit the Taylor & Francis Web site at
http://www.taylorandfrancis.com

and the CRC Press Web site at
http://www.crcpress.com

Contents

Preface

This book on hydrophilic interaction liquid chromatography (HILIC) applications is a result of the increasing interest among the separation science community in the latest chromatography technologies for isolation and purification of various chemicals. HILIC is a variation of normal-phase chromatography with the added advantage that organic solvents that are miscible in water can be used. It uses polar materials, such as amino, cyano, diol, and silanol, as its stationary phase and, thus, is sometimes referred to as "reverse reversed-phase" or "aqueous normal phase" chromatography. HILIC provides the following benefits compared to other chromatographic separation modes:

- High organic content mobile phases result in a lower operating back pressure, allowing higher flow rates to be used in high-throughput analysis.
- The high organic solvent concentration in the mobile phase leads to a higher sensitivity for LC/MS analyses due to significantly increased ionization.
- Significant improvement of peak shape and sensitivity results in more accurate and precise quantitation of polar compounds such as peptides and nucleic acids.
- Polar analytes that would be un-retained by reversed-phase chromatography are retained by HILIC.
- Capability of direct injection of solid phase extraction (SPE) or liquid–liquid extracts into the HILIC columns increases analytical throughput and facilitates sample preparation.

The HILIC concept was first introduced by Dr. Andrew Alpert in his 1990 paper. A large number of scientific papers on this subject have been published since then. This book comprehensively and systematically describes the new technology and provides detailed information and discussion on the most advanced HILIC applications in the fields of environmental sciences, food analysis, clinical chemistry, pharmaceutical research, and biotechnology discovery. Although the theory behind HILIC is not fully understood and commercial HILIC columns have limited availability, the extensive applications that we witness today have already made HILIC a unique device in the chromatography toolbox for most separation scientists. We hope that our readers will find this book to be a valuable resource for their projects ranging from academic research to industrial applications.

We are indebted to several people who assisted us during the preparation and editing of this book. We were extremely fortunate to have had the cooperation of dedicated, well-known contributing authors. Their relentless efforts and sincere scientific drive have made this book possible.

Editors

Dr. Perry G. Wang is a research chemist at the Office of Regulatory Science, Center for Food Safety and Applied Nutrition, Food and Drug Administration (FDA). His interests include analytical method development and validation for drugs and constituents in foods and cosmetic products using advanced instrumentation. His expertise focuses on high-throughput drug analysis by LC/MS/MS for the pharmaceutical industry.

Dr. Wang has recently published two books: *High-Throughput Analysis in the Pharmaceutical Industry* (published by CRC Press in October 2008) and *Monolithic Chromatography and Its Modern Applications* (published by ILM Publications in October 2010). He is currently coediting another book entitled *Identification and Analysis of Counterfeit and Substandard Pharmaceuticals* with ILM Publications, which is scheduled for publication in May 2011.

Dr. Wang has been invited to prepare, organize, and preside over symposia for the Pittsburg Conference (PittCon) and American Chemistry Society (ACS) annual meetings. He has been an invited speaker at numerous international conferences including the PittCon, the Federation of Analytical Chemistry and Spectroscopy Societies (FACSS), the Beijing Conference and Exhibition on Instrumental Analysis (BCEIA), and the International Symposium on Chemical Biology and Combinatorial Chemistry (ISCBCC). He has also been invited to teach short courses on high-throughput method development for drug analysis by LC/MS/MS at the PittCon, ACS, and at the Eastern Analytical Symposium (EAS). His current research focuses on developing analytical methods for the determination of constituents in cosmetics and dietary supplements.

Dr. Wang received his BS in chemistry from Shandong University, and his MS and PhD in environmental engineering from Oregon State University, Corvallis, Oregon.

Dr. Weixuan He is currently an associate director, Product and Process Development, at Meda Pharmaceuticals (formerly MedPointe Pharmaceuticals) located in North Brunswick, New Jersey. He is responsible for designing and conducting analytical investigations and experiments to support Meda Pharmaceuticals' product development, as well as providing technical services to ensure commercial manufacturing processes are in compliance with regulatory requirements. He also focuses on developing analytical methods using a variety of new technologies to support pharmaceutical product development.

Prior to joining Meda Pharmaceuticals in 2002, Dr. He was a senior research investigator at Bristol Myers Squibb, specializing in process development for isolation and purification of active pharmaceutical ingredient (API) from reaction mix, natural products from fermentation or biotransformation media, and diastereomer from stereoisomer mixtures. He successfully isolated and characterized various process impurities, degradants, metabolites, and natural products using modern isolation technologies and spectroscopic methods.

Before starting his career in the pharmaceutical industry in 1994, Dr. He earned his doctorate degree in organic chemistry from Oregon State University in Corvallis, Oregon, and completed his postdoctoral training in biomedical research at The Johns Hopkins University in Baltimore, Maryland.

Contributors

Mohammed Shahid Ali
Jamjoom Pharmaceuticals Company Limited
Jeddah, Saudi Arabia

M-Concepción Aristoy
Consejo Superior de Investigaciones Científicas
Instituto de Agroquímica y Tecnología de Alimentos
Valencia, Spain

Cosima Damiana Calvano
Department of Chemistry
University of Bari Aldo Moro
Bari, Italy

Allan J. Cessna
Agriculture and Agri-Food Canada

and

Environment Canada
National Hydrology Research Centre
Saskatoon, Saskatchewan, Canada

Yuming Chen
Forest Laboratories, Inc.
Hauppauge, New York

G. Corso
Department of Biomedical Sciences
University of Foggia
Foggia, Italy

Adrian Covaci
Toxicological Centre
University of Antwerp
Antwerp, Belgium

O. D'Apolito
Department of Biomedical Sciences
University of Foggia
Foggia, Italy

Carmela Dell'Aversano
Dipartimento di Chimica delle Sostanze Naturali
Università di Napoli Federico II
Napoli, Italy

Yannis Dotsikas
Department of Pharmaceutical Chemistry
University of Athens
Athens, Greece

Kasper Engholm-Keller
Department Biochemistry and Molecular Biology
University of Southern Denmark
Odense, Denmark

Yong Guo
The Janssen Pharmaceutical Companies
Johnson & Johnson
Raritan, New Jersey

Zhigang Hao
Global Analytical Science Department
Colgate-Palmolive Company
Piscataway, New Jersey

John V. Headley
Environment Canada
Water Science and Technology Directorate
Aquatic Ecosystem Protection Research Division
Saskatoon, Saskatchewan, Canada

Aleida S. Hernández-Cázares
Consejo Superior de Investigaciones Científicas
Instituto de Agroquímica y Tecnología de Alimentos
Valencia, Spain

Juris Hmelnickis
Grindeks

and

Faculty of Chemistry
University of Latvia
Riga, Latvia

Shaoxiong Huang
Covidien, Inc.
St. Louis, Missouri

Xinqun Huang
Covidien, Inc.
St. Louis, Missouri

Eugene P. Kadar
Pfizer, Inc.
Groton, Connecticut

Goran Karlsson
Octapharma AB
Stockholm, Sweden

Olga Kavetskaia
Abbott Laboratories
Abbott Park, Illinois

Aamer Roshanali Khatri
Jamjoom Pharmaceuticals Company Limited
Jeddah, Saudi Arabia

Kaspars Kokums
Grindeks
Riga, Latvia

Sandra L. Kuchta
Health Canada
Ottawa, Ontario, Canada

David Q. Liu
GlaxoSmithKline
King of Prussia, Pennsylvania

Min Liu
Amgen, Inc.
Thousand Oaks, California

Xiaodong Liu
Dionex Corporation
Sunnyvale, California

Yannis L. Loukas
Department of Pharmaceutical Chemistry
University of Athens
Athens, Greece

Minhui Ma
Amgen, Inc.
Thousand Oaks, California

Michael Matchett
Covidien, Inc.
St. Louis, Missouri

Maria T. Matyska
Department of Chemistry
San Jose State University
San Jose, California

Leticia Mora
Consejo Superior de Investigaciones Científicas
Instituto de Agroquímica y Tecnología de Alimentos
Valencia, Spain

Hugo Neels
Toxicological Centre
University of Antwerp
Antwerp, Belgium

Joanne E. Nettleship
Oxford Protein Production Facility
Wellcome Trust Centre for Human Genetics
University of Oxford
Oxford, United Kingdom

Hien P. Nguyen
Department of Chemistry and Biochemistry
The University of Texas at Arlington
Arlington, Texas

G. Paglia
Department of Biomedical Sciences
University of Foggia
Foggia, Italy

Kerry M. Peru
Environment Canada
Water Science and Technology Directorate
Aquatic Ecosystem Protection Research Division
Saskatoon, Saskatchewan, Canada

Joseph J. Pesek
Department of Chemistry
San Jose State University
San Jose, California

Christopher A. Pohl
Dionex Corporation
Sunnyvale, California

Milagro Reig
Instituto de Ingeniería de Alimentos para el Desarrollo
Universidad Politécnica de Valencia
Valencia, Spain

Peter Roepstorff
Department of Biochemistry and Molecular Biology
University of Southern Denmark
Odense, Denmark

Kevin A. Schug
Department of Chemistry and Biochemistry
The University of Texas at Arlington
Arlington, Texas

Yoichi Shibusawa
Division of Pharmaceutical and Biomedical Analysis
Tokyo University of Pharmacy and Life Sciences
Tokyo, Japan

Mingjiang Sun
GlaxoSmithKline
King of Prussia, Pennsylvania

Igors Susinskis
Grindeks
Riga, Latvia

Isabela Tarcomnicu
Toxicological Centre
University of Antwerp
Antwerp, Belgium

Morten Thaysen-Andersen
Department of Chemistry and Biomolecular Sciences
Macquarie University
Sydney, New South Wales, Australia

Fidel Toldrá
Consejo Superior de Investigaciones Científicas
Instituto de Agroquímica y Tecnología de Alimentos
Valencia, Spain

L. Trabace
Department of Biomedical Sciences
University of Foggia
Foggia, Italy

F. Tricarico
Department of Biomedical Sciences
University of Foggia
Foggia, Italy

Alexander L.N. van Nuijs
Toxicological Centre
University of Antwerp
Antwerp, Belgium

Ping Wang
The Methodist Hospital
Houston, Texas

and

Weill Cornell Medical College
New York, New York

Chad E. Wujcik
Monsanto
St. Louis, Missouri

Akio Yanagida
Division of Pharmaceutical and Biomedical Analysis
Tokyo University of Pharmacy and Life Sciences
Tokyo, Japan

Samuel H. Yang
Department of Chemistry and Biochemistry
The University of Texas at Arlington
Arlington, Texas

M. Zotti
Department of Biomedical Sciences
University of Foggia
Foggia, Italy

1 Aqueous Normal-Phase Chromatography: The Bridge between Reversed-Phase and HILIC

Joseph J. Pesek and Maria T. Matyska

CONTENTS

1.1 FUNDAMENTAL FEATURES OF AQUEOUS NORMAL-PHASE RETENTION

Reversed-phase (RP) high-performance liquid chromatography (HPLC) has been the predominant method for a vast array of separations over the last 30 years. Its principles have been applied to the analysis of compounds that include small organic and inorganic molecules as well as high molecular weight species such as proteins and peptides. However, RP HPLC is not a universal approach for all separation problems. While many molecules have sufficient hydrophobicity to result in retention on typical RP materials such as octyl (C8) or octadecyl (C18), others are too hydrophilic so that under most circumstances they elute at or near the column void volume.[1] This is especially the case when molecules are very polar or ionic. In order to enhance the retention of hydrophilic analytes, several alternative chromatographic methods have been developed. The simplest is to make a derivative of the hydrophilic compound so that the product is more hydrophobic in nature.[2,3] This can often be a time-consuming process that has the potential for error if the reaction is not quantitative. Another technique is to add ion-pairing reagents to the mobile phase in the case of ionic species so that the solutes being analyzed are neutral.[4] Both of these approaches utilize RP as the chromatographic technique since the resultant analyte becomes more hydrophobic and thus can be retained on stationary phases like C8 or C18. Another strategy for retention of polar/ionic compounds is ion-exchange.[5–7] In this case, the stationary phase retains a charge that can provide retention capabilities for hydrophilic species. Unfortunately, these methods often result in the use of mobile phases that are not compatible with mass spectrometry, a detection method that is becoming increasingly more prevalent.

These restrictions and inconveniences have led to the development of hydrophilic interaction liquid chromatography (HILIC) as a possible solution.[8–14] Since most highly polar and ionic species are generally soluble only in solvents such as water or perhaps methanol, a chromatographic method must be developed that can utilize these solvents. The retention of the analytes must be accomplished by utilizing a polar stationary phase. Thus HILIC encompasses mobile phases with some polar solvent (usually water) component and stationary phases that also are polar in nature such as silica, amino, diol, or cyano-bonded organic compounds, and zwitterionic materials that are bonded to support surfaces or are part of a polymer. Attraction and hence retention is facilitated by using low amounts of the polar solvent and a high amount of the organic component (typically acetonitrile [ACN]) in the mobile phase. However, in many cases, additives which are not compatible with mass spectrometry must be included in the mobile phase, retention times between runs are not reproducible, or the equilibration time between runs for gradient analyses (usually necessary for complex mixtures) is excessively long.

A new and even more promising approach to solving the problem of hydrophilic retention in HPLC is obtained with silica hydride-based materials.[15] The essential differences between ordinary silica (used for all commercial silica-based separation materials) and silica hydride is shown in Figure 1.1. The presence of Si–H moieties on the surface instead of silanols leads to a fundamentally different material with chromatographic properties that result in a range of selectivities and capabilities not

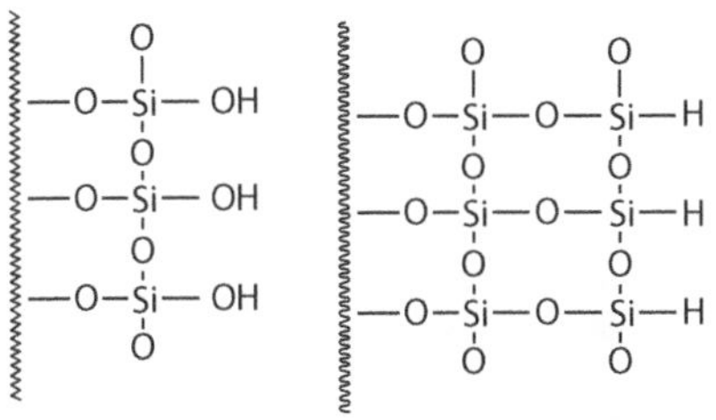

(A) (B)

FIGURE 1.1 Surface structures of ordinary silica (A) and silica hydride (B) materials used for HPLC stationary phases.

available on other modified silicas. Among the most important features of hydride-based phases are the absence of strong water adsorption on the surface leading to rapid equilibration of the separation media when used in gradient applications; the ability of all phases synthesized to date to operate in RP, aqueous normal phase (ANP), and organic normal phase modes; retention of both polar and nonpolar compounds simultaneously; operation in 100% aqueous mobile phases without any dewetting (formerly known as phase collapse) resulting in subsequent loss of retention; and a high level of reproducibility for both the retention of hydrophobic (RP) and hydrophilic (ANP) compounds. Some of these significant differences are described later as they relate to the retention and selectivity of polar compounds, the focus of this review.

1.2 MODES OF ANP

In order to distinguish between the fundamental properties and retention behavior of silica hydride-based materials in comparison to HILIC phases, the term aqueous normal phase (ANP) retention is used. The term ANP has been used occasionally in the past in the same context as HILIC. Some recent examples include the separation of polar pharmaceutical compounds,[16] the analysis of drugs in human plasma,[17] the determination of acrylamide in food samples,[18] and a general review on the analysis of a wide range of basic compounds.[19] In reality, these terms should not be used interchangeably. There are several types of ANP behavior observed with the silica hydride stationary phases which clearly distinguish this mode from HILIC. These will be briefly described with respect to the retention of hydrophilic compounds.

1.2.1 ANP 1

Figure 1.2 illustrates the retention map for two solutes (one retained by RP and the other by a normal phase mechanism) run on a stationary phase that possess ANP properties.[20] In this case, the two mechanisms are clearly evident but there is no region on the retention map where the two solutes both have significant retention. Under these circumstances, this phase can be used interchangeably for either RP or normal phase retention, depending on whether a high or low aqueous mobile phase composition is selected. With a high water percentage, the hydrophobic compound(s) will be retained while the hydrophilic analyte(s) will elute at the void volume. For high organic content mobile phases, the opposite occurs with the polar compound(s) being retained

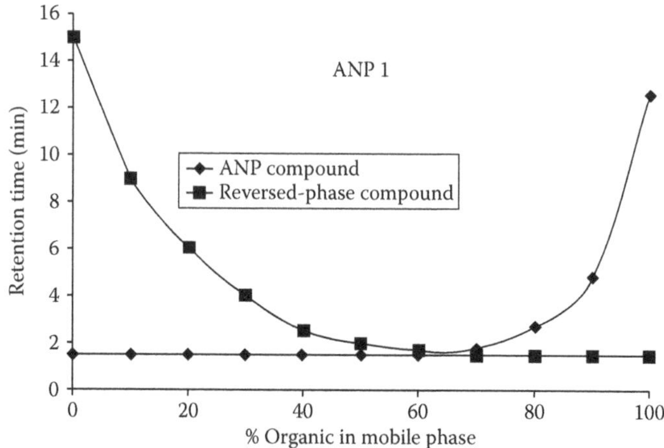

FIGURE 1.2 Retention map (t_R vs. % organic in mobile phase) for both a hydrophobic compound (■) and a hydrophilic compound (♦) on a silica hydride-based stationary phase with no overlap of RP and ANP mechanisms. (Adapted from Pesek, J.J. and Matyska, M.T., *LCGC*, 25, 480, 2007.)

and the nonpolar specie(s) eluting at or near the dead volume. In contrast, a typical HILIC material has only the behavior shown by the ANP compound (♦) in Figure 1.2. Most HILIC phases show little or no RP behavior (■) under any high aqueous mobile phase conditions. This example provides the first clear distinction between silica hydride-based stationary phases and HILIC materials. With water being the common component in the mobile phase, it can be seen how a transition can be made from RP to normal phase behavior by simply varying the aqueous content of the eluent.

1.2.2 ANP 2

Figure 1.3A shows a second type of retention map that can be obtained for two compounds (one retained by RP and the other by a normal phase mechanism) on a silica hydride phase having ANP properties. In this situation, there is a distinct range of mobile phase compositions where both compounds display measurable retention. Under these conditions, it is possible to separate mixtures of polar and nonpolar compounds, often under isocratic conditions. The other possibility suggested by this retention map is the analysis of hydrophobic and hydrophilic compounds using gradient elution with the option of running in either direction, i.e., from low to high organic content (standard gradient) or from high to low percent organic (inverse gradient). In contrast to a HILIC phase (only polar compounds retained at high organic compositions) or an ANP 1 situation (retention of the type of compound determined by % organic in mobile phase), the ANP 2 scenario offers unique separation capabilities that are available from only a limited number of commercial products, mainly those based on silica hydride.

Figure 1.3B presents an example of the separation of a mixture that illustrates the ANP 2 elution pattern. In this chromatogram, one polar (1) and one nonpolar (2) compound are tested on a silica hydride-based C18 column. In the top chromatogram at 40:60 ACN/water, the RP mechanism is dominant and the hydrophobic solute is retained more strongly than the hydrophilic compound. The middle chromatogram

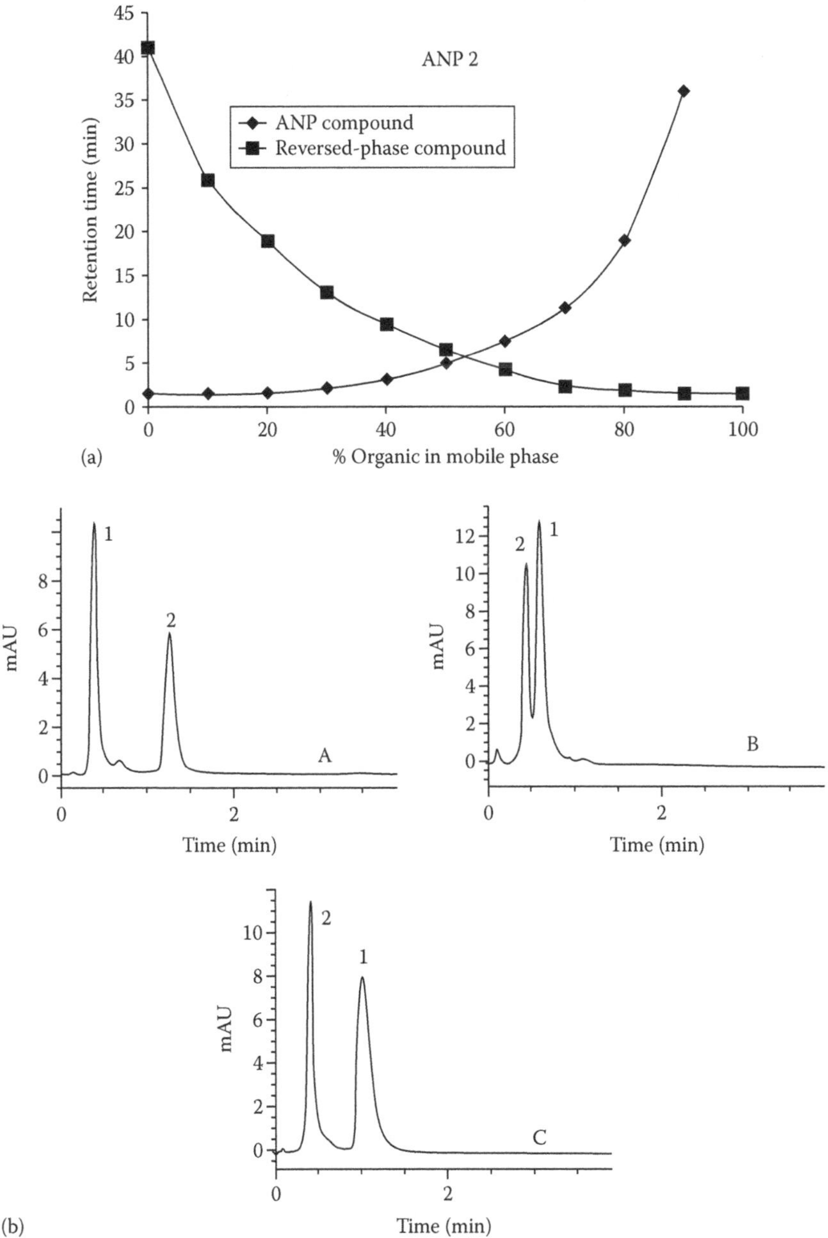

FIGURE 1.3 (a) Retention map (t_R vs. % organic in mobile phase) for both a hydrophobic compound (■) and a hydrophilic compound (♦) on a silica hydride-based stationary phase with overlap of RP and ANP mechanisms. (b) Chromatograms for a polar (1) and a nonpolar (2) compound on a silica hydride-based stationary phase at (A) 40:60 ACN/water; (B) 60:40 ACN/water and (C) 70:30 ACN/water. (Adapted from Pesek, J.J. and Matyska, M.T., *LCGC*, 25, 480, 2007.)

is obtained when the mobile phase has been changed to 60:40 ACN/water. Under these conditions, the normal-phase mechanism has become slightly stronger than the RP mode and compound 1 elutes just after compound 2. If the ACN content is increased further (70:30), then the normal-phase mechanism becomes very dominant and 1 is then retained significantly longer than 2.

1.2.3 ANP 3

A third possibility can also be obtained on a true ANP stationary phase and it also provides a clear distinction in comparison to a HILIC material. This situation is illustrated by the retention map shown in Figure 1.4A. This elution pattern occurs for certain types

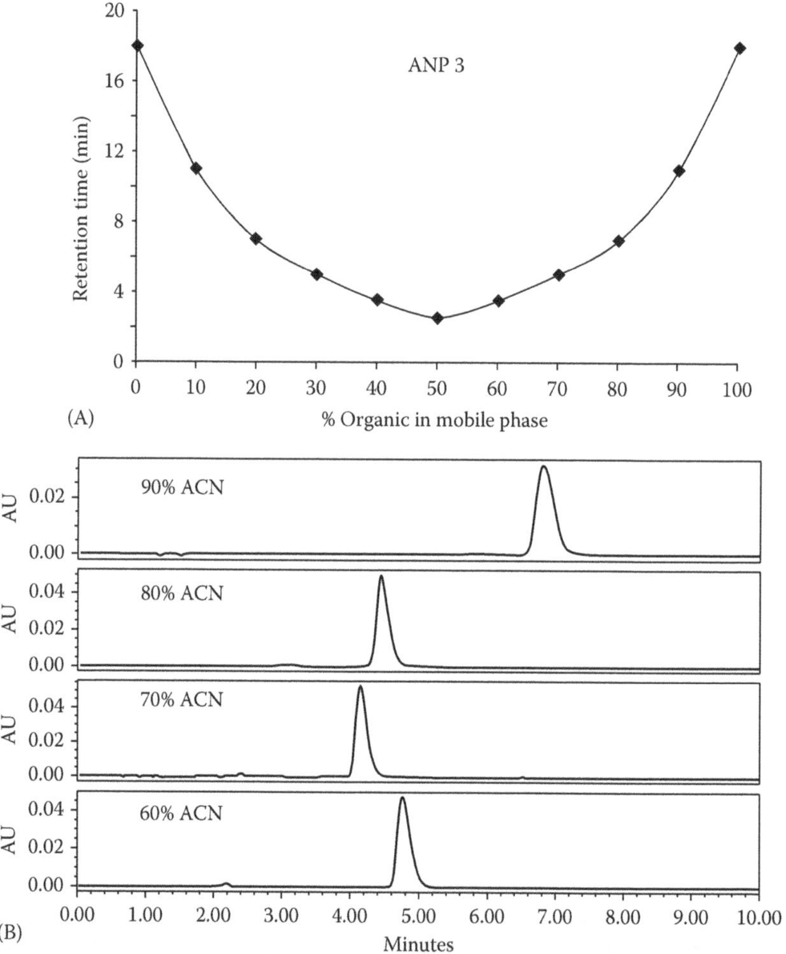

FIGURE 1.4 (A) Retention map (t_R vs. % organic in mobile phase) for a compound with both hydrophobic and hydrophilic components on a silica hydride-based stationary phase showing both RP and ANP mechanisms. (B) Chromatograms for such a compound at various amounts of ACN in the mobile phase. (Adapted from Pesek, J.J. and Matyska, M.T., *LCGC*, 25, 480, 2007.)

of compounds; those that have both hydrophobic and hydrophilic properties.[21] The solutes that result in this elution behavior are typically larger molecules with one or more polar functionalities as well as a significantly hydrophobic portion. Some peptides and proteins are examples of molecules that fit this description but some compounds with molecular weights below 1000 having both polar and nonpolar regions also behave in a similar manner. Under these circumstances, the chromatographer has a choice of operating in the RP or ANP mode, depending on the type(s) of other molecules in the mixture, sample compatibility with the mobile phase, or the means of detection utilized. This unusual capability provides experimental flexibility that is not available on typical stationary phases fabricated through the use of organosilane chemistry.

Figure 1.4B presents a series of chromatograms for a compound that is retained in both the RP and the normal-phase modes, depending on the mobile phase composition selected. The chromatograms shown follow the retention profile predicted by the map in Figure 1.4A. As expected retention goes through a minimum as a function of ACN concentration, indicating that the mechanism is changing from RP to ANP. This pattern is in clear contrast to a typical HILIC material where retention would be observed only when the mobile phase contained a high proportion of ACN.

1.3 CHROMATOGRAPHIC PROPERTIES IN ANP

1.3.1 Efficiency

Since the exact mechanism of retention for polar compounds on the hydride-based phases has yet to be elucidated, a variety of factors might affect efficiency, particularly pH. Figure 1.5 shows some examples of efficiency data obtained on diamond hydride (DH) columns.[22] Figure 1.5A is a plot of flow rate vs. height equivalent to a theoretical plate (HETP) for two carbohydrates on a 2.1 mm × 150 mm hydride-based column in the ANP mode. The interesting feature here is that efficiency is only slightly lower (~15%) while the flow rate has increased by a factor of 2.5. When making the same comparison for these solutes on a 1 mm column over the same flow rates (linear flow rates more than four times higher), the decrease in efficiency was only around 20%. Another interesting result is shown in Figure 1.5B where the efficiency of uracil under ANP conditions is measured as a function of flow rate on a 4.6 mm × 150 mm hydride-based column. In this case, there is an increase in efficiency as the flow rate increases from 0.3 to 1.0 mL/min. This result is in contrast to a commercial HILIC column of the same dimensions where efficiency drops by more than 30% over this same range of flow rates. It should be noted that linear flow rates are more than six times greater in the 2.1 mm column than in the 4.6 mm column, but direct comparison was limited due to restrictions of flow rates for the mass spectrometer used for detection. These results indicate that the silica hydride-based phases operating in the ANP mode can be used at high flow rates with little or no deterioration in column efficiency in contrast to the HILIC phase tested.

1.3.2 Temperature Behavior

Figure 1.6 shows the retention behavior of 19 amino acids as a function of temperature (log k vs. $1/T$, van't Hoff plot) in a mobile phase containing 25% water/75% ACN with 0.1% formic acid. In all cases, retention increased as the temperature was

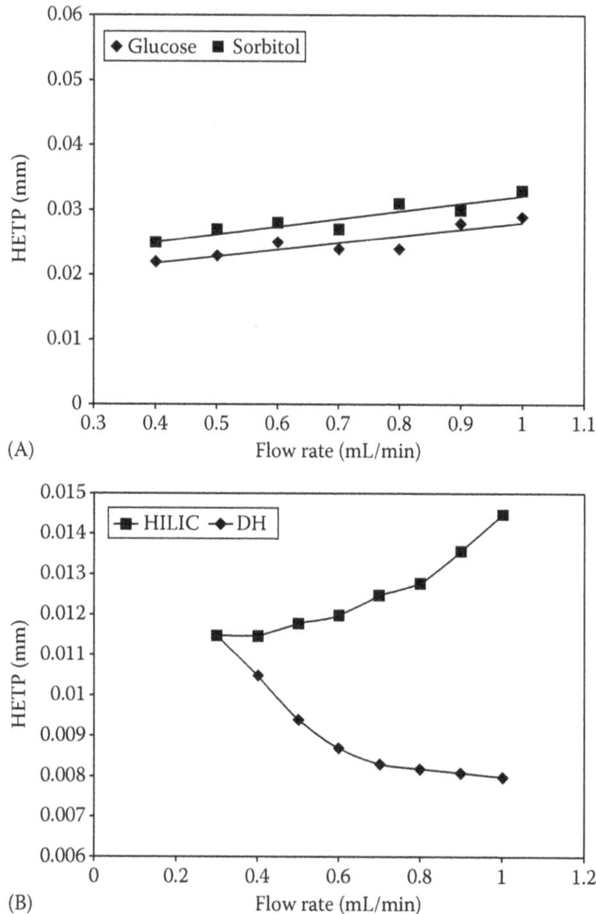

FIGURE 1.5 Plots of efficiency (HETP) vs. flow rate. (A) Glucose and sorbitol on DH column (2.1 mm × 150 mm, particle size 4 μm) in a 80:20 ACN/DI water + 0.1% formic acid mobile phase. (B) Comparison of commercial HILIC (4.6 mm × 150 mm, particle size 3.5 μm) and DH (4.6 mm × 150 mm, particle size 4.0 μm) columns for ANP retention of uracil. (Adapted from Pesek, J.J. et al., *J. Sep. Sci.*, 2200, 32, 2009.)

increased, indicating either a positive enthalpy for interaction of the solute with the stationary phase and/or substantial entropy contributions.[23] This temperature effect is opposite of what is typically observed under RP conditions in HPLC, i.e., decreasing retention with increasing temperature. This result provides other routes for improving selectivity via temperature control. Increasing temperature can lengthen retention, thus improving *R* values (resolution). Also, higher temperatures result in lower viscosity and faster mass transfer which can decrease peak widths to provide another means for improving resolution, depending on the extent of retention and diffusion in the stationary phase. However, for a few solutes such as some carbohydrates, retention decreased with temperature. Thus, a single mechanism for the ANP behavior on the hydride phases is probably not likely.

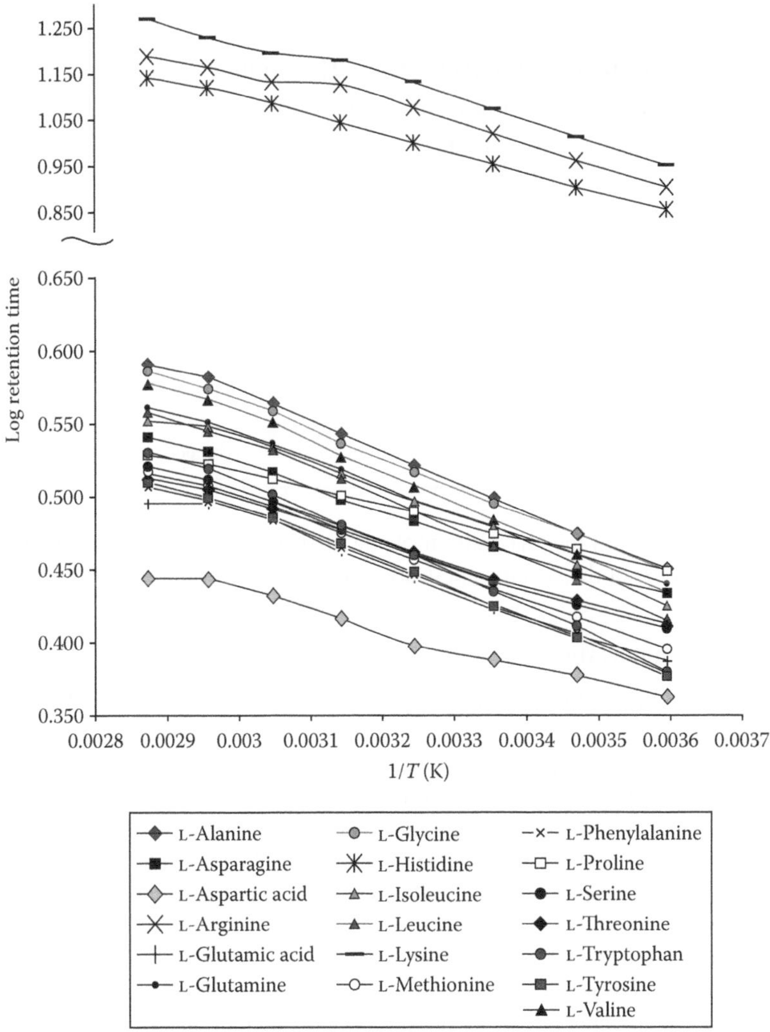

FIGURE 1.6 Plot of log *k* vs. 1/*T* (K) for 19 amino acids on the silica hydride column. (Adapted from Pesek, J.J. et al., *J. Chromatogr. A*, 1204, 48, 2008.)

1.3.3 Repeatability

Another advantage of the silica hydride phases is the small variation in retention (as measured by % relative standard deviation [RSD]) that occurs when successive injections are run for the same analysis in the ANP mode. This consistency is observed for both isocratic and gradient elution methods.[22–24] Some examples of repeatability for several types of small molecules are shown in Table 1.1. This is a representative sampling from a larger group of amino acids, carbohydrates, and organic acids that were reported in a study of hydrophilic metabolites on a silica hydride stationary phase.[23] The % RSD values are typical for a broad range of hydrophilic compounds that have

TABLE 1.1
Repeatability Results for Some Hydrophilic Compounds on the Hydride Stationary Phase[a]

Compound	Retention Time (min)	No. of Injections	%RSD
Alanine	3.72	50	0.53
Arginine	11.32	50	0.44
Glutamine	6.37	50	0.62
Histidine	12.22	50	0.20
Proline	4.69	50	0.43
Glucose	3.11	54	0.53
Fumaric acid	6.64	10	0.08
Citric acid	9.75	10	0.10

[a] Mobile phases and gradients vary from compound to compound.

been tested to date on the DH column. Some further examples of repeatability will be shown in some of the specific applications described later in this review.

1.3.4 Re-Equilibration

An important feature of hydride-based materials is their ability to re-equilibrate rapidly after changing mobile phase compositions, whether to new conditions for an isocratic run or between runs in a gradient method.[23] This feature is almost assuredly a result of the hydride surface in contrast to silanol functional groups on ordinary silica. Extensive investigations of gradient re-equilibration have been done in the RP mode and the results of a typical study are shown in Table 1.2. It can be clearly seen that with the proper experimental setup, the column has attained equilibrium within 1 min of the end of the gradient for all of the solutes tested. Such short times between runs are neither typical nor practical for most methods. All of the gradient ANP methods cited in this review have an equilibrium time of 5 min. This is considerably less than many HILIC methods reported which often have re-equilibration

TABLE 1.2
Equilibration Times for a Hydride-Based Stationary Phase

	Retention Time (min)		
Solute/equilibration time	25	10	1
Benzene	7.30	7.35	7.25
Naphthalene	11.10	1.07	11.01
Phenanthrene	14.39	14.37	14.37
Anthracene	14.81	14.80	14.80
Pyrene	16.52	16.51	16.56

times of 30–40 min. Thus, ANP utilizing hydride-based separation materials often can shorten the analysis time in comparison to other hydrophilic separation methods such as HILIC or methods that require additives or derivatization in order to achieve retention in the RP mode.

1.4 CONDITIONS FOR AQUEOUS NORMAL PHASE RETENTION

The presence and extent of ANP behavior is controlled by a number of chromatographic conditions. These features are reviewed next in order to provide a better understanding of the ANP phenomena and as a guide for developing and optimizing ANP methods.

1.4.1 Column

A number of silica hydride-based columns have been tested for ANP retention capabilities. It has been demonstrated that each one (silica hydride without any modification, silica hydride with a very low percentage of bonded organic moiety referred to as DH, C8, C18, cholesterol, and undecenoic acid [UDA]) can function in the ANP mode.[20–27] Figure 1.7 shows the retention for the amino sugar glucosamine on a silica hydride stationary phase with a bonded C18 moiety. This column is generally one that would be used for RP applications, and as expected it has substantial retention for a broad range of hydrophobic compounds as well. All three of the ANP retention scenarios described earlier have been observed on the hydride-based C18 column. The exact characteristics of retention as a function of the amount of organic constituent in the mobile phase depend on the specific solute(s) tested.

1.4.2 Organic Component of Mobile Phase

The most common solvent used in ANP, and frequently in HILIC as well, is ACN. However, it has been demonstrated that other solvents can function in the ANP

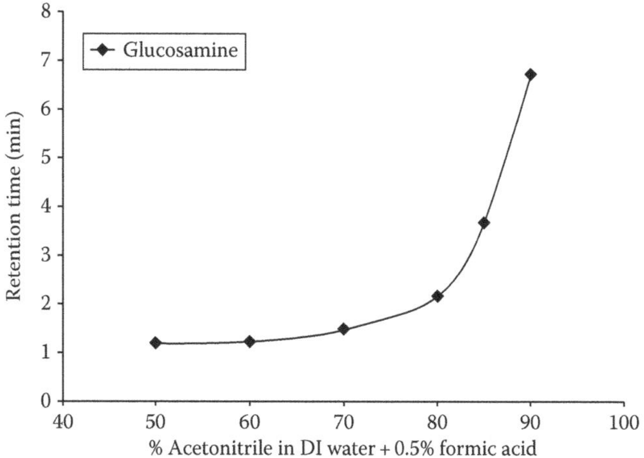

FIGURE 1.7 Retention map for the amino sugar glucosamine on a silica hydride-based C18 stationary phase. Mobile phase: ACN/water with 0.1% formic acid added.

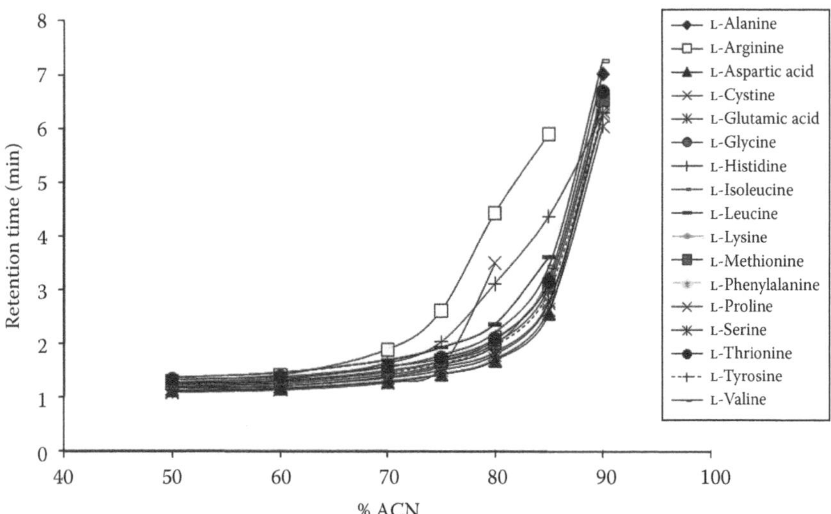

FIGURE 1.8 Retention map for 17 common amino acids on a silica hydride-based phase with minimal carbon on the surface (DH) using acetone as the organic component in the mobile phase. Additive: 0.1% formic acid. (Adapted from Pesek, J.J. et al., *J. Chromatogr. A*, 1204, 48, 2008.)

mode under certain conditions.[23] For example, with methanol being a more polar solvent than ACN, it would be expected to induce less ANP retention. This has been observed for all solutes' test. Many compounds which can be retained very well in the range of 70%–90% ACN show little or no retention with methanol. Only very basic analytes, such as the drug tobramycin, have shown any appreciable retention in methanol. In addition, the amount of methanol required to observe ANP retention is always considerably higher than the percentage of ACN needed for a comparable k value. Thus in most cases, methanol is of limited use in ANP.

In contrast, acetone, which has approximately the same polarity as ACN, is a very useful mobile phase component in ANP. Figure 1.8 shows the retention map for a number of amino acids as a function of the amount of acetone in the mobile phase. Unfortunately, for many applications using UV detection, acetone has significant absorption that precludes its use. However, when using MS or MS/MS, this is not a restriction. Thus, when ACN is not a suitable mobile phase or cannot be easily obtained, acetone may be a viable substitute for many applications.

1.4.3 Mobile Phase Additive

In most ANP applications, the use of an additive in the mobile phase enhances the retention of analytes under conditions of high organic content. When using mass spectrometry for detection, these additives must be sufficiently volatile to be compatible with the MS system.[22–24] For basic compounds, weakly acidic species usually function best. Both formic acid and acetic acid, usually at 0.1% or 0.2%, are the most often used. These additives have the proton donating ability to create a positively charged species that provides both ANP retention and a suitable ion for MS detection. For acidic compounds, it is generally required to have a higher pH additive in the mobile phase. Compounds that fulfill this requirement and are compatible with MS detection include ammonium formate and ammonium acetate usually at a concentration level of about 10–20 mM, concentrations compatible with MS detection. Under these conditions, the mobile phase pH is usually higher than the pK_a of most organic acid compounds, resulting in the formation of a negatively charged species to promote ANP retention and a suitable ion for MS detection. Other polar compounds such as carbohydrates can be retained by ANP on a silica hydride-based stationary phase even though they are not charged. For these analytes, the pH is not as crucial and in most cases any of the additives discussed earlier are suitable. To enhance MS response, a small amount (~20 µM) of a second additive such as sodium acetate can be used. The presence of this additive results in the formation of a sodium adduct which provides the ionic species analyzed by the mass spectrometer.

A more universal additive would be desirable such that hydrophilic compounds ranging in properties from acidic to basic as well as neutral could be analyzed with a single mobile phase. Such a situation is desirable for the analysis of metabolites in physiological samples. One possible choice for this additive is pyridine. Pyridine containing formic or acetic acid will result in an effective hydrogen ion concentration that will protonate most basic species and result in anions for most acidic compounds. A further refining of the hydrogen activity is achieved by adding either formic acid or acetic acid to the aqueous component of the mobile phase. Gradients can then be designed which will result in the retention of a broad range of hydrophilic compounds.

In some cases, either strong interactions between the analyte and the stationary phase or slow mass transfer effects can lead to tailing peaks. It has been demonstrated that the addition of very small amounts (0.0001%–0.001%) of trifluoroacetic acid (TFA) to the mobile phase can greatly improve peak symmetry.[23] Such concentrations of TFA are low enough so that signal suppression is not a significant problem when using mass spectrometry for detection.

1.4.4 Sample Solvent Additive

Just as the composition of the mobile phase has a profound effect on elution and peak shape characteristics in ANP, the sample solvent can also influence the chromatographic results obtained for hydrophilic compounds.[23,24] In some cases, it is not necessary or even advantageous to add TFA to the mobile phase. However, peak shape can be improved by adding a small amount of TFA (0.001%–0.005%) in the sample solvent. This amount of TFA is not sufficient to cause any changes in retention and does not suppress the signal when using MS detection. Another additive that has been identified as being beneficial for the improvement of peak shape for certain compounds is ammonia. Typically 1–2 µL of concentrated ammonia per mL of sample was shown to improve peak symmetry for nucleotides.[24] Ammonia is added to the sample when the pH of the mobile phase is above 6. At this point, it is not certain what causes the improved peak shape for either the addition of TFA or ammonia.

An even more substantial improvement in peak shape can be obtained by adjusting the ratio of organic to aqueous components in the sample solvent. Virtually all ANP methods will start at 70% ACN or higher. For gradient methods, many begin at 90%–95% ACN. Therefore, using a sample composition that is 50:50 ACN/DI water results in the sample being in an excess of the strong solvent with respect to the mobile phase into which it is being injected. Under these circumstances, a focusing effect occurs that results in both narrower and more symmetrical peak shapes.[23] This effect has been demonstrated for many RP applications so it is not a unique feature of ANP but it can be used as a means of providing the best resolution and efficiency for the separation of hydrophilic compounds.

1.4.5 Gradients

As in other modes of HPLC separation, gradients can be used to shorten the analysis time when a broad range of chromatographic retention factors (k) are present in a single sample.[22–24] In ANP, the gradient profile used to achieve a reduction in analysis time for samples with a wide variation in k values is opposite in direction to an RP gradient, i.e., the mobile phase starts at high organic content and then the amount of water is increased to shorten the elution of the most highly retained species (most polar). Water is the strong solvent in ANP and, thus, reduces the retention of hydrophilic compounds. In addition, the presence of a gradient usually leads to improved peak shape as well as enhanced efficiency. A good example of the efficiencies that can be achieved in ANP analysis under gradient conditions is shown in Figure 1.9 for the separation

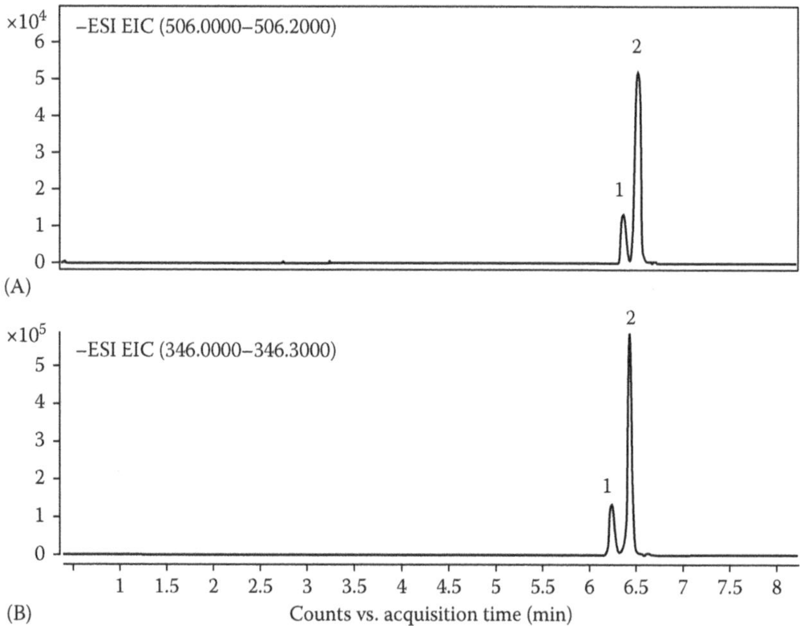

FIGURE 1.9 LC-MS EIC for isobaric nucleotide species on the DH column: (A) m/z = 506 solutes: 1 = adenosine triphosphate and 2 = deoxycytidine-5′-monophosphate; (B) m/z = 346 solutes: 1 = adenosine-3′-monophosphate and 2 = adenosine-5′-monophosphate. Column: 2.1 mm × 150 mm. Mobile phase: (A) DI water + 15 mM ammonium acetate and (B) 90% ACN + 10% water + 15 mM ammonium acetate. Gradient: 0.00 min 95% B; 0.00–1.00 min to 90% B; 1.00–3.00 min to 80% B; 3.00–4.00 min hold 80% B; 4.00–5.00 min to 50% B; 5.00–6.00 min hold 50% B; 6.00–7.00 min to 20% B. (Adapted from Pesek, J.J. et al., *J. Chromatogr. A*, 1216, 1140, 2009.)

of isobaric nucleotides. In this example, the efficiency for each of the four compounds shown in these two gradient analyses is in excess of 200,000 plates/m. Similar sharp symmetrical peaks have been obtained for many amino acids, carbohydrates, and small organic acids by selecting the appropriate additives and gradient conditions.

1.5 APPLICATIONS OF AQUEOUS NORMAL PHASE RETENTION

ANP chromatography has some features in common with HILIC, particularly those that are related to the retention of polar compounds, but as explained earlier, there are some notable differences. Because in all cases the separation material itself is different, it has been found in many cases that the capabilities of the silica hydride-based stationary phases in ANP offer a number of distinct advantages for practical applications. Some examples of analyses involving a variety of polar compounds are described next to illustrate how the fundamental aspects and conditions of ANP, as described previously, result in these new capabilities. In these applications, many of the solute molecules belong to the same general category of molecules as those used in the previous section.

1.5.1 Amino Acids

One of the more challenging analyses that has importance in many metabolomic studies is the separation of underivatized amino acids. Many methods have been developed over the years for amino acids. The most common is the separation of derivatized amino acids which simultaneously reduces the polarity to make these compounds amenable to retention and separation by RP chromatography and also providing a chromophore for UV detection. While certainly successful, these methods add a step to the analysis that results in significantly longer analysis times than would be desired in many instances. The lack of a chromophore for most amino acids necessitates analysis using detection by mass spectrometry or by light scattering methods. However, the underivatized amino acids are too polar to be retained by RP methods on columns like C18 or C8 even at very high water content in the mobile phase. Ion exchange is another alternative but these methods usually require mobile phases that are not compatible with mass spectrometry. HILIC is thus a reasonable choice for the retention and separation of amino acids.

It has already been demonstrated that ANP is a very suitable method for amino acid analysis.[23] Figure 1.8 demonstrates the ANP capability of the moderately modified silica hydride material (diamond hydride) for the retention of amino acids. As stated, with a proper choice of gradient, a mixture of 19 amino acids can be adequately separated for use with mass spectrometry detection. An example of a method suitable for MS is shown in Figure 1.10 where all analytes with different mass as well as $M + 1$ are separated. Only the isobaric compounds leucine and isoleucine are not baseline separated with this gradient. A slight modification can separate these two compounds as shown in the inset. In general, amino acids are best separated in an acidic medium. Thus, formic acid or acetic acid provides a good choice of an additive for amino acids since they are easily compatible with mass spectrometry detection. Under acidic conditions using electrospray ionization, the highest abundance ion is

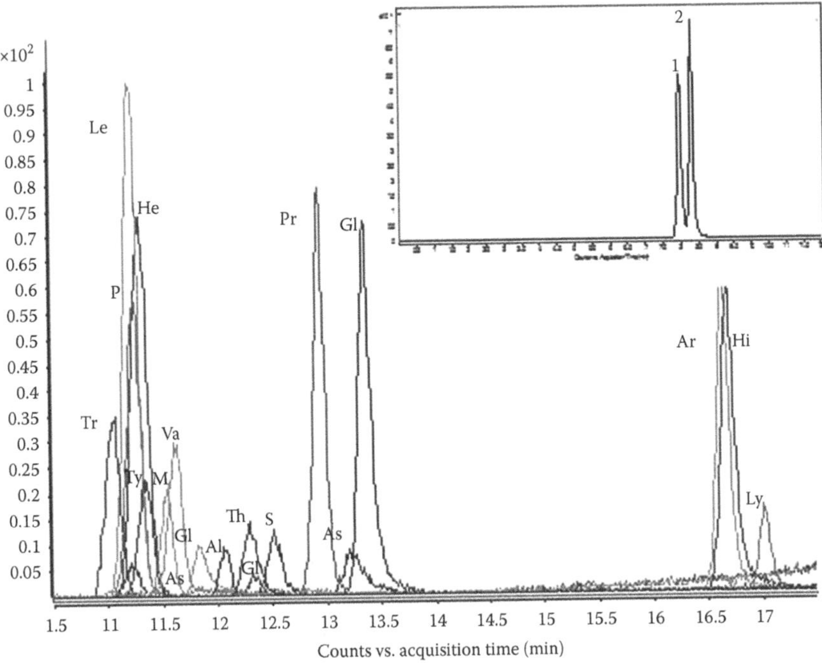

FIGURE 1.10 EICs for 19 common amino acids on silica hydride stationary phase (DH) using ACN as the organic solvent in the mobile phase. Inset: optimized separation of leucine and isoleucine under slightly different gradient conditions. (Adapted from Pesek, J.J. et al., *J. Chromatogr. A*, 1204, 48, 2008.)

the $(M + H)^+$ species. Since a weaker acid can often give better sensitivity, acetic acid will sometimes provide a higher S/N than formic acid.

While running standard samples can demonstrate the ability of the ANP method to retain amino acids, real physiological samples are more challenging and successful analyses of these mixtures prove the ultimate usefulness of this approach to hydrophilic molecule analysis. Figure 1.11 shows the analysis of several samples of human saliva for the amino acids glutamine and lysine (*m/z* 147). These composite extracted ion chromatograms (EICs) illustrate some important aspects regarding both the analysis of physiological samples and the ANP method. First, analysis of metabolites in real samples involves a range of concentrations for any particular analyte, as shown by the varying peak heights in the overlaid EICs. Second, despite the complex matrix and the variable concentration for these two compounds, the retention time is remarkably reproducible for each peak. Thus, ANP is a rugged and reliable method for the analysis of polar metabolites such as amino acids in physiological matrices.

1.5.2 Carbohydrates

Another class of solutes that can be difficult to analyze by RP as well as some methods designed for the retention of polar compounds is carbohydrates. The multiple hydroxyl groups lead to high solubility in aqueous mobile phases so that elution

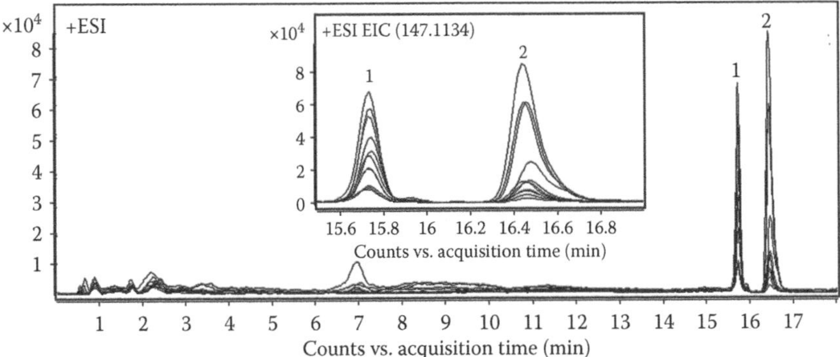

FIGURE 1.11 EICs for amino acids in saliva on the DH column: *m/z* 147, glutamine (1) and lysine (2). (Adapted from Pesek, J.J. et al., *J. Sep. Sci.*, 2200, 32, 2009.)

under typical RP conditions occurs at or near the void volume. Thus, ANP provides the ideal mode for analysis of these hydrophilic compounds utilizing its strong retention in mobile phases containing a high percentage of organic solvent such as ACN.[23]

An example of carbohydrate retention is shown in Figure 1.12 for glucose when using varying amounts of ACN in the mobile phase on a hydride-based material. These results also illustrate another important advantage of the ANP method when using mass spectrometry for detection. First, as expected, the retention of glucose increases as the amount of ACN in the mobile phase increases. Second, the S/N ratio improves as the amount of ACN in the mobile phase is increased from 60% to 90%. A decrease in S/N is seen at 95% organic. This is most likely due to a low amount of water that causes poorer ionization efficiency for the analyte. This example clearly proves the usefulness of ANP for the retention of low molecular carbohydrates which cannot be analyzed as obtained directly from metabolic samples using traditional RP methods.

The detection of carbohydrates in physiological samples (urine and saliva) using ANP has also been demonstrated.[22] The consistency of the retention times over a broad range of concentrations in these complex matrices further substantiates the ruggedness of the ANP approach using silica hydride-based stationary phases. Typical RSD values for repeated injections of the same sample were 0.5% or less for all analytes tested. Each of these analyses utilized a gradient method with re-equilibration times between runs of 5 min. Thus, the transfer of protocols from standard samples used for method development to more complex matrices generally requires little or no adjustment of the original experimental parameters.

1.5.3 Organic Acids

Organic acids are some of the more difficult metabolites to analyze when the pH of the mobile phase is above the pK_a values. Therefore, some acids have been analyzed in the RP mode at low pH. However, in the ANP better retention is obtained when the acid is ionized so the eluent is generally buffered close to neutral pH. These conditions lead to a wide range of elution times because ANP retention is strongly dependent on both the inherent acidity of the compound and the molecular

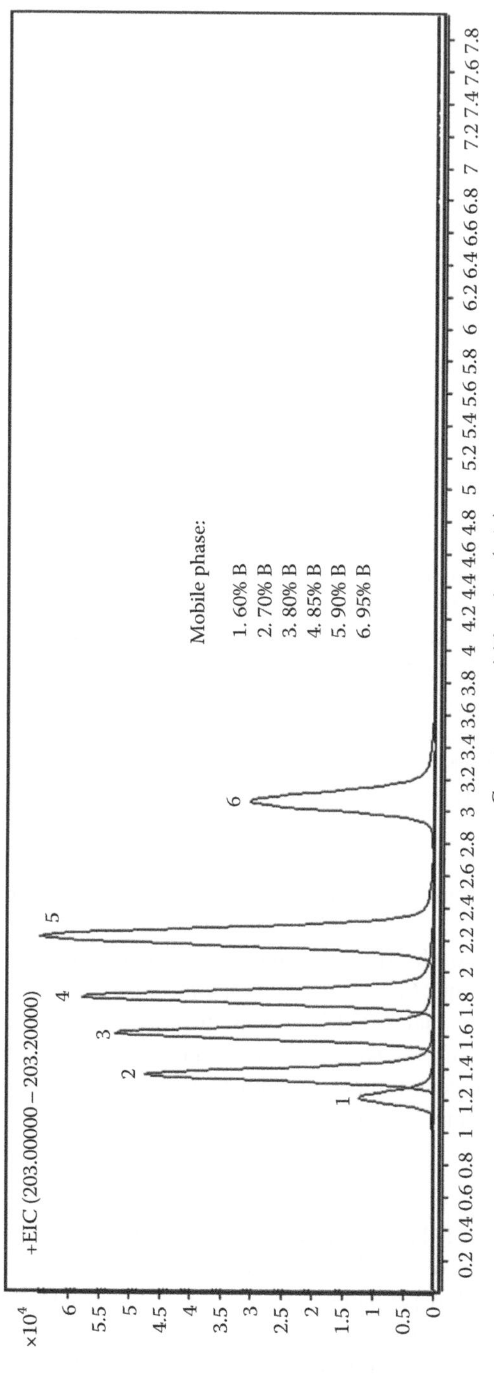

FIGURE 1.12 Comparison of the signal intensity for glucose at various compositions of ACN in the mobile phase containing 0.1% acetic acid. DH column, 2.1 mm × 150 mm. Sample: glucose 0.3 µg/mL. (Adapted from Pesek, J.J. et al., *J. Chromatogr. A*, 1204, 48, 2008.)

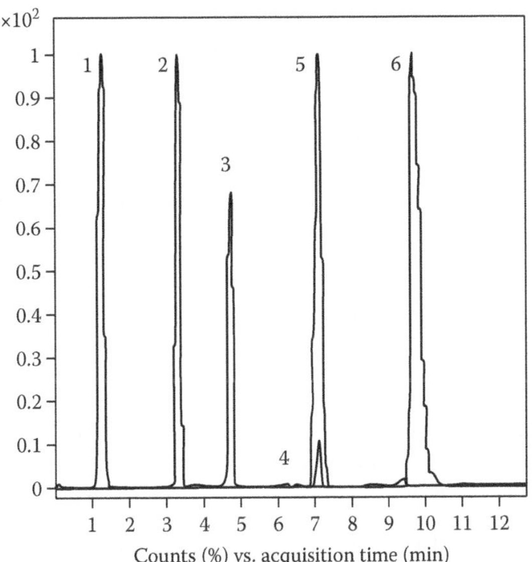

FIGURE 1.13 EICs of organic acids on a DH column. Detection in negative ion mode. Maleic acid (1), *m/z* = 115; aconitic acid (2,3,4), *m/z* = 173; fumaric acid (5), *m/z* = 115; citric acid (6), *m/z* = 191. Mobile phase: A = DI water + 0.1% ammonium formate; B = 90% ACN/10% DI water + 0.1% ammonium formate. Gradient: 0–3.0 min 90% B; 3.0–6.0 min to 70% B; 6.0–7.0 min 70% B; 7.0–7.1 min to 30% B; 7.1–8.0 min 30% B. (Adapted from Pesek, J.J. et al., *J. Chromatogr. A*, 1204, 48, 2008.)

structure.[23] Since the organic acids that have been identified as metabolites can contain one, two, or three acidic functional groups, this feature alone provides for significant differences in chemical properties that can be utilized to control their retention.

A particularly striking example of the power of the ANP method for organic acids is shown in Figure 1.13. The EICs for a number of small organic acids are presented in a single chromatogram. The mobile phase is ACN/water with ammonium formate as the additive. Thus, the pH of the mobile phase is higher than the pK_a values of the individual acids. All peak shapes are very symmetric and the efficiency is very good. The most remarkable of the separations is for compounds 1 and 5, maleic and fumaric acids. These are isobaric C4 acids with the only difference being the *cis* (maleic) and *trans* (fumaric) configuration at the double bond. Fumaric acid has pK_a values of approximately 3.0 and 4.4 while maleic acid has pK_a values of approximately 1.8 and 6.1. Thus, the longer retention for fumaric acid most likely is the result of two ionized sites at the mobile phase pH while maleic acid only has one of its acidic groups ionized under these conditions.

An example of the analysis of some small organic acids in a physiological sample is shown in Figure 1.14. The overlaid EICs in the negative ion mode are for three *m/z* values taken from several donor saliva samples. Three of the acids identified in these samples are aconitic acid (*m/z* 173), succinic acid (*m/z* 117), and adipic acid (*m/z* 145).[22] The inset shows expanded EICs for adipic acid. In each sample, the

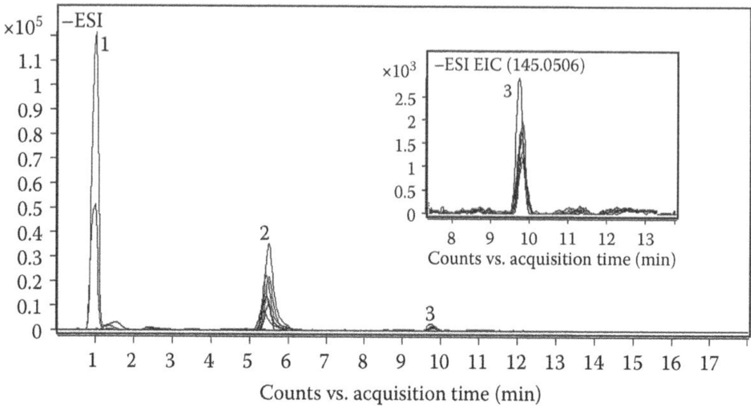

FIGURE 1.14 EICs for organic acids from saliva samples on the DH column. Peak identity: (1) aconitic acid (*m/z* 173); (2) succinic acid (*m/z* 117); and (3) adipic acid (*m/z* 145). (Adapted from Pesek, J.J. et al., *J. Sep. Sci.*, 2200, 32, 2009.)

compounds were verified from the exact mass measurement. The retention times for all three acids are very reproducible over the concentration ranges found in the saliva samples. An interesting observation was made in comparing samples from normal individuals to cancer patients. Aconitic acid (peak 1) was observed in only 2 of 10 normal samples while it was present in all of the samples obtained from cancer patients. These preliminary results indicate that further investigations, both qualitative and quantitative, could prove promising in identifying biomarkers for clinical analyses.

1.5.4 Nucleotides

Nucleotides are important phosphate-containing compounds that are found in all living cells and are involved in a broad spectrum of metabolic and biological processes. They have a key role in the synthesis of DNA and RNA, are part of signal transduction pathways, can be coenzymes in biosynthetic processes, and function as energy reservoirs in biological systems.[28–33] The development of improved methods for the analysis for these compounds is of continued interest, especially techniques which can distinguish among the nucleotides based on their degree of phosphorylation.

An example of the separation of a mixture of mono- and triphosphate nucleotides is shown in Figure 1.15. In this case, detection was done by UV absorption which is generally not possible for each of the other categories of compounds described previously. The stationary phase used in this separation consists of an 11-carbon carboxylic acid bonded to the hydride surface. The retention capabilities of the UDA column are the result of both hydrophilic (hydride surface and carboxylic acid functionality) and hydrophobic (alkyl chain of bonded acid) properties.[24] At 80% ACN in the mobile phase, good retention was obtained but one pair of analytes (thymidine triphosphate and adenosine triphosphate) was not resolved. If the organic content of the mobile phase is raised to 85% (Figure 1.15), then retention became longer and the two previously unresolved components were separated. When using the DH column,

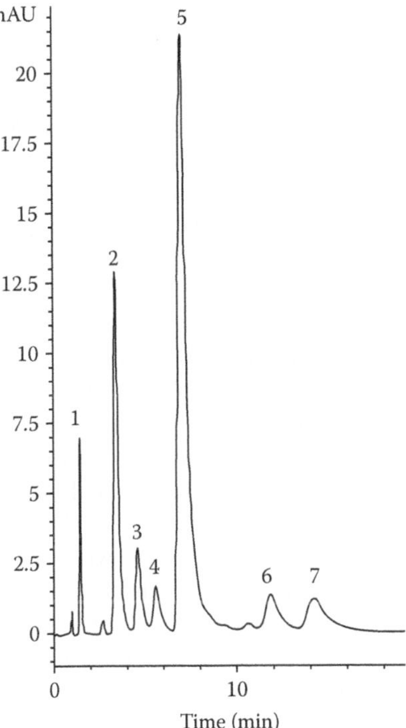

FIGURE 1.15 Gradient separation of seven-component mixture of nucleotides on a silica hydride-based UDA column. 0.0–0.5 min 95% B, 0.5–20 min to 30% B, 20.0–25.0 min hold 30% B. Mobile phase: (A) DI water + 0.1% ammonium formate; (B) 90% ACN/10% DI water + 0.1% ammonium formate. Detection at 254 nm. Solutes: 1 = adenosine-3′,5′-cyclic monophosphate; 2 = adenosine-5′-monophosphate; 3 = adenosine-5′-triphosphate; 4 = thymidine-5′-triphosphate; 5 = uridine-5′-triphosphate; 6 = cytosine-5′-triphosphate; 7 = guanosine-5′-triphosphate.

a complete resolution of this mixture was not successful under isocratic conditions and required a gradient. The improved separation capabilities of the UDA column is most likely due to the enhanced interaction of the nitrogen sites on the nucleotides with the carboxylic acid moiety on the stationary phase.

In Figure 1.9B, a challenging pair of solutes, adenosine-3′-monophosphate and adenosine-5′-monophosphate, with an *m/z* of 346 for the $(M - H)^-$ ion are separated. The identity of each species was confirmed by injection of the individual compounds since the molecular masses of these two analytes are identical and thus indistinguishable using a TOF™ analyzer. The necessity of having a suitable chromatographic method is illustrated in this example because positive identification is facilitated by their separation. These chromatograms demonstrate another advantage of MS detection: the flat baseline which is usually obtained in the EIC for a gradient elution. The experimental conditions used in the LC-MS analysis shown in Figure 1.9 result in both good peak shape and high efficiency (~200,000 plates/m). The ANP method provides high sensitivity, as demonstrated earlier, for glucose since a better S/N ratio is obtained for high percentages of organic in the mobile phase.[23]

1.5.5 Other Analytes

The above sections for specific classes of compounds are a representative cross-section of hydrophilic compounds where a significant number of challenging analyses occur, particularly in the field of metabolomics. Considerable progress has been made with the ANP method in addressing many of the issues involved in the analysis of these compounds. However, there are still improvements which can be made, especially in trying to find a method that might encompass all hydrophilic compounds, as discussed in Section 1.5.6. A number of other compounds and classes have been evaluated by the ANP procedure. A few representative examples are presented here.

A relevant and practical problem in food analysis is shown in Figure 1.16 for the analysis of melamine and cyanuric acid, two contaminants that can be difficult to analyze by RP HPLC because of their highly polar properties. In the method developed for these two compounds, the pair is easily separated and identification is facilitated through MS detection. Cyanuric acid is detected in the negative ion mode and melamine in the positive ion mode. The high degree of separation and the negligible matrix effects make quantitation relatively easy down to the ppb level or less if using an MS/MS system. The method utilized in Figure 1.16 is a simple linear ANP gradient over 15 min. The analysis time can be reduced further by increasing the steepness of the gradient in order to decrease the retention time of melamine or decreasing the column length, i.e., from 150 to 50 mm. Since the ANP column equilibrates rapidly,

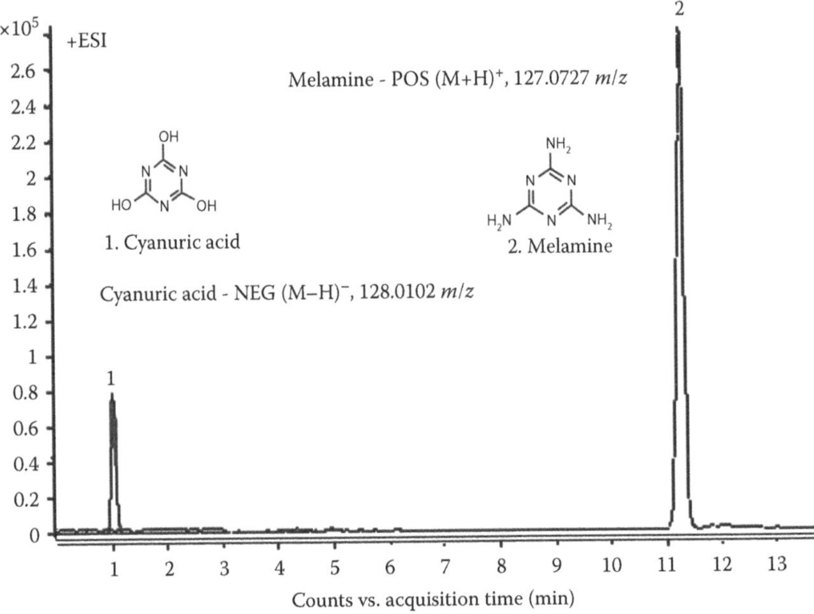

FIGURE 1.16 Analysis of melamine and cyanuric acid on the DH stationary phase. Mobile phase: (A) DI water + 0.1% ammonium acetate; (B) 90% ACN/10% DI water + 0.1% ammonium acetate.

typically no more than 5 min, laboratories could easily analyze a relatively large number of samples in a day. The column ruggedness and reproducibility add to the attractiveness of the ANP approach for this analysis. Similar analyses have been developed for herbicides such as glyphosate as well as paraquat and a number of related compounds.

A physiologically relevant pair of analytes is choline and acetylcholine. These polar compounds are difficult to analyze by RP so they are perfect candidates for the ANP method. With a mobile phase consisting of ACN/water containing 0.5% formic acid and using mass spectrometry for detection, a mixture of the two compounds was tested on two hydride-based cholesterol columns. Using 90:10 ACN/water on a 4.6 mm × 75 mm column, the selectivity factor (α) was 1.21. On a 2.1 mm × 20 mm column with a 92:8 ACN/water mobile phase, the α value was 1.28. In both cases, the two analytes were baseline separated so that good quantitative measurements could be made. Thus, quaternary amine compounds can be easily retained on the hydride-based materials. This example also illustrates the advantage for ANP with amine compounds in general. These analytes can be determined in acidic solution as opposed to the very basic conditions that are required for RP methods. The acidic experimental conditions are more favorable for long lifetimes of the column as well as the instrument.

Another interesting potential use of the ANP method is shown in Figure 1.17 where the retention of a series of hydrophilic peptides is plotted as a function of mobile phase composition.[34] These results suggest that the use of hydride-based phases in the ANP mode could be a complementary method to RP in such applications as proteomics. Just as in other applications where the sample contains a large number of compounds with a very broad range of polarities, in a typical RP analysis of a protein digest the hydrophilic peptides elute at or near the void volume while the

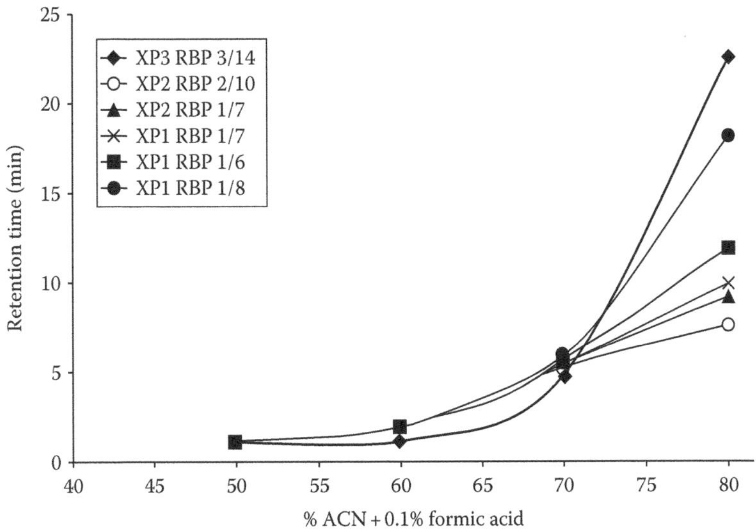

FIGURE 1.17 Retention maps for various structurally related peptides on a silica hydride-based C18 column. Mobile phase: ACN/water + 0.1% formic acid.

hydrophobic species are well retained and can be separated. In the complementary analysis by ANP, the hydrophobic species would elute at or near the void volume while the hydrophilic species would be well retained and could be separated. Thus by two analyses of a protein digest, one RP and one ANP, the complete range of peptides could be determined. A similar possibility also exists for samples such as cell extracts where a sample containing a large number of proteins could also be analyzed by the same complementary two-method approach.

A final example of the use of ANP involves the retention scenario described in Figure 1.4. Here one compound can be retained by either RP or ANP, depending on the mobile phase composition.[21] A series of cytidine-related compounds were tested for their ANP retention on a hydride-based stationary phase. Unsubstituted cytidine is sufficiently polar so that it can be easily retained with high amounts of organic in the mobile phase and an additive such as formic or acetic acid. As hydrophobic substituents are added to the molecule, a more nonpolar moiety is created and these molecules have RP retention. This effect is seen in Figure 1.18 for three substituted cytidines that have alkyl side chains of varying length. The retention map focuses on mobile phase compositions normally required for ANP behavior, i.e., >50% organic. All of the substituted cytidines studied clearly display ANP behavior. The retention of the compound shown as cytidine-R1 has the shortest alkyl chain of the three molecules tested and does not have any RP retention as low as 60% ACN in the mobile phase. Cytidine-R2 has a longer alkyl chain attached to the parent molecule and shows a slight increase in retention as the amount of organic is decreased from 70% to 60%. Cytidine-R3 has the longest alkyl chain of the three compounds and it clearly shows the greatest increase in retention as the amount of ACN is decreased from 70% to 60%. This example illustrates a unique aspect of the hydride-based phases. Typical RP materials would only have retention for this compound at high amounts of water and HILIC columns would only have retention at high organic compositions. For some analyses, the ability to utilize two modes of retention provides the greatest opportunity to find an optimal separation for a particular mixture.

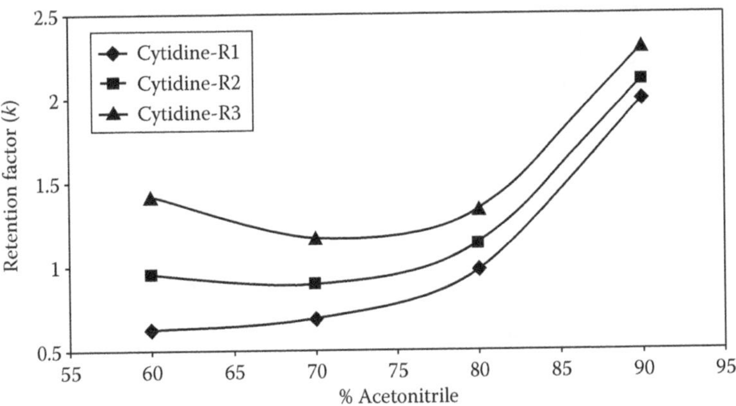

FIGURE 1.18 Retention map for substituted cytidines on a silica hydride-based cholesterol column. Mobile phase: ACN/water + 0.1% formic acid. (Adapted from Pesek, J.J. et al., *J. Sep. Sci.*, 30, 637, 2007.)

1.5.6 Universal Analytical Method for Hydrophilic Metabolites

Many of the classes of compounds described in the applications earlier are found in metabolite samples. While each of these separations typically cites the optimal conditions for separation, there is a practical need to limit the number of determinations done for each sample. Ideally, there would be one method for hydrophobic compounds based on RP conditions and another method that would encompass all of the hydrophilic metabolites. Due to its ruggedness and versatility, ANP appears an attractive choice to fill the role of analyzing for polar metabolites. The examples presented previously while covering a range of mobile phase options are not that disparate from each other. The conditions range from mildly acidic (formic acid) to near neutral pH (ammonium acetate or ammonium formate). Thus, it might be possible to conceive a gradient mobile program, either binary, tertiary, or quaternary, that could sufficiently retain a reasonable number of metabolites and separate those with identical masses or differing by only 1 amu. Somewhat relaxed separation requirements would be needed if using a mass spectrometer capable of exact mass determinations at a level of 50 ppm or less, or with an MS/MS system. The total analysis time should be minimized in developing the gradient program in order to generate as many determinations as possible within a given time frame. The hydride-based stationary phases can significantly aid in this respect since the equilibration time between runs is so short, typically no more than 5 min. At present, this goal has not yet been attained but continued investigations into understanding the exact mechanism of retention on the hydride phases may facilitate achieving such an operating format.

1.6 CONCLUSIONS

Silica hydride-based stationary phases represent a new class of materials with fundamentally different properties than ordinary silica. One of the consequences of this new surface is the ability to retain polar compounds in a manner similar to HILIC, but with other features not possessed by stationary phases operating in the HILIC format. For the stationary phases that combine a silica hydride surface with a bonded organic moiety, retention is obtained for both polar and nonpolar compounds. The hydride-based stationary phases are also characterized by rapid equilibration after changing mobile phase compositions which can be particularly important after gradient analyses. For hydride materials with a minimal amount of bonded organic moiety, the retention of hydrophilic compounds is enhanced. Separation strategies can be designed for either specific target analytes or to span a range of polarities utilizing the variety of stationary phases based on silica hydride surfaces currently available. Additional facets of these materials are being investigated and new phases are being developed which should lead to further applications utilizing all the capabilities of silica hydride, especially with respect to the ANP mode.

ACKNOWLEDGMENT

The authors would like to thank Microsolv Technology Corporation, Eatontown, NJ, for assistance in providing data and columns to obtain some of the experimental results described in this review.

REFERENCES

1. Theodoridis, G., Gika, E., and Wilson, I.D., *Trends Anal. Chem.*, 2004, 27, 251.
2. Yokoyama, T., Sokai, A., and Zenki, M., *Chromatographia*, 2008, 67, 535.
3. Concha-Herrera, V., Torres-Lapasio, J.R., Vivo-Truyols, G., and Garcia-Alvarez-Coque, M.C., *J. Liq. Chromatogr. Relat. Technol.*, 2006, 29, 2521.
4. Piraud, M., Vianey-Saban, C., Bordin, C., Acquaviva-Bordain, C., Boyer, S., Elfakir, C., and Bouchu, D., *Rapid Commun. Mass Spectrom.*, 2005, 19, 3287.
5. Tomiya, N., Alior, E., Lawrence, S.M., and Betenbaugh, M.J., *Anal. Biochem.*, 2001, 293, 129.
6. Yun, J., Shen, S., Chen, F., and Yao, K., *J. Chromatogr. B*, 2007, 860, 57.
7. Dorion, S. and Rivoal, J., *Anal. Biochem.*, 2003, 323, 188.
8. Schlichterle, H., Affolter, M., and Cerny, C., *Anal. Chem.*, 2003, 75, 2349.
9. Coulier, L., Bas, R., Jespersen, S., Verheij, E., van der Werf, M.J., and Hankemeier, T., *Anal. Chem.*, 2007, 79, 6573.
10. Bajad, S.U., Lu, W., Kimball, E.H., Yuan, J., Peterson, C., and Rabinowitz, J.D., *J. Chromatogr. A*, 2006, 1125, 76.
11. Kind, T., Tolstikov, V., Fiehn, O., and Weiss, R.H., *Anal. Biochem.*, 2007, 363, 185.
12. Tolstikov, V.V., Lommen, A., Nakanishi, K., Tanaka, N., and Fiehn, O., *Rapid Commun. Mass Spectrom.*, 2005, 19, 3031.
13. Pan, J., Song, Q., Shi, H., King, M., Junga, H., Zhou, S., and Naidong, W., *Rapid Commun. Mass Spectrosc.*, 2003, 17, 2549.
14. Cubbon, S., Bradbury, T., Wilson, J., and Thomas-Oates, J., *Anal. Chem.*, 2007, 7, 8911.
15. Pesek, J.J. and Matyska, M.T., *J. Liq. Chromatogr. Relat. Technol.*, 2006, 29, 1105.
16. Naidong, W., Shou, W.Z., Addison, T., Maleki, S., and Jiang, X., *Rapid Commun. Mass Spectrom.*, 2002, 16, 1965.
17. Shen, J.X., Xu, Y., Tama, C.I., Merka, E.A., Clement, R.P., and Hayes, R.N., *Rapid Commun. Mass Spectrom.*, 2007, 21, 3145.
18. Rosen, J., Nyman, A., and Hellenaes, K.-E., *J. Chromatogr. A*, 2007, 1172, 19.
19. McCalley, D.V., *Advances in Chromatography*, Vol. 46, CRC Press, Boca Raton, FL, 2008, pp. 305–350.
20. Pesek, J.J. and Matyska, M.T., *LCGC*, 2007, 25, 480.
21. Pesek, J.J., Matyska, M.T., and Larrabee, S., *J. Sep. Sci.*, 2007, 30, 637.
22. Pesek, J.J., Matyska, M.T., Loo, J.A., Fischer, S.M., and Sana, T.R., *J. Sep. Sci.*, 2009, 32, 2200.
23. Pesek, J.J., Matyska, M.T., Fischer, S.M., and Sana, T.R., *J. Chromatogr. A*, 2008, 1204, 48.
24. Pesek, J.J., Matyska, M.T., Hearn, M.T.W., and Boysen, R.I., *J. Chromatogr. A*, 2009, 1216, 1140.
25. Pesek, J.J. and Matyska, M.T., *LC/GC*, 2006, 24, 296.
26. Pesek, J.J., Matyska, M.T., Gangakhedkar, S., and Siddiq, R., *J. Sep. Sci.*, 2006, 29, 872.
27. Pesek, J.J., Matyska, M.T., and Sharma, A., *J. Liq. Chromatogr. Relat. Technol.*, 2008, 31, 134.
28. Sebestik, J., Hlavacek, J., and Stibor, I., *Curr. Protein Pept. Sci.*, 2005, 6, 133.
29. von Ballmoos, C., Brunner, J., and Dimroth, P., *Proc. Natl. Acad. Sci. USA*, 2004, 101, 11239.
30. Ataullakhanov, F.I. and Vitvitsky, V.M., *Biosci. Rep.*, 2002, 22, 501.
31. Begley, T.P., Kinsland, C., Mehl, R.A., Osterman, A., and Dorrestein, P., *Vitam. Horm.*, 2001, 61, 103.
32. Francis, S.H. and Corbin, J.D., *Crit. Rev. Clin. Lab. Sci.*, 1999, 36, 275.
33. Ashcroft, S.J., *Adv. Exp. Med. Biol.*, 1997, 426, 73.
34. Pesek, J.J., Matyska, M.T., Hearn, M.T.W., and Boysen, R.I., *J. Sep. Sci.*, 2007, 30, 1150.

2 Application of the Calculated Physicochemical Property, Log *D*, to Hydrophilic Interaction Chromatography in the Bioanalytical Laboratory

Eugene P. Kadar, Chad E. Wujcik, and Olga Kavetskaia

CONTENTS

2.1 INTRODUCTION

Liquid chromatography combined with tandem spectrometric detection (LC-MS/MS) is the preferred technique for quantitative bioanalysis due to its intrinsic selectivity, sensitivity, and speed.[1–4] These qualities provide optimum utility in the analysis of drugs and their metabolites in complex biological matrices such as cerebral spinal fluid, plasma, serum, tissue, urine, and whole blood in support of animal and human studies during drug development.

Hydrophilic interaction chromatography (HILIC) has become an increasingly popular choice as the chromatographic component in quantitative bioanalysis due to its unique selectivity, reduced back pressure, and enhanced mass spectrometric sensitivity relative to traditional reversed-phase chromatography.[1,2,5] In 2003, Naidong published a comprehensive literature review on the utilization of HILIC in bioanalytical LC-MS/MS.[2] This review focused on the use of unmodified silica as the stationary phase and included topics such as retention mechanism, sensitivity comparison with a reversed-phase, direct injection of SPE extracts, ultrafast LC-MS/MS, column stability, and future perspectives. A tabulation of HILIC-MS/MS methods for representative pharmaceutical compounds in various biological matrices is provided. Hsieh recently published a review article on the potential of HILIC-MS/MS in the quantitative bioanalysis of drugs and metabolites.[1] This chapter provides a tabulation of HILIC-MS/MS applications for pharmaceutical bioanalysis and presents examples of the utilization of sulfobetaine (ZIC) and amino-modified silica stationary phases for HILIC-MS/MS. Hsieh also discusses the topic of matrix effects, in the form of ionization suppression or enhancement, commonly observed in bioanalysis.

Ionization suppression or enhancement is a matrix effect in which endogenous components co-elute with the analyte of interest, causing a decrease or increase in the level of analyte ionization in the MS source and subsequent mass spectrometric signal. This phenomenon is common during the analysis of complex matrices such as biological fluids by LC-MS/MS and occurs more frequently with electropsray ionization (ESI) techniques.[1,6–10] Typically, extensive and elaborate extraction procedures, such as solid phase or liquid/liquid, are employed to alleviate matrix effects that may be encountered during LC-MS/MS using reversed-phase chromatography. For more polar compounds, the alternative selectivity of HILIC may be utilized to remediate matrix effects, allowing the analyst to use less involved and more cost-effective extraction methods like protein precipitation. The high level of orthogonality and the unique selectivity of HILIC relative to reversed-phase chromatography may also provide the means of separating isobaric compounds and other interferences that cannot be achieved by reversed-phase chromatography. This alternative selectivity is very advantageous during the early stages of drug candidate development when in vivo metabolic transformation knowledge is limited. Another factor, the reduced back pressure afforded by HILIC, allows the use of higher flow rates that significantly decreases analysis time.[1,2] The typical loss in column efficiency from increased flow rates is not as prevalent for HILIC as compared to reversed-phase chromatography.[2] This benefit is believed to be a result of the differences in diffusion and mass transfer as a result of the lower viscosity of HILIC relative to reversed-phase chromatography.[2]

HILIC typically uses 5%–50% water in the mobile phase with the remainder composed of a volatile organic solvent such as acetonitrile.[5] The relatively high organic solvent concentration is favored in the desolvation process of ESI, which is extremely beneficial considering that this ionization technique is commonly employed in the LC-MS/MS bioanalysis of polar compounds.[1,2,5] Additionally, HILIC allows the direct injection of many biological extracts obtained by protein precipitation, liquid/liquid, and solid phase extraction, without the need for further compositional modification (e.g., evaporation with reconstitution).[1–4]

Bioanalysis plays an important role in drug development, in which the speed of method development is a critical factor. The ability to predict and select appropriate separation conditions can significantly expedite method development. A consideration of the physicochemical properties of an analyte can provide great insight into this process.

The distribution coefficient, *D*, is used to quantitatively express lipophilicity under a specific pH condition. It is a physicochemical property that is defined as the equilibrium concentration ratio of a substance between an aqueous buffer at a specified pH and octanol.[11–15] The commercial software program ACD/Log*D* Suite is capable of calculating this physicochemical property from the input of the chemical structure of a compound. Recent literature has demonstrated the dependence of the HILIC capacity factor, k', on *D*.[16–18] This chapter is based on a recent publication in which a linear relationship between log *D* and the HILIC log k' was demonstrated for a series of representative pharmaceutical compounds under generic HILIC conditions.[16] This relationship was employed to rapidly assess the applicability of HILIC as well as to identify suitable internal standards for LC-MS/MS bioanalysis.

2.2 THEORETICAL INTRODUCTION AND DEFINITIONS

In 1990, Alpert coined the term "hydrophilic interaction chromatography" (HILIC) and suggested that the mechanism of HILIC involves analyte partitioning between a hydrophobic mobile phase and a partially immobilized layer of water on the stationary phase.[19] Several recent review articles have been published on the topic of HILIC, all of which conclude that the mechanism is in fact multidimensional including additional components such as adsorption and electrostatic interactions.[1,2,20–23] The following theoretical discussion is based on the hypothesis that partitioning is the primary mechanism defining retention under a generic HILIC system that includes an unmodified silica stationary phase and a relatively high mobile-phase salt concentration of 10 mM ammonium formate, pH adjusted to 3.0.

The term capacity factor (k') is used to indicate the magnitude of retention of an analyte on a chromatographic column. This term is defined as

$$k' = \frac{t_R - t_0}{t_0} \tag{2.1}$$

where

t_R is the retention time of the analyte
t_0 is the column dead time or retention time of nonretained components[24–28]

The Center for Drug Evaluation and Research (CDER) recommends a minimum k' value of >2 for separations that will be used to support regulatory submissions.[25] This criterion typically provides adequate retention, allowing a separation between the analyte of interest and the unretained hydrophobic matrix components. This separation is critical to prevent possible ionization suppression or enhancement when using atmospheric pressure ionization mass spectrometers as detectors, especially with the use of protein precipitation as the extraction technique.

The following partitioning relationship describes the dependence of the fraction of organic modifier on k' for a reversed-phase system:[20,24,26,27]

$$\log k' = \log k'_{\mathrm{w}} - S\varphi \tag{2.2}$$

where

k'_{w} is the theoretical value of k' for exclusively water as the mobile phase
φ is the volume fraction of the organic in the binary mobile phase
S is a constant representing the slope of the log k' versus the φ line[20,24,26,27]

The value of S is approximately 4 for low-molecular weight compounds (<500 Da).[24] This molecular weight range is representative of most of the drugs and drug metabolites analyzed by LC-MS/MS.

The partition mechanism observed in revered-phase chromatography is comparable to the partitioning of a compound between octanol and water.[24,26–30] Solute partitioning between octanol and water is a physicochemical property and is defined as the partition coefficient, P[30]

$$P = \frac{\gamma_{\mathrm{O}} C_{\mathrm{O}}}{\gamma_{\mathrm{W}} C_{\mathrm{W}}} \tag{2.3}$$

where

γ_{O} and γ_{W} are the activity coefficients of the compound in octanol and water, respectively
C_{O} and C_{W} are the concentrations of the compound in octanol and water, respectively

In the case of dilute solutions, such as the chromatographic mobile phase, this equation can be reduced to[30]

$$P = \frac{C_{\mathrm{O}}}{C_{\mathrm{W}}} \tag{2.4}$$

The partition coefficient is typically expressed as the logarithm and is used to represent the lipophilicity of a compound. This physicochemical property is routinely applied in the pharmaceutical industry as an indicator of the absorption and permeation characteristics of a compound.[31] This property is usually determined using the shake-flask method.[26,27,29,30,32,33] However, this method is not sensitive to impurities and degradation products and has a low throughput and limited range.[26,27,30,32,33] In order to improve these deficiencies, several laboratories have employed reversed-phase chromatography and the above-described relationship between k' and P to determine log P_{Oct} values of several test compounds.[11,26,27,29,30,32,33] Predictive commercial software programs, such as ACD/Log*P*DB, can also be used to calculate log P_{Oct} values.[34] However, the use of log P is limited, especially when applied to chromatography, because this physicochemical property considers only compounds in the unionized state.[12,13] A majority of drugs and drug metabolites are typically present in their ionized forms during the chromatographic phase of LC-MS/MS analysis.

The distribution coefficient, D, is a physicochemical property that takes into consideration the ionization state of a compound and is also typically expressed as the logarithm. This property is the equilibrium concentration ratio of a compound between the octanol and aqueous buffer at a specified pH and is expressed as[12,13,15]

$$\log D_{\text{bases}} = \log P + \log\left[\frac{1}{\left(1+10^{\mathrm{p}K_\mathrm{a}-\mathrm{pH}}\right)}\right] \tag{2.5}$$

The above equation pertains to monoprotic bases. This equation can be applied to monoprotic acids by replacing the exponent pK_a–pH with pH–pK_a.[12,13] This equation can also be expanded to include polyprotic acids, bases, and zwitterions. Lombardo et al. successfully utilized the relationship between log k' and log D at the physiological pH of 7.4 to determine $\log D_{\text{Oct}}^{7.4}$ values of a set of basic and neutral pharmaceutical compounds.[13]

Quiming et al. recently developed HILIC retention prediction models for adrenoreceptor agonist and antagonist compounds utilizing multiple linear regression (MLR) and artificial neural networks (ANN).[17,18] Although HILIC retention prediction models were determined using both unmodified silica and diol columns, the work using the unmodified silica column will be discussed. MLR is commonly used for the statistical analysis of quantitative structure–retention relationships. ANN is another method used in chromatographic optimization that has recently gained popularity. MLR was used to derive the following predictors: the percentage acetonitrile in the mobile phase, log D, the number of hydrogen bonding acceptors, the magnitude of the dipole moment, and the total absolute atomic charge. Standardization coefficients (β) of the predictors indicate the relative effect of each predictor on the logarithm of the retention factor. The percentage of acetonitrile in the mobile phase produced the largest magnitude β (positive in sign), which can be expected considering the known dependence of HILIC on organic mobile-phase concentration. Log D produced the second largest magnitude β (negative sign), representing an inversely proportional relationship to retention. The relatively high magnitude of β is indicative of the contribution of partitioning to the overall mechanism of HILIC utilizing an unmodified silica column.

ACD/LogD Suite Version 9.0 is capable of calculating log D values for polyprotic acids, bases, and zwitterions over the pH range of 0–14 in increments of 0.1 pH units.[14,15] Additionally, log D can be calculated with or without the incorporation of ion-pairing. The chemical structure of the compound of interest is entered and the software takes the proposed equilibrium scheme for all of the species in the organic and aqueous phases into account for the log D calculation.[15] Nozawa et al. successfully applied ACD Labs Log D as a measure of hydrophobicity in the advanced Marfey's method.[35] The advanced Marfey's method is a derivatization-based technique used to determine the absolute configuration of L- and D-amino acids in a peptide using LC-MS.[35] This technique is based on the hydrophobicity difference between the derivatized L- and D-amino acids.

The capability to calculate the pH-dependent distribution coefficient using ACD/LogD Suite allowed the following hypothesis to be explored: A relationship exists between the HILIC capacity factor, k', of an analyte and the calculated log $D_{pH\ 3.0}$ in a generic HILIC system. This system was designed to minimize adsorption and electrostatic interactions, thereby allowing the partition to act as the primary retention mechanism. The following assumptions were made for the evaluation of this hypothesis: (1) partitioning between the hydrophobic mobile phase and a hydrophilic partially immobilized layer of water on the stationary phase is the primary chromatographic retention mechanism of the above-defined generic HILIC system, (2) the pH of the immobilized layer of water on the silica stationary phase is 3.0, (3) ion-pairing is insignificant, and (4) the ACD/LogD Suite can provide a reasonable estimate of log $D_{pH\ 3.0}$.

2.3 EXPERIMENTAL DESIGN

A set of 30 probe compounds synthesized by Pfizer Inc., were selected to cover an extensive calculated log $D_{pH\ 3.0}$ range of −8.68 to 8.16. These probe compounds are representative of the following therapeutic areas: anti-infectives, cancer, cardiovascular and metabolic diseases, and the central nervous system. Five chemical analog pairs were included in this set based on structure and log *D* similarity, in order to determine the utility of using calculated log *D* as a criterion for analog internal standard selection. A tabulation of the probe compounds, including their physicochemical properties is presented in Table 2.1. The chemical structure of probe compound No. 1 is illustrated in Figure 2.1.

This compound was utilized as a representative basic drug in order to develop the generic HILIC system, particularly with respect to ammonium formate concentration.

ACD/LogD Suite, Version 9.0 was used to calculate the log $D_{pH\ 3.0}$ of each probe compound. Stock solutions were prepared at approximately 100 µg/mL in 50:50 (v/v) water:acetonitrile. Working solutions were prepared at approximately 100 ng/mL in acetonitrile by dilution of the corresponding stock solution, with the exception of compound No. 30. An increased concentration of this compound, approximately 10,000 ng/mL, was required to obtain an adequate peak response.

The chromatographic system included a Shimadzu SCL-10A controller, Shimadzu LC-10AD pumps, and a CTC Analytics (LEAP) HTLS PAL autosampler. An unmodified silica column, Atlantis™ HILIC silica column, 2.1 × 50 mm, 5 µm, 100 Å, P/N 186002012 from Waters Corporation (Milford, MA), was employed at ambient temperature. HPLC grade water and acetonitrile were combined with a 200 mM aqueous ammonium formate, pH 3.0 stock solution to produce three separate mobile-phase solutions with acetonitrile concentrations of 85%, 90%, and 95% (v/v) and a total ammonium formate concentration of 10 mM. These acetonitrile concentrations were chosen because they can typically provide an adequate k' range of 2–5 for HILIC applicable compounds. k' values of greater than 5 are typically not utilized in LC-MS/MS bioanalysis.

The detector was an Applied Biosystems API 4000 tandem quadrupole mass spectrometer equipped with a TurboIonspray™ source. The source was operated in the positive-ion electrospray mode except for probe compound Nos. 24 and 27, for which negative-ion electrospray was employed due to acidic functional groups. The mass spectrometer was tuned for source conditions and precursor ion/product ion

TABLE 2.1
Physicochemical Properties of Probe Compounds

Compound Identification No.	Molecular Weight (Da)	Acidic pK_a[a]	Basic pK_a[a]	Log $D_{pH\ 3.0}$[b]	Log P[b]
1	366	—	9.37, 5.65, 4.25	−8.68	−0.02
2	448	—	8.71, 7.48, 2.98	−7.44	3.09
3[c]	363	11.8	10.1, 4.11	−6.75	1.69
4[c]	377	—	10.1, 3.94	−6.36	1.64
5[d]	439	—	7.67, 5.46	−6.32	0.88
6[d]	453	—	7.67, 5.46	−5.79	1.41
7	227	—	9.95	−4.72	2.23
8[e]	322	—	9.16	−4.38	1.77
9	258	—	8.86, 2.75	−4.20	1.83
10[e]	318	—	9.35	−4.06	2.28
11	355	—	8.17, 4.83	−3.37	3.20
12	510	—	6.99, 5.60	−3.35	2.74
13[f]	378	—	9.43	−2.76	3.39
14[f]	366	—	9.41	−2.44	3.69
15	508	—	4.36, 1.31	−2.10	−0.71
16	470	12.1	8.18, 4.44	−1.08	5.12
17	300	—	9.04	−1.06	4.99
18	335	—	8.49	0.33	5.81
19	448	—	7.67, 2.61	0.67	5.14
20[g]	489	—	4.10	0.85	2.02
21[g]	474	—	4.10	0.94	2.11
22	325	9.25	—	1.51	1.51
23	382	—	5.23, 4.85	1.71	6.07
24	454	3.10	1.44	2.73	3.00
25	559	4.29	—	4.10	4.13
26	469	3.14	2.33	4.46	4.78
27	447	3.33, 11.5	—	5.88	6.08
28	639	10.5, 12.4	1.78	7.14	7.16
29	595	—	—	7.27	7.28
30	600	—	—	8.16	8.16

Source: Reproduced from Kadar, E.P. et al., *J. Chromatogr. B*, 863, 1, 2008.
[a] Calculated using ACD Labs pK_a Suite, version 9.0.
[b] Calculated using ACD Labs Log *D* Suite, version 9.0.
[c–g] Chemical analog pairs.

pair selection by the infusion of individual solutions of each probe compound. Data acquisition and chromatographic review were performed using Applied Biosystems/MDS SCIEX Analyst, Version 1.4.

The chromatographic system was allowed to equilibrate with each mobile-phase condition (10mM ammonium formate, pH 3.0 with acetonitrile concentrations of 85%, 90%, or 95% [v/v]). Exactly 3 μL of each probe compound working solution was injected in triplicate with a run time of 7 min under isocratic conditions for each of the mobile-phase compositions investigated. The retention time of compound No. 30 was utilized as the column dead time (or the retention time of nonretained components), t_0. This compound was chosen for this purpose based on its high log *P*, 8.16, and its neutral character. Additionally, experiments using the same generic HILIC system with an expanded acetonitrile concentration range of 75%–95% (v/v) produced comparable retention times for compound No. 30 (0.14–0.15 min), suggesting that it is not retained and therefore an appropriate t_0 marker.[36] The mean k' ($n = 3$) for each probe compound was calculated. Linear regression analysis of mean log k' versus calculated log $D_{\text{pH } 3.0}$ was performed for each of the three mobile-phase acetonitrile concentrations.

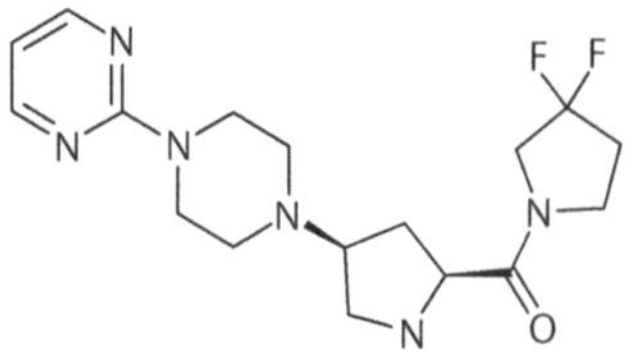

FIGURE 2.1 Chemical structure of compound No. 1. (Reproduced from Kadar, E.P. and Wujcik, C.E., *J. Chromatogr. B*, 877, 471, 2009.)

2.4 RESULTS AND DISCUSSION

2.4.1 Relationship between Experimentally Determined k' and Calculated Log $D_{\text{pH } 3.0}$

The relationship between log k' and log $D_{\text{pH } 3.0}$ was satisfactorily described by a linear function, with linear correlation coefficients of 0.751, 0.696, and 0.689, corresponding to mobile-phase acetonitrile concentrations of 85%, 90%, and 95% (v/v), respectively. Plots are presented in Figure 2.2.

Equations 2.6 through 2.8 express the linear relationship between experimentally determined log k' and calculated log $D_{\text{pH } 3.0}$ for mobile-phase acetonitrile concentrations of 85%, 90%, and 95% (v/v), respectively:

$$\log k' = -0.132\,(\log D) + 0.034 \tag{2.6}$$

$$\log k' = -0.132\,(\log D) - 0.234 \tag{2.7}$$

$$\log k' = -0.139\,(\log D) - 0.008 \tag{2.8}$$

A minimum k' value of 2 is recommended by CDER and is critical to ensure separation between the analyte of interest and unretained matrix components that could possibly produce ionization or enhancement of the analyte signal.[25] Equations 2.6 through 2.8 were utilized to produce log $D_{\text{pH } 3.0}$ cut-off ($k' \geq 2$) values of −4.0, −2.0, and −2.2 for mobile-phase acetonitrile concentrations of 85%, 90%, and 95% (v/v), respectively. The mean experimentally determined k' values are presented in Table 2.2. Additionally, the mean predicted k' values using Equations 2.6 through 2.8 are included in Table 2.2. Overall, the applicability of HILIC ($k' \geq 2$) was correctly predicted for approximately 90% of all compounds tested for each of the three generic HILIC systems evaluated.

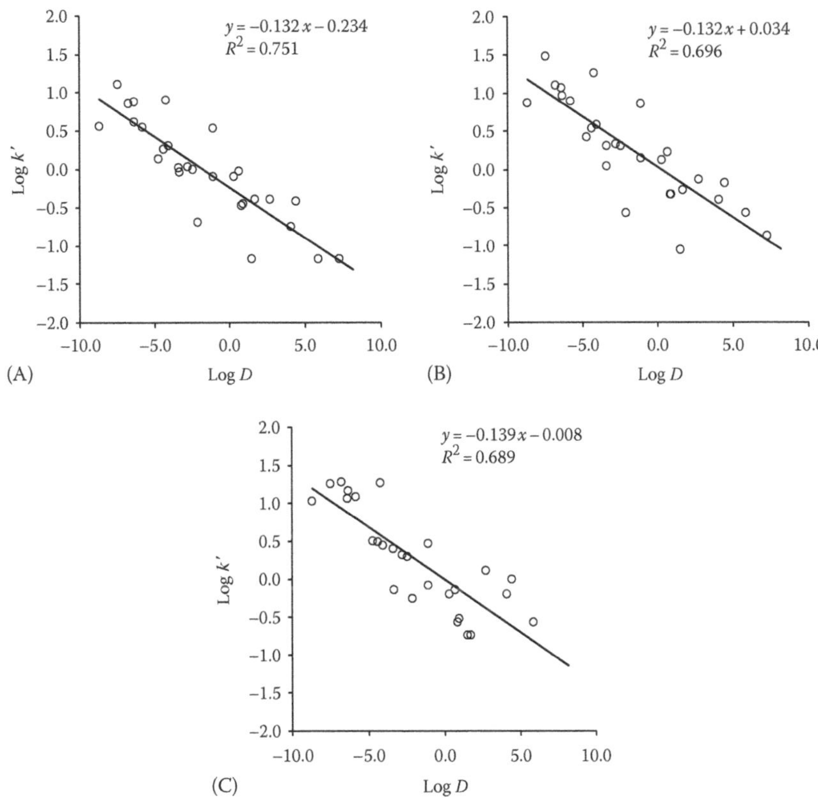

FIGURE 2.2 Experimentally determined log k' versus calculated log $D_{\text{pH 3.0}}$ obtained using generic HILIC systems (Atlantis™ HILIC silica column [5 µm], mobile phase: water and acetonitrile were combined with a 200 mM aqueous ammonium formate, pH 3.0, stock solution to produce three separate mobile-phase solutions with acetonitrile concentrations of 85%, 90%, and 95% (v/v) and a total ammonium formate concentration of 10 mM, with a flow rate of 1.0 mL/min). Linear regression analysis provided correlation coefficients (R^2) of 0.751 (A), 0.696 (B), and 0.689 (C). (Reproduced from Kadar, E.P. et al., *J. Chromatogr. B*, 863, 1, 2008.)

2.4.2 Discussion of Assumptions Made to Formulate Hypothesis

The correlation coefficients produced from linear equations (2.1 through 2.3) clearly deviate from unity. These results can be expected considering the assumptions that were made regarding the formulation of our hypothesis. With a clear understanding that the overall mechanism of HILIC is multimodal and is comprised of partitioning, adsorption, and electrostatic mechanisms, our generic HILIC systems were designed to minimize the latter two mechanisms and simplify the model.

The first assumption designates partitioning between the hydrophobic mobile phase and a hydrophilic, partially immobilized layer of water on the stationary phase as the primary chromatographic retention mechanism of our generic HILIC systems. Consideration was given to the following three principal types of intermolecular forces involved in a chromatographic separation: dispersion forces, polar

TABLE 2.2
Tabulation of Experimental k', Predicted k', and Prediction of HILIC Applicability Utilizing Log D

Mobile Phase Condition	85% Acetonitrile			90% Acetonitrile			95% Acetonitrile		
Compound ID No.	Mean (n = 3) Exptl. k'	Pred. k'	Correct Appl. Pred.[a]	Mean (n = 3) Exptl. k'	Pred. k'	Correct Appl. Pred.[a]	Mean (n = 3) Exptl. k'	Pred. k'	Correct Appl. Pred.[a]
1	3.56	8.20	Y	7.44	15.13	Y	10.73	15.92	Y
2	12.71	5.62	Y	30.64	10.38	Y	18.09	10.70	Y
3	7.18	4.56	Y	12.51	8.41	Y	18.94	8.57	Y
4	7.62	4.05	Y	11.71	7.47	Y	11.45	7.56	Y
5	4.07	4.00	Y	9.16	7.38	Y	14.67	7.47	Y
6	3.51	3.40	Y	7.89	6.28	Y	12.03	6.30	Y
7	1.36	2.46	*N*	2.67	4.54	Y	3.18	4.47	Y
8	1.80	2.22	*N*	3.40	4.09	Y	3.06	4.01	Y
9	7.98	2.10	Y	18.40	3.88	Y	18.52	3.78	Y
10	2.00	2.01	Y	3.80	3.71	Y	2.73	3.61	Y
11	1.04	1.63	Y	2.00	3.01	Y	2.52	2.90	Y
12	0.91	1.62	Y	1.11	2.99	*N*	0.73	2.88	*N*
13	1.07	1.35	Y	2.13	2.50	Y	2.06	2.38	Y
14	1.00	1.23	Y	2.04	2.27	Y	1.97	2.15	Y
15	0.20	1.11	Y	0.27	2.05	*N*	0.55	1.93	Y

16	3.40	0.81	*N*	7.31	1.50	*N*	2.91	1.39	*N*
17	0.80	0.81	Y	1.40	1.49	Y	0.82	1.38	Y
18	0.80	0.53	Y	1.33	0.98	Y	0.64	0.88	Y
19	0.93	0.48	Y	1.67	0.88	Y	0.73	0.79	Y
20	0.33	0.45	Y	0.47	0.83	Y	0.27	0.75	Y
21	0.36	0.44	Y	0.47	0.81	Y	0.30	0.73	Y
22	0.07	0.37	Y	0.09	0.68	Y	0.18	0.60	Y
23	0.40	0.35	Y	0.53	0.64	Y	0.18	0.57	Y
24	0.40	0.25	Y	0.73	0.47	Y	1.27	0.41	Y
25	0.18	0.17	Y	0.40	0.31	Y	0.64	0.26	Y
26	0.38	0.15	Y	0.67	0.28	Y	1.00	0.23	Y
27	0.07	0.10	Y	0.27	0.18	Y	0.27	0.15	Y
28	0.00	0.07	Y	0.00	0.12	Y	0.00	0.10	Y
29	0.07	0.06	Y	0.13	0.12	Y	0.00	0.10	Y
30	0.00	0.05	Y	0.00	0.09	Y	0.00	0.07	Y
% Correct applicability predictions			90			90			93

Source: Reproduced from Kadar, E.P. et al., *J. Chromatogr. B*, 863, 1, 2008.

[a] Correctly predicted if a $k' \geq 2$ would be achieved by HILIC.

forces, and ionic forces.[37] The unmodified silica stationary phase utilized in this work can exhibit all three of these intermolecular forces to a degree. Dispersion forces, also referred to as "London dispersion forces," are due to charge fluctuations from electron/nuclei vibrations. This type of intermolecular force is responsible for hydrophobic interactions. Unmodified silica columns contain siloxane bonds that are hydrophobic and may give rise to dispersive interactions with relatively hydrophobic analytes.[38] However, any dispersive interactions should be minimized by the high acetonitrile content of the generic HILIC system mobile phase.

Polar forces are due to dipole–dipole or dipole-induced dipole interactions.[37] Polar interaction between an analyte and the unmodified stationary phase should be very minimal, considering the shielding effect of the partly immobilized water layer on the stationary phase. Conversely, polar interactions, such as hydrogen bonding, occur between the analyte and the water layer. This type of interaction is incorporated into the log $D_{\text{pH 3.0}}$ value of a compound and is therefore integrated into our hypothesis.

Ionic forces are interactions between molecules having opposite net charge.[37] Possible interaction between positively charged analytes and negatively charged silanols can occur in our generic HILIC system.[20,38] This undesired interaction gives rise to enhanced retention relative to partitioning alone. Additionally, this interaction is responsible for peak tailing. The majority of the probe compounds used for this work were protonated basic amines at the mobile-phase pH of 3.0. Refer to Table 2.1 for a tabulation of the physicochemical properties of the probe compounds, including pK_a. Previous work performed utilizing compound No. 1 demonstrated the influence of mobile-phase ammonium formate concentration on retention as well as peak shape.[36] Refer to Figure 2.3 for a chromatographic illustration of the influence of ammonium formate concentration retention time and peak shape.

Figure 2.4 presents a graphical representation of the dependence of the tailing factor calculated as

$$T = \frac{A+B}{2A} \tag{2.9}$$

where

A is the width from the leading edge of the analyte peak to the peak maximum at 5% of peak height
B is the width from the peak maximum to the tailing edge of the analyte peak at 5% of peak height[24]

The tailing factor decreased from a mean value of 1.95 to 0.939 as the total ammonium formate concentration of the mobile phase was increased from 1 to 20 mM. It is apparent that a total ammonium formate mobile-phase concentration of at least 10 mM is required to achieve adequate peak symmetry. It was concluded that the ammonium formate was responsible for the alleviation of electrostatic interactions between the protonated drug, compound No.1, and the ionized silanol groups of the unmodified silica stationary phase.[36] Another experiment employing compound No. 1, in which the percentage of acetonitrile was varied from 75% to 95% (v/v) while maintaining a total ammonium formate concentration of 10 mM, was performed in

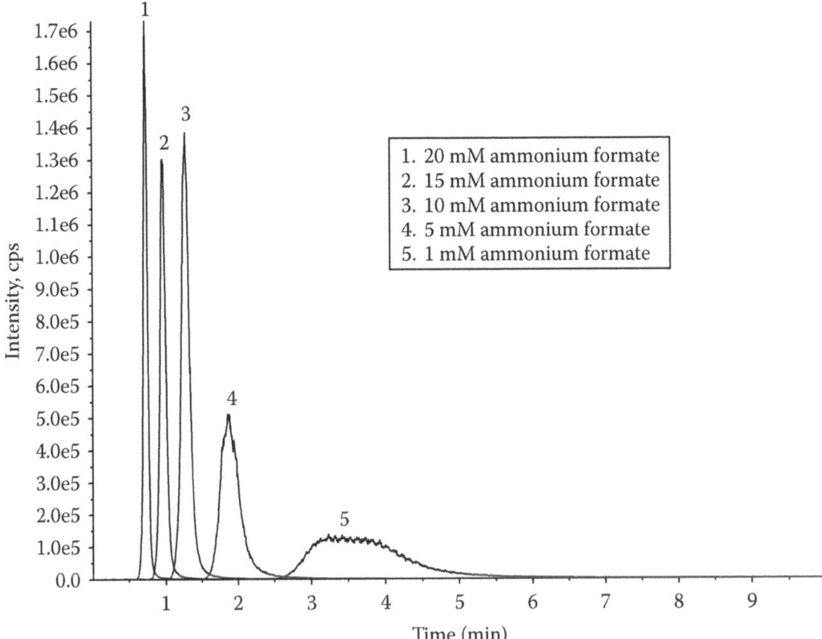

FIGURE 2.3 Influence of buffer concentration on retention time and peak shape. Chromatography was performed using an Atlantis™ HILIC silica column (5 µm) at a flow rate of 1.0 mL/min. Pump A delivered the required make-up volume of water, Pump B provided acetonitrile at a constant percent of 90% (v/v), and Pump C delivered the appropriate molarity ammonium formate, pH 3, stock buffer (i.e., 10, 50, or 200 mM) to produce a net mobile-phase concentration of 1, 5, 10, 15, and 20 mM. An ammonium formate concentration of 10 mM or greater produced symmetrical peaks. (Reproduced from Kadar, E.P. and Wujcik, C.E., *J. Chromatogr. B*, 877, 471, 2009.)

order to determine if electrostatic ionic interactions can be adequately remediated, allowing partitioning to act as the primary mechanism under an elevated ammonium formate concentration.[36] A plot of the logarithm of the mean experimentally determined k' versus the acetonitrile percentage is provided in Figure 2.5. A linear relationship between log k' and the percentage of acetonitrile in the mobile phase (correlation coefficient of 0.996) was observed. This linear relationship is suggestive of a partitioning mechanism.[20]

Based on these results, a total of 10 mM ammonium formate concentration was employed in our generic HILIC system. An examination of Table 2.2 reveals that the predicted k' of compound Nos. 2–5, 9, and 16 were significantly underestimated, which can possibly be attributed in part to secondary interactions since these compounds all contained ionized amines at the mobile-phase pH of 3.0. Consideration must be made that the scope of this work was to develop a generic HILIC system in which the applicability of HILIC ($k' \geq 2$) could readily be determined. It should also be mentioned that compounds that possess a negative charge may experience electrostatic repulsion with the ionized silanols of the unmodified silica stationary phase.[20]

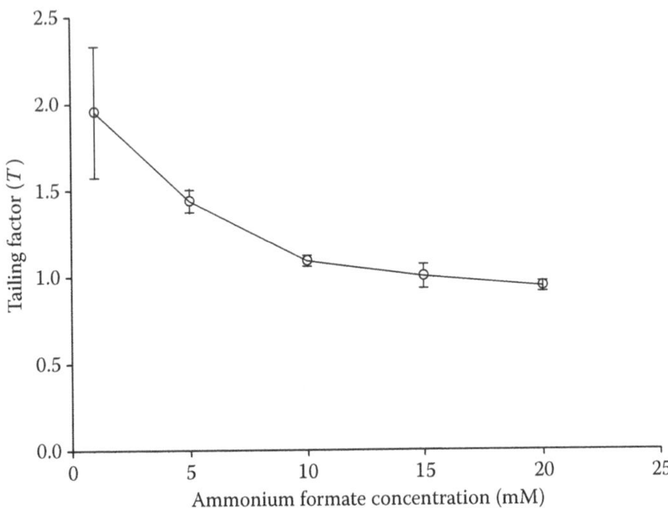

FIGURE 2.4 Influence of ammonium formate concentration on tailing factor (T). Error bars represent ±1 standard deviation (n = 3). Chromatography was performed using an Atlantis™ HILIC silica column (5 μm). Pump A delivered the required make-up volume of water, Pump B provided acetonitrile at a constant percent of 90% (v/v), and Pump C delivered the appropriate molarity ammonium formate, pH 3, stock buffer (i.e., 10, 50, or 200 mM) to produce a net mobile-phase concentration of 1, 5, 10, 15, and 20 mM. An ammonium formate concentration of 10 mM or greater produced symmetrical peaks. (Reproduced from Kadar, E.P. and Wujcik, C.E., *J. Chromatogr. B*, 877, 471, 2009.)

The second assumption is that the pH of the immobilized layer of water on the silica stationary phase is in fact 3.0. ACD/LogD Suite calculates log D values over the pH range of 0–14 in 0.1 unit increments. The log D value used for each compound in the above relationship is specific for pH 3.0. Rosés and Bosch published a review on the influence of mobile-phase acid-base equilibria on the chromatographic behavior of protolytic compounds.[39] Their article provides insight into the effect of an organic solvent on mobile-phase pH and concludes that an organic solvent, such as acetonitrile, in the mobile phase can change the pK_a of the buffer as well as that of the analyte. In their review, Rosés and Bosch present a thorough discussion of the relationship between the pK_a of a solute measured in pure water versus the pK_a of a solute in mobile phase containing an organic modifier.[39] They found the pK_a deviation to be compound specific, which can be explained by a preferential solvation of the compound by the components of the solvent mixture.[39] This finding is important with respect to our generic HILIC system, considering the relatively high acetonitrile concentration (85%–95% [v/v]) of the mobile phase. The following generalizations can be applied to the relationship between the experimentally determined k' and the calculated log $D_{\text{pH 3.0}}$ for the generic HILIC system: (1) acids generally demonstrate increasing pK_a with increasing organic modifier concentration, thereby increasing the pH of the buffer; (2) the formate, the buffer utilized in this work, would relate to this trend; and (3) the pK_a values of the protonated basic compounds, including those investigated in this chapter, decrease with the addition of an organic solvent

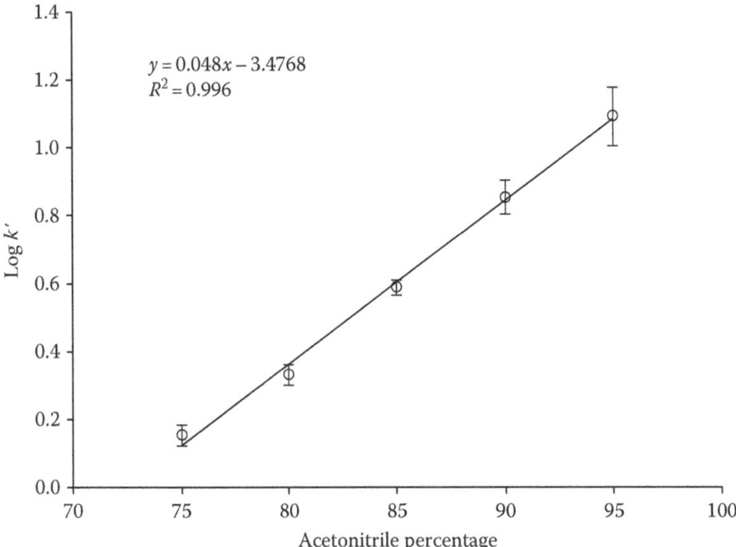

FIGURE 2.5 Influence of acetonitrile mobile-phase concentration on capacity factor, k'. Error bars represent ±1 standard deviation ($n = 3$). Chromatography was performed using an Atlantis™ HILIC silica column (5 µm). The mobile phase was prepared by combining water and acetonitrile with a 200 mM aqueous ammonium formate, pH 3.0, stock solution to produce five separate mobile-phase solutions with acetonitrile concentrations of 75%, 80%, 85%, 90%, and 95% (v/v) and a total ammonium formate concentration of 10 mM, with a flow rate of 1.0 mL/min. The linear relationship demonstrates that a partitioning mechanism is present. (Reproduced from Kadar, E.P. and Wujcik, C.E., *J. Chromatogr. B*, 877, 471, 2009.)

(minimum pK_a obtained at approximately 60% acetonitrile) and then increase to a pK_a value for a pure organic solvent higher than for water.[39] Currently, information with respect to pK_a values at the acetonitrile concentration used in generic HILIC systems are unavailable in the literature. However, it is assumed that the buffer capacity of the 10 mM ammonium formate employed in the mobile phase is adequate to provide a consistent overall pH, thereby providing a minimum impact on the utility of the relationship to rapidly predict the applicability of HILIC.

The third assumption is that ion-pairing does not make a significant contribution to the relationship between the experimentally determined k' and the calculated log $D_{\text{pH 3.0}}$ for our generic HILIC system. Formate has been demonstrated to serve as the counterion for protonated amines in some reversed-phase chromatography applications.[40,41] The total formate concentration of the mobile phase in our generic HILIC system was approximately 40 mM, considering the formic acid used to adjust the pH of the 200 mM ammonium formate stock solution to 3.0. The extent of ion-pair formation or salt precipitation is unknown for our generic HILIC system, especially given the high acetonitrile concentration of the mobile phase. However, it can be assumed to be related to the pK_a of a specific compound.[42] The log $D_{\text{pH 3.0}}$ calculations were made with the ion-pairing function of ACD/LogD Suite disabled because this function is designed to simulate physiological conditions in which the

TABLE 2.3
Chemical Analog Internal Standard Selection Evaluation Results

	85% Acetonitrile			90% Acetonitrile			95% Acetonitrile		
Compound No.	**Exptl. k'^a**	**Pred. k'^b**	**% Diff. k' Ratio[c]**	**Exptl. k'^a**	**Pred. k'^b**	**% Diff. k' Ratio[c]**	**Exptl, k'^a**	**Pred. k'^b**	**% Diff. k' Ratio[c]**
3	7.18	4.56		12.51	8.41		18.94	8.57	
4	7.62	4.05		11.71	7.47		11.45	7.56	
k' ratio	0.94	1.13	−19.5	1.07	1.13	−5.4	1.65	1.13	31.5
5	4.07	4.00		9.16	7.38		14.67	7.47	
6	3.51	3.40		7.89	6.28		12.03	6.30	
k' ratio	1.16	1.18	−1.5	1.16	1.18	−1.2	1.22	1.19	2.8
8	1.80	2.22		3.40	4.09		3.06	4.01	
10	2.00	2.01		3.80	3.71		2.73	3.60	
k' ratio	0.90	1.10	−22.7	0.89	1.10	−23.2	1.12	1.11	0.6
13	1.07	1.35		2.13	2.50		2.06	2.38	
14	1.00	1.23		2.04	2.27		1.97	2.15	
k' ratio	1.07	1.10	−2.6	1.04	1.10	−5.5	1.05	1.11	−5.9
20	0.33	0.45		0.47	0.83		0.27	0.75	
21	0.36	0.44		0.47	0.81		0.30	0.73	
k' ratio	0.92	1.02	−11.6	1.00	1.02	−2.5	0.90	1.03	−14.2

Source: Reproduced from Kadar, E.P. et al., *J. Chromatogr. B*, 863, 1, 2008.

[a] Experimentally determined k' (mean of $n = 3$).

[b] Predicted k' utilizing respective Equations 2.6 through 2.8.

[c] Percent difference between ratios of experimentally determined and predicted k'.

ionic strength is 0.15 M.[15] The ionic strength of the mobile phase used in our generic HILIC system was considerably less (approximately 0.025 M).

The fourth assumption is that ACD/LogD Suite can provide accurate calculations of log $D_{pH\,3.0}$. This calculation incorporates calculated log P and aqueous pK_a values. The error assumed for the calculated log P and aqueous pK_a is ±0.3 and ±0.2 units, respectively.[15] It is possible that the calculated pK_a error is actually greater since it is calculated based on an aqueous environment and the generic HILIC system has a relatively high acetonitrile content. Consideration should also be given to the impact of the mobile-phase pH with respect to the high acetonitrile content that is discussed in the review article by Rosés and Bosch.[39] In our work, the calculated log $D_{pH\,3.0}$ has been demonstrated to be adequate by providing clear linear relationships with the experimentally determined k'.

From the assumptions made to formulate our hypothesis, it must be realized that successful HILIC applicability predictions utilizing our generic HILIC system rely on consistency and control of the solvent system to drive partitioning as the primary mechanism.

2.4.3 Identification of Internal Standards

Chemical analog internal standards are typically utilized in LC-MS/MS bioanalysis when stable label internal standards are unavailable. Matching the internal standard and analyte retention time is desirable in order to minimize matrix effects such as ionization suppression or enhancement. Typically, analyte/analog pairs are chosen based on structural similarity without consideration for physicochemical properties, such as log D. The selection of an analog with a similar structural character can be beneficial by providing similar ionization efficiency. The analyte/analog pairs evaluated in this work were selected based on both a similar structure and log D. Table 2.3 provides a tabulation of mean experimentally determined and predicted k' ratios for chemical analog pairs. Percent difference values between the experimentally determined and predicted k' ratios ranged from −1.5% to −22.7%, −1.2% to −23.2%, and −14.2% to 31.5% for acetonitrile mobile-phase concentrations of 85%, 90%, and 95%, respectively. These relatively minor differences demonstrate the utility of using log $D_{pH\,3.0}$ similarity as an internal standard selection criterion for HILIC.

2.5 CONCLUSIONS

A relationship between the experimentally determined k' of a compound and its calculated log $D_{pH\,3.0}$, utilizing a generic HILIC system, was established. The minimization of possible secondary interactions such as adsorption and ionic interaction was achieved by the careful selection of the stationary phase and buffer concentration. This work was in agreement with the HILIC partition mechanism theory of Alpert.[19] The above discussion of the assumptions made to formulate our hypothesis clearly indicates the complexity of a HILIC system. However, the goal of this work was to provide a tool to rapidly predict the applicability of a generic HILIC system by simply entering the chemical structure of interest into the ACD/LogD Suite. The accuracy rate of approximately 90% for the HILIC applicability predictions clearly demonstrates the utility of the above-defined relationship between the experimentally determined k' and calculated log $D_{\mathrm{pH}\ 3.0}$ under a generic HILIC system. Additionally, the identification of an appropriate chemical internal standard was found to be predicted by a log $D_{\mathrm{pH}\ 3.0}$ similarity. The practical application of the calculated physicochemical property, log D, as a predictive "in-silico" tool allows a reduction in the bioanalytical method development time without compromising quality and, hence, improves the cost and resource efficiency of the bioanalytical laboratory. This approach adds considerable value to the area of LC-MS/MS bioanalysis.

ACKNOWLEDGMENT

The authors would like to acknowledge Douglas M. Fast, Covance Inc., Madison, WI, for his support and review of this chapter.

REFERENCES

1. Hsieh, Y., *J. Sep. Sci.*, **31**, 1481–1491, 2008.
2. Naidong, W., *J. Chromatogr. B.*, **796**, 209–224, 2003.
3. Xu, R.N., Fan, L., Rieser, M.J., El-Shourbagy, T.A., *J. Pharm. Biomed. Anal.*, **44**, 342–355, 2007.
4. Zhou, S., Song, Q., Tang, Y., Naidong, W., *Curr. Pharm. Anal.*, **1**, 3–14, 2005.
5. Nguyen, H.P., Schug, K.A., *J. Sep. Sci.*, **31**, 1465–1480, 2008.
6. Avery, M., *Rapid Commun. Mass Spectrom.*, **17**, 197–201, 2003.
7. Brown, S.D., White, C.A., Bartlett, M.G., *Rapid Commun. Mass Spectrom.*, **16**, 1871–1876, 2002.
8. Cappiello, A., Famiglini, G., Pierangela, P., Pierini, E., Termopoli, V., Trufelli, H., *Anal. Chem.*, **80**, 9343–9348, 2008.
9. Matuszewski, B.K., Constanzer, M.L., Chavez-Eng, C.M., *Anal. Chem.*, **75**, 3019–3030, 2003.
10. Sangster, T., Spence, M., Sinclair, P., Payne, R., Smith, C., *Rapid Commun. Mass Spectrom.*, **18**, 1361–1364, 2004.
11. Mannhold, R., *Molecular Drug Properties—Measurement and Prediction*, Wiley-VCH Verlag GmbH & Co. KGaA, Weinheim, Germany, 2008.
12. Scherrer, R.A., Howard, S.M., *J. Med. Chem.*, **20**, 53–58, 1977.
13. Lombardo, F., Shalaeva, M.Y., Tupper, K.A., Gao, F., *J. Med. Chem.*, **44**, 2490–2497, 2001.
14. ACD/LogD, ACD/Labs Release: 9.00, Product Version: 9.03, Advanced Chemistry Development, Inc., Toronto, Ontario, Canada, Copyright © 1997–2005.
15. ACD/LogD, *Suite Version 9.0 Reference Manual*, Advanced Chemistry Development, Inc., Toronto, Ontario, Canada, Copyright © 1994–2005.
16. Kadar, E.P., Wujcik, C.E., Wolford, D.P., Kavetskaia, O., *J. Chromatogr. B*, **863**, 1–8, 2008.
17. Quiming, N.S., Denola, N.L., Siato, Y., Jinno, K., *Anal. Bioanal. Chem.*, **388**, 1693–1706, 2007.
18. Quiming, N.S., Denola, N.L., Ueta, I., Siato, Y., Tatematsu, S., Jinno, K., *Anal. Chim. Acta*, **598**, 41–50, 2007.
19. Alpert, A.J., *J. Chromatogr.*, **499**, 177–196, 1990.
20. Hemström, P., Irgum, K., *J. Sep. Sci.*, **29**, 1784–1821, 2006.

21. Ikegami, T., Tomomatsu, K., Takubo, H., Horie, K., Tanaka, N., *J. Chromatogr. A*, **1184**, 474–503, 2008.
22. Hao, Z., Xiao, B., Weng, N., *J. Sep. Sci.*, **31**, 1449–1464, 2008.
23. Jandera, R., *J. Sep. Sci.*, **31**, 1421–1437, 2008.
24. Snyder, L.R., Kirkland, J.J., Glajch, J.L., *Practical HPLC Method Development*, 2nd edn., John Wiley & Sons, Inc., New York, 1997, pp. 29, 237.
25. Center for Drug Evaluation and Research, U.S. Food and Drug Administration, *Reviewer Guidance, Validation of Chromatographic Methods*, FDA, Rockville, MD, November 1994.
26. Liu, X., Tanaka, H., Yamauchi, A., Testa, B., Chuman, H., *Helv. Chim. Acta*, **87**, 2866–2876, 2004.
27. Liu, X., Tanaka, H., Yamauchi, A., Testa, B., Chuman, H., *J. Chromatogr. A*, **1091**, 51–59, 2005.
28. Vailaya, A., *J. Liq. Chromatogr. Relat. Technol.*, **28**, 965–1054, 2005.
29. Kaliszan, R., *Quant. Struct.-Act. Relat.*, **9**, 83–87, 1990.
30. Piraprez, G., Herent, M., Collin, S., *Flavour Fragrance J.*, **13**, 400–408, 1998.
31. Lipinski, C., Lombardo, F., Dominy, B.W., Feeney, P.J., *Adv. Drug. Deliv. Rev.*, **23**, 3–25, 1997.
32. Lombardo, F., Shalaeva, M.Y., Tupper, K.A., Gao, F., Abraham, M.H., *J. Med. Chem.*, **43**, 2922–2928, 2000.
33. Stella, C., Galland, A., Liu, X. et al., *J. Sep. Sci.*, **28**, 2350–2362, 2005.
34. ACD/LogP, ACD/Labs Release 11.00, Product Version 11.01, Advanced Chemistry Development, Inc., Toronto, Ontario, Canada, Copyright © 1994–2007.
35. Nozawa, Y., Kawashima, A., Hashimoto, E.H., Kato, H., Harada, K., *J. Chromatogr. A*, **1216**, 3807–3811, 2009.
36. Kadar, E.P., Wujcik, C.E., *J. Chromatogr. B*, **877**, 471–476, 2009.
37. Cazes, J., Scott, R.P.W., *Chromatography Theory*, Marcel Dekker, Inc., New York, 2002, pp. 62–73.
38. Nawrocki, J., *J. Chromatogr. A*, **779**, 29–71, 1997.
39. Rosés, M., Bosch, E., *J. Chromatogr. A*, **982**, 1–30, 2002.
40. Bianchi, F., Careri, M., Corradini, C., Elviri, L., Mangia, A., Zagnoni, I., *J. Chromatogr. A*, **825**, 193–200, 2005.
41. Roberts, J.M., Diaz, A.R., Fortin, D.T., Friedle, J.M., Piper, S.D., *Anal. Chem.*, **74**, 4927–4932, 2002.
42. Stahl, P.H., Wermuth, C.G., *Handbook of Pharmaceutical Salts: Properties, Selection, and Use*, VHCA, Verlag Helvetica Chimica Acta, Zürich, Switzerland and Wiley-VCH, Weinheim, Federal Republic of Germany, 2002.

3 HILIC Behavior of Reversed-Phase/Ion-Exchange Mixed-Mode Stationary Phases and Their Applications

Xiaodong Liu and Christopher A. Pohl

CONTENTS

3.1 INTRODUCTION

Reversed-phase/ion-exchange mixed-mode chromatography has gained increasing attention in recent years. The main driver is that such techniques provide control of selectivity through changes in the mobile-phase ionic strength, the pH, the organic solvent content, and/or the salt additives. As a result, options are now available to achieve optimal separations when compared to other separation modes, namely, reversed-phase liquid chromatography (RPLC), normal-phase liquid chromatography (NPLC), hydrophilic interaction liquid chromatography (HILIC), and ion-exchange liquid chromatography (IEX-LC). The adjustable selectivity feature results from the fact that mixed-mode chromatography combines multiple retention modes in a single stationary phase, including RP/IEX, RP/HILIC, and IEX/HILIC combinations, each of which can be controlled independently.

Analytes in a mixture can be different in hydrophobicity, polarity, charge, and size. RPLC is the most commonly used separation mode, accounting for approximately 80%–90% of all HPLC analytical separations. A major limitation of RPLC is its lack of adequate retention for polar molecules. HILIC is widely used for the separation of highly polar substances including biologically active compounds, such as pharmaceutical drugs, neurotransmitters, nucleosides, nucleotides, amino acids, peptides, proteins, oligosaccharides, carbohydrates, etc. HILIC is an important separation mode complementary to RPLC. In addition, IEX-LC is often employed to separate charged molecules. Although HILIC and IEX-LC can retain polar molecules, their low hydrophobicity makes them unsuitable for separating hydrophobic molecules. RPLC, HILIC, and IEX-LC combined together provide complementary solutions to most HPLC separations. But very often, more than one separation mode is needed to achieve desirable results. By comparison, RP/IEX mixed-mode columns can operate in various separation modes and provide broader application ranges.

This chapter discusses RP/IEX mixed-mode stationary phases, including column chemistry, chromatographic features, and applications. The HILIC behavior of these stationary phases is also discussed in detail. Some applications of RP/IEX mixed-mode columns in the HILIC mode are provided. As HILIC is thoroughly discussed in other chapters, only a brief review of this subject is given in this chapter.

3.2 HYDROPHILIC INTERACTION LIQUID CHROMATOGRAPHY

3.2.1 INTRODUCTION

While RPLC is the most powerful separation technique for a wide range of molecules, a major limitation of RPLC is its lack of adequate retention of polar molecules. Normal-phase liquid chromatography (NPLC), often with silica as the stationary phase, is used to separate compounds according to their polar features, usually with nonaqueous mobile phases, such as hexane, ethyl acetate, chloroform, etc. Therefore, NPLC is a useful separation technique as it provides selectivity complementary to RPLC. However, it is difficult to dissolve hydrophilic compounds, such as peptides, in nonaqueous mobile phases, thus limiting its application. In 1975, Linden reported the work on carbohydrate separations using an unmodified silica stationary phase in the NPLC mode but in an aqueous-containing mobile phase.[1] In these separations, the mobile phase had a relatively high water content, which helps polar molecules dissolve into the mobile phase. This technique was later named as hydrophilic-interaction liquid chromatography (HILIC) by Alpert in his 1990 paper on the separation of hydrophilic substances, such as proteins, peptides, and nucleic acids.[2] Since then, HILIC has gained much attention because of the increasing need to analyze polar compounds in complicated mixtures, as well as the widespread use of LC-MS.[3,4]

3.2.2 Retention Mechanism

HILIC is a type of NPLC that uses mobile phases typically containing 5%–40% water. In HILIC, the retention increases with an increasing polarity of the stationary phase and the analyte, and with the decreasing polarity of the predominantly organic solvent in the mobile phase, a trend opposite to that of RPLC.

The retention mechanism of HILIC is postulated as an analyte partitioning between a water-rich layer of stagnant solvent sequestered on a hydrophilic stationary phase and a mobile phase rich in organic solvent, with the main components typically composed of 5%–40% water in acetonitrile. The more polar the analyte, the more it associates with this stagnant aqueous phase and the later it elutes—a typical NPLC behavior. This partitioning mechanism can satisfactorily explain the retention of very hydrophilic and uncharged analytes. The mobile phase comprises a high percentage of an organic solvent (typically acetonitrile), which is modified by a small percentage of the water/volatile buffer. The water-rich liquid layer is established within the stationary phase, thus partitioning solutes from the mobile phase into the hydrophilic layer. The primary mechanism of separation is partitioning based on hydrophilic interaction, and a secondary mechanism, which also influences selectivity, is the electrostatic interactions between charged analytes and charged stationary phases. Elution is enabled by increasing the mobile-phase polarity (i.e., the content of the water component). Several studies on the HILIC retention mechanism have been published.[2,3,5–7]

3.2.3 Column Chemistry

In principle, any stationary phase that has the polar functionality to form a water-rich layer on its surface could be used for HILIC applications. Suitable polar functionalities include silanols on unmodified silica, neutral hydrophilic moieties (e.g., hydroxyl, cyano, amide), as well as anionic (e.g., amino), cationic (e.g., carboxylate), zwitterionic (e.g., sulfoalkylbetaine and phosphorylcholine), and amphoteric moieties. A wide variety of column chemistries are available for HILIC separations, which have been summarized in a recent review.[4]

3.2.4 Applications

HILIC is a complementary technique to RPLC with several benefits over RPLC. First, it retains polar analytes that cannot be retained on an RP column. For electrospray LC/MS applications with very polar compounds, the organic solvent-rich mobile phase used in HILIC provides a 10- to 20-fold sensitivity improvement. Moreover, by eliminating the need for an evaporation and reconstitution of a sample dissolved in a nonaqueous solvent, the sample analysis throughput can be greatly increased.

HILIC is useful for an analysis of polar solutes as RPLC is for nonpolar solutes. Since 1990, HILIC has been applied to a wide variety of applications including carbohydrates,[8,9] peptides,[6,10–15] proteins,[16–19] pharmaceuticals,[20–25] as well as other highly polar molecules, such as urea,[26,27] biurea,[28] choline and butyrobetaine,[29] tromethamine,[30] ascorbic acid and related compounds,[31,32] folic acid and its metabolites,[33] nicotine and its metabolites,[34] saponins,[35] aminoglycoside antibiotics,[36] glucosinolates,[37] ionic liquids,[38] organophosphonate nerve agent metabolites,[39] etc.

3.3 REVERSED-PHASE/ION-EXCHANGE MIXED-MODE STATIONARY PHASES

3.3.1 Introduction

RPLC is most commonly used for analyzing molecules of intermediate or higher hydrophilicity, but it often fails to retain highly hydrophilic analytes, such as catecholamines, organic acids, and inorganic ions (e.g., Na^+ and Cl^-). Although ion-pairing chromatography improves the retention and selectivity of highly hydrophilic ionizable analytes, it requires a long equilibration time and a column dedicated to use with ion-pairing mobile phases. Additionally, the mobile phase is not compatible with MS. Ion-exchange (IEX) liquid chromatography is used to separate charged molecules, but it is not generally suitable for separating neutral molecules because IEX columns normally provide inadequate hydrophobic retention for neutral compounds. HILIC complements the aforementioned techniques and may help overcome retention and selectivity problems for polar acidic solutes. However, this approach has limitations when compounds with a wide range of physicochemical properties have to be analyzed within the same analysis. Other drawbacks include low solubility of analytes in a high organic solvent and greater organic solvent consumption.

Mixed-mode columns come in various forms depending on different combinations of interactive mechanisms and different geometrical and spatial arrangements of functional groups. The reversed-phase/ion-exchange (RP/IEX) mixed-mode is the most common and effective combination because of the orthogonality of the hydrophobic and ion-exchange retention mechanisms. Such stationary phase materials separate analytes by hydrophobicity and charge differences.

RP/IEX mixed-mode liquid chromatography combines both RPLC and IEX-LC. A typical RP/IEX mixed-mode stationary phase has both reversed-phase and ion-exchange properties, and facilitates the independent control of retention for both ionizable and neutral molecules. Therefore, many application challenges involving hydrophilic ionizable compounds that are difficult for an RP column can be easily resolved on a mixed-mode column. In method development, column selectivity is one of the most important factors determining the success of a separation. Therefore, it is highly desirable to have a column that has adjustable selectivity so that an optimal separation can be easily achieved. This contrasts with conventional RPLC columns (i.e., C18 or C8) where the extent of selectivity control is rather limited. Moreover, with such a mixed-mode column, it is possible to separate several different types of molecules in a single run.

3.3.2 Column Chemistry

RP/IEX mixed-mode chromatography has been known for more than 20 years and this concept has also been applied to solid-phase extraction-based sample preparation.[40] Its use as an HPLC column material has been widely reported.[41–62] Depending on the combination of functional groups, they can be grouped into RP/anion-exchange (AEX) and RP/cation-exchange (CEX) bimodal, as well as RP/AEX/CEX trimodal materials. According to the arrangement of functional groups in the packing materials, bimodal columns may be classified into four categories (Figure 3.1). Type I materials contain a mixed-mode bed by blending two different stationary phases, such as RP and IEX (either AEX or CEX materials) in a single column. Such columns are commercially available from Thermo-Fisher under the trade name Hypersil Duet.[63] Type II materials consist of bonded phases modified with a mixture of RP and IEX ligands. It is likely that the C8/Anion and C8/Cation mixed-mode columns offered by Grace belong to this category.[64] It should be noted that both Type I and Type II materials have the inherent drawback of undesirable selectivity drift due to the difference in hydrolytic stability between the RP ligand-bonded sites and the IEX ligand-bonded sites.[42]

Both Type III and Type IV mixed-mode materials are prepared by functionalizing the silica substrate with single silyl ligands containing both RP and IEX domains. Type III materials are modified with ion-exchange functionality embedded in the alkyl ligands. Type IV materials are bonded with ion-exchange functionality at the terminus of the alkyl ligand. As a result, Type III materials are generally more hydrophobic and can be viewed as IEX-modified RP packings. Type IV materials exhibit unique IEX property supplemented by RP characteristic so that it may be considered as RP-modified IEX materials. Both Type III and Type IV packings represent state-of-the-art mixed-mode column technology, and provide improved selectivity ruggedness because of the constant ratio between the RP and IEX bonded sites. Various Type III

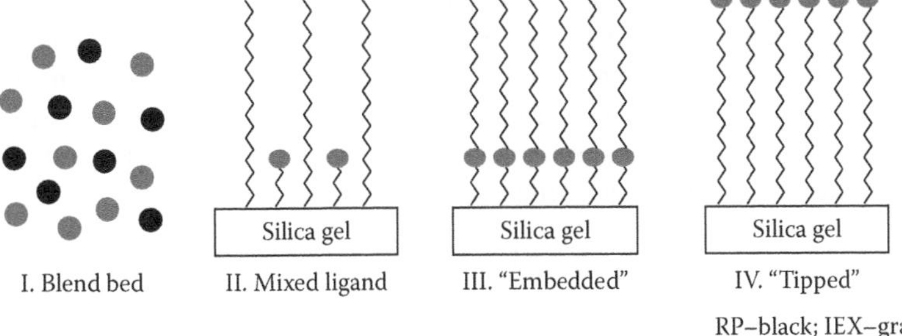

FIGURE 3.1 Four types of reversed-phase/ion-exchange bimodal mixed-mode columns classified by the arrangement of functional groups. (Courtesy of Marketing Communication Group at Dionex Corporation, Sunnyvale, CA.)

mixed-mode columns have been commercialized by SIELC under the trade name Primesep.[65] Recently, Type IV mixed-mode stationary phases have received considerable attention by both academia and industrial research organizations. Several RP/weak anion-exchange (WAX) materials consisting of a selector immobilized onto thiol-modified silica gel have been reported.[53,54,56–59] In these materials, the WAX site is located on the outer surface of the lipophilic layer and is linked to the hydrophilic silica support via a lipophilic spacer with polar-embedded amide and sulfide groups. Meanwhile, Dionex has introduced two Type IV mixed-mode columns—Acclaim® Mixed-Mode WAX-1[61,66] and Acclaim Mixed-Mode WCX-1.[62,67]

Because of the complexity and variety of HPLC analytes in terms of hydrophilicity and ionization, it is highly desirable to separate anionic, cationic, and neutral molecules within a single analysis. This necessitates a trimodal stationary phase that can provide cation-exchange, anion-exchange, and reversed-phase retention simultaneously. Packing materials with amphoteric or zwitterionic functionality have been developed. However, because anion-exchange and cation-exchange functionalities on these materials neutralize one another due to their proximity if both functional groups are ionized, they may retain and separate both anions and cations via salt exchange rather than anion-exchange and cation-exchange processes. Consequently, anion-exchange and cation-exchange retentions cannot be independently controlled. Recently, a trimodal column based on Nanopolymer Silica Hybrid (NSH) technology was commercialized by Dionex under the trade name of Acclaim Trinity™ P1.[68] This material consists of high-purity porous spherical silica gel coated with charged nanopolymer particles. The inner-pore area of the silica gel is modified with an organic layer that provides both reversed-phase and anion-exchange properties. The outer surface is modified with cation-exchange functionality (Figure 3.2). This chemistry ensures the distinctive spatial separation of the anion-exchange and cation-exchange regions, which allows both retention mechanisms to function simultaneously and be controlled independently.

A number of commercial silica-based[63–68] as well as polymer-based RP/IEX mixed-mode columns[69] are available from various suppliers and are listed in Table 3.1.

3.3.3 Separation Mechanism

The retention properties of IEX-based mixed-mode stationary phases are the result of the complex interactions of reversed-phase, ion-exchange, and ion-exclusion retention mechanisms. The relative contribution of each mechanism depends on the hydrophobicity and charge character of solutes and chromatographic conditions, such as mobile phase ionic strength, pH, and organic solvent composition. A simple empirical stoichiometric displacement model can be used for discussion.[70–72] According to this model, plots of log k versus log counterion concentration (C) are linear, and according to the following equation:[70]

$$\log k = \log Kz - Z \log C$$

where in

Kz is a constant related to the ion-exchange equilibrium constant and the ion-exchange capacity

Z is the ratio of the valencies of the solute ion (s) and counterion (c) ($Z = s/c$)

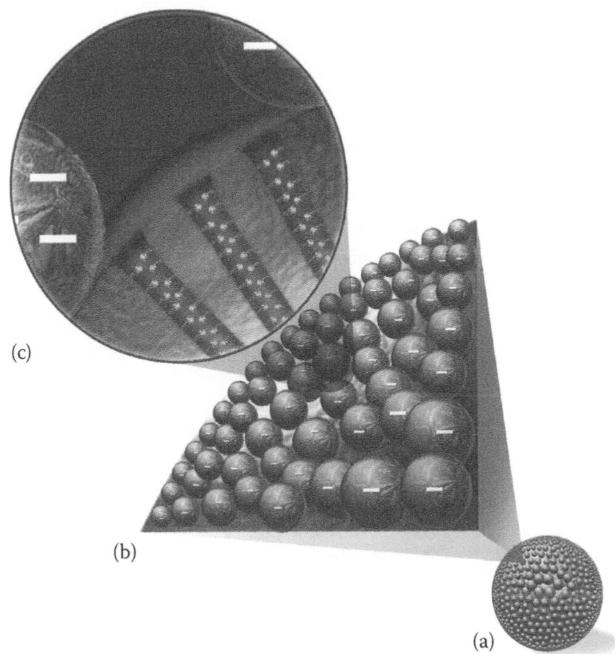

FIGURE 3.2 Column chemistry of Acclaim Trinity P1—the first reversed-phase/anion-exchange/cation-exchange trimodal mixed-mode column based on NSH technology: (a) overview of a silica particle coated with nanopolymer beads, (b) enlarged view of the silica surface coated with negatively charged nanopolymer beads, (c) inner-pore area consists of reversed-phase and weak anion-exchange functionalities and the outer surface provides strong cation-exchange interaction. (Courtesy of Marketing Communication Group at Dionex Corporation, Sunnyvale, CA.)

The empirical linear relationship between log k and log C is useful to assess the existence of an ion-exchange process. An examination of the slope Z gives the number of charges involved in the ion-exchange process, where, for a monovalent counterion, the slope is representative of the charge on the solute. In general, all mechanisms function independently, and can be modified as needed by adjusting the mobile phase ionic strength, the salt type, the pH, and the organic solvent content.[53,56–59]

3.3.4 Applications

RP/IEX mixed-mode liquid chromatography is not a new concept. However, the fact that multiple retention mechanisms provide a valuable alternative or complement to RP, IEX, and HILIC phases in a variety of HPLC applications has not been fully recognized until recently. These stationary phases are especially useful for retaining highly hydrophilic-charged analytes that are difficult for an RP column to retain. Furthermore, the increasing necessity of complementary analysis methods for validation of RPLC assay specificity required by regulatory authorities for pharmaceutical impurity profiling and stability-indicating assays has driven the need for the use of new stationary phases with complementary selectivity. Moreover, to analyze

TABLE 3.1
Commercial Reversed-Phase/Ion-Exchange Mixed-Mode Columns

Column Name	Manufacturer	Type of Column Chemistry	Substrate	References
Hypersil™ Duet C18/SAX	Thermo-Fisher	Type I, RP/SAX, bimodal mixed-mode column	Silica gel, 5 μm, 120 Å, 170 m²/g	[63]
Hypersil Duet C18/SCX	Thermo-Fisher	Type I, RP/SCX, bimodal mixed-mode column	Silica gel, 5 μm, 120 Å, 170 m²/g	[41,42,63]
Mixed-Mode C8/anion	Grace	Type II, RP/anion-exchange, bimodal mixed-mode column	Silica gel, 7 μm, 100 Å, 350 m²/g	[47,64]
Mixed-Mode C8/cation	Grace	Type II, RP/cation-exchange, bimodal mixed-mode column	Silica gel, 5 μm, 100 Å, 350 m²/g	[47,64]
Mixed-Mode C18/cation	Grace	Type II, RP/cation-exchange, bimodal mixed-mode column	Silica gel, 5 or 7 μm, 100 Å, 350 m²/g	[64]
Primesep™ 100	SIELC	Type III, RP/SCX, bimodal mixed-mode column	Silica gel, 5 or 10 μm, 100 Å	[65]
Primesep B	SIELC	Type III, RP/SAX, bimodal mixed-mode column	Silica gel, 5 or 10 μm, 100 Å	[65]
Acclaim® Mixed-Mode WAX-1	Dionex	Type IV, RP/WAX, bimodal mixed-mode column	Silica gel, 3 or 5 μm, 120 Å, 300 m²/g	[61,66]
Acclaim Mixed-Mode WCX-1	Dionex	Type IV, RP/WCX, bimodal mixed-mode column	Silica gel, 3 or 5 μm, 120 Å, 300 m²/g	[62,67]
Obelisc™ R	SIELC	Amphoteric, RP/anion-exchange/cation-exchange, mixed-mode column	Silica gel, 5 or 10 μm, 100 Å	[65]
Acclaim Trinity P1	Dionex	RP/WAX/SCX, trimodal mixed-mode column based on NSH technology	Silica gel, 3 μm, 300 Å, 100 m²/g	[68,80]
OmniPac™ PAX	Dionex	RP/SAX, mixed-mode column by coating the polymer beads with alkanol quaternary amine-functionalized latex	Ethylvinylbenzene/divinylbenzene polymer beads (55% cross-linked), 8.5 μm	[69]
OmniPac PCX	Dionex	RP/SCX, mixed-mode column by coating the polymer beads with sulfonic acid-functionalized latex	Ethylvinylbenzene/divinylbenzene polymer beads (55% cross-linked), 8.5 μm	[69]

complex samples successfully, multidimensional separation and analysis methods are often needed, which requires that the columns used in different multidimensional LC assays are orthogonal or complementary, and that the mobile phases in these dimensions are compatible. In this sense, RP/IEX mixed-mode columns may be a viable alternative or replacement to the conventional columns for established multidimensional LC methods.

Mixed-mode stationary phase materials separate analytes by both hydrophobicity and charge differences. The combination of orthogonal retention mechanisms of mixed-mode columns gives rise to a host of advantages, such as adequate retention of charged analytes, selectivity complementary to RP, IEX, and/or HILIC columns, and adjustable selectivity. As a result, greater flexibility for retention and selectivity tuning during method development and an expanded application range can be achieved.

Applications that have been reported in literature using RP/IEX mixed-mode columns include separations of biomolecules (e.g., proteins,[73–75] peptides,[52,53,76–78] oligonucleotides,[43,44] and RNAs[45,46]), metabolism studies (e.g., organophosphate[54] and catecholamines[79] in urine), ethanol consumption marker analysis in urine,[56] chromatographic profiling with mycotoxins,[59] separation of anthocyanin compounds from grapes,[48] as well as various pharmaceuticals applications.[41,42,49–51]

Compared to RP/AEX or RP/CEX bimodal columns, an RP/AEX/CEX trimodal column should provide a broader application range since anionic, cationic, and neutral analytes could be separated within a single analysis due to the combined effects of reversed-phase, anion-exchange, and cation-exchange retention mechanisms. This feature is important for the determination of active pharmaceutical ingredients (APIs) and counterions. Traditionally, APIs and counterions are analyzed in two separate assays using different LC columns and even different instruments. For example, Na^+ and Cl^- are most commonly used counterions in drug formulation and are unretained by any RP columns. It has been demonstrated that both commonly used pharmaceutical counterions and their respective APIs can be separated within a single analysis on a commercial RP/AEX/CEX trimodal column.[80]

3.4 HILIC BEHAVIOR OF REVERSED-PHASE/ION-EXCHANGE MIXED-MODE STATIONARY PHASES

RP/IEX mixed-mode stationary phases promise great versatility and capability for retaining and separating a variety of charged polar compounds in addition to more apolar ionic and nonionic analytes. The combination of an alkyl chain with embedded or terminus ion-exchange functionality generates hydrophobic, electrostatic, as well as hydrophilic interactive domains. Because of their unique column chemistries, these mixed-mode materials can be operated in a variety of chromatographic modes, including the RP mode (e.g., for neutral molecules), the IEX mode (for solutes bearing the opposite charge), the ion exclusion chromatography mode (for solutes having the same charge), and the HILIC mode (polar neutral, basic, amphoteric, and acidic compounds), depending on the chromatographic conditions and the characteristics of the analytes. While a number of studies on retention behavior and applications in RP and IEX modes have been reported, investigations of RP/IEX mixed-mode columns in the HILIC mode have been scarce. Recently, retention properties of mixed-mode columns under HILIC conditions (mobile phases with high acetonitrile content) have been the subject of research of both academia and industries.[53,56,58,59,81]

3.4.1 Retention Behavior of Xanthins, Nucleosides, and Water-Soluble Vitamins on RP/WAX Mixed-Mode Columns in HILIC Mode

Lämmerhofer et al. reported a comparative HILIC evaluation[58] of several in-house developed RP/WAX mixed-mode phases along with commercial RP/WAX mixed-mode columns, as well as a reference HILIC column (TSK-Gel Amide-80) having a neutral surface. Column chemistries of selected stationary phases in this study are illustrated in Figure 3.3. Three different commonly used HILIC test mixtures were used in the study, including xanthins (caffeine, theobromine, and theophylline), nucleosides (adenosine, cytidine, guanosine, thymidine, and uridine), and water-soluble vitamins (ascorbic acid, nicotinic acid, pyridoxine, riboflavin, and thiamine). HILIC mode separation was demonstrated for xanthins because caffeine (trimethyl-xanthine) eluted before the more polar 1,3-dimethyl xanthins (theobromine and theophylline) on all of the stationary phases tested. However, while theobromine eluted before theophylline on the TSK-Gel Amide-80 phase, the elution order of these two analytes was reversed on the basic RP/WAX phases, presumably because theophylline (pK_a, theophylline = 8.6) is more acidic than theobromine (pK_a, theobromine = 9.9) so that electrostatic interaction was superimposed upon HILIC interaction.

For nucleosides, none of the columns followed the elution order predicted based on log D values that should be in the order of adenosine < thymidine < uridine < guanosine < cytidine. This deviation, including the elution order and the elution profiles cannot be explained by a partition mechanism alone, thus specific interactions between analytes and ion-exchange functional groups of the stationary phase are most likely to be involved in the retention.

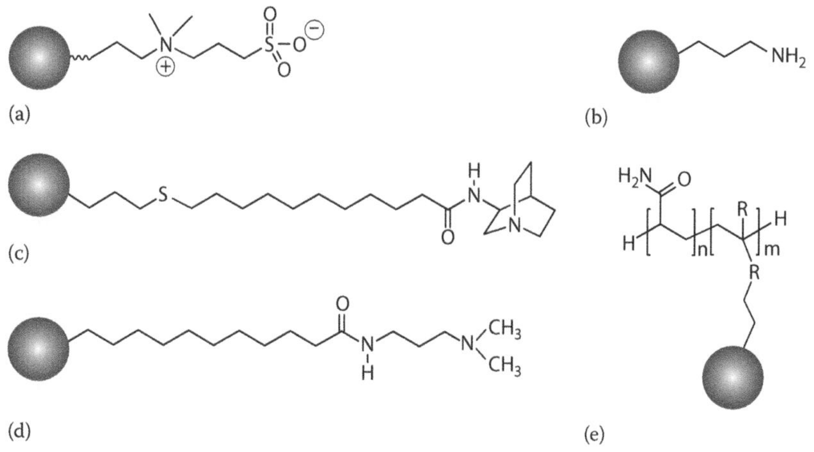

(a) (b) (c) (d) (e)

FIGURE 3.3 Column chemistries of stationary phases in the study of Ref. [58]. (a) ZIC-HILIC (Merck), (b) Luna NH_2 (Phenomenex), (c) RP/WAX-AQ, (d) Acclaim Mixed-Mode WAX-1 (Dionex), and (e) TSK-Gel Amide-80 (Tosoh).

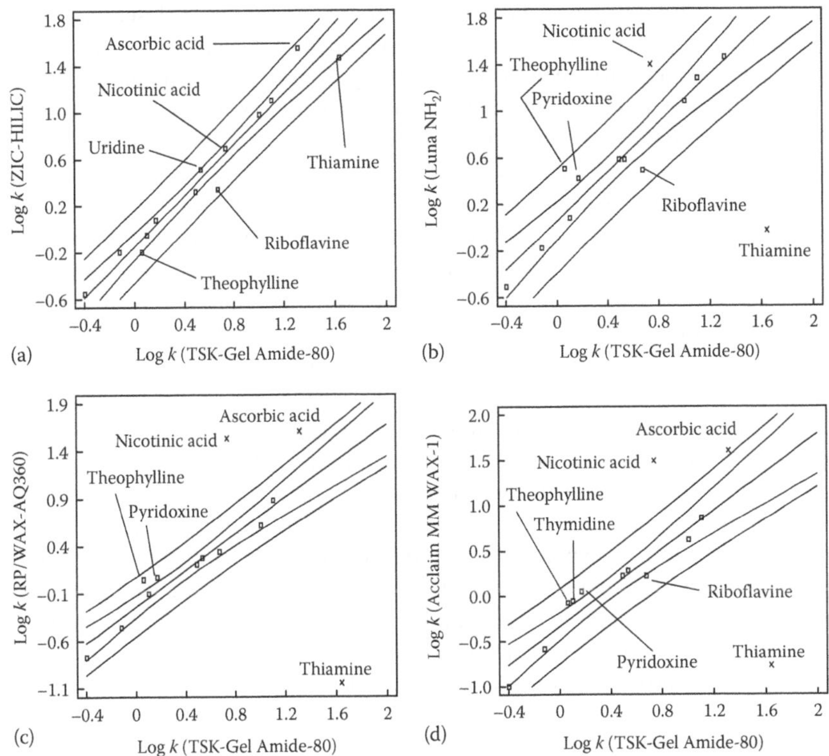

FIGURE 3.4 Orthogonality plots of log k values from HILIC tests of TSK-Gel Amide-80 versus (a) ZIC-HILIC, (b) Luna NH_2, (c) RP/WAX-AQ360, and (d) Acclaim Mixed-Mode WAX-1. Besides the fitted model, the plots show the confidence (inner lines) and prediction limits (outer lines). (Reproduced from Lämmerhofer, M. et al., *J. Sep. Sci.* 31, 2572, 2008. Copyright Wiley-VCH Verlag GmbH & Co. KGaA. With permission.)

The retention profile of the vitamins test mix was even more varied among different column chemistries because the ionic analytes in this test mix were expected to show strongly deviating retention profiles on stationary phases with cationic, anionic, zwitterionic, and neutral chromatographic bonding chemistries (Figure 3.4). Nevertheless, it appears that the retention profile differences observed are an indication of the importance of specific interaction forces between ion-exchange sites of the stationary phase and complementary solute functional groups. A partition mechanism alone fails to explain the altered retention profiles and shifts in elution orders with different stationary phases.

3.4.2 Retention Behavior of Fungal Metabolites on RP/WAX Mixed-Mode Columns in HILIC Mode

Apfelthaler et al. investigated the retention pattern profiling of fungal metabolites using the in-house developed Type IV RP/WAX mixed-mode phases.[59] In their study, retention factors k of a neutral solute (aflatoxinB1), a hydrophilic acid (moniliformin), a lipophilic acid (gibberellic acid), a base (lysergol), and an amphoteric compound (lysergic acid) were mapped in dependence of the organic modifier percentage (between 40% and 80% acetonitrile or methanol) using 3D plots (Figure 3.5). It was found that while nonionic aflatoxin B1 ($\log P = 0.45$) followed a typical RP retention mechanism, retention of the hydrophilic acid moniliformin ($\log D_{pH3} = -3.75$; $\log D_{pH7} = -4.22$; $pK_a = -0.21$) increased slightly with the mobile-phase organic solvent increase, suggesting a HILIC mechanism. In addition, a strong dependency of retention on mobile phase pH indicated the presence of anion-exchange stationary phase characteristics. For gibberellic acid ($\log D_{pH3} = -0.02$; $\log D_{pH7} = -2.81$; $pK_a = 4.13$), a slight U-shaped organic solvent dependency with a minimum of around 50%–60% was observed for acetonitrile-based mobile phases, indicating an anion-exchange process along with an RP mechanism below 50%–60% acetonitrile together with a HILIC mechanism at higher solvent percentages. Retention maps for the amphoteric lysergic acid ($\log D_{pH3} = -0.60$; $\log D_{pH7} = -0.37$; $pK_{a1} = 3.09$; $pK_{a2} = 9.18$) generally revealed a maximum retention at pH 7.5 for both acetonitrile and methanol, approximately 1 pH unit above the calculated isoelectric point of the solute in an aqueous medium. Loss of retention at a higher pH may be explained by a reduced ionization of the stationary phase ion-exchange groups, and at a lower pH by a diminished anionic character of the lysergic acid. Increasing retention with an increasing acetonitrile content at high mobile phase acetonitrile percentages indicated a HILIC-type retention mechanism. Retention profiles of the basic lysergol ($\log D_{pH3} = -1.34$; $\log D_{pH7} = 0.41$; $pK_{a1} = 8.33$; $pK_{a2} = 14.61$) can be largely explained by RP behavior, along with repulsive electrostatic interactions.

3.4.3 HILIC Behavior Comparison of RP/IEX Bimodal Mixed-Mode Columns

A recent study on the retention behavior of hydrophilic neutral (uracil), anionic (nitrate), and cationic (1,1-dimethylbiguanide) probes for both Type III (ion-exchange embedded alkyl chain) and Type IV (alkyl chain with an ion-exchange terminus) RP/IEX mixed-mode columns (column chemistries illustrated in Figure 3.6) showed that at 90% acetonitrile, the retention of uracil was in the order of Acclaim Mixed-Mode WAX-1 > Lichrosorb Diol > Acclaim Mixed-Mode WCX-1> Primesep B > Primesep 100.[81] It is clear that both Type III and Type IV mixed-mode columns possess HILIC properties, and the latter are generally more hydrophilic because of their more hydrophilic surface. Therefore, Type IV mixed-mode columns are better suited to HILIC applications.

For ionic or ionizable polar analytes, such as nitrate and 1,1-dimethylbiguanide, the retention behavior is more complicated. For an inorganic anion, nitrate, two RP/AEX mixed-mode columns (Mixed-Mode WAX-1 and Primesep B) show strong electrostatic attractions and retention decreases continuously with an increasing solvent content in the mobile phase, up to approximately 80%–90% acetonitrile. A further increase in the solvent content results in a retention increase on Mixed-Mode WAX-1 column—a typical HILIC behavior, but virtually no change for the Primesep B column (Figure 3.7). Under the same chromatographic

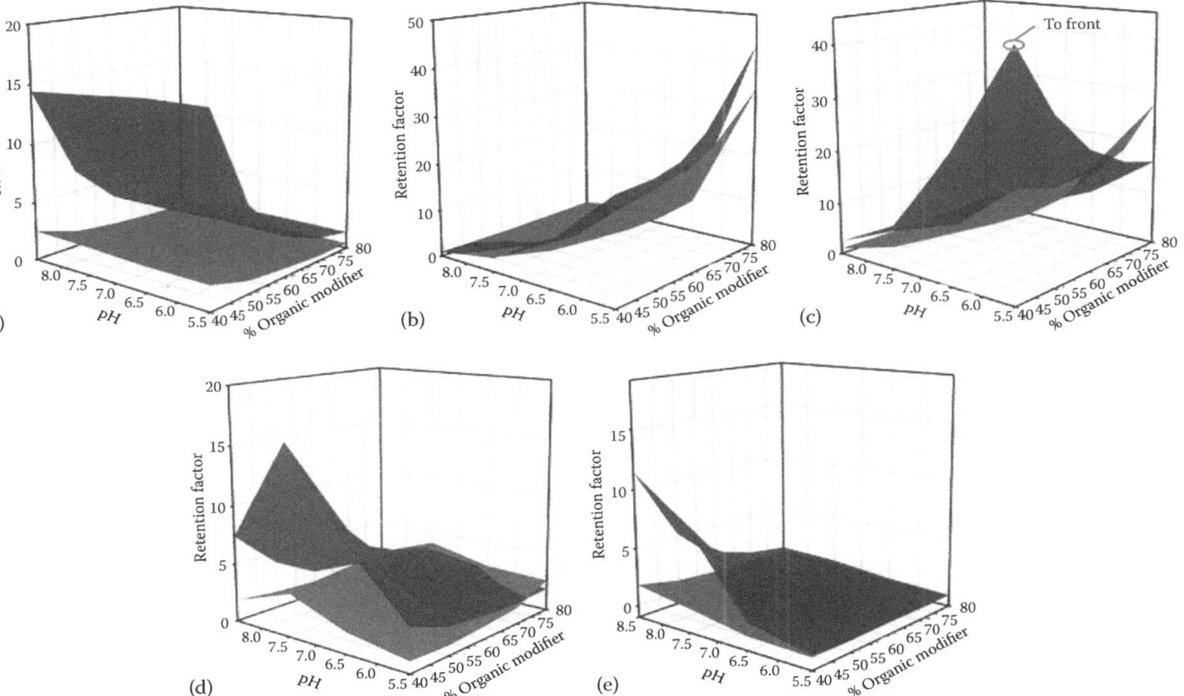

FIGURE 3.5 Retention pattern of (a) neutral (aflatoxin B1), (b) acidic (moniliformin), (c) acidic with hydrophobic properties (gibberellic acid; *note*: 40% MeOH spike pointing to front), (d) amphoteric (lysergic acid), and (e) basic (lysergol) analytes on AQ-RP/WAX as a function of organic modifier content and pH of the eluent under isocratic elution conditions (constant amount of 25 mmol L^{-1} HOAc). Darker shade (semitransparent): methanol, lighter shade: acetonitrile. (Reproduced from Apfelthaler, E. et al., *J. Chromatogr. A*, 1191, 171, 2008. Copyright Elsevier. With permission.)

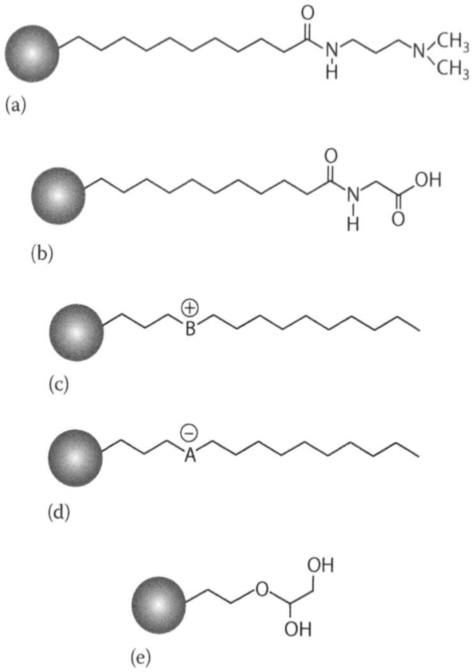

FIGURE 3.6 Column chemistries of stationary phases in the study of Ref. [81]. (a) Acclaim Mixed-Mode WAX-1 (Dionex), (b) Acclaim Mixed-Mode WCX-1 (Dionex), (c) Primesep B (SieLC), (d) Primesep 100 (SieLC), (e) Lichrosorb Diol (Phenomenex).

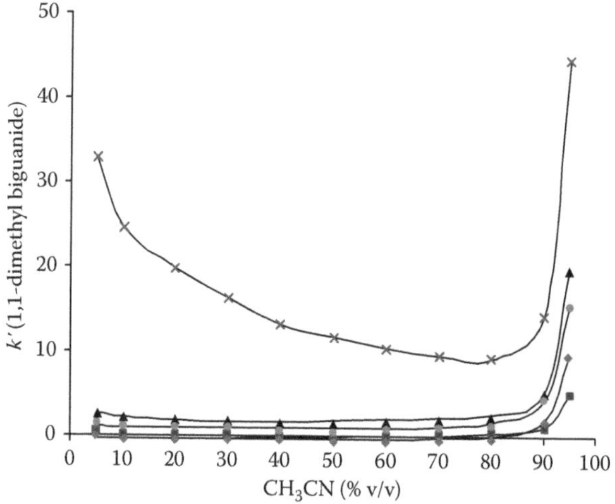

FIGURE 3.7 Comparison of the dependency of the retention factor of 1,1-dimethylbiguanide on the acetonitrile content in the mobile phase among Acclaim Mixed-Mode WAX-1 (square), Acclaim Mixed-Mode WCX-1 (triangle), Primesep B (diamond), Primesep 100 (cross), and Lichrosorb Diol (circle). (Courtesy of Marketing Communication Group at Dionex Corporation, Sunnyvale, CA.)

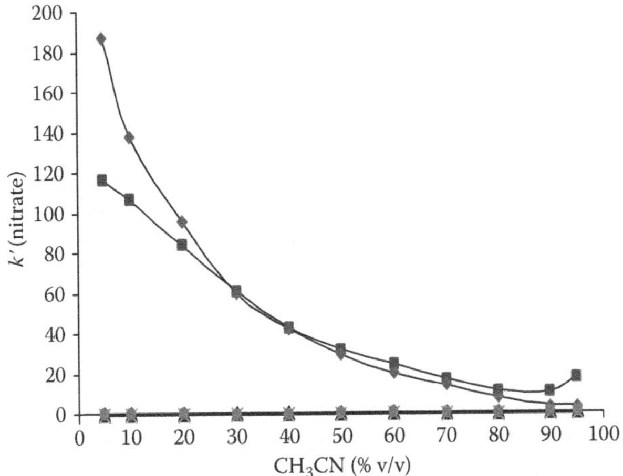

FIGURE 3.8 Comparison of the dependency of the retention factor of nitrate ion on the acetonitrile content in the mobile phase among Acclaim Mixed-Mode WAX-1 (square), Acclaim Mixed-Mode WCX-1 (triangle), Primesep B (diamond), Primesep 100 (cross), and Lichrosorb Diol (circle). (Courtesy of Marketing Communication Group at Dionex Corporation, Sunnyvale, CA.)

conditions, RP/CEX mixed-mode columns (Mixed-Mode WCX-1, Primesep 100) exhibit little retention, presumably due to the electrostatic repulsion between the stationary phase and the nitrate ion.

As shown in Figure 3.8, for the cationic probe, 1,1-dimethylbiguanide, Primesep 100 (Type III RP/SCX mixed-mode column) exhibited high retention throughout the whole solvent range indicating strong electrostatic attraction, which decreased with an increasing acetonitrile content to a minimum at 80% acetonitrile. Meanwhile, compared to Primesep 100, the Acclaim Mixed-Mode WCX-1, a Type IV RP/WCX packing material, showed weaker electrostatic retention in the 5%–80% solvent region (k' 1.5–2.6). Above 80% acetonitrile, the retention of 1,1-dimethylbiguanide sharply increased with a solvent content for both Primesep 100 and Acclaim Mixed-Mode WCX-1 due to a combination of HILIC and electrostatic interaction mechanisms. Under high organic conditions (acetonitrile >80%), retention of the basic probe increases with an increasing solvent content in the mobile phase, as the result of the combined effects of electrostatic attraction and HILIC retention. By comparison, Acclaim Mixed-Mode WAX-1 and Primesep B showed little or no retention from 5% to 80% solvent content due to electrostatic repulsion. However, above 80% solvent, retention increased with solvent content, a clear indication of HILIC interaction overcoming electrostatic repulsion.

The relative contributions of HILIC interaction and electrostatic interaction to the retention of a charged analyte on an RP/IEX mixed-mode stationary phase in the HILIC mode (in the high organic solvent mobile phase) are complex. On the weak cation-exchange surface, the ionization of the stationary phase's weak cation-exchange sites and the analyte, 1,1-dimethylbiguanide, weakens with the increase in the mobile-phase solvent content, resulting in a reduced electrostatic attraction between the stationary phase and the analyte, or lower retention. However, since the ionization of the mobile phase ions, ammonium and acetate, also decreases with an increasing solvent content, this effect weakens the mobile phase ionic strength and increases retention time. Conversely, for the weak anion-exchange surface, the ionization of weak anion-exchange sites and 1,1-dimethylbiguanide weakens with an increasing solvent content resulting in reduced electrostatic repulsion between the stationary phase and the analyte, or the retention increase. Meanwhile, the ionization of the mobile phase ions, ammonium and acetate, also decreases as solvent levels increase, reducing the suppression of electrostatic repulsion between the stationary phase and the basic analyte, thereby reducing retention. For an anionic analyte, the opposite observations can be made, but with the same reasoning.

3.4.4 HILIC Behavior of RP/WAX/SCX Trimodal Mixed-Mode Columns

For a better understanding of the contribution of hydrophilic interaction and electrostatic interaction of RP/IEX mixed-mode materials in HILIC conditions, methanol was used to replace acetonitrile. As methanol is a strong solvent for HILIC and minimizes the hydrophilic interaction between the stationary phase and the analyte, hydrophilic retention is considered negligible in a methanol/aqueous mobile phase. Thus, comparing the retention–solvent dependency for uracil, Na^+, and NO_3^- with that generated from acetonitrile will provide a better understanding of its retention mechanism in HILIC conditions. Liu and Pohl in their recent work investigated the retention behavior of a commercial trimodal mixed-mode column—Acclaim Trinity P1, using three model probes—uracil (neutral), sodium ion (cationic), and nitrate (anionic), in both acetonitrile/ammonium acetate and methanol/ammonium acetate buffer systems.[81] Because Acclaim Trinity P1 column simultaneously provides RP, AEX, and CEX functionalities on the same material, it is an ideal model to study the HILIC behavior of mixed-mode stationary phases for highly hydrophilic neutral, anionic, and cationic analytes. As shown in Figure 3.9, at higher solvent levels (>80%), acetonitrile exhibited significantly higher retention for uracil and sodium ion compared to methanol. For nitrate ion, a similar observation was noted at a 90% solvent level. Because no HILIC interaction is expected in a methanol/aqueous mobile phase, the observed retention increase with increasing acetonitrile in the mobile phase is a clear indication of a HILIC effect. At 90% acetonitrile, the presence of an ion-exchange process was also demonstrated, along with the HILIC mechanism, noting that retention decreased for charged analytes, such as sodium and nitrate, with increasing mobile-phase buffer concentrations.

3.5 APPLICATIONS OF REVERSED-PHASE/ION-EXCHANGE MIXED-MODE COLUMNS IN THE HILIC MODE

3.5.1 Pharmaceutical Active Ingredients and Counterions

Determinations of active pharmaceutical ingredients (APIs) and counterions are important assays in pharmaceutical drug development. Due to the wide variety of charges and hydrophobicities of these pharmaceutical-related molecules, it is highly challenging to analyze them via a simultaneous separation of the API and its

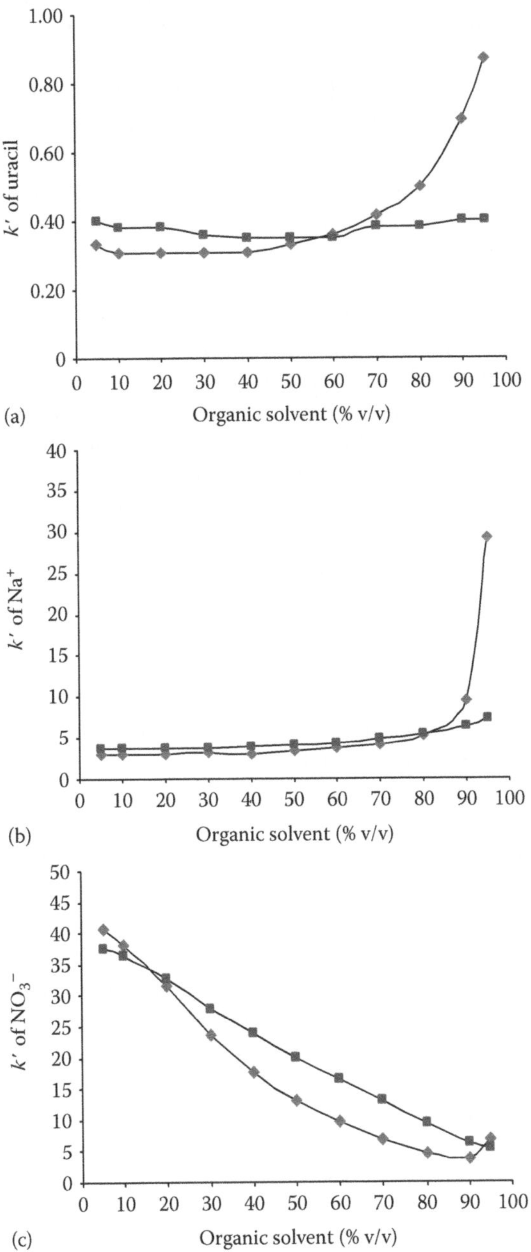

FIGURE 3.9 Comparison of the dependency of the retention factor of uracil (a), Na^+ ion (b), and NO_3^- ion (c) on the mobile phase solvent content [acetonitrile (diamond) versus methanol (square)] on the Acclaim Trinity P1—a RP/WAX/SCX trimodal column. (Courtesy of Marketing Communication Group at Dionex Corporation, Sunnyvale, CA.)

respective counterion. The most common pharmaceutical salt forms are sodium salts of acidic APIs and hydrochloride salts of basic APIs. Although RPLC is often used for analyzing drug molecules with intermediate or higher hydrophobicity, it often fails to retain highly hydrophilic analytes, such as catecholamines, organic acids, and inorganic ions (e.g., Na^+ and Cl^- ions). Ion-pairing chromatography improves the retention and selectivity of highly hydrophilic ionizable analytes, but it requires long equilibration times and a dedicated column. Furthermore, ion-pairing mobile phases are generally MS-incompatible. HILIC complements the aforementioned techniques and is suitable for analyzing highly polar analytes. However, it faces challenges, such as inadequate retentions for APIs with lower hydrophilicity and poor solubility for some highly polar analytes.

Penicillin G is an antibiotic medicine belonging to the beta-lactam family and is often formulated in the potassium salt form. Because of the highly hydrophilic nature of both the API and the counterion, it is impossible to separate them within the same analysis on any RP column. A commercial RP/WAX/SCX trimodal column (Acclaim Trinity P1) provides baseline separation of Penicillin G and K^+ ion in both the RP/IEX mode and the HILIC mode with an excellent peak shape and adequate retention (Figure 3.10).[81] The "U"-shaped retention time versus the solvent dependency of Penicillin G with a minimum at approximately 80% acetonitrile level indicates an anion-exchange process accompanied by an RP mechanism below 80% acetonitrile and by a HILIC mechanism at higher solvent levels. By comparison, the retention of K^+ increases with the mobile phase acetonitrile increase from 60% to 90% solvent, suggesting a cation-exchange mechanism superimposed with a HILIC mechanism. It should be noted that compared to the RP/IEX mode, an elution order reversal is observed in the HILIC mode separation. Therefore, depending on the specific application requirements, a suitable separation mode can be chosen for optimal results. In this case, 70% acetonitrile provided a fast separation with the smaller K^+ peak eluting before the larger Penicillin G peak. It is always desirable to avoid elution of small peaks after large peaks, which is prone to quantification inaccuracies due to the tailing of the larger peak.

1,1-Dimethylbiguanide hydrochloride (metformin), a highly hydrophilic basic drug formulated in the chloride salt form, is an antidiabetic agent that reduces blood glucose levels and improves insulin sensitivity. Figure 3.11 illustrates separations of the drug substance and its counterion using Acclaim Trinity P1 at four acetonitrile levels. Due to both the hydrophilic nature of the analytes and the multiple retention mechanisms facilitated by this stationary phase, separation in both the RP/IEX mode and HILIC mode provides baseline resolution, good peak shape, and adequate retention.[80]

3.5.2 Melamine and Cyanuric Acid

Melamine is an organic base and a trimer of cyanamide, with a 1,3,5-triazine skeleton. It is widely used in plastics, adhesives, countertops, dishware, fire retardant, etc. Cyanuric acid is a structural analogue of melamine, and may be found as an impurity of melamine. Cyanuric acid is an FDA-accepted component of feed-grade biuret, a ruminant feed additive. Melamine alone is known to form bladder stones in animal

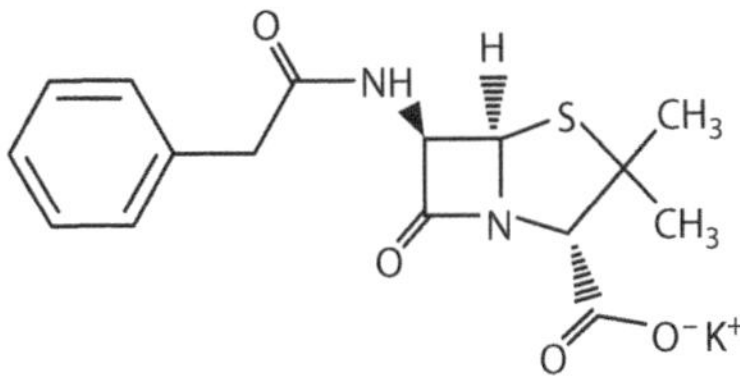

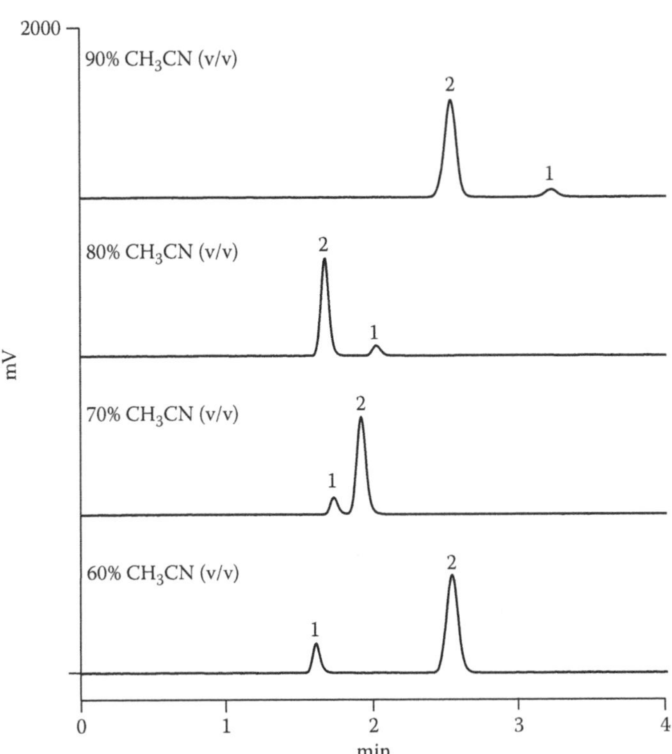

FIGURE 3.10 Analysis of Penicillin G potassium salt on a RP/WAX/SCX trimodal column. Column, Acclaim Trinity P1, 3.0 × 50-mm format; mobile phase, acetonitrile/ammonium acetate buffer, pH 5.2 (20 mmol L^{-1} total concentration) (90:10, 80:20, 70:30, and 60:40 v/v); flow rate, 0.6 mL min^{-1}; injection volume, 2.0 μL; temperature, 30°C; and detection, evaporative light scattering detector. Sample (0.5 mg mL^{-1}): (1) K^+, (2) Penicillin G. (Courtesy of Marketing Communication Group at Dionex Corporation, Sunnyvale, CA.)

tests, and when combined with cyanuric acid, it can form crystals that can give rise to kidney stones. Structures of melamine and cyanuric acid are given in Figure 3.12.

Recent investigations of adulterated pet foods and infant formula products demanded a reliable chromatographic method for monitoring the presence of melamine and cyanuric acid in food products and raw materials as well as in the tissues of animals suspected of ingesting melamine-tainted feeds. Both gas chromatography (GC) and liquid chromatography (LC) methods have been developed for the quantitative determination of melamine and cyanuric acid. The GC method requires

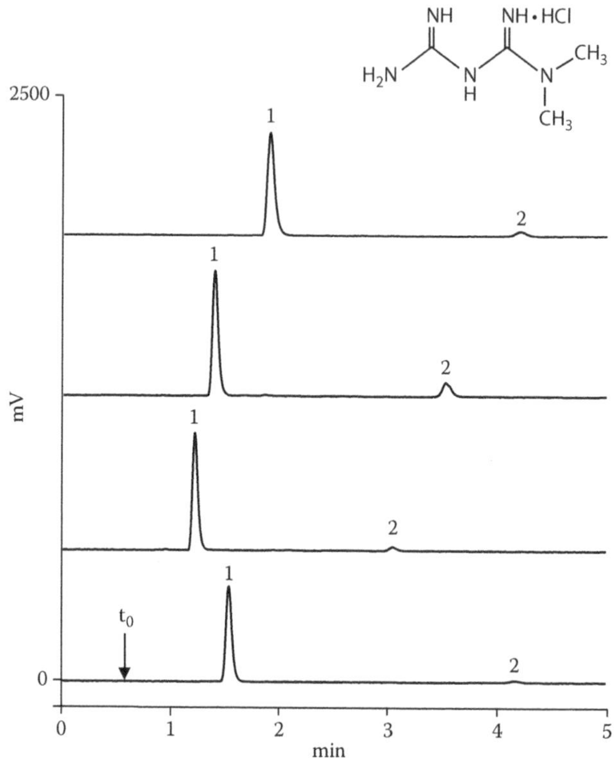

FIGURE 3.11 Analysis of 1,1-dimethylbiguanide hydrogen chloride on an RP/WAX/SCX trimodal column. Column, Acclaim Trinity P1, 3.0 × 50 mm format; mobile phase, acetonitrile/ammonium acetate buffer, pH 5.2 (20 mmol L^{-1} total concentration) (90:10, 70:30, 50:50, and 30:70 v/v); flow rate, 0.5 mL min^{-1}; injection volume, 2.5 μL; temperature, 30°C; and detection, ELS detector. Sample (0.2 mg mL^{-1}): (1) 1,1-DimethylbiguanideH^+, (2) Cl^-. (Courtesy of Marketing Communication Group at Dionex Corporation, Sunnyvale, CA.)

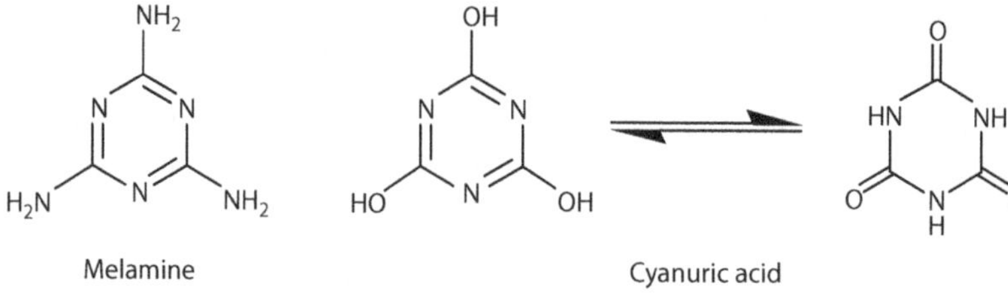

Melamine

Cyanuric acid

FIGURE 3.12 Structures of melamine and cyanuric acid.

derivatization and is thus labor intensive. LC methods, sometimes combined with mass spectrometer (MS), are more often used. RP and cation-exchange LC columns are commonly used for melamine analysis. However, both types of columns are unsuitable for cyanuric acid, due to either inadequate retention of the high hydrophilicity analyte or electrostatic repulsion between the analyte and the stationary phase. Recently, an LC-MS method was reported for the simultaneous separation of melamine and cyanuric acid on a commercial Type IV RP/WAX mixed-mode column (Acclaim Mixed-Mode WAX-1, Dionex).[82] Unlike RP or cation-exchange stationary phases on which cyanuric acid often elutes in or close to the void, this RP/WAX mixed-mode phase offers good retention with a k' greater than 3. It is expected that melamine cannot be adequately retained on an anion-exchanger with a positively charged surface due to electrostatic repulsion between the analyte and the stationary phase surface. However, when operated at a high acetonitrile level (90% v/v), the retention generated by the HILIC mechanism outstrips electrostatic repulsion, enhancing melamine retention for improved quantification accuracy. In addition, the RP/WAX mixed-mode column offers unique selectivity (different elution order) compared to its RP and cation-exchange counterparts, which may be adjusted by altering the mobile phase pH (Figure 3.13a) and the buffer concentration (ionic strength) (Figure 3.13b)—behavior typical of ion-exchangers.

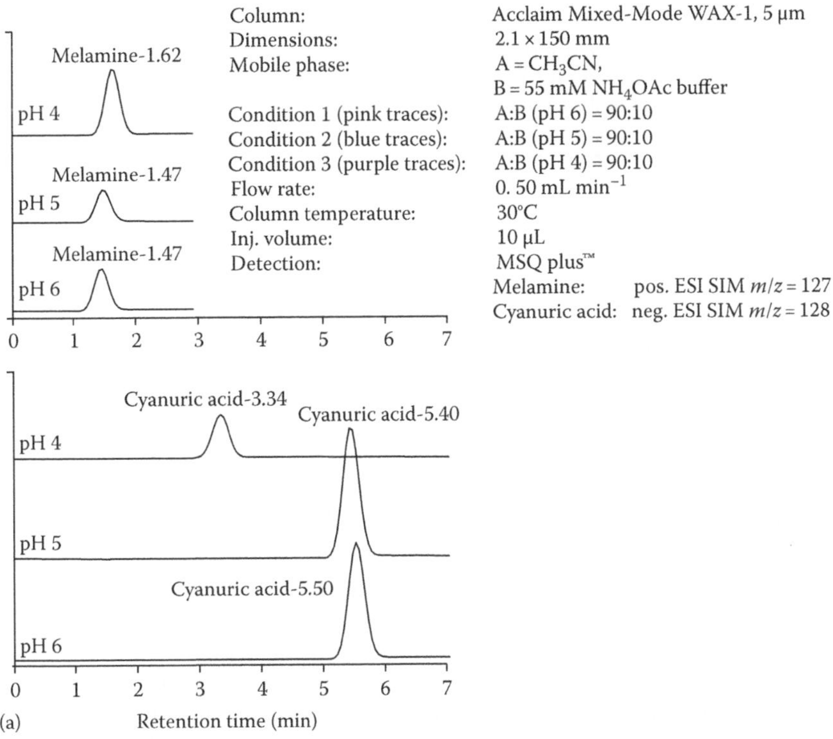

FIGURE 3.13 LC-MS analysis of melamine and cyanuric acid on Acclaim Mixed-Mode WAX-1—a Type IV RP/WAX mixed-mode column. Effects of mobile phase pH (a) and ionic strength (b). (Reproduced from Wang, L. et al., *Simultaneous Determination of Melamine and Cyanuric Acid Using LC-MS with the Acclaim Mixed-Mode WAX-1 Column and Mass Spectrometric Detection*, http://www.dionex.com/en-us/webdocs/62283_LPN%201991-01%20N_Melamine.pdf. With permission; Courtesy of Marketing Communication Group at Dionex Corporation, Sunnyvale, CA.)

(continued)

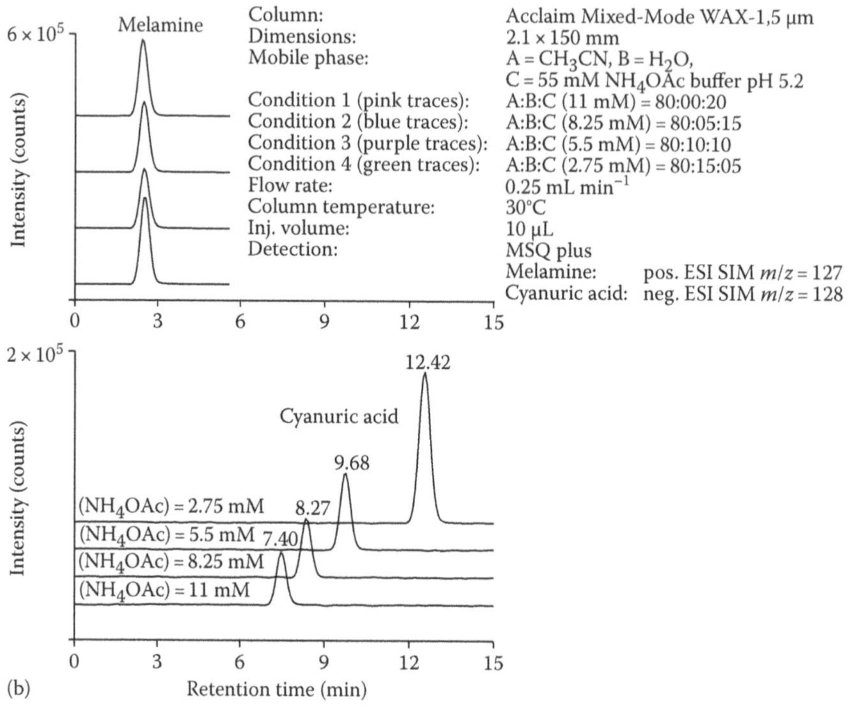

FIGURE 3.13 (continued)

3.5.3 Buffer Salts for Biopharmaceutical Applications

Good's buffers are a group of buffers first described in the research of Dr. Norman Good et al. in 1966.[83] These buffers were selected because they display characteristics that were valuable to research in biology and biochemistry. The characteristics associated with these buffers include a pK_a value between 6.0 and 8.0, high solubility, low toxicity, limited effect on biochemical reactions, very low absorbance between 240 and 700 nm, enzymatic and hydrolytic stability, minimal pH changes due to temperature and concentration, limited effects due to ionic strength or salt composition of the solution, limited interaction with mineral cations, and limited biological membrane permeability.[83]

Cell culture media is a complex blend containing amino acids, vitamins, carbohydrates, and salts with a biological buffer. All these components need to be assayed in order to ensure proper media formulation. Most of the buffers used are either inorganic buffers or Good's buffers. Tests for inorganic buffers by ICP or Ion Chromatography already exist. The conventional method for determination of Good's buffers is titration, which is inapplicable for a complex biological matrix. Due to the hydrophilic nature of Good's buffer salts, RPLC is not an appropriate approach for this separation. Figure 3.14 illustrates structures of seven commonly used Good's buffer salts, and Figure 3.15 demonstrates

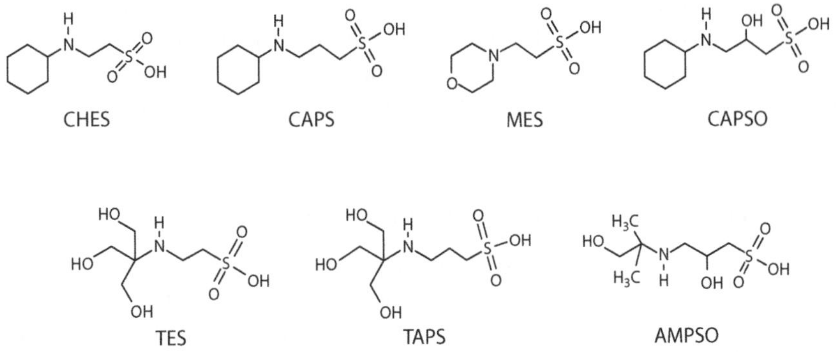

CHES CAPS MES CAPSO

TES TAPS AMPSO

FIGURE 3.14 Structures of selected Good's buffer salts. (1) 2-(Cyclohexylamino) ethanesulfonic acid (CHES), (2) 3-(cyclohexylamino)-1-propanesulfonic acid (CAPS), (3) 3-(cyclohexylamino)-2-hydroxy-1-propanesulfonic acid (CAPSO), (4) *N*-(1,1-dimethyl-2-hydroxyethyl)-3-amino-2-hydroxypropanesulfonic acid (AMPSO), (5) 2-(*N*-morpholino) ethanesulfonic acid (MES), (6) *N*-[*Tris*(hydroxymethyl)methyl]-2-aminoethanesulfonic acid (TES), and (7) *N*-[*Tris*(hydroxymethyl)methyl]-3-aminopropanesulfonic acid (TAPS).

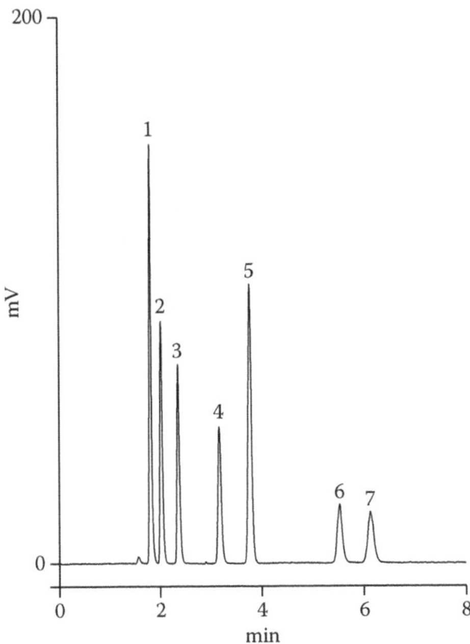

FIGURE 3.15 Separation of Good's buffer salts on the RP/WAX/SCX trimodal column. Column, Acclaim Trinity P1, 3.0 × 100 mm format; mobile phase, acetonitrile/ammonium acetate buffer, pH5.2 (12 mmol L^{-1} total concentration) (85:15 v/v); flow rate, 0.6 mL min^{-1}; injection volume, 2.5 μL; temperature, 30°C; and detection, ELS detector. Sample (0.15 mg mL^{-1}): (1) CHES, (2) CAPS, (3) CAPSO, (4) AMPSO, (5) MES, (6) TES, and (7) TAPS. (Courtesy of Marketing Communication Group at Dionex Corporation, Sunnyvale, CA.)

the separation of these salts on a commercial RP/WAX/SCX trimodal column (Acclaim Trinity P1 in 3 × 50 mm format) under the HILIC mode (85% acetonitrile/15% ammonium acetate buffer, v/v). This separation can be further "fine-tuned" by adjusting the buffer concentration as well as the organic solvent level in the mobile phase.[81]

3.5.4 Determination of Ethanol Consumption Markers

Confirmation of ethanol intake is of great importance in forensic toxicology, psychiatry, and workplace monitoring. A number of different markers, from short term to long term, are available, each of which with a specific time window of detection. Among all, the ethanol phase II metabolites ethyl glucuronide (EtG) and ethyl sulfate (EtS) have attracted particular attention lately because both compounds are sensitive, specific, and stable intermediate-term markers of acute ethanol consumption in urine and other matrices. Bicker et al. developed an LC–MS/MS method using an in-house developed Type IV RP/WAX mixed-mode column for the simultaneous analysis of EtG and EtS as well as ethyl phosphate (EtP) (see structures in Figure 3.16),[56] a potential new marker with distinct nonspecific or nonenzymatic formation in human urine. A reliable quantification of these highly polar acidic metabolites in the polar urinary matrix poses a unique challenge, particularly considering the direct injection methodology chosen for diluted urine. While an optimized RP/WAX separation was developed in which the two singly charged analytes EtG and EtS eluted before the doubly charged EtP (which was eluted in the back-flush mode to speed up the analysis), as expected from the primary anion-exchange process, the elution order of EtG and EtS, in contrast, follows the order of increasing hydrophilicity. Moreover, the retention properties of EtG and EtS as a function of acetonitrile content and the pH of the mobile phase showed that as the acetonitrile fraction increases, the retention factors increase and the more hydrophilic EtG responds more strongly to the change in solvent content than the less hydrophilic EtS (Figure 3.17). These results strongly suggest that in an acetonitrile-rich mobile phase, a HILIC mechanism is superimposed upon the primary anion-exchange process.

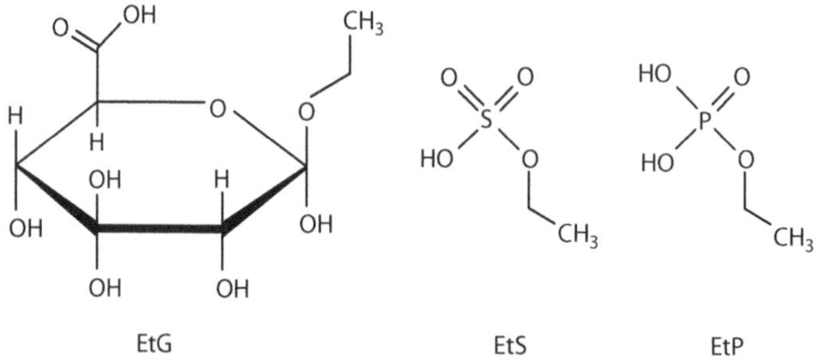

FIGURE 3.16 Structures of ethyl glucuronide (EtG), ethyl sulfate (EtS), and ethyl phosphate (EtP).

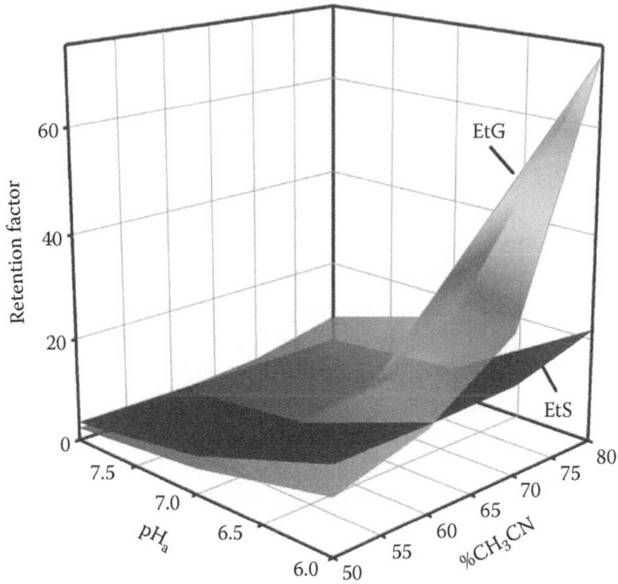

FIGURE 3.17 Dependency of the retention factor of EtG and EtS on the acetonitrile content and pH of the mobile phase using a Type IV RP/WAX stationary phase. (Reproduced from Bicker, W. et al., *Anal. Chem.*, 78, 5884, 2006. Copyright American Chemical Society. With permission)

3.6 CONCLUDING REMARKS

Reversed-phase/ion-exchange mixed-mode stationary phases promise great versatility and capability for retaining and separating a variety of charged polar compounds in addition to more apolar ionic and nonionic analytes. Compared to Type I and Type II materials, Type III and Type IV bimodal (reversed-phase/anion-exchange or reversed-phase/cation-exchange) stationary phases represent the state-of-the-art in mixed-mode column technology because of a better-defined chemistry and a more rugged selectivity of the resulting products. A reversed-phase/anion-exchange/cation-exchange trimode column based on NSH technology, a recent development in mixed-mode chromatography column technology, further extends the application range of mixed-mode phases to an even broader base.

The combination of a hydrophobic moiety with ion-exchange functionality generates hydrophobic, electrostatic, as well as hydrophilic interactive domains. Because of their unique column chemistries, these mixed-mode materials can be operated in a variety of chromatographic modes, including the RP mode (e.g., for neutral molecules), the IEX mode (for solutes bearing the opposite charge), the ion-exclusion chromatography mode (for solutes having the same charge), and the HILIC mode (polar neutral, basic, amphoteric, and acidic compounds), depending on the chromatographic conditions and the characteristics of the analytes. These stationary phases are especially useful for retaining highly hydrophilic charged analytes, which would otherwise be difficult to retain with an RP column. Furthermore, the increasing need for complementary analysis methods for validation of the RPLC assay specificity required by regulatory authorities in pharmaceutical impurity profiling and stability indicating assays has driven the need for new stationary phases with complementary selectivity. Moreover, to analyze complex samples successfully, multidimensional methods are often needed for greater method development flexibility. Consequently, facile adjustment of selectivity as a result of multi-mode retention mechanisms is highly beneficial. It might be interesting to note that for those who do not have access to MS/MS systems, retention behavior as a function of mobile phase changes can be used to help confirm peak identity from complex samples.

Due to the presence of polar ion-exchange functionality, modern reversed-phase/ion-exchange mixed-mode columns are capable of operating in high solvent (acetonitrile) conditions (in HILIC mode). Compared to their Type III counterparts, Type IV bimodal mixed-mode columns are better suited to HILIC applications. Although more studies are needed to thoroughly understand the retention mechanism of reversed-phase/ion-exchange mixed-mode columns in HILIC conditions, the existing evidence supports a retention mechanism consisting of an electrostatic interaction superimposed on a HILIC mechanism for highly hydrophilic ionic analytes and a purely HILIC retention for highly hydrophilic neutral molecules. Owing to the presence of a hydrophobic moiety, the polarity of the reversed-phase/ion-exchange surface is of intermediate reversed-phase interaction strength. However, substantial retention for charged analytes can be realized via ion-exchange interaction. As the result, the retention of charged analytes can be tuned by adjusting the mobile phase pH and ionic strength relative to neutral analytes whose retention is governed solely by the mobile phase solvent content. Such flexible selectivity can be achieved not only in reversed-phase conditions, but also in the HILIC mode. HILIC applications of mixed-mode columns offer high-sensitivity LC-MS capability with greater flexibility in method development.

Although not fully explored, it is evident that reversed-phase/ion-exchange mixed-mode columns have great potential for a variety of HPLC applications, under both reversed-phase and HILIC conditions. These stationary phases bridge the gaps between traditional reversed-phase, ion-pairing, ion-exchange, and HILIC chromatography. From the chromatographers' perspective, the unique chromatographic features of such materials stimulate innovative application method development. Besides the applications described in this chapter, many other applications will benefit from what reversed-phase/ion-exchange mixed-mode columns have to offer, including in-line SPE, multi-dimensional-LC, preparative-LC, etc. We expect that these interesting stationary phases will find their way to an increasingly broad range of applications in the coming years.

ACKNOWLEDGMENTS

The authors are thankful to Dr. Jeff Rohrer (Dionex Corporation) for the technical review and edits, Dr. Wolfgang Bicker and Dr. Michael Lämmerhofer (both from the University of Vienna) for providing Figures 3.4 and 3.5, respectively, and Scott Lefferts and Lynn Sison for preparing all the figures in this chapter.

REFERENCES

1. Linden, J. C. and Lawhead, C. L., *J. Chromatogr.*, 1975, *105*, 125.
2. Alpert, A. J., *J. Chromatogr.*, 1990, *499*, 177.
3. Hemström, P. and Irgum, K., *J. Sep. Sci.*, 2006, *29*, 1784.
4. Ikegami, T., Tomomatsu, K., Takubo, H., Horie, K. and Tanaka, N., *J. Chromatogr. A*, 2008, *1184*, 474.
5. Zhu, B. Y., Mant, C. T. and Hodges, R. S., *J. Chromatogr.*, 1991, *548*, 13.
6. Yoshida, T., *Anal. Chem.*, 1997, *69*, 3038.
7. Churms, S. C. In: *Journal of Chromatography Library*, El Rassi, Z., Ed.; Vol. 58. New York: Elsevier; 1995, Chap. 3.
8. Alpert, A. J., Shukla, M., Shukla, A. K., Zieske, L. R., Yuen, S. W., Ferguson, M. A. J., Mehlert, A., Pauly, M. and Orlando, R., *J. Chromatogr. A*, 1994, *676*, 191.
9. Churms, S. C., *J. Chromatogr. A*, 1996, *720*, 75.
10. Yoshida, T. *J. Chromatogr. A,* 1998, *808*, 105.
11. Yoshida, T. *J. Biochem. Biophys. Methods*, 2004, *60*, 265.
12. Zhang, J. and Wang, D., *J. Chromatogr. B*, 1998, *712*, 73.
13. Furuya, K., Schegg, K. M., Wang, H., King, D. S. and Schooley, D. A., *Proc. Natl Acad. Sci. USA*, 1995, *92*, 12323.
14. Boutin, J. A., Ernould, A.-P., Ferry, G., Genton, A. and Alpert, A. J., *J. Chromatogr.*, 1992, *583*, 137.
15. Faull, K. F., Feistner, G. J., Conklin, K., Roepstorff, P. and Andrews, P. C., *Neuropeptides*, 1998, *32*, 339.
16. Jenö, P., Scherer, P. E., Manning-Krieg, U. and Horst, M., *Anal. Biochem.*, 1993, *215*, 292.
17. Schmerr, M. J., Cutlip, R. C. and Jenny, A., *J. Chromatogr. A*, 1998, *802*, 135.
18. Schmerr, M. J. and Alpert, A. In: *Prions and Mad Cow Disease*, Nunnally, B. K. and Krull, I. S., Eds.; Marcel Dekker: New York, 2004, pp. 359–377.
19. Carroll, J., Fearnley, I. M. and Walker, J. E., *Proc. Natl Acad. Sci. USA*, 2006, *103*, 16170.
20. Strege, M. A., *Anal. Chem.*, 1998, *70*, 2439.
21. Strege, M. A., *Am. Pharm. Rev.*, 1999, *2*(3), 53.
22. Olsen, B. A., *J. Chromatogr. A*, 2001, *913*, 113.
23. Wang, X., Li, W. and Rasmussen, H. T., *J. Chromatogr. A*, 2005, *1083*, 58.
24. Ali, M. S., Ghori, M., Rafiuddin, S. and Khatri, A. R., *J. Pharm. Biomed. Anal.*, 2007, *43*, 158.
25. Hmelnickis, J., Pugovičs, O., Kažoka, H., Viksna, A., Susinskis, I. and Kokums, K., *J. Pharm. Biomed. Anal.*, 2008, *48*, 649.
26. Dallet, Ph., Labat, L., Kummer, E. and Dubost, J. P., *J. Chromatogr. B*, 2000, *742*, 447.
27. Doi, T., Kajimura, K., Takatori, S., Fukui, N., Taguchi, S. and Iwagami, S., *J. Chromatogr. B*, 2009, *877*, 1005.
28. Mulder, P. P. J., Beumer, B. and Van Rhijn, J. A., *Anal. Chim. Acta.*, 2007, *586*, 366.
29. Wang, Y., Wang, T., Shi, X., Wan, D., Zhang, P., He, X., Gao, P., Yang, S., Gu, J. and Xu, G., *J. Pharm. Biomed. Anal.*, 2008, *47*, 870.
30. Guo, Y. and Huang, A., *J. Pharm. Biomed. Anal.*, 2003, *31*, 1191.
31. Tai, A. and Gohda, E., *J. Chromatogr. B*, 2007, *853*, 214.
32. Nováková, L., Solich, P. and Solichová, D., *Trends Analyt. Chem.*, 2008, *27*, 942.
33. Garbis, S. D., Melse-Boonstra, A., West, C. E. and van Breemen, R. B., *Anal. Chem.*, 2001, *73*, 5358.
34. Moyer, T. P., Charlson, J. R., Enger, R. J., Dale, L. C., Ebbert, J. O., Schroeder, D. R. and Hurt, R. D., *Clin. Chem.*, 2002, *48*, 1460.
35. Soltysik, S., Bedore, D. A. and Kensil, C. R., *Ann. N.Y. Acad. Sci.*, 1993, *690*, 392.

36. McGrane, M., Keukens, H. J., O'Keeffe, M., van Rhijn, J. A. and Smyth, M. R. In: *EuroResidue IV* Veldhoven, The Netherlands; van Ginkel, L. A., Ruiter, A., Eds., 2000, pp. 765–770.
37. Troyer, J. K., Stephenson, K. K. and Fahey, J. W., *J. Chromatogr. A*, 2001, *919*, 299.
38. Le Rouzo, G., Lamouroux, C., Bresson, C., Guichard, A., Moisy, P. and Moutiers, G., *J. Chromatogr. A*, 2007, *1164*, 139.
39. Mawhinney, D. B., Hamelin, E. I., Fraser, R., Silva, S. S., Pavlopoulos, A. J. and Kobelski, R. J., *J. Chromatogr. B*, 2007, *852*, 235.
40. Fontanals, N., Marce, R. M. and Borrull, F., *J. Chromatogr. A*, 2007, *1152*, 14.
41. Walshe, M., Kelly, M. T., Smyth, M. R. and Ritchie, H., *J. Chromatogr. A*, 1995, *708*, 31.
42. Bergqvist, Y. and Hopstadiua, C., *J. Chromatogr. B*, 2000, *714*, 189.
43. Crowther, J. B., Fazio, S. D. and Hartwick, R. A., *J. Chromatogr.*, 1983, *282*, 619.
44. Bischoff, R. and McLaughlin, L. W., *J. Chromatogr.*, 1983, *270*, 117.
45. Bischoff, R. and McLaughlin, L. W., *J. Chromatogr.*, 1984, *296*, 329.
46. Bischoff, R. and McLaughlin, L. W., *J. Chromatogr.*, 1984, *317*, 251.
47. Muenter, M. M., Stokes, K. C., Obie, R. T. and Jezorek, J. R., *J. Chromatogr. A*, 1999, *844*, 39.
48. McCallum, J. L., Yang, R., Young, J. C., Strommer, J. N. and Tsao, R., *J. Chromatogr. A*, 2007, *1148*, 38.
49. Hsieh, Y., Duncan, C. and Liu, M., *J. Chromatogr. B*, 2007, *854*, 8.
50. Li, J., Shao, S., Jaworsky, M. S. and Kurtulik, P. T., *J. Chromatogr. A*, 2008, *1185*, 185.
51. Liu, X., Fang, J., Cauchon, N. and Zhou, P., *J. Pharm. Biomed. Anal.*, 2008, *46*, 639.
52. Abbood, A., Smadja, C., Herrenknecht, C., Alahmad, Y., Tchapla, A. and Taverna, M., *J. Chromatogr. A*, 2009, 1216, 3244.
53. Nogueira, R., Lämmerhofer, M. and Lindner, W., *J. Chromatogr. A*, 2005, *1089*, 158.
54. Bicker, W., Lämmerhofer, M. and Lindner, W., *J. Chromatogr. B*, 2005, *822*, 160.
55. Nogueira, R., Lubda, D., Leitner, A., Bicker, W., Maier, N. M., Lämmerhofer, M. and Lindner, W., *J. Sep. Sci.*, 2006, *29*, 966.
56. Bicker, W., Lämmerhofer, M., Keller, T., Schuhmacher, R., Krska, R. and Lindner, W., *Anal. Chem.*, 2006, *78*, 5884.
57. Bicker, W., Lämmerhofer, M. and Lindner, W., *Anal. Bioanal. Chem.*, 2008, *390*, 263.
58. Lämmerhofer, M., Richter, M., Wu, J., Nogueira, R., Bicker, W. and Lindner, W., *J. Sep. Sci.*, 2008, *31*, 2572.
59. Apfelthaler, E., Bicker, W., Lämmerhofer, M., Sulyok, M., Krska, R., Lindner, W. and Schuhmacher, R., *J. Chromatogr. A*, 2008, *1191*, 171.
60. Liu, X., Pohl, C. and Weiss, J., *J. Chromatogr. A*, 2006, *1118*, 29.
61. Liu, X. and Pohl, C., *Am. Lab.*, 2007, *39*, 22.
62. Liu, X. and Pohl, C., *Am. Lab.*, 2009, *41*, 26.
63. Thermo-Fisher website on Duet columns: http://www.selectscience.net/products/hypersil-duet-c18+sax-columns/?prodID=11596 (accessed on December 9, 2010) and http://www.selectscience.com/products/hypersil-duet-c18+scx-columns/?prodID=7515 (accessed on December 9, 2010)
64. Grace website on mixed-mode columns: http://www.discoverysciences.com/uploadedFiles/Site_for_Catalog_2008/hplcic/HPLC_Columns/Small_Molecule/alltech_specialty.pdf (Accessed on December 9, 2010)
65. SIELC website on Primesep columns: http://www.sielc.com/Technology_NovelStationaryPhases.html (accessed on December 9, 2010)
66. Dionex website on Acclaim Mixed-Mode WAX-1: http://www.dionex.com/en-us/webdocs/48829-DS_Acclaim_Mixed-Mode_WAX1_released021407.pdf (Accessed on December 9, 2010)
67. Dionex website on Acclaim Mixed-Mode WCX-1: http://www.dionex.com/en-us/columns-accessories/specialty/cons65659.html (accessed on December 9, 2010)

68. Dionex website on Acclaim Trinity P1: http://www.dionex.com/en-us/webdocs/70761-DS-Acclaim-Trinity-12Feb2010-LPN2239-02.pdf (Accessed on December 9, 2010)
69. Dionex website on OmniPac: http://www.dionex.com/en-us/products/columns/lc/mixed-mode/omnipac/lp-71739.html (Accessed on December 9, 2010)
70. Kopaciewicz, W., Rounds, M. A., Fausnaugh, F. and Regnier, F. E., *J. Chromatogr.*, 1983, *266*, 3.
71. Sellergren, B. and Shea, K. J., *J. Chromatogr. A*, 1993, *654*, 17.
72. Millot, M., Debranche, T., Pantazaki, A., Gherghi, I., Sebille, B. and Vidal-Madjar, C., *Chromatographia*, 2003, *58*, 365.
73. Hansen, M. B., Lihme, A., Spitali, M. and King, D., *Bioseparation*, 1999, *8*, 189.
74. Gao, D., Lin, D.-Q. and Yao, S.-J., *J. Chromatogr. B*, 2007, *859*, 16.
75. Rios, M., *Pharm. Technol.*, 2007, 31(5), 40, 42, 44, 46, 48.
76. Mant, C. T., Kondejewski, L. H. and Hodges, R. S., *J. Chromatogr. A*, 1998, *816*, 79.
77. Mant, C. T., Litowski, J. R. and Hodges, R. S., *J. Chromatogr. A*, 1998, *816*, 65.
78. Zhu, B. Y., Mant, C. T. and Hodges, R. S., *Pept. Chem. Biol., Proc. Am. Pept. Symp.*, 1992, *12*, 546.
79. Mashige, F., Ohkubo, A., Matshushima, Y., Takano, M., Tsuchiya, E., Kanazawa, H., Nagata, Y., Takai, N., Shinozuka, N. and Sakuma, I., *J. Chromatogr. B*, 1994, *658*, 63.
80. Liu, X., Pohl, C. and Woodruff, A. (manuscript in preparation).
81. Liu, X. and Pohl, C. *J. Sep. Sci.*, 2010, *33*, 779.
82. Wang, L., Henday, S. M., Liu, X., Tracy, M. and Schnute, W. C., *Simultaneous Determination of Melamine and Cyanuric Acid Using LC-MS with the Acclaim Mixed-Mode WAX-1 Column and Mass Spectrometric Detection* (http://www.dionex.com/en-us/webdocs/62283_LPN%201991–01%20N_Melamine.pdf).
83. Good, N. E. et al., *Biochemistry*, 1966, *5*(2), 467.

4 Mechanisms Controlling the Separation and Retention of Proanthocyanidins and Other Plant Polyphenols in Hydrophilic Interaction Chromatography

Akio Yanagida and Yoichi Shibusawa

CONTENTS

4.1 INTRODUCTION

Since the term "hydrophilic interaction chromatography (HILIC)" was proposed by Alpert in 1990,[1] a large number of highly polar (or hydrophilic) substances have been separated efficiently using the HILIC approach. HILIC requires an aqueous organic mobile phase and a polar stationary phase, such as bare silica or silica modified with ionic, neutral, or complex functional groups. HILIC is a variant of normal-phase liquid chromatography and is suitable for the separation of water-soluble polar substances that are poorly retained on conventional reversed-phase stationary phases, such as ODS-silica. Most recent HILIC applications have focused on the analyses of biologically active polar compounds such as saccharides (in glycoproteins), amino acids, peptides, nucleotides, and their pharmaceutical derivatives. HILIC is a powerful method for separating compounds that contain multiple hydrogen-bonding groups. Hydrogen bonding between the solute and the polar stationary phase is the essential mechanism for differential retention in most HILIC separations. Plant secondary metabolites, and in particular polyphenolic compounds and their glycosides, are excellent candidates for HILIC due to the presence of multiple hydroxyl groups in their structures. In this chapter, we review our recent experimental data obtained from investigating the mechanisms underlying the retention and separation of plant proanthocyanidins (oligomeric catechins) and other polyphenols by HILIC.

4.2 POLYPHENOLIC COMPOUNDS

Polyphenols are a group of secondary metabolites widely distributed throughout the plant kingdom. Almost all foods derived from edible plants contain polyphenolic compounds. Plant polyphenols are well-known natural antioxidants, and recent studies have suggested that the polyphenols in food play an important role as functional food factors in the prevention of human diseases and cancers. The molecular mass of polyphenols generally ranges from 100 to 10,000 Da. In the plant cell, the majority of polyphenols are below 500 Da and exists in the form of glycosides and/or organic acid esters, rather than as free aglycones. The separation and identification of these low-molecular-mass polyphenols are easily performed by conventional reversed-phase HPLC (high-performance liquid chromatography), because the aglycones usually contain hydrophobic aromatic groups. On the other hand, the majority of high-molecular-mass polyphenols (over 500 Da) are present as condensed or hydrolysable tannins. Condensed tannins are oligomeric flavan-3-ols, so-called proanthocyanidins (PAs), while the hydrolysable tannins are polygalloyl and related hexahydroxydiphenoyl esters. In general, the high-molecular-mass polyphenols are highly hydrophilic due to their unique 3D structures associated with the degree of polymerization (DP). These compounds are not only the water-soluble antioxidants but also strong inhibitors of numerous physiological enzymatic reactions because of their protein binding ability (often nonspecific). However, the separation and identification of these high-molecular-mass polyphenols by conventional reversed-phase HPLC are very difficult because of the heterogeneity of the monomeric units and the different polymerization linkages in their structures.

Among the high-molecular-mass polyphenols, PAs are the common macromolecules biosynthesized by enzymes in the plant cell. The monomeric units in the PA structure are neutral flavan-3-ols (catechins). Because the general structure of flavan-3-ol has two asymmetrical centers at C-2 and C-3 on the heterocyclic ring, two pairs of diastereoisomers are theoretically present (see Figure 4.1). In addition to these isomers, analogous compounds with a galloyl ester (gallate) at the C-3 hydroxyl group of each flavan-3-ol isomer have also been identified. In most plants, these isomeric flavan-3-ols and their gallates constitute the monomeric units of PAs. Plant PAs are a mixture of oligomers consisting of different chain lengths of flavan-3-ol units linked most commonly through C-4 to C-8 (or C-6) interflavan bonds (type-B PA). For example, the structures of six types of monomeric flavan-3-ols (1a–1f) and nine types of oligomeric PAs (2a–5a) are shown in Figure 4.1. The numeric number assigned to each compound

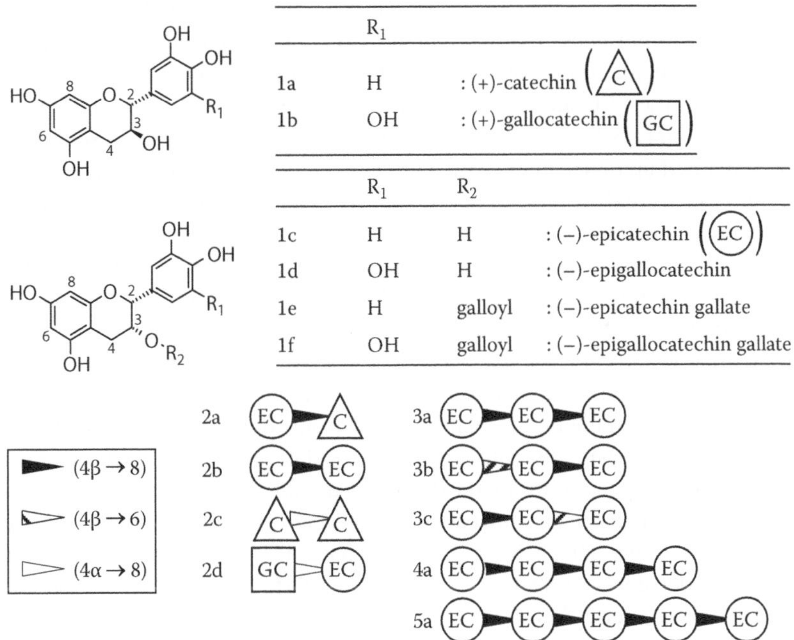

FIGURE 4.1 Structures of monomeric flavan-3-ols and oligomeric PAs used as experimental standards. (Reprinted from Yanagida, A. et al., *J. Chromatogr. A*, 1143, 153, 2007. With permission from Elsevier.)

is identical to its DP. The oligomeric PAs from dimer (except 2c and 2d) to pentamer shown in Figure 4.1 were isolated from an immature apple, as described elsewhere.[2–5] Furthermore, the MALDI-TOF MS study in our laboratory confirmed that the PA-rich fraction obtained from immature apples contained oligomeric PAs ranging from dimers (2-mer) to pentadecamers (15-mer).[2] A number of PAs with similar monomeric units but different DPs have also been identified from various plant sources, such as fruit, seeds, bark, and leaves, by other MALDI or ESI MS analyses.

4.3 CHROMATOGRAPHY CHALLENGES FOR SEPARATION OF VARIOUS PAs

As mentioned above, reversed-phase LC is not the best separation method for plant PAs. HPLC using an ODS column eluted with an aqueous acetonitrile (or methanol) mobile phase has been the general method used for the separation of flavan-3-ols and small DP oligomers such as dimers and trimers.[3,6–8] However, the elution order of these monomers and small DP oligomers was not based on their DPs, and the broad unresolved peaks derived from coexisting highly oligomeric PAs overlapped the separation profiles of these monomers and small DP oligomers.[3] For a more efficient separation of highly oligomeric PAs, several useful methods based on other LC modes have recently been reported. Size-exclusion chromatography (SEC) using dimethylformamide[9–12] or aqueous-organic mobile phases,[3,13,14] and high-speed counter-current chromatography (HSCCC) using immiscible two-phase solvent systems[15–17] enable the partial separation of oligomeric PAs based on their DP. Furthermore, normal-phase HPLC using the combination of a bare silica column and a nonaqueous organic mobile phase resulted in a more efficient DP-based separation compared with the SEC and HSCCC methods. Using an optimized HPLC method with a normal-phase silica column, oligomeric PAs from different plant sources were clearly separated up to decamers with a gradient elution of dichloromethane–methanol (containing a small volume of acidic water)[18–22] and PAs from an apple were also separated up to pentamers with a gradient elution of hexane–acetone.[4,5]

However, the above nonaqueous organic mobile phases of silica HPLC are associated with some practical drawbacks. Mobile phases containing dichloromethane must be handled carefully and need to be disposed of appropriately (dichloromethane is an organic solvent regulated and monitored by the Water Pollution Prevention Law in Japan). In the case of hexane–acetone mobile phases, highly oligomeric PAs (larger than pentamers) are barely eluted from the silica column due to their low solubility and strong adsorption to the silica stationary phase. Safer and water-compatible mobile phases are therefore necessary for these silica HPLC methods. In addition to these practical drawbacks, the mechanisms underlying the separation and retention of oligomeric PAs in this mode have not been adequately elucidated. We therefore decided to investigate HILIC using more aqueous conditions for the separation of oligomeric PAs.[23]

4.4 HILIC CONDITIONS FOR SEPARATION OF PAs

A PA-rich mixture from immature apples (apple PA mixture) was prepared, and consisted of monomeric flavan-3-ols (1a and 1c in Figure 4.1) and oligomeric PAs ranging in size from dimers to pentadecamers. The contents of this mixture were divided into two fractions by methyl acetate extraction.[4] The majority of monomers and small DP oligomers (below hexamers) were selectively extracted with methyl acetate, while highly oligomerized PAs remained in the residue. The methyl acetate extracts and the residue fraction were each subjected to HILIC separation, and in addition, the 15 different standards of flavan-3-ols and oligomeric PAs shown in Figure 4.1 were also analyzed by HILIC.

HILIC separation of these polyphenolic samples was conducted on a TSK-Gel Amide-80 column (250 mm × 4.6 mm ID, 5 μm particle size, Tosoh, Japan).[24] This silica-based stationary phase was chosen for its acrylamide nonionic (neutral) carbamoyl termini, since the solute PAs are neutral polyphenols and lack ionic groups. The HPLC system consisted of a Model L-7100 pump (Hitachi, Tokyo, Japan), a Rheodyne 7166 injector, and a Model L-7455 diode-array detector (Hitachi). Chromatograms were recorded on a PC using Model D-7000 chromatography data station software (Hitachi).

HILIC separation of oligomeric PAs in the methyl acetate extracts and residue fraction of the apple PA mixture were carried out using a linear gradient. The mobile phases were acetonitrile–water mixtures at volume ratios of (a) 90:10 and (b) 50:50, respectively. The Amide-80 column was initially conditioned with solvent A at a flow rate of 1.0 mL/min. Lyophilized methyl acetate extracts or apple PA residue mixture were dissolved in solvent A (final conc.: 10 mg/mL) and a portion (20 μL) was injected and eluted with a linear gradient from 100% A (0% B) to 0% A (100% B) over 60 min at 1.0 mL/min at ambient temperature. The absorbance of the effluent was monitored using a diode-array detector.

The retention characteristics on the Amide-80 column of the 15 standards of flavan-3-ols (1a–f) and PA oligomers (2a–d, 3a–c, 4a, and 5a) as shown in Figure 4.1 were also examined by the isocratic elution of acetonitrile–water (84:16, v/v). A 10 μL aliquot of each sample (100 μg/mL in the mobile phase) or the mixed sample solution (100 μg each/mL) was injected and eluted at a flow rate of 1.0 mL/min. The retention factor (k) of each compound was calculated using the following equation:

$$k = \frac{(t_R - t_0)}{t_0}$$

where

t_R is the retention time (min) of the solute

t_0 is the elution time of the non-retained hydrophobic compound (anthracene) through the column

4.5 HILIC SEPARATION PROFILES OF APPLE PA MIXTURES ELUTED USING A GRADIENT METHOD

HILIC chromatograms of the apple PA mixture of methyl acetate extracts and residue fraction are shown in Figure 4.2A and B, respectively. The major peaks in both chromatograms are numbered according to their elution order. The separation profile of the apple PA mixture of methyl acetate extracts (Figure 4.2A) was very similar to that obtained previously by silica HPLC using an organic hexane–acetone mobile phase.[4] The DPs of each oligomeric PA in the chromatogram were characterized by off-line MALDI-TOF-MS analysis and revealed that the oligomeric PAs in the extract eluted in the order of monomer to pentamer, in accordance with the increase in the number of their DPs (data not shown). Compared with Figure 4.2A of methyl acetate extracts, the separation profile of the residue from the apple PA mixture (Figure 4.2B) showed more separated peaks (up to 10). An LC-ESI-MS study using a semimicro Amide-80 column (2.0 mm ID) confirmed the DP-based separation of oligomeric PAs by HILIC, i.e., the number of separated peaks (up to 10) in Figure 4.2B directly corresponds to the DP of each oligomer (up to decamers).[23]

4.6 RETENTION BEHAVIOR OF OLIGOMERIC PA STANDARDS IN HILIC

Since the first report by Regaud et al. in 1993,[18] several PA separation methods using silica HPLC with organic mobile phases have been published. Although the separation methods were described as "normal-phase HPLC" in these reports, the separation and retention mechanisms of oligomeric PAs have not been elucidated. To clarify the precise mechanism of PA separation in these LC modes, HILIC studies of the 15 standards shown in Figure 4.1 were carried out using isocratic elution

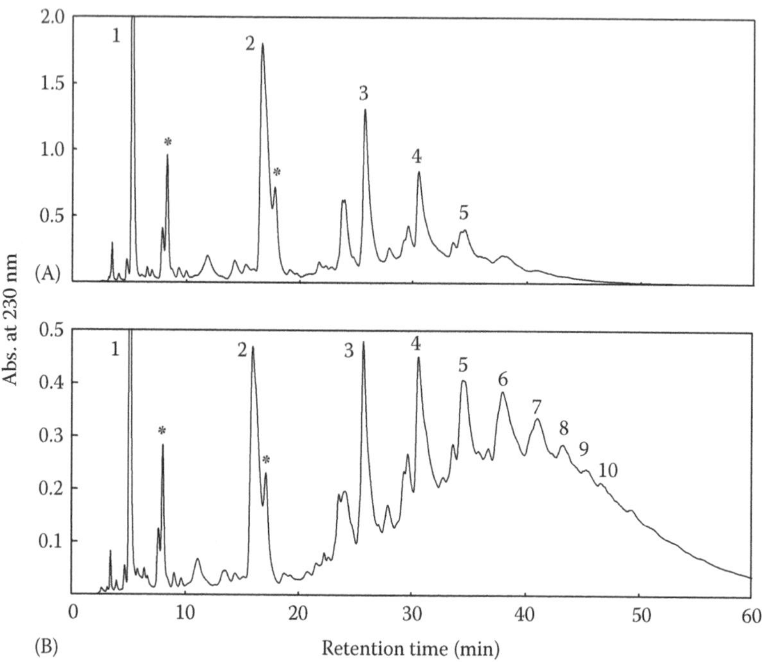

FIGURE 4.2 HILIC separation profiles of an apple PA mixture. (A) Chromatogram of oligomeric PAs in the methyl acetate extract. (B) Chromatogram of oligomeric PAs in the extraction residue. Individual peak in both chromatograms is numbered in accordance with their elution order (up to 10). The two peaks marked (*) are from impurities (caffeoylquinic acid and phloretin glycoside). (Reprinted from Yanagida, A. et al., *J. Chromatogr. A*, 1143, 153, 2007. With permission from Elsevier.)

(acetonitrile–water 84:16), and the retention factor (k) of each standard was calculated from its retention time.

The isocratic HILIC elution profile of the 15 standards is shown in Figure 4.3, and their k values are listed in Table 4.1. The chromatogram confirmed the trend of the separation of these standard oligomers based on their DP. However, the retention times of oligomers with the same DP did not match perfectly. We initially believed that this variation was due to differences in hydrophobicity. In general, the hydrophobicity of an analyte is evaluated on a logarithmic scale of its octanol–water partition coefficient, log $P_{o/w}$. Thus, we measured log $P_{o/w}$ values for the 15 different standards by HSCCC using an octanol–water two-phase solvent system.[23,25] A scatter diagram of the DP values of the 15 standards against their log $P_{o/w}$ values is shown in Figure 4.4. The diagram revealed a tendency for log $P_{o/w}$ values to decrease significantly with increasing DP values. A minus log $P_{o/w}$ value indicates that the compound is highly hydrophilic. Because these compounds are strongly partitioned into the water-enriched stationary phase in general normal-phase partition chromatography, differences in the hydrophilicity of the standards shown in Figure 4.4 actually contribute to the DP-based separation as shown in Figure 4.3. However, a more detailed analysis showed that the log $P_{o/w}$ values of compounds eluting at approximately the

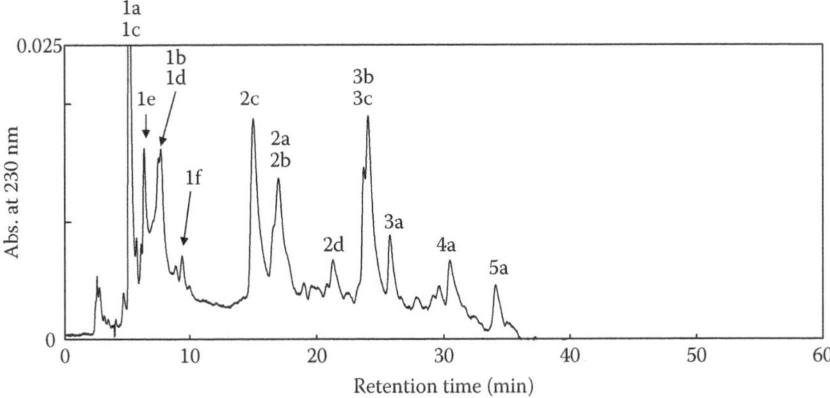

FIGURE 4.3 HILIC separation profile of 15 standards of flavan-3-ols and oligomeric PAs, eluted isocratically. The numbering scheme for the standards is identical to that in Figure 4.1. (Reprinted from Yanagida, A. et al., *J. Chromatogr. A*, 1143, 153, 2007. With permission from Elsevier.)

TABLE 4.1
Comparison of Structural and Chromatographic Parameters of 15 Standard Flavan-3-Ols and Oligomeric PAs

		Structural Parameters		Chromatographic Parameters	
Compounds		Degree of Polymerization	Number of OH Groups	log $P_{o/w}$	*k* in HILIC
1a	C	1	5	0.31	0.89
1b	GC	1	6	−0.31	1.82
1c	EC	1	5	0.10	0.89
1d	EGC	1	6	−0.53	1.82
1e	ECg	1	7	1.10	1.33
1f	EGCg	1	8	0.50	2.44
2a	EC(4b-8)C	2	10	−1.35	5.24
2b	EC(4b-8)EC	2	10	−0.90	5.24
2c	C(4a-8)C	2	10	−0.91	4.51
2d	GC(4a-8)C	2	11	−1.52	6.80
3a	EC(4b-8)EC(4b-8)EC	3	15	−1.32	8.43
3b	EC(4b-8)EC(4b-6)EC	3	15	−1.49	7.81
3c	EC(4b-6)EC(4b-8)EC	3	15	−1.13	7.81
4a	EC(4b-8)EC(4b-8)EC(4b-8)EC	4	20	−1.79	10.12
5a	EC(4b-8)EC(4b-8)EC(4b-8)EC(4b-8)EC	5	25	−2.21	11.50

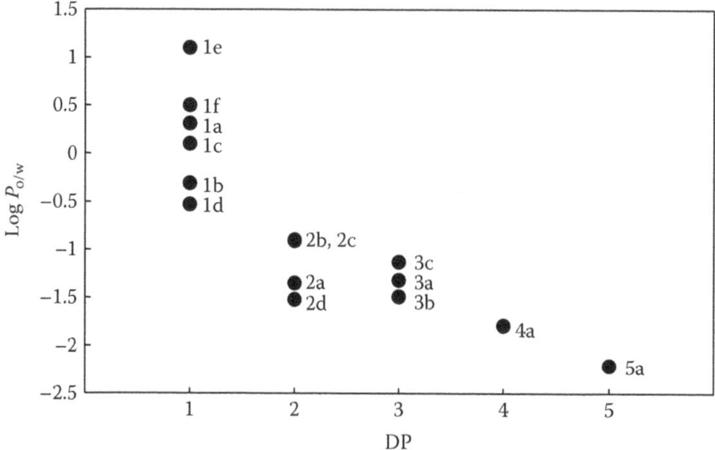

FIGURE 4.4 Scatter diagram of the degree of polymerization (DP) of the 15 standards against their log $P_{o/w}$ values. (Reprinted from Yanagida, A. et al., *J. Chromatogr. A*, 1143, 153, 2007. With permission from Elsevier.)

ttsame time in Figure 4.3 were not identical in Figure 4.4 (e.g., 1c and 1a and 1e, or 2c and 2b and 2a). This inconsistency could not be explained by a normal-phase partitioning mechanism based on the hydrophilicity of the solute PA. Therefore, we believed that another important separation factor (or interaction) contributes to the PAs separation by HILIC.

4.7 RETENTION MECHANISM OF OLIGOMERIC PAs IN HILIC

Based on the structural attributes (DP and the number of hydroxyl groups) and the chromatographic parameters (log $P_{o/w}$, t_R, k, and log k in the HILIC) listed in Table 4.1, we examined the relationship between the log k and log $P_{o/w}$ values of the 15 different standards. As shown in Figure 4.5, the relationship between log k and log $P_{o/w}$ was expressed as the linear regression equation "$y = -1.8177x - 0.4767$." This relationship revealed that the hydrophilicity (i.e., log $P_{o/w}$ values) of the oligomer standards actually contributes to DP-based separation. However, the square of the correlation coefficient of this relationship is not very high ($r^2 = 0.7949$). To gain insight into another factor affecting separation by HILIC, we examined the relationship between the log k values of the15 standards and their number of hydroxyl groups (OH). A high correlation was observed, as shown in Figure 4.6. The relationship was expressed as the linear regression equation "$y = 12.341x + 9.2541$," and the square of the correlation coefficient ($r^2 = 0.9501$) was larger than the value (0.7949) obtained from the correlation between log k and log $P_{o/w}$ values. This high correlation between log k and the number of OH groups revealed that hydrogen bonding between a carbamoyl terminal on the Amide-80 column and the OH group of the solute strongly contributes to PA separation by HILIC. The data shown in Figures 4.5 and 4.6 confirmed that two effects, hydrogen bonding and solute hydrophilicity, both contribute to DP-based PA separation, but the hydrogen bonding effect dominates HILIC separations using an aqueous acetonitrile mobile

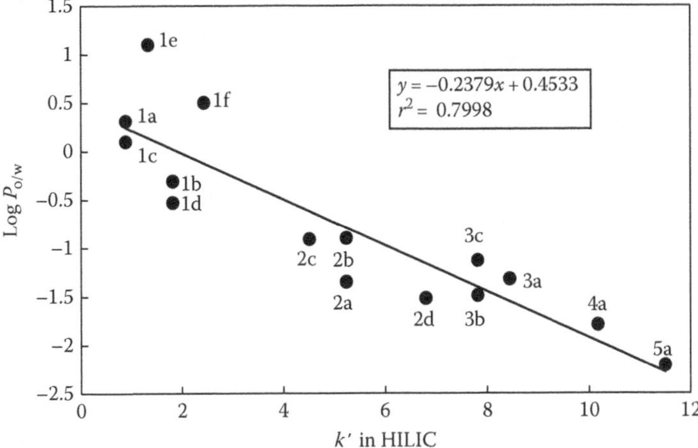

FIGURE 4.5 Relationship between log k of 15 standards in HILIC and their log $P_{o/w}$ values. The inset regression equation was calculated from the linear least-square fit of all data ($n = 15$). (Reprinted from Yanagida, A. et al., *J. Chromatogr. A*, 1143, 153, 2007. With permission from Elsevier.)

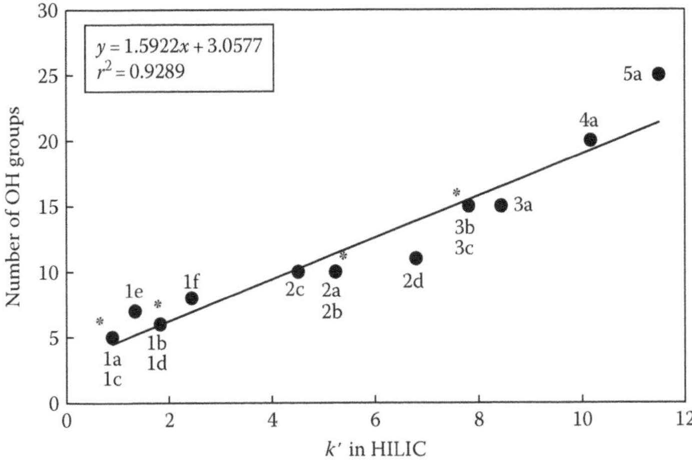

FIGURE 4.6 Relationship between log k of 15 standards in HILIC and the number of hydroxyl groups in their structures. The inset regression equation was calculated from the linear least-square fit of all data ($n = 15$). *—two plots are stacked up. (Reprinted from Yanagida, A. et al., *J. Chromatogr. A*, 1143, 153, 2007. With permission from Elsevier.)

phase. Hydrogen bond formation between two compounds is generally enhanced under hydrophobic conditions compared to that in aqueous conditions. Therefore, in previous LC methods using a bare silica column with nonaqueous organic mobile phases,[4,5,18–22] strong hydrogen bonding between the silanol groups of the stationary phase and the hydroxyl groups of the solute PAs might have been the major factor favoring retention. If this is correct, this mode of PA separation on silica columns should be referred to as "adsorption." On the other hand, in PA separations using an amide–silica column with an aqueous mobile phase, solute hydrophilicity is an important separation factor. In this case, "hydrophilic interaction chromatography (HILIC)" is the most appropriate term to describe mixed-mode separations relying on adsorption (due to hydrogen bonding) and normal-phase partitioning (due to solute hydrophilicity). Recently, another HILIC method using a diol-silica column has been used for the DP-based separation of oligomeric PAs (up to decamers) in cacao and related foods.[26] At present, HILIC methods using either amide or diol columns are the best LC approach for the DP-based separation of oligomeric PAs.

4.8 HILIC SEPARATION OF OTHER POLYPHENOLIC COMPOUNDS

For most researchers, reversed-phase HPLC using an ODS column is the first choice for chromatographic separations of plant polyphenols. HILIC has only been used in a few examples mentioned in this chapter for the separation of oligomeric PAs. However, HILIC is a potentially useful LC mode for the separation of plant polyphenols since these compounds have multiple hydroxyl and/or carboxyl groups that can interact with the HILIC stationary phase by hydrogen bonding. In this section, several examples of HILIC separation of low-molecular-mass polyphenols are discussed.

The HILIC separation profiles of six neutral phenolics (Figure 4.7A) and five phenolic acids (Figure 4.7B) on a TSK-Gel Amide-80 column (250 mm × 4.6 mm ID, 5 μm particle size) were obtained with an aqueous acetonitrile mobile phase.

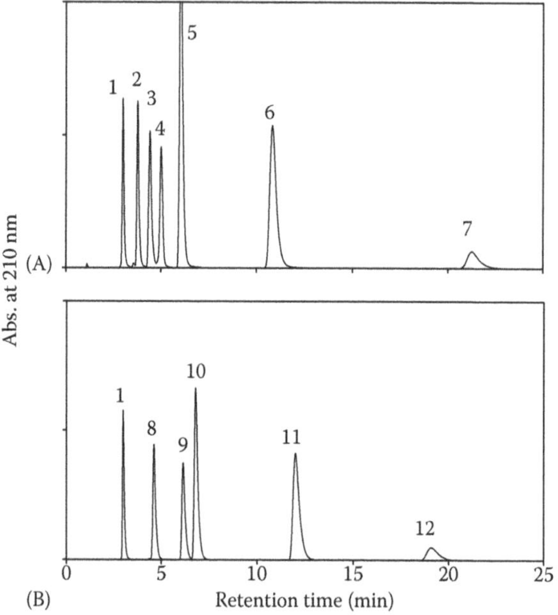

FIGURE 4.7 HILIC separation profiles of low-molecular-mass polyphenols. (A) Chromatogram of six neutral phenolics. (B) Chromatogram of five phenolic acids. Peak number 1: (not retained) anthracene; 2: catechol; 3: kaempferol; 4: phloretin; 5: phloroglucinol; 6: epicatechin; 7: phloridzin; 8: *p*-coumaric acid; 9: caffeic acid; 10: 3,4-dihydroxybenzoic acid; 11: gallic acid; and 12: chlorogenic acid.

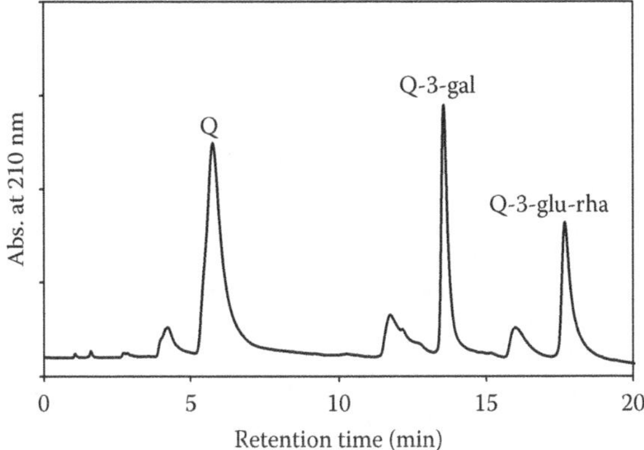

FIGURE 4.8 HILIC separation profile of quercetin (Q), its monoglycoside (Q-3-gal), and diglycoside (Q-3-glu-rha).

Both separations were achieved isocratically using acetonitrile-2.5 mM phosphoric acid (95:5, v/v) at a flow rate of 1.0 mL/min. In both chromatograms, unretained hydrophobic anthracene eluted in the void volume (t_0 about 3 min). On the other hand, all the solute phenolics and phenolic acids were retained and separated on the column, and their retention time (and thus the elution order) generally corresponded with the number of hydrogen bond donors (hydroxyl and carboxyl groups) in their structures.

The HILIC method is an effective method for the separation of polyphenolic glycosides. The HILIC separation profile of a flavonol, quercetin (Q), and its glycosides (Q-3-gal and Q-3-glu-rha) using an Amide-80 column (250 mm × 4.6 mm ID, 5 μm particle size) is shown in Figure 4.8. The separation was achieved with a linear gradient elution of acetonitrile–water at ratios of 95:5 to 80:20 (v/v) over 20 min at a flow rate of 1.0 mL/min. In the chromatogram, the most hydrophobic aglycone (Q) eluted first, its monoglycoside (Q-gal) eluted second, and its diglycosides (Q-3-glu-rha) eluted last. This elution order corresponds to the increasing number of glycoside moieties in the structure of the solute. Therefore, HILIC may be useful for determining the number of glycosides in an unknown flavonoid structure and may represent a general approach for the analysis of sugar chains in glycoproteins. Furthermore, similar separation profiles were observed from HILIC analyses of some phenolic acids and their organic acid esters (data not shown).

The separation of the polyphenolic compounds as demonstrated in Figures 4.7 and 4.8 can also be achieved by conventional reversed-phase HPLC using an ODS column. However, the elution order of the HILIC separation differs from that of a reversed-phase HPLC separation. The HILIC profile provides additional information on solute structure, such as molecular hydrophilicity, and the number of functional groups in the structure. Another advantage of HILIC is the mobile-phase composition. The acetonitrile-rich mobile phase used for HILIC can effectively solubilize moderately hydrophobic solutes (such as quercetin) compared with the water-rich mobile phase used for reversed-phase HPLC. The efficient and effective use of the above advantages will lead to the wide application of HILIC for separations and structural determination of polyphenols and other plant secondary metabolites, such as alkaloids, glycosides of steroids, or terpenes.

REFERENCES

1. Alpert, A.J., *J. Chromatogr.*, 1990, 499, 177.
2. Ohnishi-Kameyama, M., Yanagida, A., Kanda, T., and Nagata, T., *Rapid Commun. Mass Spectrom.*, 1997, 11, 31.
3. Yanagida, A., Kanda, T., Shoji, T., Ohnishi-Kameyama, M., and Nagata, T., *J. Chromatogr. A*, 1999, 855, 181.
4. Yanagida, A., Kanda, T., Takahashi, T., Kamimura, A., Hamazono, T., and Honda, S., *J. Chromatogr. A*, 2000, 890, 251.
5. Shoji, T., Mutsuga, M., Nakamura, T., Kanda, T., Akiyama, H., and Goda, Y., *J. Agric. Food Chem.*, 2003, 51, 3806.
6. Lea, A.G.H., *J. Chromatogr.*, 1982, 238, 253.
7. Delage, E., Bohuon, G., Baron, A., and Drilleau, J.-F., *J. Chromatogr.*, 1991, 555, 125.
8. Rohr, G.E., Meier, B., and Sticher, O., *J. Chromatogr. A*, 1999, 835, 59.
9. Bae, Y.S., Foo, L.Y., and Karchecy, J.J., *Holzforschung*, 1994, 48, 4.
10. López-Serrano, M. and Barceló, A.R., *J. Chromatogr. A*, 2001, 919, 267.
11. Kennedy, J.A. and Taylor, A.W., *J. Chromatogr. A*, 2003, 995, 99.
12. Kurumatani, M., Fujita, R., Tagashira, M., Shoji, T., Kanda, T., Ikeda, M., Shoji, A., Yanagida, A., Shibusawa, Y., Shindo, H., and Ito, Y., *J. Liq. Chromatogr. Relat. Technol.*, 2005, 28, 1971.
13. Yanagida, A., Shoji, T., and Kanda, T., *Biosci. Biotechnol. Biochem.*, 2002, 66, 1972.
14. Bourvellec, C.Le., Picot, M., and Renard, C.M.G.C., *Anal. Chim. Acta.*, 2006, 563, 33.
15. Shibusawa, Y., Yanagida, A., Isozaki, M., Shindo, H., and Ito, Y., *J. Chromatogr. A*, 2001, 915, 253.
16. Shibusawa, Y., Yanagida, A., Shindo, H., and Ito, Y., *J. Liq. Chromatogr. Relat. Technol.*, 2003, 26, 1609.
17. Köhler, N., Wray, V., and Winterhalter, P., *J. Chromatogr. A*, 2008, 1177, 114.
18. Rigaud, J., Escribano-Bailon, M.T., Prieur, C., Souquet, J.-M., and Cheynier, V., *J. Chromatogr. A*, 1993, 654, 255.
19. Hammerstone, J.F., Lazarus, S.A., Mitchell, A.E., Rucker, R., and Schmitz, H.H., *J. Agric. Food Chem.*, 1999, 47, 490.
20. Natsume, M., Osakabe, N., Yamagishi, M., Takizawa, T., Nakamura, T., Miyatake, H., Hatano, T., and Yoshida, T., *Biosci. Biotechnol. Biochem.*, 2000, 64, 2581.
21. Kennedy, J.A. and Waterhouse, A.L., *J. Chromatogr. A*, 2000, 866, 25.
22. Gu, L., Kelm, M., Hammerstone, J.F., Beecher, G., Cunningham, D., Vannozzi, S., and Prior, R.L., *J. Agric. Food Chem.*, 2002, 50, 4852.
23. Yanagida, A., Murao, H., Ohnishi-Kameyama, M., Yamakawa, Y., Shoji, A., Tagashira, M., Kanda, T., Shindo, H., and Shibusawa, Y., *J. Chromatogr. A*, March 2, 2007, 1143, 153.
24. Yoshida, T., *Anal. Chem.*, 1997, 69, 3038.
25. Shibusawa, Y., Shoji, A., Yanagida, A., Shindo, H., Tagashira, M., Ikeda, M., and Ito, Y., *J. Liq. Chromatogr. Relat. Technol.*, 2005, 28, 2819.
26. Kelm, M.A., Johnson, J.C., Robbins, R.J., Hammerstone, J.F., and Schmitz, H.H., *J. Agric. Food Chem.*, 2006, 54, 1571.

5 Applications of Hydrophilic Interaction Chromatography to Food Analysis

Leticia Mora, Aleida S. Hernández-Cázares, M-Concepción Aristoy, Fidel Toldrá, and Milagro Reig

CONTENTS

5.1 INTRODUCTION

There are many substances in foods that must be analyzed because of their contribution to the final quality (i.e., protein compounds, lipid compounds, nucleotides), nutrition (i.e., vitamins, essential amino acids, minerals, etc.), or safety (i.e., growth promoters or antimicrobial residues). Different methodologies are being used for their analysis. High-performance liquid chromatography (HPLC) is one of the most widely used analytical techniques due to its versatility, speed, and high resolving power. Reversed-phase high-performance liquid chromatography (RP-HPLC) with and without ion-pairing agents and ion-exchange HPLC have been the methods of choice for the determination of these compounds during the last decades. However, limitations of RP-HPLC, such as the poor retention of polar molecules, show the need for alternative separation modes. Hydrophilic interaction chromatography (HILIC) represents a viable alternative for the separation of polar compounds in foods for many reasons. This separation mode uses mobile phases with similar organic solvents as those used in RP-HPLC with the consequent advantage of analyte solubility. Due to good retention of polar analytes, neither derivatization nor ion-pair reagents are necessary for the analysis of polar or ionic compounds. An additional important advantage of HILIC is its compatibility with mass spectrometric detection, eliminating the desalting step needed when using ion-exchange chromatography. Furthermore, HILIC has proven to be a simple, fast, and reliable alternative to the existing methods.

This chapter focuses on the recent applications of HILIC to the analysis of compounds that are present in foods and are relevant for either food quality or safety. Examples where HILIC is used for food quality testing including, among others, small peptides and key compounds like creatine or nucleotides are detailed. Examples using HILIC for the analysis of residues of compounds affecting the safety of foods such as relevant growth promoters and polar antimicrobials in foods of animal origin are also presented.

5.2 APPLICATION OF HYDROPHILIC INTERACTION CHROMATOGRAPHY TO THE ANALYSIS OF QUALITY-RELATED COMPOUNDS IN FOODS

5.2.1 HILIC with Conventional Detectors

The analysis of nucleotides and nucleosides in meat is very important to follow the postmortem process just after slaughter. In fact, the evolution of such compounds at early postmortem can give a good indication of the final quality of meat.[1–3] The simultaneous analysis of adenosine triphosphate (ATP), adenosine diphosphate (ADP), adenosine monophosphate (AMP), inosine monophosphate (IMP), inosine (Ino), hypoxanthine (Hx), nicotinamide adenine dinucleotide (NAD^+), and creatinine (Cn) in pork loin muscle was developed using a zwitterionic polymeric column (ZIC-pHILIC from SeQuant). A comparison between the pork meat concentrations of the compounds analyzed by HILIC and by ion-pair reversed-phase chromatography (IP-RP-HPLC), using a Zorbax Eclipse XDB-C18 column (Agilent), was performed.[4] The good agreement obtained between the two sets of data shows that HILIC is well suited to the separation and quantification of nucleotides in complex matrices such as meat and provides an interesting alternative to other methodologies (see Figure 5.1A and B). This same method, with conditions as listed in Table 5.1, also showed good performance in the separation of guanosine (G), guanosine monophosphate (GMP), guanosine diphosphate (GDP), guanosine triphosphate (GTP), and uric acid (UA).[4]

Other compounds present in skeletal muscle and with relevance in postmortem meat have also been analyzed with a ZIC-pHILIC column. This is the case for creatine that provides the necessary energy for muscle contraction. Its amount in muscle is kept in equilibrium with phosphocreatine since creatine is reversibly converted into phosphocreatine by creatine kinase. This reaction involves the transfer of the γ-phosphate group of ATP onto the guanidine group of creatine. In postmortem meat, creatine is converted into creatinine through a nonenzymatic reaction that is enhanced at increased temperature.

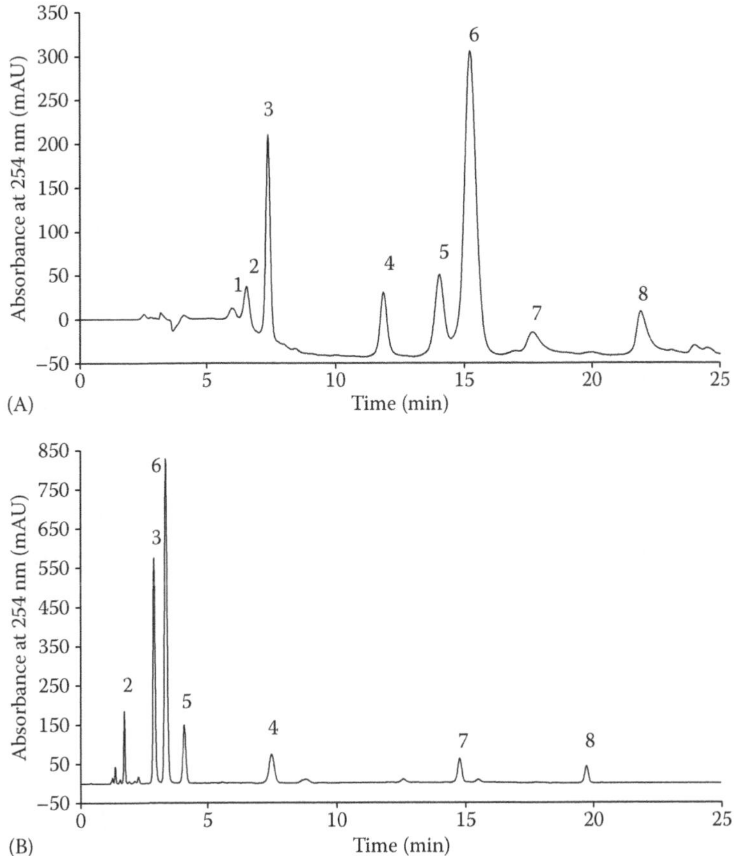

FIGURE 5.1 Chromatogram of pork meat sampled at 9 h postmortem. The column was (A) ZIC-pHILIC, 150 mm × 4.6 mm, 5 µm (SeQuant) and (B) Zorbax Eclipse XDB-C18, 150 mm × 4.6 mm, 5 µm (Agilent). Numbers 1 to 8 correspond to Cn, Hx, Ino, AMP, NAD$^+$, IMP, ADP, and ATP, respectively.

Other compounds of interest are the histidine-containing dipeptides—carnosine, anserine, and balenine which are extensively distributed in skeletal muscle. They have been reported to act as intracellular buffers, modulators for some enzymes, neurotransmitters, metal–ion chelators, antioxidants, and free-radical scavengers and are thus very important functional compounds in meat.[5] The simultaneous determination of carnosine, anserine, balenine, creatine, and creatinine in pork loin and chicken breast meat was recently developed using an Atlantis silica column (Waters).[6] Samples were extracted, deproteinized, and centrifuged and were then injected into the HPLC (see Table 5.1 for further details). The separation was monitored using a diode array detector at a wavelength of 214 nm for creatine, carnosine, anserine, and balenine, whereas a wavelength of 236 nm was used for creatinine detection (see Figure 5.2). This chromatographic method was subsequently applied in seven pork muscles of different metabolic types. The results showed that the content of creatine and creatinine were significantly higher in glycolytic muscles such as *Semimembranosus*,

TABLE 5.1
Performance of Some Recent HILIC-Based Methods for the Analysis of Compounds of Interest in Foods

Compound to Analyze	Food Matrix	Extraction	Type of Column	Elution Conditions	Detector	Reference
ATP, ADP, AMP, IMP, Ino, Hx, and NAD^+	Pork meat	0.6 M perchloric acid and neutralization with solid potassium carbonate	ZIC-pHILIC (SeQuant) 150 mm × 4.6 mm, 5 µm	Solvent A: 22.5 mM NH_4Ac, pH 3.5 in water/ACN (20:80); Solvent B: 50 mM NH_4Ac, pH 7 in water/ACN (60:40) Step gradient from 0% to 100% of B in 3 min Flow rate = 0.5 mL/min	DAD at 254 nm	[4]
Carnosine, anserine, and balenine dipeptides, creatine and creatinine	Pork loin and chicken breast	0.01N HCl and deproteinization with ACN	Atlantis silica (Waters) 150 mm × 4.6 mm, 3 µm	Solvent A: 0.65 mM NH_4Ac, pH 5.5 in water/ACN (25:75); solvent B: 4.55 mM NH_4Ac pH 5.5 in water/ACN (70:30) Linear gradient from 0% to 100% of B in 13 min Flow rate = 1.4 mL/min	DAD at 214 and 236 nm	[6]
Polar compounds contributing to umami-taste and mouth-drying oral sensation	Mushrooms	Soaking in drinking water for 12 h at room temperature	Semipreparative TSK-Gel Amide 80 (Tosoh BioSep), 300 mm × 7.8 mm	Solvent A, ACN and aqueous NH_4-formate (7 mM; pH 5.5), (80/20); solvent B, ACN and aqueous NH_4-formate (7 mM; pH 5.5) (20:80) Step gradient to 35% B Flow rate = 1.5 mL/min	ESI^+ and ESI^- LCQ MS	[11]

Free amino acids, di- and tri-peptides and glycoconjugates	Wheat gluten and Parmesan cheese	Wheat gluten was dehydrolyzed with 0.1 M HCl Parmesan cheese was defatted and extracted with water	TSK-Gel Amide 80 (Tosoh BioSep), 250 mm × 1.5 mm, 5 µm, 80 Å	In positive ionization mode: NH_4Ac buffer (pH 5.5) and an aqueous gradient (10%–40% in ACN) For analysis in the negative ionization mode, NH_4Ac buffer (pH 7.0) and same gradient Flow rate = 40 µL/min	ESI^+ and ESI^- LCQ ion trap MS	[12]
Dityrosine (DiTyr)	Milk powder	Hydrolysis with 6N HCl and recovery of DiTyr by SPE	TSK-Gel Amide 80 (Tosoh BioSep), 250 mm × 1.5 mm, 5 µm, 80 Å	Solvent A, water and solvent B, ACN Both adjusted at 2.7 with formic acid The gradient was: 20% A for 3 min, 70% A at 10 min and isocratic at 70% A Flow rate = 200 µL/min	ESI^+ triple Q	[14]
Taurine and methionine	Beverages rich in carbohydrates	Drinks were diluted with the mobile phase (methanol/water, 60:40, v/v)	Astec apHera™ NH_2 polymer, 150 mm × 4.6 mm, 5 µm	Methanol–water (60/40) Flow rate = 0.6 mL/min	ESI^- triple Q	[15]

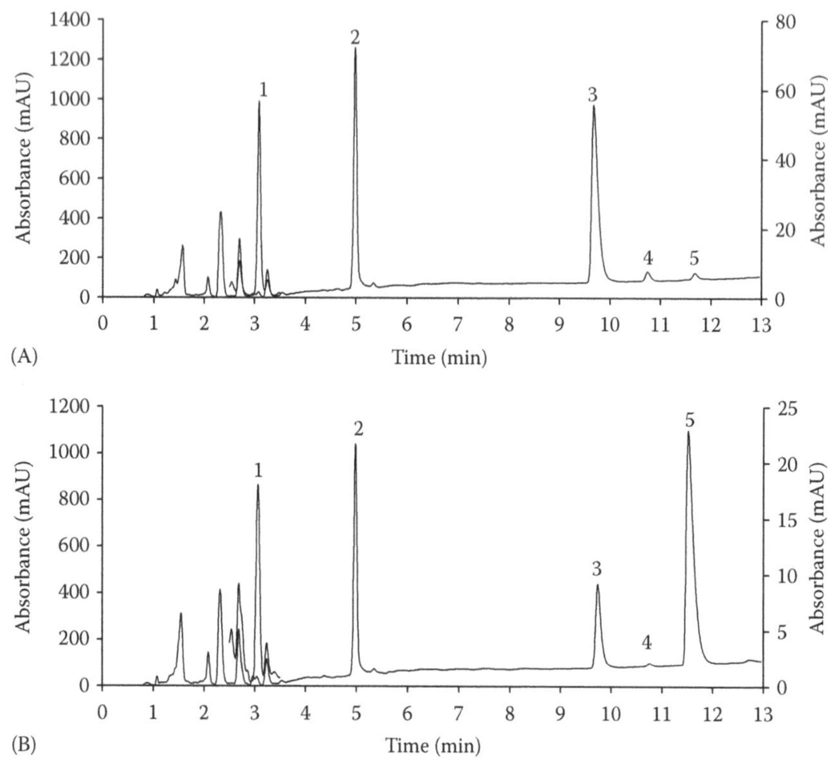

FIGURE 5.2 Chromatograms corresponding to pork loin (A) and chicken breast (B). Numbers 1, 2, 3, 4, and 5 correspond to creatinine, creatine, carnosine, baline, and anserine, respectively. The column was an Atlantis silica, 150 mm × 4.6 mm, 3 µm (water). Left scale indicates 214 nm measures for all compounds except creatinine (solid line), whereas right scale indicates 236 nm absorbance (dotted line) for creatinine.

Biceps femoris, *Gluteus maximus*, and *Longissimus dorsi*, whereas *Masseter*, a red oxidative muscle, was characterized by the lowest contents of these compounds.[7] HILIC methodology was also employed to evaluate the effect of different cooking procedures on the concentrations of creatine and creatinine in cooked ham. The results allowed the establishment of a correlation between the creatine/creatinine ratio in different sections of the ham and the effectiveness of the heat treatment in cooked ham processes.[8] The ZIC-pHILIC column has also been used by the current authors to study the evolution of creatine and creatinine during the processing of Spanish dry-cured ham obtaining good results in the isolation of these compounds (see Figure 5.3).

Another application demonstrating the benefits of HILIC separations was the detection of 1-deoxynojrimycin in mulberry leaves. This compound is a natural alkaloid with some biological activity in vivo. Mulberry leaves containing such compound are used as ingredients in functional foods in Japan.[9] Functional foods claim to have a health-promoting or disease-preventing property beyond the basic function of supplying nutrients. The method consisted of extraction with a mixture of acetonitrile and water containing 6.5 mM ammonium acetate, pH 5.5 (81:19, v/v) followed by

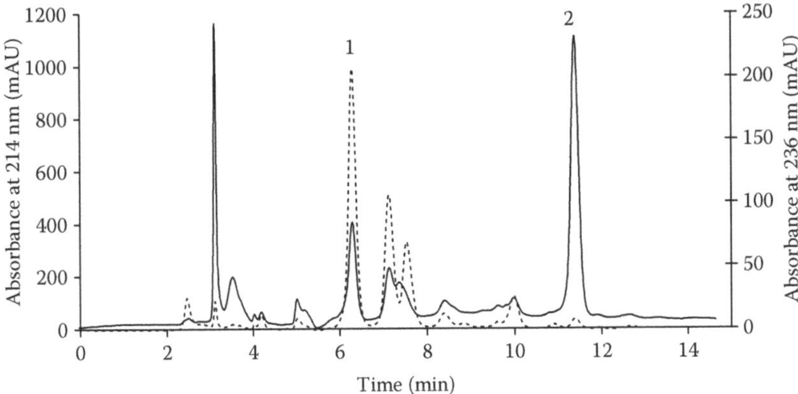

FIGURE 5.3 Chromatogram of a 50 day post-salted dry-cured ham sampled at the *semimembranosus* muscle. The column was a ZIC-pHILIC, 150 mm×4.6 mm, 5 μm (SeQuant). Creatine (1): solid line corresponding to 210 nm. Creatinine (2): dotted line corresponding to 236 nm.

separation using an amide-bonded silica column (TSK-Gel Amide-80 from Tosoh) and detection with an evaporative light-scattering detector (see Table 5.1). The detection limit was 100 ng (0.613 nmol) on the column. Thus, this analysis is useful to determine the content of mulberry leaves as ingredient in functional foods. The analysis of water soluble vitamins in energy drinks after HILIC separation with the TSK-Gel Amide-80 (Tosoh) column and UV detection has also been recently reported.[10]

5.2.2 HILIC Coupled to Mass Spectrometry

HILIC mode separations provide many advantages when coupled to mass spectrometry (MS) and this combination has been widely used in food analysis. Compounds extracted from mushroom were analyzed using HILIC-MS to determine the key compounds contributing to the umami-taste and mouth-drying oral sensation of morel mushrooms.[11] These polar compounds were analyzed using the TSK-Gel Amide-80 (Tosoh) column coupled to an electrospray ionization (ESI) quadrupole mass spectrometer (MS) as described in Table 5.1.

HILIC in combination with ESI-MS has also been employed to separate and characterize low-molecular-weight compounds in some foods. Free amino acids, glutamyl di- and tri-peptides, their glycoconjugates, and organic acids have been separated and identified without the need for derivatization in complex food matrices such as wheat gluten hydrosylate and Parmesan cheese using a TSK-Gel Amide-80 (Tosoh) column.[12] These amino acids, peptides, and glycoconjugates eluted at the void volume in an RP-HPLC but could be separated and characterized by HILIC-ESI-MS in the positive ionization mode using an ammonium acetate buffer (pH 5.5). ESI in the negative ionization mode was used for the analysis of glycoconjugates with glutamate due to the negative charge at the second carboxylic acid group.[12] A complex mixture of glycopeptides and N- and O-glycopeptides have also been efficiently separated with a capillary ZIC-HILIC (SeQuant) column coupled to ESI-MS.[13]

A TSK-Gel Amide-80 (Tosoh) column was also used for the development of an HILIC-MS method to accurately quantify dityrosine (DiTyr) in milk powder samples using d_4-DiTyr as the internal standard.[14] Milk proteins were first hydrolyzed by hydrochloric acid, and DiTyr further recovered from the amino acid hydrolyzates by a solid-phase extraction (see Table 5.1). Results demonstrated that DiTyr could be potentially used as a good chemical marker of milk protein oxidation.

HILIC-MS analysis has been also extended to the field of water and energy drinks. A validated procedure based on HILIC coupled to tandem mass spectrometry was developed for the simultaneous determination of underivatized taurine and methionine in beverages rich in carbohydrates such as energy drinks.[15] Sample preparation only required an appropriate dilution before injection, and the chromatographic separation of the compounds of interest was carried out using an Astec apHera NH_2 polymer column.

5.3 APPLICATION OF HYDROPHILIC INTERACTION CHROMATOGRAPHY TO THE ANALYSIS OF SAFETY-RELATED COMPOUNDS IN FOODS

Compounds posing health threats have received much attention in recent years. One of these substances is melamine and its determination as a contaminant in milk, fish, and other food products has to be analyzed due to its presence from adulteration or misuse.[16] Melamine is a triazine-based chemical and is typically used in the manufacture of plastics. Melamine has been detected in wheat flour, milk, fish, feed, etc. The main health concern with melamine is the formation of bladder stones with some studies reporting microcrystalluria.[17] Furthermore, insoluble melamine cyanurate salt crystals can also be formed in the kidney if cyanuric acid is present in the diet.[18] A HILIC-based method using bare silica has been recently developed for the analysis of melamine in fish and shrimp edible tissues with a detection limit as low as 3.2 ng/g.[16] Further chromatographic details are shown in Table 5.2. The authors surveyed 105 market-ready shrimp, catfish, tilapia, salmon, eel, and other fishes finding melamine concentrations exceeding 50 ng/g for those raised with feeds containing low amounts of melanine, 6.7 mg/kg. Other authors have used a ZIC-HILIC column for the analysis of melanine in milk powder with UV detection, reaching a limit of quantification of 15 ng/mL.[17]

Dichloroacetic acid was detected in drinking water using ion-exchange and HILIC on an amino column in conjunction with ESI-MS.[18] A TSK-Gel Amide-80 (Tosoh) column in combination with ESI-MS/MS was successfully employed to separate polar mushrooms toxins including amanitins, phallotoxins, and muscarine.[19] These authors studied the impact of different mobile phases on the separation, concluding that the addition of methanol as a modifier in the mobile phase (comprising acetonitrile, methanol, and ammonium formate, pH 3.5) could further enhance the HILIC separation for the studied analytes.

HILIC-ESI-MS methodology was also effective when applied to roasted coffee powder for the analysis of two polar compounds, 4(5)-methylimidazole and 2-acetyl-4(5)-1,2,3,4)-tetrahydroxybutyl-imidazole. These undesirable compounds, which are

TABLE 5.2
Performance of Some Recent HILIC-HPLC-Based Methods for the Analysis of Contaminant Residues in Foods

Compounds to Analyze	Food Matrix	Main Extraction	Type of Column	Elution Conditions	Detector	LOD[a] (ng/mL)	LOQ[a] (ng/mL)	Reference
Melamine	Fish and shrimp	Dry ice homogenization, extraction with acetonitrile, water, and HCl Oasis MCX SPE	Atlantis silica (Waters), 50 mm × 3.0 mm, 3 µm	Gradient from 95% to 50% of acetonitrile and 20 mM aqueous ammonium formate Flow rate = 0.35 mL/min	ESI$^+$ triple Q	3.2	—	[18]
Melamine	Milk powder	Aqueous solution containing 2.5% ammonia or perchloric acid Centrifugation	SeQuant ZIC-HILIC (Merck), 150 mm × 4.6 mm, 5 µm, 200 Å	Acetonitrile/ammonium acetate 25 mN (85/15%, v/v), pH 6.8 Flow rate = 2 mL/min	UV at 210 nm	—	15	[19]
Dichloroacetic acid	Drinking water	Water samples were dried under vacuum and reconstituted in 60:40 ACN:water	Luna Amino (Phenomenex) 150 mm × 2.1 mm, 5 µm	Solvent A, ACN and solvent B, 40 mM NH_4 formate in water Linear gradient from 90% to 30% of solvent A in 5 min Flow rate = 0.7 mL/min	ESI$^-$ triple Q	1	5	[20]

(continued)

TABLE 5.2 (continued)
Performance of Some Recent HILIC-HPLC-Based Methods for the Analysis of Contaminant Residues in Foods

Compounds to Analyze	Food Matrix	Main Extraction	Type of Column	Elution Conditions	Detector	LOD[a] (ng/mL)	LOQ[a] (ng/mL)	Reference
Polar toxins								
Amanitins	Mushroom	Extraction with acetonitrile and formic acid Oasis MAX anion-exchange cartridge	TSK-Gel Amide 80 (Tosoh BioSep), 250 mm × 2.0 mm, 5 µm, 80 Å	Different proportions of 2 mM NH4-formate + 5 mM formic acid at pH 3.5 acetonitrile and MeOH were tested Flow rate = 0.2 mL/min	ESI MS/MS	20	30	[21]
Phallotoxins	SAA	SAA	SAA	SAA	SAA	20	30	SAA
Muscarine	SAA	SAA	SAA	SAA	SAA	5	5	SAA
4(5)-methylimidazole (4-MeI)	Roasted coffee powder	Supercritical CO_2 modified with aqueous MeOH and SPE using SCX Disc cartridges	Atlantis silica (Waters) 150 mm × 2.1 mm, 3 µm	Mobile phase. MeOH (80%) and 0.01 M NH_4-formate in water (20%) Isocratic mode Flow rate = 0.2 mL/min	LC-DAD at 290 nm	0.0015	0.0044	[22]

2-Acetyl-4(5)-(1,2,3,4) tetrahydroxybutyl-imidazole (THI)	SAA	SAA	SAA	SAA	LC-DAD at 215 nm	0.002	0.0063	SAA
Glycoalkaloids α-chaconine and α-solanine	Potatoes	Sample extracted with chloroform:methanol: water (2:5:2, v/v/v)	Polyhydroxyethyl A (PolyLC) 100 mm × 2.1 mm, 3 µm	Solvent A, 5mM NH_4Ac pH 5.5 and solvent B, ACN Step gradient from 20% A to 100% A	ESI^+ LC MS/MS	—	20	[23]
Biogenic amines								
Cadaverine	Cheese	Hydrolysis with 0.1 M HCl C-18 SPE	Atlantis silica (Waters), 150 × 2.1 mm, 3 µm	Acetonitrile and ammonium formate 50 mM in water, pH 4.0	ESI^+ MS/MS	3.5	10	[26]
Histamine	SAA	SAA	SAA	SAA	SAA	1.5	5	SAA
Putrescine	SAA	SAA	SAA	SAA	SAA	1.2	4	SAA
Spermidine	SAA	SAA	SAA	SAA	SAA	1.0	3	SAA
Spermine	SAA	SAA	SAA	SAA	SAA	1.2	6	SAA
Tryptamine	SAA	SAA	SAA	SAA	SAA	1.3	3.5	SAA
Tyramine	SAA	SAA	SAA	SAA	SAA	1.1	3	SAA

Note: SAA, same as above.

[a] Units in ng/mL or ng/g.

neurotoxic, are formed during the coffee roasting process. The analytes were extracted with supercritical carbon dioxide modified with aqueous methanol, and subsequently analyzed using HILIC-MS with an Atlantis HILIC silica (Waters) column.[20]

HILIC has become popular recently (in the last 4–5 years) for the analysis of polar and highly hydrophilic compounds. In relation to this, a classical RP column (Hyperclone ODS C18 column, 100 × 2 mm, 3 μm, from Phenomenex) was compared to a HILIC column (polyhydroxyethyl A column, 100 × 2.1 mm, 3 μm, from PolyLC) for the detection and quantification of α-chaconine and α-solanine, the two major glycoalkaloids in potatoes, using LC-ESI-MS/MS.[21] Both methods proved to be powerful additions to standard metabolomic approaches although according to these authors, the HILIC-based method somehow disclosed its lack of ruggedness with respect to chromatographic peak shapes when hundreds of samples were analyzed for this application. In spite of this, HILIC methodology has extensively proved to be very robust in high throughput laboratories.[22,23]

The determination of biogenic amines in cheese is of interest because their presence and relative amounts give useful information about freshness, level of maturing, quality of storage, and cheese typification. A method based on HILIC MS/MS was developed for the determination of seven biogenic amines (cadaverine, histamine, putrescine, spermidine, spermine, tryptamine, and tyramine) in cheese.[24] The method used an Atlantis silica (Waters) column (see Table 5.2) and was reported to improve both the chromatographic separation and the mass spectrometry detection with respect to an HPLC-MS/MS method developed in that laboratory.

The presence of residues of growth promoters or veterinary drugs in foods of animal origin, which are generally used in farm animals for therapeutic and prophylactic purposes, makes necessary the control of these substances in order to assure consumer protection against any potentially harmful effects. Numerous analytical methods exist for determining such substances[25–28] in foods of animal origin, but there is a recent trend toward the use of HILIC to separate the more polar compounds like antibiotics. The main aspects of recent HILIC methods developed for the detection of drugs residues in foods are shown in Table 5.3.

Sulfonamide antibacterial residues have been successfully analyzed in milk and eggs using a HILIC-MS method and a Luna NH_2 (Phenomenex) column with previous polymer monolith microextraction.[29] The use of a ZIC-HILIC (SeQuant) column allowed good detection of aminoglycosides antibiotics in kidney,[30] neomycin in serum,[31] and cytostatics in wastewaters.[32] The use of an Altima HP HILIC column (Alltech) allowed the determination of the antibiotics spectinomycin and lincomycin in hog manure[33] (see Table 5.3). Carbadox and olaquindox have been analyzed in feed using an Acquity BEH HILIC (Waters) column followed by DAD.[34]

In summary, the use of HILIC coupled to DAD or MS detectors has been growing rapidly in food applications in recent years. In fact, a good number of food quality or food safety related applications are already available as reported in this chapter. Furthermore, new types of HILIC phases and columns are being developed that will facilitate the further expansion of HILIC to a wider range of applications in the field of food analysis for quality and safety purposes.

TABLE 5.3
Performance of Some Recent HILIC-HPLC-Based Methods for the Analysis of Drugs Residues in Foods

Compounds to Analyze	Food Matrix	Main Extraction	Type of Column	Elution Conditions	Detector	LOD[a] (ng/mL)	LOQ[a] (ng/mL)	Reference
Sulfonamides	Whole milk	PMME (poly(MAA-EGDMA) monolithic column	Luna NH_2 column, Phenomenex (Torrence, California) 150 mm × 2.0 mm, 3 µm	Solvent A, acetonitrile + 0.05% formic acid, and solvent B, water gradient from 100% to 95% ACN	ESI^+ MS	0.4–5.7	1.8–19.0	[31]
SAA	Eggs	SAA	SAA	SAA	SAA	0.9–9.8	3.2–32.8	SAA
Gentamicin	Swine kidney	Weak cation exchange SPE	ZIC® HILIC, Sequant (Umea, Sweden) 2.1 mm × 100 mm, 5 µm	Solvent A, 1% formic acid + 150 mM ammonium acetate and solvent B acetonitrile Gradient from 20% to 95% of solvent A Flow rate = 0.3 mL/min, 32°C	ESI^+ MS/MS	6	19	[32]
Spectinomycin	SAA	SAA	SAA	SAA	SAA	19	63	SAA
Dihydrostreptomycin	SAA	SAA	SAA	SAA	SAA	16	52	SAA
Kanamycin	SAA	SAA	SAA	SAA	SAA	6	58	SAA
Apramycin	SAA	SAA	SAA	SAA	SAA	16	52	SAA
Streptomycin	SAA	SAA	SAA	SAA	SAA	18	90	SAA
Neomycin	SAA	SAA	SAA	SAA	SAA	33	108	SAA

(continued)

TABLE 5.3 (continued)
Performance of Some Recent HILIC-HPLC-Based Methods for the Analysis of Drugs Residues in Foods

Compounds to Analyze	Food Matrix	Main Extraction	Type of Column	Elution Conditions	Detector	LOD[a] (ng/mL)	LOQ[a] (ng/mL)	Reference
Neomycin	Human serum	Oasis MCX SPE cartridges	ZIC-HILIC, Sequant (Umea, Sweden) 100 mm × 2.1 mm, 5 μm	Solvent A, mixture of acetonitrile, 10 mM ammonium acetate and formic (5/95/0.2, v/v/v), and solvent B (95/5/0.2, v/v/v) Gradient from 20% to 80% of solvent A Flow rate = 0.6 mL/min	ESI MS/MS	—	100	33
5-Fluorouracyl	Wastewater	Isolute ENV + SPE cartridge and Speedisk H_2O-philic SA-DVB disc cartridge	ZIC-HILIC in polyetheretherketone, Sequant (Umea, Sweden) 150 mm × 2.1 mm, 3.5 μm	30 mM ammonium acetate and acetonitrile (2/3, v/v) Flow rate = 0.2 mL/min	LTQ orbitrap	—	0.005	[34]
Gemcitabine	SAA		SAA		SAA	—	0.0009	SAA

Spectinomycin	Hog manure	Oasis HLB + Oasis WCX SPE cartridges	Altima HP HILIC (Alltech Assoc, IL) 150 mm × 2.1 mm, 3 µm	Solvent A, 90/10 Acetonitrile/water + 0.1% formic acid, and solvent B, 90/10 water/acetonitrile + 0.1% formic acid Isocratic conditions of 65% solvent A and 35% solvent B Flow rate = 0.2 mL/min	APCI+ triple Q	—	6	[35]
Lincomycin	SAA	Oasis HLB SPE cartridge	SAA	SAA	SAA	—	0.04	SAA
Carbadox	Feed	C18 SPE	Acquity UPLC BEH HILIC column 100 mm × 2.1 mm, 1.7 µm	Acetonitrile–water (95/5, v/v) containing 10 mM ammonium acetate Flow rate = 0.5 mL/min 30°C	DAD at 307 nm	20	70	[36]
Olaquindox	SAA	SAA	SAA	SAA	DAD at 384 nm	30	100	SAA

Note: SAA, same as above.

[a] Units in ng/mL or ng/g.

ACKNOWLEDGMENTS

Project A-001/09 from Platform on Food Safety, Conselleria de Sanitat, Generalitat Valencia (Spain) is acknowledged.

REFERENCES

1. Batlle, N. et al. *J. Food Sci.* 2000, 65: 413–416.
2. Batlle, N. et al. *J. Food Sci.* 2001, 66: 68–71.
3. Toldrá, F. In: *Handbook of Food Science, Technology and Engineering* (Y.H. Hui, Ed.), CRC Press, Boca Raton, FL, 2006, pp. 28-1–28-18.
4. Mora, L. et al. *Food Chem.* 2010, 123: 1282–1288.
5. Wu, H.C. et al. *J. Food Drug Anal.* 2003, 11: 148–153.
6. Mora, L. et al. *J. Agric. Food Chem.* 2007, 55: 4664–4669.
7. Mora, L. et al. *Meat Sci.* 2008, 79: 709–715.
8. Mora, L. et al. *J. Agric. Food Chem.* 2008, 56: 11279–11284.
9. Kimura, T. et al. *J. Agric. Food Chem.* 2004, 52: 1415–1418.
10. Roemling, R. et al. *The Applications Book* 2009, March: 21–23.
11. Rotzoll, N. et al. *J. Agric. Food Chem.* 2005, 53: 4149–4156.
12. Schlichtherle-Cerny, H. et al. *Anal. Chem.* 2003, 75: 2349–2354.
13. Tagekawa, Y. et al. *J. Sep. Sci.* 2008, 31: 1585–1593.
14. Fenaille, F. et al. *J. Chromatogr. A* 2004, 1052: 77–84.
15. De Pearson, M. et al. *J. Chromatogr. A* 2005, 1081: 174–181.
16. Food and Drug Administration. 72 FR 30014. May 30, 2007.
17. World Health Organisation, 2009. Report 1–66.
18. Andersen, W.C. et al. *J. Agric. Food Chem.,* 2008, 56: 4340–4347.
19. Jiang, W. and Ihunegoo, F.N. *The Applications Book* 2009, March: 40–41.
20. Dixon, A.M. et al. *J. Liq. Chromatogr. Relat. Technol.* 2004, 27: 2343–2355.
21. Chung, W.C. et al. *J. Chromatogr. Sci.* 2007, 45: 104–111.
22. Lojkova, L. et al. *Food Addit. Contam.* 2006, 23: 963–973.
23. Zywicki, B. et al. *Anal. Biochem.* 2005, 336: 178–186.
24. Naidong, W. *J. Chromatogr. B*, 2003, 796: 209–224.
25. Strege, M. *Anal. Chem.* 1998, 70: 2439–2445.
26. Gianotti, V. et al. *J. Chromatogr. A* 2008, 1185: 296–300.
27. Reig, M. and Toldrá, F. In: *Handbook of Muscle Foods Analysis* (L.M.L. Nollet and F. Toldrá, Eds.), CRC Press, Boca Raton, FL, 2009, pp. 837–854.
28. Reig, M. and Toldrá, F. In: *Handbook of Processed Meats and Poultry Analysis* (L.M.L. Nollet and F. Toldrá, Eds.), CRC Press, Boca Raton, FL, 2009, pp. 647–664.
29. Reig, M. and Toldrá, F. In: *Safety of Meat and Processed Meats* (F. Toldrá, Eds.), Springer, New York, 2009, pp. 365–390.
30. Verdon, E. In: *Handbook of Muscle Foods Analysis* (L.M.L. Nollet and F. Toldrá, Eds.), CRC Press, Boca Raton, FL, 2009, pp. 855–947.
31. Zheng, M.M. et al. *Anal. Chim. Acta* 2008, 625: 160–172.
32. Ishii, R. *J. Food Addit. Contam.* 2008, 20: 1–11.
33. Oertel, R. et al. *J. Pharm. Biomed. Anal.* 2004, 35: 633–638.
34. Kovalova, L. et al. *J. Chromatogr. A.* 2009, 1216: 1100–1108.
35. Peru, K.M. et al. *J. Chromatogr. A.* 2006, 1107: 152–158.
36. Kesiunaitė, G. et al. *J. Chromatogr.* 2008, 1209: 83–87.

6 Hydrophilic Interaction Liquid Chromatography–Mass Spectrometry (HILIC–MS) of Paralytic Shellfish Poisoning Toxins, Domoic Acid, and Assorted Cyanobacterial Toxins

Carmela Dell'Aversano

CONTENTS

6.1 INTRODUCTION

Marine toxins produced by harmful algae are listed among the most important causative agents of poisoning episodes involving seafood consumers.[1,2] Toxic incidents are usually associated with blooms of toxigenic plankton species in shellfish-producing regions. The phenomenon may take many forms, ranging from massive "red tides" or blooms of cells that discolor the water to dilute, inconspicuous concentrations of plankton cells that get noticed only because of the harm caused by the highly potent toxins they produce.[3] Filter-feeding mollusks consume the plankton, accumulate toxins in their edible tissues, and thus become the part of the food chain responsible for transmission of toxicity to humans.

The poisoning syndromes have been named paralytic, diarrhetic, neurotoxic, and amnesic shellfish poisoning (PSP, DSP, NSP, and ASP) based on major symptoms they induce in humans.[4] The causative toxins range from polar, low-molecular-weight (MW) compounds to high-MW lipophilic substances; most are thermally labile and some are pH, oxygen, and light sensitive. Basically, toxicity is the only common factor among different toxins. So it is not surprising that the AOAC mouse bioassay is the most widely used method for toxin detection in a regulatory setting.[5] It provides two different extraction procedures for either lipophilic or hydrophilic toxins and subsequent intraperitoneal injection of aliquots of each potentially toxic shellfish extract into three 20g mice. An observation time follows to determine symptoms and time-to-death, which correlates with the amount of toxin contained in the sample. Although this approach provides a single integrated response from all the toxins, it suffers from poor reproducibility, low sensitivity, and interferences from other components in the extract. Most importantly, it cannot single out the causative toxins and variation in toxin profiles cannot be monitored. Thus, instrumental methods are required to identify and monitor the toxins.

The analytical technique must provide low detection limits (μg/kg), high-resolution separation, and/or high detection selectivity to deal with complicated sample matrices. In addition, since a contaminated sample may contain several toxins from within a class or even different classes of toxins, analytical methods that allow the combined analysis of assorted toxins are highly desirable. The combination of liquid chromatography and mass spectrometry (LC-MS) meets all the above requirements and has shown great potential both in regulatory and research settings. It is one of the most powerful tools for detection and quantitation of toxins in plankton and shellfish samples, even at trace levels, the identification of new compounds, and the investigation of toxin production and metabolism.[6] A number of LC-MS methods have been developed for all of the known toxins.[7,8] Most of them are based on the use of reversed phase liquid chromatography (RPLC), which works well for the analysis of lipophilic toxins[9] but suffers from a number of drawbacks when very polar and hydrosoluble compounds have to be analyzed (see the following discussion). A new approach based on the use of hydrophilic interaction liquid chromatography (HILIC) coupled with electrospray ionization tandem mass spectrometry (ESI-MS/MS) has been recently proposed for sensitive and selective detection of very polar and ionic compounds belonging to the paralytic shellfish poisoning (PSP) toxin class,[10] even in combination with domoic acid (DA),[11] the causative toxin of the amnesic shellfish poisoning (ASP) syndrome, and assorted cyanobacterial toxins (Figure 6.1).[12] Details on HILIC-MS method development will be provided in the following sections.

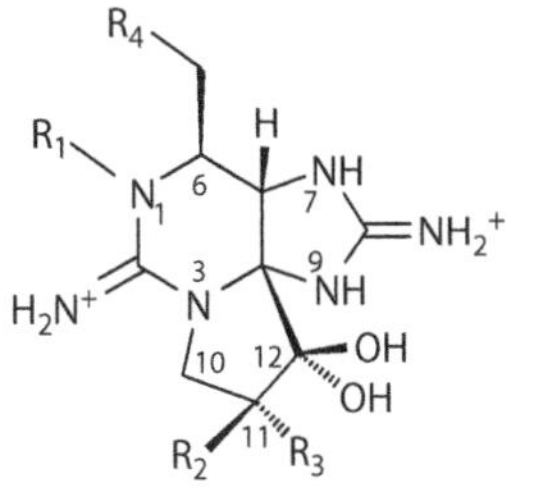

R_1	R_2	R_3	R_4	Toxin
H	H	H	$-OC(=O)NH_2$	STX
H	H	OSO_3^-		GTX2
H	OSO_3^-	H		GTX3
OH	H	H		NEO
OH	H	OSO_3^-		GTX1
OH	OSO_3^-	H		GTX4
H	H	OH		11α-OH-STX
H	OH	H		11β-OH-STX
H	H	H	$-OC(=O)NHSO_3^-$	B1 (GTX5)
H	H	OSO_3^-		C1
H	OSO_3^-	H		C2
OH	H	H		B2 (GTX6)
OH	H	OSO_3^-		C3
OH	OSO_3^-	H		C4
H	H	H	$-OH$	dcSTX
H	H	OSO_3^-		dcGTX2
H	OSO_3^-	H		dcGTX3
OH	H	H		dcNEO
OH	H	OSO_3^-		dcGTX1
OH	OSO_3^-	H		dcGTX4

STX = Saxitoxin
NEO = Neosaxitoxin
GTX = Gonyautoxins

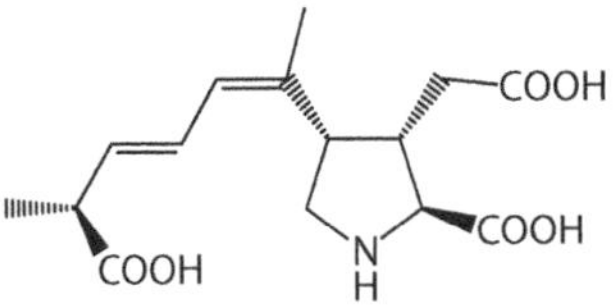

DA

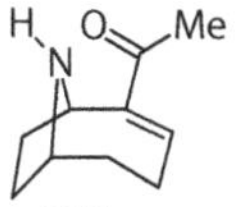

ATX-a

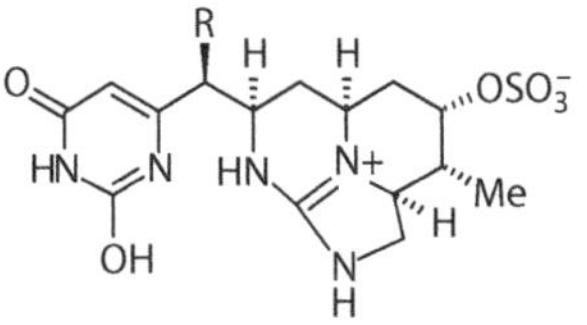

R = OH CYN
R = H doCYN

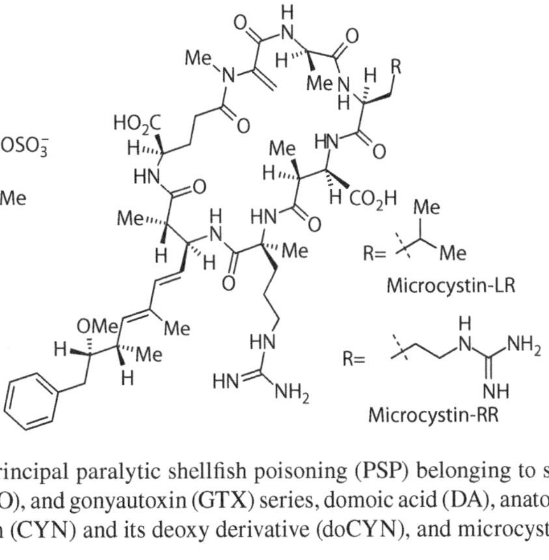

Microcystin-LR

Microcystin-RR

FIGURE 6.1 Structures of principal paralytic shellfish poisoning (PSP) belonging to saxitoxin (STX), neosaxitoxin (NEO), and gonyautoxin (GTX) series, domoic acid (DA), anatoxin-a (ATX-a), cylindrospermopsin (CYN) and its deoxy derivative (doCYN), and microcystins.

6.2 PARALYTIC SHELLFISH POISONING TOXINS

Saxitoxin (STX) and its analogues, also known as paralytic shellfish poisoning (PSP) toxins, are fast-acting neurotoxins that pose serious risks to public health and have a significant economic impact on the shellfish industry throughout the world. They are produced by marine dinoflagellates belonging to *Alexandrium*, *Pyrodinium*, and *Gymnodinium* genera[13] and, generally, cause persistent problems to humans due to their accumulation in filter-feeding mollusks. The PSP syndrome is characterized by neurological distress, which typically appears within 15–30 min following consumption of contaminated shellfish and can result in death, if the toxin concentration is sufficiently high.

Structurally, PSP toxins are tetrahydropurine derivatives and can be divided into three groups: (1) carbamoyl (STX, GTX2, GTX3, NEO, GTX1, GTX4, 11α,β-OH-STX), (2) *N*-sulfocarbamoyl (B1, C1, C2, B2, C3, C4), and (3) decarbamoyl toxins (dcSTX, dcGTX2, dcGTX3, dcNEO, dcGTX1, dcGTX4), based on the nature of the side chain (Figure 6.1). They are potent, reversible blockers of voltage-activated sodium channels on excitable cells,[14] but, due to the differences in charge state and substitution groups to the basic STX structure, they bind with different affinities to site 1 of sodium channels resulting in different toxicities. The carbamoyl toxins are the most toxic and the *N*-sulfocarbamoyl derivatives are the least toxic.[15] Thus, health risks can be reliably assessed just if the level of each toxin is individually determined.

The AOAC mouse bioassay is widely used for detecting the presence of PSP toxins in routine monitoring of shellfish.[5] In the extraction procedure used, the *N*-sulfocarbamoyl toxins are converted to the corresponding carbamoyl toxins. Alternative assays based either on in vitro cell toxicity, receptor binding, or immunological response have been developed.[8] None of the above assays provides detailed information on the toxin profile in samples, nor can they be used for the precise quantitative analysis of samples containing variable levels of individual toxins. So, positive results generally require instrumental confirmation in critical situations.

PSP toxins pose a real challenge to the development of instrumental methods for their detection and quantitation since they include a great variety of closely related structures with three different charge states (0, +1, and +2). They are nonvolatile and thermally labile and lack any useful chromophore, which eliminate most traditional chromatographic techniques such as gas chromatography (GC) and liquid chromatography (LC) with ultraviolet (UV) detection.

The most used technique for the analysis of PSP toxins is ion-pair LC coupled with postcolumn oxidation and fluorescence detection (LC-ox-FLD),[16] which provides conversion of PSP toxins into fluorescent derivatives under alkaline conditions. The technique is highly sensitive, but interfering compounds can make difficult interpretation of results.[17] An alternative method is the precolumn oxidation approach (ox-LC-FLD).[18] This method is rapid, sensitive, and fully automated but interpretation of quantitative results is complex because some toxins give the same oxidation product, while others give two or three products.[19]

Because of the efficient ionization of STX and its analogues when they are injected into a mass spectrometer (MS) equipped with an electrospray ionization (ESI) source,[20] a number of methods that use ESI-MS as detector for either a capillary electrophoresis (CE) or a liquid chromatography (LC) device have been developed.

Although CE-MS is ideally suited to the analysis of the highly charged PSP toxins,[21] it is not possible to analyze all the toxins in a single analysis due to their different charge states. In addition, the technique suffers from interference from coextracted salts.

The nature of LC stationary phase is critical for developing an efficient LC-MS method for the determination of PSP toxins. An efficient approach based on ion exchange liquid chromatography with MS detection has been recently proposed by Jaime et al.[22] but the most common reverse-phase LC provides sufficient retention of the highly polar PSP toxins only if ion-pairing reagents (heptanesulfonic acid, heptafluorobutyric acid, or tetrabutylammonium sulfate) are added to the mobile phase.[16] Such agents result in serious interference with MS detection by causing suppression of ionization and ion source contamination. In addition, the neutral C toxins have to be analyzed in a separate run from mono- and bi-charged toxins.

The combination of HILIC with ESI-MS/MS detection demonstrated to help overcome drawbacks of the above-mentioned assays and instrumental methods. The technique allows simultaneous determination of all PSP toxins with a high degree of selectivity and eliminates the need for further confirmation. The mobile phase does not use ion-pair agents, so that it does not reduce ionization efficiency. Further, the mobile phase has high amounts of organic modifier so that ionization yield is enhanced. All details of the HILIC-MS method are summarized in the following section along with its application to plankton and shellfish samples.

6.2.1 HILIC-MS Determination of PSP Toxins

A key role in setting up the HILIC-MS method for detection of PSP toxins is played by MS behavior of each saxitoxin-like compound.

In Table 6.1, MS and MS/MS data for most PSP toxins are summarized. Such data were acquired on three PE-Sciex MS instruments, namely a single and a triple quadrupole MS equipped with an ESI source (API-165 and API-III+, respectively) and a triple quadrupole MS equipped with a Turbospray® interface (API-4000).

STX, NEO, and their decarbamoyl derivatives (dcSTX and dcNEO), which exist as dications in solution, did not produce doubly charged ions in ESI-MS but gave $[M + H]^+$ ions as the base peak in their MS spectra, with a small fragment ion corresponding to the loss of a water molecule.

A different fragmentation pattern was observed with the two epimeric pairs of gonyautoxins, GTX1 and GTX4 and GTX2 and GTX3. The $[M + H]^+$ ion was still the most abundant ion in the MS spectra of GTX1 and GTX2, which present an hydroxysulfate group at C-11 in an α-orientation, whereas an $[M + H\text{-}80]^+$ fragment ion, corresponding to loss of SO_3, dominated the spectra of 11β-hydroxysulfate gonyautoxins (GTX3 and GTX4). The same behavior was observed for the decarbamoyl gonyautoxins, dc-GTX1-4.

The spectra of *N*-sulfocarbamoyl derivatives (B1, C1, C2, B2, C3, C4) varied strongly with instrument type. Orifice voltage settings of 10 and 50 V on the API-165 and API-III+ instruments, respectively, and a declustering potential of

TABLE 6.1
Mass Spectral Data of Paralytic Shellfish Poisoning (PSP) Toxins

	MS Spectra (%RI)[a]			MS/MS Spectra (%RI)[b]		
Toxin	$[M + NH_4]^+$	$[M + H]^+$	Fragment	Precursor	Product	Loss of
STX		300 (100)	282 (10)	300 (40)	204 (100)	$- 2H_2O - NH_3 - NHCO$
					138 (75)	$- H_2O - NH_3 - CO_2 - HNCNH - H_2C_2NH$
					179 (60)	$- H_2O - NH_3 - CO_2 - HNCNH$
					186 (45)	$- 3H_2O - H_3 - NHCO$
					282 (40)	$- H_2O$
					221 (25)	$- H_2O - NH_3 - CO_2$
GTX2		396 (5)	316 (100)	396 (0)	316 (100)	$- SO_3$
					298 (10)	$- SO_3 - H_2O$
GTX3		396 (100)	298 (38)	396 (0)	298 (100)	$- SO_3 - H_2O$
			316 (27)		316 (10)	$- SO_3$
			378 (23)		220 (10)	$- SO_3 - 2H_2O - NH_3 - NHCO$
NEO		316 (100)	298 (12)	316 (100)	220 (68)	$- 2H_2O - NHCO - NH_3$
					138 (65)	$- H_2O - NH_3 - CO_2 - HNCNH - H_2C_2NOH$
					298 (62)	$- H_2O$
					177 (60)	$- 2H_2O - NH_3 - CO_2 - HNCNH$
					237 (42)	$- H_2O - NH_3 - CO_2$
GTX1		412 (8)	332 (100)	412 (0)	332 (100)	$- SO_3$
					314 (15)	$- SO_3 - H_2O$
GTX4		412 (100)	394 (55)	412 (0)	314 (100)	$- SO_3 - H_2O$
			332 (12)		332 (10)	$- SO_3$
			314 (9)		253 (5)	$- SO_3 - H_2O - NH_3 - CO_2$
11OH-STX		316 (100)		316 (27)	148 (100)	Not assigned
					108 (50)	Not assigned

					220 (45)	$- 2H_2O - NHCO - NH_3$
					298 (40)	$- H_2O$
					196 (38)	$- H_2O - NHCO - NH_3 - NHCNH$
					237 (10)	$- 2H_2O - NHCO$
B1 (GTX5)		380 (100)	300 (98)	380 (0)	300 (100)	$- SO_3$
			282 (10)		282 (38)	$- SO_3 - H_2O$
			257 (8)		204 (30)	$- SO_3 - 2H_2O - NHCO - NH_3$
					221 (18)	$- SO_3 - H_2O - CO_2 - NH_3$
C1	493 (50)	476 (18)	396 (100)	493 (0)	316 (100)	$- NH_3 - 2SO_3$
			413 (38)		396 (20)	$- NH_3 - SO_3$
			316 (25)		298 (2)	$- NH_3 - 2SO_3 - H_2O$
C2	493 (45)	476 (15)	396 (100)	493 (0)	298 (100)	$- NH_3 - 2SO_3 - H_2O$
			378 (50)		316 (68)	$- NH_3 - 2SO_3$
			413 (42)		378 (30)	$- NH_3 - SO_3 - H_2O$
			316 (5)		396 (12)	$- NH_3 - SO_3$
B2 (GTX6)		396 (100)	316 (30)	396 (0)	316 (100)	$- SO_3$
					220 (5)	$- SO_3 - 2H_2O - NH_3 - NHCO$
					298 (4)	$- SO_3 - H_2O$
					237 (3)	$- SO_3 - 2H_2O - NHCO$
					177 (3)	$- SO_3 - 2H_2O - NH_3 - HNCNH - CO_2$
C3	509 (12)	492 (17)	412 (100)	509 (0)	332 (100)	$- NH_3 - 2SO_3$
			332 (11)		412 (22)	$- NH_3 - SO_3$
					314 (13)	$- NH_3 - 2SO_3 - H_2O$
					394 (2)	$- NH_3 - SO_3 - H_2O$
C4	509 (10)	492 (8)	412 (100)	509 (0)	314 (100)	$- NH_3 - 2SO_3 - H_2O$
			332 (1)		394 (95)	$- NH_3 - SO_3 - H_2O$
					332 (77)	$- NH_3 - 2SO_3$
					412 (38)	$- NH_3 - SO_3$

(continued)

TABLE 6.1 (continued)
Mass Spectral Data of Paralytic Shellfish Poisoning (PSP) Toxins

	MS Spectra (%RI)[a]			MS/MS Spectra (%RI)[b]		
Toxin	**$[M + NH_4]^+$**	**$[M + H]^+$**	**Fragment**	**Precursor**	**Product**	**Loss of**
dcSTX		257 (100)	239 (17)	257 (20)	126 (100)	Not assigned
					138 (65)	$- H_2O - NH_3 - 2NHCNH$
					222 (50)	$- H_2O - NH_3$
					180 (48)	$- H_2O - NH_3 - NHCNH$
					156 (45)	$- NH_3 - 2NHCNH$
					239 (25)	$- H_2O$
dcGTX2		353 (4)	273 (100)	353 (0)	273 (100)	$- SO_3$
					255 (15)	$- SO_3 - H_2O$
					126 (12)	Not assigned
					148 (10)	Not assigned
					238 (8)	$- SO_3 - H_2O - NH_3$
					196 (8)	$SO_3 - H_2O - NH_3 - NHCNH$
dcGTX3		353 (100)	335 (55)	353 (0)	255 (100)	$- SO_3 - H_2O$
			273 (42)		196 (25)	$- SO_3 - H_2O - NH_3 - NHCNH$
			255 (30)		273 (12)	$- SO_3$
					238 (12)	$- SO_3 - H_2O - NH_3$
					335 (7)	$- H_2O$
dcNEO		273 (100)	255 (20)	273 (40)	126 (100)	Not assigned

				225 (87)	Not assigned
				180 (60)	Not assigned
				138 (50)	Not assigned
				207 (40)	Not assigned
				255 (31)	$- H_2O$
				220 (25)	$- 2H_2O - NH_3$
dcGTX1	369 (5)	289 (100)	369 (0)	289 (100)	$- SO_3$
				271 (13)	$- SO_3 - H_2O$
				126 (7)	Not assigned
				195 (5)	Not assigned
dcGTX4	369 (100)	289 (1)	369 (0)	271 (100)	$- SO_3 - H_2O$
				195 (33)	Not assigned
				289 (15)	$- SO_3$
				178 (15)	Not assigned
				351 (10)	$- H_2O$

Source: Modified from Dell'Aversano, C. et al., *J. Chromatogr. A*, 1081, 190, 2005. With permission from Elsevier.

[a] MS spectra were obtained using a PE-SCIEX API-165 mass spectrometer. Percentage relative intensities (%RI) are reported in brackets.

[b] MS/MS spectra were obtained using a PE-SCIEX API-4000 mass spectrometer. Percentage relative intensities (%RI) are reported in brackets.

50 V on the API-4000 instrument provided minimum fragmentation of analytes while still maintaining a good background spectrum. The API-165 instrument gave a higher degree of fragmentation than the API-III+ and the API-4000, which resulted in a lower sensitivity. The Turbospray interface, which applies heated nitrogen to assist nebulization, gave more fragmentation if too high temperatures were used. Thus, a minimum temperature (275°C) had to be used to minimize temperature-induced fragmentation. Under such conditions, B1 gave a spectrum with abundant $[M + H]^+$ and $[M + H - 80]^+$ ions, while C1-4 gave both protonated and ammoniated ions, as well as associated fragment ions due to loss of SO_3 and/or H_2O molecules.

The protonated molecules or adduct ions and the main fragment ions of each toxin were selected for select ion monitoring (SIM) experiments, taking into account that, for some toxins, fragment ions caused by elimination of SO_3 from $[M + H]^+$ ion could potentially interfere with protonated molecules of other compounds (e.g., B1 with STX; C1, C2, GTX2, GTX3 and B2 with each other and NEO). This actually reduced the number of ions required for SIM, thus increasing sensitivity, but it also posed the need for a good chromatographic separation of the toxins.

Since multiple reaction monitoring (MRM) experiments can provide more selective detection of compounds, MS/MS spectra of PSP toxins were also examined. The spectra of selected toxins are shown in Figure 6.2 and the results are summarized in Table 6.1 for all compounds tested.

Significant differences were observed between fragmentation patterns of the epimeric pairs GTX2 and GTX3, GTX1 and GTX4, dcGTX2 and dcGTX3, dcGTX1 and dcGTX4, C1 and C2, and C3 and C4. Thus, more than one transition for each of these toxins had to be selected for MRM experiments. The recommended MRM transitions for all compounds are shown in Table 6.2.

The HILIC behavior of PSP toxins was investigated by analyzing a standard mixture containing STX, GTX2, GTX3, NEO, GTX1, GTX4, B1, B2, C1, C2, C3, C4, 11-OH-STX, dcSTX, dcGTX2, dcGTX3, dcNEO, dcGTX1, and dcGTX4, in comparable amounts under various conditions. The most abundant ions (Table 6.1) and transitions (Table 6.2) were selected for SIM or MRM experiments, respectively.

Two HILIC stationary phases were considered, namely TSK-Gel® Amide-80 and polyhydroxyethyl aspartamide (PHEA), since they had been reported to retain guanine, a compound very similar to PSP toxins, at various buffer concentrations.[23] The PHEA column provided a good separation between epimeric pairs (C1 and C2, GTX2 and GTX3, GTX1 and GTX4) when using water as eluent A and acetonitrile/water (95:5) as eluent B, both containing 2.0 mM ammonium formate, 3.6 mM formic acid, pH 3.5. However, selective detection of the major PSP toxins was possible only over a 120 min analysis time with poor peak shape (gradient: 90% B for 80 min, 90%–65% B over 15 min and hold 25 min). Shorter analysis time (35 min) resulted in poor separation between potentially interfering compounds, namely C1/GTX2 and STX/B1 (gradient: 80%–65% B over 20 min and hold 15 min). Several mobile phase systems were tried but separation selectivity could not be improved. Furthermore, extreme column bleed was observed at pH 2.5 to 3.5 in the mass range 200–500, which greatly hampered the determination of PSP toxins. Due to background and selectivity problems, no further development on the PHEA column was considered.

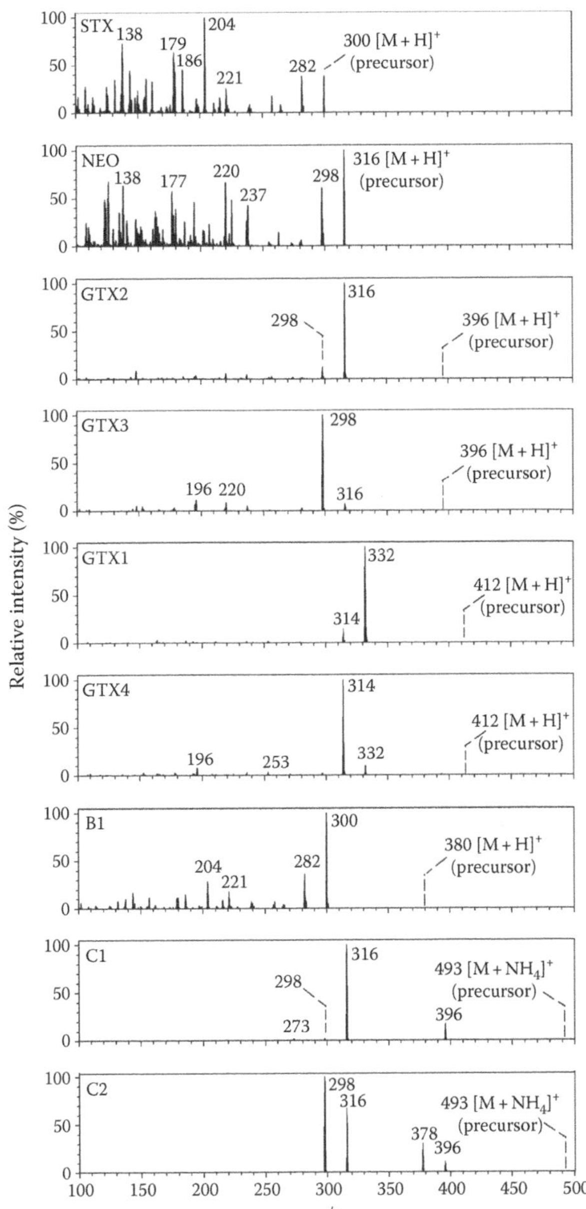

FIGURE 6.2 MS/MS spectra of selected PSP toxins obtained on the API-4000 system using a collision energy of 35 V. The $[M + H]^+$ ions of STX, NEO, GTX2, GTX3, GTX1, GTX4, B1, and the $[M + NH_4]^+$ ions of C1 and C2 were used as precursor ions. Assignments of labeled product ions are shown in Table 6.1. (Modified from Dell'Aversano, C. et al., *J. Chromatogr. A*, 1081, 190, 2005. With permission from Elsevier.)

TABLE 6.2
Recommended MRM Transitions for HILIC-MS Analysis of Major PSP Toxins[a]

Toxin	RT (min)[b]	RRT (min)[b]	*m/z > m/z*	*m/z > m/z*	*m/z > m/z*	*m/z > m/z*	*m/z > m/z*
STX	20.3	2.82	300 > 282 (100)	300 > 204 (80)			
GTX2	9.6	1.33	396 > 316 (100)	316 > 298 (30)	316 > 220 (10)	396 > 298 (5)	
GTX3	10.7	1.49	396 > 298 (100)	396 > 316 (20)	316 > 298 (10)		
NEO	21.0	2.91	316 > 298 (100)	316 > 220 (50)			
GTX1	9.8	1.36	412 > 332 (100)	412 > 314 (5)			
GTX4	10.9	1.51	412 > 314 (100)	412 > 332 (10)			
11OH-STX	24.9	3.46	316 > 298 (100)	316 > 220 (30)			
B1 (GTX5)	13.1	1.82	380 > 300 (100)	300 > 282 (5)	300 > 204 (5)		
C1[c]	7.2	1.00	493 > 316 (100)	396 > 316 (65)	396 > 298 (15)	316 > 298 (5)	493 > 298 (3)
C2[c]	8.0	1.11	396 > 298 (100)	493 > 298 (99)	493 > 316 (45)	316 > 298 (5)	396 > 316 (3)
B2 (GTX6)	14.6	2.03	396 > 316 (100)	396 > 298 (20)	316 > 298 (5)	316 > 220 (5)	
C3[c]	7.9	1.10	509 > 332 (100)	412 > 332 (50)	412 > 314 (13)	509 > 314 (8)	
C4[c]	8.8	1.22	412 > 314 (100)	509 > 314 (46)	509 > 332 (19)	412 > 332 (1)	
dcSTX	21.1	2.93	257 > 239 (100)				
dcGTX2	10.2	1.33	353 > 273 (100)	273 > 255 (30)	353 > 255 (15)		
dcGTX3	11.3	1.57	353 > 255 (100)	353 > 273 (15)	273 > 255 (5)		
dcNEO	20.8	2.89	273 > 255 (100)				
dcGTX1	10.1	1.40	369 > 289 (100)	369 > 271 (15)			
dcGTX4	11.4	1.58	369 > 271 (100)	369 > 289 (10)			

[a] Reported data for most PSP toxins were obtained using a PE-SCIEX API-III+ MS. Percentage relative intensity is provided for each transition. The most abundant transitions (100) are recommended for quantitative studies.

[b] Retention times (RT) and retention time relative to C1 (RRT) are referred to the optimized chromatographic conditions (Table 6.3).

[c] Data for C1-4 toxins were obtained using a PE-SCIEX API-4000 mass spectrometer. The $[M + H]^+$ or $[M + NH_4]^+$ ions for C toxins were not sufficiently abundant on the API-III+ MS system, in which case transitions associated with the $[M - SO_3 + H]^+$ ions were used for quantitation.

The 5 μm Amide-80 column was reported to have stability from pH 2 to 7.5 and, most importantly, showed no significant bleed at low pH. This column provided superior retention within the chromatographic window and was therefore employed for further optimization studies. The eluting system proposed by Strege[23] for the Amide-80 column was used initially (gradient: 90%–60% B over 20 min, hold for 60 min, with A being water and B acetonitrile/water (95:5), both containing 6.5 mM ammonium acetate, pH 5.5). Under these conditions, the early eluting peaks of GTX1-4 showed tailing and later eluting peaks of STX and NEO showed fronting. A number of eluting systems were tested on this column by paying particular attention to organic modifier character and percentage, buffer character and percentage, and pH.

As found by Yoshida for the separation of peptides,[24] the percentage of organic modifier in the mobile phase played a key role for the absolute retention time of PSP toxins. HILIC behaves like normal phase chromatography, so retention times increase proportional to the percentage of organic modifier and to the polarity of the solute. The neutral C toxins eluted first, followed by the single-charged gonyautoxins, then the double-charged STX and NEO, and finally the decarbamoyl derivatives.

Both acetonitrile and methanol were tested as possible organic modifiers. Acetonitrile provided sharper peaks than methanol and was thus preferred. Methanol resulted in a dramatic change in the relative order of elution and poorer separation.

The ammonium formate buffer concentration significantly affected retention time in the range 0–2 mM (>3 min) while about 1 min shifts were observed for buffer concentration 2–10 mM. Generally, in agreement with findings by Strege for the retention of guanine,[23] retention times decreased as the concentration of buffer increased. However, when no aqueous buffers were used, PSP toxin retention times exceeded 180 min and the peak widths were unacceptably broad. Therefore, an aqueous buffer was required to modify the mobile phase. Both ammonium acetate and formate were tested. The latter resulted in better peak shape and was thus preferred.

The pH of the mobile phase had the greatest influence on the separation of PSP toxins. As pH increased, retention times and separation selectivity increased. At low pH, the peaks were broader. The neutral C toxins did not change their absolute retention time when the pH was changed whereas the other toxins showed large changes in retention time (1–2 min) for relatively small changes in pH (0.1–0.2 pH units). Therefore, the retention times relative to C1 were also calculated (Table 6.2).

The HILIC column efficiency was dependent on flow rate and varied between toxins. Initial work was performed on a 250 × 4.6 mm i.d. Amide 80 column. The highest plate number was generally obtained at flow rates 0.8–1 mL/min. The later eluting toxins, namely STX and NEO, showed higher performance at 0.6–0.8 mL/min but the greatly increased analysis time was considered a serious drawback. Therefore, a flow rate of 1 mL/min was selected. For the 250 × 2 mm i.d. column, a 0.2 mL/min flow was selected.

The effect of column temperature on retention of PSP toxins in the range 10–45°C was slight. In particular, retention times of monocharged (GTX1-4 and their decarbamoyl derivatives, B1 and B2) and bicharged (STX, NEO, and their decarbamoyl derivatives) toxins increased as a function of the temperature. No effect was observed upon retention of the neutral C toxins. Similarly, selectivity increased linearly with

TABLE 6.3
Optimized Chromatographic Conditions for the HILIC-MS Analysis of PSP Toxins, DA, and Assorted Cyanobacterial Toxins[a]

Column	5 μm amide-80 column (250 × 2 mm i.d.)
Column *T* (°C)	20
Flow rate (mL/min)	0.2
Mobile phase	A = water, 2.0 mM ammonium formate, 3.6 mM formic acid (pH 3.5)
	B = acetonitrile/water (95:5), 2.0 mM ammonium formate, 3.6 mM formic acid
Elution system A	Isocratic, 65% B
Elution system B	Isocratic, 75% B
Elution system C	Gradient
	75% B for 5 min
	75%–65% B over 1 min, hold 13 min
	65%–45% B over 4 min, hold 10 min

[a] Elution system A is recommended for PSP toxins analysis, whereas elution systems B and C are indicated for determination of DA, CYN, doCYN, and ATX-a as single compounds or in combination with PSP toxins, respectively.

temperature for all PSP toxins except neutral ones. These results supported a presumed role of ion exchange in the HILIC separation of PSP toxins, as retention and selectivity are known to be temperature dependent in ion exchange chromatography.[25] In addition, as temperature increased, resolution of most peaks increased but the peak shapes of later eluting toxins broadened. A column temperature of 20°C was selected as optimum for retention, selectivity, and peak shape.

On the basis of all the above findings, the best chromatographic conditions were selected (Table 6.3). Elution system A (65%B) allowed determination of all STX-like compounds in a single chromatographic run of 30 min. HILIC-MS analyses in SIM and MRM modes of a standard mixture of PSP toxins under the optimized HILIC conditions are shown in Figure 6.3.

Some compounds could not be chromatographically resolved: GTX1 and GTX2, GTX4 and GTX3, and the associated decarbamoyl derivatives eluted in a 2 min range; C1 and C2 substantially coeluted with C3 and C4; B1 partially coeluted with B2; STX partially coeluted with NEO, and dcSTX with dcNEO. However, the additional detection selectivity provided by different channels of detection in SIM or MRM made it possible to individually detect all PSP toxins in a reasonable period of time, namely 25–30 min.

As for the mechanism of separation, HILIC is considered to behave similarly to normal phase partition chromatography, where a stagnant mobile phase (mostly aqueous) is in contact with the stationary phase and a dynamic mobile phase (mostly organic) is separated from the stationary phase.[26] The analyte partitions between the two mobile phases and may be orientated in space to interact with functions of the stationary phase. Ion exchange or electrostatic interactions can also have a role.[27,28] As for PSP toxins, the mechanism of separation appears to be primarily

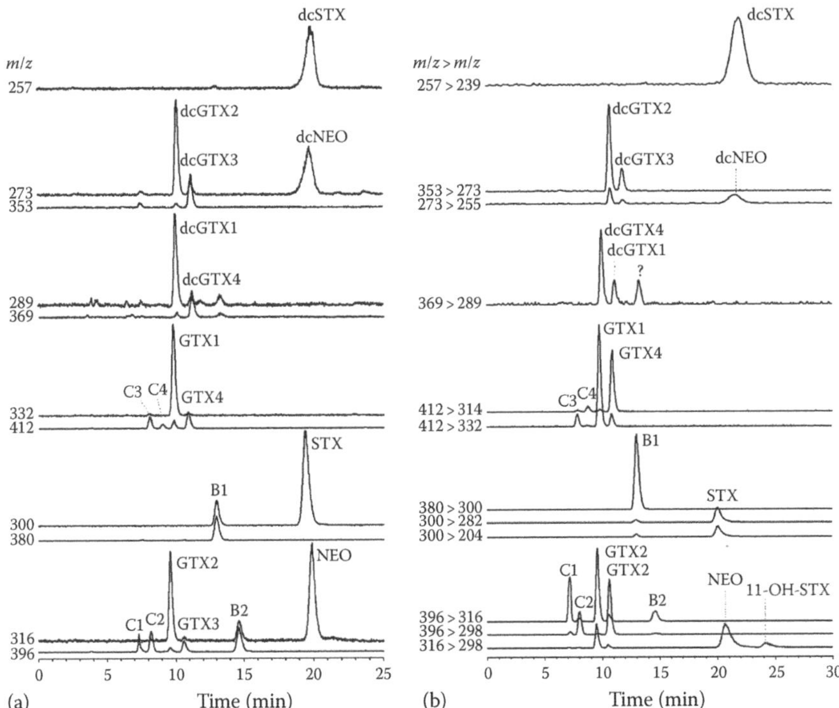

FIGURE 6.3 HILIC-MS analyses of a standard mixture of assorted PSP toxins in SIM (a) and MRM (b) modes under the optimized HILIC conditions (Table 6.3, elution system A). SIM experiment was carried on an API-165 MS system by selecting protonated and/or fragment ions (Table 6.1). MRM experiment was carried out on API-III+ MS system by selecting transitions consistent with the fragmentation pattern of each toxin (Table 6.2). (Reprinted from Dell'Aversano, C. et al., *J. Chromatogr. A*, 1081, 190, 2005. With permission from Elsevier.)

electrostatic in nature. This is supported by the short retention times of the neutral C toxins, the intermediate retention times of the single-charged gonyautoxins, and the long retention times of the double-charged toxins. The excellent separation achieved between epimeric pairs such as GTX1 and GTX4, or GTX2 and GTX3, may be explained by considering the effect that α- or β-orientation of the 11-hydroxysulfate function has on the charge states of individual functional groups. Molecular modeling showed that when the 11-hydroxysulfate group is in the α-orientation (GTX1 and GTX2), it can establish an intramolecular interaction with the guanidinium function at C-8. This would reduce the number of positively charged functions on the molecule available for interaction with the stationary phase. On the other hand, both guanidinium groups are available for interaction with the stationary phase when the 11-hydroxysulfate function is β-oriented (GTX3 and GTX4). It should, also, be noted that 11(α, β)-OH-STX do not resolve under the same conditions. This supports the argument that the 11-hydroxysulfate group and its interaction with the guanidinium function is important for separation of epimers.

TABLE 6.4
Limits of Detection (LOD, S/N = 3) for Major PSP Toxins on Different LC-MS Systems in the SIM and MRM Modes

	SIM[a]		SRM[a]				
	API-165		API-III+		API-4000		LC-ox-FLD[16]
	LOD		LOD		LOD		LOD
Toxin	*m/z*	(nM)	*m/z* > *m/z*	(nM)	*m/z* > *m/z*	(nM)	(nM)
STX	300	800	300 > 282	7000	300 > 204	20	60
GTX2	316	300	396 > 316	1000	396 > 316	20	20
GTX3	396	400	396 > 298	300	396 > 298	10	5
NEO	316	900	316 > 298	7000	316 > 220	30	60
GTX1	332	200	412 > 332	800	412 > 332	10	20
GTX4	412	800	412 > 314	400	412 > 314	5	30
B1	380	1000	380 > 300	700	380 > 300	10	100
C1	396	50	396 > 316	40	493 > 316	20	30
C2	396	60	396 > 298	50	396 > 298	10	20

Source: Modified from Dell'Aversano, C. et al., *J. Chromatogr. A*, 1081, 190, 2005. With permission from Elsevier.

[a] Optimized chromatographic conditions (Table 6.3) were used.

Limits of detection (LOD) of the HILIC-MS method for matrix-free toxins were obtained for STX, GTX2, GTX3, NEO, GTX1, GTX4, B1, C1, and C2, and ranged from 50 to 1000 nM in SIM and 40 to 7000 nM in MRM mode on the API-III+ instrument (Table 6.4). Unfortunately, a single collision energy was used for the entire group of MRM ions on the API-III+ system. The compromise value of 20 V was too low for good fragmentation of STX and NEO, resulting in poor detection limits. LOD values for such toxins could be lowered by at least fivefold by using a 30 V collision energy on $[M + H]^+$ ions of STX and NEO, which is possible if time programming of MRM transitions is used. Much better results in terms of LOD were obtained on the API-4000 MS system, which presents better sensitivity and the ability to use optimized collision energies for each ion transition. Particularly LOD values ranged from 5 to 100 nM (Table 6.4) and excellent linearity was obtained, with R^2 values of 0.999 or greater. A comparison with LOD for the LC-ox-FLD method[16] (Table 6.4) reveals that the HILIC-MS method on the API-4000 system has better sensitivity than the LC-ox-FLD method and should therefore be suitable for analysis of shellfish samples near the maximum acceptable regulatory limits for PSP toxins (0.8 mg of saxitoxin equivalent per kg of edible tissue).

6.2.2 Application to Plankton and Shellfish Samples

The HILIC-MS method for PSP toxins was tested on field plankton and mussel samples collected during an intense bloom of *Alexandrium tamarense*, which occurred in

June 2000 in a Nova Scotian harbor.[29] Extracts of the samples were analyzed in both SIM and MRM modes. Although SIM mode proved suitable for detection of most of the PSP toxins present in the samples, unambiguous interpretation of the results was hampered by the presence of many interfering peaks, a high background signal in some ion traces, and a matrix-related hump in the chromatograms at about 20 min. The higher selectivity of the MRM mode made interpretation of the results much easier due to elimination of signals from other coextractives. Anyway, a slight shift of retention times for the sample extract versus those for the standards could be observed. This was due to matrix effect that got worse as more concentrated crude extracts were used.

Figure 6.4 shows the results of the HILIC-MRM analyses of the crude extracts of the plankton sample (Figure 6.4a) and the mussel extract (Figure 6.4b). The plankton sample showed a complex array of toxins with the major toxins being GTX4, C2, C4, GTX3, B1, NEO, and STX. Only low levels of the corresponding epimeric C and GTX toxins were observed. The same toxins were observed in the mussel extract, which was not surprising, as these mussels had consumed the plankton material. Interestingly, some new saxitoxin analogues were observed in the mussel sample. The new compounds, labeled as M1, M2, and M3, were absent in the plankton sample and, thus, were likely metabolites and/or degradation products formed in the mussel. The M2 peak had an exact match of retention time and product ion spectrum for 11-hydroxy-STX. Isolation work, which included the use of preparative HILIC-MS, followed by in-depth MS and NMR investigation allowed to elucidate the structure of M1 and M3, as 11β-hydroxy-*N*-sulfocarbamoyl saxitoxin and 11,11-dihydroxy-*N*-sulfocarbamoyl saxitoxin.[30]

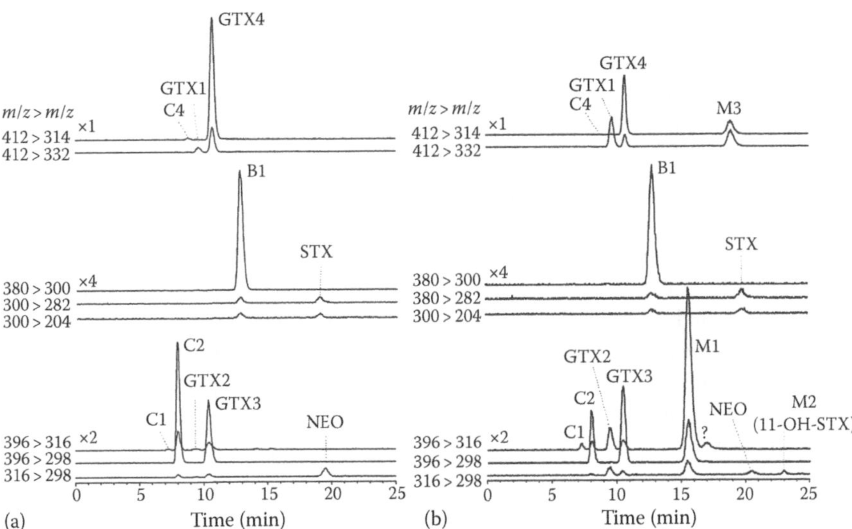

FIGURE 6.4 HILIC-MS analyses in MRM mode of an *Alexandrium tamarense* extract (a) and of a *Mytilus edulis* extract (b) containing various PSP toxins. Experiments were carried out on API-III+ MS under the optimized HILIC conditions (Table 6.3, elution system A). Some traces are plotted with expanded scale as indicated. (Reprinted from Dell'Aversano, C. et al., *J. Chromatogr. A*, 1081, 190, 2005. With permission from Elsevier.)

6.3 DOMOIC ACID

Domoic acid (DA) (Figure 6.1) is the causative toxin of amnesic shellfish poisoning (ASP) syndrome.[31] Symptoms of the intoxication include nausea, vomiting, gastroenteritis, cramps, and diarrhoea, all within 24 h after consumption of contaminated shellfish. Neurological complaints develop within 48 h and include ataxia, confusion, persistent short-term memory loss, disorientation, and even coma.

DA is produced by *Pseudo-nitzschia pungens* f. *multiseries*,[32] a diatom widely distributed in coastal waters all around the world, as well as by several species of *Pseudo-nitzschia* (*Pseudo-nitzschia australis*, *Pseudo-nitzschia pseudo-delicatissima*, *Pseudo-nitzschia galaxiae*),[33–35] although there is considerable variability in toxicity among clones and over time in culture.

The AOAC mouse bioassay for PSP toxins can be employed for detection of DA's unique toxicity,[36] since this method involves boiling shellfish tissue with an equal volume of 0.1 N HCl, which provides a common extract for screening of both DA and PSP toxins. Intraperitoneal injection of DA in mice induces a very characteristic symptomology, namely, scratching of shoulders by the hind leg, sedation-akinesia, rigidity, loss of posteral control, convulsions, and death.[37] However, such symptoms may be observed only when domoic acid is present at levels >40 μg/g edible tissue, whereas the regulatory limit for DA is 20 μg/g shellfish tissue. Therefore, the relative insensitivity of this assay precludes its use for regulatory purposes and instrumental confirmation is always needed.

Liquid chromatography with UV detection (LC-UV) is the preferred analytical technique for determination of DA in shellfish.[38,39] Mollusk tissues can be extracted by the above AOAC procedure, or by aqueous methanol (1:1, v/v),[39] which gives a better yield and a more stable extract. LOD are in the range 0.1–1 μg DA/g tissue, depending on the sensitivity of the UV detector, which is suitable for regulatory purposes. However, interferences are commonly encountered that can result in false positives. Particularly, tryptophan and its derivatives elute close to DA under the chromatographic conditions used, thus making interpretation of the results ambiguous. A strong anion exchange (SAX) solid phase extraction (SPE) cleanup is generally recommended.[39]

Alternative methods based on LC with MS detection[9,40–43] have been developed for monitoring DA in marine matrices. In this frame, the suitability of HILIC-MS technique for analysis of PSP toxins has also been examined for detection and quantitation of domoic acid.[11] The method could be used for combined analysis of DA and PSP toxins, which are coextracted with any of the extraction procedures currently used. Details on the developed method and its application to shellfish samples are provided below.

6.3.1 HILIC-MS Determination of DA

The presence of both COOH and NH functionalities in the domoic acid molecule allows MS detection of the toxin by both positive and negative ionization modes.

Full-scan mass spectra (MS) of DA were acquired on a PE-Sciex triple quadrupole MS equipped with a Turbospray source (API-2000) and showed the presence in positive ion mode of $[M + H]^+$ and $[M + Na]^+$ ions at *m/z* 312 and 334, respectively, while the $[M - H]^-$ ion at *m/z* 310 dominated the spectrum in negative ion mode. No significant fragmentation was observed under the conditions used. These ions were chosen for SIM experiments.

The $[M + H]^+$ and $[M - H]^-$ ions were selected as precursor ions for MS/MS product ion scan experiments. The fragmentation pattern produced in the MS/MS spectrum of $[M + H]^+$ consisted mainly of H_2O and CO or HCOOH losses. In particular, the following ions were observed: *m/z* 294 $[M + H - H_2O]^+$, *m/z* 266 $[M + H - HCOOH]^+$, *m/z* 248 $[M + H - HCOOH - H_2O]^+$, *m/z* 220 $[M + H - HCOOH - H_2O - CO]^+$, *m/z* 193 $[M + H - HCOOH - H_2O - CO - HCN]^+$, and *m/z* 175 $[M + H - HCOOH - 2H_2O - CO - HCN]^+$. The negative MS/MS spectrum yielded fragments due to successive losses of CO_2 and H_2O from the $[M - H]^-$ ion. The following ions were observed: *m/z* 266 $[M - H - CO_2]^-$, *m/z* 248 $[M - H - CO_2 - H_2O]^-$, *m/z* 222 $[M - H - 2CO_2]^-$, *m/z* 204 $[M - H - 2CO_2 - H_2O]^-$, *m/z* 160 $[M - H - 3CO_2 - H_2O]^-$, and *m/z* 82 $[M - H - 228]^-$.

The above MS/MS transitions of the $[M + H]^+$ and $[M - H]^-$ precursor ions were selected for MRM experiments.

Under the HILIC conditions optimized for PSP toxins (Elution system A) (Table 6.3), domoic acid eluted at the chromatographic front. A higher percentage of organic modifier, 75% B (Elution system B), was necessary to achieve effective retention of the toxin, which then eluted at 4.96 min. By way of example, MRM chromatogram in positive ion mode for a 1.1 μg/mL standard solution of domoic acid is shown in Figure 6.5a. Similar results were obtained in negative ion mode. Combined analysis of DA and PSP toxins (Figure 6.5b) was possible using the gradient elution (Elution system C in Table 6.3).

Limits of detection (LOD, S/N = 3) and of quantitation (LOQ, S/N = 10) for DA in SIM and MRM modes are reported in Table 6.5. In all cases, correlation coefficients were >0.9994, indicating a high degree of linearity of the plots within the tested concentration range.

In order to establish whether there was any ion enhancement or suppression for DA due to the presence of matrix, MRM experiments of a spiked extract and of a pure standard solution of DA (20 μg/mL) were run in triplicate. Comparison of the results showed a 53% and 3% ion suppressions in signal intensity in positive and negative ion modes, respectively. The matrix effect observed in positive ion mode was significant while the matrix interference in negative ion mode appeared negligible, suggesting that quantitation of DA in 1 g/mL crude extract was reliable using a pure standard solution as an external standard. However, the peak shape was broad and showed significant peak tailing, suggesting that some improvement was advisable, even in negative ion mode. The matrix concentration that would induce minimum suppression effect was established to be 0.0625 and 0.125 g/mL for positive and negative ion modes, respectively. Toxin-free sample extracts at such matrix concentrations were used to prepare matrix matched standards at four levels of DA concentration. Subsequently, calibration curves for pure and matrix-matched standards were generated in positive and negative MRM modes. Good linearity was observed in all cases. In negative ion mode, the slope of the curve for matrix matched standard was similar to that of pure standard, with a possible slight enhancement effect in the spiked extract. In positive ion mode, a noteworthy suppression of signal was

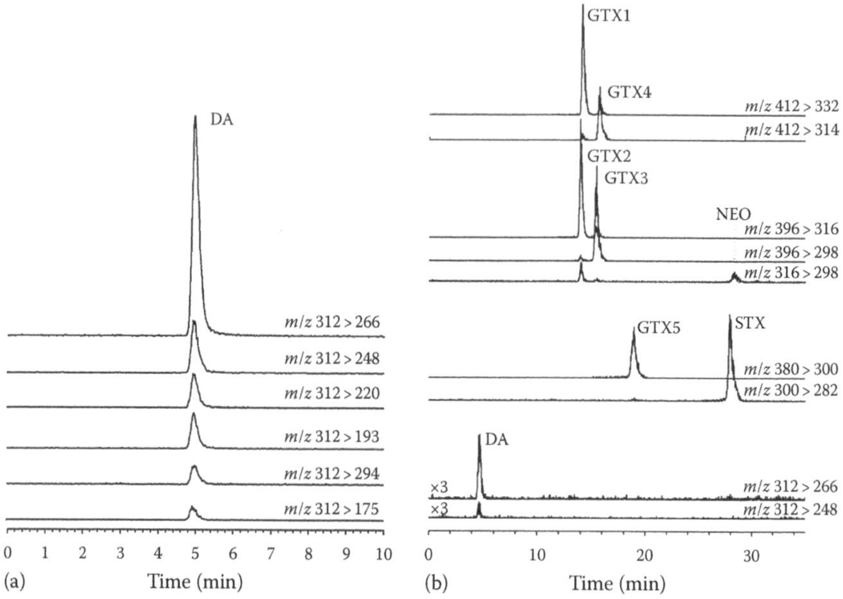

FIGURE 6.5 HILIC-MS analyses in positive MRM mode of (a) a 1.1 μg/mL standard solution of domoic acid (RT = 4.96 min) by using isocratic elution (75% B) and (b) a standard mixture containing domoic acid and selected PSP toxins (GTX2, GTX3, GTX1, GTX4, GTX5, STX, and NEO), by using gradient elution (Table 6.3, elution system C). Under the used conditions, the following toxins were determined: DA (RT = 4.69 min), GTX1 (RT = 14.17) and GTX4 (RT = 15.79), GTX2 (RT = 14.04) and GTX3 (RT = 15.51), NEO (RT = 28.33), GTX5 (RT = 19.00), and STX (RT = 27.86). (Modified from Ciminiello, P. et al., *Rapid Commun. Mass Spectrom.*, 19, 2030, 2005. With permission from Wiley.)

observed despite the additional dilution of matrix, suggesting that matrix matched standard should be used for accurate quantitation in positive mode. The minimum detection level for DA in tissue was found to be 63 and 190 ng/g in positive and negative MRM experiments, respectively.

6.3.2 Application to Shellfish Samples

In order to test suitability of the developed method for the analysis of shellfish, a number of samples of *Mytilus galloprovincialis* were collected in the period 2000–2004 in the Adriatic Sea, in coincidence with the presence in sea-water of appreciable amounts of *Pseudo-nitzschia* spp.

The samples were separately extracted with MeOH/H_2O (1:1) and with 0.1 M acetic acid; both procedures allow extraction of water-soluble PSP and DA, if they cooccur. The crude extracts were directly analyzed by HILIC-MS for both PSP toxins and DA.

As already observed for PSP toxins, SIM experiments showed a high background signal in the selected ion traces and extra peaks from other components in the crude

TABLE 6.5
Limits of Detection (LOD, S/N = 3) and Limits of Quantitation (LOQ, S/N = 10) for Domoic Acid as Determined by HILIC-MS[a] in SIM and MRM Experiments Using an API 2000 Triple Quadrupole System

Experiment	Ion	LOD (ng/mL)	LOQ (ng/mL)	R^2
SIM+	312	3	11	0.9998
	334	14	47	
MRM+	312 > 266	3	11	0.9999
	312 > 248	5	18	0.9999
	312 > 220	8		
	312 > 193	10		
	312 > 175	23		
	312 > 294	30		
SIM−	310	2	7	0.9999
MRM−	310 > 222	10	32	0.9994
	310 > 266	10	34	0.9996
	310 > 160	79		
	310 > 248	49		
	310 > 204	82		
	310 > 82	75		

Source: Modified from Ciminiello, P. et al., *Rapid Commun. Mass Spectrom.*, 19, 2030, 2005. With permission from Wiley.

[a] Optimized chromatographic conditions reported in Table 6.3 were used (Elution system B).

extract, which made interpretation of the results difficult. The more selective and sensitive (relative to S/N) MRM experiments indicated the presence of DA in some of the analyzed samples basing on peak retention time (4.90 min), the presence of six diagnostic fragments for the analyte, and the ion abundance ratios.

The domoic acid content in the *M. galloprovincialis* samples was determined by direct comparison to individual standard solutions of domoic acid at similar concentrations injected in the same experimental conditions. The most abundant transitions (*m/z* 310 > 266 and 310 > 222) were used for quantitative studies, which were carried out on the methanol/water (1:1) extract (official extraction procedure), since DA in 0.1 M acetic acid extract seemed to have poorer stability over a long storage period. Concentrations of DA in the analyzed samples were in the range 0.25–5.5 μg/g, levels well below to the regulatory limit (20 μg/g in edible tissue). However, this was the first time that domoic acid had been detected in Adriatic sea and represented a warning for domoic acid as one of the toxins that needs to be carefully monitored in Adriatic shellfish.

6.4 ASSORTED CYANOBACTERIAL TOXINS

Cyanobacteria are found in fresh and brackish water throughout the world and can present a public safety hazard through contamination of drinking water supplies. They are known to produce a number of toxins including neurotoxins, such as PSP toxins and anatoxin-a, and hepatotoxins, such as cylindrospermopsins and microcystins (Figure 6.1).

Among cyanobacteria, PSP toxins are produced by *Aphanizomenon flos-aquae*,[44,45] *Anabaena circinalis*,[46] *Lyngbya wollei*,[47] and *Cylindrospermopsis raciborskii*.[48]

Anatoxin-a (ATX-a) is a potent depolarizing neuromuscular blocking agent[49] produced by different strains of *Anabaena, Planktothrix*, and *Aphanizomenon*.[50] Cylindrospermopsin and its deoxy-derivative are alkaloids produced by *C. raciborskii*,[51] *Umezakia natans*,[52] *Aphanizomenon ovalisporum*,[53] and *Raphidiopsis curvata*.[54] Microcystins, produced mainly by *Microcystis* sp., are heptapeptides[55,56] and show tumor-promoting activity on rat liver with inhibition of protein phosphatases 1 and 2A.[57]

The monitoring of drinking water supplies for the presence of these toxins is of critical importance for the assessment of environmental and health risks. So, the HILIC-MS method for detection of PSP toxins[10] was extended to the analysis of ANTX-a, CYN, and its deoxy-derivative as well as microcystins LR and RR,[12] with the aim to develop an analytical method that could provide simultaneous detection and unambiguous identification of different cyanobacterial toxins.

6.4.1 HILIC-MS Determination of Assorted Cyanobacterial Toxins

As first step for method development, ESI MS behavior of CYN, doCYN, and ATX-a was investigated.

The MS spectrum of CYN (Figure 6.6a) contained an abundant $[M + H]^+$ ion at *m/z* 416, as well as the ammonium, sodium, and potassium adduct ions at *m/z* 433, 438, and 454, respectively. The MS/MS spectrum of the protonated ion of CYN (Figure 6.6c) showed fragment ions corresponding to the loss of SO_3 and H_2O from $[M + H]^+$, at *m/z* 336 and 318, respectively. Another fragment ion, $[M + H - 142]^+$ at *m/z* 274, resulted from loss of the [6-(2-hydroxy-4-oxo-3-hydropyrimidyl)]-hydroxymethinyl moiety of the molecule and this ion subsequently underwent loss of SO_3 and H_2O to afford ions at *m/z* 194 and 176, respectively. MS and MS/MS spectra of doCYN (Figure 6.6b and d, respectively) were very similar, the only difference being the shift of ions at *m/z* 336 and 318 in CYN down 16 mass units in doCYN. The proposed fragmentation of CYN and doCYN is summarized in Figure 6.6e.

The MS spectrum of ATX-a showed an abundant $[M + H]^+$ ion at *m/z* 166 and fragment ions at *m/z* 149 and 131 due to sequential elimination of NH_3 and H_2O. The MS/MS spectrum showed the same ions, as well as prominent ion at *m/z* 91 (not assigned) and *m/z* 43 corresponding to CH_3CO^+ from the acetyl function.

The protonated and ammonium-adduct ions of CYN and doCYN as well as the $[M + H]^+$ and the $[M + H - NH_3]^+$ ions of ATX-a were selected as ions to monitor

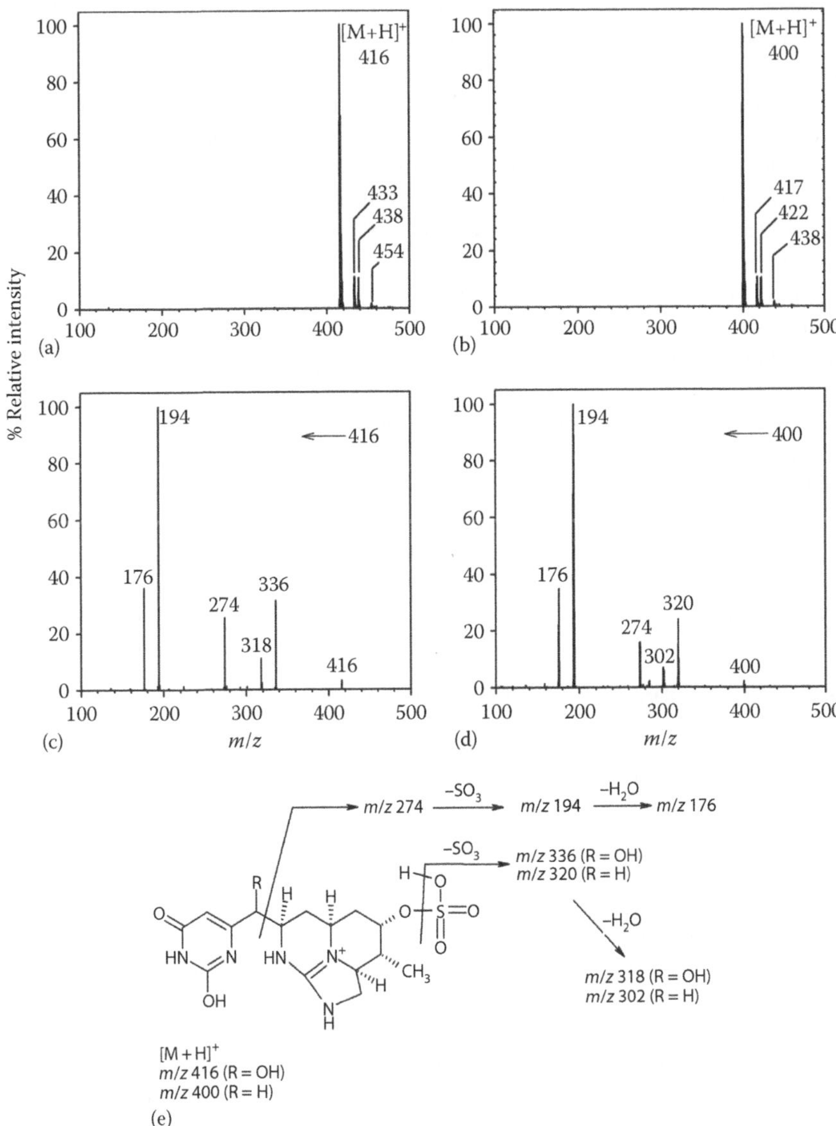

FIGURE 6.6 Electrospray mass spectra in positive ion mode of (a) cylindrospermopsin (CYN) and (b) its deoxy derivative (doCYN). MS/MS product ion spectra of the $[M + H]^+$ ions of CYN and doCYN are shown in (c) and (d), respectively. MS/MS spectra were carried out on API-III+ system using a collision energy of 30 V. Assignments of labeled fragment ions are shown (e). (Modified from Dell'Aversano, C. et al., *J. Chromatogr. A*, 1028, 155, 2004. With permission from Elsevier.)

in SIM experiments. The most abundant fragment ions contained in the MS/MS spectra of each toxin were selected for MRM experiments.

Under the same chromatographic conditions used for saxitoxin and its analogues (65% B), ATX-a, CYN and doCYN eluted too quickly. As observed for domoic acid, a higher percentage of solvent B was required for effective retention of these toxins. In particular, with 75% B isocratic (Elution system B), ATX-a, CYN, and doCYN eluted at 5.8, 7.1, and 6.2 min, respectively. The three-step gradient (Elution system C) developed for combined analysis of PSP toxins and DA allowed determination also of the above cyanobacterial toxins in combination with STX-like compounds in one 32 min run. The results are shown in Figure 6.7 for a mixture of standard compounds.

A gradient elution (90%–65% B over $T = 13$ min and hold 5 min) was required for analysis of standard solutions of microcystin-LR and -RR, which eluted at 7.9 and 13.5 min, respectively. Mono- and bi-charged protonated molecules (*m/z* 995 and 498 for microcystin-LR and *m/z* 1038 and 520 for microcystin-RR, respectively) were selected as ions to monitor in selected ion monitoring experiments. HILIC-MS technique proved to selectively detect the microcystins, but the peak shapes were not satisfactory, showing increased broadening and tailing compared to the other toxins. Likely, an adjustment of mobile phase, pH, or buffer could improve the peak shape, but many RPLC methods for the analysis of a large range of microcystins have been reported in literature,[58–60] which offer robustness and better chromatographic performance than HILIC.

6.4.2 Applications to Cyanobacterial Samples

Field and cultured cyanobacterial samples were analyzed to test suitability of the HILIC-MS method to the analysis of real samples. A simple extraction method was used and no cleanup was performed on the crude extracts in order to demonstrate rapid analysis.

A sample of *A. circinalis*, collected in October 1997 from the Coolmunda Dam, South East Queensland (Australia), was acquired and analyzed by HILIC-MS for the presence of saxitoxin-related compounds. A preliminary RPLC-ox-FLD analysis had shown that STX, GTX2, GTX3, C1, C2, and dcSTX were present in the sample at concentrations in the range of 75–1000 µg/g. HILIC-MS analysis of the crude extract was carried out in both SIM and MRM modes. The latter gave better results. All of the toxins detected by RPLC-ox-FLD, as well as GTX5, dcGTX2, and dcGTX3, were easily confirmed by matching ion transition signals and retention times with those of standards. In addition, a significant peak was present in the chromatogram for the *m/z* 380 > 300 transition at 9.7 min whose identity could not be assigned just on the basis of MS data.

The Coolmunda Dam *A. circinalis* extract was also analyzed for ATX-a. The detection limit for the ATX-a standard on the *m/z* 166 ion was estimated (S/N = 3) to be 1.4 pmol injected on column in SIM mode and the sample was found to contain no detectable ATX-a. Interestingly, European strains of *A. circinalis* produce ATX-a while Australian strains have never been observed to do so.[61]

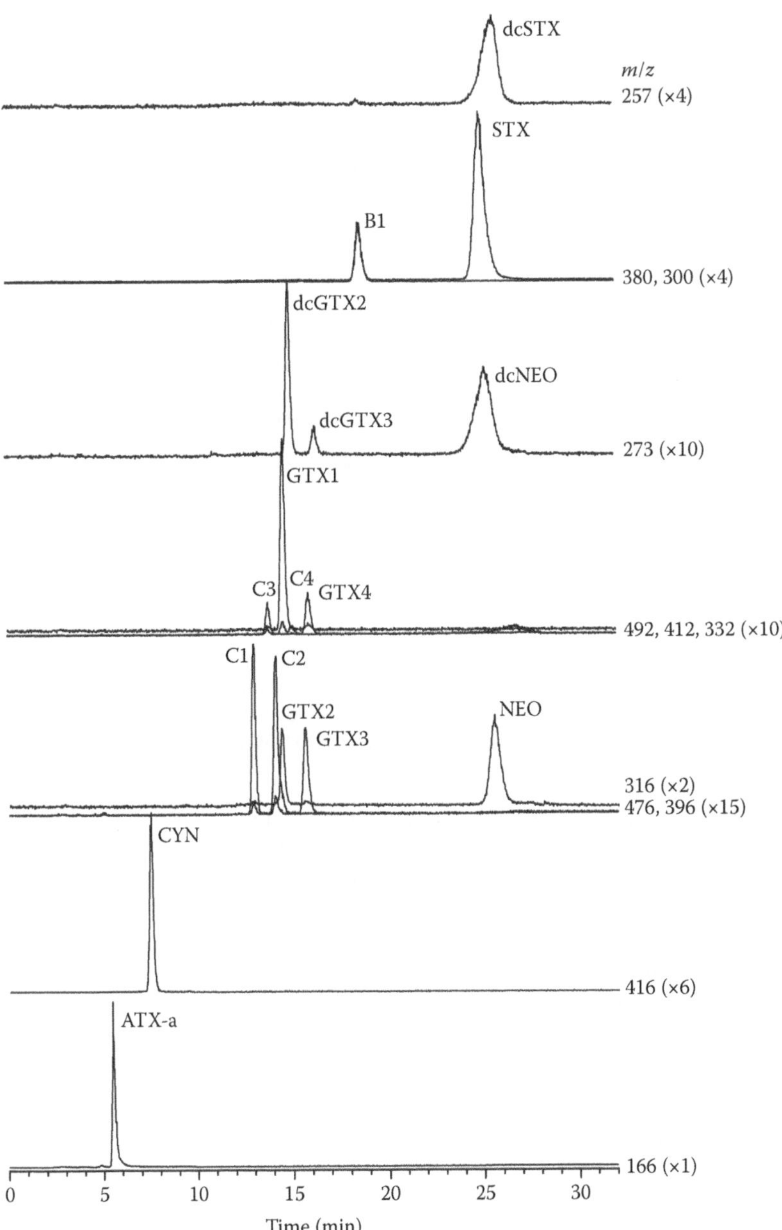

FIGURE 6.7 HILIC-MS analyses of a standard mixture containing most STX-like compounds, ATX-a and CYN. Experiments were carried out in SIM mode on the API-III+ system. Some traces are plotted with expanded scales as indicated. Gradient elution was used (Table 6.3, elution system C). (Modified from Dell'Aversano, C. et al., *J. Chromatogr. A*, 1028, 155, 2004. With permission from Elsevier.)

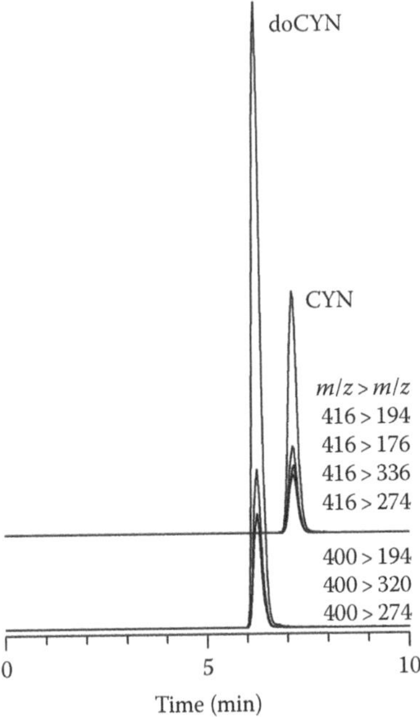

FIGURE 6.8 HILIC-MS analyses in positive MRM mode of a *Cylindrospermopsis raciborskii* extract containing cylindrospermopsin and its deoxy derivative. Experiments were carried out on the API-III+ system using isocratic elution (Table 6.3, elution system B). (Modified from Dell'Aversano, C. et al., *J. Chromatogr. A*, 1028, 155, 2004. With permission from Elsevier.)

A cultured sample of *C. raciborskii*, known to contain CYN and doCYN at levels as high as 2 mg/g, was acquired for testing. A HILIC-MS analysis of the crude extract based on SIM of the $[M + H]^+$ ions provided sufficient selectivity and sensitivity, detection limit (S/N = 3) for the CYN standard being 3.2 pmol injected on column. An MRM experiment was also performed on the sample for added confirmation. The results (Figure 6.8) confirmed the SIM results and also provided excellent sensitivity, the detection limit being (S/N = 3) 0.3 pmol injected on column. The *C. raciborskii* extract was also analyzed for saxitoxin-related toxins but none was detected in this strain.

6.5 CONCLUSIONS

HILIC-MS is a technique suitable for a number of polar hydrosoluble toxins, including saxitoxin and its analogues, most commonly known as PSP toxins, domoic acid, the causative toxin of ASP syndrome, and some assorted cyanotoxins, such as anatoxin-a, cylindrospermopsin, and its deoxy derivative. Microcystins could also be chromatographed but they are best analyzed by existing RPLC-MS methods.

For saxitoxin and its analogues as well as for domoic acid, MRM is the preferred method due to its higher selectivity. Indeed, a cleanup step is needed if SIM is the only acquisition mode available. For cylindrospermopsin and anatoxin-a, the SIM method is adequate, but MRM can provide additional selectivity for confirmatory analyses. Multitoxin determination is possible, thus allowing the rapid, simultaneous screening of an entire range of toxins.

The HILIC-MS technique provides high sensitivity which is basically due to mobile phase character, which does not use ion-pair agents and is high in organic modifier, which results in high ionization efficiency. LOD for PSP toxins in MRM mode can vary between instruments and MS tuning, but for a sensitive instrument, such as the triple quadrupole API-4000, LOD as low as 5–100 nM have been demonstrated and compete favorably with those achieved by the LC-ox-FLD method. So, the analysis of shellfish samples with a PSP toxin content near the regulatory limit is possible. LOD for domoic acid were below the regulatory limit and allow the screening of DA in shellfish tissues even at trace levels.

The developed HILIC-MS methods have shown to be sensitive, straightforward, readily automated, and robust with similar results obtained in different laboratories using different instrumentation.

REFERENCES

1. Hallegraeff, G. M., *Phycologia*, 1993, 32, 79.
2. Tibbetts, J., *Environ. Health Persp.*, 1998, 106, A326.
3. Anderson, D. M., *Red Tides: Biology Environmental Science and Toxicology*, Okaichi, T., Anderson, D. M., Nemoto, T., Eds., Elsevier, New York, 1989, p. 11.
4. Aune, T., *Seafood and Freshwater Toxins—Pharmacology, Physiology, and Detection*, Botana, L. M., Ed., CRC Press Taylor & Francis Group, Boca Raton, FL, 2nd edn., 2008, p. 21.
5. Horwitz, W., Ed., *Official Methods of Analysis of AOAC International*, 17th edn., AOAC International, Gaithersburg, MD, 2000.
6. Quilliam, M. A., *Applications of LC-MS in Environmental Chemistry*, Barcelo, D., Ed., Elsevier Science BV, Amsterdam, the Netherlands, 1996, p. 415.
7. Quilliam, M. A., *J. Chromatogr. A*, 2003, 1000, 527.
8. Botana, L. M., Ed., *Seafood and Freshwater Toxins–Pharmacology, Physiology, and Detection*, 2nd edn., CRC Press Taylor & Francis Group, Boca Raton, FL, 2008.
9. Quilliam, M. A., Hess, P., Dell'Aversano, C., *Mycotoxins and Phycotoxins in Perspective at the Turn of the Millennium*, deKoe, W. J., Samson, R. A., Van Egmond, H. P., Gilbert, J., Sabino, M., Eds., Wageningen, the Netherlands, 2001, p. 383.
10. Dell'Aversano, C., Hess, P., Quilliam, M. A., *J. Chromatogr. A*, 2005, 1081, 190.
11. Ciminiello, P. et al., *Rapid Commun. Mass Spectrom.*, 2005, 19, 2030.
12. Dell'Aversano, C., Eaglesham, G. K., Quilliam, M. A., *J. Chromatogr. A*, 2004, 1028, 155.
13. Mortensen, A. M., *Toxic Dinoflagellates*, Anderson, D. M., White, A. W., Baden, D. G., Eds., Elsevier, New York, 1985, p. 163.
14. Narahashi, T., *Fed. Proc.*, 1972, 31, 1124.
15. Genenah, A. A., Shimizu, Y., *J. Agric. Food Chem.*, 1981, 29, 1289.
16. Oshima, Y., *J. AOAC Int.*, 1995, 78, 528.
17. Sato, S., Shimizu, Y., *Harmful Algae, Proceedings of the VIII International Conference on Harmful Algae*, Reguera, B., Blanco, J., Fernandez, M. L., Wyatt, T., Eds., IOC, UNESCO, Vigo, Spain, 1997, p. 465.
18. Lawrence, J. F., Menard, C., *J. AOAC Int.*, 1991, 74, 1006.

19. Quilliam, M. A., Janecek, M., Lawrence, J. F., *Rapid Commun. Mass Spectrom.*, 1993, 7, 482.
20. Quilliam, M. A. et al., *Rapid Commun. Mass Spectrom.*, 1989, 3, 145.
21. Thibault, P., Pleasance, S., Laycock, M. V., *J. Chromatogr.*, 1991, 542, 483.
22. Jaime, E. et al., *J. Chromatogr.*, 2001, 929, 43.
23. Strege, M. A., *Anal. Chem.*, 1998, 70, 2439.
24. Yoshida, T., *Anal. Chem.*, 1997, 69, 3038.
25. Dorsey, J. G. et al., *Anal. Chem.*, 1998, 70, 591R.
26. Alpert, A. J., *J. Chromatogr.*, 1990, 499, 177.
27. Tolstikov, V. V., Fiehn, O., *Anal. Biochem.*, 2002, 301, 298.
28. Yoshida, T., *J. Biochem. Biophys. Methods*, 2004, 60, 265.
29. Cembella, A. D. et al., *Harmful Algae*, 2002, 1, 313.
30. Dell'Aversano, C. et al., *J. Nat. Prod.*, 2008, 71, 1518.
31. Quilliam, M. A., Wright, J. L. C., *Anal. Chem.*, 1989, 61, 1053.
32. Bates, S. S. et al., *Can. J. Fish. Aquat. Sci.*, 1989, 46, 1203.
33. Garrison, D. L. et al., *J. Phycol.*, 1992, 28, 604.
34. Martin, J. L. et al., *Marine Ecol.: Progress Series*, 1990, 67, 177.
35. Cerino, F. et al., *Harmful Algae*, 2005, 4, 33.
36. Lawrence, J. F. et al., *J. Chromatogr.*, 1989, 462, 349.
37. Tasker, R. A. R., Connell, B. J., Strain, S. M., *Can. J. Physiol. Pharmacol.*, 1991, 69, 378.
38. Quilliam, M. A. et al., *Int. J. Environ. Anal. Chem.*, 1989, 36, 139.
39. Quilliam, M. A., Xie, M., Hardstaff, W. R., *J. AOAC Int.*, 1995, 78, 543.
40. Lawrence, J. F. et al., *J. Chromatogr.*, 1994, 659, 119.
41. Hummert, C. et al., *Chromatographia*, 2002, 55, 673.
42. Hess, P. et al., *J. AOAC Int.*, 2001, 84, 1657.
43. Furey, A. et al., *J. Chromatogr. A*, 2001, 938, 167.
44. Ferriera, F. M. B. et al., *Toxicon*, 2001, 39, 757.
45. Li, R. et al., *Hydrobiologia*, 2000, 438, 99.
46. Humpage, A. R. et al., *Aust. J. Mar. Freshwater Res.*, 1994, 45, 761.
47. Onodera, H. et al., *Nat. Toxins*, 1997, 5, 146.
48. Lagos, N. et al., *Toxicon*, 1999, 37, 1359.
49. Carmichael, W. W., Biggs, D. F., Peterson, M. A., *Toxicon*, 1979, 17, 229.
50. Sivonen, K., Mycotoxins and phycotoxins—Developments in chemistry, toxicology and food safety, *Proceedings of the International IUPAC Symposium on Mycotoxins and Phycotoxins*, M. Miraglia, Ed., Rome, Italy, 1998, p. 547.
51. Ohtani, I., Moore, R. E., Runnegar, M. T. C., *J. Am. Chem. Soc.*, 1992, 114, 7941.
52. Harada, K.-I. et al., *Toxicon*, 1994, 32, 73.
53. Banker, R. et al., *J. Phycol.*, 1997, 33, 613.
54. Li, R. H. et al., *J. Phycol.*, 2001, 37, 1.
55. Carmichael, W. W. et al., *Toxicon*, 1988, 26, 971.
56. Rinehart, K. L. et al., *J. Am. Chem. Soc.*, 1988, 110, 8557.
57. Matsushima, R. et al., *Biochem. Biophys. Res. Commun.*, 1990, 171, 867.
58. Meriluoto, J., *Anal. Chim. Acta*, 1997, 352, 298.
59. Lawton, L., Edwards, C., Codd, G. A., *Analyst*, 1994, 119, 1525.
60. Harada, K.-I., Kondo, F., Tsuji, K., *J. AOAC Int.*, 2001, 84, 1636.
61. Fitzgerald, J., Cunliffe, D. A., Burch, M. D., *Environ. Toxicol.*, 1999, 14, 203.

7 Application of HILIC for Polar Environmental Contaminants (Including Pharmaceuticals) in Aquatic Systems

Alexander L.N. van Nuijs, Isabela Tarcomnicu, Hugo Neels, and Adrian Covaci

CONTENTS

7.1 INTRODUCTION

Reversed-phase liquid chromatography (RPLC) has been the method of choice for the analysis of various classes of compounds not amenable to gas chromatography (GC). A large number of compounds were preferably measured by RPLC and C_{18}-based silica stationary phases. Yet, highly polar compounds undergo early elution on traditional RP stationary phases, leading to lower sensitivity of the mass spectrometric (MS) detection due to high water percentage in the mobile phase at the beginning of the run. The analysis of highly hydrophilic, ionic, and polar compounds by hydrophilic interaction liquid chromatography (HILIC) coupled to MS has been demonstrated as a valuable complementary approach to RPLC.[1] The use of a low aqueous and high polar organic mobile phase in HILIC separation is almost ideal for electrospray ionization in many cases, resulting in increased sensitivity.[1]

In this review, various applications of HILIC separation coupled with MS detection recently developed for a number of environmental contaminants present in aquatic systems, such as marine toxins, pharmaceuticals, drugs of abuse, and pesticides, are discussed together with detailed description of the methodologies used.

7.2 MARINE TOXINS

7.2.1 GENERALITIES

Cyanobacteria and dinoflagellates in freshwater and brackish water produce a wide range of toxins that can accumulate in drinking water supplies or in aquatic organisms. Because of the human consumption of these contaminated aquatic organisms (shellfish, puffer fish) or this water, these compounds are of concern for public health and safety.

Paralytic shellfish poisoning (PSP) toxins can induce dangerous intoxications in humans if contaminated shellfish is consumed. PSP toxins act as potent, reversible sodium channel blockers, resulting in paralytic symptoms such as respiratory insufficiency and can even lead to death.[2]

The amnesic shellfish poisoning (ASP) syndrome causing toxin, domoic acid (DA), binds predominantly to *N*-methyl-D-aspartate receptors in the central nervous system and causes a depolarization of neurons and cell dysfunction or death.[3] The ASP syndrome is firstly characterized by gastrointestinal symptoms (nausea, diarrhea, and gastroenteritis) followed by neurological complaints (ataxia, confusion, disorientation, and coma).[4]

The toxin tetrodotoxin (TTX) and its analogs are responsible for puffer fish poisoning (PFP). TTX inhibits reversibly voltage-activated sodium channels, which results in a series of paralytic symptoms comparable with the effects of PSP toxins.[5]

The overall toxicity of TTX, PSP toxins, and DA in seafood is currently tested by an Association of Official Analytical Chemists (AOAC) official mouse bioassay method.[6] A prepared extract of shellfish or puffer fish tissue is injected in mice and the presence of the compounds is confirmed if a symptom pattern specific for these toxins is observed. Quantification is based on the time of death of the mouse. This mouse bioassay leads to severe ethical problems and does not offer very accurate and precise results. Moreover, the knowledge of a toxin profile (different analogs) is not possible with this bioassay. Because of the dangerous effects of these toxins, there is a need for techniques that allow fast and sensitive detection of these toxins in freshwater and aquatic organisms and that can offer knowledge of toxin profiles. Methods based on LC with fluorescence (FL) or MS detection have been gaining interest. These methods are often based on RPLC separations,[7–12] but because of the highly polar character of the toxins, ion-pair agents have to be used to obtain retention in RPLC. The problem with these agents is that they interfere with the MS detection (ion suppression or enhancement) and, because of the different charge of the toxin and its analogs, a single run analysis is not possible. FL detection after oxidation is an alternative, but the complex setup and the difficult interpretation of quantitative results are major drawbacks. In this part, the application of HILIC combined with FL or MS detection for the analysis of these toxins is discussed.

7.2.2 Paralytic Shellfish Poisoning Toxins

PSP toxins are produced by marine dinoflagellates belonging to the genera *Alexandrium*, *Pyrodinium*, and *Gymnodinium*, as well as by cyanobacteria present in freshwater, such as *Aphanizomenon flos-aquae*, *Anabaena circinalis*, and *Lyngbya wollei*.[13] PSP toxins all consist of a tetrahydropurine backbone and can be divided into three groups based on their side chain (carbamoyl, *N*-sulfocarbamoyl, and decarbamoyl) (Figure 7.1). Saxitoxin (STX), gonyautoxin 1, 2, 3, and 4 (GTX1, GTX2, GTX3, GTX4), neosaxitoxin (NEO), 11α-OH-saxitoxin (11α-OH-STX), and 11β-OH-saxitoxin (11β-OH-STX) are members of the carbamoyl group. The *N*-sulfocarbamoyl group consists of B1-gonyautoxin (B1), B2-gonyautoxin (B2),

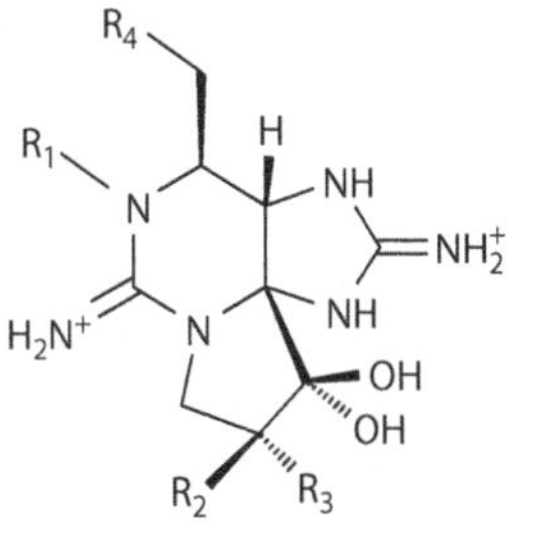

R_1	R_2	R_3	R_4	Toxin
H	H	H	$-O-C(=O)-NH_2$	STX
H	H	OSO_3^-		GTX2
H	OSO_3^-	H		GTX3
OH	H	H		NEO
OH	H	OSO_3^-		GTX1
OH	OSO_3^-	H		GTX4
H	H	OH		11α-OH-STX
H	OH	H		11β-OH-STX
H	H	H	$-O-C(=O)-NHSO_3^-$	B1(GTX5)
H	H	OSO_3^-		C1
H	OSO_3^-	H		C2
OH	H	H		B2 (GTX6)
OH	H	OSO_3^-		C3
OH	OSO_3^-	H		C4
H	H	H	$-OH$	dcSTX
H	H	OSO_3^-		dcGTX2
H	OSO_3^-	H		dcGTX3
OH	H	H		dcNEO
OH	H	OSO_3^-		dcGTX1
OH	OSO_3^-	H		dcGTX4

STX = saxitoxin
NEO = neosaxitoxin
GTX = gonyautoxin

FIGURE 7.1 Chemical structure of PSP toxins.

C1-gonyautoxin (C1), C2-gonyautoxin (C2), C3-gonyautoxin (C3), and C4-gonyautoxin (C4). Decarbamoylsaxitoxin (dcSTX), decarbamoylneosaxitoxin (dcNEO), and decarbamoylgonyautoxin 1, 2, 3, and 4 (dcGTX1, dcGTX2, dcGTX3, dcGTX4) belong to the decarbamoyl group.

The aim of the developed procedures was to analyze seafood (shellfish, puffer fish) or cell cultures for the presence of PSP toxins, DA and TTX, and analogs. Before LC analysis, extracts from the different organisms have to be prepared. For the analysis of PSP toxins in dinoflagellates and cyanobacteria, the sample preparation consists of a suspension of the organism culture in an acidic aqueous or acetonitrile solution followed by homogenization, centrifugation, and filtration.[14–17] The resulting extract is then injected in the LC system. For tissues from puffer fish and shellfish, the sample preparation consists of an acidic extraction of the toxins from the tissues followed by homogenization, centrifugation, and in most cases further cleanup with solid-phase extraction (SPE) or liquid–liquid extraction (LLE).[14,16,18–20] The resulting extract is then injected in the LC system.

To have an unambiguous and fast identification and quantification of all PSP toxins and to have knowledge of toxin profiles, methods are required that allow the analysis of all PSP toxins (Figure 7.1) in a single analysis. Chromatographic separation coupled to MS detection is the obvious choice for such analyses because of the high specificity and sensitivity of this detection. Because of the thermolabile nature of PSP toxins, LC is the preferred chromatographic technique. However, the use of RPLC reveals a major drawback in the analysis of PSP toxins: They are highly polar, resulting in poor retention. The employment of ion-pair agents can overcome the problem, but such agents interfere with the MS detection. HILIC can separate these hydrophilic toxins, while the ionization efficiency in MS detection is enhanced with HILIC due to the use of a high amount of organic solvent (methanol or acetonitrile) in the mobile phase.

Dell'Aversano et al.[16] optimized a method based on HILIC coupled to MS detection for the simultaneous analysis of all major PSP toxins in a single 30 min run with a high degree of selectivity and no need for further confirmation (Table 7.1). A TSKgel® Amide-80 stationary phase was used together with a mobile phase consisting of 65% eluent (B), where (A) was water and (B) was acetonitrile/water (95:5), both containing 2.0 mM ammonium formate and 3.6 mM formic acid (pH = 3.5). The use of a PolyHydroxyEthyl Aspartamide™ column was also tested by this group, but resulted in longer run time and bad peak shapes. Mobile phase, flow rate, and column temperature were further optimized for the amide column. The method gave excellent linearity and had a limit of detection of 30 nM, which should allow the analysis of shellfish samples near the regulatory limit of 0.8 mg STX equivalent per kg edible tissue.

Diener et al.[14] have separated PSP toxins with HILIC using two detection systems, FL and tandem MS (MS/MS), building on the HILIC separation developed by Dell'Aversano et al.,[16] which were not interested in FL detection because of co-elutions and inconstant retention times. The use of a zwitterionic ZIC®-HILIC stationary phase in combination with gradient elution with a mobile phase consisting of (A) 10 mM ammonium formate and 10 mM formic acid in water and (B) 80% acetonitrile and 20% water with a final concentration of 5 mM ammonium formate and

TABLE 7.1
Overview of Studies Concerning Marine Toxins: Studied Compounds, Chromatographic Conditions, and Analyzed Samples

Toxins Studied	Stationary Phase	Mobile Phase	Samples	References
PSP toxins				
STX, NEO, GTX1, GTX2, GTX3, GTX4, B1, B2, C1, C2, C3, C4, dcSTX, dcNEO, dcGTX1, dcGTX2, dcGTX3, dcGTX4	ZIC-HILIC (250 × 4.6 mm, 5 μm)	(A) 10 mM Ammonium formate and 10 mM formic acid in water; (B) 80% Acetonitrile and 20% water with a final concentration of 5 mM ammonium formate and 2 mM formic acid—gradient	*Alexandrium catanella* (dinoflagellate) *Gymnodinium catenatum* (dinoflagellate)	[14]
STX, NEO, GTX1, GTX2, GTX3, GTX4, B1, B2, C1, C2, C3, C4, dcSTX, dcNEO, dcGTX1, dcGTX2, dcGTX3, dcGTX4	TSK-Gel Amide-80 (250 × 2 mm, 5 μm)	(A) Deionized water; (B) 95% acetonitrile in water, both containing 2.0 mM ammonium formate and 3.6 mM formic acid—gradient	*Anabaena circinalis* (cyanobacterium) *Cylindrospermopsis raciborskii* (cyanobacterium)	[15]
STX, NEO, GTX1, GTX2, GTX3, GTX4, B1, B2, C1, C2, C3, C4, dcSTX, dcNEO, dcGTX1, dcGTX2, dcGTX3, dcGTX4, 11α-OH-STX, 11β-OH-STX	TSK-Gel Amide-80 (250 × 2 mm, 5 μm) PolyHydroxyEthyl Aspartamide™ (200 × 2 mm, 5 μm)	(A) Deionized water; (B) 95% acetonitrile in water, both containing 2.0 mM ammonium formate and 3.6 mM formic acid—isocratic	*Alexandrium tamarense* (dinoflagellate) *Mytilus edulis* (mussel)	[16]

(*continued*)

TABLE 7.1 (continued)
Overview of Studies Concerning Marine Toxins: Studied Compounds, Chromatographic Conditions, and Analyzed Samples

Toxins Studied	Stationary Phase	Mobile Phase	Samples	References
STX, NEO, GTX1, GTX2, GTX3, GTX4, B1, B2, C1, C2, C3, C4, dcSTX, dcNEO, dcGTX1, dcGTX2, dcGTX3, dcGTX4	TSK-Gel Amide-80 (250 × 2 mm, 5 μm)	(A) Deionized water; (B) 95% acetonitrile in water, both containing 2.0 mM ammonium formate and 3.6 mM formic acid—isocratic	*Alexandrium catenella* (dinoflagellate)	[17]
STX, NEO, GTX1, GTX2, GTX3, GTX4, B1, B2, C1, C2, C3, C4, dcSTX, dcNEO, dcGTX1, dcGTX2, dcGTX3, dcGTX4, 11α-OH-STX, 11β-OH-STX, DA	TSK-Gel Amide-80 (250 × 2 mm, 5 μm)	(A) Deionized water; (B) 95% acetonitrile in water, both containing 2.0 mM ammonium formate and 3.6 mM formic acid—isocratic	*Mytilus galloprovincialis* (mussel)	[19]
STX, NEO, GTX1, GTX2, GTX3, GTX4, B1, B2, C1, C2, C3, C4, dcSTX, dcNEO, dcGTX1, dcGTX2, dcGTX3, dcGTX4, 11α-OH-STX, 11β-OH-STX	TSK-Gel Amide-80 (250 × 2 mm, 5 μm)	(A) Deionized water; (B) 95% acetonitrile in water, both containing 2.0 mM ammonium formate and 3.6 mM formic acid—isocratic	*Alexandrium ostenfeldii* (dinoflagellate)	[21]
Domoic acid				
DA	TSK-Gel Amide-80 (250 × 2 mm, 5 μm)	(A) Deionized water; (B) 95% acetonitrile in water, both containing 2.0 mM ammonium formate and 3.6 mM formic acid—gradient	*Mytilus galloprovincialis* (mussel)	[18]

TTX and analogs				
STX, NEO, GTX1, GTX2, GTX3, GTX4, B1, dcSTX, dcNEO, dcGTX2, dcGTX3	ZIC-HILIC (150 × 2.1 mm, 3.5 µm) TSK-Gel Amide-80 (250 × 2.1 mm, 5 µm)	(A) Deionized water; (B) 95% acetonitrile in water, both containing 2.0 mM ammonium formate and 3.6 mM formic acid—gradient	*Mytilus edulis* (mussel)	[20]
TTX, 11-deoxyTTX, 6,11-dideoxyTTX, 5-deoxyTTX, 5,6,11-trideoxyTTX	TSK-Gel Amide-80 (150 × 2 mm, 5 µm)	16 mM Ammonium formate buffer and acetonitrile (3/7, v/v)—isocratic	*Fugu pardalis* (puffer fish)	[23]
TTX, 4-epiTTX, 11-deoxyTTX, 5-deoxyTTX, 5,6,11-trideoxyTTX, 4,9-anhydroTTX, 4-CysTTX, STX, dcSTX	TSK-Gel Amide-80 (150 × 2 mm, 5 µm)	16 mM Ammonium formate buffer and acetonitrile (3/7, v/v)—isocratic	*Fugu pardalis* (puffer fish)	[22]
TTX, 4-epiTTX, 6-epiTTX, 11-deoxyTTX, norTTX-6(S)-ol, norTTX-6(R)-ol, nor-TTX-6,6-diol, anhydroTTX, 6-epianhydroTTX, 5-deoxyTTX, trideoxyTTX	ZIC-HILIC (150 × 2.1 mm, 5 µm)	(A) 10 mM Ammonium formate and 10 mM formic acid in water; (B) 80% acetonitrile and 20% water with a final concentration of 5 mM ammonium formate and 2 mM formic acid—gradient	*Takifugu oblongus* (puffer fish)	[24]
TTX, 4-epiTTX, 6-epiTTX, 11-deoxyTTX, 5-deoxyTTX, 5,6,11-trideoxyTTX, 4,9-anhydroTTX, 6-epi-4,9-anhydroTTX, norTTX-6(S)-ol, norTTX-6(R)-ol	TSK-Gel Amide-80 (150 × 2 mm, 5 µm)	16 mM Ammonium formate buffer and acetonitrile (3/7, v/v)—isocratic	*Fugu poecilonotus* (puffer fish) *Cynops ensicauda* (newt)	[25]

2 mM formic acid was proposed (Table 7.1). The application of this gradient provided a complete separation of all regulated PSP toxins in a single run. FL detection was possible, but MS/MS detection proved to be a better choice because of two reasons: (1) the HILIC-MS/MS method led to shorter run times (40 min for MS/MS vs. 70 min for FL detection) because the FL detection required a slower gradient for the effective separation of the PSP toxins and (2) a more selective detection with MS/MS was observed compared with FL. Moreover, the HILIC-FL method needed a post-column oxidation before FL detection was possible, making this method also more complicated than the HILIC-MS/MS method. The method had good linearity and the limit of detection for the compounds ranged between 0.03 and 0.80 ng injected, depending on the compound and the type of detection.

Turrell et al.[20] evaluated two stationary phases for the separation of PSP toxins with HILIC coupled to MS and MS/MS detection (Table 7.1). Using a gradient elution with a mobile phase consisting of (A) deionized water and (B) 95% acetonitrile in water, both containing 2.0 mM ammonium formate and 3.6 mM formic acid for the TSK-Gel Amide-80 and ZIC-HILIC column, they concluded that the ZIC-HILIC column provides a faster separation (35 vs. 43 min) and better peak shape for the late eluting toxins, in accordance with the findings of Diener et al.[14]

The application of the HILIC-MS, HILIC-MS/MS, or HILIC-FLD methods to extracts of dinoflagellate cultures showed the presence of a wide range of PSP toxins in such samples (Table 7.1). Several studies revealed the presence of most of the PSP toxins in prepared extracts of different species of dinoflagellates from China, Mexico, Canada, Italy, and Australia.[14–17] Samples in Australia revealed concentrations of dcGTX2 up to 430 μg/g.[15] An accurate knowledge of the toxin profiles of these organisms could be obtained by HILIC and provided new insights into patterns and sources of toxin accumulation in marine food webs. Extracts of mussels from Canada and Scotland and extracts from puffer fish of Japan, analyzed with the developed HILIC methods, showed the presence of PSP toxins.[16,20–22] A toxicity corresponding with 178 μg STX/100 g shellfish was detected in Scottish mussels with HILIC-MS.[20]

7.2.3 Tetrodotoxin and Analogs

TTX is a toxin that is probably produced by microalgae and is considered to be the most lethal toxin coming from the marine environment.[23] It is named after the Tetraodontidae family of fish (puffer fish), but is also found in other fish and mollusks and even in amphibians and land animals.[24] An overview of the structures of TTX and its analogs is given in Figure 7.2.

Sample preparation for the analysis of TTX and its analogs in puffer fish consists of an acidic extraction of the toxins from the tissues followed by homogenization and centrifugation. Diener et al.[25] apply a direct injection of the supernatant, while other papers report further clean-up on reversed-phase resins (Table 7.1).[22,26,27]

Nakagawa et al.[27] described a method based on HILIC-MS for the detection of TTX and its analogs in puffer fish. A TSK-Gel Amide-80 column was used in combination with a mixture of 16 mM ammonium formate buffer and acetonitrile (3/7, v/v) as mobile phase. This method allowed the separation of 11 TTX analogs. A 12th

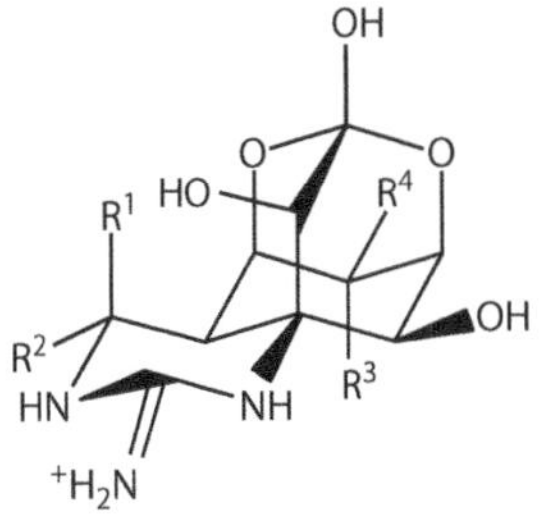

	R^1	R^2	R^3	R^4
TTX	H	OH	OH	CH_2OH
4-epiTTX	OH	H	OH	CH_2OH
6-epiTTX	H	OH	CH_2OH	OH
11-deoxyTTX	H	OH	OH	CH_3
norTTX-6(S)-ol	H	OH	OH	H
norTTX-6(R)-ol	H	OH	H	OH
norTTX-6, 6-(diol)	H	OH	OH	OH

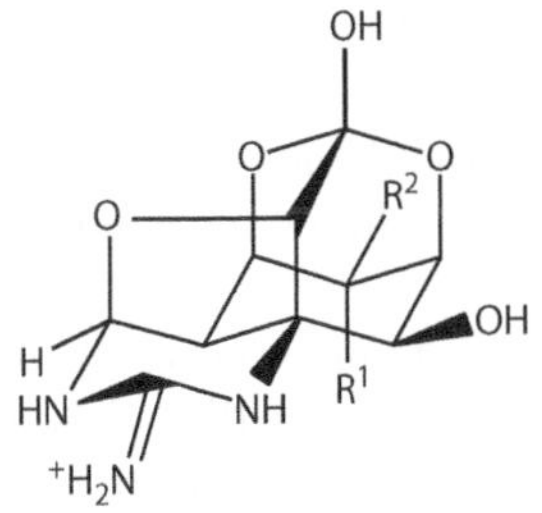

	R^1	R^2
anhydroTTX	OH	CH_2OH
6-epianhydroTTX	CH_2OH	OH

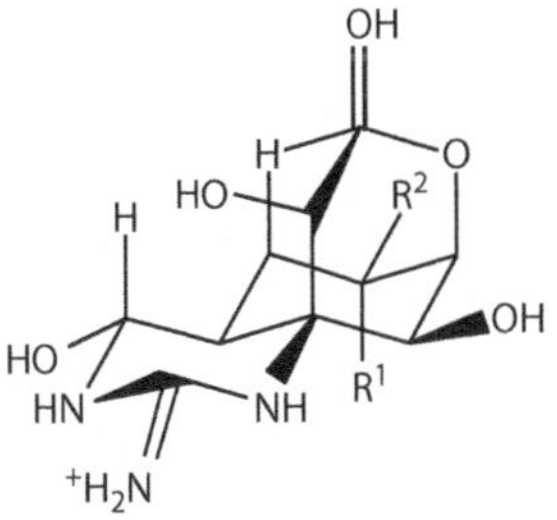

	R^1	R^2
5-deoxyTTX	OH	CH_2OH
trideoxyTTX	H	CH_3

FIGURE 7.2 Chemical structure of TTX and its analogs.

TTX analog, 6,11-dideoxytetrodotoxin, was detected and reported for the first time in puffer fish samples by applying the already optimized HILIC-MS method.

Diener et al.[25] published a HILIC-MS study for the determination of TTX and analogs in puffer fish. A ZIC-HILIC stationary phase was used for the separation of 11 toxins using a mobile phase consisting of (A) 10 mM ammonium formate and 10 mM formic acid in water and (B) 80% acetonitrile and 20% water with a final concentration of 5 mM ammonium formate and 2 mM formic acid. This setup resulted in enhanced sensitivity and accuracy compared with the previous LC-MS methods.[11,27] A further advantage of this method was that the ZIC-HILIC stationary phase and mobile phase can also be used for the analysis of PSP toxins.[27] This allowed a combined determination of TTX and analogs and PSP toxins in one single run and has major advantages over the method published by Jang and Yotsu-Yamashita,[22] which needed two separate runs for the analysis of TTX and analogs and PSP toxins in puffer fish.

Toxin profiles of TTX and its analogs could be revealed by analyzing extracts of different tissues of puffer fish from Japan and Bangladesh with HILIC-MS and HILIC-MS/MS.[22,25–27] Samples from puffer fish from Bangladesh revealed concentrations up to 2929 μg/g of 5,6,11-trideoxyTTX in ovaries[25] and up to 94 μg/g TTX could be quantified in ovaries of puffer fish from Japan.[22]

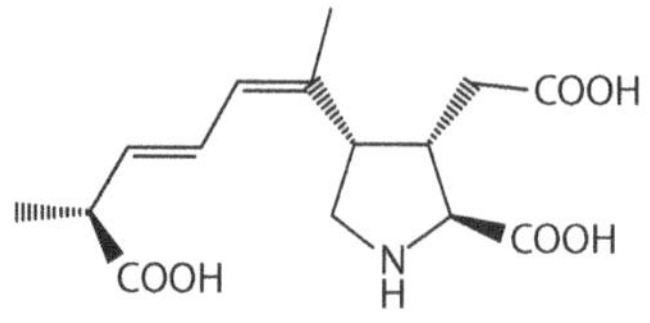

FIGURE 7.3 Chemical structure of DA.

7.2.4 Domoic Acid

DA is a naturally occurring, but rare, amino acid (Figure 7.3) produced by the diatom *Pseudo-nitzschia pungens* f. *multiseries*.[28] Other species of *Pseudo-nitzschia*, namely, *Pseudo-nitzschia australis*, *Pseudo-nitzschia pseudo-delicatissima*, and *Pseudo-nitzschia galaxiae*, are also found to produce DA.[29] This compound is toxic to marine mammals and seabirds, as well as to humans.

The sample preparation for the analysis of DA in shellfish samples is comparable with the preparation of extracts for the analysis of PSP toxins in shellfish.[18] Several efforts have been made to analyze DA in shellfish with LC-MS because of its high sensitivity and specificity. Recently, different RPLC-MS methods have been proposed for the analysis of DA in seafood.[30–32]

Ciminiello et al.[18] underlined the urgent need for multi-toxin determination methods, since shellfish can be contaminated with different classes of toxins (DA and PSP toxins). They presented a method based on HILIC-MS that simultaneously analyzes DA and PSP toxins in shellfish. A TSK-Gel Amide-80 stationary phase was used and a gradient elution with a mobile phase consisting of (A) deionized water and (B) 95% acetonitrile in water, both containing 2.0 mM ammonium formate and 3.6 mM formic acid was applied. The method was based on an earlier described procedure,[16] but in this method, a gradient elution was used, because DA was eluting at the chromatographic front if the original isocratic conditions were applied. By using a gradient elution starting with a higher amount of organic solvent, good retention for DA in combination with an efficient separation of PSP toxins was obtained. DA was detected in mussel samples from Italy (4 out of 11 samples were positive for DA) by HILIC-MS and HILIC-MS/MS.[18] Concentrations ranged from 0.25 to 2.5 μg/g, which are below the regulatory limit of 20 μg/g.

7.2.5 Concluding Remarks

The presence of PSP toxins, DA, and TTX and analogs in seafood can have severe implications for human health. Routine monitoring for these toxins is necessary to prevent human intoxications. It is clear that there is a need for replacement of the official mouse bioassay, which is now used for the evaluation of the toxicity of contaminated seafood. The application of HILIC coupled to MS detection is a good alternative and deliver simple (one analysis run), sensitive, and accurate measurements of toxins in seafood samples. In the future, methods based on HILIC-MS can serve as scientific basis for the validation and implementation of analytical techniques in (shell)fish monitoring programs to replace the traditional mouse bioassay.

7.3 PHARMACEUTICALS

Pharmaceuticals, drugs of abuse (DOA), and their metabolites are contaminants of increasing concern in the aquatic environment due to their toxicity, various endocrine-disrupting effects, and their potential accumulation in the aquatic ecosystem. Several studies pointed out that excreted human or veterinary drugs, personal care products, or DOA end up in the environment through the sewage system.[33–35] A large interest has been nowadays focused on their detection and quantification in wastewater and surface water, and HILIC is a method that is gaining interest due to the high polarity of these pollutants. Table 7.2 shows a brief list of the molecules analyzed in HILIC mode from aquatic systems.

7.3.1 Estrogens

Estrogens are steroid hormones produced by the endocrine system that are important for the human brain, reproduction, and bone development. They are also present in the environment from natural sources (human and animal excretion, plants, fungi) or as by-products of synthetic chemicals. In aquatic systems, estrogens are a risk factor as they act as endocrine disruptors producing fish feminization.[36] They may influence plant growth or even human health, and the monitoring of these compounds has therefore become imperious for more and more institutions. The analysis of estrogens (e.g., estrone) and their glucuronide and sulfate conjugates (Figure 7.4) at low ng/L levels in environmental samples can be done by immunoassays, GC/MS, or LC/MS. Immunoassay methods are sensitive and specific, but only few antibodies are available and cross-reactions remain a problem. GC/MS requires derivatization and hydrolysis of the estrogen conjugates, so the substitution group is lost before analysis. LC-MS/MS is the most advantageous technique and is widely used today. HILIC chromatography is useful for the separation of estrogen sulfates and glucuronides, which are highly hydrophilic compounds.

To detect estrogens (estrone, estriol, estradiol) and their glucuronide and sulfate conjugates, in a single run, Qin et al.[37] used a column-switching method involving a C_{18} and a HILIC stationary phase. They collected water samples from River North Saskatchewan (Canada), adjusted them to pH = 2, and stored them at 4°C prior to analysis. The analytes were isolated and pre-concentrated from 500 mL water by SPE on Oasis™ HLB cartridges. Free estrogens were eluted with ethyl acetate, while estrogen conjugates were eluted with methanol containing 2% ammonium hydroxide. After elution of the SPE cartridge, the estrogen fraction was further derivatized with dansyl chloride and the reaction mixture was purified once more on an Oasis HLB cartridge. The last methanolic eluate was mixed then with the estrogen conjugates fraction, the solvent was evaporated, and the analytes were reconstituted with mobile phase and injected in the LC/MS system. A column switching set-up, employing a binary pump connected to a quadrupole-linear ion trap mass spectrometer with an electrospray (ESI) interface operated alternatively in positive (for the detection of the dansylated estrogens) and negative (for estrogen conjugates) ion mode in the same run. A 10-port 2-positions switching valve served to link up the two columns, Luna® $C_{18}(2)$ (100 × 2.0 mm, 3 µm) for the separation of dansylated estrogens and

TABLE 7.2
Pharmaceuticals, Drugs of Abuse, and Pesticides Analyzed in HILIC Mode from Aquatic Systems—A Brief Review of Employed Columns, Mobile Phases, and Type of Samples

Compound/Class of Compounds	Matrix	Column	Mobile Phase	References
Pharmaceuticals				
Estrogen conjugates	River water	ZIC-pHILIC (100 × 2.1 mm, 5 μm)	(A) 5 mM Ammonium acetate 5 mM in water; (B) acetonitrile—gradient	[36]
Cytostatics	Wastewater	ZIC-HILIC (150 × 2.1 mm, 3.5 μm)	(A) 30 mM Ammonium acetate in water; (B) acetonitrile—gradient	[37]
Antibiotics	Liquid manure; rainfall runoff	Alltima HP HILIC (150 × 2.1 mm, 3 μm)	(A) 0.1% Formic acid in water; (B) acetonitrile—isocratic	[38]
Metformin	Wastewater; surface water	ZIC-HILIC (150 × 2.1 mm, 3.5 μm)	(A) 10 mM Ammonium formate in water (pH 3 with formic acid); (B) acetonitrile—gradient	[52]
Albuterol, cimetidine, ranitidine, metformin	Water, sludge	1. Zorbax HILIC Plus (100 × 2.1 mm, 3.5 μm)	1. (A) 10 mM Ammonium acetate in water; (B) acetonitrile—gradient	[39]
		2. Atlantis HILIC (100 × 2.1 mm, 3 μm)	2. (A) 0.1% Acetic acid/ammonium acetate in water; (B) acetonitrile—gradient	[40]

Drugs of abuse				
Cocaine and metabolites	Wastewater	Zorbax RX-Sil (150 × 2.1 mm, 5 µm)	(A) 2 mM Ammonium acetate in water (pH 4.5 with acetic acid); (B) acetonitrile—gradient	[41]
Drugs of abuse	Wastewater	Luna HILIC (150 × 3 mm, 5 µm)	(A) 5 mM Ammonium acetate in water; (B) acetonitrile—gradient	[59]
Pesticides				
Organophosphorus pesticides	Water	Atlantis HILIC (150 × 2 mm, 5 µm)	Acetonitrile/isopropanol/ammonium formate 200 mM (pH 3)—isocratic	[42]
Phenylurea pesticides and degradation products	Drinking water	Nucleosil Diol (150 × 4 mm, 5 µm)	(A) 20 mM Ammonium formate pH 3.3; (B) acetonitrile—gradient	[43]
Diquat, paraquat	Drinking water	Atlantis HILIC (150 × 2.1 mm, 3.5 µm)	(A) 250 mM Ammonium acetate pH 3; (B) acetonitrile—isocratic	[44]
		Atlantis HILIC (150 × 2.1 mm, 3 µm)	(A) 10 mM Ammonium formate pH 3.7; (B) acetonitrile—isocratic	[45]

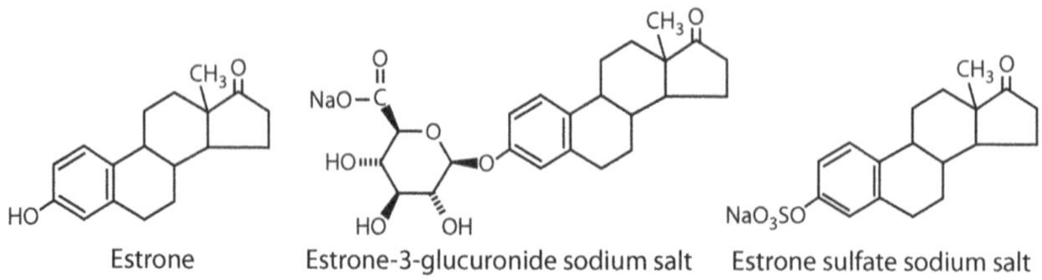

Estrone Estrone-3-glucuronide sodium salt Estrone sulfate sodium salt

FIGURE 7.4 Estrone and its conjugates.

ZIC-pHILIC (100 × 2.1 mm, 5 μm) for the estrogen conjugates. The mobile phase consisted of (A) acetonitrile/aqueous ammonium acetate 5 mM pH = 6.8 (75/25, v/v) and (B) acetonitrile/aqueous ammonium acetate 5 mM pH = 6.8 (95/5, v/v). The sample extract was injected with an initial mobile phase composition of 40% A and the valve connected both columns. In this way, estrogens are trapped on the C_{18} stationary phase and their conjugates transferred to the HILIC stationary phase. Then, with an appropriate gradient and valve program, the two groups of compounds were separated and detected. Using this approach, the method detection limits were up to 10-fold better than in the classical RP approach, and the linearity range was 0.2–200 ng/L for estrogens, and 0.5–500 ng/L for conjugates (except for estriol glucuronides with a range of 2–2000 ng/L), with an injection volume of 10 μL. However, strong matrix effects (up to 50%) were observed and these should be compensated with deuterated internal standards. Only estrone and estrone-3-sulfate were detected in river water samples.

7.3.2 Cytostatic Drugs

Cytostatic drugs are used in cancer therapy as chemotherapeutic drugs. Their action is mainly to inhibit cell growth and cell division by disrupting DNA replication. The cytotoxicity is not limited to tumor cells, but also affects other fast-dividing cells. Thus, the chemotherapeutic drugs pose an intrinsic carcinogenic potential, and their release and accumulation in the environment (they are poorly biodegraded) is a risk factor for all organisms.[38] The presence of chemotherapeutic drugs in wastewater and river water has recently received increased attention. Anti-metabolites are the most used cytostatic agents and have polar structures (5-fluorouracil, cytarabine, and gemcitabine shown in Figure 7.5). They were analyzed from biological and environmental samples, at very low concentrations, by GC/MS, often with

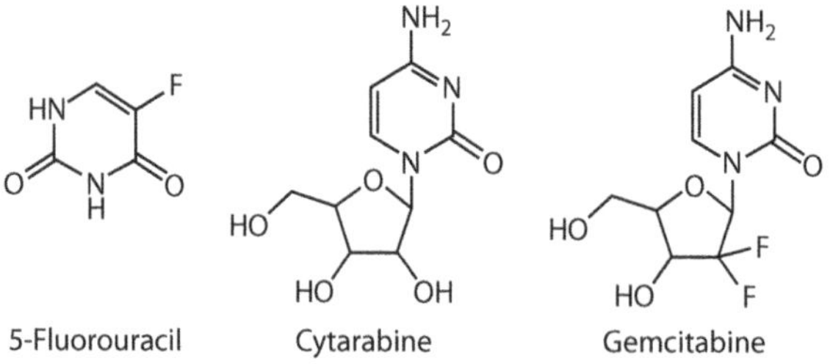

5-Fluorouracil Cytarabine Gemcitabine

FIGURE 7.5 Chemical structures of some anti-metabolites.

derivatization,[39] and by HPLC with ion-pairing on RP columns or using adequate stationary phases as porous graphite or HILIC.

Cytostatics and metabolites (5-fluorouracil, cytarabine, gemcitabine, α-fluoro-β-alanine, uracil 1-β-D-arabinofuranoside, 2,2′-dihydrodeoxyuridine) were extracted by Isolute® ENV+ cartridges.[40] Water samples (50 mL) were adjusted to pH = 6, loaded onto the cartridges and eluted with methanol. The eluate was evaporated to dryness and reconstituted with initial mobile phase before injection. For the investigation of the cytostatics in hospital wastewater, samples were collected from a Swiss hospital, filtered through glass fiber and a cellulose acetate membrane, and stored at −20°C until analysis. In the method development, wastewater samples from the University Hospital in Aachen (Germany) were used. The system employed by Kovalova et al.[40] to separate cytostatics and their human metabolites was composed of a binary pump, autosampler, and column thermostat, connected to a triple quadrupole mass spectrometer equipped with an ESI interface, operated in positive or negative ion mode (in two separate runs). The extracted compounds were chromatographed in gradient on a ZIC-HILIC column (150 × 2.1 mm, 3.5 μm), with a mobile phase consisting of (A) aqueous ammonium acetate 30 mM/acetonitrile (2/3, v/v) and (B) acetonitrile. Method limits of quantification were in the range of 0.9–9 ng/L. To confirm the masses of the analytes and elucidate the interfering peak observed in hospital wastewater, a high-resolution Orbitrap mass analyzer was used. The method was applied to hospital wastewater samples and 5-fluorouracil was detected in 76% of samples, while gemcitabine and 2,2′-difluorodeoxyuridine were present in 65% and 88% of the samples, respectively. The analysis of cytostatics in wastewater is important because of their possible negative effects on the aquatic environment. Therefore, it is necessary to have knowledge about their occurrence in wastewater and surface water and their fate during wastewater treatment.

7.3.3 Antibiotics

Antibiotics, compounds that kill or inhibit the growth of bacteria, have the potential to affect the microbial community in sewage systems. The inhibition of wastewater bacteria may seriously affect organic matter degradation and, therefore, effects of antibacterial agents on the microbial population are of great interest. Antibiotics have also been evidenced in surface water where they may affect organisms of different trophic levels.[41] Antibiotics have various (polar) structures and were determined by bioassays, GC/MS, and RPLC, often with derivatization, with ion-pair agents in the mobile phase, or even by ion-exchange techniques. Nowadays, HILIC applications are expanding and include the analysis of aminoglycoside antibiotics from biological samples,[42] analysis of lincomycin and spectinomycin in liquid hog manure supernatant, and simulated rainfall runoff from manure-treated cropland (discussed in detail in Chapters 7 and 8 of this book).[43]

7.3.4 Miscellaneous Pharmaceuticals

The high concern of measuring pharmaceuticals as an important group of emerging contaminants in the environment was expressed also by the U.S. Environmental

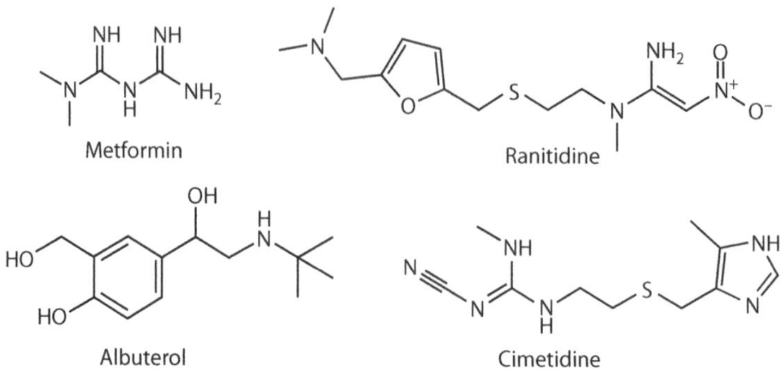

FIGURE 7.6 Chemical structures of the pharmaceuticals analyzed by HILIC with EPA 1694 method.

Protection Agency (EPA) that issued a method for the analysis of pharmaceuticals and personal care products in water, soil, sediments, and biosolids by HPLC-MS/MS.[44] Several classes of pharmaceuticals, 71 compounds in total, and 19 internal standards were determined mainly by RPLC, but for the most polar of them, albuterol, metformin, cimetidine, and ranitidine (Figure 7.6), a HILIC approach was preferred, after sample pretreatment by SPE.

Metformin is a widely used anti-diabetic drug from the biguanide class, and one of the most prescribed overall. It is not metabolized by the human body, being excreted in the urine. Albuterol (salbutamol) is also widely used as bronchodilator prescription, while ranitidine and cimetidine are histamine H_2-receptor antagonists that inhibit stomach acid production, being in the top 20 list of most prescribed drugs in different European countries.[45] Therefore, it is very important to evaluate the environmental risk of these pharmaceuticals. SPE was also involved in the extraction of water and sludge samples.[44] An ultrasonic extraction with acetonitrile was performed for the soil samples or the solid particles filtered from water. A volume of 500–1000 mL filtered water was spiked with labeled internal standards, adjusted to pH = 10, and loaded on an Oasis HLB cartridge for isolation of metformin, albuterol, ranitidine, and cimetidine. The elution was done with methanol and then with 2% formic acid solution in methanol. The extract was concentrated to nearly dryness, then 3 mL of methanol was added and spiked with instrument-labeled internal standards, and finally brought to 4 mL with 0.1% formic acid in methanol solution. For the determination of metformin in wastewater and surface water, Scheurer et al.[46] have published a method based on sample preparation with Strata™ X-CW SPE cartridges followed by a separation on a ZIC-HILIC column (150 × 2.1 mm, 3.5 μm) with a mobile phase consisting of (A) 10 mM ammonium formate in water set to pH = 3 with formic acid and (B) acetonitrile, in gradient starting with 95% B. The chromatographic separation was coupled to a 4000 Q-Trap mass spectrometer equipped with an ESI ion source for detection. Concentrations up to 129 μg/L in influent wastewater, 21 μg/L in effluent wastewater, and 1.7 μg/L in surface water were measured.

The EPA method[44] was set up using an HPLC system connected to a triple quadrupole mass spectrometer, equipped with an ESI ion source operated in positive mode. The chromatographic separation was carried out on a Waters Atlantis® HILIC column (100 × 2.1 mm, 3 μm) eluted in gradient with a mobile phase consisting of (A) ammonium acetate/acetic acid buffer 0.1% in water and (B) acetonitrile. The acquisition was done in multiple reaction monitoring (MRM) mode using one transition per compound. An alternative to the EPA setup was proposed by Agilent.[47] The sample pretreatment is the same as in the EPA method, and the separation is performed on a Zorbax® HILIC Plus column (100 × 2.1 mm, 3.5 μm) with a gradient of acetonitrile/aqueous ammonium acetate 10 mM. A second MRM transition was added for confirmation of each compound.

7.3.5 Drugs of Abuse

One of the dark sides of modern society is the spread of DOA, which are consumed to alter mood, thought, and feeling through their actions on the central nervous system (brain and spinal cord). In addition, they are capable to produce dependence, either physical or psychological.[48] Even if these substances are severely regulated and there are many efforts toward prevention, the underground consumption remains quite high and their presence in the environment has been proven.

The analysis of DOA residues in wastewater and surface water is not only a preoccupation of scientists but has become in recent years a tool to estimate the consumption of these illicit substances in different countries.[34,35,49–51] Using the excretion patterns and considering the stability of the metabolites and of the unchanged DOA in the environment, an overall estimation of the consumption can be made.[35,51]

HPLC methods for DOA (including cocaine) in RP mode have been published.[52,53] Considering the highly polar structures, a method for detection and quantification of cocaine and its principal metabolites (Figure 7.7) in wastewater and surface water using HILIC was developed, validated, and applied to samples collected from 41 wastewater treatment plants (WWTPs) and rivers across Belgium.[49–51,54] A significant increase in sensitivity for all analytes was found compared to the reversed-phase results. The target compounds were isolated by SPE on Oasis HLB cartridges. Before pretreatment, water samples (100 mL for wastewater, 500 mL for river water) were filtered through a glass microfiber filter and adjusted to pH = 6 with ammonium hydroxide. Then the deuterated internal standards were added and the samples were loaded on SPE columns preconditioned with methanol and water. After a washing step with water/methanol (95/5, v/v) and the subsequent drying, the analytes and internal standards were eluted with 2 × 4 mL methanol. The eluate was evaporated until dryness and the samples were reconstituted with 150 μL acetonitrile/methanol mixture (3/1, v/v).

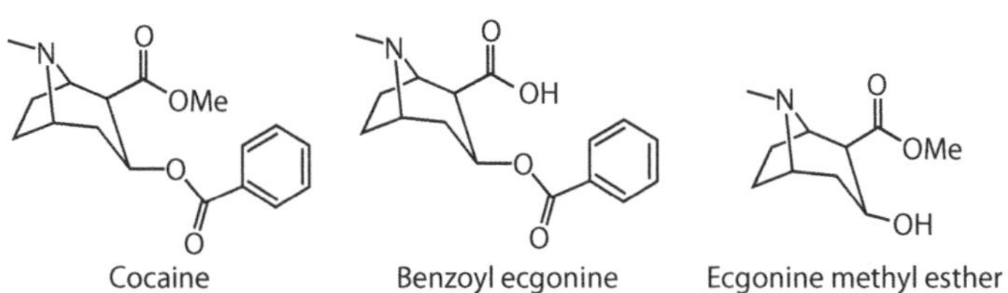

FIGURE 7.7 Cocaine and its principal metabolites.

Analysis has been carried out on a system comprised of an autosampler, binary pump, and ion trap detector with electrospray ionization (ESI).[54] The interface was operated in positive ion mode and the quantitative data were acquired in MRM mode, measuring the specific transitions for each compound and its deuterated internal standard. HILIC separation was performed on a Zorbax Rx-SIL column (150 × 2.1 mm, 5 µm) with a Zorbax Rx-Sil (12.5 × 2.1 mm, 5 µm) guard column, eluted in gradient with a mobile phase composed by (A) ammonium acetate 2 mM/acetic acid buffer (pH 4.5) and (B) acetonitrile, and the flow was 0.25 mL/min. The pump program was as follows: 0–1 min 80% B, 1–10 min linear gradient up to 40% B, 10–18 min constant on 40% B, and 21–31 min re-equilibration of the column with the initial conditions. A seven-point calibration curve based on deuterated internal standards was generated for each analyte, and linearity was obtained for the following ranges: 35–3300 pg for benzoylecgonine, 10–870 pg for cocaine, and 15–780 pg for ecgonine methyl ester injected on column. Matrix effects were also evaluated by comparing the analyte signal in samples (WWTP, river or tap water) with the analyte signal recorded for Milli-Q water. The ion suppression varied between 30% and 48% for wastewater and 12% and 22% for surface and tap water. To easily overcome these matrix effects, deuterated internal standards have to be used for quantification. With this method, 162 samples have been analyzed and the concentrations measured in Belgian WWTP samples ranged from 9 to 753 ng/L and 37 to 2258 ng/L for cocaine and benzoylecgonine, respectively.[49–51] Ecgonine methyl ester concentrations were below the method limit of quantification. A new HILIC method was recently developed for the simultaneous determination of nine DOA and metabolites (amphetamine, methamphetamine, MDMA, methadone, EDDP, 6-monoacetyl morphine, cocaine, benzoylecgonine, and ecgonine methyl ester) in wastewater, with improved quantification limits (down to 1 ng/L for all compounds, except for ecgonine methyl ester, amphetamine, and 6-monoacetyl morphine for which quantification limits were 2 ng/L).[55] This method involves a chromatographic separation on a Luna HILIC column (150 × 3 mm, 5 µm) and gradient with (A) ammonium acetate 5 mM in water and (B) acetonitrile, at a flow rate of 0.4 mL/min and starting with 95% B. The sample pretreatment was performed by SPE on Oasis MCX cartridges. All compounds, except 6-monoacetyl morphine, could be quantified in a set of 12 wastewater samples from Belgium.

7.4 PESTICIDES

On its official Web site, the European Commission states that there is a certain accepted risk associated with the use of pesticides, due to their direct benefits generated for agriculture.[56] Pesticides are employed on a large scale in modern farms, mainly for plant protection and control of harmful organisms, such as pests, but they also have unwanted dangerous effects on the human health and environment if misused, carelessness, or even if used according to label directions. Therefore, these substances are regulated for a long time in the developed countries, their environmental fate is studied, and the risks closely evaluated.[56] Some pesticides or pesticide artifacts are particularly difficult to determine by conventional analytical techniques, because of their highly hydrophilic character

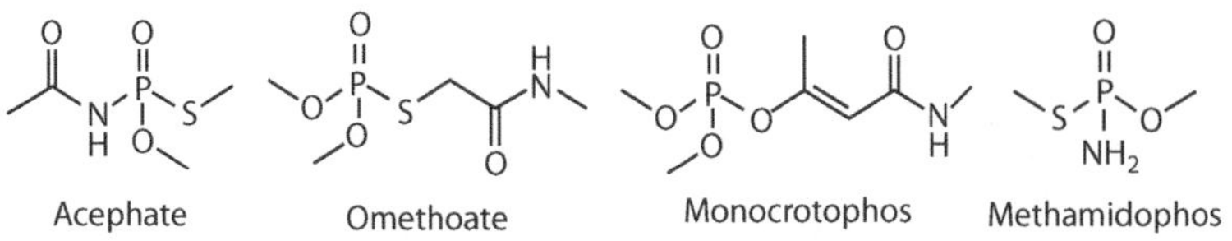

Acephate Omethoate Monocrotophos Methamidophos

FIGURE 7.8 Chemical structures of various polar organophosphorus pesticides.

(that complicates their isolation from an aqueous compatible matrix), lack of chromophore or fluorescent groups, and low molecular weights.

7.4.1 Organophosphorus Pesticides

Organophosphorus pesticides (OPPs) are cholinesterase inhibitors widely used for crop protection. The possible contamination of drinking water is monitored; in the European Union for instance, a maximum concentration of 0.1 µg/L is allowed.[57] Most OPPs are easily analyzed by gas chromatography, but a few of them are very polar and thermolabile (Figure 7.8).

Hayama et al.[58] analyzed polar OPPs from water samples by HILIC-MS/MS. Water samples (50 mL) were directly loaded to a preconditioned GL-Pak activated carbon cartridge which was back-flush eluted with 5 mL of 0.2% (v/v) formic acid in acetonitrile/isopropanol (95/5, v/v). The eluate was fortified with an internal standard ($^2H_6$-acephate) and injected into the LC-MS system. For the separation of OPPs, an Atlantis HILIC column (150 × 2.0 mm, 5 µm) was used with a mixture of acetonitrile/isopropanol/200 mM ammonium formate in water (pH = 3) (92:5:3, v/v) in isocratic conditions. Good retention times (range 3.4–4.9 min) were obtained for the target compounds. The best sensitivity was achieved in acidic conditions (buffer pH = 3), while the addition of isopropanol also improved the sensitivity. Though, only small percentages of buffer and isopropanol ensured a good retention of OPPs on the HILIC column. Hayama et al.[58] applied this method to river water samples collected from Chikugo River (Fukuoka, Japan). The spiked samples showed recoveries between 76% and 99%, similar to those obtained with distilled water.

7.4.2 Phenylurea Herbicides

Phenylurea herbicides are one of the most important classes of herbicides, mostly used in cotton, fruit, and cereal production. Diuron is also used as total herbicide in urban areas and as algicide in antifouling paints. Because of their extensive use and poor biodegradation, several phenylurea herbicides have been often detected as water pollutants.[59] This class of compounds also has high polarity and water solubility. Chemical degradation is negligible under normal pH and temperature conditions, while microbial and photodegradation may help for the decontamination of soil. Farre et al.[60] evaluated the intermediates generated during the degradation of diuron and linuron (Figure 7.9) by the photo-Fenton reaction using HILIC–MS/MS. They purified wastewater and spiked it with linuron or diuron and with Fenton reagent ($FeSO_4$ and H_2O_2), and then irradiated the mixture for 60 min with a 6W Philips black light with a measured intensity of 0.21 mW/cm^2. The photodegradation

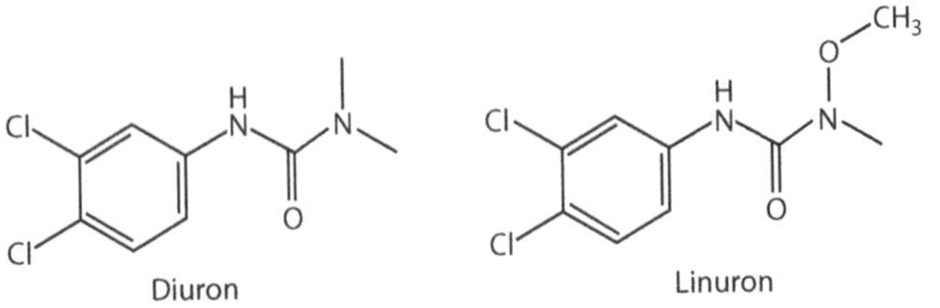

FIGURE 7.9 Chemical structures of diuron and linuron.

products of linuron and diuron were investigated on a Nucleosil® Diol (150 × 4 mm, 5 μm) column, with a mobile phase composed by (A) 20 mM ammonium formate buffer (pH = 3.3) and (B) acetonitrile, in gradient conditions. Farre et al.[60] identified by HILIC various amounts of methylurea and 1,1 dimethylurea in the photo-treated solutions of herbicides, together with chlorinated structures which were not completely elucidated. This justifies, according to the authors, the different biodegradability of the effluents.

7.4.3 Quaternary Ammonium Salt Herbicides

Diquat and paraquat, quaternary ammonium salt derivatives, are widely used herbicides. In the United States alone, the estimated use of paraquat and diquat in 1997 was over 1500 and 100 tons, respectively.[61] These compounds are particularly difficult to analyze by RPLC, due to their ionic structures (Figure 7.10), usually an ion-pair reagent being added in the mobile phase. HILIC was reported as an alternative by Makihata et al.[62] and in a Waters Application Note.[61]

The isolation procedure for paraquat and diquat presented by Waters was based on SPE using Oasis WCX cartridges.[61] After loading, the columns were washed with phosphate buffer pH = 7, water, and methanol, and then the herbicides were eluted with ACN/water/TFA 84:14:2. The dried extracts were reconstituted with mobile phase and 20 μL of the final solution were injected. The analysis of diquat and paraquat was performed on an Atlantis HILIC (150 × 2.1 mm, 3.5 μm) column, the mobile phase consisting of 40% acetonitrile, 60% aqueous buffer pH = 3.7 (250 mM ammonium formate).

Makihata et al.[62] have used a similar approach to determine diquat from Japanese surface water (tap, river, lake, shallow well, and deep well water samples) collected from Hyogo prefecture. An Atlantis HILIC column (150 × 2.1 mm, 5 μm) was used and the mobile phase was acetonitrile/10 mM ammonium formate (pH = 3.7) (50/50, v/v). In these conditions, the achieved limit of detection was 50 ng/L, 100 times lower than

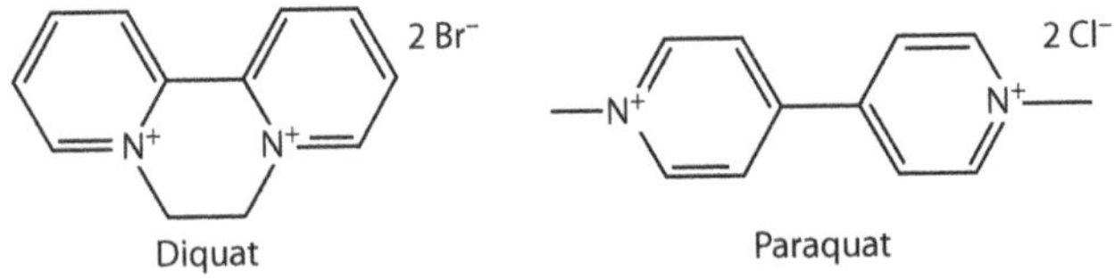

FIGURE 7.10 Chemical structures of the two common quaternary ammonium salt herbicides.

the target value for diquat residues in Japan, and the herbicide was found in one of the shallow well water samples, at a concentration of 2 μg/L.

7.5 MISCELLANEOUS

A number of other applications which cannot be categorized in the previous chapters have been found involving HILIC and contaminants of aquatic systems.

7.5.1 Dichloroacetic Acid

Dichloroacetic acid (DCA) is a small, polar compound that individuals may be exposed to as a result of drinking water consumption. DCA is found in drinking water as a disinfection by-product of chlorination or deriving from metabolization of chlorinated solvents, such as trichloroethylene. The occurrence of DCA in drinking water is of concern because DCA has been shown to be carcinogenic in laboratory animals. Most published methods use a derivatizing reagent that can artificially increase the levels of DCA during the GC-MS analysis. Recently, a LC-MS/MS method was validated for the quantitative analysis of DCA in drinking water.[63] No other sample preparation than dilution was performed with the water samples. A Luna Amino column (150 × 2.1 mm, 5 μm) was used for HILIC-ion exchange (the retention is based on the affinity of the polar analyte for the charged end group on the stationary phase) at a flow of 0.7 mL/min. The two components of the mobile phase were (A) 40 mM of ammonium formate and (B) acetonitrile. The gradient run was as follows: 90% B at 0 min, 30% B at 5 min, and 90% B at 6 min, followed by a re-equilibration period of 9 min. The triple quadrupole mass spectrometer was run in ESI using the MRM mode to monitor the transition from *m/z* 127 to 83. Method limits of quantification as low as 5 ng/mL water could be established with this method. Bottled water samples contained much less DCA than tap water samples. This is not a surprising finding, as DCA is commonly found in tap water as a disinfection by-product of chlorination.

7.5.2 Oligosaccharides

The massive accumulation of organic matter, which periodically occurs in the northern Adriatic Sea, and in other locations worldwide, is presently thought to be the result of the aggregation of dissolved organic matter (DOM) into particulate organic matter (POM). This phenomenon is the result of human activities and weather conditions. Although many aspects are well understood, the trigger mechanisms leading to mucilage formation have not been clarified yet, probably as a consequence of inadequate analytical approaches. In this context, the new advancements in LC-MS interfacing might contribute in clarifying the mechanism of mucilage formation. Recently, HILIC coupled with ESI-MS/MS was proposed as an innovative method for the investigation of underivatized oligosaccharides in mucilage samples.[64]

Although amino columns have been extensively used for derivatized sugar analysis, HILIC represents a rare application in LC-ESI-MS/MS for the analysis of underivatized oligosaccharides.[64] After defrosting, the mucilage samples were isolated from its surrounding seawater by decantation and next centrifuged at 4000rpm for 20min at ambient temperature. After centrifugation, the precipitate was immediately dialyzed against Milli-Q water overnight. The dialyzed sample was then lyophilized and subsequently extracted with water. The aqueous suspension was then centrifuged at 4000rpm for 20min, and the supernatant, containing the water-soluble fraction of the macro-aggregate, was finally subjected to LC-ESI-MS/MS analysis. Separations were performed on an Alltech Alltima™ Amino column (250 × 2.1mm, 5μm) at a flow rate of 0.150mL/min. This approach allowed the efficient separation of the oligosaccharides from the complex mucilage matrix without derivatization. A gradient elution program was applied with (A) 10mM sodium acetate and (B) acetonitrile. Solvent composition varied from 10% to 70% A in 40min. The injection volume was of 10mL. The column was connected to an ESI-quadrupole ion trap mass spectrometer for MS/MS analysis. Mass spectra of oligosaccharides sodium adduct ions were acquired in positive ion mode.

Recent findings suggest that the significant presence of these compounds in seawater can play an important role in the initial steps of the agglomeration processes forming gelatinous material. Furthermore, the presence of several maltodextrins was evidenced in the water-soluble fraction of mucilage macro-aggregates, collected in various locations of the northern Adriatic Sea.[64]

ACKNOWLEDGMENTS

Alexander van Nuijs and Dr. Adrian Covaci are grateful to the Flanders Scientific Funds for Research for financial support. Dr. Isabela Tarcomnicu acknowledges a postdoctoral fellowship from the University of Antwerp.

REFERENCES

1. Nguyen, H.P. and Schug, K.A., *J. Sep. Sci.*, 2008, 31, 1465.
2. McFarren, E.F., Schafer, M.L., Campbell, J.E., Lewis, K.H., Jensen, E.T., and Schantz, E.J., *Adv. Food Res.*, 1960, 10, 135.
3. Berman, F.W. and Murray, T.F., *J. Neurochem.*, 1997, 69, 693.
4. Quilliam, M.A. and Wright, J.L.C., *Anal. Chem.*, 1989, 61, 1053.
5. Kao, C.Y., *Pharmacol. Rev.*, 1966, 18, 997.
6. Horwitz, W., Ed., *Official Methods of Analysis of AOAC International*, AOAC International, Gaithersburg, MD, 2000.
7. Harada, K.-I., Nagai, H., Kimura, Y., and Suzuki, M., *Tetrahedron*, 1993, 49, 9251.
8. Oshima, Y., *J. AOAC Int.*, 1995, 78, 528.
9. Eaglesham, G.K., Norris, R.L., Shaw, G.R. et al., *Environ. Toxicol.*, 1999, 14, 151.
10. Pietsch, J., Fichtner, S., Imhof, L., Schmidt, W., and Brauch H.-J., *Chromatographia*, 2001, 54, 339.
11. Shoji, Y., Yotsu-Yamashita, M., Miyazawa, T., and Yasumoto, T., *Anal. Biochem.*, 2001, 290, 10.
12. Lawrence, J.F., Niedzwiadek, B., and Menard, C., *J. AOAC Int.*, 2004, 87, 83.
13. Hall, S. and Strichartz, G., Eds., *Marine Toxins: Origin, Structure, and Molecular Pharmacology*, American Chemical Society, Washington, DC, 1990.
14. Diener, M., Erler, K., Christian, B., and Luckas, B., *J. Sep. Sci.*, 2007, 30, 1821.
15. Dell'Aversano, C., Eaglesham, G.K., and Quilliam, M.A., *J. Chromatogr. A.*, 2004, 1028, 155.

16. Dell'Aversano, C., Hess, P., and Quilliam, M.A., *J. Chromatogr. A.*, 2005, 1081, 190.
17. Krock, B., Seguel, C.G., and Cembella, A.D., *Harmful Algae*, 2007, 6, 734.
18. Ciminiello, P., Dell'Aversano, C., Fattorusso, E. et al., *Rapid Commun. Mass Spectrom.*, 2005, 19, 2030.
19. Ciminiello, P., Dell'Aversano, C., Fattorusso, E. et al., *Toxicon*, 2006, 47, 174.
20. Turrell, E., Stobo, L., and Lacaze, J.-P., *J. AOAC Int.*, 2008, 91, 1372.
21. Ciminiello, P., Dell'Aversano, C., Fatturosso, E. et al., *Toxicon*, 2006, 47, 597.
22. Jang, J. and Yotsu-Yamashita, M., *Toxicon*, 2006, 48, 980.
23. Halstead, B.W., Ed., *Dangerous Marine Animals: That Bite-Sting-Shock-Are-Non-Edible*, Cornell Maritime Press, Cambridge, MD, 1959.
24. Kim, Y.H., Brown, G.B., and Mosher, F.A., *Science*, 1975, 189, 151.
25. Diener, M., Christian, B., Ahmed, M.S., and Luckas, B., *Anal. Bioanal. Chem.*, 2007, 389, 1997.
26. Jang, J. and Yotsu-Yamashita, M., *Toxicon*, 2007, 50, 947.
27. Nakagawa, T., Jang, J., and Yotsu-Yamashita, M., *Anal. Biochem.*, 2006, 352, 142.
28. Bates, S.S., Bird, C.J., De Freitas, A.S.W. et al., *Can. J. Fish. Aquat. Sci.*, 1989, 46, 1203.
29. Cerino, F., Orsini, L., Sarno, D., Dell'Aversano, C., Tartaglione, L., and Zingone, A., *Harmful Algae*, 2005, 4, 33.
30. Hess, P., Gallacher, S., Bates, L.A., Brown, N., and Quilliam, M.A., *J. AOAC Int.*, 2001, 84, 1657.
31. de Koe, W.J., Samson, R.A., van Egmond, H.P., Gilbert, J., and Sabino, M., Eds., *Mycotoxins and Phycotoxins in Perspective at the Turn of the Millenium*, Wageningen, the Netherlands, 2001.
32. Furey, A., Lehane, M., Gillman, M., Fernandez-Puente, P., and James, K.J., *J. Chromatogr. A.*, 2001, 938, 167.
33. Zuccato, E., Calamari, D., Natangelo, M., and Fanelli, R., *The Lancet*, 2000, 355, 1789.
34. Zuccato, E., Chiabrando, C., Castiglioni, S. et al., *Environmental Health: A Global Access Science Source*, 2005, 4, 14.
35. Zuccato, E., Chiabrando, C., Castiglioni, S., Bagnati, R., and Fanelli, R., *Environ. Health Perspect.*, 2008, 116, 1027.
36. Routledge, E.J., Sheahan, D., Desbrow, C., Brighty, G.C., Waldock, M., and Sumpter, J.P., *Environ. Sci. Technol.*, 1998, 32, 1559.
37. Qin, F., Zhao, Y.Y., Sawyer, M.B., and Li, X.F., *Anal. Chim. Acta*, 2008, 627, 91.
38. Johnson, A.C., Juergens, M.D., Williams, R.J., Kümmerer, K., Kortenkamp, A., and Sumpter, J.P., *J. Hydrol.*, 2008, 348, 167.
39. Tauxe-Wuersch, A., De Alencastro, L.F., Grandjean, D., and Tarradellas, J., *Int. J. Environ. Anal. Chem.*, 2006, 86, 473.
40. Kovalova, L., McArdell, C., and Hollender, J., *J. Chromatogr. A*, 2009, 1216, 1100.
41. Kümmerer, K., Antibiotics in the aquatic environment—A review—Part I, *Chemosphere*, 2009, 75, 417.
42. Oertel, R., Neumeister, V., and Kirch, W., *J. Chromatogr. A*, 2004, 1058, 197.
43. Peru, K.M., Kuchta, S.L., Headley, J.V., and Cessna, A.J., *J. Chromatogr. A*, 2006, 1107, 152.
44. EPA method 1694, Pharmaceuticals and Personal Care Products in Water, Soil, Sediment, and Biosolids by HPLC/MS/MS, 2007. http://www.epa.gov/waterscience/methods/method/files/1694.pdf (accessed on October 15, 2010)
45. Fent, K., Weston, A.A., and Caminada, D., *Aquat. Toxicol.*, 2006, 76, 122.
46. Scheurer, M., Sacher, F., and Brauch, H., *J. Environ. Monit.*, 2009, 11, 1608.
47. Ferrer, I., Thurman, E.M., and Zweigenbaum, J., Agilent's 6410A LC/MS/MS Solution for Pharmaceuticals and Personal Care Products in Water, Soil, Sediment, and Biosolids by HPLC/MS/MS, 2009. Agilent application note, http://www.chem.agilent.com/Library/applications/5989–9665EN.pdf (accessed on October 15, 2010)

48. U.S. Drug Enforcement Administration (DEA), DEA Briefs & Background, Drugs and Drug Abuse, Drug Descriptions, Drug Classes, 2010. http://www.usdoj.gov/dea/concern/drug_classes.html (accessed on October 15, 2010)
49. van Nuijs, A.L.N., Pecceu, B., Theunis, L. et al., *Environ. Pollut.*, 2009, 157, 123.
50. van Nuijs, A.L.N., Pecceu, B., Theunis, L. et al., *Water Res.,* 2009, 43, 1341.
51. van Nuijs, A.L.N., Pecceu, B., Theunis, L. et al., *Addiction*, 2009, 104, 734.
52. Castiglioni, S., Zuccato, E., Crisci, E., Chiabrando, C., Fanelli, R., and Bagnati, R., *Anal. Chem.*, 2006, 78, 8421.
53. Huerta-Fontela, M., Galceran, M.T., Martin-Alonso, J., and Ventura, F., *Sci. Total Environ.*, 2008, 397, 31.
54. Gheorghe, A., van Nuijs, A., Pecceu, B. et al., *Anal. Bioanal. Chem.*, 2008, 391, 1309.
55. van Nuijs, A.L.N., Tarcomnicu, I., Bervoets, L. et al., *Anal. Bioanal. Chem.*, 2009, 395, 819.
56. EU Commission. Sustainable Use of Pesticides, 2009. http://ec.europa.eu/environment/ppps/home.htm (accessed on October 15, 2010)
57. Strosser, P., Pau Vall, M., and Plotscher, E., Water and Agriculture: Contribution to an Analysis of a Critical but Difficult Relationship, 1999. http://ec.europa.eu/agriculture/envir/report/en/eau_en/report.htm (accessed on October 15, 2010)
58. Hayama, T., Yoshida, H., Todoroki, K., Nohta, H., and Yamaguchi, M., *Rapid Commun. Mass Spectrom.* 2008, 22, 2203.
59. Sorensen, S., Bending, G., Jacobsen, C., Walker, A., and Aamand, J., *FEMS Microbiol. Ecol.*, 2003, 453, 1.
60. Farre, M.J., Brosillon, S., Domenech, X., and Peral, J., *J. Photochem. Photobiol. A: Chem.*, 2007, 189, 364.
61. Young, M.S. and Jenkins, K.M., Waters application note, http://www.waters.com/waters/library.htm?locale = en_us&lid = 10077196
62. Makihata, N., Yamasaki, T., and Eiho, J., *Bunseki Kagaku*, 2007, 56, 579–585.
63. Dixon, A.M., Delinsky, D.C., Bruckner, V.J., Fisher, J.W., and Bartlett, M.G., *J. Liq. Chromatogr. Relat. Technol.*, 2004, 27, 2343.
64. Cappiello, A., Trufelli, H., Famiglini, G. et al., *Water Res.*, 2007, 41, 2911.

8 Hydrophilic Interaction Chromatography–Mass Spectrometry: An Environmental Application

Kerry M. Peru, Sandra L. Kuchta, John V. Headley, and Allan J. Cessna

CONTENTS

8.1 INTRODUCTION

8.1.1 Pharmaceuticals in the Environment

Pharmaceuticals are used extensively in both human and veterinary medicine. Many are excreted in the feces and urine as the nonmetabolized parent compound and, consequently, may be present in municipal biosolids and livestock manure. Because biosolids and manure are valuable sources of nutrients for crop growth, the use of these materials as a fertilizer is widespread in many agricultural areas. Consequently, use of biosolids and manure as fertilizer may be a potential source of pharmaceutical contamination of surface and ground water via surface runoff and leaching, respectively. Other sources of water contamination by pharmaceuticals include municipal sewage discharge, aquaculture, and inappropriate disposal[1] (Figure 8.1). Some pharmaceuticals detected in environmental waters are known to exhibit hormonal activity with the potential to disrupt normal endocrine function.[2] There is also concern that the presence of veterinary antimicrobials in environmental waters may accelerate the development of antimicrobial-resistant bacteria. While there is no clear evidence that the transfer of antimicrobial resistance and that the selection for resistant microbial agents occur at antimicrobial concentrations found in the environment, the spread of resistant bacteria and resistance genes by use of biosolids and manure on the agricultural landscape is not yet fully understood. Thus, in recent years, concern over the impact, fate, and transport of both human and veterinary pharmaceuticals entering the environment has gained much attention.[3] In Canada, water quality has become an important issue with the public both in terms of safety of drinking water and in terms of protecting and conserving aquatic ecosystems.

8.1.2 Antimicrobials in the Environment

One class of pharmaceuticals commonly used in both human and veterinary medicine is antimicrobials that are generally used to treat bacterial infections. However, in livestock production, antimicrobials are routinely used not only at therapeutic levels for the treatment of disease but also at subtherapeutic levels (milligrams per kilogram of feed) to improve feed efficiency and to promote growth. Increasingly, swine, poultry, and cattle are being produced in large confined animal feeding operations that are also increasing in size. The number of animals housed in these

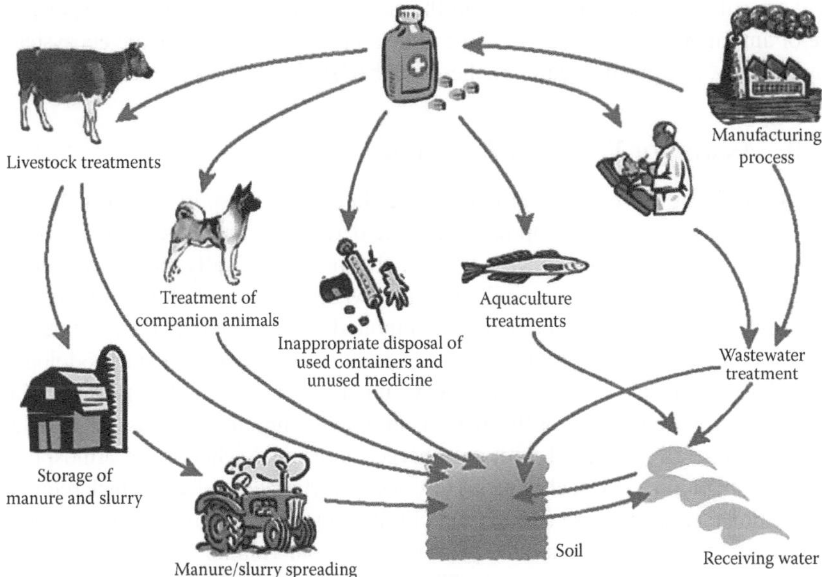

FIGURE 8.1 Pharmaceutical routes of entry into the environment. (Reprinted from European Molecular Biology Organization (EMBO), *EMBO Rep.*, 5, 12, 1110, 2004. With permission.)

facilities increased by approximately two to three times in the United States over 14 years.[4] A similar increase in swine numbers from 1990 to 2000 was observed in Canada.[5] There has been a corresponding increase on the reliance of antimicrobial therapy with animals housed under high-density confined conditions that are conducive for the transmission of infectious disease. Estimates of the total antimicrobial use in North American agriculture are highly variable, ranging from 9 to 16 million kg annually with between 10% and 70% being used subtherapeutically.[6]

Most antimicrobials are poorly absorbed through the gut and only partially metabolized in the animal and, as a consequence, can be excreted up to 75% or more in the feces and urine.[7] Excretion rates for antimicrobials commonly used as feed additives in beef cattle[8,9] and swine production[10,11] have been reported. There is also considerable evidence that antimicrobials can also persist for prolonged periods in livestock manure.[10–15] Concentrations of antimicrobials present in lagoon manure can be at milligrams per liter levels.[12,16–18]

The fact that residual antimicrobials persist in manure raises the possibility that, after land application of the manure, they may persist in soil[15,16,19–21] and potentially come in contact with surface[16,17,22–25] and ground water.[22,26,27] Antimicrobials used in the livestock industry have been detected in surface waters in Canada,[28,29] the United States,[30–32] Europe,[33,34] and Asia.[35] Veterinary antimicrobials have been detected in ground water following manure application[16] and in ground water in the vicinity of confined animal feeding operations in the United States, albeit in a limited number of samples.[12,32,36] There is limited information available on the biologic effect of antimicrobials in the environment. They have been shown to affect the composition of microbial communities in soil[37,38] and to affect bacterial populations in wetlands.[39] However, there are still major knowledge gaps in the environmental occurrence, transport, and fate of antimicrobials and their biologic effects in terrestrial and aquatic ecosystems; consequently, continued research in these areas of study is an ongoing requirement.

8.1.3 Analysis of Veterinary Antimicrobials

The study of the environmental occurrence, fate, and transport of antimicrobials and other pharmaceutical compounds and the determination of the biologic impact from short- and long-term exposure to these chemicals, generally requires the analysis of environmental matrices (soil, sediment, water, plant and animal tissues, and manure). To analyze and quantify these compounds at environmental concentrations in such complex matrices requires sensitive and robust analytical methodologies. Most pharmaceutical compounds are polar in nature and are soluble in water. Mass spectrometry is well suited for the analysis of polar compounds because greater sensitivity is generally achieved with pre-charged molecules when using atmospheric pressure ionization (API) techniques. However, pre-charged molecules lack retention on traditional reverse-phase (RP) liquid chromatography column packing materials and analytes of interest often elute in or near the void volume of the column along with interfering matrix components. This may lead to significant enhancement or suppression of ionization in the mass spectrometer source, as well as possible mass interference, by co-eluting matrix co-extractives. A great deal of time and expense by column manufacturers and researchers has been invested in developing RP packing materials to achieve adequate polar analyte retention, often in conjunction with additives (e.g., ion pairing reagents, pH adjustment) to enhance analyte retention and separation. Success has been achieved, but sometimes at the expense of reduced sensitivity due to the ionization suppression effects of these additives.[40,41]

8.1.4 Analysis of Spectinomycin and Lincomycin

Confined animal feeding operations require that infections be minimized for profitable production. *Escherichia coli* infections causing diarrhea or edema disease in weanling pigs are estimated to cause up to 50% of the economic losses.[42] Spectinomycin and lincomycin (Figure 8.2) are two commonly administered antimicrobials used to control postweaning diarrhea. Spectinomycin is a member of the aminocyclitol group of antimicrobials produced by a species of soil microorganism designated as *Streptomyces spectabilis*. It is effective against a broad spectrum of gram-positive and gram-negative bacteria by inhibiting protein synthesis through interactions with the 30S ribosomal subunit of bacterial cells.[43] Spectinomycin is excreted mostly unchanged in the feces and urine.[44,45] Lincomycin is a member of the lincosamide group of antimicrobials produced from the actinomyces *S. lincolnensis*. It is effective against a variety of anaerobic and gram-positive bacteria by also inhibiting protein synthesis.[45] Lincomycin undergoes some hepatic metabolism, however, up to 32% can be excreted in the bile and urine.[11,44,45]

Waste management in confined animal feeding operations can be daunting. It is estimated that each pig produces just over 2 m^3 of waste annually and the liquid manure management system results in a total of 2.5 m^3 of manure per pig to process.[46]

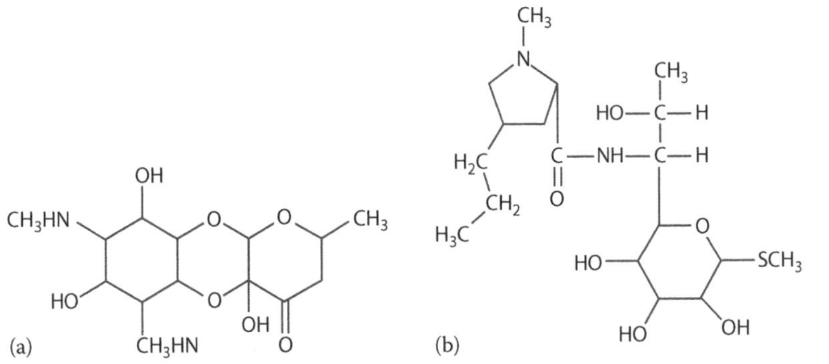

FIGURE 8.2 Chemical structures of (a) spectinomycin and (b) lincomycin. (Reprinted from Peru, K.M. et al., *J. Chromatogr. A*, 1107, 152, 2006. With permission from Elsevier.)

As an economic means to waste management, liquid manure is frequently utilized as a nutrient source on cropland. The applied manure is thus a potential source of spectinomycin and lincomycin to the environment; however, little information has been established on their environmental fate.

8.1.5 Spectinomycin

Previously, methods have been reported for the determination of spectinomycin using microbial assays,[47–52] gas chromatographic assay,[53] thin layer chromatography/densitometry,[54] high-performance liquid chromatography (HPLC) with electrochemical detection,[55–57] HPLC with amperometric detection,[58] and HPLC methods requiring either pre- or post-column derivatization.[59–65] These methods have mainly focused on residues in animal tissues and milk for human consumption. Trifluoroacetic acid (25 mM) has been used as an ion pairing agent to establish the retention of spectinomycin on an Agilent SB-C18 column.[66] Detection was with evaporative light scattering, however, the detection limit (6 mg L^{-1} in aqueous samples) was insufficient for an environmental level analysis.

Analysis of spectinomycin by LC-MS-MS has proved challenging largely because of the high polarity and basic properties of the analyte. Chromatographic retention is difficult to achieve and thus poor separation from co-eluting matrix components leads to API enhancement and/or suppression interference.[41,67] Previously, two methods have been reported for the determination of spectinomycin in tissues using ion-pairing reagents with detection by LC-MS[68] and LC-MS-MS.[69] More recently, the utility of heptafluorobutyric acid (HFBA) ion-pairing to aid in LC retention and solid-phase extraction was reported.[70,71] However, it has been observed that the HFBA concentrations required for sufficient retention of spectinomycin and separation leads to compromised sensitivity and detection limits when using API techniques.[41,67] An LC/MS/MS method utilizing a pH gradient was employed to better retain and resolve spectinomycin from interfering co-eluting compounds.[67] This method provided improved detection limits compared to the HFBA method, eliminated or significantly reduced the ionization inference in the MS source, and gave reproducible results for samples of simulated rainfall runoff. However, the method did not provide the required retention and reproducibility for analysis of the more complex matrix of liquid hog manure.

8.1.6 Lincomycin

Methods have been reported for the determination of lincomycin in animal tissues using GC with derivatization,[72,73] HPLC with electrochemical detection in milk and animal tissues;[74] ion-pair HPLC with electrochemical detection in salmon tissues,[75] and HPLC with pulsed electrochemical detection in animal feed.[76] Recently, LC-MS-MS with electrospray ionization (ESI) has been used to determine lincomycin in surface waters,[31,33] swine tissues,[77] and milk.[78] One method utilized both a radioimmunoassay and LC-MS with ESI to determine lincomycin residues in samples collected from swine lagoons.[12] Ultra performance liquid chromatography (UPLC) methods have also been reported using UV detection with success, however, mass spectrometry was additionally required to exclude the misidentification of low lincomycin levels.[79] The use of LC-MS-MS with APCI has also been described for the determination of lincomycin in honey.[80] However, applications using APCI in combination with HILIC separations are sparse.

Recently, research using HILIC separations has been increasing steadily, along with various stationary phases developed for HILIC.[40,81] HILIC was explored in this work to improve the LC retention of the two antimicrobials investigated, particularly spectinomycin. HILIC, based on silica columns, is an alternative to normal-phase chromatography but utilizes traditional RP mobile phases.[82] HILIC aids in the retention of basic, polar compounds that are not retained well on RP columns. Mobile phases containing a high organic solvent content (typically acetonitrile) and a low water content are used to retain analytes depending on their hydrophilicity. Retention is proportional to the polarity of the solute and inversely proportional to the polarity of the mobile phase.[83] Retention is therefore, the opposite of what is normally observed using RP chromatography. Several retention mechanisms, namely, hydrophilic interaction, ion-exchange, and RP retention occur simultaneously when using silica as the stationary phase. The combination of these retention mechanisms offers unique selectivity and provides retention of polar solutes. These attractive features of HILIC have been used to separate peptides[84] and histones,[85,86] and, more recently, some polar pharmaceuticals,[87–89] including the antimicrobials neomycin,[90] avoparcin,[91] and tetracycline, chlortetracycline, and oxytetracycline.[92]

8.1.7 Objective

As part of an investigation to fill knowledge gaps on the environmental fate and transport of veterinary antimicrobials, we report a robust, specific, and sensitive analytical method for the determination of spectinomycin and lincomycin in liquid hog manure, manure-treated soil, and snowmelt runoff and ground water from manure-treated cropland. The APCI LC/MS/MS method described in this work uses hydrophilic interaction chromatography (HILIC) to provide sufficient resolution to reduce interference from the complex liquid swine manure matrix while providing a baseline resolution between lincomycin and spectinomycin. This was achieved without using mobile phase additives and, consequently, without compromising sensitivity.

8.2 EXPERIMENTAL

8.2.1 Chemicals and Reagents

Spectinomycin dihydrochloride (purity ≥98%) and lincomycin hydrochloride (purity ≥90%) were obtained from Sigma-Aldrich Co., (St. Louis, Missouri). HPLC grade acetonitrile and methanol, and certified A.C.S. grade formic acid, ammonium hydroxide, trisodium citrate, citric acid, and 20- to 30-mesh Ottawa sand were obtained from Fisher Scientific (Edmonton, AB, Canada). ASE-prep diatomaceous earth was obtained from Dionex (Sunnyvale, California). Deionized water (18 MΩ) containing less than 4 μg L^{-1} total organic carbon was obtained using a Millipore Milli-Q Gradient A10 (with a total organic carbon detector) purification system (Millipore Corp., Billerica, Massachusetts).

8.2.2 Preparation of Standards and Solutions

Stock standard solutions (1000 mg L^{-1}) of spectinomycin and lincomycin were prepared by weighing and dissolving each antimicrobial in 50% aqueous acetonitrile. The stock solutions were stored in the dark at 4°C for a maximum of 1 month. Nine calibration standards were prepared (0.1–500 μg L^{-1}) in mobile phase (acetonitrile/H_2O, 75:25, v/v containing 0.1% formic acid). Due to differences in detection limits for spectinomycin and lincomycin, a subset of five calibration solutions (10–500 μg L^{-1}) was used to generate the calibration curve for spectinomycin, while another subset of five calibration solutions (0.1–10 μg L^{-1}) was used for lincomycin.

8.2.3 Environmental Samples

8.2.3.1 Sample Collection

Details of field sites, sampling procedures, and timelines are described elsewhere.[16,17]

8.2.3.1.1 Liquid Swine Manure

Liquid swine manure was collected from two commercial swine barns in Saskatchewan, Canada. The liquid manure was typically a brown, opaque slurry containing an average of 15% solids.[11] In one barn, control manure was collected from manure pits located under the pens of grower-finisher pigs that were not currently being administered antimicrobials. These samples were used as a matrix blank and for fortification experiments. As part of a study to determine the excretion of spectinomycin and lincomycin by weanling pigs, liquid swine manure was also collected from manure pits located under the pens of weanling pigs housed in the nursery area of the barn.[11] These pigs were being administered approximately equal amounts of spectinomycin and lincomycin in their feed to prevent and control postweaning diarrhea. Details of liquid manure sampling procedures are described elsewhere.[11]

Liquid swine manure from storage lagoons at both barns was applied in the fall to cropland and pasture in 2003 and/or 2004 at rates ranging from 60,000 to 110,000 L ha^{-1}.[16,17] Liquid manure samples were collected during manure application and would permit the determination of the rates at which the antimicrobials were applied to the agricultural landscape.

8.2.3.1.2 Soil

Manure-amended soil was sampled immediately after liquid manure application and then monthly after snowmelt runoff the following spring.[16] These samples would permit the determination of the half-lives of lincomycin and spectinomycin in soil and their presence in the upper layer of soil would confirm their availability for transport in surface runoff.

8.2.3.1.3 Snowmelt Runoff

Spring snowmelt runoff from manure-amended soil was sampled from ephemeral wetlands and closed basin depressions immediately after snowmelt was complete and then weekly until the snowmelt had completely infiltrated.[16,17] These samples would confirm whether the management practice of utilizing liquid swine manure as a plant nutrient source on crop and pasture land would result in the transport of lincomycin and spectinomycin to aquatic ecosystems and would permit the determination of their persistence in the water column of the ephemeral wetlands.

8.2.3.1.4 Ground Water

Ground water from piezometers installed in manure-amended fields was collected immediately after manure application and then monthly after spring snowmelt runoff.[16] These samples would confirm whether the management practice of utilizing liquid swine manure as a plant nutrient source on crop and pasture land would result in leaching of lincomycin and spectinomycin to ground water.

8.2.3.2 Sample Preparation

Liquid swine manure samples were separated into solids and liquid components as described previously.[11,41] Briefly, samples (10 mL) were centrifuged at 3000 *g* for 15 min followed by the decanting of the supernatant. This initial supernatant was centrifuged two more times to maximize solids removal. The solids components were combined for subsequent extraction.

8.2.4 Solid-Phase Extraction

8.2.4.1 Spectinomycin

8.2.4.1.1 Manure Liquid Component

A Waters Oasis HLB cartridge was stacked on top of a Waters Oasis WCX column for extraction and cleanup purposes. Cartridges were conditioned by sequentially washing with 10 mL methanol followed by 10 mL deionized water. The clear supernatant (2 mL ca) was diluted to 200 mL with deionized water and thoroughly mixed. Dilution was required to aid in the extraction process and to minimize potential losses from sample handling. A subsample (100 mL) was passed through the stacked cartridges at approximately 1 mL min^{-1} (the remaining 100 mL was used for lincomycin analysis). After sample extraction, the HLB cartridge was discarded while the WCX cartridge was washed using 5 mL of 25 mM citrate buffer (pH 5) followed by 5 mL of methanol. The cartridge was then eluted with 10 mL of acetonitrile containing 3% formic acid. The eluate was taken to dryness using a stream of N_2 gas and the extract residue dissolved in acetonitrile (1.0 mL) for LC/MS/MS analysis.

8.2.4.1.2 Manure Solids Component

The pressurized liquid extraction (PLE) procedure has been described previously.[11] Briefly, a subsample (1 g) of the air-dried manure solids component was mixed with diatomaceous earth and Ottawa sand to fill a 33 mL PLE cell. The packed PLE cell was then extracted as follows: the cell was heated to 100°C and extracted for 5 min with deionized water at a pressure of 1500 psi. The cell was then flushed with a 30% volume of deionized water and purged with nitrogen gas for 1.5 min. Optimal extraction was achieved by extracting each cell twice. The resulting aqueous extracts (~30 mL each) were combined and diluted to 200 mL with deionized water and extracted as described for the manure liquid component.

8.2.4.1.3 Manure-Amended Soil

Air-dried soil (2 g) was extracted by PLE as described for the manure solids component. The resulting aqueous extracts (~30 mL each) were combined and diluted to 200 mL with deionized water and extracted as described for the manure liquid component.

8.2.4.1.4 Surface and Ground Water

Surface or ground water (100 mL) was extracted as described for the manure liquid component.

8.2.4.2 Lincomycin

8.2.4.2.1 Manure Liquid Component

An Oasis HLB cartridge was conditioned by sequentially washing with 10 mL acetonitrile followed by 10 mL deionized water. The clear supernatant (2 mL ca) was diluted to 200 mL with deionized water, and a subsample (100 mL) was adjusted to pH 9 with 1 M ammonium hydroxide solution and passed through the cartridges at approximately 1 mL min^{-1} followed by elution with 10 mL of acetonitrile. The eluate was taken to dryness using a stream of N_2 gas and the extract residue dissolved in acetonitrile (1.0 mL) for LC/MS/MS analysis.

8.2.4.2.2 Manure Solids Component

Manure solids samples were subjected to the same PLE procedure as that used for the analysis of spectinomycin. The combined PLE extracts (~30 mL each) were diluted to 200 mL with deionized water and extracted as described for the manure liquid component.

8.2.4.2.3 Manure-Amended Soil

Air dried soil (2 g) was extracted by PLE as described for the manure solids component. The combined PLE extracts (~30 mL each) were diluted to 200 mL with deionized water and extracted as described for the manure liquid component.

8.2.4.2.4 Surface and Ground Water

Surface or ground water (100 mL) was extracted as described for the manure liquid component.

8.2.5 Fortification Studies

Fortification studies were carried out with all matrices studied (liquid manure, manure liquid component, manure solids component, soil, and water) in order to determine the reliability of the analytical methods. Liquid manure was fortified with both spectinomycin and lincomycin at 100 (n = 6) and 10 (n = 6) μg L^{-1}, manure liquid component at 100 (n = 3), 10 (n = 3), and 1 (n = 3) μg L^{-1} and manure solids component at 100 (n = 5), 50 (n = 4), and 1 (n = 5) μg kg^{-1}. Soil was fortified with both antimicrobials at 100 (n = 5) and 1 (n = 5) μg kg^{-1}. South Saskatchewan River water (as a surrogate for snowmelt runoff and ground water) and deionized water were fortified with spectinomycin at 1 μg L^{-1} (n = 9) and lincomycin at 0.01 μg L^{-1} (n = 9). The fortification procedures have been described previously.[11,16]

8.2.6 LC/MS/MS Analysis

8.2.6.1 HILIC LC System

An Alliance 2695 Separations Module (Waters Corp., Milford, Massachusetts) consisting of an LC pump, column oven, solvent degasser, and auto-sampler was used for all chromatographic separations. HILIC separations were carried out using a silica-based Altima HP hydrophilic interaction column (2.1 mm i.d. × 150 mm, 3 μm, Alltech Associates Inc., Deerfield, Illinois), maintained at 36°C. The mobile phases were, A: 90/10 acetonitrile/water + 0.1% formic acid and B: 90/10 water/acetonitrile + 0.1% formic acid. The sample injection volume was 15 μL. Separation was achieved with isocratic conditions (200 μL min^{-1}) of 65% mobile phase A and 35% mobile phase B.

8.2.6.2 Mass Spectrometer

All experiments were conducted on a Micromass Quattro Ultima (Waters Corp., Milford, Massachusetts) triple quadrupole mass spectrometer equipped with an APCI interface operated in the positive ion mode. Optimized instrumental settings were as follows: corona 0.20 mA, cone 60 V, source temperature 120°C, desolvation temperature 550°C, cone gas flow rate of 153 L h^{-1} N_2, desolvation gas flow rate of 199 L h^{-1} N_2, nebulizer gas N_2 maximum, collision energy of 30 V, and multiplier 700 V. Multiple reaction monitoring (MRM) was used for quantitative analysis with the following settings: 0.5 s dwell time and 0.1 s interchannel delay while monitoring two channels. Argon was used as the collision gas at a pressure sufficient to increase the Pirani gauge of the collision cell to a reading of 2.41×10^{-4} mbar.

8.3 RESULTS AND DISCUSSION

8.3.1 Extraction Method

Methods for the extraction of lincomycin are numerous in the literature,[31,33,72–75,80] as it is commonly included in multiresidue methods for the determination of other antimicrobials. Lincomycin is amenable to solid-phase extraction (SPE) using traditional packings (C18, C8, etc.) and good recoveries are obtained from a variety of matrices. In contrast, only one SPE extraction method along with data on the levels of spectinomycin has been reported.[70] This may be due in part to the difficulty of the analysis. The analyte is poorly retained on RP packings leading to poor SPE recoveries and inadequate chromatographic retention on traditional RP LC packing materials. Our approach was to use stacked SPE cartridges for all sample matrices, such that the Oasis HLB cartridge was used for cleanup (removal of nonpolar compounds) followed by the trapping of spectinomycin on a second cartridge containing a weak cation-exchange sorbent (Oasis WCX). The WCX sorbent contains carboxylic acid groups that allow protons to be exchanged with basic ionic compounds at pH 5 or higher. Acetonitrile containing 3% formic acid is then passed through the WCX cartridge that re-protonates the carboxylic acid groups, resulting in the elution of the sorbed basic analytes. Improved chromatography (with reproducible retention times and fewer matrix effects) was observed when the stacked cartridges were used versus use of the WCX cartridge alone. This can be attributed to the retention of nonpolar co-extractives on the HLB cartridge, resulting in less fouling of the WCX sorbent and providing a cleaner final extract.

In this study, lincomycin was extracted from snowmelt runoff, ground water, diluted liquid manure supernatant, diluted manure solids extract, and diluted soil extract using an Oasis HLB cartridge. In order to suppress protonation of the amino nitrogen and to achieve acceptable recovery, it was necessary to adjust water samples and diluted matrix extracts to pH = 9 using 25% ammonium hydroxide solution. This required step, unfortunately, circumvented the use of the HLB cartridge used for cleanup during spectinomycin analysis for the simultaneous extraction of lincomycin.

8.3.2 Recovery Studies

Method recoveries were determined by the extraction of fortified liquid swine manure (liquid and solids components and whole manure), soil, deionized water, and river water. The river water was used as a control sample because it contained no detectable levels of the antimicrobials investigated and served as a matrix similar to snowmelt runoff. Because similar recoveries of lincomycin or spectinomycin were obtained from each matrix regardless of the fortification level, the recovery for each analyte was determined as an average recovery for a given matrix.

8.3.2.1 Liquid Manure

The average recoveries of spectinomycin and lincomycin were 84 ± 2% and 78 ± 7%, respectively, from the manure liquid component (n = 9). Recovery of both analytes from the manure solids component was unreliable at the 1 μg kg^{-1} fortification level.

Average recoveries for the 100 and 10 μg kg^{-1} levels (n = 9) were 62% ± 12% and 70% ± 10%, respectively, for spectinomycin and lincomycin.[11] Average recoveries from liquid manure (liquid plus solids components, n = 12) were 69% ± 9% and 77% ± 14% for spectinomycin and lincomycin, respectively.[11]

8.3.2.2 Soil

The average recoveries of spectinomycin and lincomycin from fortified soil (n = 10) were similar to those for the manure solids component and were 66% ± 15% and 74% ± 8%, respectively.[16]

8.3.2.3 Water

Average recoveries from deionized water (n = 9) were 108% ± 7% and 103% ± 10% for spectinomycin and lincomycin, respectively,[40] whereas corresponding recoveries from river water (n = 9) were 95% ± 8% and 91% ± 4%.[16,17,41]

8.3.3 Mass Spectrometry

Prior to developing an alternative to ion-pairing for sufficient LC separation, MS parameters were evaluated to determine the most sensitive ionization technique available in our laboratory. This evaluation was performed by infusing a 1 mg L^{-1} solution of spectinomycin and lincomycin in 50:50 methanol/water, and for comparison in 50:50 acetonitrile/water. Both ESI and APCI in the positive ion mode using either solvent produced an $(M + H)^+$ ion for lincomycin (*m/z* 407) of a similar intensity (Figure 8.3). However, in contrast to previously reported ion trap data,[70] both ESI and APCI produced a very weak $(M + H)^+$ at *m/z* 333 for spectinomycin (Figure 8.4a and b) when using either solvent. Varying instrumental operating conditions and a changing solvent composition did not improve the $(M + H)^+$ abundance

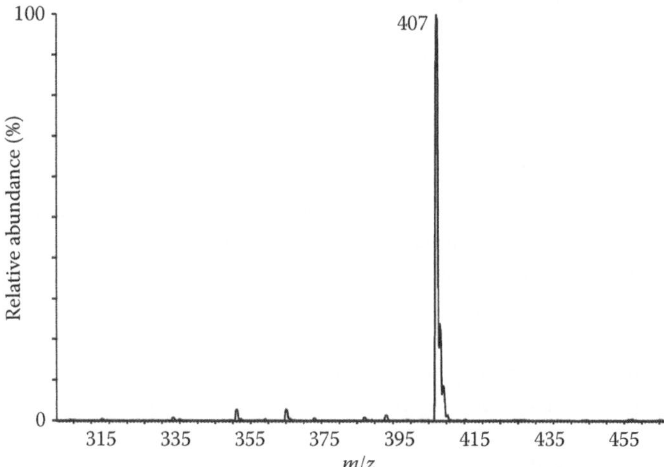

FIGURE 8.3 Mass spectrum of lincomycin obtained using APCI using acetonitrile/water as the solvent. (Reprinted from Peru, K.M. et al., *J. Chromatogr. A*, 1107, 152, 2006. With permission from Elsevier.)

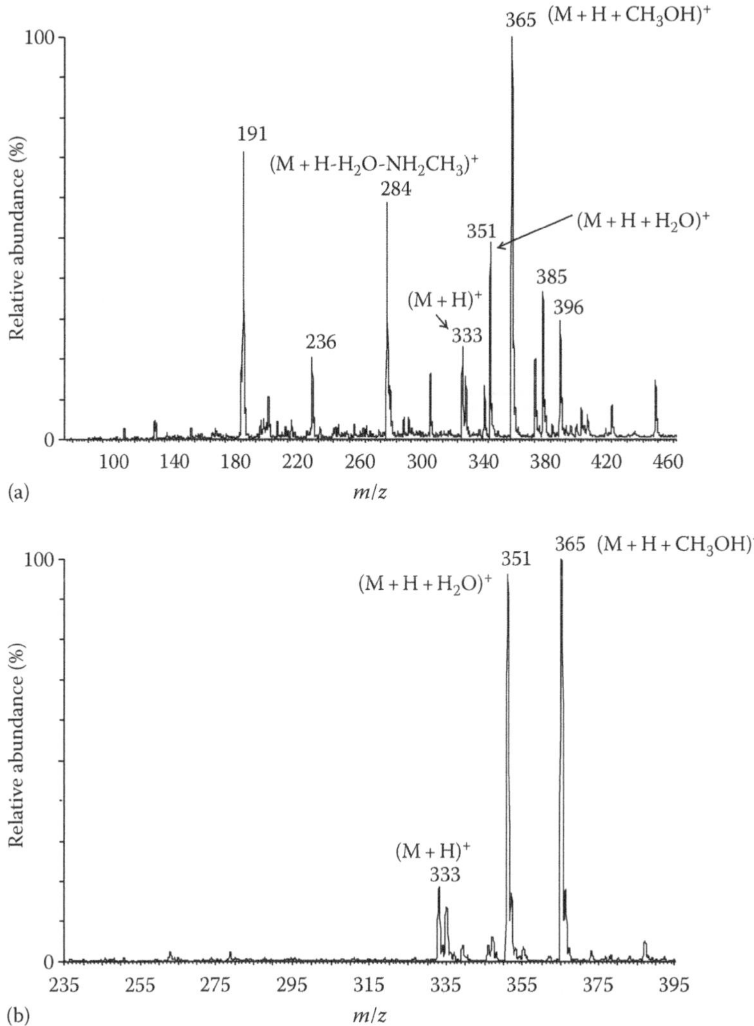

FIGURE 8.4 Mass spectrum of spectinomycin obtained using (a) ESI and (b) APCI using methanol/water as the solvent. (Reprinted from Peru, K.M. et al., *J. Chromatogr. A*, 1107, 152, 2006. With permission from Elsevier.)

for spectinomycin. The corresponding ESI mass spectrum (Figure 8.4a) was noisy with the ion current being averaged over many ions. Thus, monitoring the $(M + H)^+$ did not provide adequate sensitivity for the quantitative analysis of spectinomycin at levels present in environmental matrices. However, intense adduct ions $(M + H + H_2O)^+$ and $(M + H + CH_3OH)^+$ were observed with APCI (Figure 8.4b) when using methanol/water as the solvent and similarly $(M + H + H_2O)^+$ and $(M + H + CH_3CN)^+$ when using acetonitrile/water as the solvent. APCI was thus the ionization technique of choice for both antimicrobials because it produced the most intense signals and, consequently, the lowest detection limits. Acetonitrile/water was used as the mobile

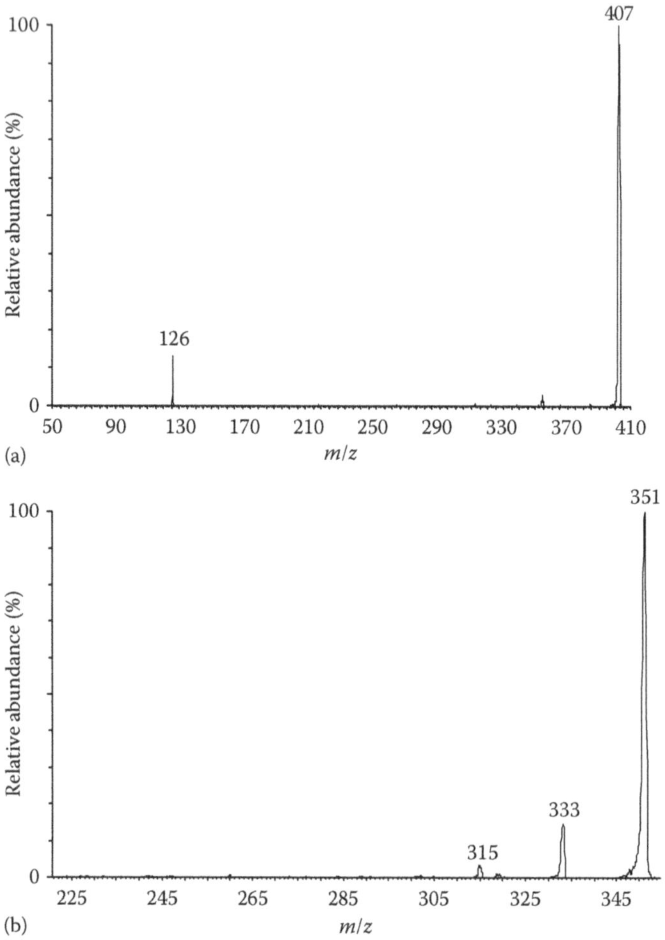

FIGURE 8.5 Product ion mass spectrum of (a) lincomycin $(M + H)^+$, *m/z* 407 and (b) water adduct of spectinomycin $(M + H + H_2O)^+$, *m/z* 351.

phase for the liquid chromatographic separation of the antimicrobials because it provided a lower chromatographic system pressure and improved separation.

Product ion scans were completed to evaluate the best MRM transitions (Figure 8.5a and b). The optimal MRM transitions monitored were *m/z* 351 > 333 for spectinomycin and *m/z* 407 > 126 for lincomycin. Although monitoring the loss of H_2O for spectinomycin is not a highly specific transition, this compromise was made to optimize detection limits.

8.3.4 Chromatography

Initial results using C8 or C18 HPLC column packing materials with a variety of mobile phase compositions generated little to no retention of spectinomycin (Figure 8.6a) and resulted in matrix components affecting the ionization within the source

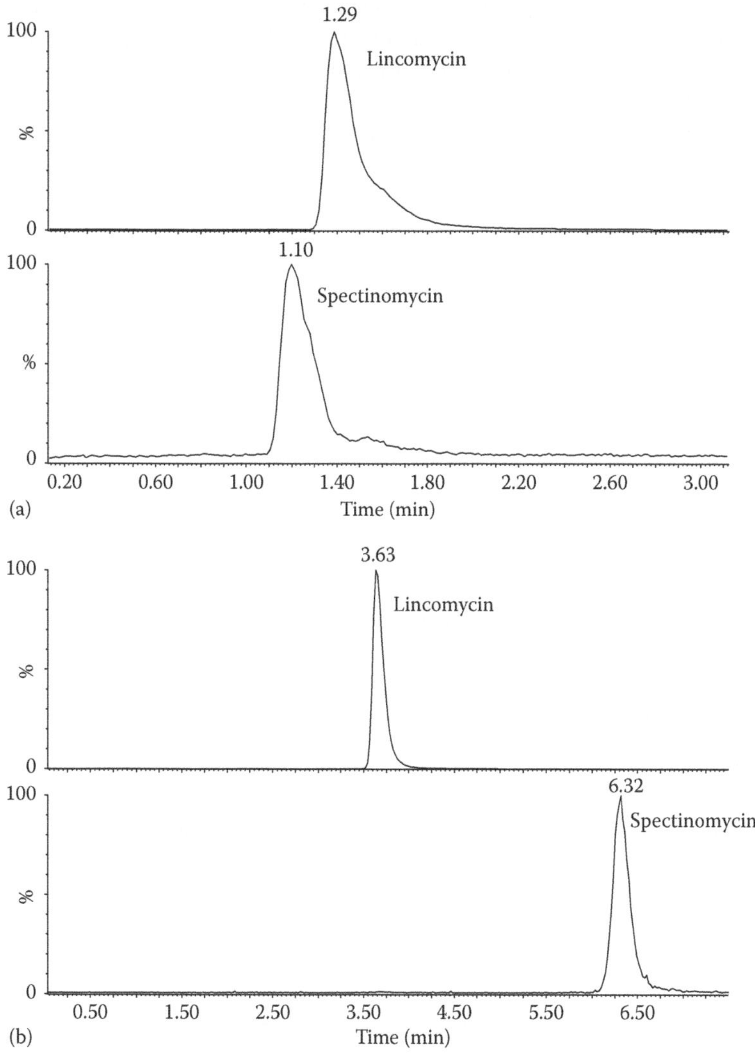

FIGURE 8.6 LC chromatograms illustrating retention of spectinomycin and lincomycin using (a) C18 analytical column and (b) HILIC analytical column. (Reprinted from Peru, K.M. et al., *J. Chromatogr. A*, 1107, 152, 2006. With permission from Elsevier.)

of the mass spectrometer. We observed up to a 200% recovery for spectinomycin due to ionization enhancement when elution of the analyte was at or near the void volume. By utilizing HILIC, analyte retention was increased (Figure 8.6b) while ionization enhancement and/or matrix interference was significantly reduced or eliminated. Even under isocratic conditions, HILIC provided very good retention and separation from matrix components while providing baseline resolution of lincomycin and spectinomycin (Figure 8.6b). Furthermore, the addition of formic acid to the mobile phase to promote better source ionization, increased the overall sensitivity, provided better peak shape, and improved the reproducibility of the retention times of the analytes. The retention order of the two antimicrobials is reversed from that on RP columns because more polar compounds are retained longer using HILIC. Figure 8.7 illustrates a chromatogram observed from each MRM channel for both the WCX extract and the HLB extract (fortified liquid hog manure supernatant).

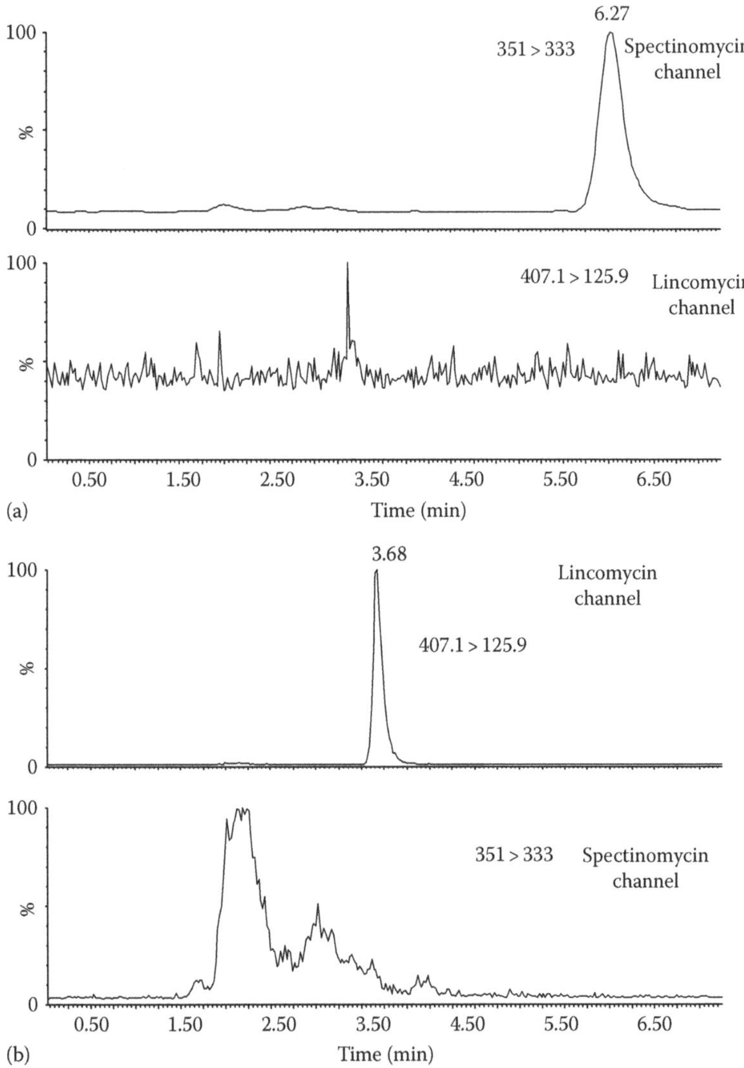

FIGURE 8.7 Chromatogram of each MRM channel for (a) WCX cartridge cleanup of a manure liquid component extract for spectinomycin analysis and (b) HLB cartridge cleanup of a manure liquid component extract for lincomycin analysis. (Reprinted from Peru, K.M. et al., *J. Chromatogr. A*, 1107, 152, 2006. With permission from Elsevier.)

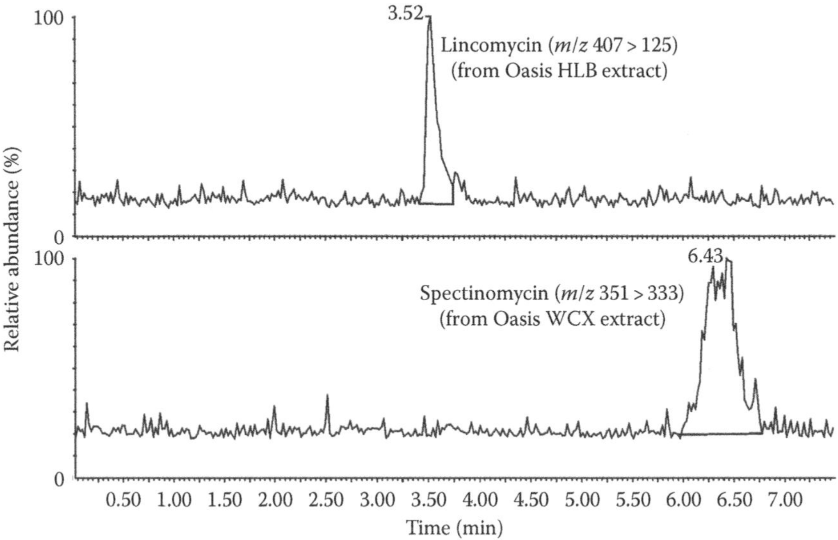

FIGURE 8.8 Typical chromatogram of the lowest fortification level (1 μg L^{-1}) of lincomycin and spectinomycin in the manure liquid component. (Reprinted from Peru, K.M. et al., *J. Chromatogr. A*, 1107, 152, 2006. With permission from Elsevier.)

8.3.5 Quantification

Five-point calibration curves were generated for both spectinomycin and lincomycin. Over the tested concentration ranges, the linear regression of observed peak areas versus concentration gave excellent linearity with R^2 values of 0.998 or greater. Limits of quantitation were determined to be 5 μg L^{-1} (whole liquid manure),[11] 0.5 μg kg^{-1} (soil),[16] 0.2 μg L^{-1} (river water; spectinomycin),[41] and 0.005 μg L^{-1} (river water; lincomycin).[16] A typical chromatogram of the lowest fortification level of liquid swine manure liquid component (1 μg L^{-1}) is illustrated in Figure 8.8.

8.3.6 Application of Method to Analysis of Environmental Matrices

Over the course of several studies, more than 300 samples (including liquid swine manure, manure-amended soil, snowmelt runoff, and ground water samples) have been analyzed.[11,16,17,41] The method has proven to be robust and required minimal ion-source cleaning. Depending on the sample type and matrix loading, slight shifts in retention times (up to ±0.4 min) for both spectinomycin and lincomycin were sometimes observed.

Both antimicrobials were excreted by weanling pigs and were detected in the liquid manure collected from the nursery area of a commercial-scale barn.[11] Concentrations ranged from 173 to 686 μg L^{-1} for spectinomycin and 2520 to 9780 μg L^{-1} for lincomycin.[11] Liquid manure from the storage lagoon associated with this barn was applied in the fall to crop and pasture land. Lincomycin was detected in manure samples (25.1–231 μg L^{-1})[16,17] collected during manure application, but spectinomycin was not. Consequently, spectinomycin was not detected in any soil, snowmelt runoff, or ground water samples. Lincomycin, however, was detected in manure-amended soil (<0.5–117 μg kg^{-1}),[16] snowmelt runoff (<0.005–3.2 μg L^{-1}),[16,17] and ground water (generally <0.005 μg L^{-1}).[16]

8.4 CONCLUSIONS

In order to overcome the effects of matrix components on analyte ionization within the source of the mass spectrometer, the separation of analytes from matrix components must be achieved. Such a separation could not be achieved with traditional reverse-phase liquid chromatographic packing materials for the analysis of spectinomycin and lincomycin without sacrificing overall detection limits. In contrast, hydrophilic interaction chromatography provided an excellent retention of spectinomycin and lincomycin resulting in separation of both antimicrobials from interfering matrix components. In addition, HILIC provided a baseline separation of the two antimicrobials without the use of ion-pairing reagents. HILIC also facilitated the use of formic acid as a mobile-phase additive that was compatible with API techniques and provided good ionization efficiency. APCI provided intense ions that were conducive to trace analysis for both antimicrobials using MS-MS. Weak cation-exchange solid-phase extraction provided good recoveries for spectinomycin from complex matrices such as liquid hog manure. Stacking an Oasis HLB cartridge on top of the WCX cartridge provided the necessary cleanup prior to analysis in order to reduce or eliminate the matrix effects and to provide reproducible chromatography.

The method was used to demonstrate that both spectinomycin and lincomycin were present in liquid swine manure collected from the nursery area of a commercial-scale barn. Likewise, it was shown that lincomycin was detectable in lagoon manure and, after fall liquid manure application, in manure-amended soil, snowmelt runoff, and ground water collected from manure-treated agricultural fields. These results indicate that liquid swine manure applied to agricultural land as a nutrient source is a potential source of veterinary antimicrobials to aquatic ecosystems.

ACKNOWLEDGMENTS

The authors thank David Gallén and Sumith Priyashantha for field technical assistance, Jonathan Bailey for laboratory technical assistance, Prairie Swine Centre Incorporated and River Lakes Stock Farm barn personnel for information on lincomycin use and times and rates of manure application, the collaborating producers for permission to sample on their farms, and the operators of the manure applicators for manure sample collection. The study was funded, in part, by the Saskatchewan Agriculture Development Fund.

REFERENCES

1. European Molecular Biology Organization (EMBO), The Environmental Side Effects of Medication, *EMBO Rep.*, 2004, 5, 12, 1110.
2. Ternes, T.A., Joss, A., and Siegrist, H., *Environ. Sci. Technol.*, 2004, 38, 393.
3. Richardson, S.D., *Anal. Chem.*, 2008, 80, 4373.
4. United States General Accounting Office (GAO), *Animal Agriculture: Information on Waste Management and Water Quality Issues*, GAO/RCED-95-200BR, Washington, DC, 1995.
5. Statistics Canada, *Number of Farms Reporting Pigs and Average Number of Pigs per Farm, 2002, Livestock Statistics*, Cat. No. 23–603-UPE, http://www.statcan.ca.

6. Sarmah, A.K., Meyer, M.T., and Boxall, A.B.A., *Chemosphere*, 2006, 65, 725.
7. Kumar, K. et al., *Adv. Agron*., 2005, 87, 1.
8. EMEA, *Tylosin Summary Report (Parts 1–5)*, 1994–2002,
   Part 1: http://www.emea.europa.eu/pdfs/vet/mrls/tylosin1.pdf;
   Part 2: http://www.emea.europa.eu/pdfs/vet/mrls/tylosin2.pdf;
   Part 3: http://www.emea.europa.eu/pdfs/vet/mrls/020597en.pdf;
   Part 4: http://www.emea.europa.eu/pdfs/vet/mrls/073200en.pdf;
   Part 5: http://www.emea.europa.eu/pdfs/vet/mrls/082902en.pdf.
9. Halling-Sørensen, B. et al., Worst-case estimates of predicted environmental soil concentrations (PEC) of selected veterinary antibiotics and residues used in Danish agriculture. In: Kűmmerer, K., Ed., *Pharmaceuticals in the Environment*. Springer Verlag, Berlin, Germany, 2001, p. 143.
10. Winckler, C. and Grafe, A., *J. Soils Sediments*, 2001, 1, 66.
11. Kuchta, S.L. and Cessna, A.J., *Arch. Environ. Contam. Toxicol*., 2009, 57, 1.
12. Campagnolo, E.R. et al., *Sci. Total Environ*., 2002, 299, 89.
13. Schlusener, M. et al., *Arch. Environ. Contam. Toxicol*., 2006, 51, 21.
14. Arikan, O.A. et al., *Bioresour. Technol*., 2006, 98, 169.
15. De Liguoro, M. et al., *Chemosphere*, 2003, 52, 203.
16. Kuchta, S.L. et al., *J. Environ. Qual*., 2009, 38, 1719.
17. Kuchta, S.L. and Cessna, A.J., *Chemosphere*, 2009, 76, 439.
18. Haller, M.Y. et al., *J. Chromatogr. A*, 2002, 952, 111.
19. Aga, D.S. et al., *J. Agric. Food Chem*., 2005, 53, 7165.
20. Jacobsen, A.M. et al., *J. Chromatogr. A*, 2004, 1038, 157.
21. Christian, T. et al., *Acta Hydroch. Hydrob*., 2003, 31, 36.
22. Dolliver, H. and Gupta, S., *J. Envrion. Qual*., 2008, 37, 1227.
23. Stoob, K. et al., *Environ. Sci. Technol*., 2007, 41, 7349.
24. Burkhardt, M. et al., *J. Environ. Qual*., 2005, 34, 1363.
25. Kay, P., Blackwell, P.A., and Boxall, A.B.A., *Chemosphere*, 2005, 59, 951.
26. Kay, P., Blackwell, P.A., and Boxall, A.B.A., *Environ. Toxicol. Chem*., 2004, 23, 1136.
27. Boxall, A.B.A. et al., *Toxicol. Lett*., 2002, 131, 19.
28. Forrest, F. et al., *Livestock Pharmaceuticals in Agricultural Streams: A Scoping Study for Alberta. Alberta Agriculture*, Food and Rural Development, Edmonton, Alberta, 2006, p. 35.
29. Lissemore, L. et al., *Chemosphere*, 2006, 64, 717.
30. Yang, S. and Carlson, K., *Water Res*., 2003, 37, 4645.
31. Kolpin, D.W. et al., *Envrion. Sci. Technol*., 2002, 36, 1202.
32. Lindsey, M.E., Meyer, M., and Thurman, E.M., *Anal. Chem*., 2001, 73, 4640.
33. Calamari, D. et al., *Environ. Sci. Technol*., 2003, 37, 1241.
34. Hirsch, R. et al., *Sci. Total Environ*., 1999, 225, 109.
35. Managaki, S. et al., *Environ. Sci. Technol*., 2007, 41, 8004.
36. Batt, A.L., Snow, D.D., and Aga, D.S., *Chemosphere*, 2006, 64, 1963.
37. Vaclavik, E., Halling-Sørensen, B., and Ingerslev, F., *Chemosphere*, 2004, 56, 667.
38. Westergaard, K. et al., *Soil Biol. Biochem*., 2001, 33, 2061.
39. Sura et al., *Presented at SETAC North America 29th Annual Meeting*, Tampa, FL, November 2008.
40. Rubio, S. and Pérez-Bendito, D., *Anal. Chem*., 2009, 81, 4601.
41. Peru, K.M. et al., *J. Chromatogr. A*, 2006, 1107, 152.
42. Stahl, C.H., Iowa State University Animal Industry Report, A.S. Leaflet R2025, Ames, IA, 2005.
43. Murray, L., Ed., *Physicians Desk Reference*, 45th edn., Mont Vale, NJ, Medical Economies, 1991, pp. 2260.
44. Burrows, G.E., *J. Am. Vet. Med. Assoc*., 1980, 176, 1072.

45. Aiello, S.E., Ed., *The Merck Veterinary Manual*, 8th edn., Merck, Whitehouse Station, NJ, 1998.
46. Health Canada, *Canadian Handbook on Health Impact Assessment—Volume 4: Health Impacts by Industry Sector*, Cat. No.: H46-2/04-363E, 2004.
47. Stahl, G.L. and Zaya, M.J., *J. Assoc. Off. Anal. Chem.*, 1991, 74, 471.
48. El-Sayed, M.G.A. et al., *Dtsch. Tierarztl. Wochenschr.*, 1995, 102, 446.
49. Tanaka, T., Ikebuchi, H., and Sawada, J.I., *J. AOAC Int.*, 1996, 79, 426.
50. Shaikh, B. et al., *J. AOAC Int.*, 1999, 8, 1002.
51. Medina, M.B. *J. Agric. Food Chem.*, 2004, 52, 3231.
52. Thacker, J.D., Casale, E.S., and Tucker, C.M., *J. Agric. Food Chem.*, 1996, 44, 2680.
53. Hoebus, J., Yun, L.M., and Hoogmartens, J., *Chromatographia*, 1994, 39, 71.
54. Krzek, J. et al., *J. AOAC Int.,* 2000, 83, 1502.
55. Elrod, L., Bauer, J.F., and Messner, S.L., *Pharm. Res.*, 1988, 5, 664.
56. Schermerhorn, P.G., Chu, P., and Kijak, P.J., *J. Agric. Food Chem.*, 1995, 43, 2122.
57. Debremaeker, D. et al., *J. Chromatogr. A*, 2002, 953, 123.
58. Phillips, J.G. and Simmonds, C., *J. Chromatogr. A*, 1994, 675, 123.
59. Tsuji, K. and Jenkins, K.M., *J. Chromatogr.*, 1995, 333, 365.
60. Myers, H.M. and Rindler, J.V., *J. Chromatogr.*, 1979, 176, 103.
61. Burton, S.D. et al., *J. Chromatogr.*, 1991, 571, 209.
62. Haagsma, N., Keegstra, J.R., and Scherpenisse, P., *J. Chromatogr.*, 1993, 615, 289.
63. Haagsma, N. et al., *J. Chromatogr. B*, 1995, 672, 165.
64. Hornish, R.E. and Wiest, J.R., *J. Chromatogr. A*, 1998, 812, 123.
65. Bergwerff, A.A., Scherpenisse, P., and Haagsma, N., *Analyst*, 1998, 123, 2139.
66. Wang, J. et al., *J. Chromatogr. B*, 2006, 834, 178.
67. Peru, K.M. et al., *Presented at 52nd ASMS Conference on Mass Spectrometry*, Nashville, TN, June 2004.
68. McLaughlin, L.G. and Henion, J.D., *J. Chromatogr.*, 1992, 591, 195.
69. McLaughlin, L.G., Henion, J.D., and Kijak, P.J., *Biol. Mass Spectrom.*, 1994, 23, 417.
70. Carson, M.C. and Heller, D.N., *J. Chromatogr. B*, 1998, 718, 95.
71. Zhu, W.X. et al., *J. Chromatogr. A*, 2008, 1207, 29.
72. Farrington, W.H. et al., *Food Addit. Contam.*, 1987, 5, 67.
73. Luo, W. et al., *J. Chromatogr. B*, 1996, 687, 405.
74. Moats, W.A., *J. Agric. Food Chem.*, 1991, 39, 1812.
75. Luo, W. et al., *J. AOAC Int.*, 1996, 79, 839.
76. LaCourse, W.R. and Dasenbrock, C.O., *J. Pharm. Biomed. Anal.*, 1999, 19, 239.
77. Sin, D.W.M. et al., *Anal. Chim. Acta*, 2004, 517, 39.
78. Crellin, K.C., Sible, E., and Van Antwerp, J., *Int. J. Mass Spectrom.*, 2003, 222, 281.
79. Olsovska, J. et al., *J. Chromatogr. A*, 2007, 1139, 214.
80. Thompson, T.S. et al., *J. Chromatogr. A*, 2003, 1020, 241.
81. Ikegami, T. et al., *J. Chromatogr. A*, 2008, 1184, 474.
82. Alpert, A.J., *J. Chromatogr.*, 1990, 499, 177.
83. Peru, K.M. et al., *Presented at 53rd ASMS Conference on Mass Spectrometry*, San Antonio, TX, June 2005.
84. Yoshida, T., *J. Biochem. Biophys. Methods*, 2004, 60, 265.
85. Lindner, H. et al., *J. Chromatogr. A*, 1996, 743, 137.
86. Lindner, H., Sarg, B., and Helliger, W., *J. Chromatogr. A*, 1997, 782, 55.
87. Strege, M.A., *Anal. Chem.*, 1998, 70, 2439.
88. Strege, M.A., Stevenson, S., and Lawrence, S.M., *Anal. Chem.*, 2000, 72, 4629.
89. Olsen, B.A., *J. Chromatogr. A*, 2001, 913, 113.
90. Oertel, R., Renner, U., and Kirch, W., *J. Pharm. Biomed. Anal.*, 2004, 35, 633.
91. Curren, M.S.S., and King, J.W., *J. Chromatogr. A*, 2002, 954, 41.
92. Valette, J.C. et al., *Chromatographia*, 2004, 59, 55.

9 Clinical Applications of Hydrophilic Interaction Liquid Chromatography

Ping Wang

CONTENTS

Hydrophilic interaction liquid chromatography (HILIC) has found many applications in clinical diagnostics and clinical research in recent years. HILIC is suitable for clinical specimens such as plasma and urine, as many metabolites and biomarkers are present in these matrices as polar forms. Using HILIC obviates the needs for ion-pairing reagents necessary for polar analytes in reverse phase liquid chromatography (RPLC). The relatively high concentration of organic solvents in HILIC mobile phase leads to increased ionization signal in the subsequent mass spectrometry analysis. The composition of HILIC mobile phases is also friendlier to the HPLC pump seals than that of the classical normal phase chromatography. Compared with gas chromatography (GC), HILIC eliminates the hydrolysis and derivatization steps, and can preserve the intactness of heat-sensitive molecules. HILIC may be interfaced with single or tandem mass spectrometry, electrochemical detectors, and nuclear magnetic resonance spectroscopy for high sensitivity and specificity. These characteristics make HILIC well suited for the detection and quantitation of low-level biomarkers for early disease detection and monitoring. HILIC may also be used for other clinical purposes, such as the speciation and quantitation of contrast agents, metabolites of abused drugs, and toxic agents.

This chapter focuses on applications of HILIC in clinical diagnostics and clinical research analyzing human clinical samples. Applications in animal studies will not be included. Use of HILIC for pharmaceutical analysis will also not be covered, although human samples may be used in these assays. The unique characteristics of clinical samples and clinical analysis will be discussed. HILIC applications in three categories will then be introduced. These include sample preparation and concentration, biomarker profiling, and quantitation of specific analytes. Examples encompassing applications in a diverse array of sample types such as plasma, urine, tissue, and breath condensate will be given. A comparison will be made between the HILIC method and other methods when applicable.

9.1 CHARACTERISTICS OF CLINICAL SAMPLES AND CLINICAL ANALYSIS

Various types of human fluid and tissue samples are analyzed in clinical laboratories for the purpose of disease diagnosis or prognosis. These include serum, plasma, cerebrospinal fluid, urine, dialysis fluid, hair, sweat, saliva, and solid tissues. Biomarkers are parent forms or metabolites of endogenous proteins or small molecules, whose levels in the above matrices change when diseases are present. Detection and quantitation of biomarkers provide valuable information for laboratorians and clinicians.

As most samples that are easy to procure for clinical analysis are aqueous, many biomarkers of interest are present as polar forms in these matrices. Historically, both GC and RPLC have been widely used in the clinical laboratories for the analysis of polar biomarkers. This is because both technologies are well developed and the expertise is readily available in the clinical laboratory environment. However, both methods have limitations when it comes to clinical sample analysis. GC requires extraction and derivatization of nonvolatile compounds, and may lead to heat-induced fragmentation of analytes. High temperature in the injection port may also cause heat-induced artifactual formation of derivatives. On the other hand, ion-pairing reagents are needed to retain polar analytes in order to be separated by RPLC. Assays are often optimized individually with the optimal ion-pairing reagent for each analyte. This may have practical concerns in clinical laboratories, especially when many different markers are analyzed on the same instrument. Switching between different ion-pairing reagents and mobile phases may cause interference between assays, increased run time, and erroneous results. Furthermore, the relatively high concentration of salt in RPLC mobile phase suppresses electrospray ionization, resulting in decreased detection sensitivity in interfaced mass spectrometry.

HILIC methods separate compounds by forming a water-rich layer on the surface of the polar stationary phase. When analytes are introduced in mobile phases rich in organic solvents, hydrophilic molecules are extracted from the mobile phase and retained in the water-rich layer due to hydrogen bonding and dipole–dipole interactions. Because of this mechanism, HILIC is well suitable for the analysis of polar compounds such as carbohydrates, amino acids, polar metabolites, and peptides.[1–3] The high organic content of the mobile phase in HILIC is favorable for electrospray ionization efficiency when HILIC is interfaced with MS, which leads to enhanced sensitivity compared to RPLC-MS.

As a result, HILIC is increasingly being used in clinical sample analysis in recent years. New methods using HILIC have been developed for analytes that have traditionally been measured using GC or RPLC. At the same time, HILIC has enabled clinical applications that have not been accomplished before.

9.2 CLINICAL APPLICATIONS OF HILIC

9.2.1 Sample Preparation and Concentration

The ability of the polar stationary phase in HILIC to retain hydrophilic molecules is exploited to concentrate hydrophilic components in human clinical samples. One example is the enrichment of glycoproteins in plasma for *N*-glycosylation pattern profiling.[4–14] Protein glycosylation is an important posttranslational modification that has diverse biological roles. Clinically, changes in glycosylation pattern have ramifications in immune system diseases, cancers, and congenital disorders.[15–17] The concentration of plasma glycoproteins is relatively low compared to other high-abundance proteins such as albumin and immunoglobulins. To enhance the detection signal of glycans, HILIC solid phase extraction columns are used to concentrate *N*-glycans, which are released from glycoproteins by peptide-*N*-glycosidase digestion. *N*-glycans, either native or fluorescent tag-labeled, bind to the HILIC stationary phase, while hydrophobic peptides and other high-abundance matrix proteins are removed in the flow-through fractions. Bound glycans are later eluted by water. A variety of stationary phases have been used in glycoprofiling, including microcrystalline cellulose and Sepharose, synthetic polymer Diacovery DPA-6S, and resin with a covalently bound zwitterionic sulfobetaine functional group. The HILIC columns demonstrate great selectivity for the hydrophilic *N*-glycans. Eluted glycans can be analyzed by matrix-assisted laser desorption ionization time-of-flight mass spectrometry (MALDI-TOF-MS), high-performance liquid chromatography (HPLC) with fluorescence detection, capillary electrophoresis mass spectrometry (CE-MS), or liquid chromatography tandem mass spectrometry (LC-MS/MS).

9.2.2 Biomarker Profiling

HILIC is a technique gaining increasing popularity for urinary metabolomics. Urine is a commonly analyzed matrix for metabolomics studies. It can be obtained non-invasively and contains a wealth of metabolites of endogenous as well as exogenous molecules. A significant portion of the metabolites are hydrophilic to be soluble in urine and excreted out of the body. This feature greatly facilitates the use of HILIC in urinary metabolomics studies.

In a comparison of HILIC with RPLC to correctly classify mass spectrometry data based on sex, diurnal variation, and age, statistically comparable results were observed.[18] To generate a complete picture, both positive and negative electrospray ionization modes were used. It was concluded that HILIC is a comparable, if not better separation method for LC-MS metabolomics studies. One should keep in mind that urine samples used in the above work were from healthy volunteers. Under clinical circumstances when the analytes of interest are likely to exist in polar forms in the urine samples, one may opt for HILIC for metabolomic analysis of these samples, especially when the volume of sample is limited. Several method development studies have demonstrated that very polar endogenous metabolites or xenobiotics not resolved in RP separations can be separated and studied using HILIC coupled to MS and nuclear magnetic resonance (NMR) spectroscopy.[19–21] The stationary phase may be either silica gel or silica modified with zwitterionic hydrocarbon chains.

The high organic solvent concentration in HILIC mobile phase is advantageous for retention of analytes introduced in organic solvents. As an example, urinary nucleosides in 100% acetonitrile were preconcentrated using an online solid phase extraction column packed with aprotic boronic acid. Eluted nucleosides were separated by HILIC to avoid the possibility that any acetonitrile entering the analytical column may compromise retention of nucleosides on RP stationary phases. A HILIC column packed with polyhydroxyethyl aspartamide provided robust separation of 22 urinary nucleosides, including structural isomers.[22]

In comprehensive urinary metabolomic analysis, HILIC, RPLC, and GC are usually used as complementary techniques to cover the majority of urine metabolites. GC-MS may generate data for unconjugated metabolites with masses up to 500 Da, while LC-MS covers conjugated metabolites up to 4000 Da.[23] Thousands of features may be identified using each of the separation method, the combination of which provides a good coverage of individual metabolomic profiles. The approach may be used for disease marker discovery or basic disease mechanism research.

9.2.3 Quantitation of Specific Analytes

9.2.3.1 Serum/Plasma

9.2.3.1.1 S-Adenosylmethionine

S-adenosylmethionine (SAM) has been suggested as a diagnostic marker for *Pneumocystis jirovecii* pneumonia (PCP) in HIV-positive patients. It is synthesized from methionine and ATP by SAM synthetase and serves as a universal methyl donor in biometabolism. SAM quantitation requires a sensitive method. Plasma or serum levels 60–160 nM (24–64 ng/mL) have been reported for healthy people; these levels are further decreased in PCP-infected patients. Plasma SAM can be measured using HPLC after derivatization with a fluorescent reagent.[24] This method has the drawback of requiring an expensive derivatization reagent and fluorescence detectors are not routinely used or available in clinical laboratories. A sensitive coulometric electrochemical method was also reported but had the same disadvantage of not being amenable to routine clinical use.[25] Methods using RPLC-MS/MS require ion-pairing reagents in order for the polar SAM molecule to be retained by C18 columns.[26–28] The presence of ion-pairing reagents can affect the performance of other LC assays. In contrast, HILIC is ideal to retain and quantify the highly polar SAM molecule.[29] In one application, serum SAM was first extracted using solid-phase extraction and separated on a silica column. The limit of quantitation (LOQ) (intraassay coefficient of variation [CV] <20%) for SAM was 10 ng/mL. Using this method, it was found that serum SAM concentration was correlated with fasting status, especially methionine intake. Consistent with previous studies, SAM concentrations in convalescent serum samples were significantly increased compared to acute levels only in patients with PCP.[29]

9.2.3.1.2 Free Metanephrines

Plasma-free metanephrines are metabolic products of catecholamines, which are excreted from pheochromocytoma cells in a continuous manner. Compared to catecholamines, metanephrines are better markers for diagnosis and follow-up of pheochromocytoma, because the plasma levels of metanephrines are less subjected to environmental changes and metanephrines are more stable. Sensitive and specific assays are demanded to quantify the low levels of plasma metanephrines and to distinguish among the structurally similar metanephrine family members. HPLC with electrochemical detection is currently most frequently used.[30–32] These methods invariably require laborious sample preparation and are subjected to interferences of coeluting compounds. A cyano column-based LC-MS/MS method increases the specificity but still requires manual offline solid-phase extraction.[33] The total analysis times for the above methods are long. Immunoassays provide rapid results but have the disadvantage of cross-reactivity and nonspecific binding.[34,35] The polar basic nature of metanephrines is well suited for HILIC separation. Using a silica stationary phase and gradient elution, a method was able to achieve lower limits of detection (LODs) and LOQs than electrochemical methods with a plasma volume of 50 μL.[36] The use of online solid-phase extraction also reduced the total analysis time compared to other assays.

9.2.3.1.3 Gadolinium-Based MRI Contrast Agents

Gadolinium-based contrast agents are used in magnetic resonance imaging to enhance the contrast of images. In these compounds, gadolinium is complexed with chelators to prevent the toxic effects of free gadolinium. The agents exhibit good safety profile in most patients. However, a recently described nephrogenic systemic fibrosis syndrome may arise in some terminally ill and renally impaired patients. Decrease in renal function leads to increased accumulation of contrast agents, which may result in skin and inner organ fibrosis. To better elucidate the pathogenesis of the disease, sensitive and specific quantification of gadolinium-based contrast agents is necessary. The total amount of contrast agents is quantifiable with element-selective techniques such as inductively coupled plasma-optical emission spectroscopy (ICP-OES), inductively coupled plasma-mass spectrometry (ICP-MS), and atomic absorption spectroscopy. The individual contrast agent has been determined using capillary electrophoresis,[37–39] size exclusion chromatography with ICP-MS,[40] ion exchange chromatography with ultraviolet detection,[41] and RPLC.[42–45] None of the above methods are able to simultaneously detect several contrast agents. Most provide only qualitative results. The ionic contrast agent species can be elegantly separated by a ZIC-HILIC column and detected by electrospray mass spectrometry, which provides good correlation with ICP-OES with regard to total amount of gadolinium.[46] Five different gadolinium-related contrast agents can be simultaneously quantitated using this method, with LODs (signal to noise ratio [S/N] = 3) in the range 0.1–1 nM and LOQs in the range 0.5–5 nM. The drawback of this method is the long chromatographic run time (gradient elution time of 30 min and column equilibration time of 20 min).

9.2.3.1.4 Methylmalonic Acid

Cobalamin (vitamin B12) deficiency is a common cause of macrocytic anemia. Methylmalonic acid is a nonvolatile aliphatic dicarboxylic acid that is related to the catabolism of valine, isoleucine, and propionic acid. The conversion of methylmalonyl coenzyme A to succinyl coenzyme A requires cobalamin. Therefore, cobalamin deficiency causes increase in the level of methylmalonic acid, which can be used as a biomarker for cobalamin deficiency. Due to the nonvolatile and polar nature of methylmalonic acid, GC-MS assays for methylmalonic acid require lengthy extraction and derivatization steps,[47] and RPLC does not show good retention. The physiological concentration of plasma methylmalonic acid is less than 0.2 μM, which is much lower than that of its structural isomer succinic acid. The low molar absorption coefficient of methylmalonic acid makes direct ultraviolet detection of physiological concentration of methylmalonic acid difficult. To increase assay sensitivity and selectivity, methylmalonic acid is derivatized with fluorescent agents in HPLC or capillary electrophoresis methods, and detected by fluorescence detectors.[48,49] Derivatization is also usually required for LC-MS/MS methods,[50–52] with some exceptions.[53] A recently developed HILIC-MS assay took advantage of the bonded zwitterionic stationary phase ZIC-HILIC to separate methylmalonic acid and succinic acid without derivatization.[54] Using an isocratic mobile phase of 4 volumes acetonitrile and 1 volume 100 mM ammonium acetate adjusted to pH 4.5 with formic acid, succinic acid eluted in the void volume in the first 0.5 min. The LOQ (10 × standard deviation [SD]) was 0.09 μM, and the LOD (3 × SD) was 0.03 μM, which was low enough for the quantification of physiological methylmalonic acid levels. The HILIC method has several advantages. First, labor-intensive and time-consuming derivatization is avoided, which makes the assay suitable for automation and decreases the labor and consumable costs. Secondly, succinic acid is eluted in the void volume, leaving only methylmalonic acid, which can be detected by simple single-ion monitoring. Finally, the high acetonitrile concentration in the mobile phase yields high electrospray ionization efficiency in the mass spectrometer, leading to high analytical sensitivity.

9.2.3.1.5 Morphine and Metabolites

Morphine is an opioid drug used for pain relief. The main metabolites are morphine-3 glucuronide and morphine-6-glucuronide, of which morphine-6-glucuronide is biologically active. For therapeutic monitoring and pharmacokinetic research purposes, sensitive and specific quantitation methods that can be applied to various biological matrices are desirable. RPLC methods with or without ion-pairing reagents have been developed for plasma and microdialysis samples, with either MS or electrochemical detection.[55–59] Most of these methods require solid-phase extraction to clean up sample matrix. Desalting is especially important when analyzing microdialysis samples using MS detectors, since the perfusion fluid contains a high concentration of salts in order to prevent fluid loss due to osmotic gradient. Electrospray ionization is known to be susceptible to the ionization suppression effect of high concentration of salts. Since morphine glucuronides are hydrophilic compounds, normal phase or HILIC-MS/MS methods have also been developed.[60–62] The comparison of three normal phase or HILIC methods is listed in Table 9.1.

9.2.3.1.6 Ornithine

Ornithine is a basic amino acid that is an important component of the urea cycle, which converts ammonia, the protein catabolic product, into the less toxic compound urea. Ornithine is synthesized from arginine by arginase, as urea is generated. Arginine can also be synthesized from ornithine and citrulline. Because arginine is

TABLE 9.1
Comparison of the Three Normal Phase or HILIC-MS Methods for Morphine and Morphine Glucuronides

Method	Sample Type	Sample Preparation	Analytical Column	Mobile Phase	Chromatographic Run Time	LOQ
Bengtsson et al.[62]	Microdialysis Fluid (5 µL), plasma (100 µL)	Online desalting with C18 column	ZIC-HILIC	70% acetonitrile + 30% ammonium acetate	10 min	M, M3, M6: 0.5, 0.22, 0.55 ng/mL (microdialysis fluid); 0.78, 1.49, 0.53 ng/mL (plasma)
Naidong et al.[60]	Plasma (1.0 mL)	Solid-phase extraction with C18 column	Silica	1% formic acid + 10% water + 90% acetonitrile	3 min	M, M3, M6: 0.5, 10, 1 ng/mL
Shou et al.[61]	Plasma (250 µL)	Solid-phase extraction with Oasis HLB plate	Silica	11% water + 89% acetonitrile, supplemented with 0.01% TFA	48 s	M, M3, M6: 1, 20, 2 ng/mL

the source of nitric oxide generated in endothelial cells, ornithine is an indirect factor that influences nitric oxide generation rate and endovascular function. Ornithine is also marketed as a nutritional supplement with the claim to increase anabolic hormone levels and expedite healing process. Ornithine is traditionally analyzed together with other amino acids. Amino acids in biological fluids are either separated by ion exchange chromatography and then detected using the ninhydrin reaction, or are first derivatized with *o*-phthalaldehyde and then separated by RPLC and detected by fluorescence detectors.[63–66] An RPLC-MS/MS method used ion-pairing reagent to help retain underivatized amino acids on an RP column.[67] The run times of chromatography in the above methods are typically 30–120 min in order to separate many different amino acids. A HILIC-MS/MS method dedicated to ornithine shortened the run time to 4 min and required only a protein precipitation step for sample preparation,[68] which made the method amenable to high-throughput analysis. The use of a silica column resulted in efficient ornithine retention. The LOQ was determined to be 7.5 μmol/L, and the LOD (S/N = 3) was 0.1 μmol/L with 100 μL of plasma. The method was used successfully to study the pharmacokinetics of ornithine after arginine supplementation.[68]

9.2.3.1.7 Quaternary Ammonium Compounds

The quanternary ammonium compounds, choline, betaine, and dimethylglycine, are important precursors of the neurotransmitter acetylcholine and membrane phospholipids. Betaine is linked to homocysteine and folate metabolism. The highly polar quaternary ammonium compounds are well suited to be detected by HILIC-MS. Methods using HILIC have been developed for acetylcholine, choline, and butyrobetaine in human liver tissues,[69] and for choline, betaine, and dimethylglycine in human plasma.[70] Comparison of the two HILIC methods is presented in Table 9.2.

Historically, radioenzymatic assays,[71,72] GC,[73] GC-MS,[74,75] HPLC with electrochemical detection (ECD)[76–82] and RP HPLC-MS[83–86] have been used to measure choline or acetylcholine. The radioenzymatic assays are cumbersome, whereas GC applications require extraction and derivatization. HPLC-ECD assays usually provide high sensitivity and specificity, but may produce unstable results from electrode to electrode. In order to detect the hydrophilic quaternary ammoniums by HPLC-MS, highly aqueous mobile phase, ion-pairing reagents, or ion exchange columns are used, which lead to reduced ionization and decreased sensitivity. An RP HPLC method for dimethylglycine and betaine requires derivatization,[87] while GC-MS for betaine necessitates an enzymatic conversion step.[88] The HILIC-MS methods described above provide simple and elegant solutions to detect the quaternary ammonium compounds simultaneously.[69,70]

9.2.3.1.8 Tobacco-Specific Nitrosamine Metabolites

The tobacco-specific nitrosamine 4-(methylnitrosamino)-1-(3-pyridyl)-1-butanone (NNK) is a suspected lung carcinogen. NNK is metabolized to 4-(methylnitrosamino)-1-(3-pyridyl)-1-butanol (NNAL), which is then conjugated and excreted in urine as NNAL-*O*-glucuronide and NNAL-*N*-glucuronide. Plasma and urine methods for determining NNAL and NNAL–glucuronide concentrations have been developed using GC-thermoelectric analysis.[89–91] The methods typically require large volumes

TABLE 9.2
Comparison of the Two HILIC Methods for Quaternary Ammonium Compounds

Method	Sample Type	Compounds Detected	HILIC Stationary Phase	Linearity	LOD	LOQ	Recovery
Wang et al.[69]	Liver tissue	Acetylcholine (Ach), choline (Ch), butyrobetaine (Bb)	Polyhydroxyethyl A	Ach: 1.0–200 ng/mL; Ch: 100–10,000 ng/mL; Bb: 20–5,000 ng/mL	S/N = 3 Ach: 0.2 ng/mL; Ch: 30 ng/mL; Bb: 2.0 ng/mL	S/N = 10 Ach: 0.6 ng/mL; Ch: 80 ng/mL; Bb: 15 ng/mL	94%–105.7%
Holm et al.[70]	Serum or plasma	Choline, betaine, dimethylglycine	Hypersil silica	0.4–400 µM (40.8–46,800 ng/mL)	S/N = 5 0.3 µM (30–35 ng/mL)	N/A	87%–105%

of plasma or urine, and have a chromatographic run time of 100 min. Extraction and derivatization are also required as sample preparation steps. An RPLC-MS/MS assay has an LOQ of 20 pg/mL using 15 mL of urine.[92] Because of the accumulation of basic compounds on the RP column, the method requires acetonitrile flushing of the analytical column every 20 injections. Using a silica column and an isocratic mobile phase, a HILIC-MS/MS assay requires only a simple liquid–liquid extraction and has a short run time of 1 min.[93] With a total plasma volume of 1.0 mL, both total and free NNAL can be quantified with an LOQ of 5 pg/mL.

9.2.3.2 Urine

9.2.3.2.1 Acrylamide Metabolites

Low concentration of acrylamide is generated during heat preparation of carbohydrate-rich foods.[94] Concerns have been raised regarding increased cancer risks caused by long-term exposure to acrylamide from food. Therefore, efforts have been devoted to quantitate acrylamide metabolites in human urine to improve risk assessment. Although RPLC-MS/MS methods have been developed,[95–98] the highly polar metabolites are not well retained on RP columns and usually elute close to or within the void volume, which contains coeluting salts and matrix proteins. This causes ion suppression when samples are introduced into the mass spectrometer. To avoid this effect, interfering substances need to be cleaned up from urine samples using solid phase extraction in RPLC-MS/MS methods.

Recently a HILIC-MS/MS method using zwitterionic sulfoalkylbetaine stationary phase was developed to bypass the problems associated with the RP methods.[99] An online trap column was used to preconcentrate the acrylamide metabolites, including glycidamide, *iso*-glycidamide, mercapturic acids, and mercapturic acid-sulfoxide. ZIC-HILIC was able to separate the above metabolites well, including the regioisomers of glycidamide and diastereomers of mercapturic acid-sulfoxide. The LOD (S/N = 3) ranged 0.1–1 μg/L and the LOQ (S/N = 10) was 0.5–2.0 μg/L.

9.2.3.2.2 1,5-Anhydroglucitol

Hemoglobin A1c and fructosamine are glycemic markers that reflect the glycemic control over a period of 2–6 weeks. 1,5-Anhydroglucitol, on the other hand, is a short-term marker that changes almost real-time with plasma glucose levels in diabetic patients. Renal tubule reabsorption of 1,5-anhydroglucitol is inhibited when glucose concentration is high in glomerular filtrate.[100] This causes decreased plasma level and increased urinary level of 1,5-anhydroglucitol.[101–103] In contrast, the plasma concentration of 1,5-anhydroglucitol is very stable in euglycemic individuals.[101,104,105] 1,5-Anhydroglucitol is an extremely polar compound that does not retain on RP columns. An enzymatic assay was developed to measure serum 1,5-anhydroglucitol, with an LOD of 0.3 mg/L based on 0.5 mL of serum.[106,107] Serum and urine assays were also developed with GC-MS, which required sample preparation and derivatization.[105,108,109] The lowest LOD in urine by GC-MS was 0.06 μg/mL with 0.1–0.2 mL of sample.[110] Ion exchange HPLC methods had detection limits between 100 and 200 ng/mL for 1,5-anhydroglucitol in serum or urine.[111,112] Cation-exchange chromatography coupled to mass spectrometry had a similar detection limit of 200 ng/mL with 0.5 mL of serum.[113] More sensitive and specific detection for 1,5-anhydroglucitol was achieved using a HILIC-MS^3 method with a linear ion trap.[114] Using this method, only minimal sample preparation was involved and an LOQ of 50 ng/mL could be achieved with 50 μL of human urine.

9.2.3.2.3 Arginine, Synthetic Dimethylarginine, and Asymmetric Dimethylarginine

The importance of asymmetric dimethylarginine (ADMA) as a marker in cardiovascular disease, renal failure, diabetes mellitus, essential hypertension, and hypercholesterolemia is gradually being revealed in recent years.[115–118] ADMA is a product of arginine methylation during protein posttranslational modification.[119] The methylation reaction is catalyzed by two types of protein arginine methyltransferases. Both types catalyze the formation of monomethylarginine. Type I enzyme forms ADMA, whereas type II enzyme also generates symmetric dimethylarginine (SDMA). Both ADMA and SDMA are released into cytosol during protein catabolism. Most of the ADMA is metabolized intracellularly and excreted into the urine. A small amount of ADMA enters the plasma and may subsequently enter the endothelial cells.[120] Recent studies suggest that ADMA is a competitive inhibitor of the endothelial nitric oxide synthase,[121,122] which generates nitric oxide (NO) from arginine. NO has important physiological functions as a powerful vasodilator and inhibitor of vascular wall adhesion of proinflammatory cells, platelet aggregation, and smooth muscle cell proliferation. Although SDMA does not inhibit endothelial nitric oxide synthase, it shares the same cell entry pathway with ADMA and therefore may influence the NO production rate indirectly.[123,124]

Various methods have been developed for the determination of arginine, ADMA, and SDMA in biological fluids. A commercial ELISA assay that measures only ADMA is available, but shows no correlation with chromatographic results.[125] Importantly, chromatographic methods offer better sensitivity, specificity, and simultaneous detection of all the three compounds. The earliest RPLC methods required derivatization with *o*-phthalaldehyde and detection by fluorescence.[66,126–128] The run times for these methods are usually long in order to separate the structurally similar ADMA and SDMA. Mass spectrometric detection obviates the need for complete chromatographic resolution of the two compounds by taking advantage of unique molecular transitions. Nonetheless, RPLC-MS methods require precolumn derivatization in order to retain the polar amino acids on RP columns.[129,130] Two HPLC-MS assays achieved complete chromatographic separation of ADMA and SDMA using a hypercarb column and an octadecylsilyl column,[131,132] respectively, but did not use isotopic labeled analogs as internal standards, which may cause incorrect quantitation due to matrix effect.

Two HILIC-MS methods have been developed to quantitate arginine, ADMA, and SDMA simultaneously.[133,134] Both methods require only protein precipitation as a minimal sample preparation step, use isotopic labeled analogs as internal standards, and employ isocratic mobile phases and silica columns for separation. A detailed parameter optimization study was carried out to explore the optimal mobile phase composition, percentage of trifluoroacetic acid, and column temperature.[135] Both methods were simple and robust enough to be applied to the analysis of clinical samples. Statistically similar results were reported for plasma arginine, ADMA, and

TABLE 9.3
Comparison of the Two HILIC-MS Methods for Arginine and Its Dimethylated Derivatives

Method	Sample Type	Mobile Phase	Internal Standards	Complete Chromatographic Resolution of ADMA and SDMA
Martens-Lobenhoffer et al.[133]	Plasma (100 μL), urine (50 μL), cell culture supernatant (100 μL)	10% water + 90% acetonitrile, supplemented with 0.025% TFA and 1% propionic acid	$^{13}C_6$-Arginine, $^{2}D_6$-ADMA	No
D'Apolito et al.[134]	Plasma (0.02 mL)	10% water + 90% acetonitrile, supplemented with 0.025% TFA and 1% acetic acid	$^{13}C_6$-Arginine	Yes

SDMA levels in healthy individuals using the two methods. One method detected statistically significant differences between arginine derivative levels in healthy subjects and those of type 2 diabetic patients.[134] The comparison of the two methods is listed in Table 9.3.

9.2.3.2.4 Cocaine and Metabolites

Cocaine (benzoylmethyl ecgonine) is one of the most often abused recreational drugs. Many cocaine-related deaths are reported every year. Cocaine is a strong central nervous system stimulant that acts by blocking the reuptake of dopamine and serotonin. Physiological responses to cocaine include constricted blood vessels, increased body temperature, increased heart rate, and blood pressure. When ingested in large doses, cocaine can cause irritability, paranoid psychosis, myocardial infarction, stroke, and sudden death.

Cocaine is rapidly hydrolyzed by plasma and liver carboxylesterases to benzoylecgonine and ecgonine methylester. If cocaine is administered by smoking, pyrolysis of cocaine results in the formation of methylecgonidine or anhydroecgonine methylester. Cocaine also undergoes metabolism in the liver by the Cytochrome P450 enzymes. *N*-demethylation of cocaine produces norcocaine, while aromatic hydroxylation produces hydroxycocaine and hydroxybenzoylecgonine. When cocaine is coingested with ethanol, cocaethylene is formed from the two compounds. Cocaethylene is further metabolized into norcocaethylene, benzoylecogonine, and ecgonine ethylester. Cocaine and its metabolites form a family of basic compounds with a wide range of polarity, which are traditionally detected using GC-MS methods with derivatization. One of the drawbacks of GC-MS methods is the possibility of artifactual formation of anhydroecgonine methylester. This possibility is avoided in RPLC-MS methods, which require solid-phase extraction of urine samples before chromatography.[136–139] A HILIC-MS method was reported that used a silica column to separate cocaine and eight of its metabolites.[140] Blood, bile, urine as well as brain tissue could be analyzed following chloroform/isoporanol extraction, evaporation, and reconstitution. The LOQ was 5 ng/mL and the LOD (S/N = 3) was below 0.5 ng/mL for most of the metabolites analyzed. This method was used successfully to detect and quantitate cocaine and metabolites in a case of fatal overdose.[140]

9.2.3.2.5 Estrogen Metabolites

Estrogen is metabolized through conjugation with sulfate and glucuronide moieties. Various estrogen conjugates are then excreted in the urine. The concentration of free estrogen in urine is usually extremely low. Estrogen metabolism varies significantly between individuals, suggesting the profile of urinary estrogen metabolites may provide important physiological and pathological information. Immunoassays are widely used in the clinical laboratory for estrogen measurement. However, these assays are usually subjected to cross-reactivity and poor precision at low concentrations. Labor-intensive sample preparation steps including hydrolysis and derivatization are needed in GC-MS assays, as estrogen metabolites are highly polar.[141–144] HPLC assays have been developed with electrochemical detection for estrogens and metabolites in human tissues,[145] but suffer from low sensitivity and background interference. Several RPLC-MS methods determine estrogens with or without derivatization with toluenesulfonhydrazide or dansyl chloride. Estrogen conjugates are hydrolyzed and converted to estrogen for quantification in these methods.[146–152] Besides time-consuming derivatization and hydrolysis, the disadvantage of these assays is the inability to quantify individual estrogen conjugates and isomers. RPLC-MS/MS methods have also been developed to quantify different estrogen conjugates.[153,154] The RPLC-MS/MS method with the highest sensitivity has an LOQ of 0.02–0.6 ng mL for seven estrogen conjugates based on 5 mL of urine.[154] A HILIC-MS/MS assay provides 2- to 10-fold higher sensitivity for the same metabolites than that of the RPLC-MS/MS method, with 1 mL of urine sample.[155] This is due to the higher content of organic solvent in the mobile phase, which enhances the ionization efficiency in the electrospray ion source. The assay uses an analytical column packed with derivatized silica (carbamoyl groups attached to silica via an aliphatic carbon chain). Urine samples undergo a solid-phase extraction using a Waters Oasis HLB cartridge. Complete chromatographic separation of all conjugates is not necessary due to the unique MRM transitions. Using this HILIC method, the unique profiles and changes of urinary estrogen conjugates during different stages of pregnancy were obtained. The assay was also able to demonstrate individual-specific urinary estrogen conjugate profiles in postmenopausal breast cancer patients before and after aromatase inhibitor treatment.[155]

9.2.3.2.6 Organophosphorus Nerve Agent Metabolites

The organophosphorus nerve agents are extremely toxic substances that could be used as terrorist weapons. The agents include sarin (GB), soman (GD), tabun (GA), cyclohexylsarin (GF), VX, and rVX (VR). These compounds act by binding and reacting with the active site of acetylcholinesterase, inhibiting the hydrolysis of acetylcholine in the neural synapse. Excess acetylcholine results in overstimulation of the acetylcholine receptors in the neuromuscular junction, leading to paralysis of muscles. Death may result from the paralysis of the respiratory muscles. It is thought that these agents are metabolized within the human body by hydrolysis breakdown. Typical metabolites analyzed to assess human exposure to the organophosphorus nerve agents are the alkyl methylphosphonic acids. Various GC-MS methods were developed to quantify the alkyl methylphosphonic acids.[156–162] Methods without chromatographic separation have also been developed.[163–165] In clinical and forensic analysis, isotope dilution GC-MS or GC/MS/MS methods are usually used, with solid-phase extraction and derivatization required. The polar acidic nature of the metabolites makes HILIC an attractive alternative. In an effort to maximize the signal-to-noise ratio generated by the mass spectrometer, HILIC was also found to be optimal in order to maintain the high acetonitrile content of the mobile phase.[166] Using a solid-phase extraction procedure with non-bonded silica and a silica analytical column, the LOD in the range 30–240 pg/mL was achieved for five alkyl methylphosphonic acid metabolites. When compared to an established GC-MS/MS method[160] using the same blind unknown urine samples distributed to six different laboratories, the HILIC-MS/MS method had comparable accuracy for three metabolites and better accuracy for the other two metabolites.[166]

9.2.3.2.7 Pheomelanin Hydrolysis Products

Pheomelanin is a skin pigment thought to be responsible for the higher incidence of malignant melanomas in individuals with light colored skin than those with dark skin. It is often excreted in the urine of human melanoma patients and can be used as a disease marker. As pheomelanin is formed by oxidative polymerization of different bezothiazine derivatives, analysis of pheomelanin is usually accomplished by measurement of the reductive hydrolysis products. Two major hydrolysis products, 4-amino-3-hydroxyphenylalanine (4-AHP) and 3-amino-4-hydroxyphenylalanine (3-AHP), are formed after hydriodic acid reduction of pheomelanin.[167] Of the two compounds, 4-AHP is a more specific indicator of pheomelanin, as 3-AHP can be generated from reduction of other compounds.[168,169] Early HPLC methods cannot resolve the peaks,[170] and therefore are not optimal for quantitation of pheomelanin. RPLC methods can separate the two peaks by using ion-pairing reagents,[168,171–173] which may cause problems in subsequent ionization in mass spectrometer. A silica-based HILIC stationary phase with covalently bonded zwitterionic sulfabetaine groups separates 4-AHP and 3-AHP well and is suitable for mass spectrometric analysis.[174] When HILIC is coupled to an electrochemical detector, the simple mobile phases, long column life, and robust performance of the method enable quantitation of pheomelanin in urine, hair and melanoma tissues, structural analysis of pheomelanin, and stability study of hydrolysis products. The LOD (S/N = 3) was as low as 73 pg for 4-AHP and 51 pg for 3-AHP, respectively.[174]

9.2.3.3 Breath Condensate

9.2.3.3.1 3-Nitrotyrosine, Tyrosine, Hydroxyproline, and Proline

Monitoring of biomarkers in breath condensate is a noninvasive method for detection of diseases related to the lung. Nonvolatile compounds such as amino acids are exhaled in droplets of the bronchoalveolar lining fluid in the airflow. As the breath condensates on a cold surface, the analytes are collected in the aqueous phase.

3-Nitrotyrosine is formed from the reaction of tyrosine with peroxynitrite, which is a reactive form of nitric oxide especially during inflammation. It has been discovered that the ratio between 3-nitrotyrosine and tyrosine is a good marker of asthmatic inflammation. On the other hand, hydroxyproline is supposed to be a marker for collagen turnover. The ratio between hydroxyproline and its precursor proline is hypothesized to be a good marker for asbestosis, fibrotic reaction of the lung. The extremely low concentration of these amino acids in breath condensate calls for a highly sensitive method. Although determination of 3-nitrotyrosine alone in exhaled breath condensate has been achieved using enzyme immunoassay,[175] GC-MS/MS,[176] GC-MS(NCI),[177] and LC-MS/MS[178,179] after extensive sample preparation, the simultaneous measurement of nitrotyrosine, tyrosine, proline, and hydroxyproline is beneficial. This is because the variation of analyte dilution by exhaled water vapor may be internally corrected for by the measurement of tyrosine and proline. The first method that simultaneously quantifies the four amino acids was developed using a ZIC-HILIC column, without any need for derivatization.[180] The condensate sample was lyophilized and reconstituted. The LOD (S/N = 3) was 5 ng/L for hydroxyproline and nitrotyrosine and 0.05 μg/L for proline and tyrosine, which was favorable compared to previous detection limits in breath condensate.

REFERENCES

1. Alpert, A. J. 1990. Hydrophilic-interaction chromatography for the separation of peptides, nucleic acids and other polar compounds. *J. Chromatogr.* 499: 177–196.
2. Ikegami, T.; Tomomatsu, K.; Takubo, H.; Horie, K.; Tanaka, N. 2008. Separation efficiencies in hydrophilic interaction chromatography. *J. Chromatogr. A* 1184: 474–503.
3. Hemstrom, P.; Irgum, K. 2006. Hydrophilic interaction chromatography. *J. Sep. Sci.* 29: 1784–1821.
4. Anumula, K. R.; Dhume, S. T. 1998. High resolution and high sensitivity methods for oligosaccharide mapping and characterization by normal phase high performance liquid chromatography following derivatization with highly fluorescent anthranilic acid. *Glycobiology* 8: 685–694.
5. Hagglund, P.; Bunkenborg, J.; Elortza, F.; Jensen, O. N.; Roepstorff, P. 2004. A new strategy for identification of N-glycosylated proteins and unambiguous assignment of their glycosylation sites using HILIC enrichment and partial deglycosylation. *J. Proteome. Res.* 3: 556–566.
6. Omaetxebarria, M. J.; Hagglund, P.; Elortza, F.; Hooper, N. M.; Arizmendi, J. M.; Jensen, O. N. 2006. Isolation and characterization of glycosylphosphatidylinositol-anchored peptides by hydrophilic interaction chromatography and MALDI tandem mass spectrometry. *Anal. Chem.* 78: 3335–3341.
7. Prater, B. D.; Anumula, K. R.; Hutchins, J. T. 2007. Automated sample preparation facilitated by PhyNexus MEA purification system for oligosaccharide mapping of glycoproteins. *Anal. Biochem.* 369: 202–209.
8. Royle, L.; Campbell, M. P.; Radcliffe, C. M. et al. 2008. HPLC-based analysis of serum N-glycans on a 96-well plate platform with dedicated database software. *Anal. Biochem.* 376: 1–12.
9. Ruhaak, L. R.; Huhn, C.; Waterreus, W. J. et al. 2008. Hydrophilic interaction chromatography-based high-throughput sample preparation method for N-glycan analysis from total human plasma glycoproteins. *Anal. Chem.* 80(15): 6119–6126.

10. Tajiri, M.; Yoshida, S.; Wada, Y. 2005. Differential analysis of site-specific glycans on plasma and cellular fibronectins: Application of a hydrophilic affinity method for glycopeptide enrichment. *Glycobiology* 15: 1332–1340.
11. Thaysen-Andersen, M.; Thogersen, I. B.; Nielsen, H. J. et al. 2007. Rapid and individual-specific glycoprofiling of the low abundance N-glycosylated protein tissue inhibitor of metalloproteinases-1. *Mol. Cell. Proteomics* 6: 638–647.
12. Wada, Y.; Tajiri, M.; Yoshida, S. 2004. Hydrophilic affinity isolation and MALDI multiple-stage tandem mass spectrometry of glycopeptides for glycoproteomics. *Anal. Chem.* 76: 6560–6565.
13. Wuhrer, M.; Koeleman, C. A.; Hokke, C. H.; Deelder, A. M. 2005. Protein glycosylation analyzed by normal-phase nano-liquid chromatography–mass spectrometry of glycopeptides. *Anal. Chem.* 77: 886–894.
14. Yu, Y. Q.; Gilar, M.; Kaska, J.; Gebler, J. C. 2005. A rapid sample preparation method for mass spectrometric characterization of N-linked glycans. *Rapid Commun. Mass Spectrom.* 19: 2331–2336.
15. Daniels, M. A.; Hogquist, K. A.; Jameson, S. C. 2002. Sweet 'n' sour: The impact of differential glycosylation on T cell responses. *Nat. Immunol.* 3(10): 903–910.
16. Dwek, M. V.; Ross, H. A.; Leathem, A. J. 2001. Proteome and glycosylation mapping identifies post-translational modifications associated with aggressive breast cancer. *Proteomics* 1: 756–762.
17. Freeze, H. H. 2001. Update and perspectives on congenital disorders of glycosylation. *Glycobiology* 11: 129R–143R.
18. Cubbon, S.; Bradbury, T.; Wilson, J.; Thomas-Oates, J. 2007. Hydrophilic interaction chromatography for mass spectrometric metabonomic studies of urine. *Anal. Chem.* 79(23): 8911–8918.
19. Godejohann, M. 2007. Hydrophilic interaction chromatography coupled to nuclear magnetic resonance spectroscopy and mass spectroscopy—A new approach for the separation and identification of extremely polar analytes in bodyfluids. *J. Chromatogr. A* 1156(1–2): 87–93.
20. Idborg, H.; Zamani, L.; Edlund, P. O.; Schuppe-Koistinen, I.; Jacobsson, S. P. 2005. Metabolic fingerprinting of rat urine by LC/MS Part 2. Data pretreatment methods for handling of complex data. *J. Chromatogr. B Analyt. Technol. Biomed. Life Sci.* 828: 14–20.
21. Idborg, H.; Zamani, L.; Edlund, P. O.; Schuppe-Koistinen, I.; Jacobsson, S. P. 2005. Metabolic fingerprinting of rat urine by LC/MS Part 1. Analysis by hydrophilic interaction liquid chromatography–electrospray ionization mass spectrometry. *J. Chromatogr. B Analyt. Technol. Biomed. Life Sci.* 828: 9–13.
22. Tuytten, R.; Lemiere, F.; Van Dongen, W. et al. 2008. Development of an on-line SPE-LC-ESI-MS method for urinary nucleosides: Hyphenation of aprotic boronic acid chromatography with hydrophilic interaction LC-ESI-MS. *Anal. Chem.* 80(4): 1263–1271.
23. Kind, T.; Tolstikov, V.; Fiehn, O.; Weiss, R. H. 2007. A comprehensive urinary metabolomic approach for identifying kidney cancer. *Anal. Biochem.* 363(2): 185–195.
24. Merali, S.; Vargas, D.; Franklin, M.; Clarkson, A. B., Jr. 2000. S-Adenosylmethionine and *Pneumocystis carinii*. *J. Biol. Chem.* 275: 14958–14963.
25. Melnyk, S.; Pogribna, M.; Pogribny, I. P.; Yi, P.; James, S. J. 2000. Measurement of plasma and intracellular S-adenosylmethionine and S-adenosylhomocysteine utilizing coulometric electrochemical detection: Alterations with plasma homocysteine and pyridoxal 5′-phosphate concentrations. *Clin. Chem.* 46: 265–272.
26. Struys, E. A.; Jansen, E. E.; de Meer, K.; Jakobs, C. 2000. Determination of S-adenosylmethionine and S-adenosylhomocysteine in plasma and cerebrospinal fluid by stable-isotope dilution tandem mass spectrometry. *Clin. Chem.* 46: 1650–1656.

27. Gellekink, H.; van Oppenraaij-Emmerzaal, D.; van Rooij, A.; Struys, E. A.; den Heijer, M.; Blom, H. J. 2005. Stable-isotope dilution liquid chromatography–electrospray injection tandem mass spectrometry method for fast, selective measurement of S-adenosylmethionine and S-adenosylhomocysteine in plasma. *Clin. Chem.* 51: 1487–1492.
28. Delabar, U.; Kloor, D.; Luippold, G.; Muhlbauer, B. 1999. Simultaneous determination of adenosine, S-adenosylhomocysteine and S-adenosylmethionine in biological samples using solid-phase extraction and high-performance liquid chromatography. *J. Chromatogr. B Biomed. Sci. Appl.* 724(2): 231–238.
29. Wang, P.; Huang, L.; Davis, J. L. et al. 2008. A hydrophilic-interaction chromatography tandem mass spectrometry method for quantitation of serum S-adenosylmethionine in patients infected with human immunodeficiency virus. *Clin. Chim. Acta* 396(1–2): 86–88.
30. Lenders, J. W.; Pacak, K.; Walther, M. M. et al. 2002. Biochemical diagnosis of pheochromocytoma: Which test is best? *JAMA* 287: 1427–1434.
31. Pallant, A.; Mathian, B.; Prost, L.; Theodore, C.; Patricot, M. C. 2000. Determination of plasma methoxyamines. *Clin. Chem. Lab. Med.* 38(6): 513–517.
32. Lenders, J. W.; Eisenhofer, G.; Armando, I.; Keiser, H. R.; Goldstein, D. S.; Kopin, I. J. 1993. Determination of metanephrines in plasma by liquid chromatography with electrochemical detection. *Clin. Chem.* 39: 97–103.
33. Lagerstedt, S. A.; O'Kane, D. J.; Singh, R. J. 2004. Measurement of plasma free metanephrine and normetanephrine by liquid chromatography–tandem mass spectrometry for diagnosis of pheochromocytoma. *Clin. Chem.* 50: 603–611.
34. Manz, B.; Kuper, M.; Booltink, E.; Fischer-Brugge, U. 2004. Development of enantioselective immunoassays for free plasma metanephrines. *Ann. N Y Acad. Sci.* 1018: 582–587.
35. Lenz, T.; Zorner, J.; Kirchmaier, C. et al. 2006. Multicenter study on the diagnostic value of a new RIA for the detection of free plasma metanephrines in the work-up for pheochromocytoma. *Ann. N Y Acad. Sci.* 1073: 358–373.
36. de Jong, W. H.; Graham, K. S.; van der Molen, J. C. et al. 2007. Plasma free metanephrine measurement using automated online solid-phase extraction HPLC–tandem mass spectrometry. *Clin. Chem.* 53: 1684–1693.
37. Vetterlein, K.; Buche, K.; Hildebrand, M.; Scriba, G. K.; Lehmann, J. 2006. Capillary electrophoresis for the characterization of the complex dendrimeric contrast agent Gadomer. *Electrophoresis* 27(12): 2400–2412.
38. Vetterlein, K.; Bergmann, U.; Buche, K. et al. 2007. Comprehensive profiling of the complex dendrimeric contrast agent Gadomer using a combined approach of CE, MS, and CE-MS. *Electrophoresis* 28(17): 3088–3099.
39. Campa, C.; Rossi, M.; Flamigni, A.; Baiutti, E.; Coslovi, A.; Calabi, L. 2005. Analysis of gadobenate dimeglumine by capillary zone electrophoresis coupled with electrospray–mass spectrometry. *Electrophoresis* 26(7–8): 1533–1540.
40. Loreti, V.; Bettmer, J. 2004. Determination of the MRI contrast agent Gd-DTPA by SEC-ICP-MS. *Anal. Bioanal. Chem.* 379(7–8): 1050–1054.
41. Vora, M. M.; Wukovnig, S.; Finn, R. D.; Emran, A. M.; Boothe, T. E.; Kothari, P. J. 1986. Reversed-phase high-performance liquid chromatographic determinationof gadolinium-diethylenetriaminepentaacetic acid complex. *J. Chromatogr.* 369: 187–192.
42. Hagan, J. J.; Taylor, S. C.; Tweedle, M. F. 1988. Fluorescence detection of gadolinium chelates separated by reversed-phase high-performance liquid chromatography. *Anal. Chem.* 60(6): 514–516.
43. Hvattum, E.; Normann, P. T.; Jamieson, G. C.; Lai, J. J.; Skotland, T. 1995. Detection and quantitation of gadolinium chelates in human serum and urine by high-performance liquid chromatography and post-column derivatization of gadolinium with Arsenazo III. *J. Pharm. Biomed. Anal.* 13(7): 927–932.

44. Behra-Miellet, J.; Briand, G.; Kouach, M.; Gressier, B.; Cazin, M.; Cazin, J. C. 1998. On-line HPLC-electrospray ionization mass spectrometry: A pharmacological tool for identifying and studying the stability of Gd3+ complexes used as magnetic resonance imaging contrast agents. *Biomed. Chromatogr.* 12(1): 21–26.
45. Arbughi, T.; Bertani, F.; Celeste, R.; Grotti, A.; Sillari, S.; Tirone, P. 1998. High-performance liquid chromatographic determination of the magnetic resonance imaging contrast agent gadobenate ion in plasma, urine, faeces, bile and tissues. *J. Chromatogr. B Biomed. Sci. Appl.* 713(2): 415–426.
46. Kunnemeyer, J.; Terborg, L.; Nowak, S. et al. 2008. Speciation analysis of gadolinium-based MRI contrast agents in blood plasma by hydrophilic interaction chromatography/electrospray mass spectrometry. *Anal. Chem.* 80(21): 8163–8170.
47. Windelberg, A.; Arseth, O.; Kvalheim, G.; Ueland, P. M. 2005. Automated assay for the determination of methylmalonic acid, total homocysteine, and related amino acids in human serum or plasma by means of methylchloroformate derivatization and gas chromatography–mass spectrometry. *Clin. Chem.* 51: 2103–2109.
48. Schneede, J.; Ueland, P. M. 1993. Automated assay of methylmalonic acid in serum and urine by derivatization with 1-pyrenyldiazomethane, liquid chromatography, and fluorescence detection. *Clin. Chem.* 39: 392–399.
49. Schneede, J.; Ueland, P. M. 1995. Application of capillary electrophoresis with laser-induced fluorescence detection for determination of methylmalonic acid in human serum. *Anal. Chem.* 67(5): 812–819.
50. Kushnir, M. M.; Komaromy-Hiller, G.; Shushan, B.; Urry, F. M.; Roberts, W. L. 2001. Analysis of dicarboxylic acids by tandem mass spectrometry. High-throughput quantitative measurement of methylmalonic acid in serum, plasma, and urine. *Clin. Chem.* 47: 1993–2002.
51. Schmedes, A.; Brandslund, I. 2006. Analysis of methylmalonic acid in plasma by liquid chromatography–tandem mass spectrometry. *Clin. Chem.* 52: 754–757.
52. Magera, M. J.; Helgeson, J. K.; Matern, D.; Rinaldo, P. 2000. Methylmalonic acid measured in plasma and urine by stable-isotope dilution and electrospray tandem mass spectrometry. *Clin. Chem.* 46: 1804–1810.
53. Blom, H. J.; van Rooij, A.; Hogeveen, M. 2007. A simple high-throughput method for the determination of plasma methylmalonic acid by liquid chromatography–tandem mass spectrometry. *Clin. Chem. Lab. Med.* 45(5): 645–650.
54. Lakso, H. A.; Appelblad, P.; Schneede, J. 2008. Quantification of methylmalonic acid in human plasma with hydrophilic interaction liquid chromatography separation and mass spectrometric detection. *Clin. Chem.* 54: 2028–2035.
55. Ekblom, M.; Hammarlund-Udenaes, M.; Paalzow, L. 1993. Modeling of tolerance development and rebound effect during different intravenous administrations of morphine to rats. *J. Pharmacol. Exp. Ther.* 266: 244–252.
56. Toyo'oka, T.; Yano, M.; Kato, M.; Nakahara, Y. 2001. Simultaneous determination of morphine and its glucuronides in rat hair and rat plasma by reversed-phase liquid chromatography with electrospray ionization mass spectrometry. *Analyst* 126(8): 1339–1345.
57. Projean, D.; Minh Tu, T.; Ducharme, J. 2003. Rapid and simple method to determine morphine and its metabolites in rat plasma by liquid chromatography–mass spectrometry. *J. Chromatogr. B Analyt. Technol. Biomed. Life Sci.* 787(2): 243–253.
58. Whittington, D.; Kharasch, E. D. 2003. Determination of morphine and morphine glucuronides in human plasma by 96-well plate solid-phase extraction and liquid chromatography–electrospray ionization mass spectrometry. *J. Chromatogr. B Analyt. Technol. Biomed. Life Sci.* 796(1): 95–103.
59. Mabuchi, M.; Takatsuka, S.; Matsuoka, M.; Tagawa, K. 2004. Determination of morphine, morphine-3-glucuronide and morphine-6-glucuronide in monkey and dog plasma by high-performance liquid chromatography–electrospray ionization tandem mass spectrometry. *J. Pharm. Biomed. Anal.* 35(3): 563–573.

60. Naidong, W.; Lee, J. W.; Jiang, X.; Wehling, M.; Hulse, J. D.; Lin, P. P. 1999. Simultaneous assay of morphine, morphine-3-glucuronide and morphine-6-glucuronide in human plasma using normal-phase liquid chromatography–tandem mass spectrometry with a silica column and an aqueous organic mobile phase. *J. Chromatogr. B Biomed. Sci. Appl.* 735(2): 255–269.
61. Shou, W. Z.; Chen, Y. L.; Eerkes, A. et al. 2002. Ultrafast liquid chromatography/tandem mass spectrometry bioanalysis of polar analytes using packed silica columns. *Rapid Commun. Mass Spectrom.* 16(17): 1613–1621.
62. Bengtsson, J.; Jansson, B.; Hammarlund-Udenaes, M. 2005. On-line desalting and determination of morphine, morphine-3-glucuronide and morphine-6-glucuronide in microdialysis and plasma samples using column switching and liquid chromatography/tandem mass spectrometry. *Rapid Commun. Mass Spectrom.* 19(15): 2116–2122.
63. Fekkes, D.; van Dalen, A.; Edelman, M.; Voskuilen, A. 1995. Validation of the determination of amino acids in plasma by high-performance liquid chromatography using automated pre-column derivatization with o-phthaldialdehyde. *J. Chromatogr. B Biomed. Appl.* 669(2): 177–186.
64. Le Boucher, J.; Charret, C.; Coudray-Lucas, C.; Giboudeau, J.; Cynober, L. 1997. Amino acid determination in biological fluids by automated ion-exchange chromatography: Performance of Hitachi L-8500A. *Clin. Chem.* 43: 1421–1428.
65. Patchett, M. L.; Monk, C. R.; Daniel, R. M.; Morgan, H. W. 1988. Determination of agmatine, arginine, citrulline and ornithine by reversed-phase liquid chromatography using automated pre-column derivatization with o-phthalaldehyde. *J. Chromatogr.* 425(2): 269–276.
66. Zhang, W. Z.; Kaye, D. M. 2004. Simultaneous determination of arginine and seven metabolites in plasma by reversed-phase liquid chromatography with a time-controlled ortho-phthaldialdehyde precolumn derivatization. *Anal. Biochem.* 326(1): 87–92.
67. Piraud, M.; Vianey-Saban, C.; Petritis, K.; Elfakir, C.; Steghens, J. P.; Bouchu, D. 2005. Ion-pairing reversed-phase liquid chromatography/electrospray ionization mass spectrometric analysis of 76 underivatized amino acids of biological interest: A new tool for the diagnosis of inherited disorders of amino acid metabolism. *Rapid Commun. Mass Spectrom.* 19(12): 1587–1602.
68. Martens-Lobenhoffer, J.; Postel, S.; Troger, U.; Bode-Boger, S. M. 2007. Determination of ornithine in human plasma by hydrophilic interaction chromatography–tandem mass spectrometry. *J. Chromatogr. B Analyt. Technol. Biomed. Life Sci.* 855(2): 271–275.
69. Wang, Y.; Wang, T.; Shi, X. et al. 2008. Analysis of acetylcholine, choline and butyrobetaine in human liver tissues by hydrophilic interaction liquid chromatography–tandem mass spectrometry. *J. Pharm. Biomed. Anal.* 47(4–5): 870–875.
70. Holm, P. I.; Ueland, P. M.; Kvalheim, G.; Lien, E. A. 2003. Determination of choline, betaine, and dimethylglycine in plasma by a high-throughput method based on normal-phase chromatography–tandem mass spectrometry. *Clin. Chem.* 49: 286–294.
71. Wang, F. L.; Haubrich, D. R. 1975. A simple, sensitive, and specific assay for free choline in plasma. *Anal. Biochem.* 63(1): 195–201.
72. Eckernas, S. A.; Aquilonius, S. M. 1977. Free choline in human plasma analysed by simple radio-enzymatic procedure: Age distribution and effect of a meal. *Scand. J. Clin. Lab. Invest.* 37(2): 183–187.
73. McMahon, K. E.; Farrell, P. M. 1985. Measurement of free choline concentrations in maternal and neonatal blood by micropyrolysis gas chromatography. *Clin. Chim. Acta* 149(1): 1–12.
74. Pomfret, E. A.; daCosta, K. A.; Schurman, L. L.; Zeisel, S. H. 1989. Measurement of choline and choline metabolite concentrations using high-pressure liquid chromatography and gas chromatography–mass spectrometry. *Anal. Biochem.* 180(1): 85–90.

75. Tsai, T. H. 2000. Separation methods used in the determination of choline and acetylcholine. *J. Chromatogr. B Biomed. Sci. Appl.* 747(1–2): 111–122.
76. Guerrieri, A.; Palmisano, F. 2001. An acetylcholinesterase/choline oxidase-based amperometric biosensors as a liquid chromatography detector for acetylcholine and choline determination in brain tissue homogenates. *Anal. Chem.* 73(13): 2875–2882.
77. Huang, T.; Yang, L.; Gitzen, J.; Kissinger, P. T.; Vreeke, M.; Heller, A. 1995. Detection of basal acetylcholine in rat brain microdialysate. *J. Chromatogr. B Biomed. Appl.* 670(2): 323–327.
78. Niwa, T. 1996. Organic acids and the uremic syndrome: Protein metabolite hypothesis in the progression of chronic renal failure. *Semin. Nephrol.* 16(3): 167–182.
79. Tsai, T. R.; Cham, T. M.; Chen, K. C.; Chen, C. F.; Tsai, T. H. 1996. Determination of acetylcholine by on-line microdialysis coupled with pre- and post-microbore column enzyme reactors with electrochemical detection. *J. Chromatogr. B Biomed. Appl.* 678(2): 151–155.
80. Yang, S.; Khaledi, M. G. 1995. Chemical selectivity in micellar electrokinetic chromatography: Characterization of solute-micelle interactions for classification of surfactants. *Anal. Chem.* 67(3): 499–510.
81. You, T.; Niwa, O.; Tomita, M.; Hirono, S. 2003. Characterization of platinum nanoparticle-embedded carbon film electrode and its detection of hydrogen peroxide. *Anal. Chem.* 75(9): 2080–2085.
82. Fossati, T.; Colombo, M.; Castiglioni, C.; Abbiati, G. 1994. Determination of plasma choline by high-performance liquid chromatography with a postcolumn enzyme reactor and electrochemical detection. *J. Chromatogr. B Biomed. Appl.* 656(1): 59–64.
83. Hows, M. E.; Organ, A. J.; Murray, S. et al. 2002. High-performance liquid chromatography/tandem mass spectrometry assay for the rapid high sensitivity measurement of basal acetylcholine from microdialysates. *J. Neurosci. Methods* 121(1): 33–39.
84. Reubsaet, J. L.; Ahlsen, E.; Haneborg, K. G.; Ringvold, A. 2003. Sample preparation and determination of acetylcholine in corneal epithelium cells using liquid chromatography–tandem mass spectrometry. *J. Chromatogr. Sci.* 41(3): 151–156.
85. Zhang, M. Y.; Hughes, Z. A.; Kerns, E. H.; Lin, Q.; Beyer, C. E. 2007. Development of a liquid chromatography/tandem mass spectrometry method for the quantitation of acetylcholine and related neurotransmitters in brain microdialysis samples. *J. Pharm. Biomed. Anal.* 44(2): 586–593.
86. Koc, H.; Mar, M. H.; Ranasinghe, A.; Swenberg, J. A.; Zeisel, S. H. 2002. Quantitation of choline and its metabolites in tissues and foods by liquid chromatography/electrospray ionization-isotope dilution mass spectrometry. *Anal. Chem.* 74(18): 4734–4740.
87. Laryea, M. D.; Steinhagen, F.; Pawliczek, S.; Wendel, U. 1998. Simple method for the routine determination of betaine and N,N-dimethylglycine in blood and urine. *Clin. Chem.* 44: 1937–1941.
88. Allen, R. H.; Stabler, S. P.; Lindenbaum, J. 1993. Serum betaine, N,N-dimethylglycine and N-methylglycine levels in patients with cobalamin and folate deficiency and related inborn errors of metabolism. *Metabolism* 42(11): 1448–1460.
89. Hecht, S. S.; Carmella, S. G.; Ye, M. et al. 2002. Quantitation of metabolites of 4-(methylnitrosamino)-1-(3-pyridyl)-1-butanone after cessation of smokeless tobacco use. *Cancer Res.* 62: 129–134.
90. Carmella, S. G.; Akerkar, S.; Hecht, S. S. 1993. Metabolites of the tobacco-specific nitrosamine 4-(methylnitrosamino)-1-(3-pyridyl)-1-butanone in smokers' urine. *Cancer Res.* 53: 721–724.
91. Hecht, S. S.; Carmella, S. G.; Chen, M. et al. 1999. Quantitation of urinary metabolites of a tobacco-specific lung carcinogen after smoking cessation. *Cancer Res.* 59: 590–596.
92. Byrd, G. D.; Ogden, M. W. 2003. Liquid chromatographic/tandem mass spectrometric method for the determination of the tobacco-specific nitrosamine metabolite NNAL in smokers' urine. *J. Mass Spectrom.* 38(1): 98–107.

93. Pan, J.; Song, Q.; Shi, H. et al. 2004. Development, validation and transfer of a hydrophilic interaction liquid chromatography/tandem mass spectrometric method for the analysis of the tobacco-specific nitrosamine metabolite NNAL in human plasma at low picogram per milliliter concentrations. *Rapid Commun. Mass Spectrom.* 18(21): 2549–2557.
94. Tareke, E.; Rydberg, P.; Karlsson, P.; Eriksson, S.; Tornqvist, M. 2000. Acrylamide: A cooking carcinogen? *Chem. Res. Toxicol.* 13(6): 517–522.
95. Bjellaas, T.; Stolen, L. H.; Haugen, M. et al. 2007. Urinary acrylamide metabolites as biomarkers for short-term dietary exposure to acrylamide. *Food Chem. Toxicol.* 45(6): 1020–1026.
96. Boettcher, M. I.; Angerer, J. 2005. Determination of the major mercapturic acids of acrylamide and glycidamide in human urine by LC-ESI-MS/MS. *J. Chromatogr. B Analyt. Technol. Biomed. Life Sci.* 824(1–2): 283–294.
97. Urban, M.; Kavvadias, D.; Riedel, K.; Scherer, G.; Tricker, A. R. 2006. Urinary mercapturic acids and a hemoglobin adduct for the dosimetry of acrylamide exposure in smokers and nonsmokers. *Inhal. Toxicol.* 18(10): 831–839.
98. Fuhr, U.; Boettcher, M. I.; Kinzig-Schippers, M. et al. 2006. Toxicokinetics of acrylamide in humans after ingestion of a defined dose in a test meal to improve risk assessment for acrylamide carcinogenicity. *Cancer Epidemiol. Biomarkers Prev.* 15: 266–271.
99. Kopp, E. K.; Sieber, M.; Kellert, M.; Dekant, W. 2008. Rapid and sensitive HILIC-ESI-MS/MS quantitation of polar metabolites of acrylamide in human urine using column switching with an online trap column. *J. Agric. Food Chem.* 56(21): 9828–9834.
100. Buse, J. B.; Freeman, J. L.; Edelman, S. V.; Jovanovic, L.; McGill, J. B. 2003. Serum 1,5-anhydroglucitol (GlycoMark): A short-term glycemic marker. *Diabetes Technol. Ther.* 5(3): 355–363.
101. Yamanouchi, T.; Akanuma, H.; Asano, T.; Konishi, C.; Akaoka, I.; Akanuma, Y. 1987. Reduction and recovery of plasma 1,5-anhydro-D-glucitol level in diabetes mellitus. *Diabetes* 36: 709–715.
102. Yamanouchi, T.; Akanuma, H.; Nakamura, T.; Akaoka, I.; Akanuma, Y. 1988. Reduction of plasma 1,5-anhydroglucitol (1-deoxyglucose) concentration in diabetic patients. *Diabetologia* 31(1): 41–45.
103. Yoshioka, S.; Saitoh, S.; Negishi, C. et al. 1983. Variations of 1-deoxyglucose(1,5-anhydroglucitol) content in plasma from patients with insulin-dependent diabetes mellitus. *Clin. Chem.* 29: 1396–1398.
104. Yamanouchi, T.; Minoda, S.; Yabuuchi, M. et al. 1989. Plasma 1,5-anhydro-D-glucitol as new clinical marker of glycemic control in NIDDM patients. *Diabetes* 38: 723–729.
105. Yamanouchi, T.; Tachibana, Y.; Akanuma, H. et al. 1992. Origin and disposal of 1,5-anhydroglucitol, a major polyol in the human body. *Am. J. Physiol.* 263: E268–E273.
106. McGill, J. B.; Cole, T. G.; Nowatzke, W. et al. 2004. Circulating 1,5-anhydroglucitol levels in adult patients with diabetes reflect longitudinal changes of glycemia: A U.S. trial of the glycomark assay. *Diabetes Care* 27: 1859–1865.
107. Fukumura, Y.; Tajima, S.; Oshitani, S. et al. 1994. Fully enzymatic method for determining 1,5-anhydro-D-glucitol in serum. *Clin. Chem.* 40: 2013–2016.
108. Yoshioka, S.; Saitoh, S.; Fujisawa, T.; Fujimori, A.; Takatani, O.; Funabashi, M. 1982. Identification and metabolic implication of 1-deoxyglucose (1,5- anhydroglucitol) in human plasma. *Clin. Chem.* 28: 1283–1286.
109. Yamanouchi, T.; Akaoka, I.; Akanuma, Y.; Akanuma, H.; Miyashita, E. 1990. Mechanism for acute reduction of 1,5-anhydroglucitol in rats treated with diabetogenic agents. *Am. J. Physiol.* 258: E423–E427.
110. Akanuma, Y.; Morita, M.; Fukuzawa, N.; Yamanouchi, T.; Akanuma, H. 1988. Urinary excretion of 1,5-anhydro-D-glucitol accompanying glucose excretion in diabetic patients. *Diabetologia* 31(11): 831–835.

111. Tanabe, T.; Tajima, S.; Suzuki, T. et al. 1997. Quantification of 1,5-anhydro-D-glucitol in urine by automated borate complex anion-exchange chromatography with an immobilized enzyme reactor. *J. Chromatogr. B Biomed. Sci. Appl.* 692(1): 23–30.
112. Tanaka, S.; Nakamori, K.; Akanuma, H.; Yabuuchi, M. 1992. High performance liquid chromatographic determination of 1,5-anhydroglucitol in human plasma for diagnosis of diabetes mellitus. *Biomed. Chromatogr.* 6(2): 63–66.
113. Niwa, T.; Tohyama, K.; Kato, Y. 1993. Analysis of polyols in uremic serum by liquid chromatography combined with atmospheric pressure chemical ionization mass spectrometry. *J. Chromatogr.* 613(1): 9–14.
114. Onorato, J. M.; Langish, R. A.; Shipkova, P. A. et al. 2008. A novel method for the determination of 1,5-anhydroglucitol, a glycemic marker, in human urine utilizing hydrophilic interaction liquid chromatography/MS(3). *J. Chromatogr. B Analyt. Technol. Biomed. Life Sci.* 873(2): 144–150.
115. Zoccali, C.; Bode-Boger, S.; Mallamaci, F. et al. 2001. Plasma concentration of asymmetrical dimethylarginine and mortality in patients with end-stage renal disease: A prospective study. *Lancet* 358(9299): 2113–2117.
116. Abbasi, F.; Asagmi, T.; Cooke, J. P. et al. 2001. Plasma concentrations of asymmetric dimethylarginine are increased in patients with type 2 diabetes mellitus. *Am. J. Cardiol.* 88(10): 1201–1203.
117. Surdacki, A.; Nowicki, M.; Sandmann, J. et al. 1999. Reduced urinary excretion of nitric oxide metabolites and increased plasma levels of asymmetric dimethylarginine in men with essential hypertension. *J. Cardiovasc. Pharmacol.* 33(4): 652–658.
118. Boger, R. H.; Bode-Boger, S. M.; Szuba, A. et al. 1998. Asymmetric dimethylarginine (ADMA): A novel risk factor for endothelial dysfunction: Its role in hypercholesterolemia. *Circulation* 98: 1842–1847.
119. Gary, J. D.; Clarke, S. 1998. RNA and protein interactions modulated by protein arginine methylation. *Prog. Nucleic Acid Res. Mol. Biol.* 61: 65–131.
120. Teerlink, T. 2007. HPLC analysis of ADMA and other methylated L-arginine analogs in biological fluids. *J. Chromatogr. B Analyt. Technol. Biomed. Life Sci.* 851(1–2): 21–29.
121. Bode-Boger, S. M.; Boger, R. H.; Kienke, S.; Junker, W.; Frolich, J. C. 1996. Elevated L-arginine/dimethylarginine ratio contributes to enhanced systemic NO production by dietary L-arginine in hypercholesterolemic rabbits. *Biochem. Biophys. Res. Commun.* 219(2): 598–603.
122. Cooke, J. P. 2004. Asymmetrical dimethylarginine: The Über marker? *Circulation* 109: 1813–1818.
123. Fleck, C.; Janz, A.; Schweitzer, F.; Karge, E.; Schwertfeger, M.; Stein, G. 2001. Serum concentrations of asymmetric (ADMA) and symmetric (SDMA) dimethylarginine in renal failure patients. *Kidney Int. Suppl.* 78: S14–S18.
124. Lluch, P.; Mauricio, M. D.; Vila, J. M. et al. 2006. Accumulation of symmetric dimethylarginine in hepatorenal syndrome. *Exp. Biol. Med. (Maywood)* 231(1): 70–75.
125. Valtonen, P.; Karppi, J.; Nyyssonen, K.; Valkonen, V. P.; Halonen, T.; Punnonen, K. 2005. Comparison of HPLC method and commercial ELISA assay for asymmetric dimethylarginine (ADMA) determination in human serum. *J. Chromatogr. B Analyt. Technol. Biomed. Life Sci.* 828(1–2): 97–102.
126. Meyer, J.; Richter, N.; Hecker, M. 1997. High-performance liquid chromatographic determination of nitric oxide synthase-related arginine derivatives in vitro and in vivo. *Anal. Biochem.* 247(1): 11–16.
127. Pettersson, A.; Uggla, L.; Backman, V. 1997. Determination of dimethylated arginines in human plasma by high-performance liquid chromatography. *J. Chromatogr. B Biomed. Sci. Appl.* 692(2): 257–262.

128. Teerlink, T.; Nijveldt, R. J.; de Jong, S.; van Leeuwen, P. A. 2002. Determination of arginine, asymmetric dimethylarginine, and symmetric dimethylarginine in human plasma and other biological samples by high-performance liquid chromatography. *Anal. Biochem.* 303(2): 131–137.
129. Martens-Lobenhoffer, J.; Krug, O.; Bode-Boger, S. M. 2004. Determination of arginine and asymmetric dimethylarginine (ADMA) in human plasma by liquid chromatography/mass spectrometry with the isotope dilution technique. *J. Mass Spectrom.* 39(11): 1287–1294.
130. Schwedhelm, E.; Tan-Andresen, J.; Maas, R.; Riederer, U.; Schulze, F.; Boger, R. H. 2005. Liquid chromatography–tandem mass spectrometry method for the analysis of asymmetric dimethylarginine in human plasma. *Clin. Chem.* 51: 1268–1271.
131. Kirchherr, H.; Kuhn-Velten, W. N. 2005. HPLC–tandem mass spectrometric method for rapid quantification of dimethylarginines in human plasma. *Clin. Chem.* 51: 249–252.
132. Huang, L. F.; Guo, F. Q.; Liang, Y. Z.; Li, B. Y.; Cheng, B. M. 2004. Simultaneous determination of L-arginine and its mono- and dimethylated metabolites in human plasma by high-performance liquid chromatography–mass spectrometry. *Anal. Bioanal. Chem.* 380(4): 643–649.
133. Martens-Lobenhoffer, J.; Bode-Boger, S. M. 2006. Fast and efficient determination of arginine, symmetric dimethylarginine, and asymmetric dimethylarginine in biological fluids by hydrophilic-interaction liquid chromatography–electrospray tandem mass spectrometry. *Clin. Chem.* 52: 488–493.
134. D'Apolito, O.; Paglia, G.; Tricarico, F. et al. 2008. Development and validation of a fast quantitative method for plasma dimethylarginines analysis using liquid chromatography–tandem mass spectrometry. *Clin. Biochem.* 41(16–17): 1391–1395.
135. Paglia, G.; D'Apolito, O.; Tricarico, F.; Garofalo, D.; Corso, G. 2008. Evaluation of mobile phase, ion pairing, and temperature influence on an HILIC-MS/MS method for L-arginine and its dimethylated derivatives detection. *J. Sep. Sci.* 31(13): 2424–2429.
136. Van Bocxlaer, J. F.; Clauwaert, K. M.; Lambert, W. E.; Deforce, D. L.; Van den Eeckhout, E. G.; De Leenheer, A. P. 2000. Liquid chromatography–mass spectrometry in forensic toxicology. *Mass Spectrom. Rev.* 19(4): 165–214.
137. Tatsuno, M.; Nishikawa, M.; Katagi, M.; Tsuchihashi, H. 1996. Simultaneous determination of illicit drugs in human urine by liquid chromatography–mass spectrometry. *J. Anal. Toxicol.* 20(5): 281–286.
138. Marquet, P. 2002. Progress of liquid chromatography–mass spectrometry in clinical and forensic toxicology. *Ther. Drug Monit.* 24(2): 255–276.
139. Weinmann, W.; Svoboda, M. 1998. Fast screening for drugs of abuse by solid-phase extraction combined with flow-injection ionspray–tandem mass spectrometry. *J. Anal. Toxicol.* 22(4): 319–328.
140. Giroud, C.; Michaud, K.; Sporkert, F. et al. 2004. A fatal overdose of cocaine associated with coingestion of marijuana, buprenorphine, and fluoxetine. Body fluid and tissue distribution of cocaine and its metabolites determined by hydrophilic interaction chromatography–mass spectrometry (HILIC-MS). *J. Anal. Toxicol.* 28(6): 464–474.
141. Ding, W. H.; Chiang, C. C. 2003. Derivatization procedures for the detection of estrogenic chemicals by gas chromatography/mass spectrometry. *Rapid Commun. Mass Spectrom.* 17(1): 56–63.
142. Gibson, R.; Tyler, C. R.; Hill, E. M. 2005. Analytical methodology for the identification of estrogenic contaminants in fish bile. *J. Chromatogr. A* 1066(1–2): 33–40.
143. Zuo, Y.; Zhang, K.; Lin, Y. 2007. Microwave-accelerated derivatization for the simultaneous gas chromatographic-mass spectrometric analysis of natural and synthetic estrogenic steroids. *J. Chromatogr. A* 1148(2): 211–218.

144. Quintana, J. B.; Carpinteiro, J.; Rodriguez, I.; Lorenzo, R. A.; Carro, A. M.; Cela, R. 2004. Determination of natural and synthetic estrogens in water by gas chromatography with mass spectrometric detection. *J. Chromatogr. A* 1024(1–2): 177–185.
145. Rogan, E. G.; Badawi, A. F.; Devanesan, P. D. et al. 2003. Relative imbalances in estrogen metabolism and conjugation in breast tissue of women with carcinoma: Potential biomarkers of susceptibility to cancer. *Carcinogenesis* 24: 697–702.
146. Draisci, R.; Palleschi, L.; Ferretti, E.; Marchiafava, C.; Lucentini, L.; Cammarata, P. 1998. Quantification of 17 beta-estradiol residues in bovine serum by liquid chromatography–tandem mass spectrometry with atmospheric pressure chemical ionization. *Analyst* 123(12): 2605–2609.
147. Xu, X.; Ziegler, R. G.; Waterhouse, D. J.; Saavedra, J. E.; Keefer, L. K. 2002. Stable isotope dilution high-performance liquid chromatography–electrospray ionization mass spectrometry method for endogenous 2- and 4-hydroxyestrones in human urine. *J. Chromatogr. B Analyt. Technol. Biomed. Life Sci.* 780(2): 315–330.
148. Xu, X.; Keefer, L. K.; Waterhouse, D. J.; Saavedra, J. E.; Veenstra, T. D.; Ziegler, R. G. 2004. Measuring seven endogenous ketolic estrogens simultaneously in human urine by high-performance liquid chromatography–mass spectrometry. *Anal. Chem.* 76(19): 5829–5836.
149. Nelson, R. E.; Grebe, S. K.; O' Kane, D. J.; Singh, R. J. 2004. Liquid chromatography–tandem mass spectrometry assay for simultaneous measurement of estradiol and estrone in human plasma. *Clin. Chem.* 50: 373–384.
150. Xia, Y. Q.; Chang, S. W.; Patel, S.; Bakhtiar, R.; Karanam, B.; Evans, D. C. 2004. Trace level quantification of deuterated 17 beta-estradiol and estrone in ovariectomized mouse plasma and brain using liquid chromatography/tandem mass spectrometry following dansylation reaction. *Rapid Commun. Mass Spectrom.* 18(14): 1621–1628.
151. Xu, X.; Veenstra, T. D.; Fox, S. D. et al. 2005. Measuring fifteen endogenous estrogens simultaneously in human urine by high-performance liquid chromatography–mass spectrometry. *Anal. Chem.* 77(20): 6646–6654.
152. Xu, X.; Roman, J. M.; Issaq, H. J.; Keefer, L. K.; Veenstra, T. D.; Ziegler, R. G. 2007. Quantitative measurement of endogenous estrogens and estrogen metabolites in human serum by liquid chromatography–tandem mass spectrometry. *Anal. Chem.* 79(20): 7813–7821.
153. Volmer, D. A.; Hui, J. P. 1997. Rapid determination of corticosteroids in urine by combined solid phase microextraction/liquid chromatography/mass spectrometry. *Rapid Commun. Mass Spectrom.* 11(17): 1926–1933.
154. D'Ascenzo, G.; Di Corcia, A.; Gentili, A. et al. 2003. Fate of natural estrogen conjugates in municipal sewage transport and treatment facilities. *Sci. Total Environ.* 302(1–3): 199–209.
155. Qin, F.; Zhao, Y. Y.; Sawyer, M. B.; Li, X. F. 2008. Hydrophilic interaction liquid chromatography–tandem mass spectrometry determination of estrogen conjugates in human urine. *Anal. Chem.* 80(9): 3404–3411.
156. Shih, M. L.; Smith, J. R.; McMonagle, J. D.; Dolzine, T. W.; Gresham, V. C. 1991. Detection of metabolites of toxic alkylmethylphosphonates in biological samples. *Biol. Mass Spectrom.* 20(11): 717–723.
157. Nakajima, T.; Sasaki, K.; Ozawa, H. et al. 1998. Urinary metabolites of sarin in a patient of the Matsumoto sarin incident. *Arch. Toxicol.* 72(9): 601–603.
158. Driskell, W. J.; Shih, M.; Needham, L. L.; Barr, D. B. 2002. Quantitation of organophosphorus nerve agent metabolites in human urine using isotope dilution gas chromatography–tandem mass spectrometry. *J. Anal. Toxicol.* 26(1): 6–10.
159. Kataoka, M.; Seto, Y. 2003. Discriminative determination of alkyl methylphosphonates and methylphosphonate in blood plasma and urine by gas chromatography–mass spectrometry after tert.-butyldimethylsilylation. *J. Chromatogr. B Analyt. Technol. Biomed. Life Sci.* 795(1): 123–132.

160. Barr, J. R.; Driskell, W. J.; Aston, L. S.; Martinez, R. A. 2004. Quantitation of metabolites of the nerve agents sarin, soman, cyclohexylsarin, VX, and Russian VX in human urine using isotope-dilution gas chromatography–tandem mass spectrometry. *J. Anal. Toxicol.* 28(5): 372–378.
161. Barr, J. R. 2004. Biological monitoring of human exposure to chemical warfare agents. *J. Anal. Toxicol.* 28(5): 305.
162. Riches, J.; Morton, I.; Read, R. W.; Black, R. M. 2005. The trace analysis of alkyl alkylphosphonic acids in urine using gas chromatography–ion trap negative ion tandem mass spectrometry. *J. Chromatogr. B Analyt. Technol. Biomed. Life Sci.* 816(1–2): 251–258.
163. Joshi, K. A.; Prouza, M.; Kum, M. et al. 2006. V-type nerve agent detection using a carbon nanotube-based amperometric enzyme electrode. *Anal. Chem.* 78(1): 331–336.
164. Mulchandani, P.; Chen, W.; Mulchandani, A. 2006. Microbial biosensor for direct determination of nitrophenyl-substituted organophosphate nerve agents using genetically engineered *Moraxella* sp. *Anal. Chim. Acta* 568(1–2): 217–221.
165. Shu, Y. R.; Su, A. K.; Liu, J. T.; Lin, C. H. 2006. Screening of nerve agent degradation products by MALDI-TOFMS. *Anal. Chem.* 78(13): 4697–4701.
166. Mawhinney, D. B.; Hamelin, E. I.; Fraser, R.; Silva, S. S.; Pavlopoulos, A. J.; Kobelski, R. J. 2007. The determination of organophosphonate nerve agent metabolites in human urine by hydrophilic interaction liquid chromatography tandem mass spectrometry. *J. Chromatogr. B Analyt. Technol. Biomed. Life Sci.* 852(1–2): 235–243.
167. Ito, S.; Fujita, K. 1985. Microanalysis of eumelanin and pheomelanin in hair and melanomas by chemical degradation and liquid chromatography. *Anal. Biochem.* 144(2): 527–536.
168. Wakamatsu, K.; Ito, S.; Rees, J. L. 2002. The usefulness of 4-amino-3-hydroxyphenylalanine as a specific marker of pheomelanin. *Pigment Cell Res.* 15(3): 225–232.
169. Crowley, J. R.; Yarasheski, K.; Leeuwenburgh, C.; Turk, J.; Heinecke, J. W. 1998. Isotope dilution mass spectrometric quantification of 3-nitrotyrosine in proteins and tissues is facilitated by reduction to 3-aminotyrosine. *Anal. Biochem.* 259(1): 127–135.
170. Ito, S.; Jimbow, K. 1983. Quantitative analysis of eumelanin and pheomelanin in hair and melanomas. *J. Invest. Dermatol.* 80(4): 268–272.
171. Kolb, A. M.; Lentjes, E. G.; Smit, N. P.; Schothorst, A.; Vermeer, B. J.; Pavel, S. 1997. Determination of pheomelanin by measurement of aminohydroxyphenylalanine isomers with high-performance liquid chromatography. *Anal. Biochem.* 252(2): 293–298.
172. Borges, C. R.; Roberts, J. C.; Wilkins, D. G.; Rollins, D. E. 2001. Relationship of melanin degradation products to actual melanin content: Application to human hair. *Anal. Biochem.* 290(1): 116–125.
173. Takasaki, A.; Nezirevic, D.; Arstrand, K.; Wakamatsu, K.; Ito, S.; Kagedal, B. 2003. HPLC analysis of pheomelanin degradation products in human urine. *Pigment Cell Res.* 16(5): 480–486.
174. Nezirevic, D.; Arstrand, K.; Kagedal, B. 2007. Hydrophilic interaction liquid chromatographic analysis of aminohydroxyphenylalanines from melanin pigments. *J. Chromatogr. A* 1163(1–2): 70–79.
175. Bodini, A.; Peroni, D. G.; Zardini, F. et al. 2006. Flunisolide decreases exhaled nitric oxide and nitrotyrosine levels in asthmatic children. *Mediators Inflamm.* 2006(4): 31919.
176. Larstad, M.; Soderling, A. S.; Caidahl, K.; Olin, A. C. 2005. Selective quantification of free 3-nitrotyrosine in exhaled breath condensate in asthma using gas chromatography/tandem mass spectrometry. *Nitric Oxide* 13(2): 134–144.
177. Celio, S.; Troxler, H.; Durka, S. S. et al. 2006. Free 3-nitrotyrosine in exhaled breath condensates of children fails as a marker for oxidative stress in stable cystic fibrosis and asthma. *Nitric Oxide* 15(3): 226–232.

178. Goen, T.; Muller-Lux, A.; Dewes, P.; Musiol, A.; Kraus, T. 2005. Sensitive and accurate analyses of free 3-nitrotyrosine in exhaled breath condensate by LC-MS/MS. *J. Chromatogr. B Analyt. Technol. Biomed. Life Sci.* 826(1–2): 261–266.
179. Baraldi, E.; Giordano, G.; Pasquale, M. F. et al. 2006. 3-Nitrotyrosine, a marker of nitrosative stress, is increased in breath condensate of allergic asthmatic children. *Allergy* 61(1): 90–96.
180. Conventz, A.; Musiol, A.; Brodowsky, C. et al. 2007. Simultaneous determination of 3-nitrotyrosine, tyrosine, hydroxyproline and proline in exhaled breath condensate by hydrophilic interaction liquid chromatography/electrospray ionization tandem mass spectrometry. *J. Chromatogr. B Analyt. Technol. Biomed. Life Sci.* 860(1): 78–85.

10 Dimethylarginine Analysis by HILIC–MS/MS and Its Applications in Clinical Chemistry and In Vivo Animal Models

O. D'Apolito, G. Paglia, M. Zotti, F. Tricarico, L. Trabace, and G. Corso

CONTENTS

10.1 INTRODUCTION

Suitable methods are needed to separate polar compounds but small polar analytes are not well retained during a conventional reversed-phase liquid chromatography (RPLC) even with high aqueous mobile phases. On the other hand, normal-phase liquid chromatography (NPLC), using nonaqueous mobile phase, is not generally suitable for routine application because of poor reproducibility and difficulty in interfacing with mass spectrometry (MS) detection.[1]

Hydrophilic interaction liquid chromatography (HILIC), a method where polar stationary phase is used in conjunction with a low aqueous/high organic mobile phase,[2] is a useful alternative to RPLC for applications involving polar compounds.[3] This feature helps to eliminate the problem associated with low solubility in high organic mobile phase of polar compounds, which are usually more soluble in acetonitrile (ACN)/water mobile phase than in hexane/acetate based mobile phase; moreover, interfacing with electrospray ionization-mass spectrometry (ESI-MS) works very well with the typical low salt acetonitrile/water HILIC eluents whereas efficient ionization is not as easily achieved with normal phase solvents.[1] HILIC is often more reproducible and eluent preparation is simpler since there is no need for controlling over a low water content in the organic solvents.

HILIC coupled with tandem MS (HILIC-MS/MS) has been utilized as a quantitative method for a number of bioanalytical applications with complex matrices.

As reported by Yong Guo and Sheetal Gaiki, HILIC applications have been initially focused on carbohydrates and peptides and then broadened to include many small polar compounds, such as pharmaceuticals, toxins, plant extracts, and other compounds important for analyses in food and pharmaceutical industries.[1] Many bioanalytical applications of HILIC with and without extraction procedures are reported in literature. In 2003, Naidong reviewed bioanalytical LC-MS/MS methods using underivatized silica columns and low aqueous/high organic mobile phases for the analysis of various types of polar analytes, extracted by using protein precipitation (PP), liquid/liquid extraction (LLE), or solid-phase extraction (SPE) and then analyzed using LC-MS/MS on the silica columns.[4] Kadar et al. reported many examples of bioanalytical applications of HILIC on different biological samples (human and animals plasma/serum, extracellular fluid of rat, mouse brains, and swine kidney) with and without extraction procedures.[5]

Here, we describe the application of a HILIC-MS/MS for separation and simultaneous quantification of arginine (Arg), asymmetric N^G, $-N^G$-dimethylarginine (ADMA), and symmetric N^G, $-N^{G\prime}$-dimethylarginine (SDMA), with a short run time (less than 5 min), using a small sample volume of human plasma (0.02 mL), useful for routine applications. Furthermore, we report the evaluation of some chromatographic aspects (mobile phase, ion pairing, and temperature) influencing the HILIC-MS/MS method for the detection and quantitation of L-arginine and its dimethylated derivatives. Finally, we describe the application of our HILIC-MS/MS method for the determination of dimethylarginine levels in rats' cerebrospinal fluid (CSF), for an in vivo microdialysis study, by a fast chromatography method using a shorter silica column.

10.2 ARGININE AND ITS DIMETHYL DERIVATIVES

As reported in literature, lysine, histidine, and arginine N-methylation of proteins is a posttranslational modification that allows the cell to expand its proteome repertoire.[6] Arginine methylation in proteins is catalyzed by two types of protein arginine methyltransferases (PRMTs) which transfer the methyl group from *S*-adenosyl-L-methionine to the guanidine group of arginine residue. Both types of enzymes catalyze the formation of N^G-monomethylarginine (MMA) in proteins.[7]

Type I PRMT enzyme produces ADMA, whereas type II enzyme produces SDMA, a constitutional isomer of ADMA.[7–9]

In 1992, Vallance et al.[10] first described ADMA as an endogenous competitive inhibitor of endothelial nitric oxide synthase (eNOS), competing with the natural substrate L-arginine (Arg), thus decreasing nitric oxide (NO) synthesis. NO is an ubiquitous intracellular messenger involved in many cellular events. It is formed by the enzymatic oxidation of L-arginine to L-citrulline in the presence of oxygen and NADPH using flavin adenine dinucleotide, flavin mononucleotide, heme, thiol, and tetrahydrobiopterin as cofactors.[11,12] In vascular endothelial cells, NO is essential for the regulation of vascular homeostasis; it is a powerful endogenous vasodilator and inhibits the adhesion of inflammatory cells to the vascular wall, the aggregation of platelets, and the proliferation of smooth muscle cells.[13] NO synthesized by the neuronal NOS (nNOS) is involved in many processes such as learning and memory.[14] The immune responses of the organism also induces an other isoform of NOS (iNOS).[13]

SDMA has no inhibitory effect on NOS but may interfere with NO synthesis indirectly by competing with arginine for transport across cell membrane.[15]

10.3 MS SETUP

The described methods used a high performance liquid chromatography (HPLC) system with an autosampler (Alliance 2695, Waters, Milford, MA). The HPLC was coupled with a Triple Quadrupole Mass Spectrometer (Micromass Quattromicro, Waters, Milford, MA) equipped with an electrospray ion source and operated in positive mode. The optimal conditions of the method were obtained using the flow injection analysis of aqueous *working solutions* directly pumped into the electrospray probe. *Working solutions* were obtained by diluting the *standard solutions* containing Arg, ADMA, and SDMA (Sigma–Aldrich-Steinheim, Germany), and $^{13}C_6$-L-Arginine (Cambridge Isotope Laboratories, Inc., Andover, MA), as internal standard (IS). *Standards solutions* were obtained by dissolving Arg, ADMA, SDMA, and $^{13}C_6$-L-Arginine ($^{13}C_6$-L-Arg) in water and diluted to 0.010 mg/mL for Arg and IS and 0.001 mg/mL for dimethylarginines.

Optimal mass conditions were set with a capillary voltage of 3.3 kV and a cone voltage of 25 V. Source temperature and desolvation gas temperature for human plasma method were set at 150°C and 450°C, respectively, while for CSF method were set at 110°C and 320°C, respectively. The optimum collision energy was observed at 18 eV for Arg, 15 eV for IS, 16 eV for ADMA, and 14 eV for SDMA.

Selected reaction monitoring (SRM) experiments were used to obtain higher specificity for the target molecules, selecting the molecular ion on the first analyzer and monitoring the specific product ion, from fragments induced in collision cell, on the second analyzer. Both ADMA and SDMA showed the same protonated molecule $[M + H]^+$ at *m/z* 203, while specific product ions of *m/z* 46 and 172, respectively, have been selected for their analysis. Arg and IS showed the protonated molecule $[M + H]^+$ at *m/z* 175 and 181, respectively; the selected product ions from the fragmentation pattern were of *m/z* 70 and 74, respectively (Figure 10.1). No interferences were observed during analytes detection and quantitation.

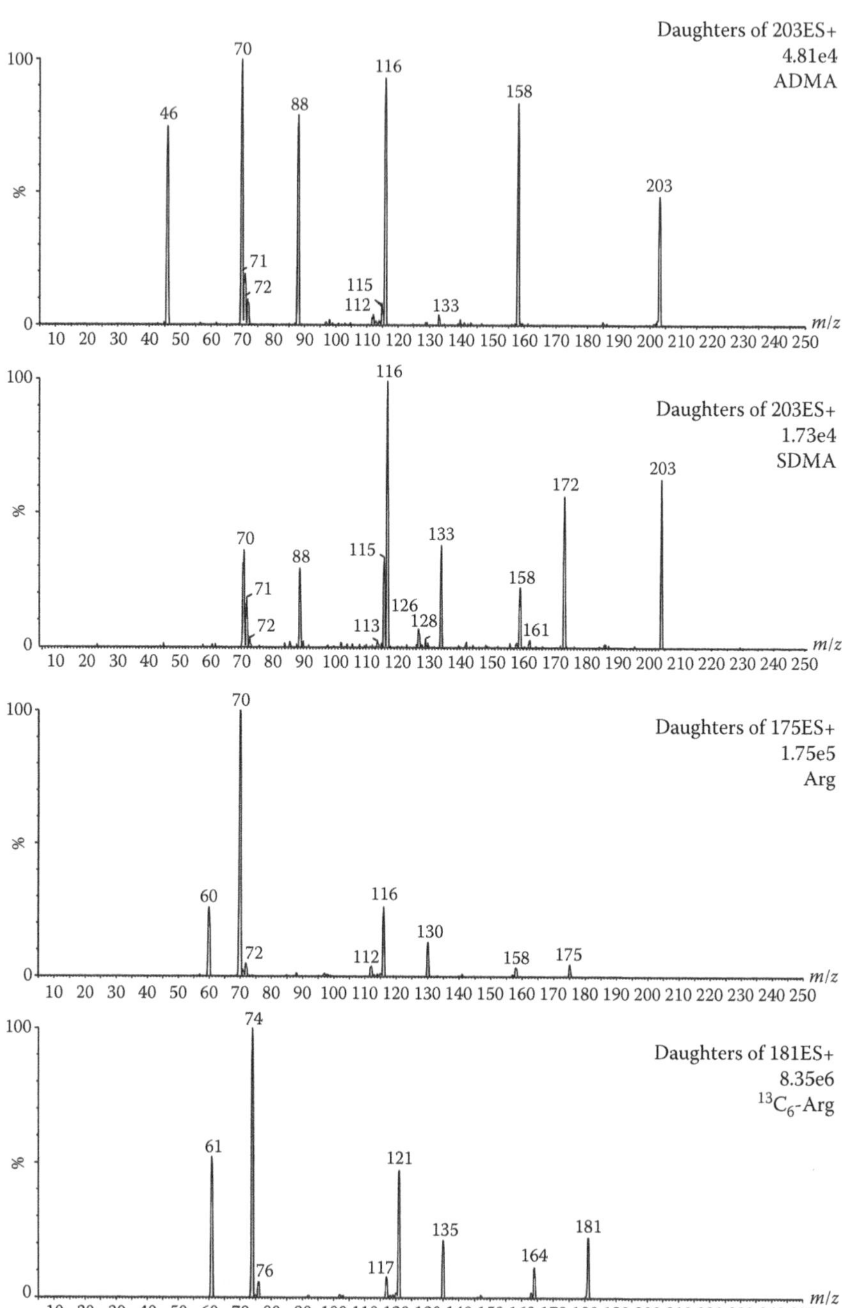

FIGURE 10.1 Full-scan positive-ion mass spectra (fragmentation spectra) of ADMA, SDMA, Arginine, and IS ($^{13}C_6$-L-Arginine).

10.4 CHROMATOGRAPHIC EVALUATIONS

10.4.1 Implementation of Two HILIC Methods

Two different chromatographic methods were evaluated in our laboratory. Firstly, the chromatographic method for the quantification of free L-Arginine and dimethylarginines in human plasma was developed and validated using a silica column (Luna Silica, 3 µm, 100 mm × 2 mm i.d., Phenomenex, Torrance, CA).

We have evaluated the impact of some experimental parameters (mobile phase, ion pairing, and temperature) that could influence the retention time and peak shape of the studied compounds in order to achieve the best compromise to obtain a robust and useful HILIC method for the routine application.

Secondly, a faster chromatographic method was developed using a shorter silica column (Zorbax RX-SIL, 1.8 µm, 50 mm × 2.1 mm i.d., Agilent Technologies, Santa Clara, CA) and applied to the protein-free fluids analysis, such as rats' CSF obtained by in vivo microdialysis.

Both HILIC methods were performed with an isocratic mobile phase of 90% Solution A (ACN/trifluoroacetic acid (TFA)/acetic acid; 1000:0.25:10) and 10% Solution B (water/TFA/acetic acid; 1000:0.25:10) at a flow rate of 0.4 mL/min. The HPLC eluent was split by a Valco valve and approximately only one-third of flow was introduced into the stainless steel capillary probe. The column temperature was kept at about 22°C for plasma analysis and at 25°C for CSF analysis.

Using these chromatographic conditions, a short run time and a good peak separation were obtained, although it was reported that it might not be necessary to completely separate all the peaks using the MS-MS technology.[16–18] In fact, Arg, ADMA, and SDMA can be unambiguously identified by their specific fragmentation pattern (Figure 10.1).

The robustness of the chromatographic method was evaluated by varying the column temperatures and the proportion of organic solvent and TFA content in the mobile phase. For chromatographic method optimization we used two *work solutions*, containing 50 ng/mL of Arg, ADMA, and SDMA; *standard solution 1* was obtained by diluting the *standard solutions* in ACN/water (90:10) and the *standard solution 2* by using ACN/water (95:5).

Column efficiency value (N) was calculated using the full width at half height of the peak. The asymmetric factor (As) was calculated using the ratio of the half widths of the rear and front sides of the peak at 10% of the peak height. SDMA/Arg and ADMA/SDMA selectivities were measured by the retention time ratios between SDMA and Arg, and between ADMA and SDMA, respectively.

10.4.2 Effect of ACN Percentage on HILIC and Comparison between Organic-Rich and Water-Rich Mobile Phase on Silica Column Chromatography

A typical HILIC application uses mobile phases containing 5%–40% of water in ACN. The chromatographic effects due to different percentages of ACN in an aqueous mobile phase, while keeping both TFA (0.025%) and acetic acid (1%) constant, were evaluated by injecting 0.01 mL of the *standard solution 1* at a flow rate of 0.4 mL/min.

As shown in Figures 10.2 and 10.3, the decrease of ACN in mobile phase causes reduction in retention time and column efficiency.[19] In fact, when ACN percentage decreased from 95% to 80%, the retention time of Arg changed from 9.56 to 2.12 min, SDMA retention time from 11.45 to 2.21 min, and ADMA retention time from 12.15 to 2.29 min. Although higher ACN content in HILIC mobile phases should improve peak shape, we observed a negative effect on peak shape, analytes retention, and selectivity.

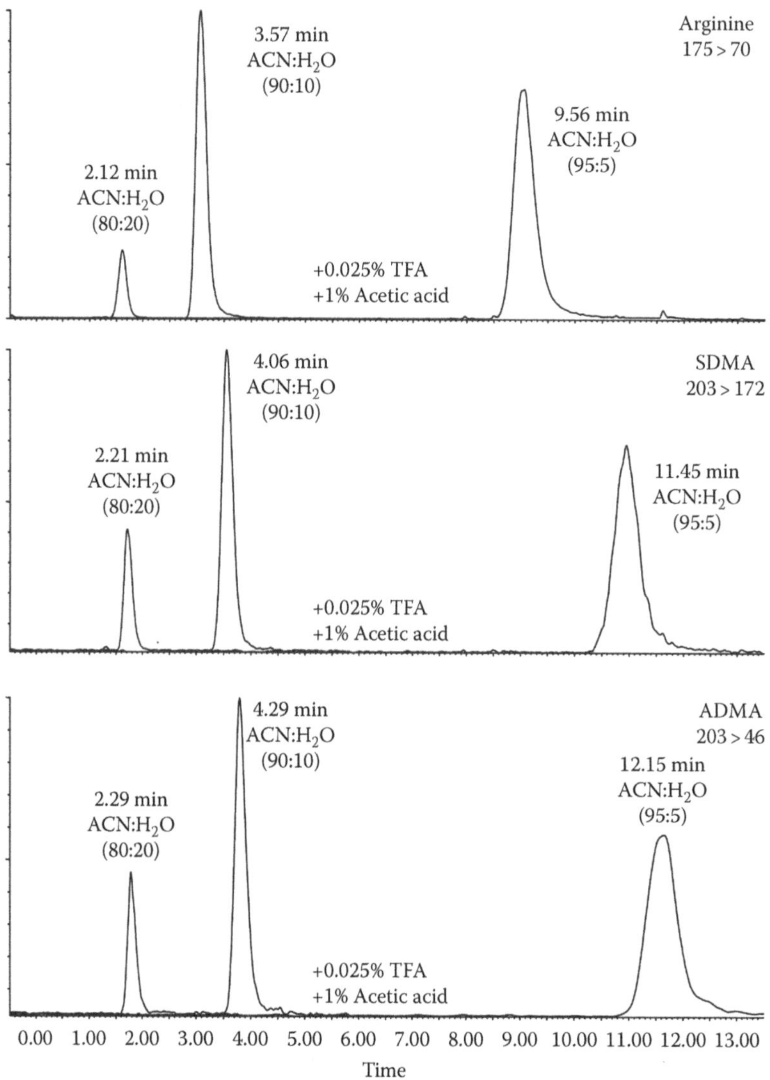

FIGURE 10.2 Effects of increasing ACN in mobile phase. (Reprinted from Paglia, G. et al., *Rapid Commn. Mass Spectrom.*, 22, 3809, 2008. With permission from Wiley-VCH Verlag Gmbh & Co. KGaA.)

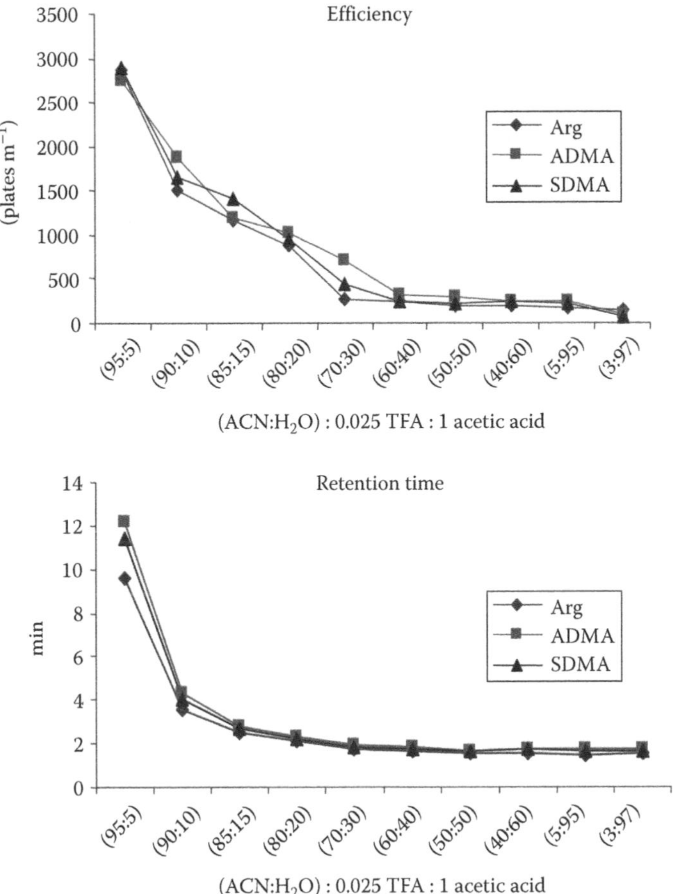

FIGURE 10.3 Effects of water percentage in mobile phase. (Reprinted from Paglia, G. et al., *Rapid Commn. Mass Spectrom.*, 22, 3809, 2008. With permission from Wiley-VCH Verlag Gmbh & Co. KGaA.)

To exclude the hypothesis of the peak broadening that was caused by sample overloading, different injection volumes and different water contents in sample diluent were tested. Volumes of 2, 5, 10, 15, 20, 25, and 30 μL of *standard solution 1* with a mobile phase containing ACN 90% were injected. No overloading effect was detected. Moreover, injections of 2, 5, and 10 μL, for both *standard solutions 1* and *2*, using 90% and 95% ACN/water mobile phases showed peak broadening when mobile phase was changed from 90:10 ACN/water to 95:5 of ACN/water.

Figure 10.4 shows that the same elution pattern was obtained using a water-rich mobile phase compared to an ACN-rich mobile phase.[19] In fact, with a mobile phase of water/ACN/methanol/formic acid (95:2.5:2.5:0.1) and 10 mM ammonium formate, a good separation of the studied analytes and high selectivity were achieved, especially between SDMA and Arg (see also Table 10.1). Nevertheless, the use of a water-rich mobile phase was not suitable to the silica columns as it would decrease the limit of

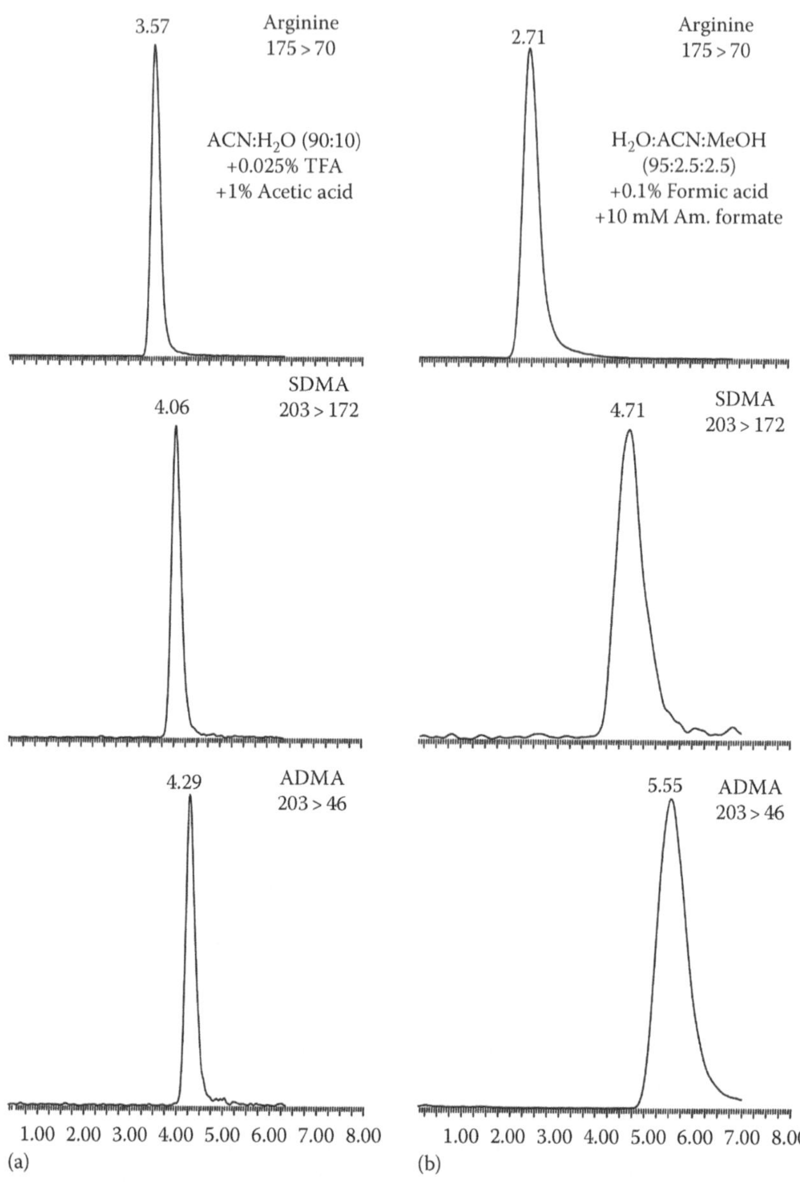

FIGURE 10.4 Combined effects by mobile phase composition. (Reprinted from Paglia, G. et al., *Rapid Commn. Mass Spectrom.*, 22, 3809, 2008. With permission from Wiley-VCH Verlag Gmbh & Co. KGaA.)

detection, efficiency, and asymmetric factor and considerably increase column back pressure (Table 10.1).[19] All of these would lead to a high stress on HPLC hardware (valve, seals, etc.), a faster impairment of the column, and a decreased ESI efficiency because of worse desolvation and increased surface tension. Alternatively, by using an organic-rich mobile phase, it is possible to inject the same organic solvent used for protein

TABLE 10.1
Comparison of Some Chromatographic Variables

	H_2O:ACN:MeOH (95:2.5:2.5) + 0.1% Formic Acid + Ammonium Formate 10 mM			ACN:H_2O (90:10) + 0.025% TFA + 1% Acetic Acid		
Flow rate (μL/min)		300			400	
Column back pressure (psi)		2900			1150	
Selectivity SDMA/ARG (α)		1.738			1.137	
Selectivity ADMA/SDMA (α)		1.178			1.057	
Retention Time (min)	**Arginine**	**SDMA**	**ADMA**	**Arginine**	**SDMA**	**ADMA**
Efficiency, N (plates m^{-1})	2.71	4.71	5.55	3.57	4.06	4.29
Asymmetric factor (As)	374	318	414	1515	1645	1863
Limit of detection (ng/mL)	1.79	1.6	1.72	1.25	1.34	1.28
	4.53	10.42	8.44	0.32	1.61	1.42

Source: Extracted from Paglia, G. et al., *Rapid Commn. Mass Spectrom.*, 22, 3809, 2008. With permission from Wiley-VCH Verlag Gmbh & Co. KGaA.

precipitation, liquid/liquid extraction, or SPE, directly in chromatographic system without evaporation and reconstitution, resulting in a reduction of sample workup.[20–23]

10.4.3 Effect of TFA Percentage on the HILIC Method

TFA is commonly used as an additive in HPLC and LC-MS for analysis of basic compounds. It is also routinely added to aqueous–organic mobile phases utilized in the HILIC-ESI/MS/MS[16,24] to improve peak shape. According to Shou and Naidong,[24] acetic acid was also added to alleviate sensitivity loss caused by TFA. The effect of different percentages of TFA and acetic acid (0.025% TFA—1% acetic acid; 0.020% TFA—0.8% acetic acid; 0.015% TFA—0.6% acetic acid; 0.010% TFA—0.4% acetic acid; 0.005% TFA—0.2% acetic acid) in the mobile phase consisting of ACN/water (90:10) on the retention time, selectivity, and efficiency for Arg, ADMA, and SDMA was investigated. Figure 10.5 shows that by decreasing the TFA percentage from 0.025% to 0.005%, and correspondingly acetic acid amounts, the retention time (min) increases from 3.57 to 5.94 for Arg, from 4.05 to 7.15 for SDMA, and from 4.28 to 7.76 for ADMA.[19] The selectivity is almost constant between ADMA and SDMA while between SDMA and Arg increases a little bit, reducing the amount of TFA and acetic acid. We have not observed significant differences of selectivity and efficiency when TFA content changed from 0.025% to 0.015% and acetic acid from 1% to 0.6%. In fact, the proper amounts of the acids used would optimize the chromatographic method.

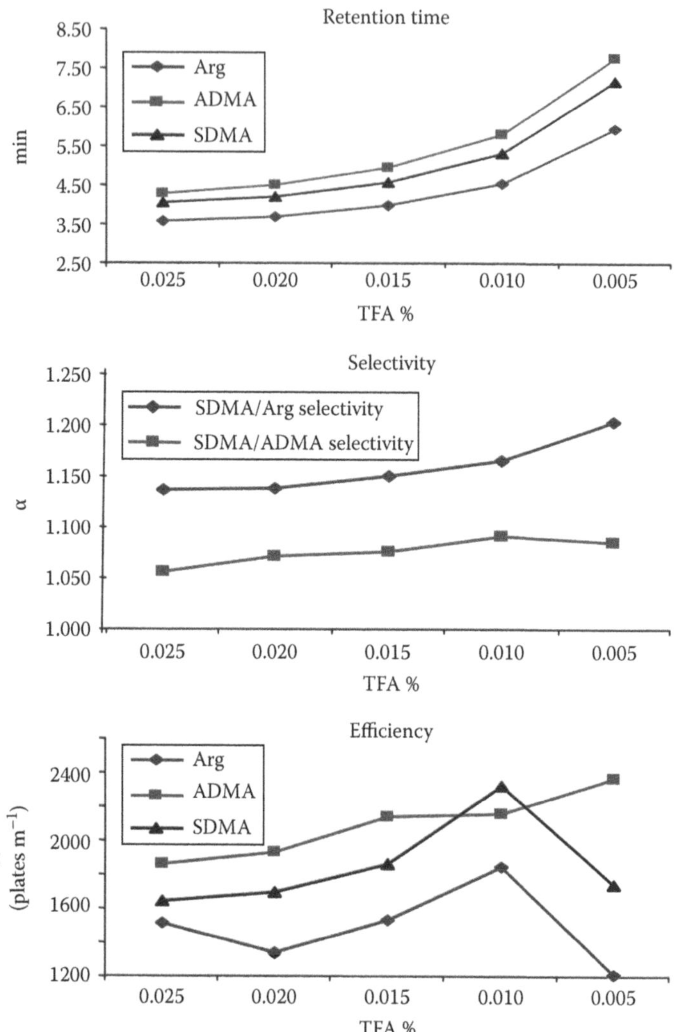

FIGURE 10.5 Effects of different TFA percentages in mobile phase. (Reprinted from Paglia, G. et al., *Rapid Commn. Mass Spectrom.*, 22, 3809, 2008. With permission from Wiley-VCH Verlag Gmbh & Co. KGaA.)

10.4.4 Effect of Column Temperature on the HILIC Method

Temperature may affect liquid chromatographic physical parameters such as solubility, vapor pressure, viscosity, and diffusivity and subsequently on retention, column efficiency, selectivity, and column back pressure. Then, the effects of different column temperatures on efficiency, retention time, and selectivity of HILIC, using ACN/water/TFA/acetate (90:10:0.025:1) as mobile phase were investigated.

As shown in Figure 10.6, the temperature increase would cause little increase in both the retention and the efficiency.[19] The selectivity remained nearly constant

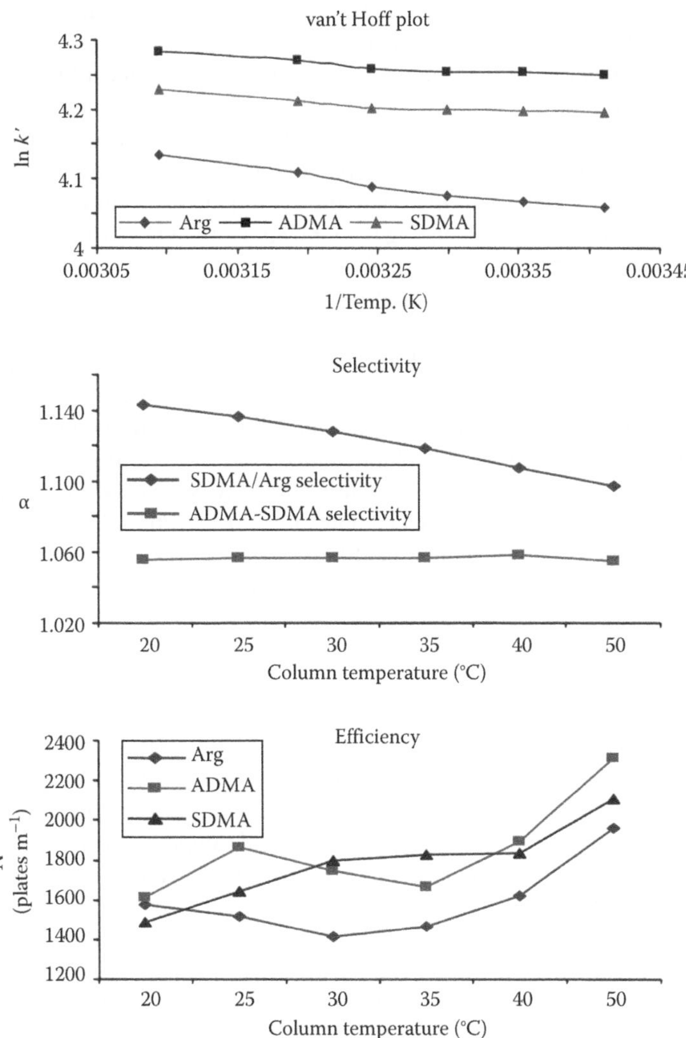

FIGURE 10.6 Effects of different column temperatures. (Reprinted from Paglia, G. et al., *Rapid Commn. Mass Spectrom.*, 22, 3809, 2008. With permission from Wiley-VCH Verlag Gmbh & Co. KGaA.)

between ADMA and SDMA, but decreases between SDMA and Arg when higher temperatures were used. The van't Hoff plot depicted in Figure 10.6 shows a remarkable endothermic adsorption process.[19] According to Dong and Huang,[25] the amount of adsorbed solvent on bare silica might change with the temperature, influencing the partitioning of the analytes between a water-enriched layer of semi-immobilized eluent on a hydrophilic stationary phase and a relatively hydrophobic bulk eluent. As expected, the column back pressure would decrease when higher column temperature was applied. Nevertheless, using a constant flow rate and an ACN-rich mobile phase, the pressure changed from 1500 to 1100 psi by varying temperature from 20°C to 50°C. Setting the column temperature between 25°C and 30°C seems to be a good compromise.

The main objective of our analytical laboratory is to perform high-throughput analyses, which requires the analytical method to be highly robust and reproducible. Thus, many potential sources of variability need to be accurately evaluated during the development. Endogenous compounds or salts always present in biological matrices together with high amount of ion pairing agent (TFA, etc.) could affect the reproducibility, influencing the ionization efficiency of ESI source. Using diluted samples (1:30–1:50) and a mobile phase with acetic acid, to alleviate the ionization loss caused by TFA, could be a solution to overcome problems due to the matrix salts.

As described in the text, Arg, ADMA, and SDMA are well retained using HILIC, and their chromatographic behavior is mainly influenced by the amount of ACN and TFA content in the mobile phase. The optimal chromatographic conditions for *standard solutions* of the method are: column temperature at 25°C, injection sample volume of 0.01 mL, and mobile phase containing 90% ACN/water, with a fixed amount of 0.025% TFA and 1% acetic acid, and flow rate of 0.4 mL/min. Nevertheless, many other parameters (different organic modifiers, stationary phases, mobile phase pH, etc.) should be evaluated as a potential source of chromatographic variability.

10.5 PLASMA METHOD

After evaluation and optimization, a simple, fast, selective, and sensitive HILIC-MS/MS method has been developed and validated. It allows the simultaneous quantitative analysis of plasma Arg and its dimethylated forms, ADMA and SDMA, in less than 5 min. This is a micromethod, with a short chromatographic run time, that allows to work with just 0.02 mL or less of biological samples. The method is suitable for routine analysis of human plasma samples for clinical studies.

To evaluate the sample preparation, six different protein precipitation solutions were tested to obtain a rapid and efficient plasma sample preparation: (a) methanol; (b) methanol:acetonitrile (50:50); (c) methanol:acetonitrile (25:75); (d) methanol:acetonitrile (25:75) to which 250 μL/L TFA and 10 mL/L acetic acid have been added; (e) acetonitrile to which 250 μL/L TFA and 10 mL/L acetic acid have been added; (f) acetonitrile. These solutions have been tested on blank and on two spiked plasma samples at Arg concentration of 105 and 128 μM, at ADMA concentration of 0.63 and 0.72 μM, and at SDMA concentration of 0.60 and 0.67 μM. Based on the best signal intensity, peak shape, and retention time, solution (d) methanol:acetonitrile (25:75) was chosen for sample preparation.

Blood samples were collected in EDTA tubes from 30 apparently healthy subjects and 33 type 2 diabetic patients, without ($n = 24$) or with ($n = 9$) renal dysfunction. Plasma samples, obtained after centrifugation at 2500 rpm for 10 min, were stored at −20°C until the analysis. Plasma (0.02 mL) was mixed with 0.02 mL of IS (0.56 nmol) and incubated at room temperature (10 min). To precipitate proteins, 0.16 mL of methanol:acetonitrile mixture (25:75) was added, vortex mixing for 3 min and centrifuging at 3000 rpm for 10 min. Then, 0.015 mL of supernatant was injected into LC-MS/MS.

The method validation study evaluated a few critical method parameters, such as calibration curves, QCs, recovery, and carryover.

Calibration standards (calibrators) were prepared by spiking blank human plasma with Arg and ADMA and SDMA *standard solutions* and diluting to give concentrations of (μM) 42, 84, 94, 105, 128, 169, and 254 for Arg, of (μM) 0.22, 0.47, 0.57, 0.63, 0.72, 0.91, and 1.29 for ADMA, and of (μM) 0.20, 0.42, 0.53, 0.60, 0.67, 0.87, and 1.27 for SDMA. They were prepared fresh for each assay and analyzed along with plasma and QCs. The concentrations of blank plasma for all analytes were calculated using the intercept value of a first calibration curve obtained on theoretical concentrations of spiked plasma with *standard solutions*. The lowest concentration point of the calibration curve was obtained diluting the blank plasma with a saline buffer (1:2). The linearity of calibration curves was estimated by the coefficients of correlation (r) which ranged from 0.9926 to 0.9984; slopes ranged from 0.0397 to 0.0685; and intercept values ranged from −0.2466 to −0.0009.

Three levels of plasma quality control (QC) samples, used in the validation study, were prepared as follows. The lowest QC level consisted of a pooled blank plasma different from that used for the preparation of the calibration curve. This pool was analyzed five times and the mean concentrations of 48.6 μM for Arg, 0.41 μM for ADMA, and 0.33 μM for SDMA were calculated using the calibration curves; this QC level has been used just for imprecision study. The other two QC levels were obtained by spiking the new pooled human plasma with diluted *stock standard solutions* to provide concentrations of (μM) 94.2 and 144.2 for Arg, of (μM) 0.88 and 1.38 for ADMA, and of (μM) 0.80 and 1.30 for SDMA. The QC samples were used to evaluate the within-day and between-day imprecision while inaccuracy and recovery were calculated using the last two spiked plasma QC levels. All QC samples were aliquoted and stored at −20°C with the clinical samples until use.

QC samples were analyzed after a sequence of unknown samples. The within-day imprecision (CV%) and inaccuracy (%) were calculated in the same analytical run, each level of QC samples 15 times. The imprecision (CV%) ranged from 2.3 to 3.9 for Arg, from 5.7 to 8.4 for ADMA, and from 5.8 to 7.5 for SDMA; the inaccuracy (%) ranged from 1.8 to 2.3 for Arg, from 6.1 to 8.7 for ADMA, and from 5.3 to 6.3 for SDMA.

The between-day imprecision (CV%) and inaccuracy (%) were calculated by analyzing each level of QC samples once a day for 15 days. The imprecision (CV%) ranged from 2.3 to 3.0 for Arg, from 6.9 to 8.4 for ADMA, and from 6.1 to 8.0 for SDMA; the inaccuracy (%) ranged from −0.1 to 0.4 for Arg, from −0.8 to 1.8 for ADMA, and from −0.1 to 0.9 for SDMA.

Recovery was calculated according to the IUPAC recommendations[25] using the following formula: $R_A = [Q_A(O + S) - Q_A(O)]/Q_A(S)$, in which R_A is the recovery; $Q_A(S)$ the quantity of analyte A added (spiked value); $Q_A(O)$ is the quantity of analyte A in the original sample; and $Q_A(O + S)$ is the quantity of analyte A recovered from the spiked sample. Recovery was obtained from three different experiments in which the two pooled blank spiked QC samples were repeated five times ($n = 15$). The mean recovery percentage for Arg, using the spiked amounts of (μM) 45.6 and 95.6, was 104% and 101%, respectively; the mean recovery, using the spiked amounts of (μM)

0.47 and 0.97, for ADMA was 106% and 102%, respectively, and for SDMA was 105% and 102%, respectively.

Carryover was evaluated as the percentage of the peak areas ratio between the mobile phase and the highest concentration of QC plasma sample injected just before the mobile phase. Each experiment was performed three times. Carryover was less than 1.2% compared to plasma sample levels. Data from calibration curves and QCs showed that the method was accurate and precise and it could be used for the determination of Arg, ADMA, and SDMA in patient samples.

A typical chromatogram obtained from human plasma is depicted in Figure 10.7.[26] The retention time was of 2.93 min for Arg and IS, 3.30 and 3.53 min for SDMA and ADMA, respectively.

Listed in Table 10.2 are the analytical results of Arg, ADMA, and SDMA from 30 plasma samples of apparently healthy subjects (group 1), and from 33 type 2 diabetic patients without kidney dysfunction (group 2, $n = 24$) and with kidney dysfunction (group 3, $n = 9$).[26] Plasma samples at concentration greater than 250 μM for Arg and 1.5 μM for ADMA and SDMA were opportunely diluted and reanalyzed.

The plasma Arg, ADMA, and SDMA concentrations calculated by the validated method are comparable to those reported in literature[17,28–30] even though our results are generally a little bit higher, most likely because of different selection of subjects and of the small number of patients. However, recent reviews suggest that the mean levels for ADMA measured in plasma of healthy subjects by HPLC or mass spectrometry-based analytical methods range between 0.4 and 0.6 μM,[31] which are in agreement with the results as we reported here.

Comparison among three groups of studied patients was performed using the analysis of variance (ANOVA) with Bonferroni correction; $p < .05$ was considered statistically significant. Arg and ADMA levels of group 2 (diabetic) and group 3 (diabetic + KD) are statistically different ($p < .05$) from group 1 (healthy subjects), and SDMA levels of group 3 are statistically different from groups 1 and 2. Besides, Arg and ADMA levels of group 2 vs. group 3 and SDMA levels of group 1 vs. group 2 were not statistically different. In particular, ADMA levels of diabetic patients are significantly elevated both in groups 2 (+11%) and 3 (+25%) compared to healthy subjects while SDMA levels of group 3 are significantly elevated (2.25-fold) compared to groups 2 and 1 (Table 10.2). These values are in agreement with those of other studies,[32,33] given the same general observation that plasma SDMA is a sensitive marker of reduced renal function.[33]

In conclusion, a practical method is developed for routine analytical applications and it allows a short analytical cycle of 5 min or less and requires minimum sample volume and preparation.

In literature are reported many chromatographic–mass spectrometric methods for the measurement of L-Arginine and its dimethyl derivatives in biological fluids; they have been discussed in detail by Martens-Lobenhoffer and Bode-Böger who stated that there is an uncertainty in the measurement of Arg, ADMA, and SDMA due to a few factors.[34] The implementation of this method was aimed to overcome some of the analytical problems linked to these published methods. As regards sample preparation, most methods include sample extraction by ion-exchange solid phase, but, as known, small changes in conditioning protocol of the extraction columns

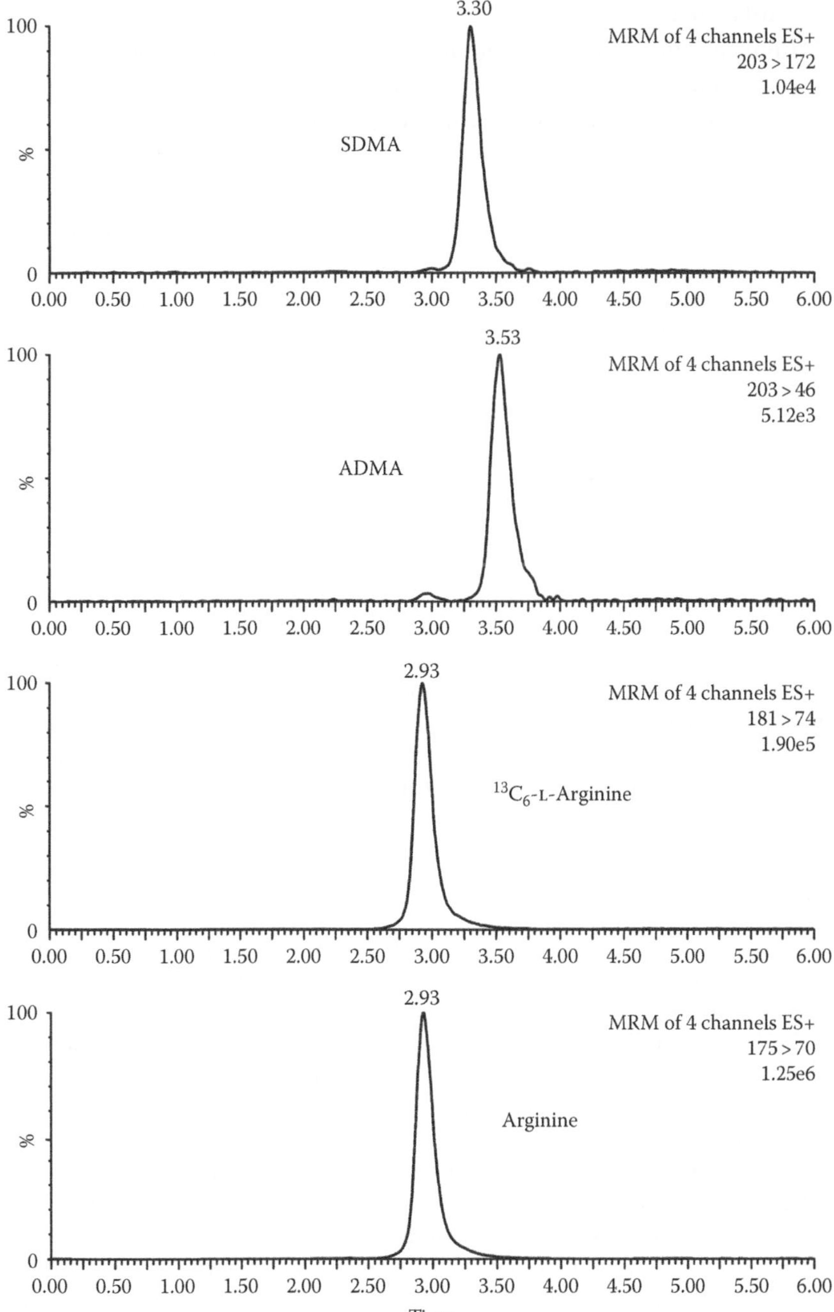

FIGURE 10.7 The figure depicts a chromatography result of a human plasma sample. (Reprinted from D'Apolito, O. et al., *Clin. Biochem.*, 41, 1391, 2008. With permission from Elsevier.)

TABLE 10.2

Concentrations (μM) of Arginine, ADMA, and SDMA in Plasma from Apparently Healthy Subjects and Diabetic (Type 2) Patients with and without Kidney Dysfunction (KD) Reported as Mean Values (SD)

Subjects	*n*	ARG Mean (SD)	ADMA Mean (SD)	SDMA Mean (SD)
Apparently healthy (group 1)	30	49.9 (19.0)	0.57 (0.11)	0.52 (0.10)
Diabetic without KD (group 2)	24	64.2 (20.5)[a]	0.63 (0.10)[a]	0.53 (0.12)
Diabetic with KD (group 3)	9	66.2 (17.3)[a]	0.71 (0.24)[a]	1.17 (0.68)[b]

Source: Extracted from D'Apolito, O. et al., *Clin. Biochem.*, 41(16–17), 1391, 2008. With permission from Elsevier.

[a] $p < .05$ group 2 vs. group 1; group 3 vs. group 1 (Arg and ADMA of group 2 vs. group 3 were not statistically significant).

[b] $p < .05$ group 3 vs. group 1 and group 2 (SDMA of group 1 vs. group 2 were not statistically significant).

(e.g., pH values of the solutions utilized, extraction and elution speed, and other factors) can affect the extraction efficiency.[28] The sample preparation procedure presented here consisting of a fast and simple protein precipitation could be completed in 13 min; this procedure is also well tolerated to MS detection and together with a fast chromatographic separation produces a very short total analytical run time.[26] To evaluate protein precipitation procedure, we tested six different solutions, including the one used by Martens-Lobenhoffer and Bode-Böger[16] (acetic acid is used rather than propionic acid with no remarkable differences in pH value). The methanol:acetonitrile (25:75) solution was chosen because it gave best peak intensity and peak shape.

As shown in Figure 10.7, using the HILIC-MS/MS method with SRM quantitative analysis and with $^{13}C_6$-L-Arginine as an internal standard, we did not observe endogenous interferences in the separation of the analytes, as reported by others.[34] Indeed, the mass spectrometric detection provides a high specificity for qualitative and quantitative analyses despite the fact that the sample composition after a simple extraction step remains very complex.[17]

In addition, the SRM analysis generally does not require high chromatographic resolution. Using an HPLC separation of underivatized analytes, the peaks were unaffected by a noisy baseline as occurs in the methods reported by others in which the peaks were also not well separated from endogenous interferences.[34,35]

The compounds were analyzed in their underivatized forms,[16,29,30,35] using an isocratic mobile phase for separation rather than a gradient elution, as reported by others.[29,30] In addition, the best chromatographic performances were obtained by a controlled low column temperature.

This method could present some minimal limitations, mainly the mobile phase which has pH 1.38, whereas the suggested optimal working pH range of the column was from 2 to 8; thus, we would expect that the acidity of the mobile phase could, over time, impair the efficiency of the column. However, we did not observe any alterations of the chromatographic reproducibility during all steps of this study. Another consideration is that we used just $^{13}C_6$-L-Arginine as an internal standard. As known, the isotope dilution analysis with labeled analogue is undoubtedly the gold standard and the labeled ADMA and/or SDMA should be recommended for the analysis.

10.6 COLLECTION OF CSF SAMPLES FROM RAT BRAIN BY MICRODIALYSIS

Microdialysis technique is one of the best methodological approaches to evaluate extracellular concentrations of several endogenous compounds. It is based on the diffusion of these substances down a concentration gradient from the extracellular fluid compartment to the dialysis fluid compartment, separated through a semi-permeable microdialysis membrane, of the microdialysis probe.[36,37] Experiments were conducted in conscious, freely moving Wistar rats (Harlan, S., Pietro al Natisone, Udine, Italy) weighing 250–300 g, housed at a constant room temperature (22°C ± 1°C) and relative humidity (55% ± 5%) under a 12 h light/dark cycle. Food and water were freely available. For the experiments, the fibers of microdialysis probe, inserted in the prefrontal cortex (PFC), were perfused, at a constant flow rate of 2 μL/min, with an artificial CSF, buffered at pH 7.3 with a 0.6 mM NaH_2PO_4, containing 145 mM NaCl, 3 mM KCl, 1.26 mM $CaCl_2$, 1 mM $MgCl_2$, and 1.4 mM Na_2HPO_4 in distilled water. The microdialysis membrane was allowed to stabilize for 2 h, collecting samples at the end of the stabilization period at 20 min intervals for 1 h for the evaluation of basal levels. The method reported here is an extension of the above-described HILIC-MS/MS method adapted to the new matrix (CSF). The aim was to determine the endogenous ADMA and SDMA in CSF samples and if their basal levels are stable over time.

10.7 METHOD DEVELOPMENT FOR THE ANALYSIS OF CSF

This method was optimized using *working solutions* prepared fresh daily by diluting aqueous *stock standard solutions* (1 g/L) of all analytes with water to provide concentrations of 57.4 μM for IS and 4.9 μM for ADMA and SDMA.

Calibration standards were prepared by diluting the *stock standard solutions* with a Krebs–Ringer buffer containing CSF-like electrolyte solution of (mM) 138 NaCl, 11 KCl, 1.5 $CaCl_2$, 1.0 $MgCl_2$, 11 $NaHCO_3$ at pH 7.4, to obtain concentrations of (nM) 256, 128, 64, 32, 16 for ADMA and SDMA before the analysis. The calibration solutions were prepared fresh for each assay and analyzed twice along with CSF and QC samples. Correlation coefficients of calibration curves (r) were greater than 0.998 in all cases for the concentration range chosen.

QC samples were prepared for the validation study by spiking the Krebs–Ringer solution with *working standard solutions* to give two levels of concentrations (20 and 100 nM) both for ADMA and SDMA, then aliquoted and stored at −20°C until use. The imprecision of the method was evaluated by analyzing daily the two levels of QC samples in triplicate for 3 days, together with calibration and CSF samples.

The imprecision (CV%, $n = 9$) ranged from 10% to 16% for ADMA and from 13% to 14% for SDMA. The inaccuracy (%) was less than 15% on average for ADMA and SDMA.

Microdialysis samples were stored at −20°C until the analysis. The analytical samples were prepared by mixing 0.03 mL of CSF sample with 0.03 mL of mobile phase containing $^{13}C_6$-L-Arginine (15 nmol) as labeled internal standard. The mix was vortexed vigorously and incubated for 10 min and further diluted with 0.12 mL of acetonitrile. An aliquot (0.02 mL) of this solution was automatically injected into the LC-MS/MS system. CSF samples with concentrations over the range of calibration curves were opportunely diluted and re-assayed.

Compared to the other approaches, the MS detector is a highly selective tool due to its ability of a structure-based analysis. The direct separation of analytes will enhance the efficiency of analysis and will also reduce the time needed for the sample preparation and ultimately enhances the analytical samples' throughput. Due to the advantage of HILIC separation and MS detection, the coupling between these two tools starts to seem attractive for the analyses of neurotransmitters and related compound.

Figure 10.8 depicts the selected ion chromatograms acquired for a typical in vivo CSF sample in MRM mode for ADMA, SDMA, and IS.[38] The peak at retention time of 1.87 min was for IS, 1.98 min for SDMA, and 2.03 min for ADMA. To avoid ion suppression, a switch valve was used to divert the eluent to waste so that signals acquisition occurs only from 1.3 to 2.8 min. Moreover, ion suppression was strongly minimized by diluting the sample and using the chromatographic parameters and column, as reported in Section 10.3.

Analytes values (mean ± SD) measured in CSF samples of rats PCF ($n = 5$) collected over time intervals of 20, 40, and 60 min were of (nmol/L) 72 ± 9, 74 ± 40, and 55 ± 35, respectively, for ADMA and of (nmol/L) 54 ± 6, 55 ± 30, and 40 ± 16, respectively, for SDMA. Data were analyzed by using one-way ANOVA for repeated measures. Statistical computations were performed with Sigma Stat software package version 3.1.

These results showed that both ADMA and SDMA levels were detectable in the PFC of freely moving rats by combining microdialysis technique and HILIC-MS/MS. Furthermore, the present study showed that ADMA and SDMA extracellular concentrations did not differ among the three time points considered. Since NO is a cell signaling substance of paramount importance in various tissues and its altered generation has been implicated in the pathogenesis of several diseases, it is of great importance to determine the concentration of circulating methylated arginine analogues that occur free in the brain and can exert a control on signal transduction through the nitrergic system. As mentioned earlier, ADMA and SDMA are the major circulating forms of methylarginines in human plasma and the elevated ADMA concentrations are an established risk marker for various diseases. However, little information is available on the levels of this molecule in CSF. Direct CSF determinations are essential for the estimation of neuronal NO production and the involvement of oxidative stress processes. These experiments showed that ADMA and SDMA levels were detectable in vivo in the PFC of rats by using a method which allows the simultaneous determination of both endogenously produced compounds using only a fast dilution as sample

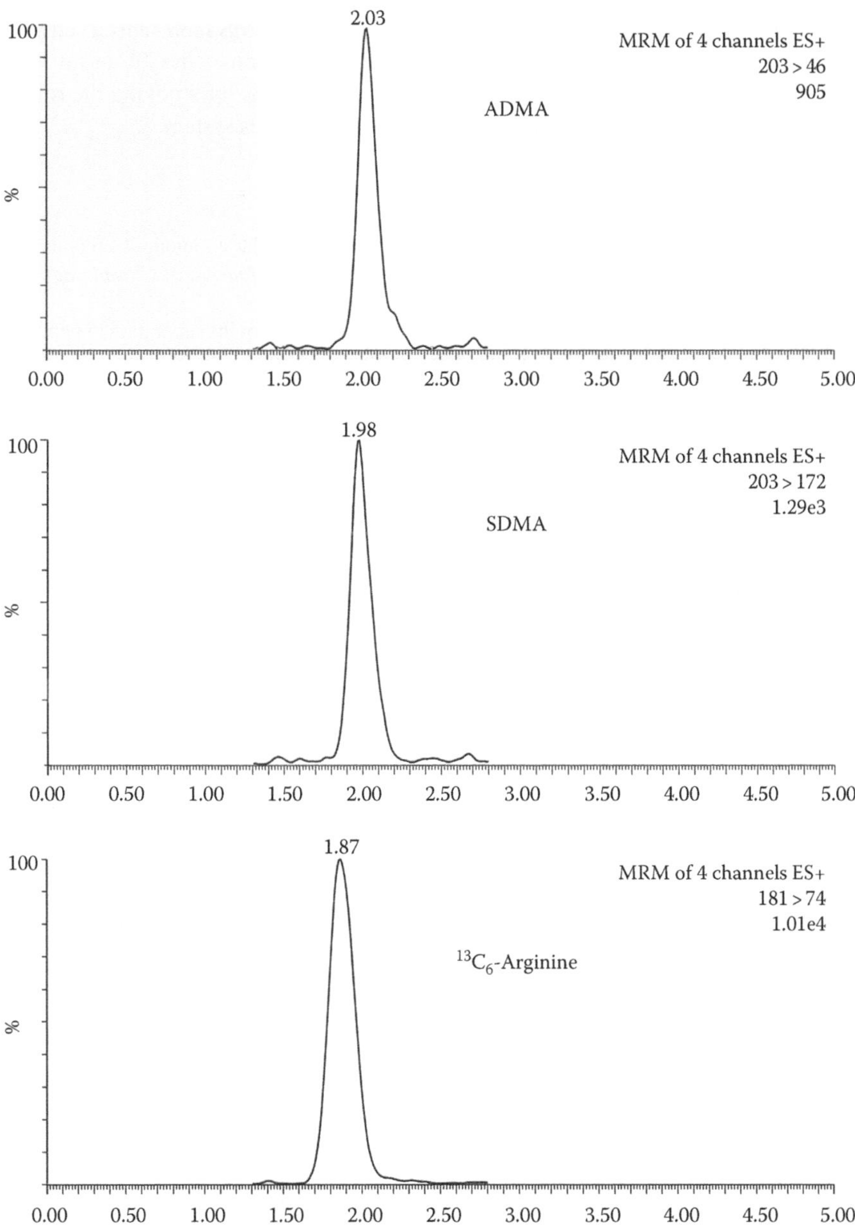

FIGURE 10.8 Chromatograms of a CSF sample. (Reprinted from Zotti, M. et al., *J. Sep. Sci.*, 31, 2511, 2008. With permission from Wiley-VCH Verlag Gmbh & Co. KGaA.)

preparation and with a total chromatographic run time of 3 min or less.[38] In this study, the combined advantages of this technique with microdialysis is useful to realize an in vivo approach; the present findings suggest that HILIC-MS/MS represents a sensitive technique for the simultaneous monitoring of dimethylarginines in CSF of rat PFC. Interestingly, these results could have important implications for studying the role of ADMA and SDMA in several diseases of the central nervous system.

REFERENCES

1. Guo, Y. and Gaiki, S. 2005. Retention behavior of small polar compounds on polar stationary phases in hydrophilic interaction chromatography. *Journal of Chromatography A* 1074:71–80.
2. Alpert, A.J. 1990. Hydrophilic-interaction chromatography for the separation of peptides, nucleic acids and other polar compounds. *Journal of Chromatography* 499:177–196.
3. Olsen, B.A. 2001. Hydrophilic interaction chromatography using amino and silica columns for the determination of polar pharmaceuticals and impurities. *Journal of Chromatography A* 913:113–122.
4. Naidong, W. 2003. Bioanalytical liquid chromatography tandem mass spectrometry methods on underivatized silica columns with aqueous/organic mobile phases. *Journal of Chromatography B* 796:209–224.
5. Kadar, E.P. et al., 2008. Rapid determination of the applicability of hydrophilic interaction chromatography utilizing ACD labs log D suite: A bioanalytical application. *Journal of Chromatography B* 863:1–8.
6. Teerlink, T. 2007. HPLC analysis of ADMA and other methylated L-arginine analogs in biological fluids. *Journal of Chromatography B* 851:21–29.
7. Gehrig, P.M. et al. 2004. Fragmentation pathways of N(G)-methylated and unmodified arginine residues in peptides studied by ESI-MS/MS and MALDI-MS. *Journal of American Society of Mass Spectrometry* 15:142–149.
8. Bedford, M.T. and Richard, S. 2005. Arginine methylation an emerging regulator of protein function. *Molecular Cell* 18:263–272.
9. Valtonen, P. et al. 2005. Comparison of HPLC method and commercial ELISA assay for asymmetric dimethylarginine (ADMA) determination in human serum. *Journal of Chromatography B* 828:97–102.
10. Vallance, P. et al. 1992. Accumulation of an endogenous inhibitor of nitric oxide synthesis in chronic renal failure. *Lancet* 339:572–575.
11. Giraldi-Guimarães, A. Bittencourt-Navarrete, R.E., and Mendez-Otero, R. 2004. Expression of neuronal nitric oxide synthase in the developing superficial layers of the rat superior colliculus. *Brazilian Journal of Medical and Biological Research* 37:869–877.
12. Akyol, O. et al. 2004. Nitric oxide as a physiopathological factor in neuropsychiatric disorders. *In Vivo* 18:377–390.
13. Cooke, J.P. and Dzau, V.J. 1997. Nitric oxide synthase: Role in the genesis of vascular disease. *Annual Review of Medicine* 48:489–509.
14. Fedele, E., and Raiteri, M. 1999. In vivo studies of the cerebral glutamate receptor/NO/cGMP pathway. *Progress in Neurobiology* 58:89–120.
15. Lluch, P. et al. 2006. Accumulation of symmetric dimethylarginine in hepatorenal syndrome. *Experimental Biology and Medicine (Maywood)* 231:70–75.
16. Martens-Lobenhoffer, J. and Bode-Böger, S.M. 2006. Fast and efficient determination of arginine, symmetric dimethylarginine, and asymmetric dimethylarginine in biological fluids by hydrophilic-interaction liquid chromatography-electrospray tandem mass spectrometry. *Clinical Chemistry* 52(3):488–493.

17. Kushnir, M.M. et al. 2005. Assessing analytical specificity in quantitative analysis using tandem mass spectrometry. *Clinical Biochemistry* 38:319–327.
18. Schwedhelm, E. et al. 2005. Liquid chromatography-tandem mass spectrometry method for the analysis of asymmetric dimethylarginine in human plasma. *Clinical Chemistry* 51:1268–1271.
19. Paglia, G., D'Apolito, O., and Corso, G. 2008. Precursor ion scan profiles of acylcarnitines by atmospheric pressure thermal desorption chemical ionization tandem mass spectrometry. *Rapid Communication in Mass Spectrometry* 22(23):3809–3815.
20. Song, Q. and Naidong, W. 2006. Analysis of omeprazole and 5-OH omeprazole in human plasma using hydrophilic interaction chromatography with tandem mass spectrometry (HILIC-MS/MS)—Eliminating evaporation and reconstitution steps in 96-well liquid/liquid extraction. *Journal of Chromatography B* 830:135–142.
21. Naidong, W. et al. 2004. Direct injection of 96-well organic extracts onto a hydrophilic interaction chromatography/tandem mass spectrometry system using a silica stationary phase and an aqueous/organic mobile phase. *Rapid Communication in Mass* Spectrometry 18:2963–2968.
22. Li, A.C. et al. 2004. Direct injection of solid-phase extraction eluents onto silica columns for the analysis of polar compounds isoniazid and cetirizine in plasma using hydrophilic interaction chromatography with tandem mass spectrometry. *Rapid Communication in Mass Spectrometry* 18:2343–2350.
23. Naidong, W. et al. 2002. Liquid chromatography/tandem mass spectrometric bioanalysis using normal-phase columns with aqueous/organic mobile phases—A novel approach of eliminating evaporation and reconstitution steps in 96-well SPE. *Rapid Communication in Mass Spectrometry* 16:1965–1975.
24. Shou, W.Z. and Naidong, W. 2005. Simple means to alleviate sensitivity loss by trifluoroacetic acid (TFA) mobile phases in the hydrophilic interaction chromatography-electrospray tandem mass spectrometric (HILIC-ESI/MS/MS) bioanalysis of basic compounds. *Journal of Chromatography B* 825:186–192.
25. Dong, L. and Huang, J. 2007. Effect of temperature on the chromatographic behavior of epirubicin and its analogues on high purity silica using reversed-phase solvents. *Chromatographia* 65:519–526.
26. D'Apolito, O. et al. 2008. Development and validation of a fast quantitative method for plasma dimethylarginines analysis using liquid chromatography–tandem mass spectrometry. *Clinical Biochemistry* 41:1391–1395.
27. Burns, D.T., Danzer, K., and Townshend, A. 2002. Use of the terms "recovery" and "apparent recovery" in analytical procedures (IUPAC recommendations 2002). *Pure and Applied Chemistry* 74:2201–2205.
28. Martens-Lobenhoffer, J. and Bode-Böger, S.M. 2006. Measurement of asymmetric dimethylarginine (ADMA) in human plasma: From liquid chromatography estimation to liquid chromatography–mass spectrometry quantification. *European Journal of Clinical Pharmacology* 62:61–68.
29. Kirchherr, H. and Kühn-Velten, W.N. 2005. HPLC–tandem mass spectrometric method for rapid quantification of dimethylarginines in human plasma. *Clinical Chemistry* 51:249–252.
30. Bishop, M.J. et al. 2007. Direct analysis of un-derivatized asymmetric dimethylarginine (ADMA) and L-arginine from plasma using mixed-mode ion-exchange liquid chromatography–tandem mass spectrometry. *Journal of Chromatography B* 859:16416–16419.
31. Bode-Böger, S.M., Scalera, F., and Ignarro, L.J. 2007. The L-arginine paradox: Importance of the L-arginine/asymmetrical dimethylarginine ratio. *Pharmacology & Therapeutics* 114:295–306.
32. Fleck, C. et al. 2001. Serum concentrations of asymmetric (ADMA) and symmetric (SDMA) dimethylarginine in renal failure patients. *Kidney International* 59:S14–S18.

33. Fleck, C. et al. 2003. Serum concentrations of asymmetric (ADMA) and symmetric (SDMA) dimethylarginine in patients with chronic kidney diseases. *Clinica Chimica Acta* 336:1–12.
34. Martens-Lobenhoffer, J. and Bode-Böger, S.M. 2007. Chromatographic–mass spectrometric methods for the quantification of L-arginine and its methylated metabolites in biological fluids. *Journal of Chromatography B* 851:30–41.
35. Vishwanathan, K. et al. 2000. Determination of arginine and methylated arginines in human plasma by liquid chromatography–tandem mass spectrometry. *Journal of Chromatography B* 748:157–166.
36. Benveniste, H. et al. 1989. Determination of brain interstitial concentrations by microdialysis. *Journal of Neurochemistry* 52:1741–1750.
37. Lindefors, N., Amberg, G., and Ungerstedt, U. 1989. Intracerebral microdialysis: I. Experimental studies of diffusion kinetics. *Journal of Pharmacological Methods* 22:141–156.
38. Zotti, M. et al. 2008. Determination of dimethylarginine levels in rats using HILIC-MS/MS: An in vivo microdialysis study. *Journal of Separation Science* 31:2511–2515.

11 Polar Functional Groups for HILIC Method

Zhigang Hao

CONTENTS

11.1 INTRODUCTION

Using polar stationary phases such as bare silica paired with aqueous-organic mobile phases like water-acetonitrile (H_2O-ACN) to separate hydrophilic ingredients can be traced back to the 1970s.[1,2] The name hydrophilic interaction chromatography (HILIC) was coined by Alpert in 1990.[3] Dr. Weng initiated the first HILIC review on bioanalytical application with underivatized silica column and mass spectrometer (MS) detection in 2003.[4] Bare silica has no secondary modification on its surface, and it is still the most popular stationary phase in HILIC applications due to column stability and low back-pressure. Different types of silica materials (types A, B, and C) have been developed for method selectivity. The rapid growth of HILIC chromatography began around 2003 and since then hundreds of applications with many new stationary phases have been successfully applied to polar analyte separation. Hemstrom and Irgum constructed review of the HILIC field in 2006 and attempted to ascertain the extent to which partition or adsorption accounted for the separation mechanism.[5]

In 2008, Prof. Laemmerhofer edited a special HILIC issue in *Journal of Separation Science*. Prof. Jandera reviewed HILIC stationary phases and differentiated the selectivity between HILIC and reverse phase modes.[6] It was postulated that the mobile phase similarity in both HILIC and reverse phase systems provided an ideal combination for two-dimensional (2D) chromatographic separation. Polar analytes are separated in HILIC mode, usually in the first dimension, whereas less hydrophilic ones in reverse phase mode. Univariate and multivariate approaches in HILIC method development were summarized by Prof. Vander Heyden's group.[7] Parameters including stationary phase, column temperature, mobile phase composition, and flow rate for the detection of biological and nonbiological samples were discussed in this review. Our group discussed the importance of column temperature and mobile phase for HILIC selectivity.[8] Drs. Nguyen and Schug highlighted the advantages of HILIC separation coupled with ESI-MS detection.[9] The HILIC-MS method application in biological assays was reviewed by Dr. Hsieh.[10] Several new HILIC studies were also included in this special issue. These articles gave readers an updated and representative overview of what can be accomplished using HILIC mode, the typical application fields, and the benefits for polar analytes as compared to traditional reverse phase chromatography.

The need to identify novel analytes such as those found in biological fluid metabolites and synthetic mixtures leads to analytical method development. With research advances in genomics, proteomics, and metabolomics, more and more small polar components have appeared in the analytical laboratory. These compounds are usually not retained well enough in a reverse phase high performance liquid chromatography (HPLC) system to satisfy the recommendation from the Center for Drug Evaluation and Research (CDER), which suggests a minimum capacity factor (k') value of >2.0 to ensure adequate separation of the analytes from un-retained matrix components.[11] However, these analytes are retained more efficiently in HILIC mode, and the use of a high ratio of organic solvent such as ACN in the mobile phase is compatible with MS detection.

For reverse phase chromatography, retention is considered to be (ideally) controlled by partition since most functional groups such as silanols are covered/shielded by carbon chain materials (C_8 or C_{18}) in the stationary phase. Compared to reverse phase HPLC, columns for HILIC have a much wider variety of functional groups such as silanol, amine, amide, diol, cyano, poly-succinimide, sulfoalkylbetaine, and zirconia. All polar functionalities including anionic and cationic moieties on the HILIC packing material surface can absorb some water (0.5%–1.0%) to form a stagnant water-enriched layer between the mobile and stationary phases, especially when the water ratio is low (usually less than 40%) in the mobile phase. This layer is immobilized and can be considered as a portion of the stationary phase. The transition between adsorption and partition mechanisms is probably continuous as the water content in the mobile phase gradually increases.[6,12] The term "HILIC" refers to practical application possibilities rather than to a special retention mechanism. Of course, this is the simplest way to consider the HILIC separation compared to reverse phase chromatography. The more hydrophilic an analyte, the more it associates with the stagnant water-rich layer and the later it elutes out. The functional moieties in the structure that convey this property to highly polar compounds are either charged groups or groups capable of entering strong dipolar or hydrogen bonds. An empirical formula can be used to describe analyte polarity:

$$\text{Polarity} = \frac{(\text{number of polar group})}{(\text{number of carbon})} \tag{11.1}$$

where the polar group can be ionic, protic, CN, C=O, and CONH. The analyte can become very hydrophilic (the polarity is high) when a large number of polar groups relative to the number of carbons are present in a structure. A significant characteristic of polar analytes is that they contain multiple polar groups. If only partition dominates the HILIC separation, we can use the empirical formula (11.1) to predict the elution order, which will be reversed compared to that in reverse phase chromatography. However, the HILIC stationary phase is different from reverse phase materials in that polar functional groups bound on a HILIC column are not usually highly covered or shielded. They can be exposed directly to the mobile phase and even to the polar functional groups on analyte structures inside the stagnant liquid layer. The variety of polar functional groups bound on a HILIC phase exhibit more significant, strong, and different interactions with the hydrophilic functional groups on analyte structures. Some authors consider these secondary interactions undesirable,[13] whereas others utilized them for better method selectivity.[14,15] The structural diversity of HILIC surface chemistries provides analysts with a lot of opportunity for better separation of polar analytes in terms of column selection and the manipulation of the experimental conditions. So far, less attention has been paid to electrostatic (or called as ionic), hydrogen-bond, and hydrophobic interactions for HILIC retention capability. More importantly, these interactions specifically correspond to the different functional group moieties in both analyte and HILIC phase structures. An electrostatic interaction (attraction or repulsion) is produced primarily by either cations or anions from both analytes and stationary phases. Experimentally, mobile phase pH, buffer species, concentration, and column temperature can impact upon this type of interaction. The hydrogen bond usually can be generated by protic groups, and its interaction strength can be affected by varying the mobile phase composition (either protic or aprotic organic solvents). The hydrophobic interaction comes mainly from nonpolar moieties in an analyte structure, and it becomes relatively strong only when a high ratio of water content (>40%) is present in the mobile phase. It is not a typical HILIC separation condition and will not be discussed in this chapter.

When polar analytes from different projects are sent to an analytical laboratory, columns are selected based on analyte structures. When an ionic group (either positive or negative) is present in an analyte, logically the anionic or cationic stationary phase under a designed experimental condition should be selected to enhance or reduce the analyte retention due to electrostatic attraction or repulsion. If multiple protic groups are presented in an analyte such as a carbohydrate, a protic stationary phase such as an amide column might be used for retention on the column due to strong hydrogen-bond interaction. Different polar functional groups from both analytes and stationary phases can exhibit varying levels of interaction strength in a HILIC separation. The most polar analytes usually contain more than one polar functional group and each group can contribute to some degree. The individual interaction strength can be manipulated with different experimental conditions such as column selection, pH, buffer concentration, mobile phase composition, and column temperature. This chapter focuses on specific interactions between polar functional groups in both analytes and the stationary phase. For better understanding, ionic and nonionic groups will be reviewed and discussed in some detail separately.

11.2 IONIC FUNCTIONAL GROUPS IN HILIC SEPARATION

An ionic compound is a chemical compound in which ions are held together by the strong electrostatic force between oppositely charged groups. Small ionic compounds tend to dissolve in polar solvents like water and form strong electrostatic bonds. Usually, the positive-to-positive or negative-to-negative ions exhibit an electrostatic repulsive interaction. In a HILIC system, the ionic functional groups, like other polar groups, prefer to stay inside the stagnant water-rich layer and can interact with ions either inside the stagnant layer (usually from buffer) or bound on the stationary phase by either electrostatic attraction or repulsion. The partition retaining power of an analyte can be determined by the empirical formula (Equation 11.1), as described earlier. The more polar groups an analyte possesses, the stronger its retaining power. The retention strength from ionic interaction can be determined by several factors. The first factor is the presence of ionic functional group(s) (either cationic or anionic) in the analyte structures or bound on the stationary phase under a specific pH condition. Certainly, the silanol, amine, and zwitterion chromatography (ZIC) stationary phases should be considered first in the HILIC separation of ionic analytes. The secondary factor is inclusion of buffer salts in the separation system. Different buffer species and their concentration could affect the ionic interaction between analytes and the stationary phase. The third factor is column temperature when a significant analyte transferring enthalpy is involved between the dynamic mobile phase and the stagnant water-rich layer. These three factors for ionic analyte separation will be reviewed and discussed in the following sections. The organic solvent compositions of the mobile phase can impact on ionic group retention in a HILIC system. However, this is more relevant to the hydrogen-bonding retention mechanism and will be reviewed in the section of nonionic functional groups in HILIC separation.

11.2.1 Effect of Stationary Phase Composition

The role and classification of the stationary phase in HILIC separation efficiencies have been reviewed by Prof. Tanaka's group.[16] Numerous ionic analytes, including acids, bases, amino acids, peptides, nuclei bases, and nucleosides, have been separated on bare silica, amino-silica, amide-silica, poly(succinimide), sulfoalkylbetaine, diol, cyano, cyclodextrin, and triazol columns. The application of the cyano-silica column to the HILIC mode is still limited probably because the cyano group does not have sufficient association with water molecules to generate an effective water-rich layer for partitioning. The residual silanol groups under the cyano-group perhaps are unable to stimulate accumulation of water due to the surface shielding by organic ligands.[16–18]

In general, different HILIC stationary phases exhibit enough variability to retain a wide spectrum of analytes. For example, bare silica columns have been successfully applied for the separation of many nitrogen-containing analytes such as atenolol,[19] carvedilol,[20] levofloxacin,[21] doxazosin,[22] donepezil,[23] and glycyl-sarcosine.[24] These nitrogen-containing analytes usually provide a positive charge under an acidic condition. However, electrostatic interaction between basic analytes and the different bare silica materials has not been discussed intensively enough to make an informed decision on column selection. Type-A silica materials usually have a lower average pK_a value and should provide a stronger electrostatic attraction to the positive charged group on polar analytes under most pH conditions. When this attraction is the secondary retaining power in the HPLC separation, tailing phenomenon can be observed. To eliminate interaction with the strong acidic silanol groups on the silica surface, newer silica materials, so-called types B and C, were developed. Silanol groups on type-B and type-C silica surfaces usually have a higher average pK_a value. However, when the primary partition in HILIC separation

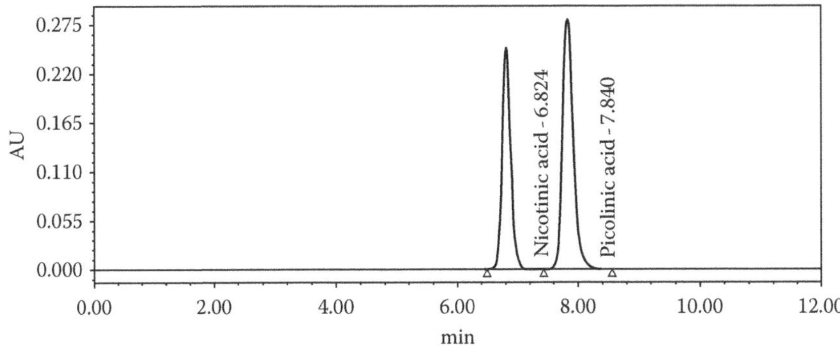

FIGURE 11.1 Representative chromatogram of nicotinic and picolinic acids at 70:30 ACN/buffer with a Phenomenex Luna NH_2 column. The buffer was made by equal volume of 200 mM ammonium acetate and 100 mM glacial acetic acid in water. (Adapted from Christopherson, M.J. et al., *J. Liq. Chromatogr. Relat. Technol.*, 29, 2545, 2006. With permission.)

cannot provide enough retention capability, the electrostatic attraction from type-A silica materials can provide a very powerful retention for basic analytes compared with type-B and type-C silica materials. Of course, type-A silica may provide a stronger electrostatic repulsion to acidic functional groups, resulting in very weak analyte retention.

A comparison of nicotinic acid separation on different columns is a good example of electrostatic interaction in HILIC chromatography. Different types of stationary phases, Thermo Hypersil silica column and Phenomenex Luna NH_2, were used to separate nicotinic acid. Nicotinic acid retention on the amino phase (Luna NH_2) column shown in Figure 11.1 is much stronger than its retention on the Hypersil silica column shown in Figure 11.2, even though a much stronger mobile phase containing 30% water was used with the amino phase column. The different retentions from Figures 11.1 and 11.2 can be explained by electrostatic interaction of functional groups between analytes and the stationary phase. In Figure 11.2, the carboxyl group (RCOOH) on analytes was deprotonated and the amine group (RNH_2) on column was protonated under the experimental pH condition.[15] The negative ion ($RCOO^-$) can be electrostatically attracted by the positive ion (NH_3^+). Nicotinic acid and picolinic acid can be retained longer, even though a high ratio of water (30%) was presented in the mobile phase. However, Hypersil silica column in Figure 11.2 is a more acidic silica type-A material. The silanol group (Si-OH) bound on this material surface can be deprotonated into a negative ion ($Si\text{-}O^-$) under the pH condition employed. The negative silanol ions give a repulsive interaction to the negative carboxyl group ($RCOO^-$) on analytes and result in very short retention time even though a low ratio of water (2%) was presented in the mobile phase.

The repulsive interaction between silanol and carboxyl groups is further elucidated by the data in Table 11.1.[14] A Hypersil silica column was used. The chemical structures of glycine (G), diglycine (DG), triglycine (TG), *N*-[1-deoxy-D-glycose-1-yl]-glycine (GG), *N*-[1-deoxy-D-glycose-1-yl]-diglycine (GDG), and *N*-[1-deoxy-D-glycose-1-yl]-triglycine (GTG) are shown in Figure 11.3. The equilibrium reactions for ionic functional groups in both analytes and stationary phase are shown in the following formula:

$$Si^{\ominus} \rightleftarrows Si\text{-}OH \qquad RCOO^{\ominus} \rightleftarrows RCOOH \qquad RNH_2 \rightleftarrows RNH_3^+$$

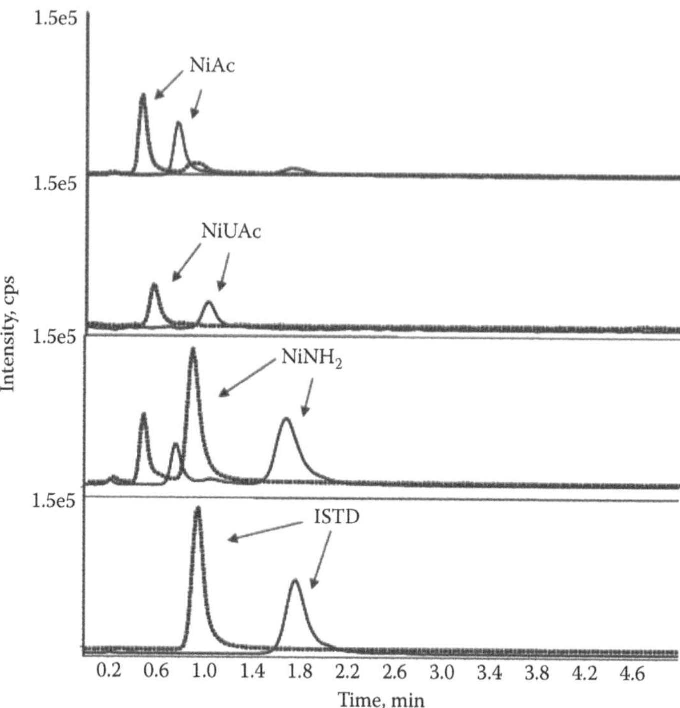

FIGURE 11.2 HILIC-APCI-MS/MS chromatograms for nicotinamide ($NiNH_2$), nicotinic acid (NiAc), 6-methylnicotinic acid (ISTD), and nicotinuric acid (NiUAc), using high organic mobile phase containing 2% (solid line) or 5% (dotted line) water with a Thermo Hypersil silica column. (Adapted from Hsieh, Y. and Chen, J., *Rapid Commun. Mass Spectrom.*, 19, 3031, 2005. With permission.)

All equilibrium reactions will move to the right-hand side when mobile phase acidity increases. Amine $NH_{(1-2)}$ groups in the analyte structures should be fully protonated to positive ammonium $NH_{(2-3)}^+$ groups based on the pH values of different formic acid concentrations in the mobile phase (from 0.1% to 0.7%) and the pK_a values of G, DG, TG, GG, GDG, and GTG listed in Table 11.1, even though the real pH values in aqueous organic mobile phase are slightly different from those in water media. Because their pK_a values are 3.5, the neutral form of carboxyl groups in the DG, TG, GDG, and GTG structures will be dominant under mobile phase pH conditions from 2.20 to 2.68. The ionic form of silanol groups on the silica surface will decrease when the formic acid concentration increases because these groups usually have a broad range of pK_a values. As expected, DG, TG, and their Amadori compounds, GDG and GTG, are retained less when formic acid content is increased because less ionic interaction between Si-O$^-$ and $NH_{(2-3)}^+$ results in a weaker retention. It was interesting to note that retention times of G and GG were completely different from the other four analytes. They were retained longer when the formic acid content was increased from 0.1% to 0.7% in the mobile phase. The observation can be explained by the equilibrium reactions described earlier. The pK_a of silanol groups

TABLE 11.1
Retention Times of Six Analytes, G, DG, TG, GG, GDG, and GTG under the HPLC Condition of Hypersil Silica Column, 100 × 1 mm, Particle Size: 3 μm with Mobile Phase 2% Water and 98% Methanol and the Variable Formic Acid Contents in Both Solvents from 0.1% to 0.7% and Column Temperature at 30°C

	0.1%FA (pH:2.68)[a]	0.2%FA (pH:2.50)	0.3%FA (pH:2.41)	0.4%FA (pH:2.34)	0.5%FA (pH:2.28)	0.6%FA (pH:2.25)	0.7%FA (pH:2.20)
G (pK_a: 2.34, 9.6)	3.92	4.74	5.15	5.46	6.69	6.90	6.80
DG (pK_a: 3.5, 8.0)	10.62	10.52	10.31	9.90	8.87	8.67	8.26
TG (pK_a: 3.5, 8.1)	13.62	13.11	12.49	11.87	10.43	10.12	9.61
GG (pK_a: 3.5, 8.5)[b]	3.05	3.15	3.35	3.35	4.38	4.69	4.79
GDG (pK_a: 3.5, 8.0)[b]	8.20	7.59	7.48	7.28	6.87	6.56	6.35
GTG (pK_a: 3.5, 8.0)[b]	10.28	9.15	8.63	8.22	7.29	7.09	6.99

Source: Adapted from Hao, Z. et al., *J. Chromatogr. A*, 1147, 165, 2007. With permission.

[a] The pH values in this row were measured in water media and the real pH values in aqueous organic mobile phase could be slight different in water media.

[b] The pK_a data for GG, GDG, and GTG were not measured and the data were just assumed to be similar to G, DG, and TG, respectively.

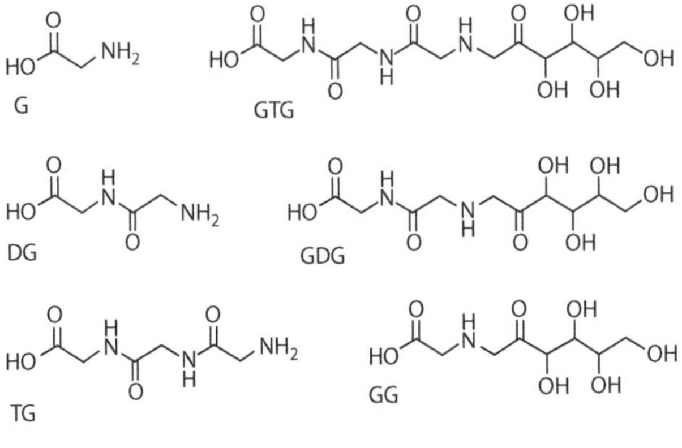

FIGURE 11.3 Structures of G, DG, TG, GG, GDG, and GTG.

is the average value of all silanols present on the silica surface. When the pH falls below 2.1, the ionization of silanols is suppressed with the exception of the most acidic ones.[25] The silanol groups on type-A silica provide a lower average pK_a value. A significant amount of the ionic form, $Si\text{-}O^-$, is presented on the stationary phase surface under the pH conditions listed in Table 11.1. At a pH of 2.68, the ionic form of carboxyl group, COO^-, is dominant in the G and GG structures due to their pK_a values of 2.34 and 2.5, respectively. The electrostatic repulsion between COO^- and $Si\text{-}O^-$ results in less retention. When formic acid content is increased, both carboxyl acid and silanol ionizations were suppressed. The decreased repulsive force provided a longer retention time. Indeed, the retention times were increased from 3.92 and 3.05 to 6.80 and 4.79 min for G and GG, respectively, when the formic acid concentration was changed from 0.1% to 0.7% (Table 11.1). A similar situation was found in sodium cromoglicate separation on an Atlantis HILIC-Si column. Its retention time became longer when the mobile phase pH increased from 4.2 to 5.8.[26] Usually, a bare silica column cannot provide a very strong retention capability to acidic groups in polar analytes under an acidic condition without buffer applications.[27,28]

The repulsive interaction is not only present in the negative-to-negative groups but also in the positive-to-positive groups between analytes and the stationary phase. Dr. Liu et al. found that only a Zorbax-NH_2 column caused four hydrazines to be eluted out before the void volume compared to three other columns (YMC-Pack Diol-120-NP, Amide-80, and ZIC-HILIC; see Figure 11.4).[29] The Zorbax-NH_2 column also showed a weak retention to basic 4-(aminomethyl)pyridine and related compounds when compared with six other columns.[18]

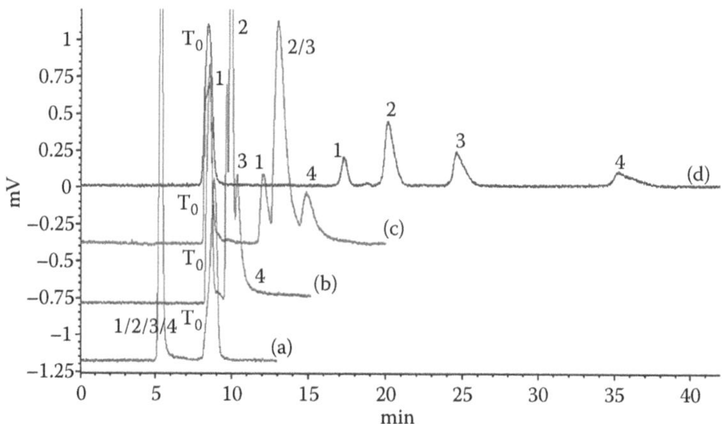

FIGURE 11.4 The separation of hydrazines on different columns. Parameters for the studies included 30°C column temperature, 0.4 mL/min flow rate with splitter, CLND detector system set at 10, 50°C combustion furnace, 50 mL/min argon, 280 mL/min oxygen, 75 mL/min makeup (argon), 30 mL/min ozone, 5°C cooler, gain x1, and 750 V on PMT. The analyte concentrations were about 30–70 μg/mL in water/ethanol (20/80, v/v). Injection volume was 10 μL. Mobile phase was formic acid/water/ethanol (0.5/20/80, v/v/v). 0.1% acetonitrile (v/v) in ethanol was used as a void volume marker. 1: 1,2-dimethylhydrazine, 2: 1,1-dimethylhydrazine, 3: methylhydrazine, and 4: hydrazine. (a) Zorbax NH_2, (b) Diol, (c) Amide-80, and (d) ZIC HILIC. (Adapted from Liu, M. et al., *J. Chromatogr. A*, 1216, 2362, 2009. With permission.)

The secondary retaining force can be an undesired interference in HPLC separation causing tailing or broad peak issues.[13] But it can also be a very important retention mechanism if we understand which column is able to provide such a retaining force for our analytes of interest. Switching from bare silica to an amino-phase for acidic or even basic functional groups should introduce a significant separation capability. More recently, Alpert has coined another chromatographic acronym, ERLIC, for electrostatic repulsion hydrophilic interaction chromatography.[30] The HILIC mobile phase (a high ratio of organic solvents) is combined with an ion-exchange column (a charged stationary phase) in this methodology. The separation strategy in ERLIC superimposes a second mode of chromatography that selectively reduces the retention of analytes that are usually the most strongly retained. Actually, many HILIC columns have this ion-exchange capability. For example, negative residual silanol groups are present in all silica-based HILIC columns,[31] which not only provide an extra retention capability to the basic or positive-charged functional groups but can also generate an electrostatic repulsion to acidic or negative-charged functional groups. The negative residual silanol groups are more significant in type-A silica material. For the amine column, the positive charge on the outer layer of the stationary phase can give a strong electrostatic attraction to acidic analytes, even though the residual silanols are present in the inner layer of the stationary phase. In contrast, the ZIC-HILIC stationary phase has an inner positive quaternary ammonium ion and an outer negative sulfonate ion separated by a three-methylene group, which has been reported for the separation of basic analytes such as hydrazines,[29] mildronates,[17] and acetylcholine.[32] In addition, some degree of electrostatic interaction was found between ionizable analytes and the residual silanol groups under nonionic (neutral) polar stationary phase surfaces such as diol,[33] sorbitol methacrylate,[34] and amide-silica.[35] These initial studies are promising, but more studies are required to better understand the role of electrostatic interaction in HILIC separation.

11.2.2 Effect of Buffer Concentration (Ionic Strength)

The buffer is a very important component when considering ionic interaction. The buffer can be used to control mobile phase pH, and pH can modify the ionization status not only for analytes but for stationary phases as well. Optimizing buffer concentration can also generate the ion strength necessary to mediate the electrostatic interaction between analytes and stationary phase.[36] In addition, buffer composition can affect analyte retention or transition enthalpy between mobile and stationary phases. Retention enthalpy is an important parameter if column temperature needs to be considered for improved analyte separation, which will be discussed in next section.

In general, at a low buffer concentration or a buffer-free condition, ionic analytes are effectively retained on stationary phases containing counter ionic groups. As the buffer concentration increases, a high level of organic composition in the mobile phase could force the buffer ions inside the stagnant water-rich liquid layer. Higher buffer ion concentration would drive more solvated salt ions into this liquid layer and result in an increase in volume or hydrophilicity of the liquid layer. The strength of the electrostatic interaction between analytes and the stationary phase would be weakened with an increase of the water-rich layer volume, thus resulting in a weak electrostatic interaction. If the electrostatic interaction is with counter ions, the electrostatic attraction would be weakened, resulting in a longer retention times. If the charge–charge interaction is with co-ions, the electrostatic repulsion would also be weakened, resulting in shorter retention times. Most experimental data, especially with ammonium acetate or formate buffer, support this rationale.[29,37–39] For example, room temperature ionic liquid (RTIL) imidazolium cations were not eluted (a very strong capacity factor) on diol stationary phase with a salt-free mobile phase due to their strong electrostatic interactions with the negative residual silanols under the diol stationary phase surface.[40] A decrease in capacity factor was observed when the ammonium acetate concentration was raised from 5 to 20 mmol/L (shown in Figure 11.5). The negative residual silanols on a diol phase column was also reported for a uric acid separation.[41] The negative silanol groups exert electrostatic repulsion on the negatively charged acids under the experimental conditions. Increasing the concentration of salt in the mobile phase would reduce this electrostatic repulsion leading to stronger retention.

The retention of anionic nicotinic and picolinic acids on an amine-silica column were reduced by increasing the buffer concentration, as shown in Figure 11.6.[15] In another example, the retention time of acidic analytes (aspirin and salicylic acid) increased by about 20%–40% on amide-silica, HILIC silica, and ZIC-HILIC columns but decreased sharply on the amine-silica column when the ammonium acetate concentration was increased from 5 to 20 mM (shown in Table 11.2).[36] The buffer concentration can contribute to both electrostatic repulsion and attraction between analytes and stationary phases. The ionized residual silanol groups on amide-silica and silica columns can repulse the ionized carboxyl groups on analytes under pH 6.9.

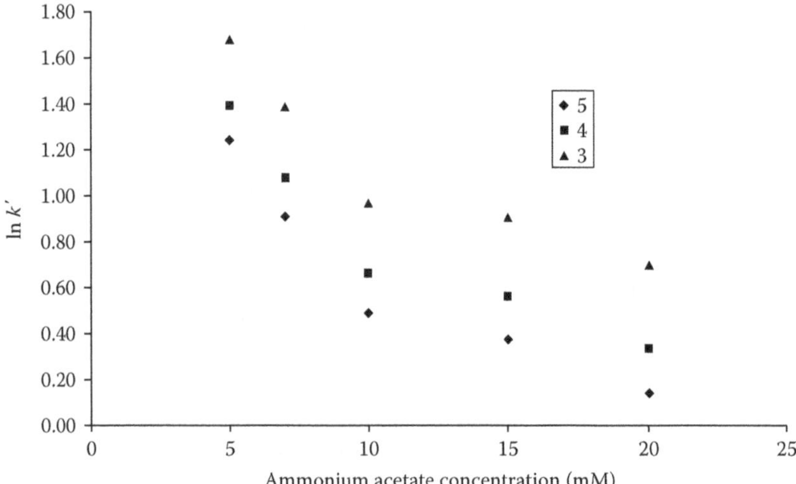

FIGURE 11.5 Plots of ln k' vs. ammonium acetate concentration in the mobile phase for three RTILs: (3) 1-butyl-2-methyl-3-methyl imidazolium bis-(trifluoromethylsulfonyl)-imide, (4) 1-hexyl-3-methyl imidazolium chloride, and (5) 1-methyl-3-octyl imidazolium chloride. Stationary phase: Uptisphere OH; mobile phase: ACN/H_2O with 10 mM ammonium acetate at 0.2 mL/min. (Adapted from Rouzo, G.L. et al., *J. Chromatogr. A*, 1164, 139, 2007. With permission.)

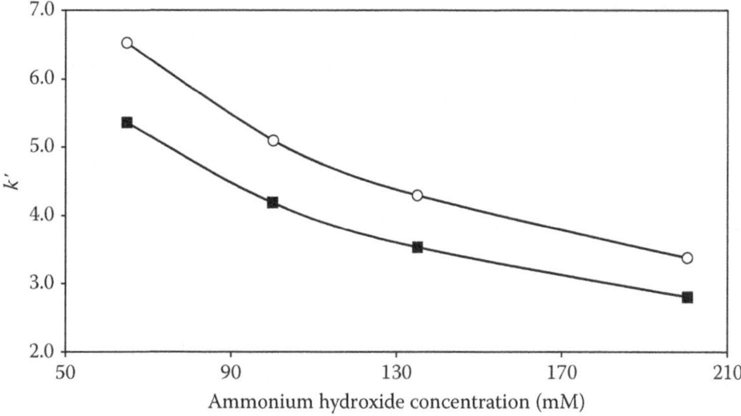

FIGURE 11.6 Effect of buffer concentration on the retention of nicotinic acid (■) and picolinic acid (○). (Adapted from Christopherson, M.J. et al., *J. Liq. Chromatogr. Relat. Technol.*, 29, 2545, 2006. With permission.)

TABLE 11.2
Retention Time of the Model Compounds at Different Ammonium Acetate Concentrations in the Mobile Phase[a]

Column	Concentration (mM)	Salicylic Acid	Aspirin	Cytosine
TSK-Gel Amide-80	5	2.07	3.06	6.84
	10	2.39	3.65	7.19
	20	2.61	4.14	8.01
YMC-Pack NH_2	5	7.59	20.21	6.03
	10	4.72	11.50	6.10
	20	3.56	7.17	6.45
HILIC Silica	5	1.78	2.94	5.51
	10	2.06	3.51	5.78
	20	2.49	4.21	6.62
ZIC-HILIC	5	2.16	2.78	5.52
	10	2.44	3.22	5.59
	20	2.64	3.55	5.98

Source: Adapted from Guo, Y. and Gaiki, S., *J. Chromatogr. A*, 1074, 71, 2005. With permission from Elsevier.

[a] Mobile phase: acetonitrile/ammonium acetate solution (85/15, v/v). Column temperature: 30°C. Flow rate: 1.5 mL/min.

An increase in buffer ions would eliminate such repulsive interaction, resulting in a longer retention time. The ZIC-HILIC stationary phase has an inner positive quaternary ammonium ion and an outer negative sulfonate ion separated by a three-methylene group.[29] The ascorbic acid (vitamin C) retention time on this column became longer when the ammonium acetate concentration increased from 10 to 100 mM.[42]

The electrostatic repulsion between the analyte carboxyl group and sulfonate group on the outer layer of the stationary phase seems to be predominant. Analyte size may block or reduce the electrostatic attraction between analyte carboxyl ions and quaternary ammonium ions on the inner layer of the stationary phase. The ZIC-HILIC column has been reported for the separation of inorganic anions by ion chromatography.[43] The retention behavior of four hydrazine analogues in different ammonium formate concentrations is shown in Figure 11.7. Since the HCl salts of hydrazine and 1,2-dimethylhydrazine were used and RI is a universal detector, the chloride peak was also observed. The retention of positively charged hydrazines decreased with increasing ionic strength, whereas the retention behavior of negatively charged chloride was just the opposite (Figure 11.7a).

In conventional ion-exchange chromatography (IEC) with water media, the relationship of the retention factor k' of an analyte and the buffer concentration $[C]$ is as described below[44,45]

$$\log k' = -s \cdot \log [C] + \text{constant}$$

where s is a constant slope, which is dependent on the overall charge of the analytes and counter-ions. For a singly charged analyte and univalent counterion, the slope should be −1. Ammonium formate concentrations in the range of 5–30 mM were used for drawing the plots of log k' vs. log $[C]$ for hydrazines on a ZIC-HILIC column. The linear relationships observed for all hydrazines and for chloride (coefficient of determination $r^2 \geq 0.97$) are shown in Figure 11.7b. All hydrazines in this

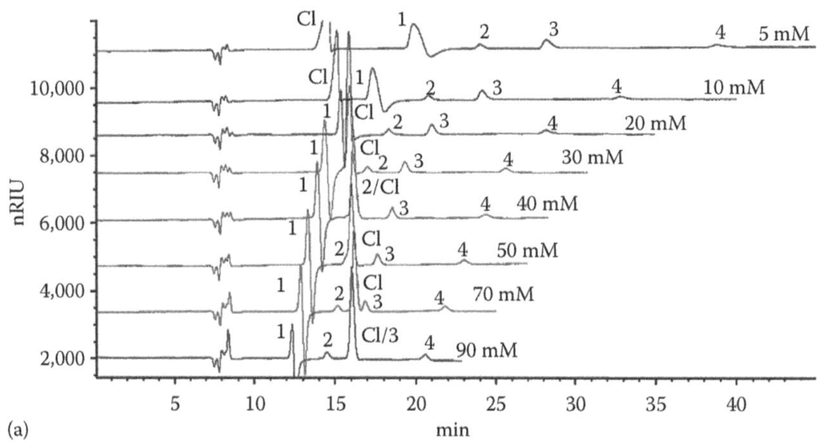

FIGURE 11.7 (a) The effect of ionic strength on the retention of hydrazines on a ZIC-HILIC column. Column temperature was set at 30°C. Isocratic runs with a mobile phase of 5–90 mM ammonium formate buffer pH 3.0/ethyl alcohol (20/80, v/v) (the buffer concentrations refer to the concentration before mixing with organic solvent). Flow rate was 0.4 mL/min with RI detection. The analyte concentrations were about 0.8–1.2 mg/mL in water/ethyl alcohol (20/80, v/v). The injection volume was 1 μL. 0.5% toluene (v/v) in ethyl alcohol was used as a void volume marker. Chloride is from the HCl salts of hydrazine and 1,2-dimethylhydrazine. 1: 1,2-dimethylhydrazine, 2: 1,1-dimethylhydrazine, 3: methylhydrazine, and 4: hydrazine.

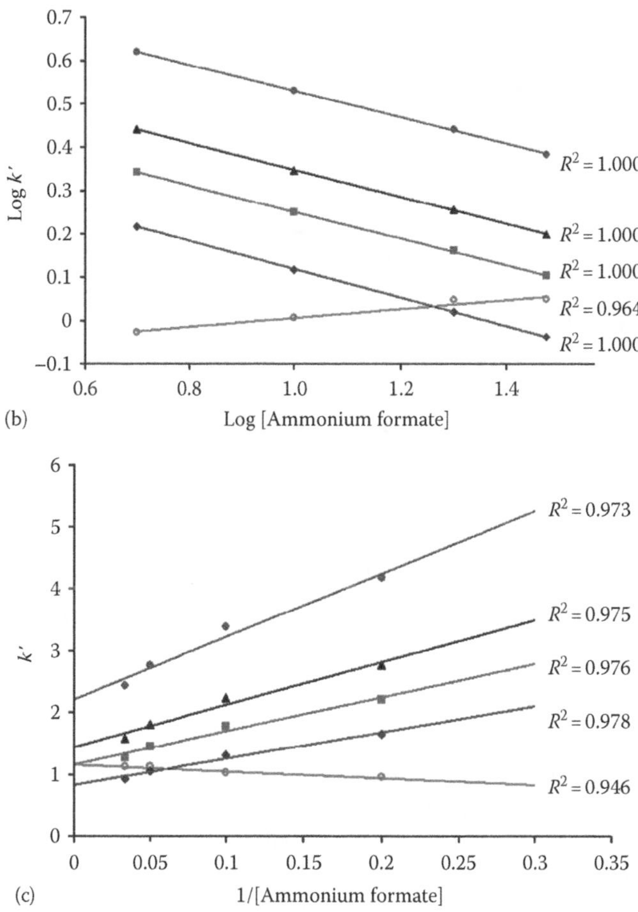

FIGURE 11.7 (continued) (b) Plot of logarithm of retention factor (k') against logarithm of buffer concentration (5–30mM). (c) Plot of k' against the inversed buffer concentration (5–30mM). Hydrazine (•), methylhydrazine (▲), 1,1-dimethylhydrazine (▪), 1,2-dimethylhydrazine (♦), and chloride (○). (Adapted from Liu, M. et al., *J. Chromatogr. A*, 1216, 2362, 2009. With permission.)

study have negative slopes while chloride has a positive slope, which are indicative of the net electrostatic attraction and repulsion with the stationary phase, respectively. The absolute slope values in Figure 11.7b (~0.3 for hydrazines and 0.1 for chloride) indicated other interactions, and mobile phase composition could also contribute to analyte retention. Positive intercepts were observed for all components, which provide evidence for the existence of additional retention mechanisms at "infinite" buffer concentration. Cumulatively, data in Figure 11.7 indicate that both ionic interaction and hydrophilic interaction were involved in the separation of hydrazine analogues and their counterions under an aqueous organic mobile phase condition.

An exceptional example was found with the positive analytes (metformin hydrochloride [MFH], cyanoguanidine [CGD], and melamine [MLN]) separated on an Atlantis HILIC-Si column.[46] Ammonium acetate and ammonium formate buffers proved to

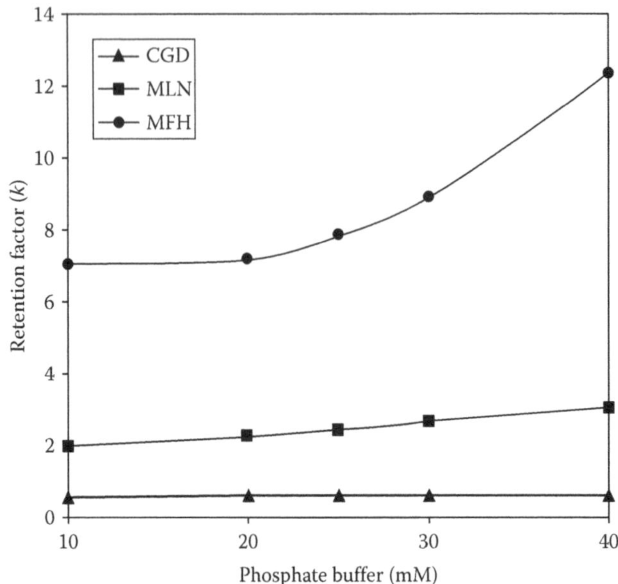

FIGURE 11.8 Effect of sodium buffer strength on the retention of MFH, CGD, and MLN on a 5 μm, 250 × 4.6 Atlantis HILIC-Si column. Acetonitrile/buffer (84/16, v/v) with a pH of 3. Flow rate: 2.0 mL/min. (Adapted from Ali, M.S. et al., *Chromatographia*, 67, 517, 2008. With permission from Elsevier.)

be unsuitable for UV detection at 218 nm. The retention of metformin increased when the concentration of sodium phosphate increased from 10 to 40 mM (Figure 11.8). It is not entirely clear whether sodium phosphate buffer has an impact upon ionic analyte separation in HILIC since the use of this buffer is limited with HILIC method.

Determining how buffer type modifies electrostatic interaction between analytes and stationary phases can be complicated. When ammonium acetate was replaced with ammonium formate in Drs. Guo and Gaiki study described previously,[36] no significant change was found for the neutral analyte, cytosine. Little retention change was observed on amide-silica, HILIC silica, and ZIC-HILIC columns for acidic analytes (aspirin and salicylic acid). A significant retention increase was obtained on an amine-silica column for acidic analytes, which might be related to formic acid size since these smaller sized ions might penetrate through the outer amine layer to suppress silanol ionization in the inner layer of stationary phase. Surprisingly, when ammonium acetate was replaced with ammonium bicarbonate, the latter caused drastic decreases in retention on all four HILIC columns and destroyed the separation of the acids.

The study of buffer species is limited because some buffers that are typically used in reverse phase HPLC may not be suitable for HILIC due to their poor solubility in mobile phase containing a high level of ACN.[36] In addition, the use of buffers is relatively limited when HILIC separation is coupled with mass spectrometry detection. Most nonvolatile buffers produce a very strong ionic suppression on MS detection signals. Therefore, the most common buffer solutions used in HILIC-MS methods are ammonium acetate or formate. Practical HILIC applications coupled with MS detection only use organic acid to control the mobile phase pH because most HILIC stationary phases are stable under an acidic condition.

11.2.3 Effect of Column Temperature

Column temperature has long been recognized as an important parameter in HPLC separation, especially after column and mobile phase selections have been made. Column temperature has a significant impact on analyte diffusivity, mobile phase viscosity, and analyte transferring enthalpy between mobile and stationary phases. In general, temperature elevation increases the diffusion coefficient and decreases the mobile phase viscosity, resulting in narrower peaks. But its drawback is shorter retention times. The total resolution could not be improved much under such a scenario. Only if increased column temperature would simultaneously retain various analytes longer on the column, then retention differences between individual analytes with the narrowed peaks would be enlarged and overall resolution could be improved. To achieve such a scenario, a positive transferring enthalpy (ΔH value), or the negative slope in the van't Hoff plot, needs to be present. Small ionic analytes combined with a counterionic stationary phase in HILIC has provided such an opportunity. When ionic functional groups are present in small polar analytes, column temperature should therefore be considered in HILIC separations.

For example, when anionic aspirin (acetylsalicylic acid) was separated on four different HILIC columns, YMC-pack NH_2, TSK-Gel Amide-80, Atlantis HILIC silica (type-B), and SeQuant ZIC, only the positive amino stationary phase (YMC-pack NH_2) obtained negative slopes or positive ΔH values (Figure 11.9).[36] The positive retention enthalpy indicated an endothermic process of transferring analytes from the mobile phase to the stationary phase. Guo et al. further investigated six organic acids, acetylsalicylic acid, salicylic acid, gentisic acid, hippuric acid, salicyluric acid, and α-hydroxyhippuric acid, on the five columns (Figure 11.10b), and found

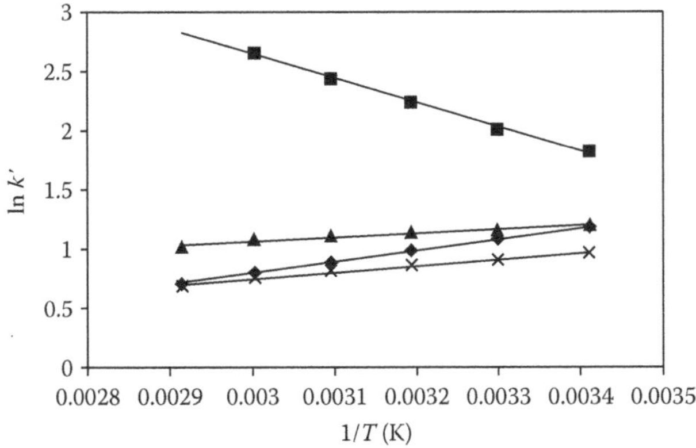

FIGURE 11.9 The van't Hoff plots for aspirin on (♦) TSK-Gel Amide-80, (■) YMC-Pack NH2, (▲) HILIC Silica, and (×) ZIC HILIC columns. Mobile phase: ACN/water (90:10, v/v) containing 10 mM ammonium acetate. (Adapted from Guo, Y. and Gaiki, S., *J. Chromatogr. A*, 1074, 71, 2005. With permission.)

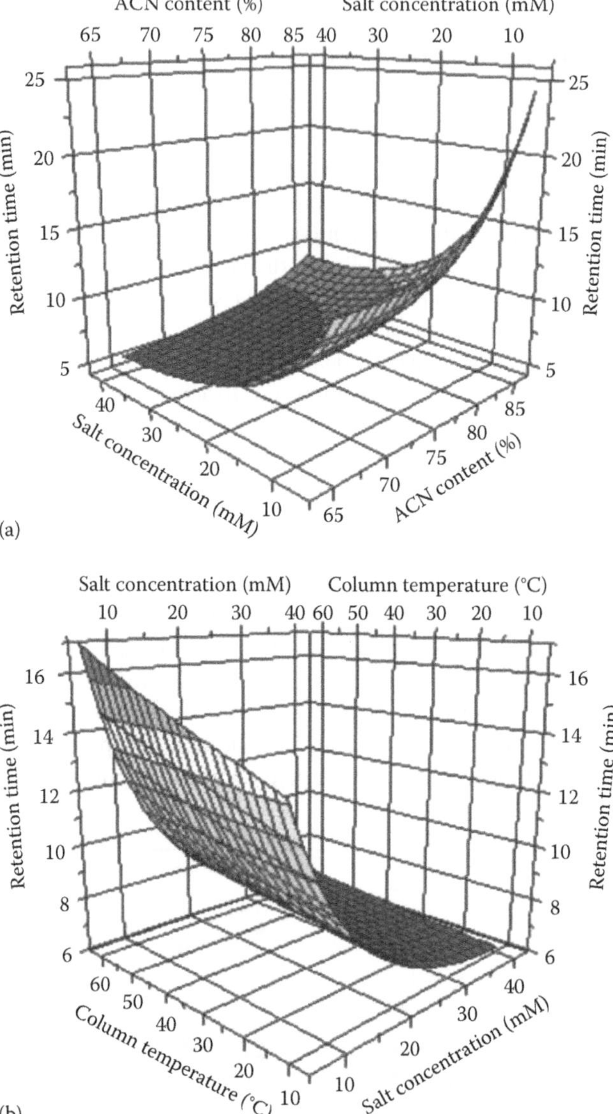

FIGURE 11.10 3D response surfaces for the amino phase (YMC-pack NH_2 column) generated by DOE software. (a) Influence of ACN and ammonium acetate concentration on acidic analyte retention. (b) Influence of ammonium acetate concentration and column temperature on acidic analyte retention. (Adapted from Guo, Y. et al., *Chromatographia*, 66, 223, 2007. With permission.)

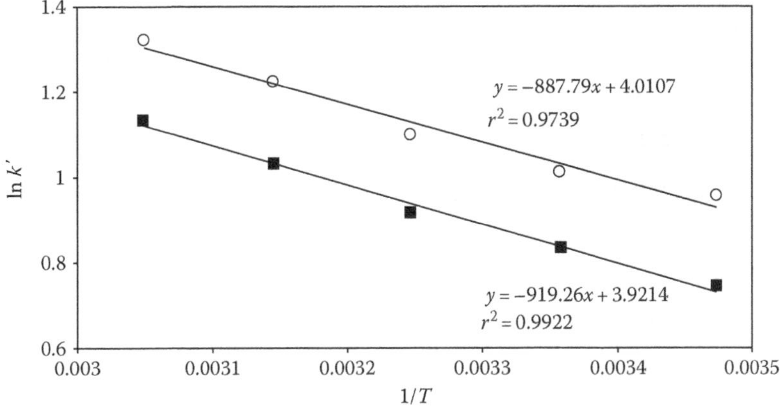

FIGURE 11.11 The van't Hoff curves for nicotinic acid (■) and picolinic acid (○) on a Phenomenex Luna NH_2 column using 70:30 (acetonitrile:buffer). (Adapted from Christopherson, M.J. et al., *J. Liq. Chromatogr. Relat. Technol.*, 29, 2545, 2006. With permission.)

the amino phase had very different response surfaces. The retention time increased when the column temperature was elevated from 10°C to 60°C. Similarly, when picolinic and nicotinic acids were separated using the positive Phenomenex Luna NH_2 stationary phase shown in Figure 11.1, negative slopes from both anionic analytes were obtained in the van't Hoff plots (shown in Figure 11.11).

In contrast, when cationic analytes are separated with an anionic stationary phase, the analytes which transfer enthalpy might also exhibit a better separation. The cationic analytes, G, DG, TG, GG, GDG, and GTG (their structures are in Figure 11.3), were separated by four silica type-A columns (Figure 11.12). Their retention times increased when the column temperature was elevated no matter if methanol or ACN was used in the mobile phase.

When these six analytes were separated on bare silica type-B and type-C columns, much weaker retentions were observed for all six analytes (Figure 11.13). More importantly, the positive slopes or negative transferring enthalpies were obtained in van't Hoff plots when a column temperature program from 5°C to 80°C was applied for analyte separation. A possible rationale is higher average pK_a values attributed to silanol groups on type-B and type-C silica material surfaces.

However, negative slopes were reported for the separation of the basic analogue epirubincin on a Kromasil KR100-5SIL bare silica column (Figure 11.14).[47] More importantly, the column plate number (N) was increased from 32 to 55K plates/m by elevating the column temperature from 25°C to 40°C (about 70% improvement in column efficiency).

For comparison, we also used this type of column to test the impact of column temperature on the separation of six analytes in our laboratory. In contrast, slopes for all six analytes were positive in the van't Hoff plots (Figure 11.15). Buffer solution was not used in our experiments. It is not clear whether the different outcomes are the result of a difference in analyte characteristics or the sodium formate buffer used in epirubicin analogue separation. More studies with Kromasil KR100-5SIL are needed for a more definitive conclusion.

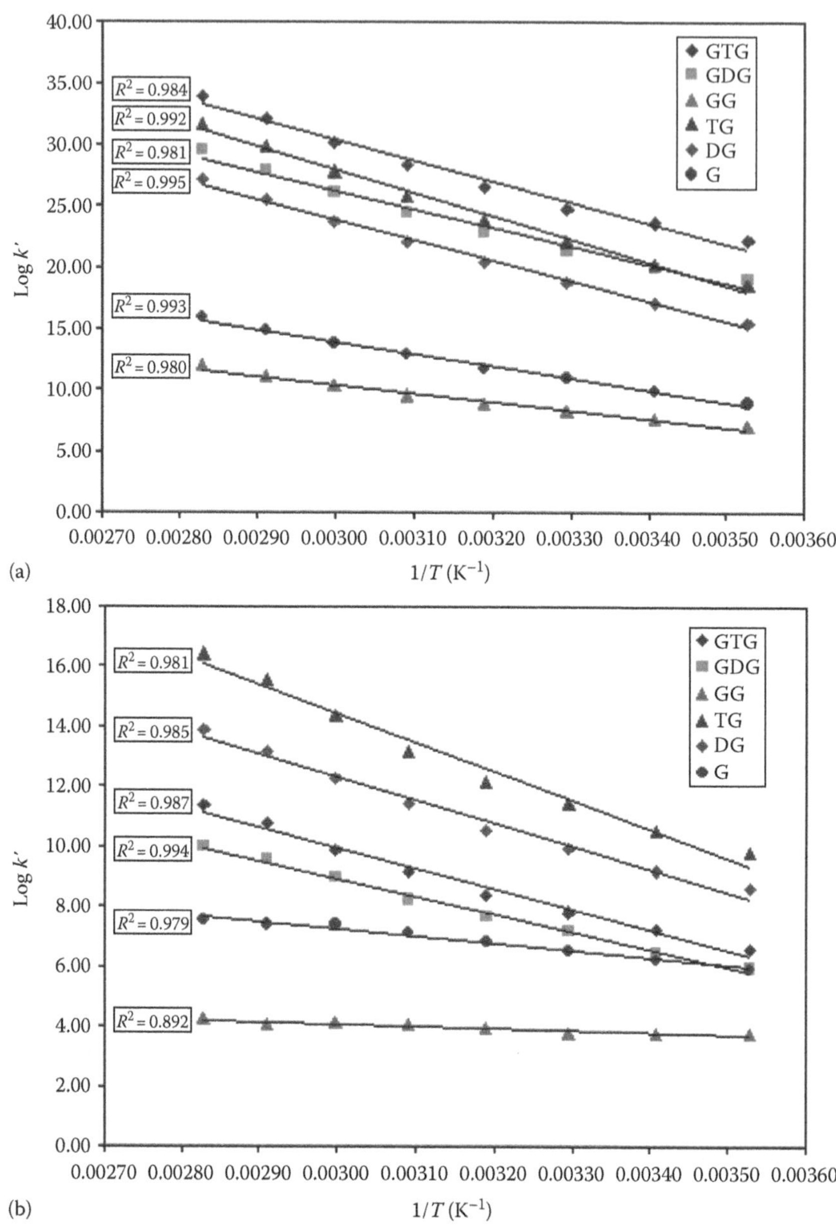

FIGURE 11.12 The van't Hoff plots for G, DG, TG, GG, GDG, and GTG under HILIC condition of Hypersil silica column, 100 × 1.0 mm, particle size of 3 µm with a flow rate of 100 µL/min, the column temperature varying from 5°C to 80°C, mobile phase for top: water/ACN (25:75, v/v) containing 0.4% formic acid and mobile phase for bottom: water/MeOH (2:98, v/v) containing 0.4% formic acid. (Adapted from Hao, Z. et al., *J. Sep. Sci.*, 31, 1449, 2008. With permission from Wiley-VCH Verlag Gmbh & Co. KGaA.)

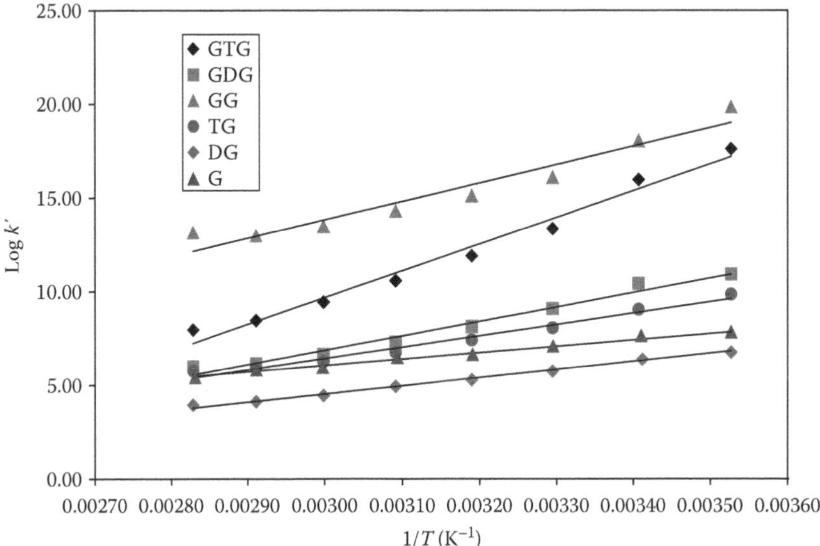

FIGURE 11.13 The van't Hoff plots for six analytes, G, DG, TG, GG, GDG, and GTG under HILIC condition of Atlantis silica column, 50 × 2.1 mm, particle size of 5 μm with a flow rate of 100 μL/min, column temperature varying from 5°C to 80°C, mobile phase of water/ACN (10:90, v/v) containing 0.4% formic acid. (Adapted Hao, Z. et al., *J. Sep. Sci.*, 31, 1449, 2008. With permission from Wiley-VCH Verlag Gmbh & Co. KGaA.)

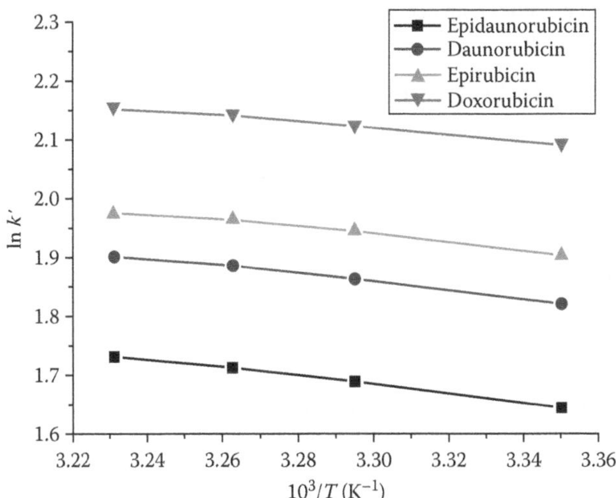

FIGURE 11.14 Effect of temperature upon retention with acetonitrile as modifier. Conditions: Kromasil KR100-5SIL (5 μm); mobile phase: 90% (v/v) ACN in sodium formate buffer (30 mM, pH 2.9) and UV detection at 254 nm for epirubicin and its analogues; flow rate: 1.0 mL/min; injection volume: 20 μL. (Adapted from Dong, L. and Huang, J., *Chromatographia*, 65, 519, 2007. With permission.)

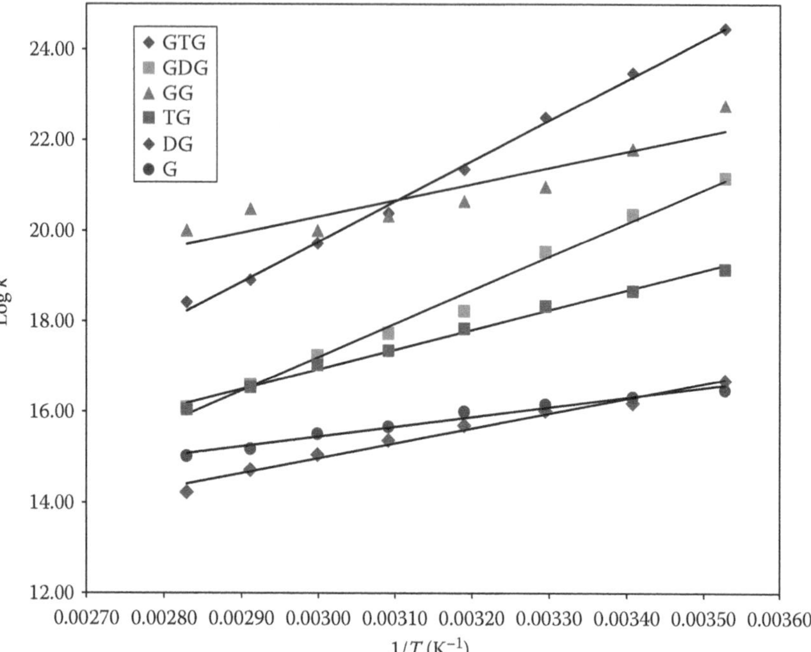

FIGURE 11.15 The van't Hoff plots for six analytes, G, DG, TG, GG, GDG, and GTG, under HILIC condition of Kromasil KR100-5SIL silica column, 100 × 2.0 mm, particle size of 5 μm with a flow rate of 100 μL/min, the column temperature varying from 5°C to 80°C, mobile phase of water/acetonitrile (10:90, v/v) containing 0.4% formic acid. (Adapted Hao, Z. et al., *J. Sep. Sci.*, 31, 1449, 2008. With permission from Wiley-VCH Verlag Gmbh & Co. KGaA.)

11.3 NONIONIC POLAR FUNCTIONAL GROUPS IN HILIC SEPARATION

The nonionic polar functional group has a permanent dipole moment but lacks a complete electric charge. The permanent dipole occurs when two atoms in a functional group have substantially different electronegativity, one atom attracts an electron more than the other becoming more negative, while the other atom becomes more positive. The major and strongest interaction between nonionic polar functional groups is hydrogen bonding. The hydrogen bond can also provide powerful retaining power in HILIC, even though it is weaker than ionic interaction. Hydrogen-bond energy is within 5–30 kJ/mol while ionic bond energy is usually larger than 100 kJ/mol. Mobile phase composition and column temperature can dramatically impact upon hydrogen bonding between analytes and the stationary phase.

11.3.1 Effect of Mobile Phase Composition

In HILIC, the secondary hydrogen-bond retention mechanism can be distinguished from the primary partition retention mechanism. Hydrogen-bond strength can be explained by the neutral polar functional groups, including protic or aprotic. The polar protic groups can be both donors and acceptors of hydrogen bonds whereas aprotic solvents can be only hydrogen-bond acceptors. Partition is usually determined by the entire analyte polarity or hydrophilicity, represented by the log *P* value, which is the logarithm of the octanol–water partition coefficient. Log *P* values are unavailable and an extra independent experiment is needed to determine them. For practical HILIC method development, the hydrogen-bond retention mechanism is more suitable to predict the retention capability of neutral functional groups within analytes and stationary phase. A separation study of neutral oligomeric proanthocyanidins (structures in Figure 11.16) on an amide-silica column (TSK-Gel Amide-80) is a good example showing the contributions of hydrogen bonding and partitioning in a HILIC separation.[48] The correlation between the logarithm of retention factors (log *k*) and the number of hydroxyl groups in Figure 11.17 is even better than between log *k* and log *P* values in Figure 11.18 ($r^2 = 0.9501$ vs. 0.7949, respectively).[48]

When hydrogen bonding is strongly involved in HILIC retention, switching organic components in the mobile phase can be a very important strategy to improve analyte separation. These organic components can be subdivided into polar protic and aprotic solvents. Methanol, ethanol, isopropanol, and acetic acid are representative polar protic solvents. The typical polar aprotic solvents are ACN and tetrahydrofuran (THF). Because of their strong ability to hydrogen bond, polar protic solvents can more effectively compete for polar active sites on the HILIC phase surface, perturbing the formation of water layers by replacing water molecules, thus

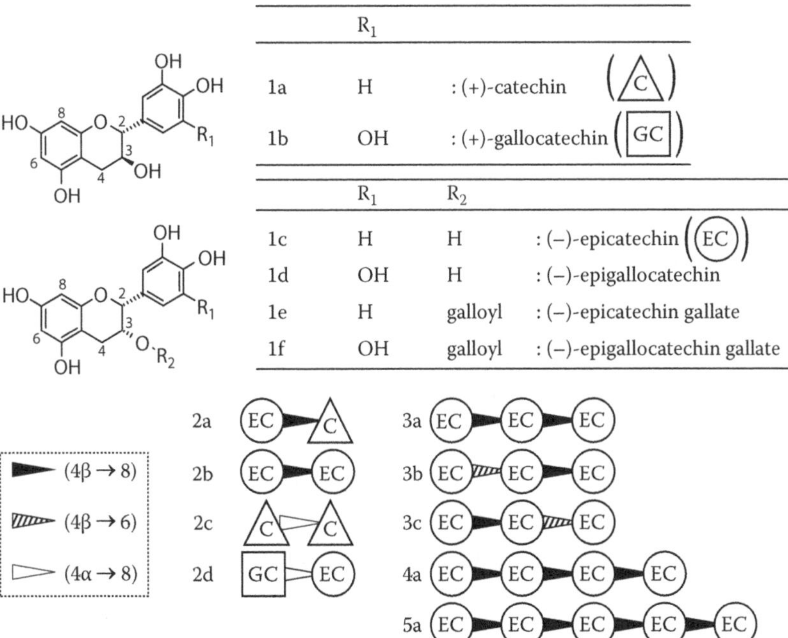

FIGURE 11.16 Chemical structures of monomeric flavan-3-ols and oligomeric proanthocyanidins. (Adapted from Yanagida, A. et al., *J. Chromatogr. A*, 1143, 153, 2007. With permission.)

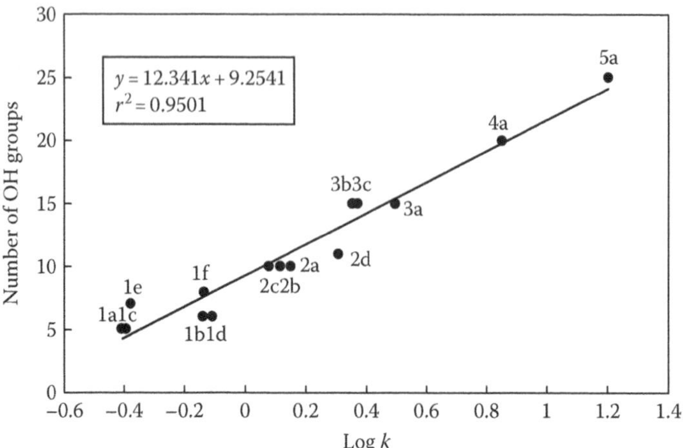

FIGURE 11.17 Relationship between log k' of 15 standards in amide HILIC column (TSK-Gel Amide-80) and the number of hydroxyl groups (–OH) in their structures. Inset regression formula was calculated from the linear least-square fit of all data (n = 15). (Adapted from Yanagida, A. et al., *J. Chromatogr. A*, 1143, 153, 2007. With permission.)

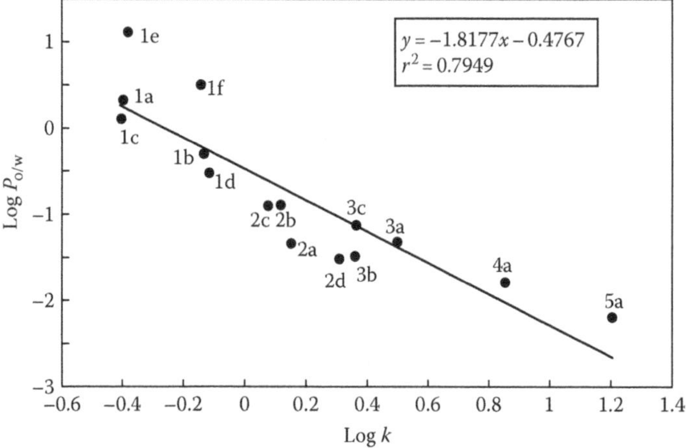

FIGURE 11.18 Relationship between log k′ of 15 standards in amide HILIC column (TSK-Gel Amide-80) and log P values measured by HSCCC. Inset regression formula was calculated from the linear least-square fit of all data (n = 15). (Adapted from Yanagida, A. et al., *J. Chromatogr. A*, 1143, 153, 2007. With permission.)

producing a more hydrophobic stationary phase.[49] As a consequence, analytes with strong hydrogen-bond capability are poorly retained.

Li and Huang compared the ability of various organic modifiers to retain epirubicin analogues using a Kromasil KR100-5SIL column (Figure 11.19).[50] Methanol caused all four analytes to be eluted with no retention. Isopropanol has a longer alkyl chain and less hydrophilic character. It competes less strongly for the active

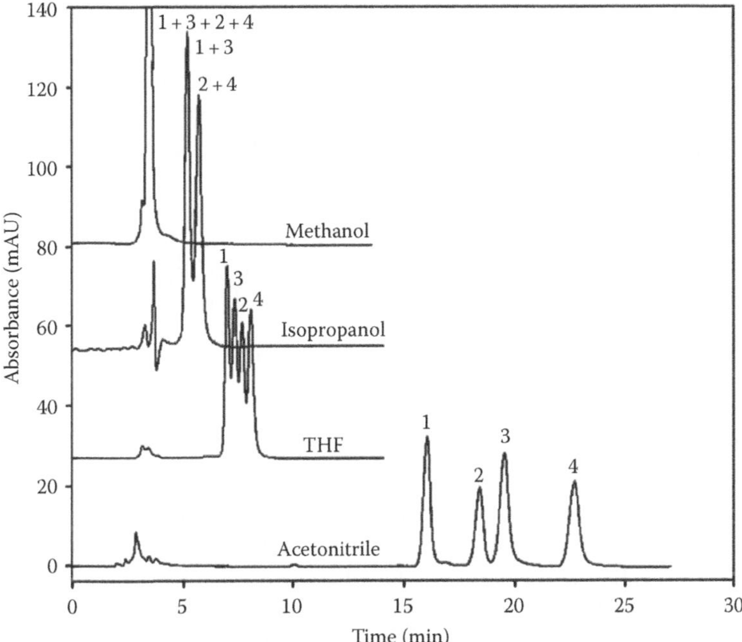

FIGURE 11.19 Effect of organic modifier on separation of epirubicin and its analogues. Conditions: Kromasil KR100-5SIL (5 μm); mobile phase: sodium formate buffer (20 mM, pH 2.9) modified with various organic solvents (10:90, v/v). Peaks: (1) epidaunorubicin, (2) daunorubicin, (3) epirubicin, (4) doxorubicin. (Adapted from Li, R.P. and Huang, J.X., *J. Chromatogr. A*, 1041, 163, 2004. With permission.)

sites and left more time for analytes to be retained. In contrast to methanol and isopropanol, the aprotic solvents ACN and THF provided more effective retention of the analytes on the column. The analytes were retained more strongly with ACN than THF because the latter is a better hydrogen-bond acceptor. Interestingly, the elution order of epirubicin and daunorubicin with ACN is different from THF and isopropanol (Figures 11.19 and 11.20). The major differences between these two structures are an extra hydroxyl group by the keto group in epirubicin and an inner hydrogen bond between the hydroxyl group and amine group in daunorubicin. Epirubicin was retained longer in the mobile phase containing ACN, where stronger hydrogen bonding occurs between analyte and the stationary phase. When ACN was replaced by THF or isopropanol, the retention contribution from such hydrogen bonding became weaker and ion-exchange interactions between the analyte and the stationary phase became stronger. Daunorubicin, with a higher pK_a due to such inner hydrogen bonding, was retained longer. The four hydrazine analogues mentioned earlier were successfully separated by an aqueous ethanol mobile phase on a ZIC-HILIC column because ACN cannot be used with a chemiluminescent nitrogen detector (CLND).[36]

A systematic comparison of selectivity (log α values) between G vs. GG, DG vs. GDG, and TG vs. GTG was investigated on a Hypersil silica column with

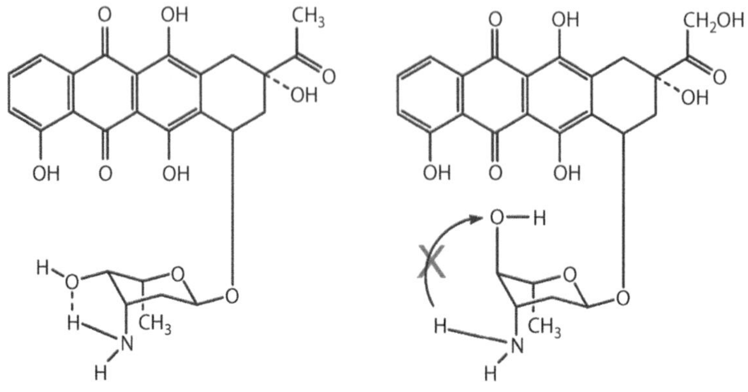

(2) (3)

FIGURE 11.20 Chemical structures of daunorubicin (2) and epirubicin (3).

different organic solvents used in the mobile phase (Figure 11.21).[14] When the methanol content was low and log $\alpha < 0$, G, DG, and TG eluted before GG, GDG, and GTG, respectively. When the methanol content was increased and log $\alpha = 0$, no separation occurred for each of the individual pairs. When the methanol content was further increased and log $\alpha > 0$, G, DG, and TG eluted after GG, GDG,

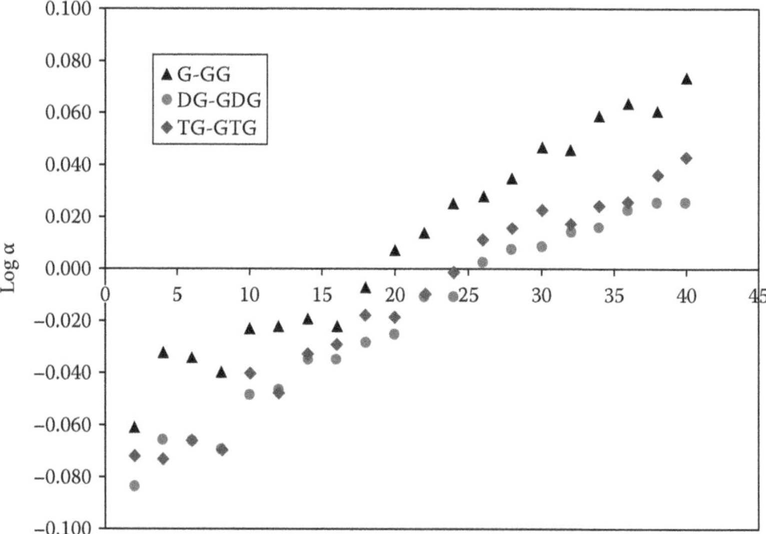

FIGURE 11.21 The plots of log α (selectivity) vs. %methanol relative to acetonitrile in the mobile phase for three pairs of compounds, G vs. GG, DG vs. GDG, and TG vs. GTG, under HILIC condition on a Hypersil silica column, 100 × 1 mm, particle size of 3 μm, 0.1% formic acid in all mobile phase solvents used, and column temperature at 30°C. Total organic content (acetonitrile + methanol) in the mobile phase remained at 75%. (Adapted from Hao, Z. et al., *J. Chromatogr. A*, 1147, 165, 2007. With permission.)

and GTG, respectively. The active silanol groups on the stationary phase surface can effectively be competed with either the hydroxyl groups from methanol or the neutral polar hydroxyl groups from Amadori compounds. The more active sides are occupied by methanol, the less are left for Amadori compounds, thus resulting in a reversed elution order. A similar phenomenon was found in an analysis of morphine and its metabolite, morphine-3-glucuronide, and their elution order was switched when ACN was replaced by methanol in the mobile phase on the Inertsil silica and ZIC-HILIC columns.[39,51]

Switching a protic organic solvent with aprotic can improve separation selectivity not only for similar but also for completely different structures. For example, 50% ACN-H_2O containing 0.1% formic acid as mobile phase was used to separate choline and arginine on a Hypersil bare silica column (Figure 11.22a). If the aprotic organic solvent ACN was replaced by the protic organic solvent methanol, the elution order in Figure 11.22a was switched to those shown in Figure 11.22b.[8] The rationale behind this switching is hydrogen-bond competition. Abundant protic functional groups like

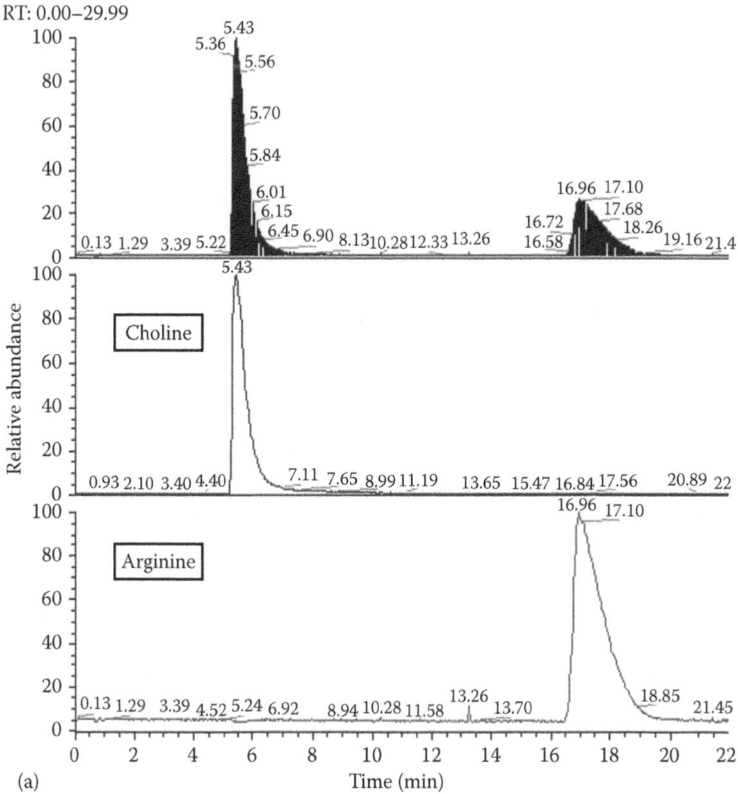

FIGURE 11.22 HILIC-MS/MS chromatograms of choline and arginine under HILIC condition on a Hypersil silica column, 100 × 1 mm, particle size of 3 µm. (a) The mobile phase is consistent of 50% water and 50% ACN and 0.4% formic acid in both solvents.

(continued)

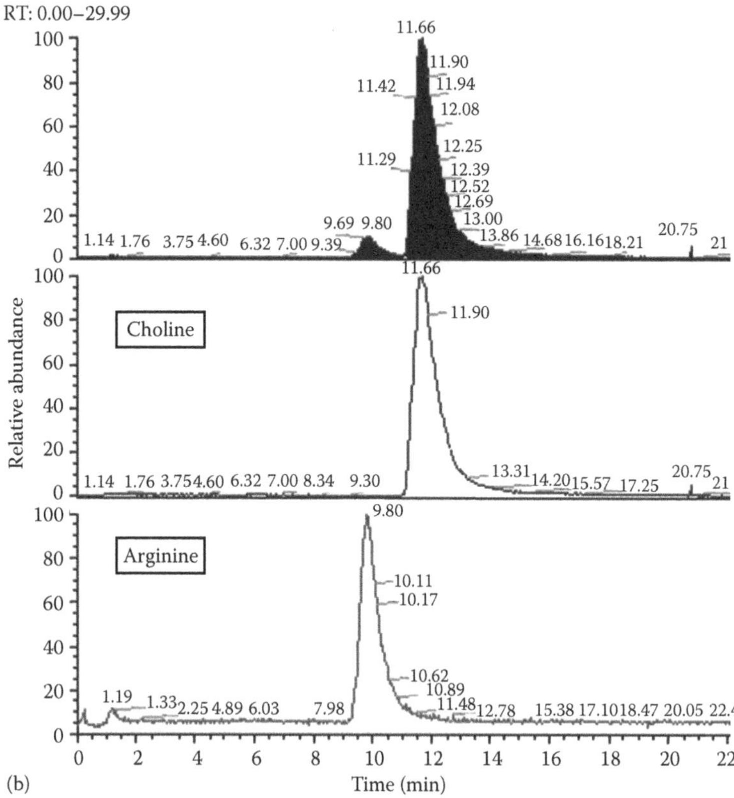

FIGURE 11.22 (continued) (b) The mobile phase is consistent of 30% water and 70% methanol and 0.4% formic acid in both solvents. (Adapted from Hao, Z. et al., *J. Sep. Sci.*, 31, 1449, 2008. With permission from Wiley-VCH Verlag Gmbh & Co. KGaA.)

–$NH_{(1-2)}$ and –OH in the arginine structure can form hydrogen bonds with silanol groups on the silica surface or hydroxyl groups from the immobilized water-rich layer on the silica surface, contributing to arginine retention on the stationary phase. However, this hydrogen bond could be eliminated by a protic solvent such as methanol, thus resulting in a decrease in retention of arginine. The choline polarity may be higher or lower than arginine but its retention was less affected by ACN replacement with methanol compared to arginine.

Methanol was also preferred for the separation of taurine and methionine in a beverage matrix relative to ACN. Many carbohydrates such as glucose, fructose, and saccharose present in the beverage solution could be strongly retained on a HILIC column through a hydrogen-bonding retention mechanism if an ACN–water mobile phase was used, whereas with a methanol–water mobile phase they are not retained (Figure 11.23).[52]

Methanol has not always been successfully used to replace ACN for selectivity improvement. In general, ACN is often selected over methanol since it has many advantages including nearly ideal spectroscopic qualities, low viscosity, and

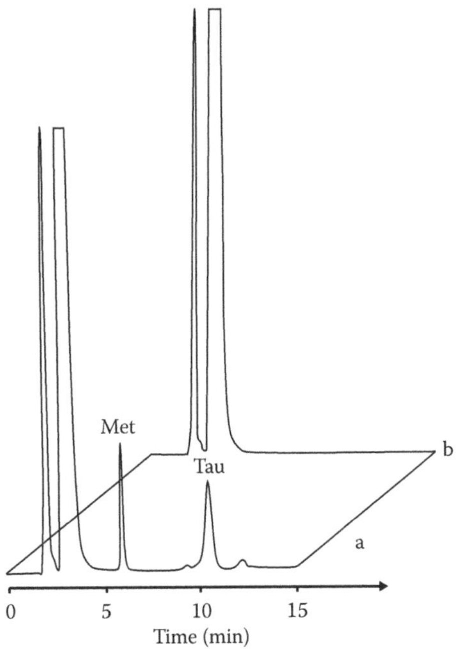

FIGURE 11.23 HILIC-ELSD analysis of a beverage diluted to 1/10 with mobile phase. (a) Beverage containing Met and Tau, (b) amino acid-free beverage. Column: apHera NH_2 (150 mm × 4.6 mm I.D., 5 μm). Column temperature is at 37°C. Mobile phase: MeOH/H_2O (60:40, v/v) under isocratic elution mode; flow rate 0.6 mL/min; injection volume: 10 μL. (Adapted from de Person, M. et al., *J. Chromatogr. A*, 1081, 174, 2005. With permission.)

unique chromatographic properties. Acetonitrile is basically a polar solvent and it is miscible with water in all proportions. More importantly, ACN does not associate strongly with water and ACN–water mixtures remain binary in character. This simplifies the interactive theory and allows a more simple prediction of retention based on the ACN concentration in the partition retention mechanism. Methanol, on the other hand, not only interacts with analytes and the stationary phase but also forms a strong association with water so that at a high concentration of water, the mobile phase behaves as a binary mixture of water and water–methanol associate. At high concentrations of methanol, the converse applies, that is, the mobile phase consists of a mixture of methanol and water–methanol associate. Between these extremes the mobile phase consists of a complex ternary mixture of methanol, water–methanol associate, and water. A review by Scott showed that there was some association between water with ACN and water–THF, but not nearly to the same extent as water with methanol.[53] At the point of maximum association in water/methanol mixtures, the solvent contained nearly 60% of the water–methanol associate. In contrast, the maximum amount of water–THF associate formed was only 17%, and that for water–ACN as little as 8%. That is why methanol was considered to be too strong an elution solvent, leading to poor retention,[54] and ACN usually provided sharper peaks than methanol.[55]

The hydrogen-bonding interaction between methanol and analytes may introduce extra resonance structures and cause broad or tailing peaks.[56,57]

It is worthy to note that hydrogen bonding can become unstable at elevated temperatures.[58] The broad and tailing peaks with a methanol mobile phase could be improved if the column is running at an elevated temperature, which will be discussed in the next section. The effects of mobile phase composition and temperature are often complementary and a simultaneous optimization of these parameters could become a useful approach to control analyte retention and improve selectivity and peak shape.[59] One should be aware, however, that polar analytes are poorly soluble in mobile phases with high concentrations of ACN.[37]

11.3.2 Effect of Column Temperature

Column temperature can change not only the analyte transferring enthalpy from mobile phase to stationary phase but can also have an effect on analyte structure. For example, when crystalline glucose, which is a single compound, is dissolved in an aqueous solution, tautomerization occurs. Ultimately an equilibrium mixture of at least five compounds is formed: the α-pyranose, the β-pyranose, the α-furanose, the β-furanose, and the aldehyde form.[60] The transformation rate between different isomers is temperature dependent. The isomers α- and β-glucopyranoses have been separated at lower temperature on Ca^{2+}-form Aminex HPX-87C column.[61] Partially resolved double peaks were also observed for L-fucose on a carbamoyl-silica HILIC column (TSK-Gel Amide-80).[35] Elevated temperature can preclude the existence of specific isomers of carbohydrate analytes. When isomers of the same analyte cannot be distinguished at higher temperatures, broad or split peaks will be narrowed down into a single peak.[35,62]

Amadori compounds can also exist as tautomers due to the carbohydrate portion of their structures. The different α-, β-, and acyclic anomers of Amadori compounds in Figure 11.24 have been confirmed by nuclear magnetic resonance.[63]

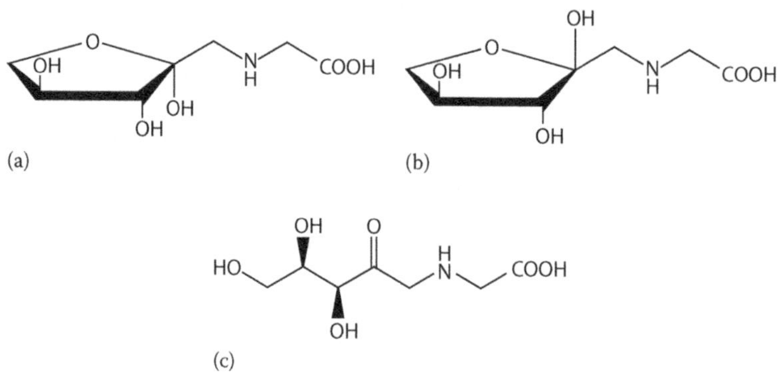

(a) (b) (c)

FIGURE 11.24 Chemical structures of open-chain and cyclic forms of *N*-(1-deoxy-D-xylose-1-yl)-glycine: (a), α-anomer; (b), β-anomer; (c), acyclic form. (Adapted from Davidek, T. et al., *Anal. Chem.*, 77, 140, 2005. With permission.)

Within the six structures of G, DG, TG, GG, GDG, and GTG seen in Figure 11.3, only three Amadori compounds contain a carbohydrate (glucose) portion. At a low temperature of 5°C, broad/split peaks were observed when these compounds were separated on bare silica columns and became narrower single peaks at an elevated column temperature (Figure 11.25).[14]

Elevated column temperatures can narrow down the peak widths of analytes containing nonionic polar functional groups, especially for carbohydrate rings, but it also shortens their retention times. The overall resolution of these types of compounds is still primarily analyte dependent.

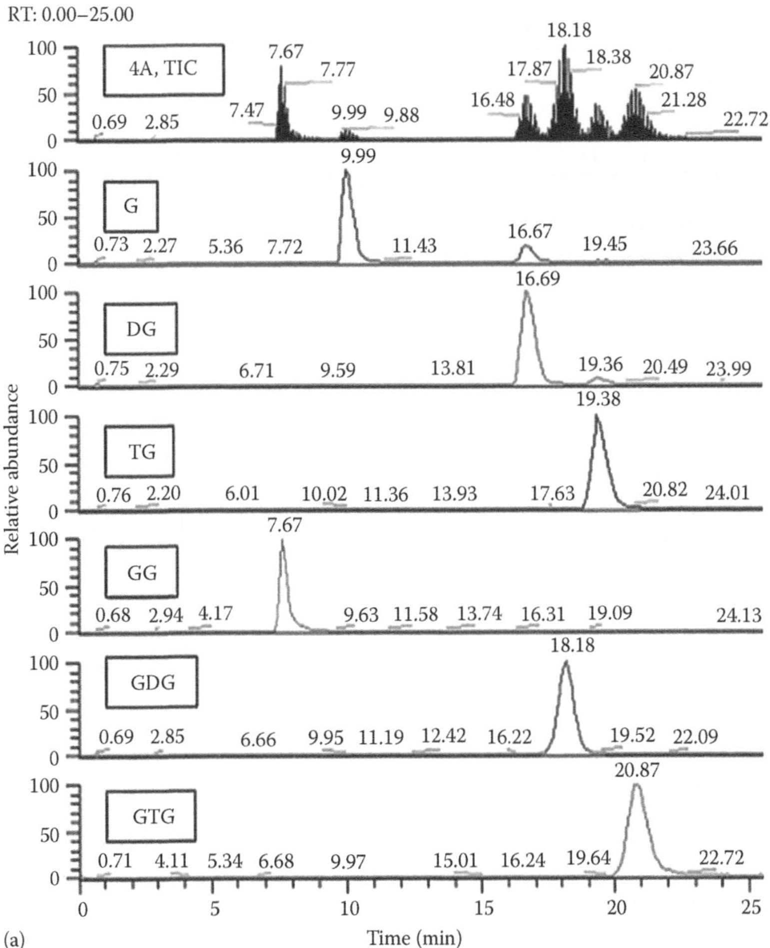

FIGURE 11.25 HILIC-MS/MS chromatograms of G, DG, TG, GG, GDG, and GTG under HILIC condition on a Hypersil silica column, 100 × 1 mm, particle size of 3 μm with mobile phase of 25% water and 75% ACN and 0.4% formic acid in both solvents. Column temperature was at 80°C for (a), 30°C for (b), and 5°C for (c). (Adapted from Hao, Z. et al., *J. Chromatogr. A*, 1147, 165, 2007. With permission.)

(continued)

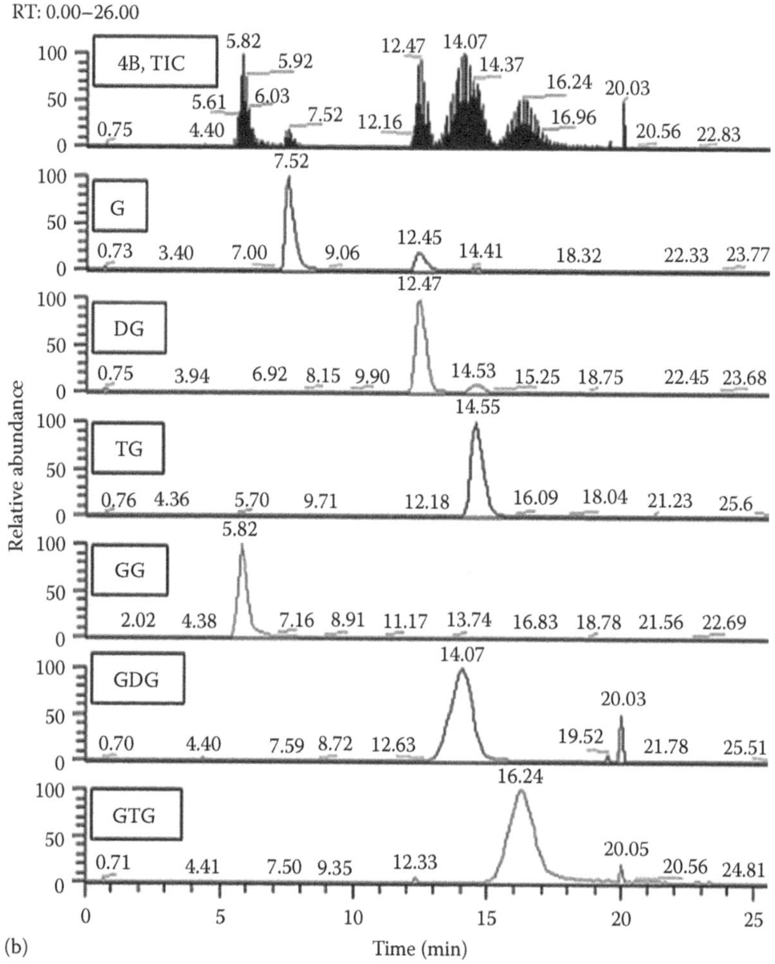

(b)

FIGURE 11.25 (continued)

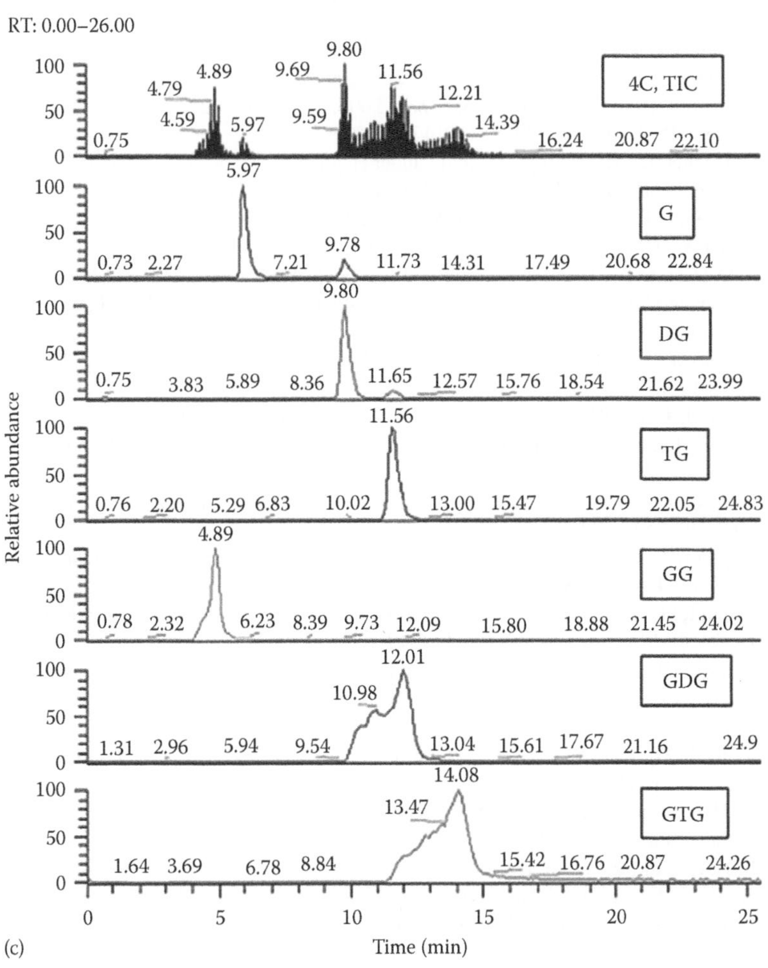

FIGURE 11.25 (continued)

ACKNOWLEDGMENTS

The author would like to thank Mark Storton and Kate Jackson from our Global Analytical Science Department for their helpful comments on my manuscript!

REFERENCES

1. Linden, J.C. and Lawhead, C.L., *J. Chromatogr.*, 1975, 105, 125–133.
2. Palmer, J.K., *Anal. Lett.*, 1975, 8, 215–224.
3. Alpert, A.J., *J. Chromatogr.*, 1990, 499, 177–196.
4. Naidong, W., *J. Chromatogr. B*, 2003, 796, 209–224.
5. Hemstron, P. and Irgum, K., *J. Sep. Sci.*, 2006, 29, 1784–1821.
6. Jandera, P., *J. Sep. Sci.*, 2008, 31, 1421–1437.
7. Dejaegher, B., Mangelings, D., and Vander Heyden, Y., *J. Sep. Sci.*, 2008, 31, 1438–1448.
8. Hao, Z., Xiao, B., and Weng, N., *J. Sep. Sci.*, 2008, 31, 1449–1464.
9. Nguyen, H.P. and Schug, K.A., *J. Sep. Sci.*, 2008, 31, 1465–1480.
10. Hsieh, Y., *J. Sep. Sci.*, 2008, 31, 1481–1491.
11. FDA. Center for Drug Evaluation and Research, U.S. Food and Drug Administration, Reviewer Guidance, Validation of Chromatographic Method, FDA, Rockville, MD, November, 1994.
12. Kovalova, L., McArdell, C.S., and Hollender, J., *J. Chromatogr. A*, 2009, 1216, 1100–1108.
13. Kadar, E.P. and Wujcik, C.E., *J. Chromatogr. B*, 2009, 877, 471–476.
14. Hao, Z., Lu, C. Y., Xiao, B., Weng, N., Parker, B., Knapp, M., and Ho, C.T., *J. Chromatogr. A*, 2007, 1147, 165–171.
15. Christopherson, M.J., Yoder, K.J., and Hill, J.T., *J. Liq. Chromatogr. Relat. Technol.*, 2006, 29, 2545–2558.
16. Ikegami, T., Tomomatsu, K., Takubo, H., Horie, K., and Tanaka, N., *J. Chromatogr. A*, 2008, 1184, 474–503.
17. Hmelnickis, J., Pugovics, O., Kazoka, H., Viksna, A., Susinskis, V., and Kokums, K., *J. Pharm. Biomed. Anal.*, 2008, 48, 649–656.
18. Liu, M., Chen, E.X., Ji, R., and Semin, D., *J. Chromatogr. A*, 2008, 1188, 255–263.
19. Li, W., Li, Y., Francisco, D.T., and Naidong, W., *Biomed. Chromatogr.*, 2005, 19, 385–393.
20. Jeong, D.W., Kim, Y.H., Ji, H.Y., Youn, Y.S., Lee, K.C., and Lee, H.S., *J. Pharm. Biomed. Anal.*, 2007, 44, 547–552.
21. Ji, H.Y., Jeong, D.W., Kim, Y.H., Kim, H.H., Sohn, D.R., and Lee, H.S., *J. Pharm. Biomed. Anal.*, 2006, 41, 622–627.
22. Ji, H.Y., Park, E.J., Lee, K.C., and Lee, H.S., *J. Sep. Sci.*, 2008, 31, 1628–1633.
23. Park, E.J., Lee, H.W., Ji, H.Y., Kim, H.Y., Lee, M.H., Park, E.S., Lee, K.C., and Lee, H.S., *Arch. Pharm. Res.*, 2008, 31, 1205–1211.
24. Sun, Y., Sun, J., Liu, J., Yin, S., Chen, Y., Zhang, P., Pu, X., Sun, Y., and He, Z., *J. Chromatogr. B*, 2009, 877, 649–652.
25. Nawrocki, J., *J. Chromatogr. A*, 1997, 779, 29–71.
26. Ali, M.S., Rafiuddin, S., Al-Jawi, D.A., Al-Hetari, Y., Ghori, M., and Khatri, A.R., *J. Sep. Sci.*, 2008, 31, 1645–1650.
27. Hsieh, Y. and Chen, J., *Rapid Commun. Mass Spectrom.*, 2005, 19, 3031–3036.
28. Chen, C.Y., Chang, S.N., and Wang, G.S., *J. Chromatogr. Sci.*, 2009, 47, 67–74.
29. Liu, M., Ostovic, J., Chen, E.X., and Cauchon, N., *J. Chromatogr. A*, 2009, 1216, 2362–2370.
30. Alpert, A.J., *Anal. Chem.*, 2008, 80, 62–76.

31. Neue, U.D., VanTran, K., Iraneta, P.C., and Alden, B.A., *J. Sep. Sci.*, 2003, 26, 174–186.
32. Schebb, N.H., Fischer, D., Hein, E.M., Hayen, H., Krieglstein, J., Klumpp, S., and Karst, U., *J. Chromatogr. A*, 2008, 1183, 100–107.
33. Wu, J.Y., Bicker, W., and Linder, W., *J. Sep. Sci.*, 2008, 31, 1492–1503.
34. Persson, J., Hemstron, P., and Irgum, K., *J. Sep. Sci.*, 2008, 31, 1504–1510.
35. Karlsson, G., Winge, S., and Sandberg, H., *J. Chromatogr. A*, 2005, 1092, 246–249.
36. Guo, Y. and Gaiki, S., *J. Chromatogr. A*, 2005, 1074, 71–80.
37. Ali, M.S., Ghori, M., Rafiuddin, S., and Khatri, A.R., *J. Pharm. Biomed. Anal.*, 2007, 43, 158–167.
38. Guo, Y., Srinivasan, S., and Gaiki, S. *Chromatographia*, 2007, 66, 223–229.
39. Vikingsson, S., Kronstrand, R., and Josefsson, M., *J. Chromatogr. A*, 2008, 1187, 46–52.
40. Rouzo, G.L., Lamouroux, C., Bresson, C., Guichard, A., Moisy, P., and Moutiers, G., *J. Chromatogr. A*, 2007, 1164, 139–144.
41. Quiming, N.S., Denola, N.L., Saito, Y., Catabay, A.P., and Jinno, K., *Chromatographia*, 2008, 67, 507–515.
42. Novakova, L., Solichova, D., Pavlovicova, S., and Solich, P., *J. Sep. Sci.*, 2008, 31, 1634–1644.
43. Jiang, W. and Irgum, K., *Anal. Chem.*, 1999, 71, 333–344.
44. Walton, H.F. and Rochlin, R.D., *Ion Exchange in Analytical Chemistry*, CRC Press, Boca Raton, FL, 1990.
45. Yang, X.Q., Dai, J., and Carr, P.W., *J. Chromatogr. A*, 2003, 996, 13–31.
46. Ali, M.S., Rafiuddin, S., Ghori, M., Rafiuddin, S., and Khatri, A.R., *Chromatographia*, 2008, 67, 517–525.
47. Dong, L. and Huang, J., *Chromatographia*, 2007, 65, 519–526.
48. Yanagida, A., Murao, H., Ohnishi-Kameyama, M., Yamakawa, Y., Shoji, A., Tagashira, M., Kanda, T., Shindo, H., and Shibusawa, Y., *J. Chromatogr. A*, 2007, 1143, 153–161.
49. Scott, R.P.W., *Adv. Chromatogr.*, 1982, 20, 169–187.
50. Li, R.P. and Huang, J.X., *J. Chromatogr. A*, 2004, 1041, 163–169.
51. Naidong, W., Lee, J.W., Jiang, X., and Wehling, M., *J. Chromatogr. B*, 1999, 735, 255–269.
52. de Person, M., Hazotte, A., Elfakir, C., and Lafosse, M., *J. Chromatogr. A*, 2005, 1081, 174–181.
53. Scott, R.P.W., *J. Chromatogr. A*, 1993, 656, 51–68.
54. Tai, A. and Gohda, E., *J. Chromatogr. B*, 2007, 83, 214–220.
55. Aversano, C.D., Hess, P., and Quillian, M.A., *J. Chromatogr. A*, 2005, 1081, 190–201.
56. Olsen, B.A., *J. Chromatogr. A*, 2001, 913, 113–122.
57. Valette, J.C., Demesmay, C., Rocca, J.L., and Verdon, E., *Chromatographia*, 2004, 59, 55–60.
58. Linden, H.V.D., Herber, S., Olthuis, W., and Bergveld, P., *Sens. Mater.*, 2002, 14, 129–139.
59. Guillarme, D. and Heinisch, S., *Sep. Purif. Rev.*, 2005, 34, 181–216.
60. Angyal, S.J., *Angew. Chem. Int. Ed. Engl.*, 1969, 8, 157–166.
61. Nishikawa, T., Suzuki, S., Kubo, H., and Ohtani, H., *J. Chromatogr. A*, 1996, 720, 167–172.
62. Alpert, A.J., Shukla, M., Shukla, A.K., Zieske, L.R., Yuen, S.W., Ferguson, M.A.J., Mehlert, A., Paulv, M., and Orlando, R., *J. Chromatogr. A*, 1994, 676, 191–202.
63. Davidek, T., Kraehenbuehl, K., Devaud, S., Robert, F., and Blank, I., *Anal. Chem.*, 2005, 77, 140–147.

12 Analysis of Pharmaceutical Impurities Using Hydrophilic Interaction Liquid Chromatography

Mingjiang Sun and David Q. Liu

CONTENTS

12.1 INTRODUCTION

Organic impurities in active pharmaceutical ingredients (APIs) are derived from many different sources during manufacturing and storage. They can be starting materials,[1] intermediates,[2] reagents, side-reaction products,[3] packaging extractables, degradants,[4] etc. The presence of such impurities could pose risks to the safety of the general public, and thus their levels need to be controlled according to International Conference on Harmonization (ICH) guidelines.[5–8] Impurity analysis and control are the key activities in process analytical development, and thus adequate analytical methods are needed for the accurate determination of those impurities.[9] Some API impurities, however, are so polar that they cannot be retained on the conventional reversed-phase high-performance liquid chromatographic (RP-HPLC) columns; therefore their analyses constitute a real challenge to the analytical community. Research has been devoted to the development of general strategies for the accurate analysis of very polar analytes. For instance, some polar compounds containing reactive functional groups such as hydroxyl and amino can be analyzed after chemical derivatization to reduce the polarity.[10] Some polar analytes are charged, therefore ion chromatography or capillary electrophoresis has been used for their analysis. Nonetheless, the application of these methods has been limited because of the relatively poor reproducibility, complexity in method development, and the limitation in choosing detection methodologies.

The polar pharmaceutical impurities discussed in this chapter refer to those that are poorly retained on conventional RP-HPLC columns even if high aqueous or pure aqueous mobile phases are used. It is well known that typical RP-HPLC columns such as C18 columns may collapse in the high aqueous mobile phase by de-wetting, resulting in poor retention, low selectivity, and irreproducibility.[11] Newer RP-HPLC columns that are more amenable to high aqueous mobile phases are commercially avaliable.[12] For the use of such columns, high aqueous content in sample diluent is required in order to prevent peak fronting caused by solvent mismatch. Therefore, its application in pharmaceutical analysis is constrained by the low aqueous solubility of samples. Poor separation efficiency and reproducibility are also potential issues for this approach. Furthermore, some compounds are so polar that they do not retain on conventional RP-HPLC columns at all, even when the pure aqueous mobile phase is used. Alternatively, ion pairing RP-HPLC employs ion pairing agents to improve retention of polar compounds. Ion pairing reagents, however, are generally not compatible with mass spectrometry due to their ion suppression effects.[13] Polar compounds very often have low molecular weight and lack a chromophore for UV detection. Therefore, universal detectors such as evaporative light scattering (ELSD), conduct activity (CAD), and reflective index (RID) detectors have been explored. Nonetheless, issues such as low specificity and narrow dynamic ranges were experienced.[14–16] Because of superior sensitivity, specificity, and versatility, LC/MS has become the technique of choice for the low-level detection of many types of pharmaceutical compounds. As such, the use of mass spectrometry compatible mobile phases has become an important attribute of HPLC methods.

Hydrophilic interaction liquid chromatography (HILIC) is a complementary tool for the analysis of polar compounds.[17] HILIC uses polar stationary phases including bare silica, polar functional group modified silica, or modified polymers, while the mobile phases contain water and an organic solvent, mostly acetonitrile. The interaction between water and the polar stationary phase forms a water-enriched layer on the surface of the polar stationary phase. The partition of analytes between the bulk of the organic-enriched mobile phase and the water-enriched layer on the surface of the stationary phase results in their chromatographic separation. Polar compounds will partition between the two phases, while nonpolar compounds may run through the column in the mobile phase directly. As a consequence, polar compounds are better retained than less polar ones. To date, various types of stationary phases including bare silica, amino, amide, diol, cyano, poly(succinamide), sulfoalkylbetaine, and cyclodextrin-modified silica have been developed and commercialized. Ikegami et al. provided an excellent review on such column types and their applications.[18] The stationary phases can be acidic, basic, or neutral, which may interact with analytes by ionic interaction and/or hydrogen bonding. Thus, the separation of analytes on HILIC columns could be a mix of several different mechanisms.[19–22] The applications of HILIC in the analysis of amino acids, peptides, proteins, sugars, toxins, and drugs have been demonstrated in the literature.[17] The recent development in column technology has facilitated the rapid growth of HILIC applications in various industries, such as pharmaceutical, food, and environmental analyses. In the pharmaceutical industry, the technique has been used in drug discovery,[23,24] pharmacokinetics,[13] quality and process controls of APIs and drug products.[25,26] The fact that many pharmaceutical APIs are highly soluble in organic solvents makes HILIC a viable option for the determination of polar impurities in APIs.

The aim of this chapter is to provide a review on the applications of HILIC in the analysis of polar impurities in APIs. In particular, the strategies to convert genotoxic alkylating agents to positively charged quaternary amines to enhance mass spectrometry detection sensitivity are highlighted. Practical examples from both unpublished data from the authors' laboratory and recent literature will be presented.

12.2 TRACE ANALYSIS OF GENOTOXIC IMPURITIES

Potential genotoxic impurities are defined as impurities that may react with DNA and potentially increase cancer risk in patients. The European Medicines Agency (EMEA) issued guidelines regarding the control of potential genotoxic impurities in drug substances or drug products.[8] The U.S. Food and Drug Administration (FDA) also drafted a guidance document and made it available online recently. Pharmaceutical manufacturers are actively seeking strategies to monitor and control impurity levels in APIs according to the regulatory requirements.[27] When a new manufacturing process is developed, one of the goals is to avoid the usage and generation of genotoxic impurities.[28] Nonetheless, it is almost impossible to completely eliminate the risk of introducing genotoxic impurities into a particular manufacturing process. Therefore, the levels of genotoxic impurities in APIs are managed by a controlled risk approach. The acceptable risk caused by pharmaceutical products for patients was defined as no greater than 10^{-5} based on lifetime exposure. As such, the daily intake of a specific genotoxin for a person cannot be greater than 1.5 μg. Consequently, the concentration limits of genotoxic impurities in drug substances or drug products can be derived based on the maximum daily dose. For instance, if the

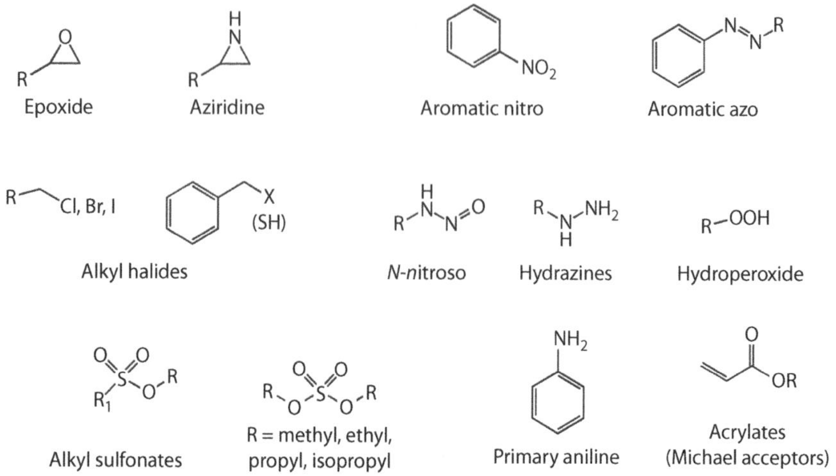

FIGURE 12.1 Structures with high genotoxic concerns.

daily dose of a drug is 1 g, the allowable concentration of a genotoxic impurity in the drug substance should be no more than 1.5 ppm (ppm, μg genotoxic impurity per g API). Regulatory agencies also advocate applying the limit to the total of all genotoxic impurities in an API that act through the same mechanism. Given such a low ppm concentration limit, developing sensitive and robust methodology for their detection poses great challenges on analytical technologies.

The EMEA issued a list of alerted structures for potential genotoxins, as illustrated in Figure 12.1, which must be controlled in APIs at a trace level unless they are demonstrated to be nongenotoxic by other experiments such as the Ames test.[8] Alkyl sulfonates and alkyl halides are among the alerted structures of potential genotoxic impurities.[29,30] They could be reagents used during manufacturing processes, or the by-products from side reactions. For example, common alcohols such as methanol, ethanol, and isopropanol can be converted to alkyl chlorides or alkyl sulfonates in the presence of HCl or sulfonic acid, respectively, during API salt formation processes.[31,32] They should be analyzed and controlled at trace levels depending on the daily doses. The challenges of developing robust analytical methods with desired low ppm sensitivity for those genotoxins are well known.[33] The instability of some of the analytes poses additional difficulties to method development, and thus the sensitivity and reproducibility of direct analysis methods of the unstable compounds are often unsatisfactory.[34–37]

Both alkyl sulfonates and alkyl halides are strong alkylators, which can readily react with electron rich nucleophiles. This property can be advantageous for derivatization reactions. By reacting with alkyl amines, permanently positively charged quaternary amines could be produced (Scheme 12.1), which are excellent candidates for electrospray (ES) mass spectrometry detection. Therefore, converting to quaternary amines followed by LC/MS detection seems to be a very effective strategy for analyzing alkyl sulfonates and alkyl halides. However, HPLC separation of the resulting very polar quaternary amines is a new challenge. Previous reports demonstrate that some quaternary amines can be analyzed by HPLC after derivatization or by ion

$$\text{LG-CH(R}_1\text{)-R}_2 + \text{N(R}_3)_3 \longrightarrow \text{R}_2\text{-CH(R}_1\text{)-}\overset{\oplus}{\text{N}}\text{(R}_3)_3 + \text{LG}$$

LG: Leaving group

SCHEME 12.1 Alkylators react with trialkyl amines producing quaternary ammoniums.

chromatography directly.[10,38] The advances in HILIC made it possible to directly analyze these compounds by HPLC. The high organic mobile-phase compositions of HILIC methods are ideal for electrospray ionization (ESI), resulting in improved ionization effficiency.[39] Guo et al. explored the analysis of several quaternary amines using amino or amide HILIC columns. It was found that these compounds were better retained on an amide column than on an amino column.[40] The interfacing of HILIC with mass spectrometry lowered the limit of detection (LOD) of choline by 75 times in comparison to the use of an RP-HPLC method.[40] Mckeown et al. analyzed several basic compounds with silica HILIC columns using a mobile phase with a pH from 2 to 9.[41] The quaternary amines are permanently positively charged, while the surface silanol groups of the stationary phase of silica HILIC columns are negatively charged in the neutral and even weak acidic mobile phase.[42] Thus, silica HILIC columns may offer an ion-exchange separation mechanism in addition to the simple partition mechanism of HILIC, and thus the quaternary amines may be better retained and resolved on silica columns.[43] Aqueous ammonium formate buffers are typically used as the strong solvent for HILIC silica columns. The peak shape and retention factor of an analyte are affected by the concentration of ammonium ions, and higher buffer concentrations reduce peak broadening and tailing.[41] The retention factors are also strongly affected by the mobile phase pH values. At pH 6, the percentage of the aqueous solvent in the mobile phase may need to be as high as 30% to elute the quaternary amines derivatives with a desirable retention time. Since a higher percentage of the organic mobile phase is desirable for better ionization (sensitivity) in ESI, the aqueous mobile phase should be adjusted to pH 4 by adding formic acid for optimal detection. The analytes can be eluted by using less than 20% of the aqueous mobile phase typically. The nonpolar APIs, on the other hand, are generally eluted in the solvent front and diverted to waste stream before entering the mass spectrometer to avoid interfering with the detection of the target analytes (Figure 12.2). The isocratic gradient can often be implemented to reduce turn round time by eliminating the extra column equilibrium time. The modified conditions are optimal for mass spectrometry detection, thus providing superior method specificity and sensitivity. Integrating HILIC/MS with derivatization offers a generic approach for the analysis of the class of genotoxins, alkyl sulfonates, and alkyl halides.

12.2.1 Alkyl Sulfonates

The derivatization-HILIC/MS strategy was initially demonstrated for the determination of a group of 16 commonly encountered alkyl esters of sulfonates or sulfates (Figure 12.3) in drug substances at low ppm levels by An et al.[43] The method uses trimethylamine as the derivatization reagent for ethyl, propyl, and isopropyl esters, and triethylamine for the methyl esters. The sulfonates readily react with the derivatization

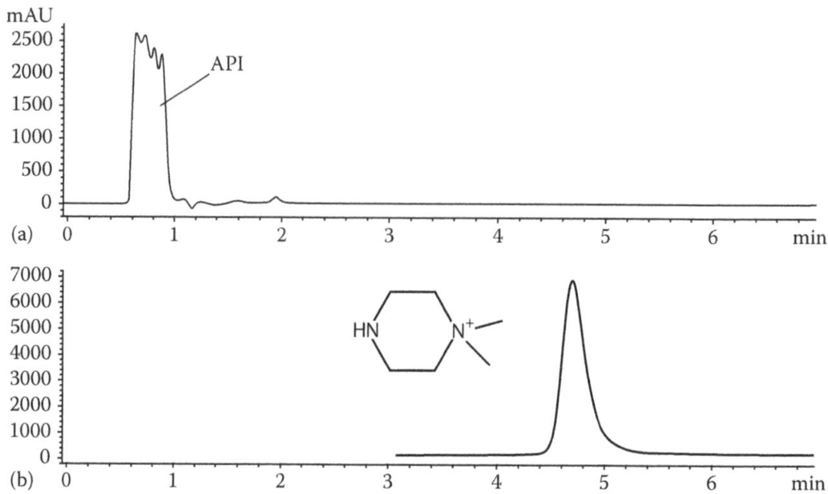

FIGURE 12.2 Separation of derivatized bis(2-chloroethyl)amine from API by HILIC. (a) UV trace, (b) MS trace. The first 3 min of the LC flow was diverted to waste after UV detection. See Tables 12.1 and 12.2 for experimental details.

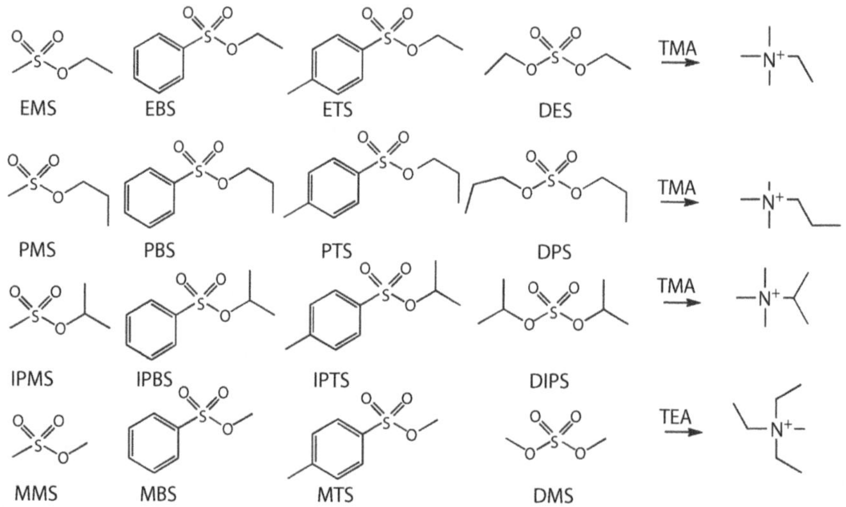

FIGURE 12.3 Structures of sixteen alkyl sulfonates and their trialkylamine derivatives.

reagents, producing four common derivatives: *N,N,N*-triethylmethanaminium, *N,N,N*-trimethylethanaminium, *N,N,N*-trimethylpropanaminium, and *N,N,N*-trimethylisopropanaminium. The resulted quaternary ammonium derivatization products are highly polar (ionic) and can be retained by a silica HILIC column and readily separated from the main interfering peak, e.g., the less polar API peak, which is usually present at a very high concentration. The mobile phase used 15% of 50 mM ammonium formate buffer (pH 4) in acetonitrile (Table 12.1). The chromatograms of all four

TABLE 12.1
HILIC Separation Conditions of Pharmaceutical Compounds

Matrices	Impurities/Analytes	Method Details	References
	Choline, chlormequat, Mepiquat, acetylcholine, betaine	HPLC-MS with an amide HILIC column (TSK-Gel Amide-80, 250 mm × 2.0 mm, 3 µm), mobile phase consisting of a mixture of water (10 mM ammonium acetate) and acetonitrile (19/81). Flow rate 1 mL/min	[40]
Seven APIs	Sixteen sulfonates	Analytes were converted to quaternary amine before analysis. HPLC-MS with a silica HILIC column (Atlantis, 50 mm × 2.1 mm, 3 µm), mobile phase consisting of a mixture of water (50 mM ammonium formate and 0.1% formic acid, pH 4) and acetonitrile (15/85/v/v). Flow rate 0.3 mL/min. SIM: *m/z* 88, *m/z* 102, or *m/z* 116	[43]
Six APIs	Seven alkyl halides	Analytes were converted to quaternary amine before analysis. HPLC-MS with a silica HILIC column (Atlantis, 50 mm × 2.1 mm, 3 µm), mobile phase consisting of a mixture of water (50 mM ammonium formate and 0.1% formic acid, pH 4) and acetonitrile (with varied ratio). Flow rate 0.3 mL/min	[44] and [45]
API	Imidazole	HPLC with a silica HILIC column (Atlantis, 100 mm × 2.1 mm, 3 µm), mobile phase consisting of a mixture of water (50 mM ammonium formate and 0.1% formic acid) and acetonitrile (1/99). Flow rate 1 mL/min. λ: 220 nm	
Ethanol cleaning solution	(3*S*)-3-Morpholinemethanol	HPLC-MS with a silica HILIC column (Atlantis, 50 mm × 2.0 mm, 3 µm), mobile phase consisting of a mixture of water (20 mM ammonium formate) and acetonitrile (15/85). Flow rate 0.3 mL/min. SIM: *m/z* 118	
Intermediate	Acetamide	HPLC with an amino HILIC column (Zorbax NH_2, 250 mm × 4.6 mm, 5 µm), mobile phase acetonitrile-water (90/10); Flow rate 1.0 mL/min. λ: 205 nm	[49]
API	Oxamide, oxamic acid, oxalic acid	HPLC with an amino HILIC column (Zorbax NH_2, 250 mm × 4.6 mm, 5 µm), mobile-phase acetonitrile-water (60/40), 50 mM potassium phosphate, pH 7.0; Flow rate 1.0 mL/min. λ: 205 nm	[49]

(continued)

TABLE 12.1 (continued)
HILIC Separation Conditions of Pharmaceutical Compounds

Matrices	Impurities/Analytes	Method Details	References
5-Fluorocytosine	5-Fluorouracil	HPLC with a silica HILIC column (Zorbax SIL, 250 mm × 4.6 mm, 5 μm), mobile-phase acetonitrile-water (75/25), 5 mM phosphoric acid; Flow rate 1.0 mL/min. λ: 275 nm Or HPLC with an amino HILIC column (Zorbax NH_2, 250 mm × 4.6 mm, 5 μm), mobile-phase acetonitrile-water (80/20), 25 mM potassium phosphate pH 6.5; Flow rate 1.0 mL/min. λ: 275 nm	[49]
Acyclovir	Guanine	HPLC with a silica HILIC column (Zorbax SIL, 250 mm × 4.6 mm, 5 μm), mobile-phase acetonitrile-water (70/30), 5 mM phosphoric acid; Flow rate 1.0 mL/min. λ: 275 nm	[49]
Epirubicin	Epidaunorubicin Daunorubicin Doxorubicin	HPLC with a silica HILIC column (Kromasil KR100-5SIL, 250 mm × 4.6 mm, 5 μm), mobile-phase acetonitrile, 30 mM sodium formate, pH 2.9 (70/30); Flow rate 1.0 mL/min. λ: 254 nm	[55]
LY293558	LY293559	HPLC with a Chirobiotic T™ column (250 mm × 4.6 mm), mobile-phase acetonitrile-water (65/35); Flow rate 1.0 mL/min. ELSD detection	[57]
API	Three polar carbamate impurities	HPLC with a diol column (YMC-pack Diol-120-NP, 250 mm × 4.6 mm, 5 μm), mobile phase 10 mM NH_4Cl in acetonitrile-water (95/5); Flow rate 1.5 mL/min. λ: 215 nm	[2]
Sodium cromoglicate	Unspecified degradants	HPLC with a silica HILIC column (Atlantis, 250 mm × 4.6 mm, 5 μm), mobile phase 30 mM NH_4Ac in acetonitrile-water (86/14, pH 5.6); Flow rate 1.5 mL/min. λ: 326 nm	[60]
Cytosine	Uracil 7-Aminopyrido[2,3-*d*] pyrimidin-2-ol	HPLC with an amide column (TSK-Gel Amide-80, 250 mm × 4.6 mm, 5 μm), mobile phase A: sodium hydroxide/formic acid (pH 3.0), mobile phase B: acetonitrile; Gradient: 5% A for 5 min, then linear increase to 25% A in 10 min, and hold for 5 min; Flow rate 1.0 mL/min	[61]

TABLE 12.1 (continued)
HILIC Separation Conditions of Pharmaceutical Compounds

Matrices	Impurities/Analytes	Method Details	References
10 APIs	Inorganic ions: Na^+, K^+, Zn^{2+}, Ca^{2+}, Mg^{2+}, Cl^-, Br^-, NO^{3-}, SO_4^{2-}, PO_4^{3-}; Organic ions: lysine, diethanolamine, trizma, piperazine, choline, esylate, mesylate, edisylate, isethionate, citrate, glucuronate, mandelate, tartrate, fumarate, glycolate, glutarate, maleate, malate, tosylate, napadisylate, benzylamine, arginine	HPLC with a zwitterionic column (ZIC-HILIC, 250 mm × 4.6 mm, 5 µm), mobile phase A: acetonitrile/buffer (85/15), mobile phase B: acetonitrile/buffer (10/90); Buffer: ammonium acetate, pH adjusted by acetic acid; Gradient: 0% B for 2 min, then linear increase to 100% A in 20 min, and hold for 5 min; Flow rate 1.0 mL/min. Detector: ELSD	[68]
API	Tromethamine	HPLC with an amino HILIC column (Zorbax NH_2, 250 mm × 4.6 mm, 5 µm), mobile-phase acetonitrile-water (80/20); Flow rate 1.0 mL/min. Detector: RID	[67]

derivatives are depicted in Figure 12.4. The cationic analytes can be retained on a silica HILIC column through the interaction of the cations with the ionized silanol groups. The method gave excellent sensitivity for all the alkyl esters at the targeted analytical level of 1–2 ppm when the API samples were prepared at 5 mg/mL with a single quadrupole mass detector. The recoveries at 1–2 ppm were generally above 85% for all the alkyl esters in various APIs. The injection precisions of the lowest standards ranged from 0.4% to 4% (%RSD). A linear range from 0.2 to 20 ppm concentrations was established, with $R^2 \geq 0.99$. The generic method has been applied to the development of several investigational drugs, and there was no need for method alteration for different APIs. However, the method is unable to distinguish the alkyl donors if multiple sources coexist in the sample. On the other hand, the method can be used to quantify the total methyl donors when multiple reactive methyl impurities coexist, and this also applies to the ethyl, propyl, or isopropyl alkylating impurities.

12.2.2 Alkyl Halides

Alkyl halides are another group of common alkylators that need to be controlled in APIs. Low molecular weight volatile alkyl halides are typically analyzed by headspace gas chromatography (HSGC) with mass spectrometry or an electron capture

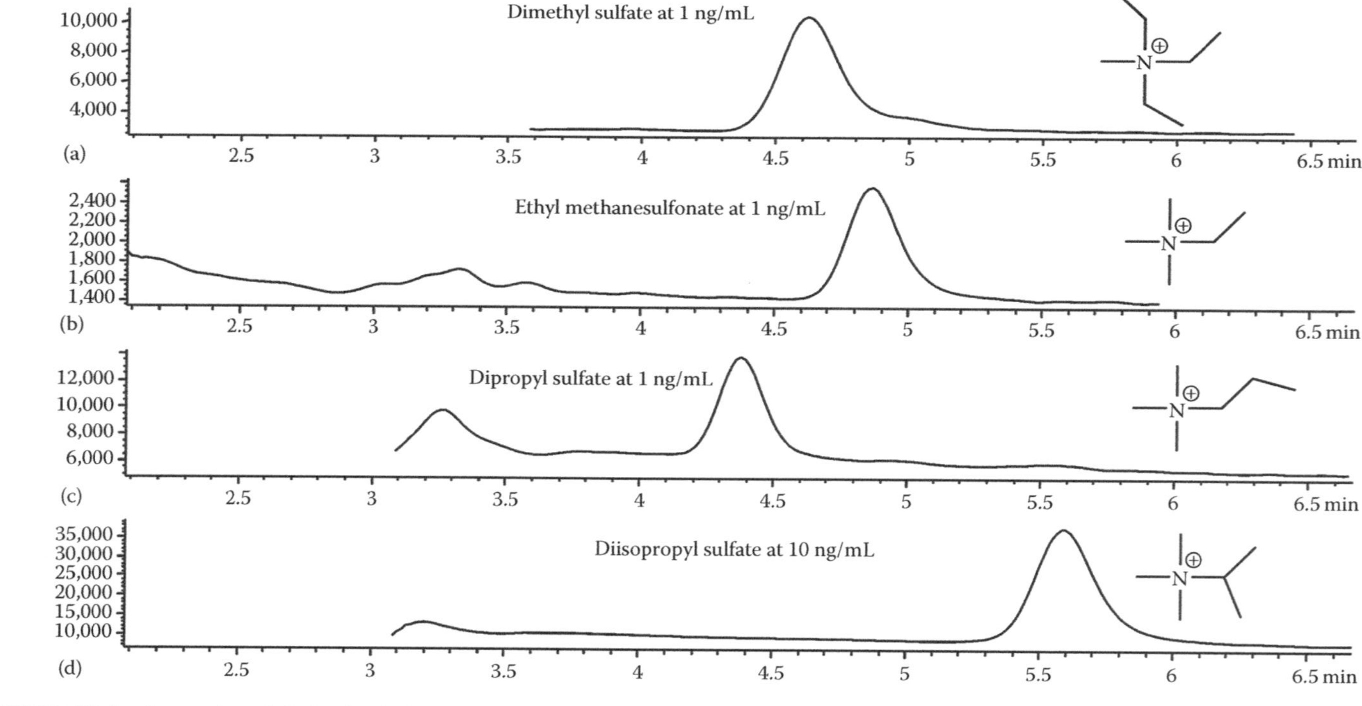

FIGURE 12.4 Separation of derivatized alkyl sulfonates on a silica HILIC column. See Table 12.1 for experimental details.

detector (ECD).[29] Direct injection GC methods are generally not desirable because of potential instrument contamination issues due to the presence of a large amount of API. Furthermore, for some unstable or nonvolatile polar alkyl halides, the HSGC approach is not an option because of the long incubation duration (typically 10 min) at the elevated temperature (near boiling point of analytes). And these compounds may not survive such conditions especially in the presence of API. A group of alkyl halides that fall into this category is listed in Table 12.2. They are either unstable or not volatile enough for HSGC analysis, nor suited for direct HPLC analysis.[44,45] For example, bis(2-chloroethyl)amine seems a good candidate for direct LC/MS analysis. However, this compound is readily hydrolyzed to a series of degradants in an aqueous solution.[46,47] Therefore, an analysis of this compound after derivatization not only stabilized the compound but also improved its detectability.[45] 3-Chloropropane-1,2-diol, has been analyzed by headspace solid-phase microextraction (HS-SPME) coupled with the GC/MS method following derivatization with phenylboronic acid.[48] The detection limit of the method was 3.78 ppm, which was insufficient for the application in the analysis of pharmaceutical genotoxins.

Following the same strategy developed for the analysis of alkyl sulfonates, alkyl halides can also be converted to polar amines that can be readily separated from APIs by HILIC and detected by electrospray mass spectrometry.[44,45] Based on the reactivity of the alkyl halides, they can be derivatized with either dimethylamine (DMA) or trimethylamine (TMA). The derivatization conditions and structures of derivatives are listed in Table 12.2. For the majority of the compounds, the derivatization reactions proceed quickly. Alkyl chlorides, on the other hand, are generally not as reactive as their bromide and iodide analogues. 4-Chlorobutan-1-ol was derivatized with dimethylamine catalyzed by NaI. Its derivative is not permanently charged, but is protonated in the slightly acidic mobile phase. A separation of the derivatives from API and other interferences was achieved on a silica HILIC column with the mobile phase containing acetonitrile and an ammonium formate buffer of pH 4 (Table 12.1). Because of the diverse structures of the alkyl halides, a single gradient may not be sufficient to cover the separation of all the analytes. For optimal separation and detection sensitivity, the percentage of the aqueous buffer in the mobile phase is adjusted in a typical range from 13% to 30% (Table 12.2). The retention of the derivatives appears to be related to both their polarity and their ability to generate ionic interactions with the stationary phase (Figure 12.5). Nonetheless, further investigations seem warranted to understand how the molecule structures and chromatographic conditions impact the retention factor.

Excellent sensitivity was achieved for all these trace levels of analytes in the presence of API. The limits of quantitation (LOQs) of the various analytes in the range of 0.1–0.2 ppm (ca) were successfully achieved. In the case of 3-chloropropane-1,2-diol,[44] the HILIC/MS approach was about 20 times more sensitive than the derivatization-HS-SPME-GC/MS method, which had an LOQ of 3.78 ppm.[48] The recovery tests were performed by spiking alkyl halides into various investigational APIs where their manufacturing process involved the use of the corresponding alkyl halides. It was observed that the recovery of the analytes ranged from 50% to 109% depending on the API sample matrices (Table 12.2). Since these methods are limit tests only, the

TABLE 12.2
Structures of Seven Alkyl Halides, and Key Analytical Validation Results of Derivatization HILIC/MS Methods

Genotoxic Impurity	Reagent	Derivative	Reaction Temp/ Time (°C)/h	Diluent	Mobile Phase A/B	LOQ (ppm) (ca)	Recovery (%)
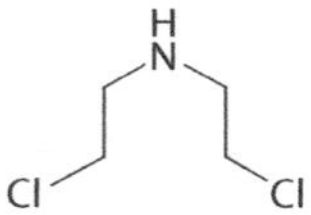	DMA	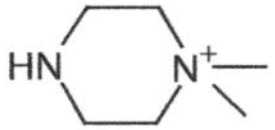 *m/z* 115	60/1	ACN/acetic acid (99/1)	30/70	0.2	98% at 1 ppm with %RSD 1.6%
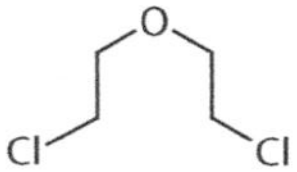	DMA	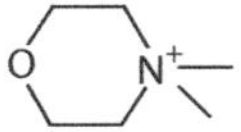 *m/z* 116	35/1	ACN/H_2O (85/15)	13/87	0.1	89% at 2 ppm with %RSD 0.4%
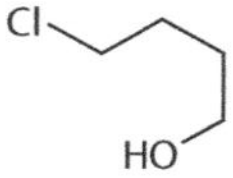	DMA	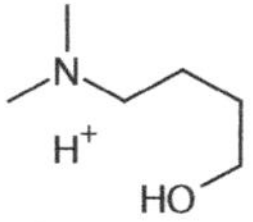 *m/z* 118	60/24	2 mg/mL NaI in ACN/H_2O (9/1)	14/86	0.2	50% at 2 ppm with %RSD 5.1%
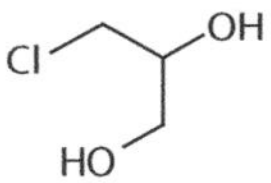	DMA	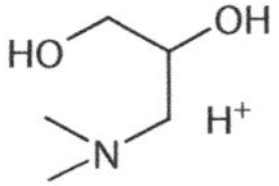 *m/z* 120	75/2	ACN/H_2O (4/1)	17/83	0.2	58% at 3.5 ppm with %RSD 3.0%

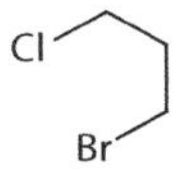 Cl Br	TMA	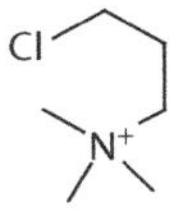 Cl N+ *m/z* 136	60/1.5	ACN/H_2O (4/1)	15/85	0.1	87% at 5 ppm with %RSD 3.3%
 Br OH	TMA	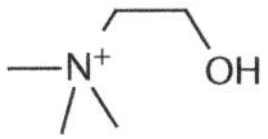 N+ OH *m/z* 104	60/1	ACN/H_2O (4/1)	15/85	0.1	56% at 2 ppm with %RSD 2.1%
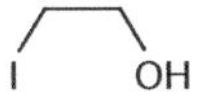 I OH	TMA	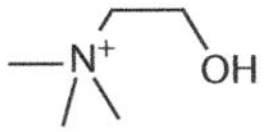 N+ OH *m/z* 104	70/4	ACN/H_2O (4/1)	15/85	0.1	109% AT 3 ppm with %RSD 3.2%

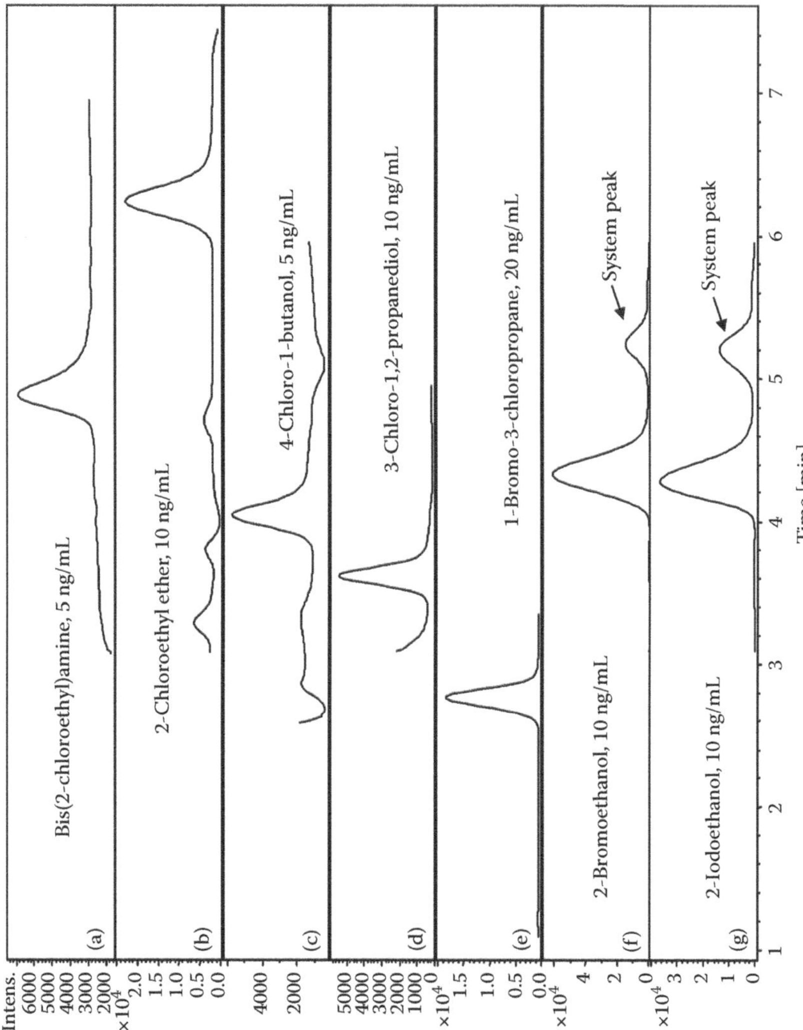

FIGURE 12.5 Separation of derivatized alkyl halides by HILIC. See Tables 12.1 and 12.2 for details.

relatively low recoveries are acceptable per regulatory method validation guidelines. The relatively low recovery of certain analytes in certain sample matrices requires further investigation. One possible reason could be that the derivatization reaction conditions for those compounds were not optimal and APIs may have interfered with the derivatization reaction. In summary, the HILIC/MS approach provides an alternative strategy for the sensitive detection of some unstable alkyl halides.

12.3 ANALYSIS OF POLAR IMPURITIES IN NONPOLAR APIs OR INTERMEDIATES

Certain polar impurities in nonpolar APIs or intermediates cannot be quantified using a single impurity profile method because they may not be retained on the RP-HPLC columns. In such a situation, a second method is usually required to quantify the specific polar impurity. HILIC has been successfully applied to the analysis of polar impurities in the presence of APIs or intermediates, of which the polar impurities are retained on HILIC columns, while the nonpolar components are eluted earlier than the targeted polar impurities.

12.3.1 IMIDAZOLE

N,N′-Carbonyldiimidazole is a common coupling reagent for the synthesis of amides and carbamates. Imidazole (Figure 12.6) is a by-product of the coupling reactions, which needs to be controlled as low as 0.05% w/w. Imidazole was not retained on the C18 column that was used for the impurity profile analysis. An ion pairing chromatographic method was also explored. However, imidazole was barely retained by a C8 column with 1-hexanesulfonic acid as the ion pairing agent in a high aqueous mobile phase. Ultimately, a HILIC method was developed for the accurate determination of imidazole using a silica HILIC column. The mobile phase was a combination of the ammonium formate buffer and acetonitrile. It was observed that the retention factor of the imidazole peak was greatly affected by the percentage of the aqueous buffer, as expected for HILIC. The desired retention of the analyte was achieved with an isocratic elution using an ammonium formate buffer in acetonitrile (Table 12.1). The API peak eluted in the column void for about 0.5 min while the imidazole peak had a retention time of about 2.0 min with a good peak shape and a tailing factor of 1.2 (Figure 12.7). The method was validated according to ICH guidelines. The %RSD of injection precisions is less than 0.01%. LOQ can be as low as 0.0004% (w/w) based on a 50 mg/mL sample.

12.3.2 (3*S*)-3-MORPHOLINEMETHANOL

In regulated API manufacturing facilities, reaction vessels or drying equipment must be cleaned and verified before being released for producing a different drug or intermediate. Depending on the dose of the compounds, the levels of impurities in the cleaning solution are usually controlled at ppm (μg/mL) levels, and must be verified by validated limit test analytical methods. (3*S*)-3-Morpholinemethanol (Figure 12.6) is a building block of an experimental drug in early development. After production

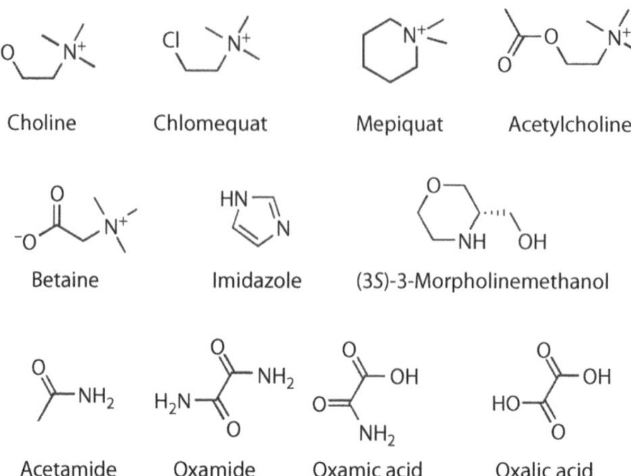

FIGURE 12.6 Structures of representative pharmaceutical compounds and impurities analyzed by HILIC.

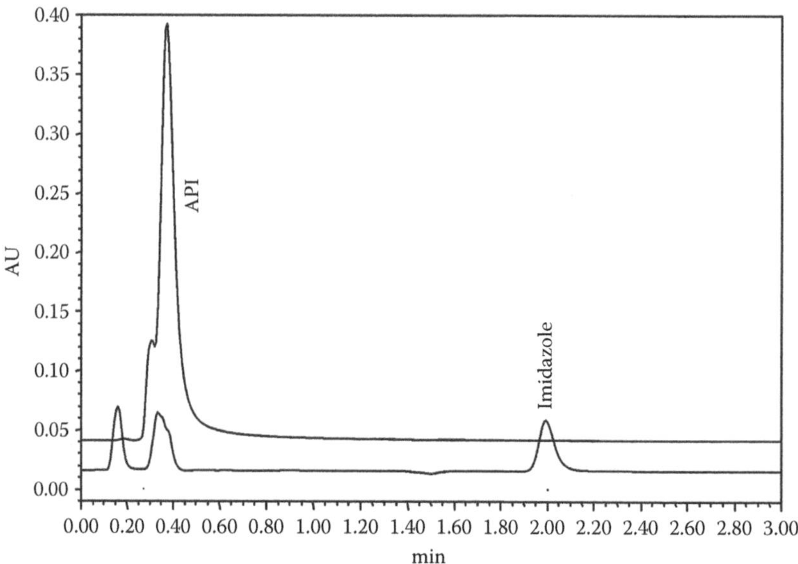

FIGURE 12.7 Separation of imidazole from API on an Atlantis Silica HILIC column. See Table 12.1 for details.

of the API, the equipment was cleaned with 200 proof ethanol. The levels of the compound need to be controlled to less than 8 ppm. The process monitoring GC method cannot provide sufficient sensitivity. Considering the polar nature of this compound, a HILIC method was developed. A silica HILIC column was used for the separation with 15% of 20 mM ammonium formate buffer (pH 6) in acetonitrile as the mobile phase. The separation of the peaks is illustrated in Figure 12.8.

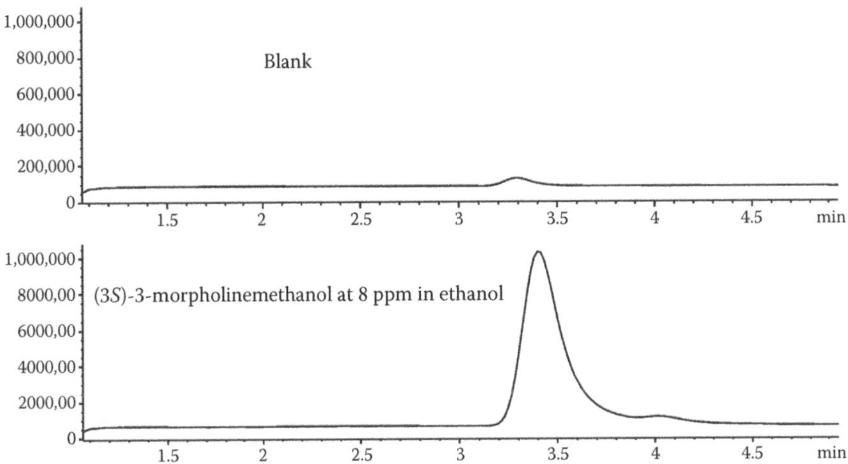

FIGURE 12.8 Analysis of (3*S*)-3-morpholinemethanol by HILIC/MS. See Table 12.1 for details.

An Agilent LC/MSD was used, and the analyte was detected in the SIM mode by monitoring the ion at *m/z* 118. The S/N at 8 ppm was better than 50. The injection precision was 8% RSD. The relatively high value is because *m/z* 118 is one of the ions used for instrument calibration.

12.3.3 Acetamide

Olsen explored the application of silica and amino columns for the analysis of low molecular weight acetamide (Figure 12.6).[49] Both the columns seemed able to retain the compound. To determine the acetamide level in a pharmaceutical intermediate, an amino HILIC column with the acetonitrile/water (90/10) mobile phase was used. As expected, the bulk intermediate peak eluted earlier than the acetamide peak (Figure 12.9). The acetamide was sufficiently retained on the column to allow detection at a 0.05% level. However, the sensitivity was undermined by the tailing of the large intermediate peak due to the insufficient specificity of UV detection.

12.3.4 Organic Acids

When neutral or acidic mobile phases are used, the stationary phase of amino columns are protonated. Anionic analytes such as organic acids can be retained via ion-exchange mechanisms in addition to hydrophilic partition. Oxamic acid, oxalic acid, and neutral oxamide, the possible impurities in an API (Figure 12.6), could not be resolved by RP-HPLC but were separated by an ion-exclusion chromatography previously.[50] On an amino column with acetonitrile/phosphate buffer at pH 7 as the mobile phase, the compounds were retained in the following order: di-anionic oxalic acid > mono-anionic oxamic acid > neutral oxamide. The elution order appears to be in correlation with acidity.[49] Their retentions seem dependent on the buffer concentrations. Increasing the buffer concentration resulted in less retention of the two

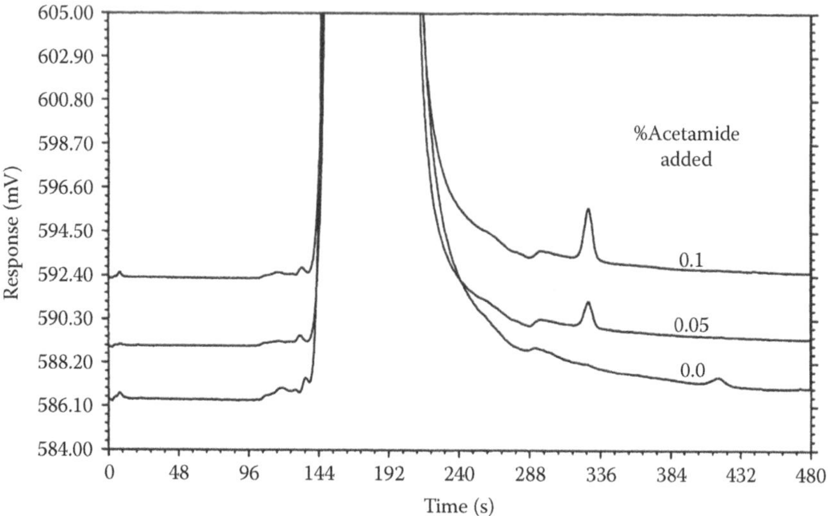

FIGURE 12.9 Determination of acetamide in a pharmaceutical intermediate. Conditions: Zorbax NH_2 column; acetonitrile–water (90:10) mobile phase; 205 nm detection; sample concentration: 4 mg/mL in tetrahydrofuran. (Reprinted from Olsen, B.A., *J. Chromatogr. A*, 913, 113, 2001. With permission from Elsevier.)

acids but did not affect the retention of the neutral oxamide. The experiments demonstrated that the ion-exchange mechanism is responsible for the retention of the acids on the HILIC columns. A successful determination of the three compounds in a nonpolar API was achieved on an amino column using acetonitrile and 50 mM potassium phosphate pH 7 as the mobile phase (Table 12.1). The two acids were well separated from the API peak while oxamide eluted at the tail of the API peak. Oxalic acid gave a relatively broad peak so its sensitivity was no better than that of the ion-exclusion chromatography method.[50] Other factors such as the pH of the mobile phases could be further optimized to improve the separation.

12.4 PURITY AND IMPURITY ANALYSIS OF POLAR APIs AND INTERMEDIATES

12.4.1 5-Fluorouracil in 5-Fluorocytosine

5-Fluorouracil (Figure 12.10), an impurity in 5-fluorocytosine, needs to be controlled below the limit of 0.1%.[49] Both silica and amino columns were explored for the determination of 5-fluorouracil in 5-fluorocytosine. The less polar 5-fluorouracil was eluted in front of 5-fluorocytosine on both columns with good resolution (Figure 12.11). Two limit test methods were developed using the two columns respectively. One utilized a silica HILIC column eluted with an acidic mobile phase and 5 mM phosphoric acid (25%) in acetonitrile. Another used the amino HILIC column with a near neutral mobile phase and 25 mM potassium phosphate in acetonitrile with a pH of 6.5 (Table 12.1). Recently, other studies suggest that a silica-based amino stationary phase is not

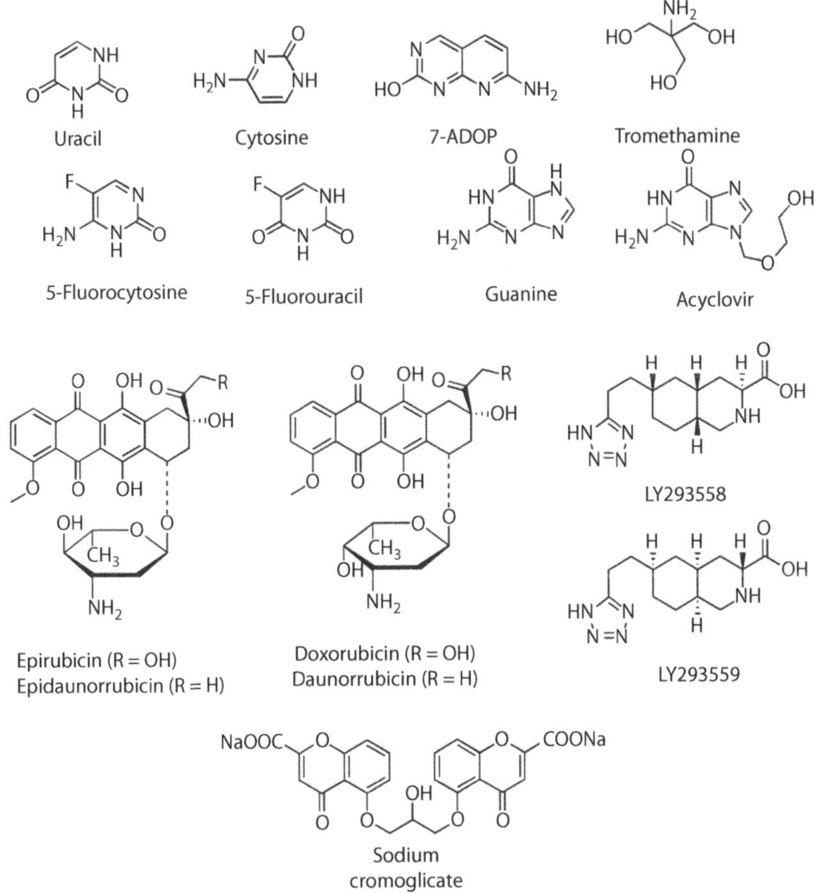

FIGURE 12.10 Structures of representative pharmaceutical compounds and impurities analyzed by HILIC continued.

stable under these chromatographic conditions, and a polymer-based amino column is a better choice in terms of chromatographic stability and column bleeding.[51,52]

12.4.2 Guanine in Acyclovir

A similar limit test method was developed for the determination of guanine (Figure 12.10), an impurity in acyclovir with an LOQ of 0.7%.[49] The mobile phase contained acetonitrile and 5 mM phosphoric acid (Table 12.1). Both guanine and acyclovir were well retained and resolved with the guanine peak eluting after the acyclovir peak. The silica HILIC column performed better than the amino column in terms of method selectivity, and thus it was selected over the amino column where guanine was not baseline resolved from the acyclovir peak. Furthermore, a zwitterionic HILIC column may be potentially evaluated as an alternative approach for the similar purposes.[53]

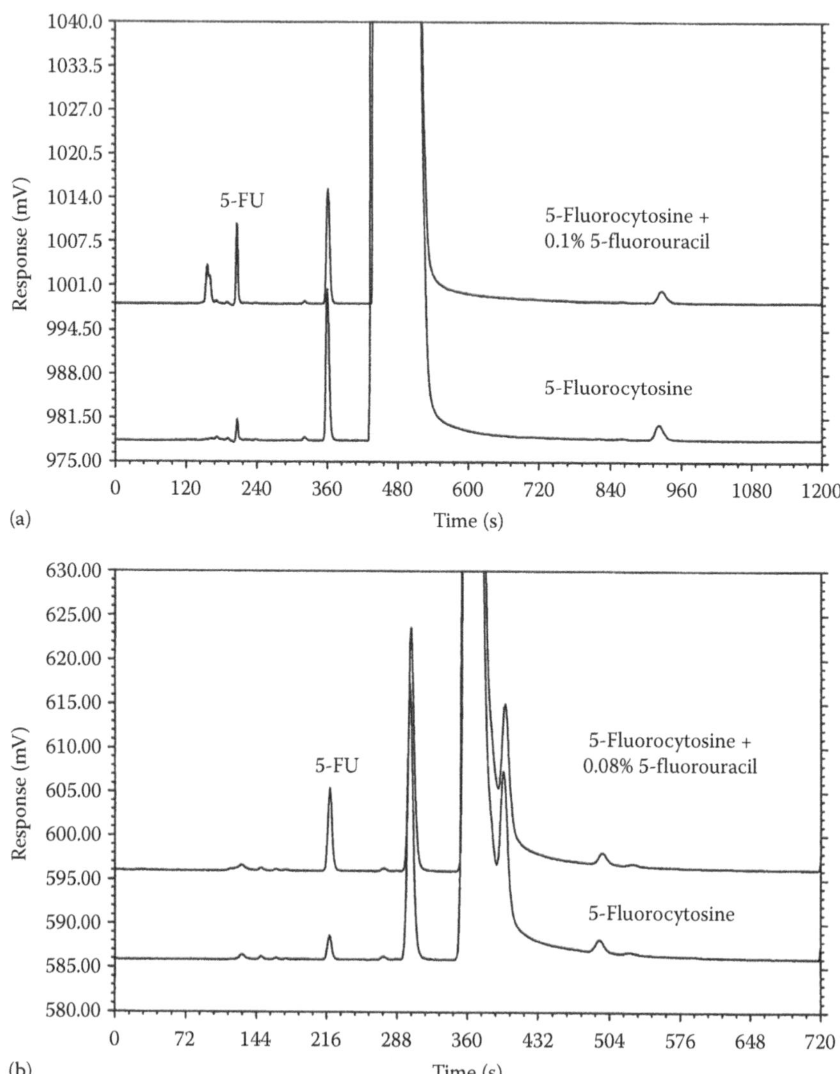

FIGURE 12.11 Determination of 5-fluorouracil in 5-fluorocytosine. (a) Zorbax SIL column; mobile-phase acetonitrile–water (75:25), 5 mM phosphoric acid; 275 nm detection; 5-fluorocytosine (1.0 mg/mL). (b) Zorbax NH_2 column; mobile-phase acetonitrile–25 mM potassium phosphate, pH 6.5 (80:20); 275 nm detection; 5-fluorocytosine (1.0 mg/mL). (Reprinted from Olsen, B.A., *J. Chromatogr. A*, 913, 113, 2001. With permission from Elsevier.)

12.4.3 Epirubicin

Epirubicin, an oncology drug, contains doxorubicin, daunorubicin, and epidaunorubicin as the major impurities (Figure 12.10). They all contain a very polar daunosamine sugar moiety and have been analyzed by ion pairing HPLC.[54] Li and Huang studied the separation of these compounds on a bare silica HILIC column and obtained

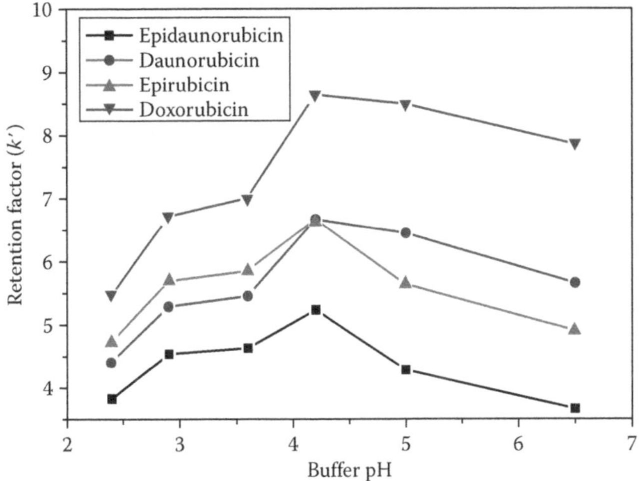

FIGURE 12.12 Effect of buffer pH on the retention factor (k') of epirubicin and its analogues. Conditions: Kromasil KR100-5SIL (5 μM); mobile phase: sodium formate buffer (20 mM) modified with acetonitrile (10:90, v/v). Peaks: (■) epidaunorubicin, (•) daunorubicin, (▲) epirubicin, (▼) doxorubicin. (Reprinted from Li, R. and Huang, J., *J. Chromatogr. A*, 1041, 163, 2004. With permission from Elsevier.)

promising results.[55] A mixture of acetonitrile and sodium formate buffer was used as the mobile phase. The pH value of the mobile phase was screened between pH 2.4 and 6.5. It was observed that the pH value of the buffer had a significant effect on the retention factors of the analytes (Figure 12.12). At a low pH, the retention of these molecules increased with the increase of the mobile phase pH. The best retention was obtained at about pH 4.2 for all compounds. However, a small change of the mobile phase pH around pH 4.2 may change the elution order of daunorubicin and epirubicin. A further increase of the mobile phase pH caused a decrease of the retention factors, and the analytes displayed strong irreversible binding to the silica column at pH 6.5 or higher, which was consistent with the instability of the dihydroquinone type of compounds at higher pH. Based on the experimental results, the optimal pH value of the mobile phase should be in the range from 2.9 to 3.6 or 5 to 6.5. In these pH ranges, the compounds were well resolved and the pH effects on the retention factor and elution order were relatively small. A higher pH should be avoided because of the irreversible adsorption of the analytes on the column. A sodium formate buffer (30 mM) at pH 2.9 was selected for the batch analysis of epirubicin containing the related impurities (Table 12.1). All impurities were resolved from the major peak with resolutions better than 2.0 (Figure 12.13). Under similar conditions, the temperature effects on the separation were studied separately.[56] A higher column temperature at 40°C improved column efficiency resulting in a slightly better separation of the analytes.

12.4.4 Chiral Compounds

An experimental drug LY293558 cannot be easily separated from its enantiomeric impurity LY293559 (Figure 12.10) by a normal-phase chiral method without involving

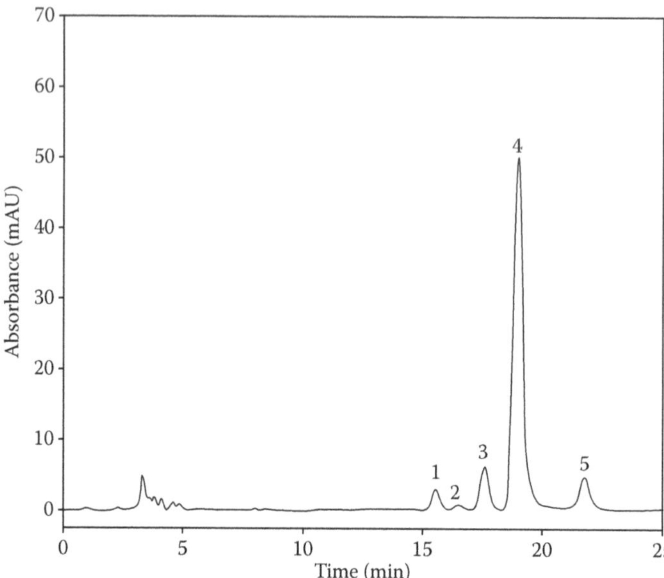

FIGURE 12.13 Chromatogram of the analysis of epirubicin hydrochloride in a raw material. Conditions: Kromasil KR100-5SIL (5 μm); mobile phase: sodium formate buffer (30 mM, pH 2.9)–acetonitrile (10:90, v/v). Peaks: (1) epidaunorubicin, (2) unknown compound, (3) daunorubicin, (4) epirubicin, (5) doxorubicin. (Reprinted from Li, R. and Huang, J., *J. Chromatogr. A*, 1041, 163, 2004. With permission from Elsevier.)

deravitization.[57,58] Guisbert et al. reported the direct separation of the enantiomers using a Chirobiotic T™ column, which displayed classic HILIC characteristics when the water–acetonitrile mobile phase was used. The retention factor of LY293558 was improved from 2.0 to 5.2 when the acetonitrile content was increased from 40% to 80% (Figure 12.14). At a higher composition of acetonitrile (80%) in the mobile phase, the resolution of the two enantiomers was improved to 7.3 from 3 at a lower organic composition (5%). The pH value in a narrow range (4.0–6.5) did not seem to have significant effects on the retention and resolution of the two enantiomers. The optimized separation of the two enantiomers was accomplished with a mobile phase of aqueous acetonitrile (65%) with an ELSD for detection (Figure 12.15). To overcome the narrow dynamic range of the detector, a "high–low" injection strategy was applied. That is, high concentration samples were injected for the determination of the low level impurity, while a further diluted sample was injected for the determination of the major peak. The method was validated in terms of linearity, precision, accuracy, recovery, and sensitivity. At a sample concentration of 10 mg/mL of LY293558, the quantification limit of LY293559 could be as low as 0.1%.

12.4.5 Carbamates

Wang et al. developed a stability-indicating method for the analysis of a proprietary experimental drug.[2] Three impurities or degradants were structurally closely related to the API. The structures of the analytes were not disclosed. Compound 1 has amino and

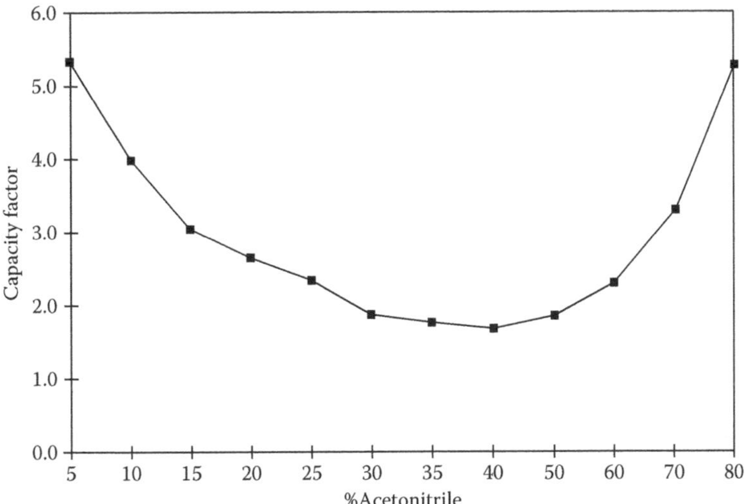

FIGURE 12.14 Effect of increasing the organic modifier on the capacity factor. (Reprinted from Guisbert, A.L. et al., *J. Liq. Chromatogr. Relat. Technol.*, 23, 1019, 2000. With permission from Elsevier.)

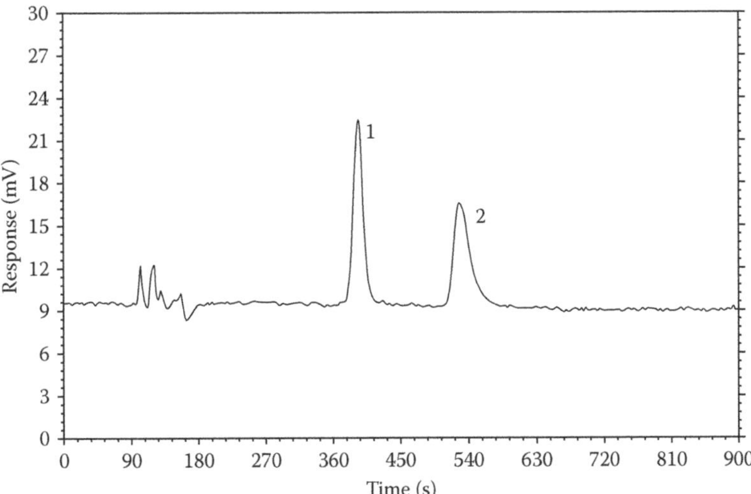

FIGURE 12.15 Sample chromatogram for the separation of LY293558 (1) and LY293559 (2). (Reprinted from Guisbert, A.L. et al., *J. Liq. Chromatogr. Relat. Technol.*, 23, 1019, 2000. With permission from Elsevier.)

hydroxyl functional groups. The polar API (Compound 2) is aromatic, and contains amino and carbamate functional groups. Compound 3 has urea and alcohol functional groups. Compound 4 contains a carbamate functional group. The polarity and basicity of Compound 1 and API were obviously higher than that of compounds 2 and 3 because of their amino functional groups. In an RP-HPLC analysis, their retention factor was in the order of compound 1 < compound 2 < compound 3 < compound 4

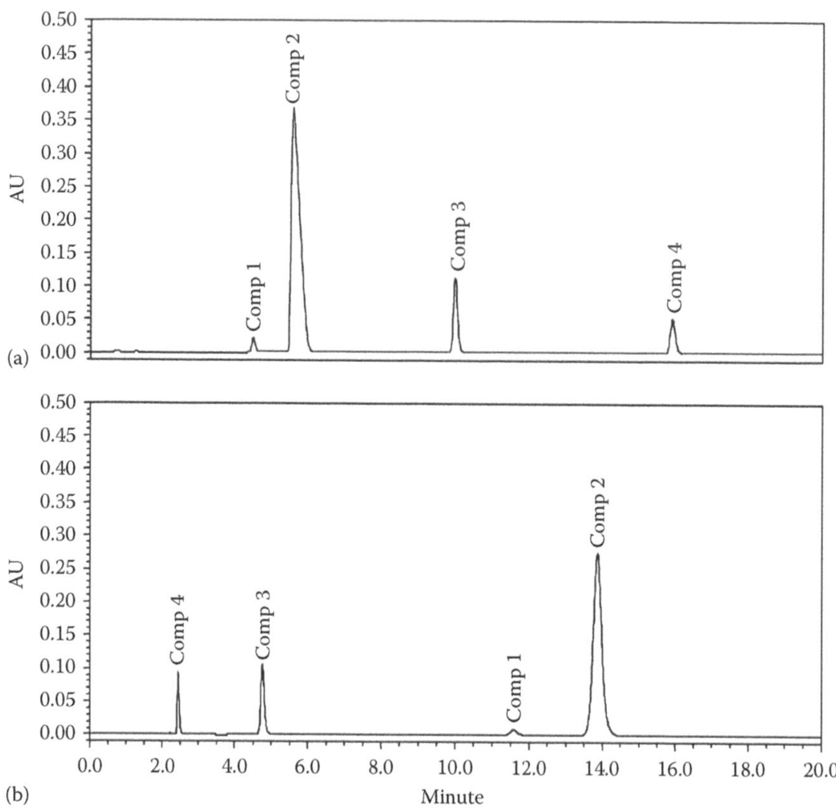

FIGURE 12.16 Chromatograms of the specificity solution on (a) RP-HPLC and (b) HILIC. RP-HPLC conditions: column 100 mm × 4.6 mm 5 μm Xterra MS C18; gradient elution with (A) 0.09% phosphoric acid and (B) acetonitrile (B, 2%–25% in 20 min); column temperature 35°C; flow rate 1 mL/min; UV detection 215 nm. HILIC conditions: column 250 mm × 4.6 mm 5 μm YMC-pack Diol-120 NP; mobile phase 10 mM NH_4Cl in acetonitrile/water (95:5, v/v); column temperature 30°C; flow rate 1.5 mL/min; UV detection 215 nm. (Reprinted from Wang, X. et al., *J. Chromatogr. A*, 1083, 58, 2005. With permission from Elsevier.)

(Figure 12.16a). An orthogonal HILIC method using a diol column was also developed. Under HILIC conditions, the retention factor was changed to the order of compound 4 < compound 3 < compound 1 < compound 2 (Figure 12.16b). The addition of ammonium chloride to the water/acetonitrile mobile phase did not assert significant effects on the separation. A mobile phase containing acetonitrile and 10 mM NH_4Cl was selected for the final method. The method was validated according to ICH guidelines, and LOQs of the impurities were estimated to be 0.05%. The HILIC method served as an orthogonal method for the RP-HPLC method.

12.4.6 Sodium Cromoglicate

Sodium Cromoglicate (SCG, Figure 12.10), a polar drug for the treatment of allergy, had been analyzed by RP-HPLC with a high aqueous mobile phase.[59] SCG (pK_a 1.9) is negatively charged under basic or weakly acidic conditions. Alternatively, the compound could be analyzed by HILIC. Ali et al. developed a stability-indicating impurity profile method for the compound.[60] SCG was analyzed by a silica HILIC column using ammonium acetate buffered acetonitrile/water as the mobile phase (Table 12.1). Increasing the pH of the mobile phase from 4.0 to 5.8 reduced its retention factor from 8.7 to 3.7, probably due to the increased repulsion between the negatively charged silanol and the negatively charged analyte. Lowering the mobile phase pH increased the retention of SCG, and thus a buffer of pH 3.0 was selected. At pH 3.0, increasing the concentration of ammonium acetate slightly improved the retention from 3.1 to 4.4, presumably because the ammonium ion would interact with the negative charges of the silanols and thus reduce the repulsion between SCG and ionized silanol. The method was validated with respect to linearity, specificity, accuracy, precision, and sensitivity. Forced degradation samples of SCG were analyzed by the method, and the degradants were well separated from the API peak (Figure 12.17). A limited method robustness test was performed. Furthermore, as demonstrated by Yang et al., amino columns might be an alternative for acidic compounds.[50]

12.4.7 Cytosine

Cytosine (Figure 12.10) is a common building block for the synthesis of many drug substances. 7-ADOP and uracil are two major impurities observed in commercial supplies.[61] Cytosine and other pyrimidines have been analyzed either by GC after derivatization or ion-pairing HPLC methods.[62,63] One amide and two amino columns were screened for the separation of the three compounds using a gradient mobile phase from 5% to 25% aqueous sodium formate (pH 3.5)/acetonitrile in 10 min. Cytosine and 7-ADOP were retained on both columns, while uracil was only retained on amide columns. Thus, a HILIC method using an amide column was developed. The method was fully validated, demonstrating acceptable linearity, precision, repeatability, and sensitivity. During robustness evaluation, a range of parameters including flow rate, column temperature, injection volume, mobile phase pH, buffer concentration, values of retention factor, tailing factor, and resolution were shown to be satisfactory. Intermediate precision and sample stability were also evaluated. Should there be a need to convert the method to a mass spectrometry compatible method, ammonium formate buffer could be used.

12.5 COUNTERION ANALYSIS

In order to improve physicochemical properties, drug substances are often produced as salts. The counterions could be inorganic ions, polar organic bases, or acids. The stoichiometry of the salt form is a critical attribute of APIs and thus must be determined as part of batch release tests before the usage. These counterions are generally not retained on conventional RP-HPLC columns, and had been analyzed by several special techniques including inductively coupled plasma-atomic emission spectrometry (ICP-AES),[64] capillary electrophoresis (CE),[65] and ion chromatography (IC).[66] While IC is the most commonly used approach in the pharmaceutical industry, individual IC method developments are necessary for each assay. HILIC-based HPLC

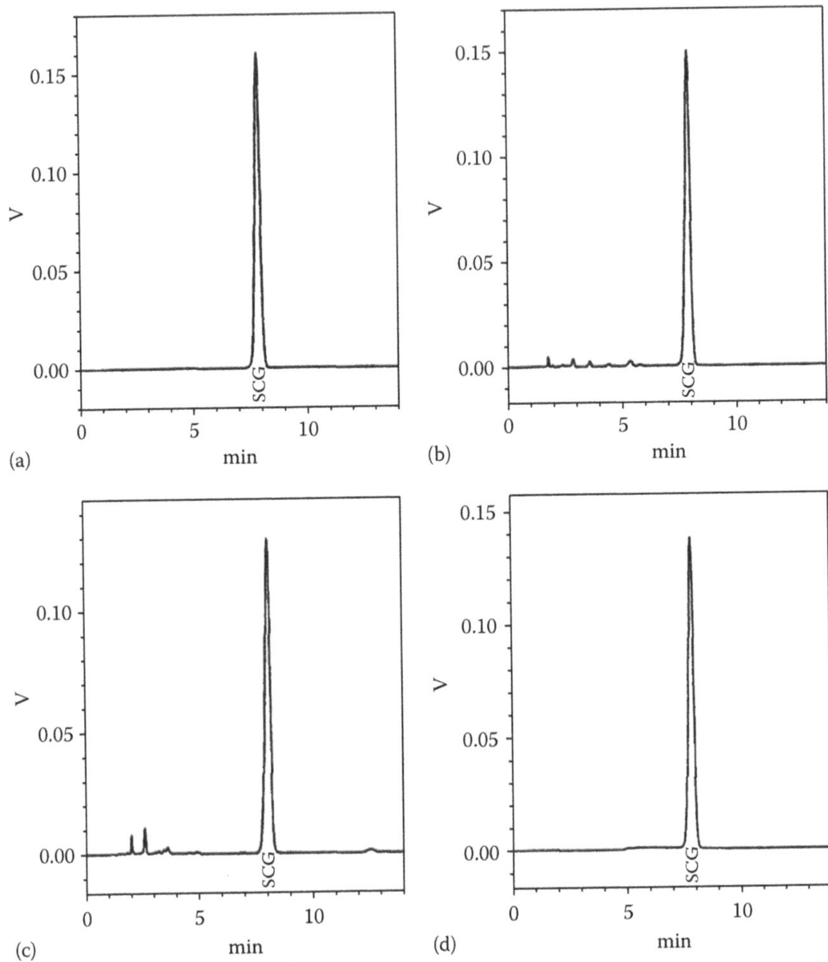

FIGURE 12.17 Chromatograms of SCG obtained at various stress conditions. (a) Standard (no stress), (b) Oxidative (5% H_2O_2, 5 mL, 80°C, 5 h), (c) Photolytic-254 nm (24 h) and (d) Basic 0.1N NaoH, 5 mL, RT, 2 h). (Reprinted from Ali, M.S. et al., *J. Sep. Sci.*, 31, 1645, 2008. With permission from Wiley-VCH Verlag GmbH & Co. KGaA.)

methods have been shown as a promising approach to simplify the method development.[67–69] Risley and Pack developed a method for the simultaneous analysis of 12 cations and 21 anions using HILIC interfaced with an ELSD (Table 12.1).[68] The method used a zwitterionic HILIC column with a mixture of acetonitrile and an ammonium formate buffer as the mobile phase, which gave satisfactory separation of all the ions. It was believed that the analytes were separated by an ion-exchange mechanism. The mobile-phase pH had significant effects on the retention of the ions in the range from pH 3.1 to 6.6. The ionic interaction between the analytes and the stationary phase dominated the separation mechanism; thus, increasing the pH resulted in a decreased retention of anions and an improved retention of cations. For the same reason, buffer concentrations affected the retention and the peak shape and a buffer concentration ranging from 50 to 100 mM was recommended. The low buffer concentration caused peak tailing, while the high concentration resulted in peak fronting. Satisfactory validation data including linearity, precision, and accuracy were obtained. The accuracy for the majority of the ions was within ±2.5% of the theoretical values except for PO_4^{-3}, which was believed to be impaired by the limited solubility in the organic diluent. Limited solubility of some API salts in the organic-rich diluent is the limiting factor for the application of the method. Huang et al. explored charged aerosol detection (CAD) for the analysis of inorganic counterions.[69]

A group of amino HILIC columns was also explored for the analysis of organic counterions. Guo and Huang[67] determined tromethamine (Figure 12.10) in an API using an amino column interfaced with a refractive index (RID) detector. The effects of ammonium acetate buffer on the separation were examined. The presence of ammonium acetate increases the retention, probably due to the interaction between the amino groups of the stationary phase and the negatively charged acetate ions that reduce the repulsion between the analytes and the stationary phase. However, increasing the buffer concentration caused the fronting of the analyte peak. The authors attributed the phenomenon to the potential changes of the water-enriched layer on the surface of the stationary phase that may be disturbed by the higher salt concentration. As demonstrated in the analysis of other basic compounds, ZIC or silica HILIC columns appear to be better choices for the task.[4,68] The method was validated with regard to the specificity, repeatability, linearity, and sensitivity. The %RSD of six injections was 1.9%, and the detection limit was 0.03 mg/mL. The application was demonstrated with an actual formulated API (Figure 12.18). The method robustness was satisfied by slightly changing the mobile phase composition, flow rate, and column temperature. Three columns from the same manufacturer were used for testing the repeatability, with about 10% variation observed in the retention time and the separation efficiency.

12.6 CONCLUSIONS AND FUTURE PROSPECTS

The above examples demonstrate the representative applications of HILIC in the analysis of polar impurities in various pharmaceutical products in pharmaceutical R&D. When compared to its applications in the bioanalysis of drugs and drug metabolites, HILIC is to some extent underutilized in pharmaceutical analysis.[13] The majority of the published applications appears to be in the R&D phases rather than in the commercial production stage. As described earlier, HILIC has been a useful tool in the analysis of polar impurities in polar APIs and intermediates as well as in the analysis of polar impurities in nonpolar pharmaceuticals. In some cases, HILIC methods were developed as orthogonal methods to the most popular RP-HPLC methods, either as a substitution for ion-pairing methods or as mass spectrometry compatible methods.

Most small molecule drugs in pharmaceutical pipelines are nonpolar, and thus RP-HPLC methods are well suited for their analyses. Typically, their impurity profile methods are modified to accommodate all the impurities including the polar analytes whenever possible to minimize the need for an additional method. Thus, HILIC is only used as a supplement method for the determination of very polar compounds

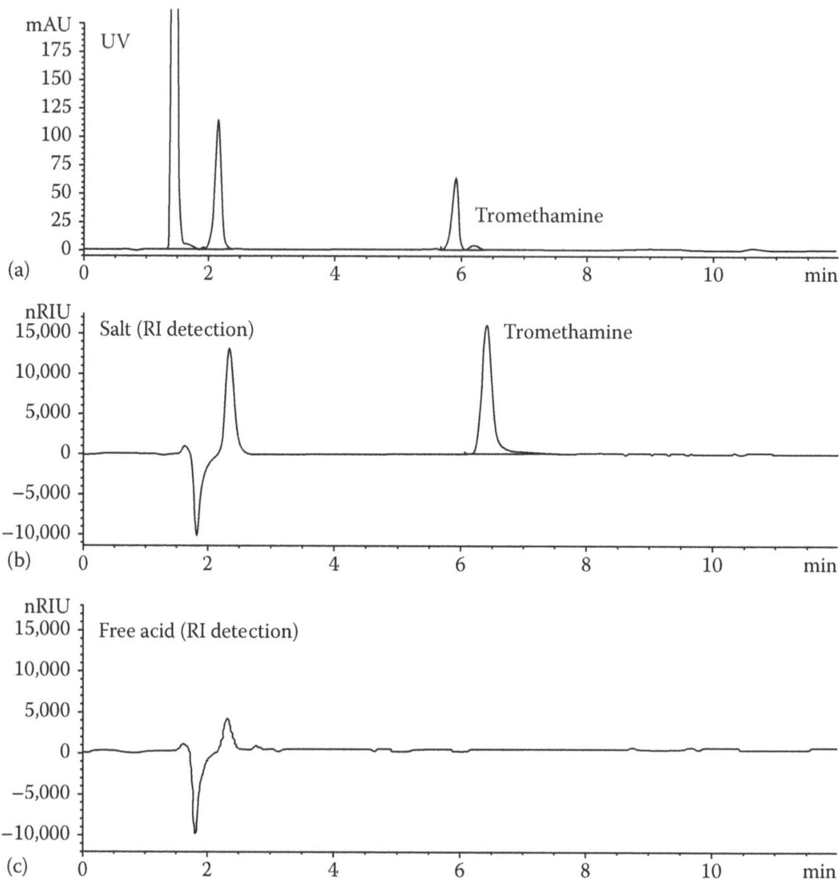

FIGURE 12.18 Chromatograms for the tromethamine salt (a and b) and free acid (c) forms of investigational drug, from UV (a) and RI detectors (b and c). Column: Zorbax NH_2, 4.6 × 150 mm, 5 mm particle size. Column temperature: 25°C. Mobile phase: acetonitrile/water (80/20, v/v). Flow rate: 1 mL/min. Sample: the investigational API (salt form/2 mg/mL in the mobile phase). Injection volume: 50 mL. (Reprinted from Guo, Y. and Huang, A., *J. Pharm. Biomed. Anal.*, 31, 1191, 2003. With permission from Elsevier.)

when conventional RP-HPLC approaches fail to deliver the desired separation or when facing other considerations such as mass spectrometry detection.[4] Among the process analytical methods to be transferred to manufacturing facilities, UV detection is the most popular detection method for HPLC analysis, while advanced techniques such as mass spectrometry are undesirable. Therefore, mass spectrometry compatible advantages of HILIC would not be a key factor regarding choosing either an HILIC method or an ion-pairing RP-HPLC method, unless mass spectrometry detection is used.

In light of the increasing attention on trace level genotoxic impurities in pharmaceuticals, analyte-specific trace analysis methods are being developed regularly. In order to achieve the required sensitivity and specificity, they are often mass spectrometry-based approaches.[70] HILIC/MS-based methods for the determination of genotoxic alkyl sulfonates and alkyl halide have been successfully developed. Furthermore, some of the low molecular weight polar genotoxic impurities such as small aromatic amines (Figure 12.1) can be potentially analyzed by HILIC. Therefore, increasing the usage of HILIC in pharmaceutical analytical development and even in commercial settings can be foreseen.

HILIC stationary phases are of different types of chemical structures, and the concerning polar analytes have diverse structures that could be acidic, basic, or neutral molecules. Depending on the mobile-phase composition, the analytes not only partition between the mobile phase and the water-rich layer on the surface of the stationary phase, but also interact with the stationary phase directly.[18] Thus, the retention mechanism of analytes on stationary phases could be a combination of hydrophilic, ionic, hydrogen bonding, and even hydrophobic factors.[20] Depending upon the conditions, one mechanism could dominate the others. Selecting the stationary phase and mobile phase for the target analytes is critical to the separation and could be challenging during method development. A systematic understanding of the retention mechanisms of analytes on different stationary phases has gained increased attention recently.[71] Hao et al. suggested taking advantage of the ion-exchange mechanism to separate ionizable analytes: selecting acidic columns for cationic analytes and basic columns for anionic analytes, respectively.[20] For example, the silica stationary phase can be used for separating bases, while the amino column can be used for separating acids, although this may not always work. Understanding of factors such as mobile-phase composition, buffer pH and concentration, column temperature that affects the retention factor, efficiency, resolution, peak shape, and peak tailing has been reviewed in the literature.[17,18,20,41,71] Knowledge in terms of structure and retention relationship, however, remains very limited. Column selectivity-based systematic column screening strategies that have been adopted in pharmaceutical analysis for RP-HPLC method development have not been reported for HILIC method development.[72] Further understanding of retention mechanisms and factors are imperative, which should help facilitate HILIC method development.

ACKNOWLEDGMENTS

The authors would like thank Dr. Jianguo An, Lin Bai, and Josephine Vega for providing analytical data and thoughtful discussions.

REFERENCES

1. Liu, M. et al., *J. Chromatogr. A*, 2009, 1216, 2362.
2. Wang, X., Li, W., and Rasmussen, H.T., *J. Chromatogr. A*, 2005, 1083, 58.
3. Hmelnickis, J. et al., *J. Pharm. Biomed. Anal.*, 2008, 48, 649.
4. Liu, M. et al., *J. Chromatogr. A*, 2008, 1188, 255.
5. International Conference on Harmonisation (ICH) Guidance for Industry: Impurities in New Drug Substances Q3A (R2); U.S. Department of Health and Human Services, Food and Drug Administration, Center for Drug Evaluation and Research (CDER): Rockville, MD, June 2008, http://www.fda.gov/RegulatoryInformation/Guidances/ucm127942.htm

6. International Conference on Harmonisation (ICH) Guidance for Industry: Impurities in New Drug Products Q3B (R2); U.S. Department of Health and Human Services, Food and Drug Administration, Center for Drug Evaluation and Research (CDER): Rockville, MD, July 2006, http://www.fda.gov/RegulatoryInformation/Guidances/ucm128032.htm
7. International Conference on Harmonisation (ICH) Guidance for Industry: Impurities: Guideline for Residue Solvents Q3C and Q3C(M); U.S. Department of Health and Human Services, Food and Drug Administration, Center for Drug Evaluation and Research (CDER): Rockville, MD, July 1997 and September 2002, http://www.fda.gov/downloads/RegulatoryInformation/Guidances/ucm128317.pdf
8. Guideline on the Limits of Genotoxic Impurities, CPMP/SWP/5199/02, EMEA/CHMP/QWP/251344/2006; Committee for Medicinal Products (CHMP), European Medicines Agency (EMEA): London, 28 June 2006, http://www.emea.europa.eu/pdfs/human/swp/519902en.pdf
9. Argentine, M.D., Owens, P.K., and Olsen, B.A., *Adv. Drug Deliv. Rev.*, 2007, 59, 12.
10. Tsai, T.H., *J. Chromatogr. B: Biomed. Sci. Appl.*, 2000, 747, 111.
11. Przybyciel, M. and Majors, R.E., *LCGC North Am.*, 2002, 20, 516.
12. Majors, R.E., and Przybyciel, M., *LCGC North Am.*, 2002, 20, 584.
13. Hsieh, Y., *J. Sep. Sci.*, 2008, 31, 1481.
14. Ramos, R.G. et al., *J. Chromatogr. A*, 2008, 1209, 88.
15. Vervoort, N., Daemen, D., and Török, G., *J. Chromatogr. A*, 2008, 1189, 92.
16. de Villiers, A. et al., *J. Chromatogr. A*, 2007, 1161, 183.
17. Hemstroem, P. and Irgum, K., *J. Sep. Sci.*, 2006, 29, 1784.
18. Ikegami, T. et al., *J. Chromatogr. A*, 2008, 1184, 474.
19. Ali, M.S. et al., *Chromatographia*, 2008, 67, 517.
20. Hao, Z., Xiao, B., and Weng, N., *J. Sep. Sci.*, 2008, 31, 1449.
21. Jin, G. et al., *Talanta*, 2008, 76, 522.
22. Strege, M.A., Stevenson, S., and Lawrence, S.M., *Anal. Chem.*, 2000, 72, 4629.
23. Strege, M.A., *Anal. Chem.*, 1998, 70, 2439.
24. Boyman, L. et al., *Biochem. Biophys. Res. Commun.*, 2005, 337, 936.
25. Beilmann, B. et al., *J. Chromatogr. A*, 2006, 1107, 204.
26. Risley, D.S., Yang, W.Q., and Peterson, J.A., *J. Sep. Sci.*, 2006, 29, 256.
27. Genotoxic and Carcinogenic Impurities in Drug Substances and Products: Recommended Approaches; U.S. Department of Health and Human Services, Food and Drug Administration, Center for Drug Evaluation and Research (CDER): Silver Spring, MD, December 2008; http://www.fda.gov/downloads/Drugs/Guidance-Compliance RegulatoryInformation/Guidances/ucm079235.pdf
28. Delaney, E.J., *Regul. Toxicol. Pharmacol.*, 2007, 49, 107.
29. Elder, D.P., Lipczynski, A.M., and Teasdale, A., *J. Pharm. Biomed. Anal.*, 2008, 48, 497.
30. Elder, D.P., Teasdale, A., and Lipczynski, A.M., *J. Pharm. Biomed. Anal.*, 2008, 46, 1.
31. Snodin, D.J., *Regul. Toxicol. Pharmacol.*, 2006, 45, 79.
32. Teasdale, A. et al., *Org. Process Res. Dev.*, 2009, 13, 429.
33. Jacq, K. et al., *J. Pharm. Biomed. Anal.*, 2008, 48, 1339.
34. Li, W., *J. Chromatogr. A*, 2004, 1046, 297.
35. Ramjit, H.G., Singh, M.M., and Coddington, A.B., *J. Mass Spectrom.*, 1996, 31, 867.
36. Raman, N.V.V.S. et al., *J. Pharm. Biomed. Anal.*, 2008, 48, 227.
37. Ramakrishna, K. et al., *J. Pharm. Biomed. Anal.*, 2008, 46, 780.
38. Salamoun, J., Nguyen, P.T., and Remien, J., *J. Chromatogr.*, 1992, 596, 43.
39. Nguyen, H.P. and Schug, K.A., *J. Sep. Sci.*, 2008, 31, 1465.
40. Guo, Y., *J. Liq. Chromatogr. Relat. Technol.*, 2005, 28, 497.
41. McKeown, A.P. et al., *J. Sep. Sci.*, 2001, 24, 835.
42. Weng, N., *J. Chromatogr. B*, 2003, 796, 209.

43. An, J. et al., *J. Pharm. Biomed. Anal.*, 2008, 48, 1006.
44. Bai, L. et al., *J. Chromatogr. A*, 2010, 1217, 302.
45. Sun, M. et al., *J. Pharm. Biomed. Anal.*, 2010, 52, 30.
46. Hemminki, K. et al., *Arch. Toxicol.*, 1987, 61, 126.
47. Watson, E., Dea, P., and Chan, K.K., *J. Pharm. Sci.*, 1985, 74, 1283.
48. Huang, M. et al., *Anal. Sci.*, 2005, 21, 1343.
49. Olsen, B.A., *J. Chromatogr. A*, 2001, 913, 113.
50. Yang, L. et al., *J. Pharm. Biomed. Anal.* 2000, 22, 487.
51. Kosovec, J.E. et al., *Rapid Commun. Mass Spectrom.*, 2008, 22, 224.
52. Pisano, R. et al., *J. Pharm. Biomed. Anal.*, 2005, 38, 738.
53. Kamleh, M.A. et al., *FEBS Lett.*, 2008, 582, 2916.
54. Nicholls, G., Clark, B.J., and Brown, J.E., *J. Pharm. Biomed. Anal.*, 1992, 10, 949.
55. Li, R. and Huang, J., *J. Chromatogr. A*, 2004, 1041, 163.
56. Dong, L. and Huang, J., *Chromatographia*, 2007, 65, 519.
57. Guisbert, A.L. et al., *J. Liq. Chromatogr. Relat. Technol.*, 2000, 23, 1019.
58. Risley, D.S. and Strege, M.A., *Anal. Chem.*, 2000, 72, 1736.
59. Ozoux, M.L. et al., *J. Chromatogr. B: Biomed. Sci. Appl.*, 2001, 765, 179.
60. Ali, M.S. et al., *J. Sep. Sci.*, 2008, 31, 1645.
61. Strege, M. et al., *LCGC North Am.,* 2008, 26, 632.
62. Schram, K.H., Taniguchi, Y., and McCloskey, J.A., *J. Chromatogr.*, 1978, 155, 355.
63. Schilsky, R.L. and Ordway, F.S., *J. Chromatogr. Biomed. Appl.*, 1985, 337, 63.
64. Wang, L. et al., *J. Pharm. Biomed. Anal.*, 2003, 33, 955.
65. Wiliams, R.C. et al., *J. Pharm. Biomed. Anal.*, 1997, 16, 469.
66. Kotinkaduwe, R.P. and Kitscha, R.A., *J. Pharm. Biomed. Anal.*, 1999, 21, 105.
67. Guo, Y. and Huang, A., *J. Pharm. Biomed. Anal.*, 2003, 31, 1191.
68. Risley, D.S. and Pack, B.W., *LCGC North Am.*, 2006, Suppl., 82, 776–785.
69. Huang, Z. et al., *J. Pharm. Biomed. Anal.*, 2009, 50, 809.
70. Borman, P.J. et al., *J. Pharm. Biomed. Anal.*, 2008, 48, 1082.
71. Dejaegher, B., Mangelings, D., and Vander Heyden, Y., *J. Sep. Sci.*, 2008, 31, 1438.
72. Krisko, R.M. et al., *J. Chromatogr. A*, 2006, 1122, 186.

13 Fast In-Process Method for the Determination of Ioversol and Related Polar Compounds by Hydrophilic Interaction Chromatography (HILIC) and UPLC

Yuming Chen, Xinqun Huang, Shaoxiong Huang, and Michael Matchett

CONTENTS

13.1 OVERVIEW OF HILIC/UPLC

13.1.1 HILIC AND HILIC STATIONARY PHASES

HILIC is used primarily for the separation of very polar compounds that are poorly retained by reversed-phase liquid chromatography (RPLC).[1–23] Alpert[24] described HILIC as a variant of normal-phase liquid chromatography (NPLC), where the stationary phase is more polar than the mobile phase, and analytes typically elute in an order opposite to that of RPLC. Different from traditional NPLC, a HILIC system allows for water as well as buffering agents to be present in the mobile phase. It is commonly believed that in HILIC separations, water in the mobile phase is preferentially adsorbed to the polar stationary phase, resulting in a layer of liquid enriched in water near the stationary phase surface vs. a layer of liquid enriched in organic in the bulk mobile phase. Analytes can then partition between these two layers to achieve chromatographic separation. Polar compounds have higher solubility in the water layer and are thus retained longer than less polar compounds. The retention mechanism may include hydrogen bonding and electrostatic interactions between the analytes and functional groups on the surface of the stationary phase. Therefore, the separation mechanisms in HILIC may be characterized as mixed modes of partition and absorption.

Compared with RPLC, HILIC provides an orthogonal separation that can be valuable in many applications, particularly for complex mixtures. In addition, HILIC often uses more volatile mobile phases relative to RPLC, so it offers higher sensitivity when combined with mass spectrometer detection.[5,25,26]

In recent years, an increasing number of HILIC stationary phases (SPs) have become commercially available. The selection of a proper column is the first step to assure a successful analytical method. A brief overview is given here to categorize the most widely used HILIC SPs.

Bare silica: Bare silica is one of the classical and still often used HILIC SPs. Zorbax SIL,[27] Supelcosil LC-Si,[28] Hypersil Silica,[29,30] Nucleosil SIL, and Atlantis HILIC[27,31–35] all belong to this category. The siloxane and silanol groups, in some cases with trace amounts of metal ions on the surface of the silica particles, provide a polar surface to retain polar compounds. The operating pH range for such SPs is usually 2–6. These SPs may offer unique selectivity,[36] but in some cases bear the common problem of poor reproducibility. Columns with 5,[37] 4,[38] 3,[31,32,39–43] and 1.7 μm[2–4,9,14,44] particle sizes are commercially available. Bare silica columns are mainly used for the separation of small polar compounds.

Polar neutral bonded silica: The most commonly used SPs in this group include diol- and amide-modified silica.

Diol SPs usually provide improved reproducibility and different selectivity when compared with bare silica (e.g., YMC-Pack Diol,[45] Uptisphere Diol, Nucleosil Diol). Because hydrogen bonding with the diol functional groups is not as strong as with the silanol groups on a bare silica surface, the retention is usually less with diol SPs. Diol columns are mostly used for the separation of small organic compounds.[46,47]

Amide-bonded SPs are one of the most popular SPs in HILIC separation (e.g., TSK-Gel Amide 80). The columns are packed with spherical silica-based material to which carbamoyl functional groups are covalently bonded. These SPs have been used to separate oligosaccharides,[48–51] peptides,[52–55] and small organic molecules.[18,56–58] The TSK-Gel Amide 80 is also available in a 3 μm particle size.[59]

Silica-based ion-exchange SPs: Ion-exchange SPs can be further categorized into positively charged and negatively charged SPs.

Amine-bonded silica columns, such as YMC-Pack NH_2, Luna Amino, Alltima Amino, Zorbax NH_2, and Hypersil NH_2 columns, are typical positively charged columns and have been used for the separation of carbohydrates.[60,61] They are not recommended for peptide separations due to low analyte recovery.[62] In some cases, these SPs demonstrated poor reproducibility and short column life because of chemical instability of the amine group.

Negatively charged cation-exchange columns have been successfully used for peptide and protein separations. PolyCAT A column, a weak cation-exchange column formed by poly(aspartic acid) bonded to silica, was reported for the separation of phosphorylated H1 histone.[63] Synchropak CM 300, a weak cation-exchange column with carboxymethyl functionalities, was reported for the successful separation of histone with different degrees of acetylation.[64] Polysulfoethyl A, a strong cation-exchange column, was used for peptide separations.[65–67]

Zwitterionic SPs: Over the past few years, HILIC columns have progressed to second- and third-generation embodiments, most of which involve mixed or multiple-interaction modes. A good example is the zwitterion SPs, which has sulfoalkylbetaine functional groups that strongly absorb water by hydrogen bonding. This type of SPs offers multiple analyte-SP interactions including hydrogen bonding, dipole–dipole interactions, and electrostatic interactions for the separation of charged and neutral compounds. Zwitterionic SPs are commercially available either on silica gel support (e.g., ZIC HILIC) or polymer support (e.g., ZIC pHILIC). ZIC HILIC columns have demonstrated impressive effectiveness in separations of complicated protein[68] and peptide samples.[69–71] They have also become one of the most popular choices for small organic molecules.[72–75]

13.1.2 Overview of UPLC and HILIC-UPLC Applications

There is always a drive to pursue high efficiency in the field of liquid chromatography (LC). The underlying principle of LC can be explained by the van Deemter equation, which describes the kinetics of a chromatographic separation[19] using three terms:

$$H = A + \frac{B}{u} + C \times u$$

where

H is the plate height (column efficiency)
A is the Eddy diffusion
B is the longitudinal diffusion
C is the mass transfer kinetics of the analyte between the mobile and stationary phase
u is the linear velocity

Among these three terms, the A term is proportional to d_p (particle size of packing material), whereas the C term is proportional to d_p^2. Based on the van Deemter equation, high column efficiency can be attained over a wide range of linear velocity if the particle size of the packing is smaller than 2 μm (Figure 13.1). Ultra-performance

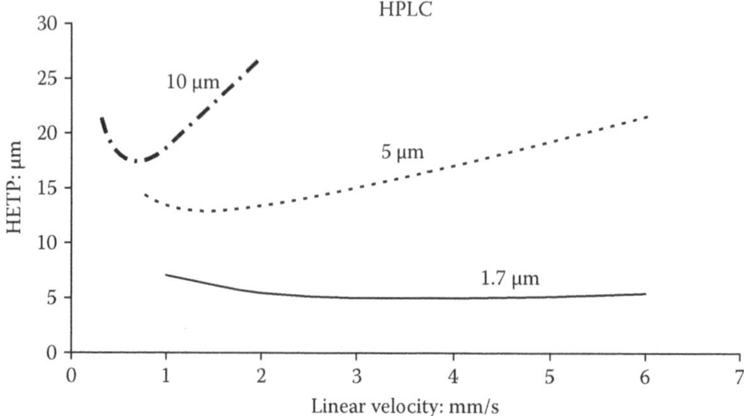

FIGURE 13.1 van Deemter plots showing relationships between HETP and the linear velocity of the mobile phase for columns with stationery phases of different particle sizes.

liquid chromatography (UPLC) has been developed to take advantage of these chromatographic principles, so the separation can be conducted using sub-2 µm columns at higher linear velocity, which results in increased speed for the separation with superior resolution and higher sensitivity.

Currently, commercially available UPLC systems include the Waters Acquity UPLC system, the Agilent 1200 series Rapid Resolution liquid chromatographic system, the Shimadzu UFLC system, and the Thermo Fisher Accela UHPLC system. All these systems are built specifically for delivering accurate flow at very high back pressure (up to 18,000 psi).

Literature reports on HILIC-UPLC applications are limited simply because of the limited availability of suitable columns. HILIC-UPLC methods so far are based on the bridged ethylene hybrid (BEH) SP.[2–4,9,14,20] The BEH HILIC column is packed with underivatized 1.7 µm BEH particles. Because the ethylene bridged groups are embedded in the silica matrix, nearly one-third of the surface silanols are removed, therefore causing decreased retention attributed to the reduced number of silanols.[76]

Speed is one of the major drivers for HILIC-UPLC applications. Kesiunaite et al.[9] reported the separation of cabadox and olaquindox, two synthetic antibacterial drugs for animal feeding, within 1 min on a 1.7 µm BEH HILIC column by isocratic elution with a mobile phase of 10 mM ammonium acetate in acetonitrile-water (95:5, v:v). The resolution was comparable to that obtained by an HPLC method,[77] but the separation was 10 times faster. Demacker et al.[3] developed an UPLC-HILIC-MS method to analyze citrulline, a nonessential amino acid in plasma samples. With a simple protein precipitation/extraction step, 12 samples could be analyzed in an hour by the method.

HILIC-UPLC has also been utilized for the mapping of complex biosamples. Gika et al.[4] evaluated a HILIC-UPLC-MS method for profiling metabolites in Zucker rat urine samples. Both HILIC-UPLC and RP-UPLC methods were used to generate complementary profiles. HILIC conditions were optimized to separate highly polar metabolites found in these samples, which included simple organic acids, amino acids, and amphoteric basic compounds. Compared to the RPLC profiles, the HILIC profiles showed higher signal intensity in total ion current (TIC) chromatograms, presumably due to higher ionization efficiency. The HILIC-UPLC-MS results revealed different biomarker information from the RP-UPLC data and these biomarkers were from polar metabolites of the urine samples. Such findings demonstrate the value of HILIC as an orthogonal method to RPLC by providing critical information in biomarker identification and characterization for complex biosamples.

In other HILIC-UPLC applications, Chen et al.[2] compared HPLC with a BetaMax acid column and HILIC-UPLC with a BEH HILIC column for the detection of 10 haloacetic acids in drinking water. The HILIC-UPLC method provided lower on-column detection limits. New and Chan[14] evaluated BEH C18, BEH HILIC, and the high-strength silica (HSS) T3 C18 column for the analysis of glutathione, glutathione disulfide, and ophthalmic acid. A gradient elution was used along with MS/MS detection. However, BEH HILIC did not show an advantage over the other SPs.

13.2 CASE STUDY: FAST IN-PROCESS ANALYSIS OF IOVERSOL BY HILIC-UPLC

13.2.1 Introduction

Ioversol is an x-ray imaging enhancing reagent widely used in clinical diagnostics, such as angiocardiography,[78] urography,[79] herniography,[80] and arthrography.[81] The chemical structure of Ioversol is shown in Figure 13.2. Each Ioversol molecule contains three iodine atoms, which are responsible for x-ray opacity. Ioversol is highly water-soluble, and its safety has been demonstrated in a number of animal and clinical studies.[82–85] In clinical applications, a single dose of Ioversol can be as high as 40 g,[86] and the same dosage can be repeated if needed.[80] Because of the high dosing amount, a stringent control of the impurity levels is required to ensure the safety of the product.

During the large-scale production of Ioversol, up to seven process impurities were monitored and controlled. Speed and specificity are the two most important attributes

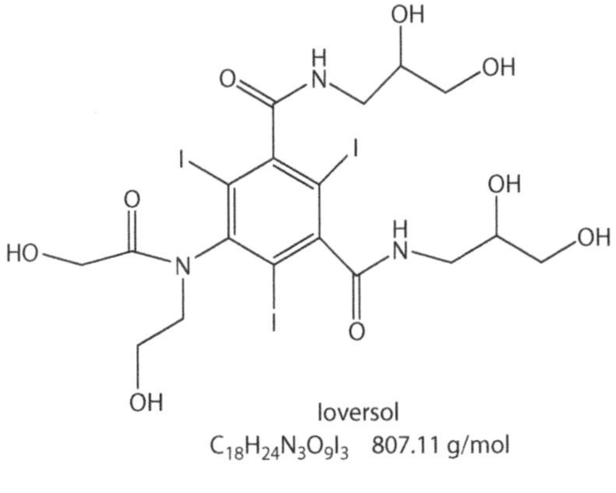

FIGURE 13.2 Structure of Ioversol.

for an in-process method. Ioversol and its process impurities are highly polar analogues, which are difficult to separate by RPLC. Assay[87] and inprocess[88] methods for Ioversol based on RPLC have been reported, but proved to be unsuitable for resolving all the impurities in our case. So a method using HILIC-UPLC was developed and optimized to separate Ioversol and its process impurities in less than 10 min. To the author's knowledge, the separation of Ioversol and its related impurities using a HILIC-UPLC method has not been attempted before.

The HILIC-UPLC method was developed and optimized by both univariate and multivariate approaches. Method parameters including buffer concentration, acetonitrile percentage, temperature, and flow rate were investigated in univariate ways. Based on the results of univariate studies, the multivariate approach was used for the first time[89] for a HILIC method to demonstrate the correlations among different parameters.

13.2.2 Experimental

Reagents and materials: HPLC grade acetonitrile was purchased from Fisher Scientific (Fair Lawn, NJ). Acetic acid (Glacial) was purchased from J. T. Baker (Philipsburg, NJ). HPLC grade water was obtained from an in-house Milli-Q water purification system (Millipore, Bedford, MA).

Standards of Ioversol and seven process impurities (RS-1 to RS-7) were supplied by Covidien's Department of Imaging Research and Development. Ioversol consists of several stereo-isomers[90] and often shows as a group of unresolved peaks in chromatographic analysis. The process impurities are structural analogs of Ioversol. Each of them also contains three iodine atoms, and some of them also have one or more structural isomers. The functional groups and physical properties of these compounds are listed in Table 13.1.

Preparation of solutions: The stock solution for each process impurity was prepared at a concentration of 0.5 mg/mL in 95% acetonitrile. About 0.2 mg of Ioversol standard and 1 mL of each impurity stock solution were added into the same 50 mL

TABLE 13.1
Ioversol and Its Process Impurities

Name	Molecular Weight	Functional Group	pK_a
Ioversol	807	–OH (6)	11.35
RS-1	763	–OH (5)	11.48
RS-2	705	–OH (4) Aromatic –NH_2 (1)	11.58
RS-3	791	–OH (5)	11.35
RS-4	807	–OH (5)	11.58
RS-5	851	–OH (7)	N/A
RS-6	747	–OH (4)	N/A
RS-7	781	–OH (4) –Cl (1)	N/A

volumetric flask and diluted to volume with 95% acetonitrile. This resolution solution contained about 0.4 mg/mL of Ioversol and 0.01 mg/mL of each impurity.

UPLC conditions: An Acquity UPLC system (Waters, Milford, MA) equipped with a binary solvent delivery system, an autosampler, and an Acquity photodiode array detector was used. System control and data processing were performed by the Empower 2 software (Waters).

An Acquity BEH HILIC column (50 mm × 2.1 mm I.D., 1.7 µm particle size) was used for the separation. The optimized method has a column temperature of 35°C, a flow rate of 0.4 mL/min, and a mobile phase composed of acetonitrile:water (95:5, v/v) with 10 mM ammonium acetate. The injection volume was 2 µL using a partial loop with the needle overfill injection mode. Data was collected at 254 nm.

13.2.3 Results and Discussion

In the method development stage, a one-variable-at-a-time approach was used. Four parameters including acetonitrile concentration, buffer concentration, column temperature, and flow rate were investigated to identify a suitable range for each of the parameters. In the method optimization stage, three parameters including buffer concentration, flow rate, and column temperature were selected for further optimization using a multivariate design-of-experiment approach.

13.2.3.1 Univariate Method Development

Acetonitrile concentration: The effect of acetonitrile concentration on the retention of Ioversol and its impurities was investigated by varying the percentage of acetonitrile in the mobile phase while keeping the ammonium acetate concentration (10 mM), flow rate (0.5 mL/min), and column temperature (25°C) constant. As can be seen in Figure 13.3, all analytes exhibited the typical retention behavior of a HILIC system, i.e., decreasing retention with increasing water content in the mobile phase. The retention was

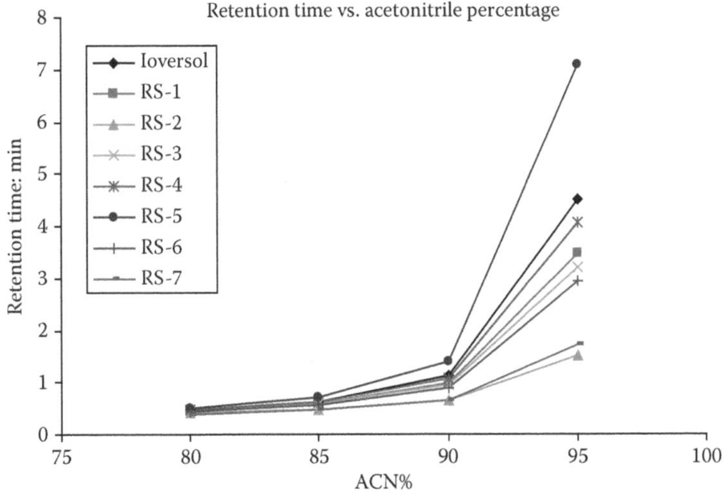

FIGURE 13.3 Retention time vs. acetonitrile concentration in the mobile phase.

poor for all components in the acetonitrile range of 80%–90%. Significant increases in retention were observed as the acetonitrile concentration increased from 90% to 95%. Similar retention behavior was also observed by Grumbach et al.[5] who evaluated the performance of BEH HILIC columns. The mobile phases with greater than 95% acetonitrile were not tested because about 5% of water was needed to maintain the desired buffer concentration. The best selectivity was also observed with 95% acetonitrile in the mobile phase.

Buffer concentration: Buffer (or salt) is an essential component in a HILIC method. A buffer solution is needed even though the analyte(s) of interest may not have an ionizable functional group. A suitable buffer is also critical for achieving desirable efficiency and sometimes selectivity. In some cases, the choice of a buffer or salt is limited to its solubility in the high-organic-content mobile phases. In developing the HILIC method for Ioversol and its impurities, ammonium acetate was selected for its relatively high solubility in 95% acetonitrile. To study the effect of buffer concentration on retention, the ammonium acetate concentration was varied from 2.5 to 15 mM while holding the flow rate (0.5 mL/min), column temperature (25°C), and acetonitrile concentration (95%) constant. Buffer concentration was found to have a significant impact on both the retention and selectivity of the analytes. As shown in Figure 13.4, the logarithms of retention for Ioversol and its impurities vs. buffer concentration show a linear relationship. More interestingly, there is a difference in slope among these linear plots, indicating that the variation in buffer concentration causes a change in selectivity. Ioversol, RS-1, and RS-5 show steeper slopes compared with the rest. Based on the hypothesis that ammonium acetate prefers the water-enriched surface layer of the polar stationary phase, increasing the ammonium acetate levels thereby promotes the formation of the water-rich layer, thus increasing the retention of polar analytes. Because Ioversol and its impurities did not retain well at low buffer concentrations, higher levels of 10, 12.5, and 15 mM ammonium acetate were later investigated for method optimization.

Flow rate: The effect of flow rate on column efficiency for this group of compounds is presented in Figure 13.5. Compared with the van Deemter plot in Figure 13.1, the

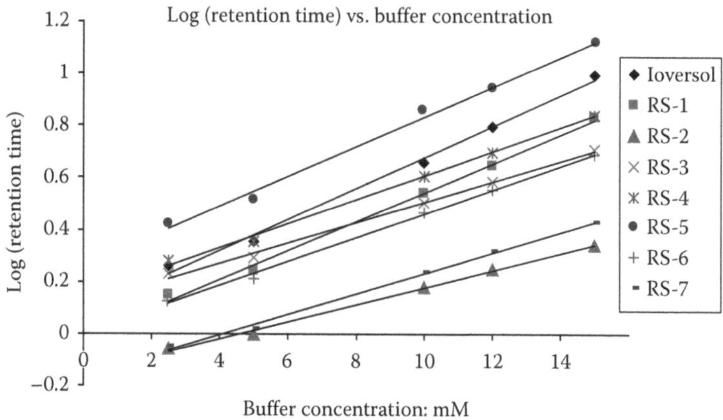

FIGURE 13.4 Effect of buffer concentration on retention.

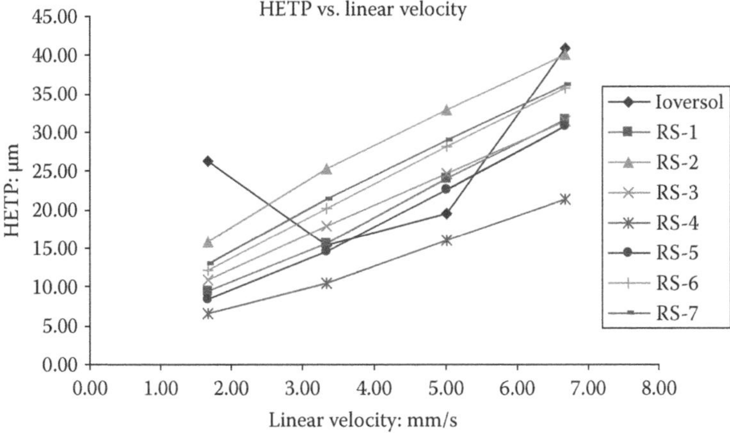

FIGURE 13.5 Effect of linear velocity on column efficiency.

abnormal shape and trend of the Ioversol curve were due to the partial separation of Ioversol stereo-isomers. Curves of six out of the seven impurities show characteristics of HETP-flow relationships typically observed with 5 µm particle size HPLC columns. The curve of RS-4 is closer to the 1.7 µm curve in Figure 13.1 with a steeper slope. This observation may be related to the presence of multiple retention mechanisms in the HILIC mode, i.e., partition as well as hydrogen bonding interactions, which may slow down mass transfer between the SP and the mobile phase.

Column temperature: The effect of column temperature on retention and column efficiency was explored across the range of 15°C–55°C while holding the other conditions constant. Column temperature had a significant impact on both retention and selectivity (Figure 13.6). RS-1 and RS-3 reversed their elution order when the

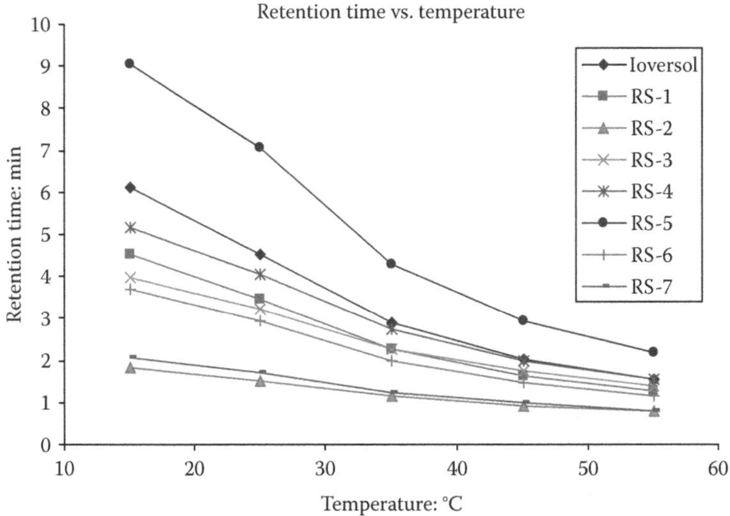

FIGURE 13.6 Effect of column temperature on retention time.

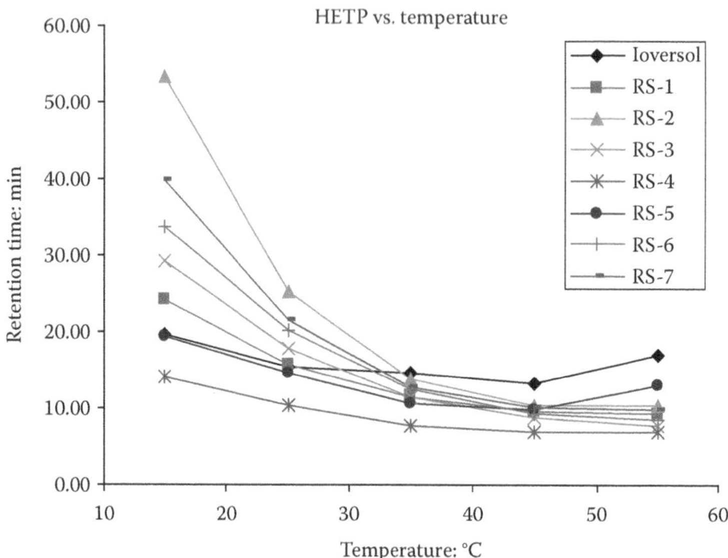

FIGURE 13.7 Relationship between column temperature and column efficiency.

temperature increased from 15°C to 25°C. Ioversol and RS-4 co-eluted at 55°C but were well separated at lower temperatures. Temperature may affect the separation in multiple ways. Higher temperature facilitates mass transfer, which increases column efficiency. On the other hand, higher temperature increases longitudinal diffusion, which decreases column efficiency. Secondary interactions, such as hydrogen bonding and electrostatic interactions between analytes and the stationary phase may be reduced at high temperature, which then changes the selectivity. Figure 13.7 shows that maximum column efficiency was achieved around 45°C for Ioversol and some of the impurities. The impact of column temperature on column efficiency was further analyzed using multivariate analysis.

13.2.3.2 Multivariate Method Optimization

The one-variable-at-a-time studies were necessary to identify suitable operating ranges for the four method parameters investigated. Figure 13.4 implies that acetonitrile concentration in the mobile phase should be kept at 95% to obtain sufficient retention for this group of compounds. The buffer concentration may be varied from 10 to 15 mM, whereas column temperature showed a promising range between 25°C and 35°C based on Figures 13.6 and 13.7 considering both selectivity and column efficiency. On the other hand, these results did not demonstrate relationships among those parameters, and the relative importance of these factors for the method was also unknown. Therefore, a multivariate method optimization study was conducted to further optimize the method. A simple three-factor by two-level design with an additional center point (Table 13.2) was used. After execution of the designed experiments, multiple responses were collected, which included the resolution of all peaks, the column efficiency for all components, and the tailing factor for all components. To limit the discussions within the scope of this article, only two examples regarding the data analysis are given here.

TABLE 13.2
Factors in Designed Experiments and Responses

	Factors			Response 1	Response 2—Peak Width (min)							
Run	Temperature (°C)	Buffer (mM)	Flow Rate (mL/min)	Resolution (Peaks Resolved)	RS-2	RS-7	RS-6	RS-3	RS-1	RS-4	Ioversol	RS-5
1	25	10	0.4	Most	0.095	0.098	0.163	0.168	0.171	0.158	0.241	0.323
2	35	10	0.4	Some	0.049	0.051	0.091	0.091	0.086	0.165	0.129	0.165
3	25	15	0.4	Some	0.133	0.133	0.228	0.235	0.287	0.225	1.030	0.490
4	35	15	0.4	All	0.064	0.068	0.111	0.114	0.119	0.113	0.243	0.228
5	25	10	0.6	Some	0.068	0.070	0.116	0.120	0.122	0.112	0.153	0.228
6	35	10	0.6	Some	0.037	0.039	0.060	0.068	0.063	0.064	0.093	0.123
7	25	15	0.6	Some	0.094	0.100	0.171	0.167	0.197	0.173	0.691	0.367
8	35	15	0.6	All	0.049	0.051	0.082	0.084	0.088	0.084	0.119	0.164
9	30	12.5	0.5	Most	0.068	0.071	0.117	0.119	0.126	0.115	0.190	0.232

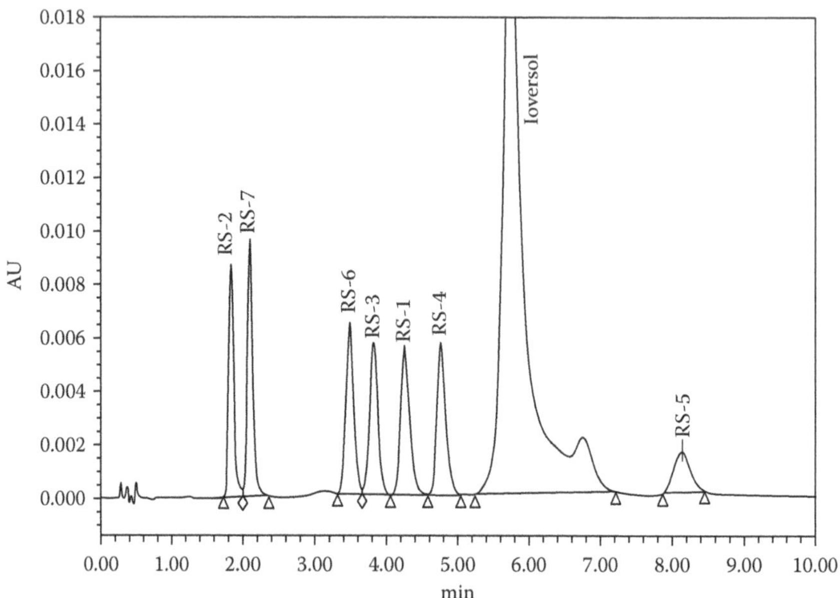

FIGURE 13.8 Typical chromatogram showing the separation of Ioversol from seven process impurities.

Data analysis by visual inspection of results: The first example involved tabulating the resolution results and analyzing them by visual inspection. The resolution results under different experimental conditions are summarized in Table 13.2. A complete resolution of all components was achieved in two out of nine runs. A chromatogram showing the separation is presented in Figure 13.8.

Multivariate data analysis: The second example of data analysis utilized a partial least squares (PLS) algorithm. In this case, experimental condition values were used as x-variables ($n = 3$) and the peak width values (Table 13.2) for all components measured at half peak height were used as y-variables ($n = 8$). The data was normalized using the equation of 1/(standard deviation). Results of the PLS modeling are graphically presented in Figure 13.9. The PLS algorithm uses latent variables to describe variances within the two sets of experimental data. A latent variable may represent a particular correlation that can have specific physical meanings. The residual validation variance plot (bottom-left) in Figure 13.9 indicates that only one latent variable is needed to account for 87% of the variances in the y-variables. The correlation loadings plot (top-right), characterized by two ellipses, demonstrates the correlation between the x- and y-variables. The outer ellipse is the unit circle and indicates 100% explained variance. The inner ellipse indicates 50% of explained variance. The x- and y-variables that are closest to +1 (positive correlation) or −1 (negative correlation) along the PC1 axis have the strongest correlation. The most significant correlations are observed between temperature (in blue) and the eight y-variables (in red). In other words, column temperature plays the most important role in defining peak width among the studied parameters and within the experimental conditions. The other two plots in Figure 13.9 also support this conclusion. The scores

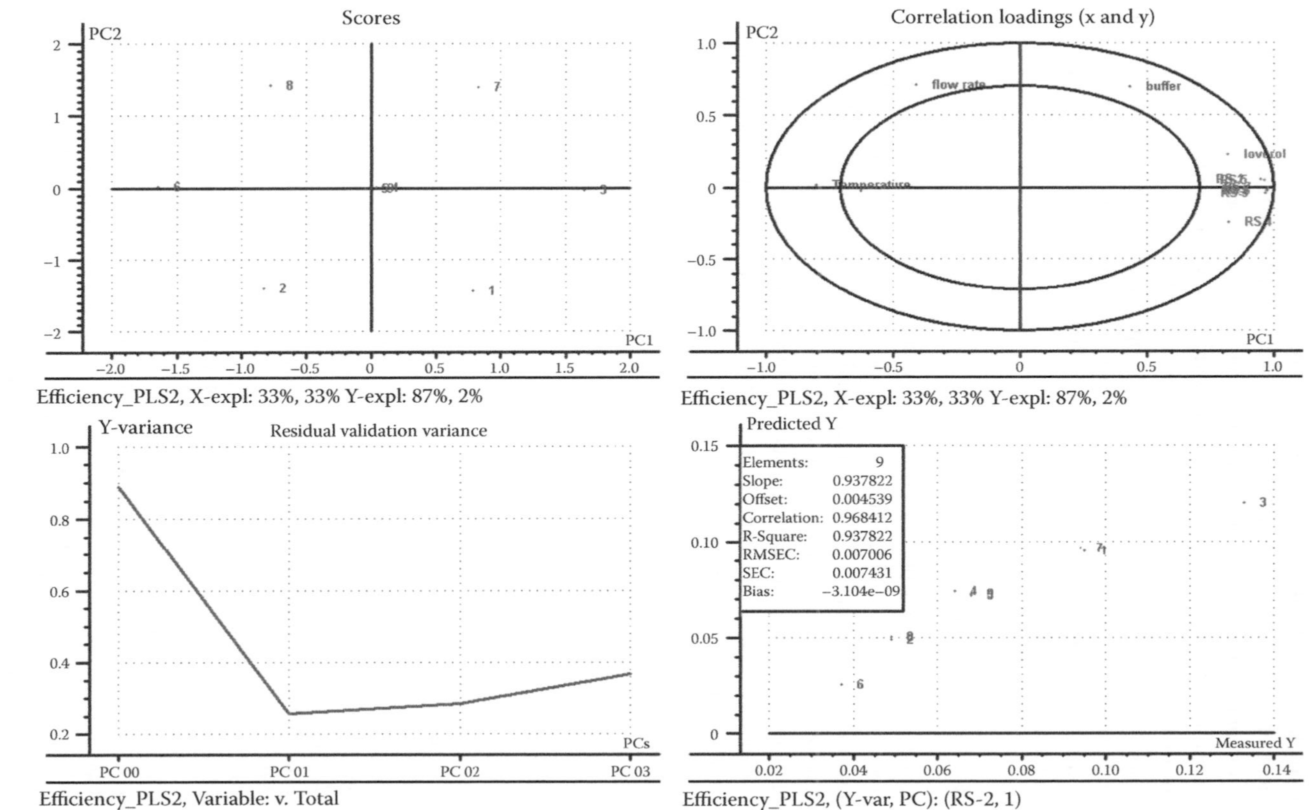

FIGURE 13.9 Multivariate data analysis using a PLS model.

plot (top-left) shows a pattern that is highly column-temperature dependent along the PC1 axis. The measured vs. predicted peak width plot (bottom-right) demonstrates a quantitative relationship between the measured and predicted peak widths as a function of column temperature. Even though the plot shows a curvature, a correlation coefficient of 0.968 and an R^2 of 0.938 were obtained for impurity RS-2. Similar correlation and trends were observed for the other seven compounds (not presented here). For simplicity of data analysis, the peak width data was not corrected for retention time differences, which may cause the curvature in the plots.

The above analyses effectively identified the optimized HILIC conditions, and revealed the controlling factors with regard to one study outcome. The approach can also be used for studying method robustness, which is sometimes a more serious concern for a HILIC method.

13.2.4 Conclusions

Method development and optimization were conducted by a two-stage process. The univariate approach defined the suitable operation ranges for four of the most important method parameters. The experimental results showed that small changes in any one of the parameters (acetonitrile concentration, buffer concentration, or column temperature) would cause a significant change in retention, column efficiency, and resolution. The design-of-experiment approach was effective for the identification of the conditions to achieve complete separation of all components. The chromatographic conditions were optimized using a flow rate of 0.4 mL/min, a column temperature at 35°C, and with a mobile phase of 15 mM ammonium acetate in 95% acetonitrile.

REFERENCES

1. L.T. Andersen, H. Schlichtherle-Cerny, Y. Ardo, *Dairy Science & Technology* 88 (2008) 467.
2. C.-Y. Chen, S.-N. Chang, G.-S. Wang, *Journal of Chromatographic Science* 47 (2009) 67.
3. P.N.M. Demacker, A.M. Beijers, H. van Daal, J.P. Donnelly, N.M.A. Blijlevens, J.M.W. van den Ouweland, *Journal of Chromatography, B: Analytical Technologies in the Biomedical and Life Sciences* 877 (2009) 387.
4. H.G. Gika, G.A. Theodoridis, I.D. Wilson, *Journal of Separation Science* 31 (2008) 1598.
5. E.S. Grumbach, D.M. Diehl, U.D. Neue, *Journal of Separation Science* 31 (2008) 1511.
6. J. Guitton, S. Coste, N. Guffon-Fouilhoux, S. Cohen, M. Manchon, M. Guillaumont, *Journal of Chromatography, B: Analytical Technologies in the Biomedical and Life Sciences* 877 (2009) 149.
7. D.N. Heller, C.B. Nochetto, *Rapid Communications in Mass Spectrometry* 22 (2008) 3624.
8. E.P. Kadar, C.E. Wujcik, *Journal of Chromatography, B: Analytical Technologies in the Biomedical and Life Sciences* 877 (2009) 471.
9. G. Kesiunaite, E. Naujalis, A. Padarauskas, *Journal of Chromatography, A* 1209 (2008) 83.
10. L. Kovalova, C.S. McArdell, J. Hollender, *Journal of Chromatography, A* 1216 (2009) 1100.

11. H.-A. Lakso, P. Appelblad, J. Schneede, *Clinical Chemistry* (Washington, DC) 54 (2008) 2028.
12. S.-M. Liang, D.-M. Liu, W.-J. Zeng, Y. Zeng, Y.-H. Luo, *Xiandai Shipin Keji* 24 (2008) 1180.
13. K.T. Myint, K. Aoshima, S. Tanaka, T. Nakamura, Y. Oda, *Analytical Chemistry* (Washington, DC) 81 (2009) 1121.
14. L.-S. New, E.C.Y. Chan, *Journal of Chromatographic Science* 46 (2008) 209.
15. R. Oertel, N. Arenz, J. Pietsch, W. Kirch, *Journal of Separation Science* 32 (2009) 238.
16. M. Strege, C. Durant, J. Boettinger, M. Fogarty, *LC-GC Europe* 21 (2008) 608.
17. Y. Sun, J. Sun, J. Liu, S. Yin, Y. Chen, P. Zhang, X. Pu, Y. Sun, Z. He, *Journal of Chromatography, B: Analytical Technologies in the Biomedical and Life Sciences* 877 (2009) 649.
18. E. Turrell, L. Stobo, J.-P. Lacaze, S. Piletsky, E. Piletska, *Journal of AOAC International* 91 (2008) 1372.
19. J.J. van Deemter, F.J. Zuiderweg, A. Klinkenberg, *Chemical Engineering Science* 5 (1956) 271.
20. J.-H. Wang, X.-Y. Lu, M. Huang, Z. Wu, G.-P. Ma, C.-Y. Xu, *Fenxi Huaxue* 35 (2007) 1509.
21. D. Xie, J. Mattusch, R. Wennrich, *Engineering in Life Sciences* 8 (2008) 582.
22. L. Yan, M. Wu, Z. Zhang, Y. Zhou, L. Lin, E. Fang, D. Xu, L. Chen, *Sepu* 26 (2008) 759.
23. N.L. Young, M.D. Plazas-Mayorca, B.A. Garcia, *American Biotechnology Laboratory* 27 (2009) 23.
24. A.J. Alpert, *Journal of Chromatography* 499 (1990) 177.
25. E.S. Grumbach, D.M. Diehl, J.R. Mazzeo, *LC-GC Europe* 22 (2004) 1010.
26. F. Qin, Y.-Y. Zhao, M.B. Sawyer, X.-F. Li, *Analytical Chemistry* (Washington, DC) 80 (2008) 3404.
27. V. Goekmen, H.Z. Senyuva, J. Acar, K. Sarioglu, *Journal of Chromatography, A* 1088 (2005) 193.
28. M.S. Ali, S. Rafiuddin, M. Ghori, A.R. Khatri, *Chromatographia* 67 (2008) 517.
29. Z. Hao, C.-Y. Lu, B. Xiao, N. Weng, B. Parker, M. Knapp, C.-T. Ho, *Journal of Chromatography, A* 1147 (2007) 165.
30. Y. Hsieh, J. Chen, *Rapid Communications in Mass Spectrometry* 19 (2005) 3031.
31. T.M. Baughman, W.L. Wright, K.A. Hutton, *Journal of Chromatography, B: Analytical Technologies in the Biomedical and Life Sciences* 852 (2007) 505.
32. D.W. Jeong, Y.H. Kim, H.Y. Ji, Y.S. Youn, K.C. Lee, H.S. Lee, *Journal of Pharmaceutical and Biomedical Analysis* 44 (2007) 547.
33. T. Ohmori, M. Nakamura, S. Tada, T. Sugiyama, Y. Itoh, Y. Udagawa, K. Hirano, *Journal of Chromatography, B: Analytical Technologies in the Biomedical and Life Sciences* 861 (2008) 95.
34. H. Zhang, Z. Guo, F. Zhang, Q. Xu, X. Liang, *Journal of Separation Science* 31 (2008) 1623.
35. Y. Iwasaki, M. Hoshi, R. Ito, K. Saito, H. Nakazawa, *Journal of Chromatography, B: Analytical Technologies in the Biomedical and Life Sciences* 839 (2006) 74.
36. B.A. Olsen, *Journal of Chromatography, A* 913 (2001) 113.
37. W. Naidong, *Journal of Chromatography, B: Analytical Technologies in the Biomedical and Life Sciences* 796 (2003) 209.
38. M. Godejohann, *Journal of Chromatography, A* 1156 (2007) 87.
39. H.Y. Ji, D.W. Jeong, Y.H. Kim, H.-H. Kim, Y.-S. Yoon, K.C. Lee, H.S. Lee, *Rapid Communications in Mass Spectrometry* 20 (2006) 2127.
40. H.Y. Ji, D.W. Jeong, Y.H. Kim, H.-H. Kim, D.-R. Sohn, H.S. Lee, *Journal of Pharmaceutical and Biomedical Analysis* 41 (2006) 622.

41. S. Cohen, F. Lhuillier, Y. Mouloua, B. Vignal, P. Favetta, J. Guitton, *Journal of Chromatography, B: Analytical Technologies in the Biomedical and Life Sciences* 854 (2007) 165.
42. J. Vacek, B. Klejdus, J. Petrlova, L. Lojkova, V. Kuban, *Analyst* (Cambridge, U.K.) 131 (2006) 1167.
43. T. Kawamoto, M. Yano, N. Makihata, *Analytical Sciences* 22 (2006) 489.
44. E.S. Grumbach, T.E. Wheat, J.R. Mazzeo, *LCGC North America* 24 (2006) 70.
45. X. Wang, W. Li, H.T. Rasmussen, *Journal of Chromatography, A* 1083 (2005) 58.
46. P. Uutela, R. Reinila, P. Piepponen, R.A. Ketola, R. Kostiainen, *Rapid Communications in Mass Spectrometry* 19 (2005) 2950.
47. A.E. Karatapanis, Y.C. Fiamegos, C.D. Stalikas, *Journal of Separation Science* 32 (2009) 909.
48. G. Karlsson, S. Winge, H. Sandberg, *Journal of Chromatography A* 1092 (2005) 246.
49. V.V. Tolstikov, O. Fiehn, *Analytical Biochemistry* 301 (2002) 298.
50. R.R. Towsend, P.H. Lipniumas, C. Bigge, A. Ventom, R. Parekh, *Analytical Biochemistry* 239 (1996) 200.
51. S. Inagaki, J.Z. Min, T. Toyo'oka, *Biomedical Chromatography* 21 (2007) 338.
52. T. Yoshida, *Journal of Chromatography A* 811 (1998) 61.
53. T. Yoshida, *Journal of Chromatography A* 808 (1998) 105.
54. T. Yoshida, *Analytical Chemistry* 69 (1997) 3038.
55. D.E. McNulty, R.S. Annan, *Molecular and Cellular Proteomics* 7 (2008) 971.
56. T. Kimura, K. Nakagawa, Y. Saito, K. Yamagishi, M. Suzuki, K. Yamaki, H. Shinmoto, T. Miyazawa, *BioFactors* 22 (2004) 341.
57. D.S. Risley, W.Q. Yang, J.A. Peterson, *Journal of Separation Science* 29 (2006) 256.
58. C. Dell'Aversano, G.K. Eaglesham, M.A. Quilliam, *Journal of Chromatography A* 1028 (2004) 155.
59. C. Tomasek, *LCGC North America: The Application Notebook* 25 (2007) 21.
60. B. Beilmann, P. Langguth, H. Haeusler, P. Grass, *Journal of Chromatography A* 1107 (2006) 204.
61. S.-C. Lin, W.-C. Lee, *Journal of Chromatography A* 803 (1998) 302.
62. T. Yoshida, T. Okada, *Journal of Chromatography A* 840 (1999) 1.
63. H. Lindner, B. Sarg, W. Helliger, *Journal of Chromatography A* 782 (1997) 55.
64. H. Lindner, B. Sarg, C. Meraner, W. Helliger, *Journal of Chromatography A* 743 (1996) 137.
65. C.T. Mant, J.R. Litowski, R.S. Hodges, *Journal of Chromatography A* 816 (1998) 65.
66. C.T. Mant, L.H. Kondejewski, R.S. Hodges, *Journal of Chromatography A* 816 (1998) 79.
67. R.S. Hodges, Y. Chen, E. Kopecky, C.T. Mant, *Journal of Chromatography A* 1053 (2004) 161.
68. C. Viklund, A. Sjoegren, K. Irgum, I. Nes, *Analytical Chemistry* 73 (2001) 444.
69. M.J. Omaetxebarria, P. Haegglund, F. Elortza, N.M. Hooper, J.M. Arizmendi, O.N. Jensen, *Analytical Chemistry* 78 (2006) 3335.
70. W. Jiang, G. Fischer, Y. Girmay, K. Irgum, *Journal of Chromatography A* 1127 (2006) 82.
71. Y. Takegawa, K. Deguchi, H. Ito, T. Keira, H. Nakagawa, S.-I. Nishimura, *Journal of Separation Science* 29 (2006) 2533.
72. P. Appelblad, P. Abrahamsson, *LCGC Europe*, Suppl., Mar. (2005) 47.
73. J. Bengtsson, B. Jansson, M. Hammarlund-Udenaes, *Rapid Communications in Mass Spectrometry* 19 (2005) 2116.
74. H. Idborg, L. Zamani, P.-O. Edlund, I. Schuppe-Koistinen, S.P. Jacobsson, *Journal of Chromatography, B: Analytical Technologies in the Biomedical and Life Sciences* 828 (2005) 9.
75. K.L. Wade, I.J. Garrard, J.W. Fahey, *Journal of Chromatography A* 1154 (2007) 469.

76. K.D. Wyndham, J.E. O'Gara, T.H. Walter, K.H. Glose, N.L. Lawrence, B.A. Alden, G.S. Izzo, C.J. Hudalla, P.C. Iraneta, *Analytical Chemistry* 75 (2003) 6781.
77. G. Kesiunaite, A. Padarauskas, *Chemija* 18 (2007) 30.
78. S. Morimoto, T. Kozuka, M. Takamiya, K. Kimura, S. Matsuyama, S. Kuribayashi, A. Shigeta, J. Umemura, J. Harada, Y. Yamada, *Nihon Igaku Hoshasen Gakkai Zasshi. Nippon Acta Radiologica* 50 (1990) 1087.
79. H. Jahn, R. Muller-Spath, *Annales de Radiologie* 35 (1992) 297.
80. http://www.drugs.com/mmx/ioversol.html
81. http://www.drugs.com/mmx/optiray-350.html
82. W.H. Ralston, M.S. Robbins, P. James, *Investigative Radiology* 24 (1989) S16.
83. R.A. Wilkins, J.R. Whittington, G.S. Brigden, A. Lahiri, M.E. Heber, L.O. Hughes, *Investigative Radiology* 24 (1989) 781.
84. M. Akagi, S. Masaki, K. Kitazumi, M. Mio, K. Tasaka, *Methods and Findings in Experimental and Clinical Pharmacology* 13 (1991) 449.
85. M. Akagi, K. Tasaka, *Methods and Findings in Experimental and Clinical Pharmacology* 13 (1991) 377.
86. http://www.rxlist.com/optiray-injection-drug.htm, The Internet Drug Index.
87. P. Zou, S.-N. Luo, Y.-L. Liu, M.-H. Xie, Y.-J. He, Z. Yang, *Zhongguo Xiandai Yingyong Yaoxue* 22 (2005) 68.
88. E. Collins, S. Muldoon, B. O'Callaghan, D. Carolan, Mallinckrodt Medical Imaging-Ireland, Ire., Application: WO (1997) p. 25.
89. B. Dejaegher, D. Mangelings, Y. Vander Heyden, *Journal of Separation Science* 31 (2008) 1438.
90. J. Dunn, Y. Lin, D. Miller, M. Rogic, W. Neumann, S. Woulfe, D. White, *Investigative Radiology* 25 (1990) S102.

14 Retention Behavior of Hydrazine Derivatives in HILIC Mode

Juris Hmelnickis, Igors Susinskis, and Kaspars Kokums

CONTENTS

14.1 INTRODUCTION

In pharmaceutical research and development, separation and analysis of a variety of raw materials, intermediates, and excipients from the active pharmaceutical ingredient (API) is a hard task, and often demands unusual tools. The nature of impurities is difficult to predict and analytical methods must be able to detect a wide range of physicochemical properties, depending on the type of analysis.

A good separation method should combine sufficient retention, selectivity, and sensitivity. Reversed-phase liquid chromatography (RPLC) is currently the most popular method in the field of HPLC. Although RPLC is a powerful separation mode, it has a major limitation: the lack of adequate retention for polar molecules.

A suitable analytical method should avoid both too weak retention that usually leads to poor separation and too strong retention that results in the longer run and often is accompanied with unsatisfactory peak shape.

It is hypothesized that analytes are not retained by traditional RPLC on C18 packing with aqueous mobile phase, containing ammonium acetate pH 5.5 buffer and 2% ACN, then these compounds would represent highly polar analytes.[1]

Some chromatographic methods, such as size-exclusion chromatography (SEC), ion-exchange chromatography (IEC), hydrophobic-interaction chromatography (HIC), and immobilized metal-affinity chromatography (IMAC), are not compatible for mass-spectrometric detection due to high salt concentrations in the mobile phases. Traditional RPLC can work without salts, but separation of small hydrophilic compounds is generally poor, because of weak retention. However, derivatization of those compounds is laborious and prone to introduce errors in the overall analytical scheme. Side reactions and slow kinetics at trace concentrations are also the factors limiting the usefulness of derivatization. Hydrophilic-interaction chromatography (HILIC) can solve this problem by retaining polar compounds because such separation mode is based on the hydrophilicity of the analyte.

HILIC as an analytical tool for the analysis of polar compounds has been studied extensively in the past few years. HILIC mode is based on the combination of hydrophilic stationary phases and hydrophobic, mostly organic, mobile phases. HILIC is effective for separation of low molecular weight polar and ionizable compounds.[2–5] This mode was first discussed in detail in the works of Alpert.[6] It should be pointed out that chromatography with hydrophilic sorbents has been routinely used for analysis of sugars since 1975, and the use of low amounts of water in organic solvents for chromatographic elution was described by Samuelson in the 1950s—but Alpert was the first who used abbreviation "HILIC."[7,8] A comprehensive review on the scope and limitations of HILIC approach for analysis of polar analytes was published recently by Hemström and Irgum.[9] HILIC method has been used for pharmaceutical product analysis and is more suitable to mass-spectrometric detection and since it gives better mass-spectrometric sensitivity compare to the RPLC.[1,10–18]

In HILIC mode, the mobile phase usually is an aqueous/organic mixture with the water content in less than 60%. Polar compounds always have good solubility in aqueous mobile phase, which overcome poor solubility of these compounds in normal-phase chromatography. Polar analytes usually are retained stronger than nonpolar ones. The retention mechanism for HILIC is a partitioning process between the bulk mobile phase and a partially immobilized water-rich layer on the stationary phase. Some researchers call such conditions as "the aqueous normal phase" (pointing out that "traditional" normal-phase chromatography is based on adsorption onto a polar stationary phase, not a water layer); others state that HILIC and the "aqueous normal phase" are the same. But "pure HILIC" separation mode appears, however, quite rare.[9] Depending on the surface chemistry of the stationary phase, ion exchange in particular can contribute significantly to the retention of ionic analytes.[2,3,9,19–23]

Yoshida, same as Alpert, considered that HILIC retention encompassed both hydrogen bonding (which depends on Lewis acidity/basicity) and dipole–dipole interactions (dependent on dipole moments and polarizability of molecules).[24] Clearly, the mechanism of HILIC separations—partition, ion exchange, and hydrophobic retention—contribute to various degrees, depending on the particular conditions employed.[25]

In this chapter, we describe the retention behavior of a hydrazine-based pharmaceutical compound mildronate and its analogs (see Table 14.1) under HILIC mode on various polar stationary phases (silica, amino, cyano, amide, and zwitterionic sulfobetaine). Mildronate or 3-(2,2,2-trimethylhydrazinium)propionate dihydrate (also

TABLE 14.1
Studied Compounds

Formula	Compound No.
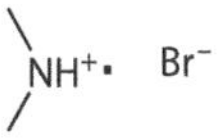 $NH^{+} \cdot Br^{-}$	1
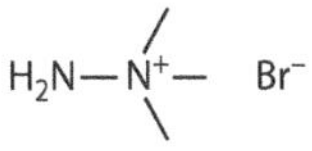 $H_2N—N^{+}—$ Br^{-}	2
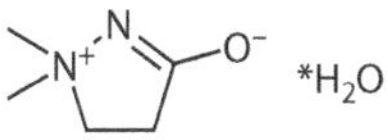 N^{+} N O^{-} $*H_2O$	3
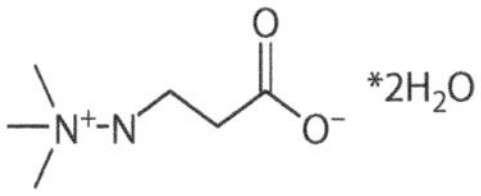 —N^{+}-N O O^{-} $*2H_2O$	4
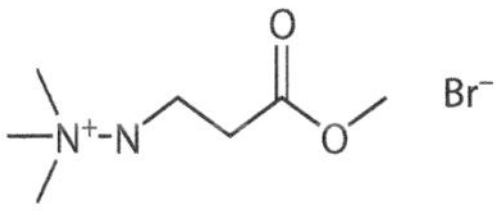 —N^{+}-N O O Br^{-}	5
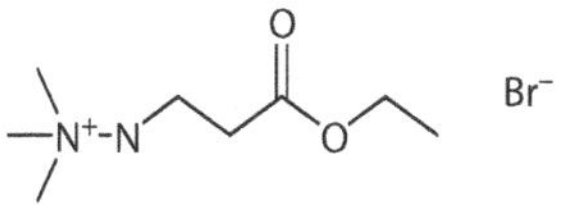 —N^{+}-N O O Br^{-}	6
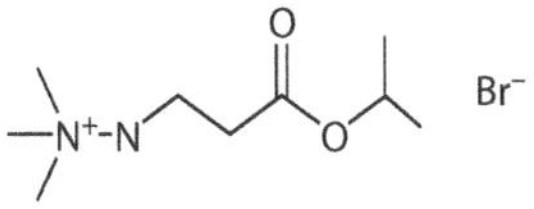 —N^{+}-N O O Br^{-}	7

known as Mildronatum, Meldonium hydrate, MET-88, 3-TMHP) is an anti-ischemic drug developed at the Latvian Institute of Organic synthesis over 20 years ago. The impact of separation conditions (organic solvent content, buffer pH, temperature, salt concentration in mobile phase) on retention has been systematically studied.

As mildronate and its related substances are small polar molecules, weak retention of these substances under the reversed-phases mode can be expected, and application of ion-pair chromatography (on RPLC sorbents) seems to be the easiest alternative. However, there is another restriction for the above compounds—the lack of distinguished chromophores, which makes mass spectrometry (MS) the appropriate detection alternative. Unfortunately, ion-pair agents as well as a high water content in the mobile phase usually reduce the sensitivity of mass-spectrometric detection.[26,27]

HPLC method with evaporative light-scattering detection (ELSD) was used for the determination of mildronate previously.[28,29] It is known that stripping voltammetry, HPLC-MS-MS, or UV detection (with derivatization) were used for mildronate detection in the plasma and urine.[30–32] In all these methods, polar stationary phases (columns Silasorb 600 Silica, Inertsil NH_2, Inertsil CN-3) and water-rich (more than 60% water) mobile phases were used. In our opinion, such chromatographic conditions suffer from insufficient sensitivity, even HPLC-MS being employed. Impurities mentioned in Table 14.1 have not been studied chromatographically before, except for a brief discussion by the same researchers as this article (determination of these compounds is important for purity control of the active pharmaceutical compound).[33]

Latest works described the detection of 1,1-dimethylhydrazine (this compound, as well as hydrazine are classified as a carcinogen requiring control) in HILIC mode by chemiluminescent nitrogen detector (CLND). Acetonitrile in mobile phase was replaced by ethanol.[34] The same scientists refer to other studies, mentioning ion chromatography with conductivity detection.[35] Other researchers used derivatization followed by GC or HPLC.[36–42] But derivatization is complicated, time consuming, and sometimes not suitable for detection of hydrazine in compounds with amine functional groups.

14.2 EFFECT OF ORGANIC SOLVENT

The selectivity and sensitivity of an LC-MS method depends not only on the ionization technique but also on the LC technique. One of the most important parameters is the mobile phase composition. Variation of organic solvent (acetonitrile, methanol, ethanol, isopropanol, acetone, tetrahydrofuran) concentration is commonly used for optimizing retention, selectivity, and sensitivity.

Typical HILIC mobile phases consist of 40%–97% acetonitrile in water (with volatile buffer, if necessary). At least 3% of water (some researchers talk about "non-water HILIC"—using another polar solvent instead) is necessary to ensure sufficient hydration of surface of stationary phase.

Acetonitrile is the most popular organic solvent for HILIC; nevertheless it is possible to use several other polar, water-miscible solvents. Solvent strength is inverted to what is observed in RPLC, and relatively:

Tetrahydrofuran < acetone < acetonitrile < isopropanol < ethanol < methanol < water[43] (other sources mention the sequence "acetonitrile < tetrahydrofuran < isopropanol < methanol"[19]).

The organic components of the HILIC mobile phase can be subdivided into polar protic and aprotic solvents. Methanol, ethanol, and isopropanol represent polar protic solvents. Examples of polar aprotic solvents are acetonitrile and tetrahydrofuran (THF). Protic solvents can be both donors and acceptors of hydrogen bonds, and aprotic solvents can be only hydrogen-bond acceptors. Due to their ability to make hydrogen bond, protic solvents can compete for polar active sites on the surface of the stationary phase, affecting the formation of water layer on the surface, thus changing stationary phase to less hydrophilic. As a consequence, the analytes with strong hydrogen-bonding potentials are poorly retained.

According to Snyder, there are three major contributions to solvent selectivity, i.e., solvent strength, solvent–analyte localization, and solvent–analyte hydrogen bonding.[44] The solvent strengths (ε°) is following the order of: methanol (0.73), isopropanol (0.60), THF (0.48), acetonitrile (0.50), and acetone (0.50).

Because of its strong hydrogen-bonding ability, methanol can perturb the formation of water layer on silica surface, replacing water molecules and producing a more hydrophobic stationary phase.[45] As a consequence, the ionic analytes are poorly retained. Compared to methanol, a decrease of the hydrophilic character of isopropanol is caused by the longer alkyl chain, which competes less strongly for the active sites, and leaves more space for analyte molecules to be absorbed. The hydrogen-bonding interactions between analytes and solvent molecules arise in both the mobile phase and the stationary phase, significantly affecting the analytes' retention and solvent selectivity. In contrast to methanol and isopropanol, the hydrogen-bond donor strengths (denoted α) of acetonitrile and THF are weak, and their α values are 0.15 and 0.00, respectively. The elution strengths of acetonitrile and THF are approximate equal, but the analytes are retained more strongly with acetonitrile than THF. This is because THF is a better hydrogen-bond acceptor (the corresponding strength being denoted β) than acetonitrile, and the solvent–analyte localization effect is correlated with β of the solvent.[44] The higher β of the solvent, the lower the pK_a value of the basic analyte.[46] The β values of THF and acetonitrile are 0.49 and 0.25, respectively.

It was found that retention factors of analytes follow the values of pK_a of amino groups of these analytes when methanol is used in mobile phase.[20] The results have shown again that the ion-exchange interactions dominate the retention of the analytes because of the ionization of silica silanols at elevated pH values, causing weakened adsorption interactions between methanol and silica surface.

Some researchers stated that even sample diluent can influence the peak shape of the analytes.[47,48] The diluent should be as close to the initial mobile phase as possible. The peak shape improves as the aqueous content of the diluent is decreased. However, if water is removed completely from diluent and the sample is prepared in, for example, 100% methanol, wide peaks are observed and the resolution of the compounds decreases. Castells et al. determined that the best compromise for solubility and peak shape would be 75:25 (v/v) mixture of acetonitrile and methanol. It should be noted that the solubility in various organic solvent combinations is analyte dependent and should be optimized accordingly.

It is important to be aware that organic solvents in the mobile phase can change pK_a of a buffer as well as that of the analyte. Consequently, that shifts protolytic equilibrium and changes conditions of chromatography.

Other researchers tested replacing water in the mobile phase by other polar solvents, mainly alcohols.[2] In such conditions, hydrophilic interaction was still observed, but a greater content (approximately doubled) of methanol was required to keep retention the same as for a water-containing mobile phase. Some compounds were not eluted by 100% *n*-propanol at all, apparently due to the decreasing hydrophilic character caused by the longer alkyl chain.

Ideally, samples should be dissolved in the same solvent as the mobile phase. Differences in viscosity between the sample and mobile phase can lead to peak distortion, while differences in the solvent strength can lead to peak broadening. The strong injection solvent prevents the analyte from absorbing at the head of the column. As the solvent injected is diluted by the mobile phase, the analyte retention gradually approaches that characteristic of the mobile phase. However, as the rate of dilution is finite, additional broadening occurs.[49]

All experiments were carried out with mobile phases containing from 60% to 95% of the organic part, with addition of 0.1% of formic acid in aqueous part, and Atlantis HILIC column.

First, retention behavior of the above hydrazine derivatives in various organic solvents was compared by varying the amount of water in mobile phase and changing different organic modifiers as well. The main tendency is obvious: retention is constantly rising with the increase of the organic part of the mobile phase. It seems that mobile phases with a lower content of organics are too strong: no compounds were retained, separation did not occur.

Retention becomes significant only when the mobile phase contains at least 80%–85% of organics. On the other hand, there was no use to employ more than 95% of organics: retention was too strong, and one chromatography run took more than 2 h. It was not surprising that in the line of homologs of methanol, ethanol, and isopropanol (as shown in Figures 14.1 through 14.3), the greatest retention of analytes was provided by the one with the lowest polarity—isopropanol, and other members of this line presented predictable change of properties. It is found no clear connection

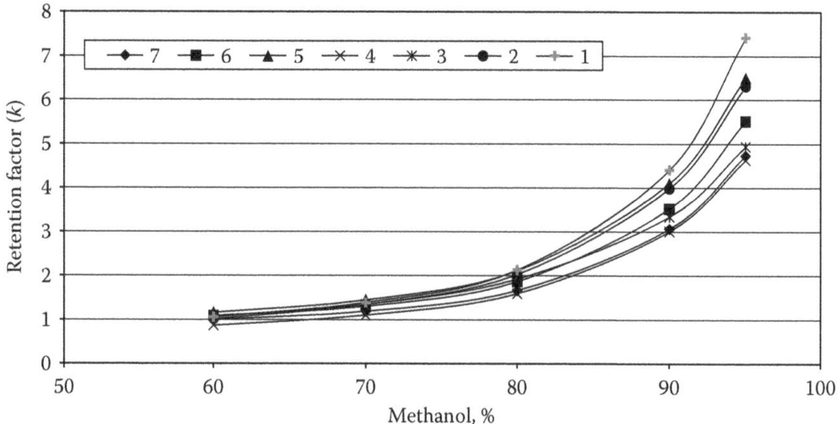

FIGURE 14.1 Retention of hydrazine derivatives in methanol-containing mobile phase.

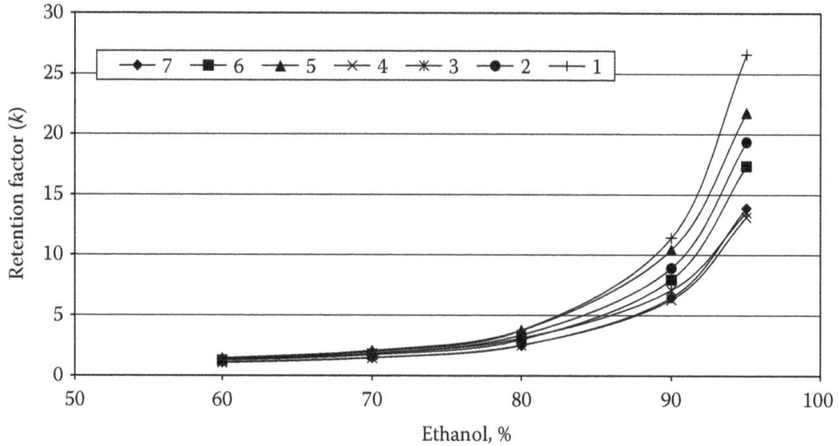

FIGURE 14.2 Retention of hydrazine derivatives in ethanol-containing mobile phase.

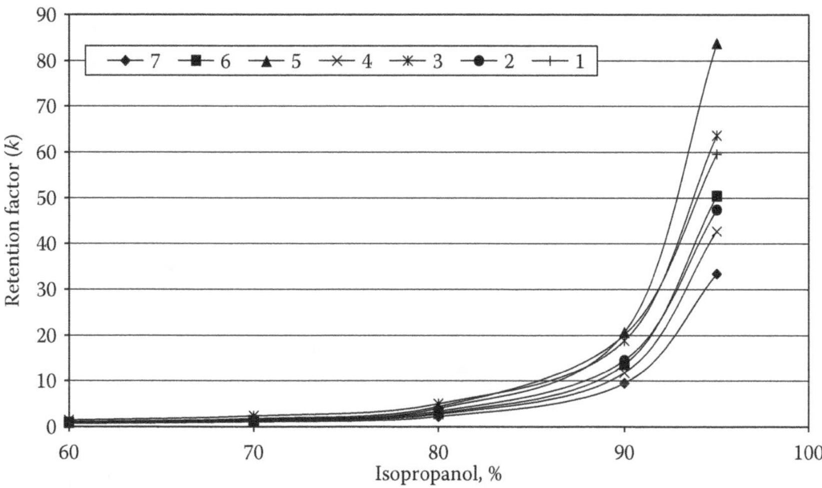

FIGURE 14.3 Retention of hydrazine derivatives in isopropanol-containing mobile phase.

between the retention of analytes and any of organic solvent's properties; probably there is not such a simple law or formula due to the complex mechanism of retention on HILIC sorbents.

Acetone was chosen for its low viscosity and good miscibility with water. As predicted, acetone turned out as weak mobile phase. There was no literature found about acetone as the mobile phase—maybe due to its UV-absorbing properties (UV cutoff at 340 nm). But the authors hope that this example will encourage other researchers to look for unusual mobile phases—suitable for their specific applications.

Acetonitrile—a solvent very popular in HPLC—showed a behavior (see Figure 14.4) similar to that of ethanol (Figure 14.2). THF—a solvent sometimes used in RP-HPLC—showed a behavior similar to that of acetone (see Figures 14.5 and 14.6).

The solvent strength for hydrazine compounds is as follows:

Tetrahydrofuran < acetone < isopropanol < acetonitrile < ethanol < methanol < water

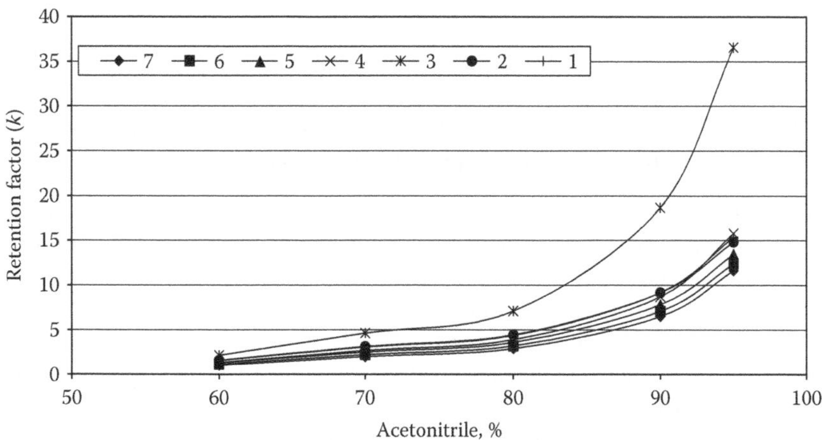

FIGURE 14.4 Retention of hydrazine derivatives in acetonitrile-containing mobile phase.

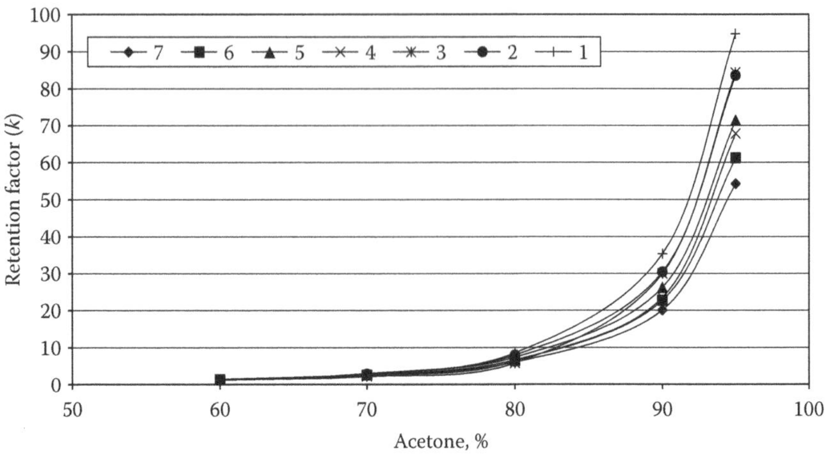

FIGURE 14.5 Retention of hydrazine derivatives in acetone-containing mobile phase.

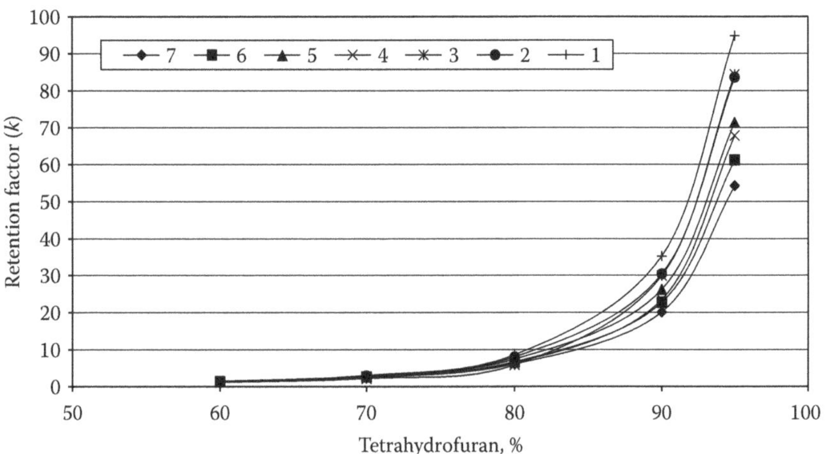

FIGURE 14.6 Retention of hydrazine derivatives in tetrahydrofuran-containing mobile phase.

14.3 EFFECT OF ACETONITRILE CONTENT ON RETENTION WITH DIFFERENT STATIONARY PHASES

Structurally, the studied substances (see Table 14.1) can be divided into three different groups: compounds 1–2 (cations), compounds 3–4 (zwitterions), and compounds 5–7 (cationic esters).

The columns 1–6 (Table 14.2) were tested with respect to their ability of retaining the analytes 1–7 in typical HILIC conditions. The percentage of acetonitrile in the mobile phase was studied in the range between 60% and 95%, with deionized water and 0.1% FA (formic acid).

It was found that the polar bonded phase Discovery Cyano appeared to be the least retentive (Figure 14.7). At acetonitrile concentrations up to 90%, the retention was

TABLE 14.2
Columns Used and Properties of Stationary Phases

No.	Column Name	Phase Type	Pore Size (Å)	Particle Size (μm)	Surface Area (m^2/g)	Dimension (mm)
1	Discovery® Cyano	Cyano	180	5	200	2.1 × 100
2	Hypersil APS-1	Amino	120	3	170	3.2 × 100
3	Atlantis HILIC Silica	Silica	100	3	330	2.1 × 150
4	Alltima HP Silica	Silica	100	3	450	2.1 × 150
5	Spherisorb® Silica	Silica	80	3	220	2.1 × 100
6	ZIC®-HILIC	Sulfobetaine	200	5	135	2.1 × 100
7	Spherisorb® NH2	Amino	80	3	220	2.1 × 100
8	TSK-Gel Amide-80	Amide	80	5	313	2.0 × 100

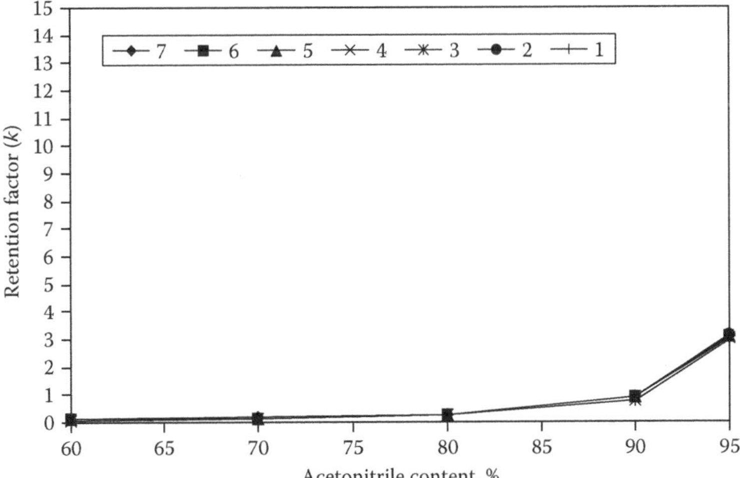

FIGURE 14.7 Relationship between the retention factor (k) and the acetonitrile content. Column: Discovery Cyano. Mobile phases: 0.1% FA in water and acetonitrile.

negligible. Even at 95% of acetonitrile, the retention factors (k) were only about 3. This indicates that under the conditions used, the volume of dynamically generated water-rich stationary phase is small. Probably, the residual silanol groups are unable to accumulate water because of surface shielding by organic ligands.

An interesting phenomenon was the fact that there was almost complete absence of structural selectivity on this cyano column. All analytes irrespective of their molecular size, functionalities, and ionized sites present elute within an extremely narrow k range. Obviously, such behavior is not attractive from a purely analytical point of view. Still, its mechanism might be a subject of a more thorough investigation in the future.

The amino-modified stationary phase (Hypersil APS-1 column) showed a different behavior (Figure 14.8). Cationic substances were retained very weakly in the entire range of acetonitrile concentrations studied (60%–95%). On the other hand,

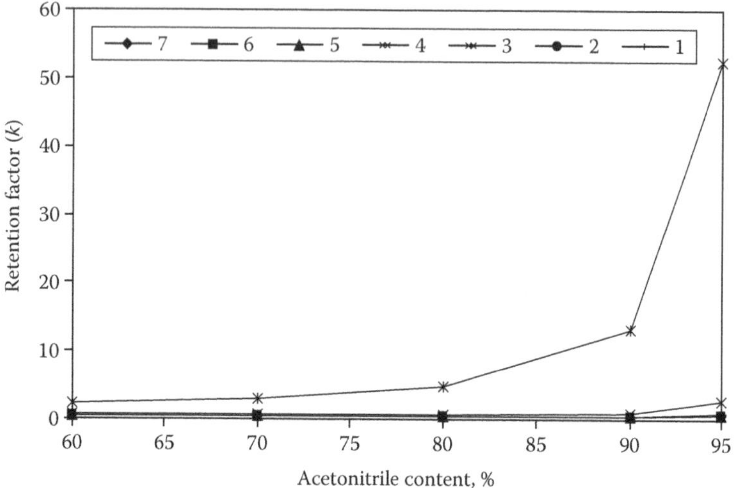

FIGURE 14.8 Relationship between the retention factor (k) and the acetonitrile content. Column: Hypersil APS-1. Mobile phases: 0.1% FA in water and acetonitrile.

zwitterions 3–4 were strongly retained which indicates electrostatic interaction between the protonated amino groups of the stationary phase and the negatively charged moieties of zwitterions.

Specialized HILIC silica (Atlantis HILIC Silica column) showed an increase of retention with the increase of acetonitrile content in the mobile phase (Figure 14.9). The cationic esters 5–7 differ by one carbon atom while having identical functionality. Retention behavior within such a structurally close group of compounds is mainly defined by the balance of hydrophilic and hydrophobic parts. The linear

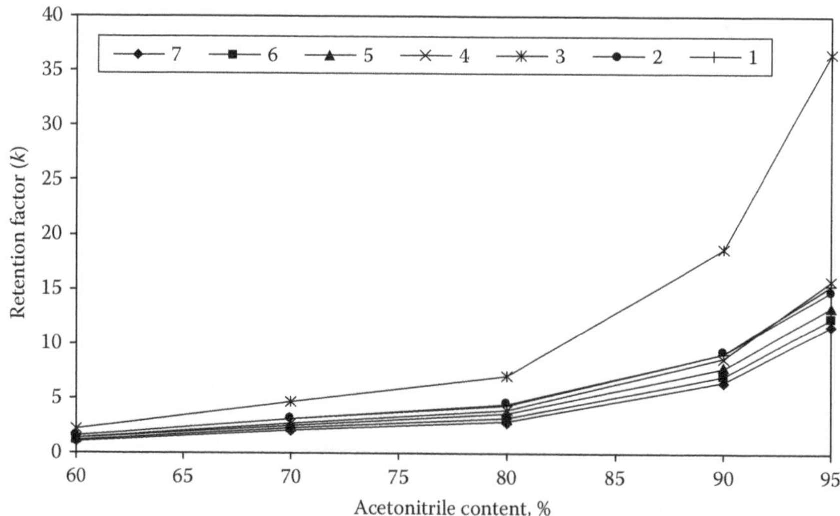

FIGURE 14.9 Relationship between the retention factor (k) and the acetonitrile content. Column: Atlantis HILIC Silica. Mobile phases: 0.1% FA in water and acetonitrile.

relationships between the number of carbon atoms and retention is a common phenomenon in such cases, indicative of the role of partition mechanism between the stationary and mobile phases. The cationic compounds 1–2 cannot be separated on this column, even when retention is strong, at 95% of acetonitrile. One could expect that compound 4 should be retained much more strongly than cationic ester 5. In fact, data show only slight difference between these two compounds. Most likely, in the acidic conditions used in these experiments, the carboxylic group does not influence partition values. The role of this group seems to be so insignificant that compounds 5–7 and 4 belonging to esters and a betaine behave like homologs (Figure 14.9). Compound 3 is retained much stronger than 4–7 and falls out of the common line. This can be explained by a more pronounced contribution of ion exchange in the case of this cyclic structure.

It was observed that the common silica (Alltima HP Silica column) when used with 70%–80% acetonitrile mobile phase showed similar retention (Figure 14.10) and similar elution order to the specialized HILIC silica (Atlantis HILIC Silica column). This suggests the retention mechanisms are similar on both silica packings. On the other hand, at higher (90%–95%) acetonitrile concentrations, the retention on the common silica is much stronger than that on HILIC silica. A possible explanation for this phenomenon is the larger surface area of Alltima HP Silica (see Table 14.2). It can retain a larger volume of a water-rich layer acting as the stationary phase. Spherisorb silica column generally is about three times (Figure 14.11) more retentive than Alltima HP silica if the mobile phase with 70% acetonitrile is used. At higher concentrations of ACN, the difference is even more pronounced. Retention behaviors within homolog series 5–7 are similar to what was observed on Alltima HP and Atlantis HILIC silica. At the same time, the profile of structural selectivity is different. For example, zwitterions are retained relatively weak on this stationary phase.

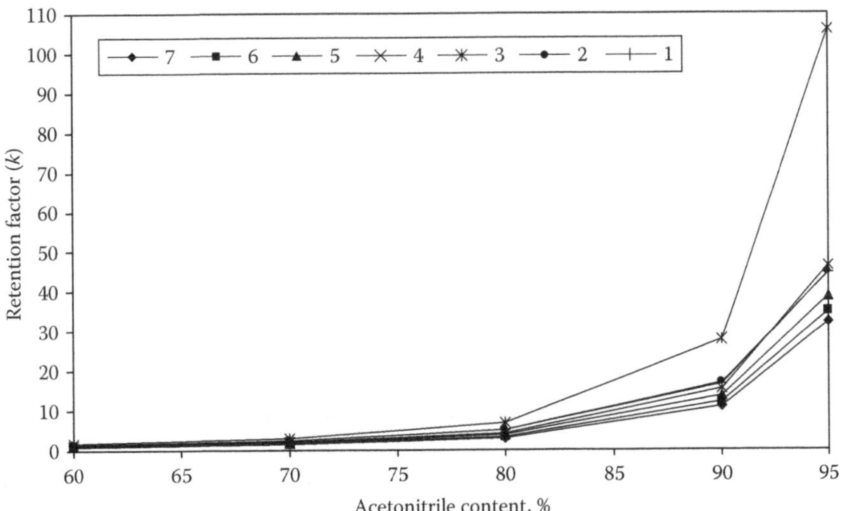

FIGURE 14.10 Relationship between the retention factor (k) and the acetonitrile content. Column: Alltima HP Silica. Mobile phases: 0.1% FA in water and acetonitrile.

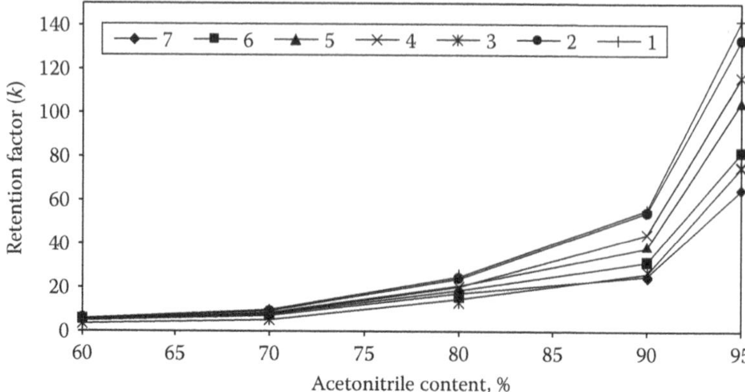

FIGURE 14.11 Relationship between the retention factor (*k*) and the acetonitrile content. Column: Spherisorb Silica. Mobile phases: 0.1% FA in water and acetonitrile.

Compound 3 is the least retained, and the behavior of mildronate (compound 4) is different as compared to esters (compounds 5–7). A possible explanation lies in the properties of Spherisorb silica. It is one of the earliest types of HPLC silica (type A silica), most likely manufactured according to an old technology and containing metal impurities, capable of secondary interactions with nitrogen-containing compounds.

Type A silica gels, prepared by precipitation from solutions of silicates, are acidic because of contamination with certain metals, which activate surface silanol groups and can form complexes with some chelating analytes, causing strong retention of asymmetrical peaks. Newer, highly purified and less-acidic spherical silica particles, "sil-gel" (type B silica) are formed by the aggregation of silica sols in the air, contain very low amounts of metals and are more stable at intermediate and higher pH than the xerogel-type materials, to at least pH 9. Atlantis HILIC Silica column is made out of type B silica. At higher pH, silanol groups are ionized and cation exchange plays a significant role in retention, especially of positively charged basic compounds. Suppression of silanol ionization by the addition of trifluoroacetic acid (TFA) may promote the ion-pairing mechanism. Occasionally, irreversible adsorption may occur on bare silica in the HILIC mode. Silica gel of type C having hydrosilated surface populated with nonpolar silicon hydride (Si-H) groups instead of silanol groups may have up to 95% of original silanols removed, so that it is less polar than silica gels with higher population of silanol groups. An important characteristic of the hydride phases is little attraction for water and consequently improved reproducibility of retention in both the organic and aqueous normal phases (NPs).[50]

Some researchers found that columns of ultra-high purity silica gave lower retention and less selectivity than the columns made from metal containing silica.[49] This was caused by higher acidity of the silanols activated by metal impurities in their vicinity, mainly aluminum and iron. McKeown et al.[51] explained that as combined RP and ion exchange interactions, but Hemström and Irgum insist that these findings might be a mixed mode of cation exchange and HILIC, as well.[9]

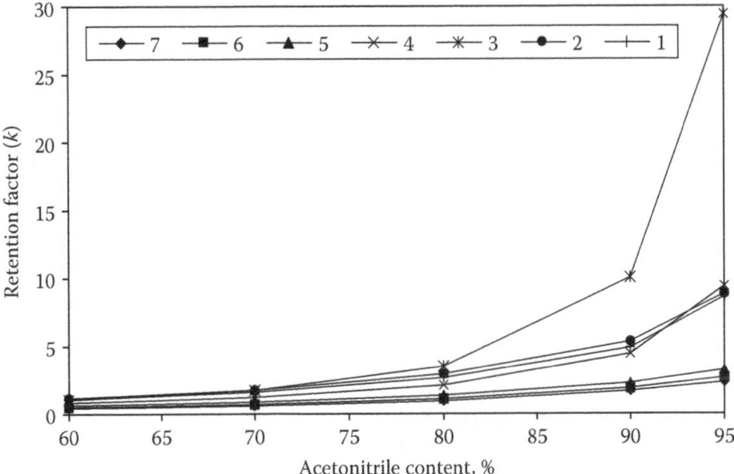

FIGURE 14.12 Relationship between the retention factor (k) and the acetonitrile content. Column: ZIC-HILIC. Mobile phases: 0.1% FA in water and acetonitrile.

The fact that considerable retention is observed even at 60% of acetonitrile (when the volume of the aqueous-rich stationary phase is relatively low) suggests a direct interaction with surface groups of the stationary phase.

Retention of test analytes under HILIC conditions on ZIC-HILIC column is represented in Figure 14.12. Average retention values on ZIC-HILIC are lower than those observed for silica columns, but the general character of plots is similar. On the other hand, the structural factors are more visible on ZIC-HILIC column. First, like it was on silica columns, the retention factors of cations 1 and 2 show very similar values. On the other hand, the retention of cationic esters 5–7 is much lower than that observed on silica. Zwitterions behave differently. At intermediate content of acetonitrile (70%–80%), their behavior is similar to that of other analytes, but at above 80% they show a more rapid increase in retention than other compounds do. A similar, still less pronounced effect was observed on HILIC silica.

14.4 EFFECT OF pH

The concept of electrostatic ion chromatography (EIC), or zwitterionic ion chromatography (ZIC) as it was named later, with a zwitterionic stationary phase for the separation of ions, was first proposed by Hu et al. in 1993.[52] The separation uses zwitterionic stationary phase that maintains a fixed positive and negative charge in close proximity to each other. Analyte ions can access both the fixed positive charge, in the case of anions, and the fixed negative charge, in the case of a cation. As a result of the proximity of the charges, the analyte ions will be repulsed and attracted at the same time. Thus, a unique and sometimes complicated selectivity is obtained. Many studies have been performed to attempt outlining the charge interactions on a molecular level. Hu and Haddad reported the formation of an electrical double layer to explain retention mechanisms.[53,54] Okada and Patil modeled zwitterionic retention based upon Poisson–Boltzmann theory.[55] The formation of a Donnan membrane combined the previous theories of Hu (electric double layer) and Patil (charged surfaces) to explain both elution order and the effect that mobile composition has no retention.[56,57] There have been just a few applications reported that take advantage of the separating power of this unique stationary phase.[58] Jonsson and Appelblad demonstrated the separation of polar and hydrophilic compounds with a sulfobetaine-type zwitterionic stationary phase. The article focused on the selectivity from a HILIC perspective, where the effect of acetonitrile and methanol was evaluated for the retention of RNA–DNA bases in an ammonium formate buffer system.[59]

The combination of separation mechanisms between HILIC and EIC can theoretically be complicate and difficult to understand; however, the performance of zwitterionic column is greatly enhanced with the addition of organic to the mobile phase to take advantage of the HILIC effect. It already has been demonstrated that acetonitrile will promote the HILIC effect more than methanol.

Across the pH range 3.1–6.6, sulfobetaine-type zwitterionic stationary phase retains its permanent positive and negative charges. Increasing pH (within this range), the retention of all the cations increases and the retention of the anions decreases. The effect on cation retention is presumably due to the H^+ interacting with the negatively charged part of the zwitterions (SO_3^-), which ultimately shields the cation from having a strong interaction at a lower pH. The anions are following standard ion-exchange theory. The same studies showed that while increasing the buffer concentration from 10 to 200 mM ammonium acetate both the peak shape and retention times of the ions are drastically affected. The peaks of cations had tailing at lower buffer concentration, then peak fronting at higher buffer concentrations. From these experiments, a range of 50–100 mM buffer concentration was recommended for the experiments. Such concentration of the buffer is doubtful for mixtures of aqueous buffer with organic part (acetonitrile) of the mobile phase: the solubility of ammonium salts in a mobile phase containing high concentration of organics is limited. Besides, high buffer concentration is unacceptable for MS detection.

Change of retention can be explained by a two-part mechanism. The main mechanism of interaction in the HILIC mode is based on a partitioning of the ions into stagnant layer of water on stationary-phase surface, so the decrease of the retention time might be best understood by a shift of equilibrium concentrations. As the buffer cation's (NH_4^+) concentration increases preferentially in the aqueous layer, there is less opportunity for the analyte counterions to partition into the aqueous layer. Thus, the ions are swept through the column (mainly in the organic layer) with less interaction with the column and the aqueous phase. In addition, as the NH_4^+ interacts strongly with the SO_3^- fixed negative charges as the buffer concentration increases, access to these fixed charges is diminished. As a result, cations do not interact with SO_3^- and are not significantly retained; anion retention is affected in the opposite manner. The anions do not experience the typical repulsion forces of the SO_3^- functionality and can then access the tertiary amine for ion exchange. This ion-exchange interaction causes the anions to be retained more strongly.[58]

Peptides are amphoteric molecules whose charges change with pH of the surrounding medium. Their net charge is zero at pH = pI and increases with decreasing pH of buffer solution and vice versa, owing to protonation and dissociation of weekly basic and acidic side chains of the peptide, and of the amino and carboxyl terminals. In this study, hydrophilic interaction and elution power of ionic interaction were kept constant (changing pH of buffers, while maintaining the acetonitrile content and buffer concentration constant). The retention will mainly depend on the charge change of peptides and stationary phase.[60] Thus, the retention data can be correlated with the estimated pH-dependent charge of the peptides. The k of small peptides decreased slightly when pH was increased from 3 to 7. It can be explained as increased positive charge accompanied by an increased hydrophilicity of the peptides at lower pH, which in turn increased retention factors in the HILIC separation mode. An opposite corresponding dependence has been found by Guo et al., i.e., that the positive charge at lower pH results in large decrease in retention times of peptides in RP-HPLC.[60,61] Small peptides were well separated based on a mix of hydrophilic interaction and ion interaction between the stationary phase and analyte. Compared to native silica, zwitterionic sorbent showed more stable retention for basic peptides within the pH range 3–7 and higher retention capacity for acidic peptides. The contribution from ionic interaction from ionic interaction on zwitterionic sorbent was also lower than on silica at pH > 5, and can be eliminated when the buffer concentration in mobile phase is above 20 mM.

Some researchers were surprised when they encountered dramatic loss of resolution or theoretical plates while significantly increasing mobile-phase flow rate (as described, from 1 to 5 mL/min).[62] It seems they had forgotten conclusions from van Deemter's equation—that there is an optimal flow rate with minimal theoretical plate height, and at conditions far away from such plates would not be so ideal. The same researchers, working with a silica column in HILIC mode, came to a conclusion that increasing organic content in the mobile phase substantially increases retention times of cations. In addition, this column could exhibit ion-exchange characteristics as a part of the retention mechanism; thus, the retention of the cations should be affected by a change in buffer concentration and pH. It is experimentally proved: increasing buffer concentration from 10 to 200 mM ammonium acetate (pH 6.65), the retention times of cations monitored dropped about 2–4 times. When pH decreased from 6.65 to 3.55 at constant buffer concentration of 50 mM ammonium acetate, a decrease of retention times of the cation was observed—for about one-third. All these observations can be attributed to the active silanol sites and ion exchange.

Effect of pH on retention and selectivity of seven compounds was examined in pH range of 2.6–5.8 for Atlantis HILIC Silica column (Figure 14.13) and in pH 2.5–7.6 for ZIC-HILIC column (Figure 14.14).

The retention factor of cations 1–2 and cationic esters 5–7 on Atlantic HILIC Silica column increased steadily in the pH range between 2.8 and 5.8. The most probable reason of this is ion exchange. At higher pH, the degree of ionization of surface silanol group increases, thus creating more favorable conditions for such interactions. Mildronate (4) behaves in a similar way as cations 1–2. Another zwitterion 3 shows a completely different pattern: its retention is independent of pH.

While changes of retention on Atlantis HILIC are monotonous, similar series of experiments on ZIC-HILIC column, performed at a wider pH range, showed maxima

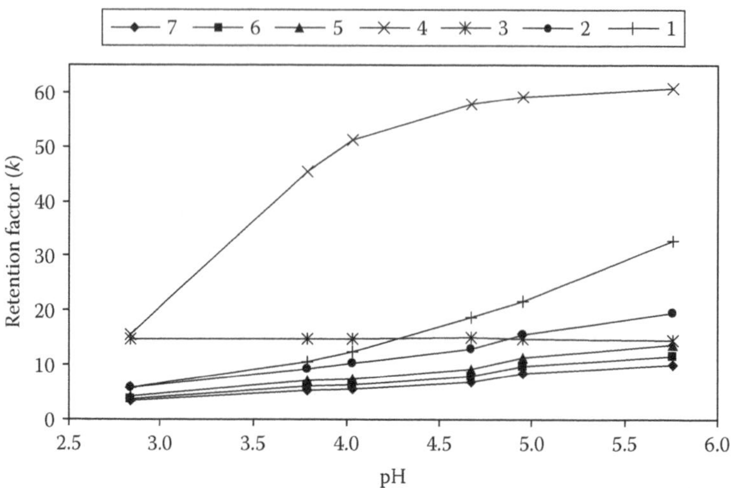

FIGURE 14.13 Effect of buffer pH on the retention factor (*k*). Column: Atlantis HILIC. Mobile phase: acetonitrile–5 mM ammonium formate (85/15, v/v).

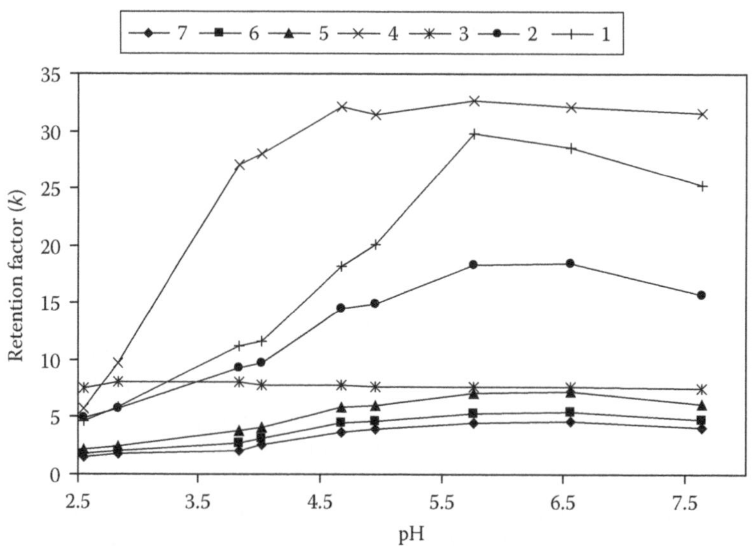

FIGURE 14.14 Effect of buffer pH on the retention factor (*k*). Column: ZIC-HILIC. Mobile phase: acetonitrile–5 mM ammonium formate (85/15, v/v).

at pH about 6. Again, the compound 3 is different, its retention steadily decreasing in the whole pH range studied.

Chromatograms of the seven test compounds are shown in Figure 14.15a for Atlantic HILIC Silica column, pH 5.0; and Figure 14.15b for ZIC-HILIC column at pH 5.0. ZIC-HILIC column shows the best selectivity for the seven test compounds.

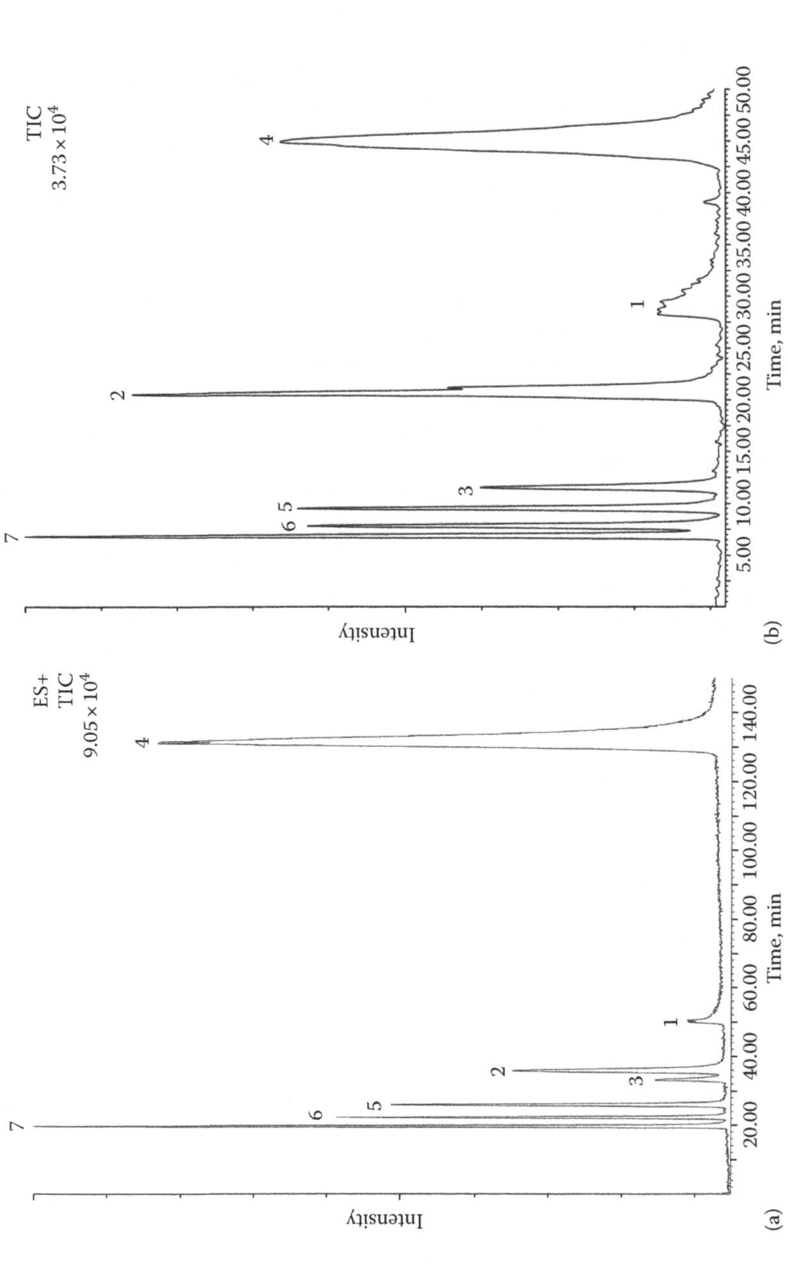

FIGURE 14.15 Separation of test compounds 1–7 (see Table 14.1) under HILIC mode: (a) Atlantis HILIC silica; (b) ZIC HILIC column. Mobile phase: acetonitrile–5 mM ammonium formate pH 5.0 (85:15, v/v).

14.5 EFFECT OF BUFFER CONCENTRATION WITH VARIOUS STATIONARY PHASES

Many salts typically used in reversed-phase (RP) chromatography may not be suitable for HILIC due to poor solubility in the mobile phases containing high level of acetonitrile. In addition to ammonium formate, other salts with relatively high solubility at high organic levels have also been used for HILIC, such as ammonium acetate and bicarbonate salts, triethylamine phosphate and sodium perchlorate, but the last two are not compatible with MS detection.

It is reported in the literature that low concentration of buffer salts in the mobile phase improves peak shape and separation efficiency in HILIC.[1] The use of buffered mobile phase is crucial for achieving reproducible chromatographic separations of charged species, since electrostatic interactions between the analyte and stationary phase impacting retention are influenced and controlled by the buffer. However, although volatile buffering agents are compatible with ESI-MS, it is known that electrospray ionization is adversely affected by buffer salts due to ionization suppression. It has also been determined that HILIC retention is inversely proportional to mobile phase ionic concentration, since increases in salt concentration result in increases of solution hydrophilicity. Therefore, an investigation of the effects of volatile buffer salt concentration at low levels required for optimal retention and ESI-MS sensitivity needs to be performed. Even the presence of 3.3 mM of ammonium acetate (pH 5.5) significantly improves the peak shape and efficiency. The factors influencing sample HILIC retention in extremely low buffer salt environments are complex and may include effects such as ion exclusion and/or ion exchange, in addition to the impact of the salt upon mobile phase hydrophilicity. Other set of experiments proved that the increase in both sample solution pH and ionic strength can by minimized through the use of the ammonium acetate buffer.[1]

Mobile phases containing TFA have been successfully employed for the normal phase HILIC chromatography of polar compounds. But at the conditions of this study, TFA provided unacceptably weak retention, probably due to ion-pairing effects which serve to decrease the hydrophilicity of the analytes. In general, in the buffered mobile phases retention of the test compounds appeared to follow the pattern of "acetate ≈ formate > bicarbonate," which corresponds to a relationship similar to the differences in anion polarity. Similar results have been reported for the HILIC separation of peptides, where retention in the presence of TFA, formic acid, and acetic acid was compared and the differences were attributed to the effectiveness of these acids in minimizing ion-exchange interactions. Since both the acidic and basic test compounds behaved in a relatively similar fashion from one buffer system to another, the results suggested that the effect of the pH upon analyte retention were minimal in comparison to the changes occurring within the stationary base directed adsorption/desorption processes, as influenced by the character of the buffer anion.

HILIC columns contain hydrophilic and often charged stationary phases at lower pH range. The main part of chromatographic separation occurs in water layer on the surface of the stationary phase. But when the stationary phase is charged, it is necessary to keep in mind electrostatic or ionic interactions—or to use salts or buffers in the mobile phase to eliminate such interactions.

Three types of mobile phases are commonly used in the HILIC mode, including acetonitrile/water, acetonitrile/salt solution, and acetonitrile/buffer. The mobile phase should be chosen on the basis of chemical properties of the analytes. For example, for neutral molecules, the acetonitrile/water mobile phase is preferred but any of the three types can be used. For organic salts of strong bases, the acetonitrile/salt solution mobile phase can be used. The acetonitrile/buffer type of the mobile phase is the best for weak acids or weak bases, as well as for samples containing unknown components. If the mobile phase with salt solution is used, the observed effects are consistent with the HILIC retention mechanism. It has been proposed that there may be a layer of stagnant water on the surface of the polar stationary phase to facilitate the partition interaction for analytes. In the absence of a salt or buffer, the charged analytes have a much greater tendency to reside in the water layer. Therefore, the analytes have long retention time or are permanently retained on the column. As the salt concentration increases, the presence of excess amount of counter ions will promote the formation of ion pairs for the charged analytes. The formed ion pairs will have better solubility in the mobile phase, which results in shorter retention times.[63]

There are a number of publications dealing with studies of salt-concentration effects in HILIC, but no information was found on compounds shown in Table 14.1.[1,3,4,62,63] Six commercially available silica-based columns 3–8 (see Table 14.2) representing different polar stationary phases were selected for this study. The effect of ammonium formate concentration in the mobile phase of acetonitrile/pH 5.0 ammonium formate buffer (85/15, v/v) on the retention of the seven test compounds (see Table 14.1) was investigated. For all seven compounds, samples of 0.1 ng/mL concentrations were prepared. The sample solvent was a binary mixture of acetonitrile–water (90:10, v/v). The concentration range of ammonium formate was from 1 to 30 mM (the buffer concentration and pH values refer to the aqueous part only). A further increase in the salt concentration was not possible due to solubility limitations in the mobile phase.

As shown in Figure 14.16a through c, there are significant retention differences between silica columns of different manufacturers. The Alltima HP stationery phase yielded the highest retention while both Atlantis HILIC and Waters Spherisorb show similar retention factor values. Although purity of the silica used in the preparation of the stationary phase may cause such differences,[16] one cannot exclude other reasons like pore size, surface area, and even different manufacturing technologies.

As shown in Figure 14.16a through c, the increase of ammonium formate concentration from 1 to 30 mM, on all silica phases results in a significant decrease of retention for compounds 1 and 2, as well as 5, 6, and 7. Retention of betaine-type substances 3 and 4 seems not affected by the change of ammonium formate concentration.

We found that the retention factor of unmodified silica sorbents is proportional to the relative surface area of stationary phase (see Figures 14.17 and 14.18). Sorption capacity per unit of surface is constant; the same is the percentage of ion exchange and partition.

One can speculate that the increase in ammonium formate concentration may cause a certain saturation of the active silanol sites by excess of ammonium ions thus decreasing the ability of the positively charged analytes to interact with these sites.[14] Our opinion is that the presence of a salt in the mobile phase suppresses the solution

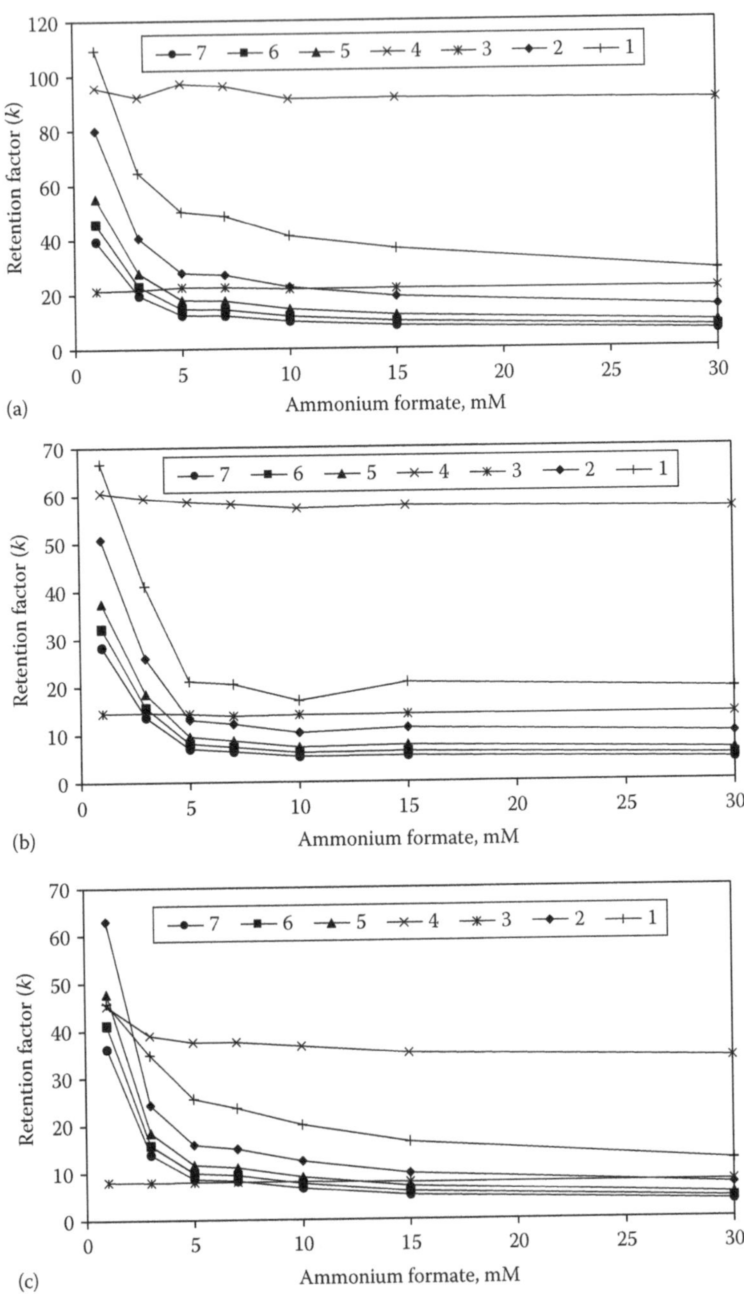

FIGURE 14.16 Effect of ammonium formate buffer solution concentration in mobile phase on retention of mildronate and related substances in sorbents. (a) Alltima HP silica, (b) Atlantis HILIC silica, (c) Spherisorb silica.

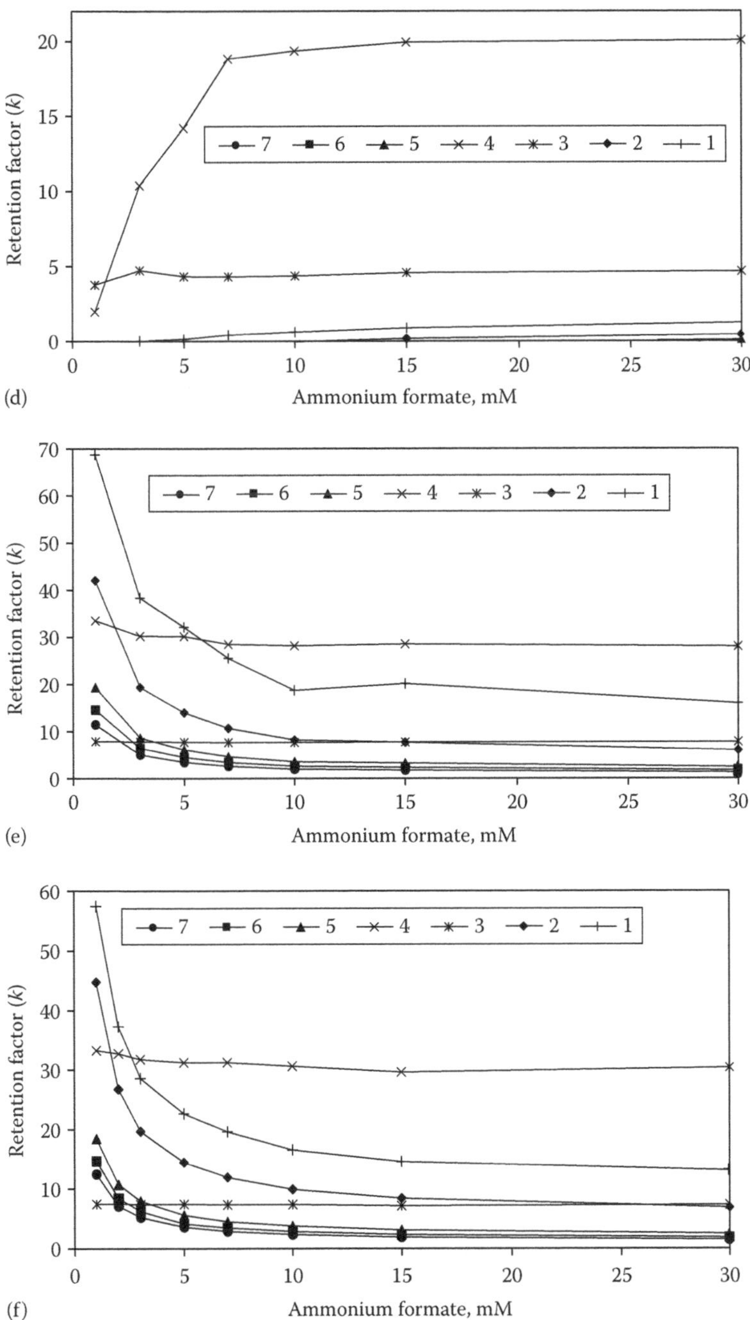

FIGURE 14.16 (continued) (d) Spherisorb NH_2, (e) TSK-Gel Amide-80, (f) ZIC-HILIC.

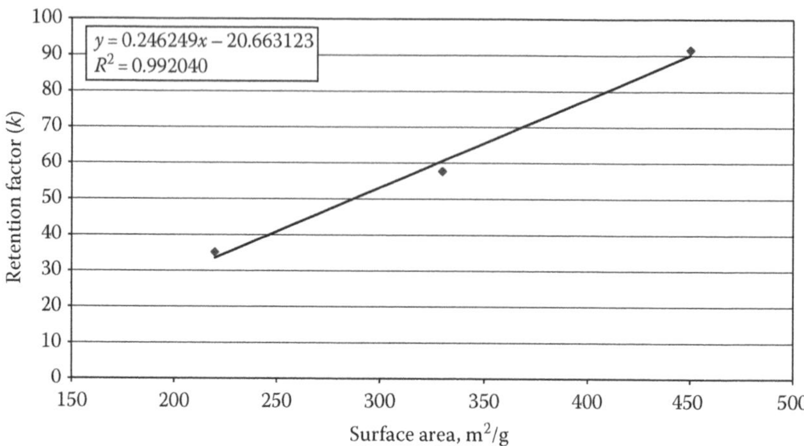

FIGURE 14.17 Retention factor (k) depending on surface area of silica packing. Measured compound: 4. Mobile phase: acetonitrile–15 mM ammonium formate pH 5.0 (85/15, v/v).

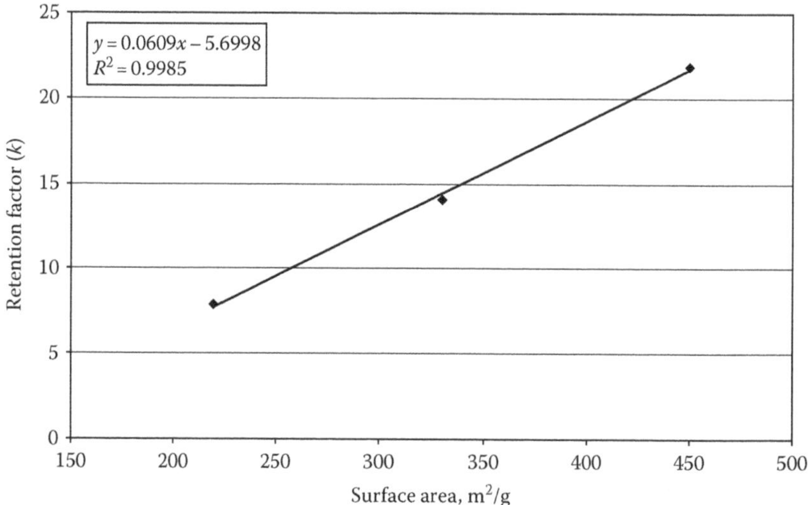

FIGURE 14.18 Retention factor (k) depending on surface area of silica packing. Measured compound: 3. Mobile phase: acetonitrile–15 mM ammonium formate pH 5.0 (85/15, v/v).

of analytes. The retention of the betaine-like compounds 3 and 4 then is determined by a mechanism different from the affinity of positively charged particles toward silanol groups of the stationary phase.

The average retention values are the highest for Alltima column, followed by Atlantis and Spherisorb ones. It correlates the surface areas of these stationary phases and suggests that the surface chemistry factor plays a similar role on these three silicas.

The Waters Spherisorb NH_2 column has an aminopropyl group without endcapping. Aminopropyl-bonded stationary phases are protonated at neutral and acidic mobile-phase pH values. For compounds that are anions in the mobile phase, this offers the possibility of an anion-exchange mechanism in addition to hydrophilic interactions. The ion-exchange nature of the separation was demonstrated in Figure 14.16d, the aqueous–acetonitrile ratio of the mobile phase was constant but the buffer concentration was varied.

The formate counter ions might be adsorbed onto the positively charged amino phase via electrostatic interaction. This could reduce electrostatic repulsion of the positively charged analytes from the positively charged stationary phase, thus causing an increased retention.

TSK-Gel Amide-80 stationary phase bears carbamoyl groups attached to the silica surface through an aliphatic carbon chain. The amide group is less reactive compared to the amine and lacks its basicity.[64]

As shown in Figure 14.16e, the increase of ammonium formate concentration from 1 to 5 mM results in a significant decrease of retention of compounds 1 and 2 as well as 5, 6, and 7. The retention of betaine-type substances 3 and 4 has little or not affected by the change of ammonium formate concentration with Amide-80 stationary phase.

Similar to a silica phase, the decrease of retention time for compounds 1–2 and 5–7 could be attributed to the ion-exchange effect of surface silanol groups.[6] Retention of betaine-like compounds 3 and 4 then is determined by a mechanism different from the affinity of positively charged particles toward silanol groups of the stationary phase.

ZIC-HILIC phase includes a $–CH_2N^+(Me)_2CH_2CH_2CH_2SO_3^-$ ligand covalently attached to porous silica. This ligand includes two oppositely charged functional groups. Close proximity of these two charged moieties preclude their independent interaction with analytes.

Zwitterionic stationary phase has the negatively charged sulfonic acid group exposed to the mobile phase and the positively charged quaternary ammonium group "hidden" inside the alkyl chain. Hence, the attractive interaction of any positive charge of the analyte with the sulfonic acid group of the zwitterionic stationary phase is stronger than the respective interaction between a negative charge of the analyte with the positively charged group of the zwitterionic stationary phase. Similarly, negatively charged analytes are more strongly repelled by the sulfonic acid groups than the "hidden" ammonium groups are attracted. Altogether, this leads to higher retention times for positively charged analytes, and shorter retention times for negatively charged analytes.[18]

The dependence of retention on the ammonium formate concentration is shown in Figure 14.16f. The retention of cations 1–2 and cationic esters 5–7 follows a very similar pattern: an increase in buffer concentration leads to a decrease of retention. Usually, it is explained by increasing competition for sorption sites when the number of counterions in the mobile phase increases.

On the other hand, zwitterions behave differently. Retention of these compounds is slightly influenced by the salt concentration. This indicates that on conditions of the experiment ion-exchange role in the retention mechanism of zwitterions is insignificant.

The buffer concentration is of primary importance for adjusting retention of cations and cationic esters, but has little effect on zwitterions. The retention of the cations and the cationic esters follows a dual (partition and ion exchange) mechanism. Retention data of zwitterionic compounds suggest that the role of ion exchange is less significant for this group of compounds. Concentration of ammonium formate buffer makes selective impact on retention of the compounds, and the most significant change is observed at lower buffer concentrations of less than 10 mM.

14.6 TEMPERATURE EFFECT WITH VARIOUS STATIONARY PHASES

Mobile phase composition and column temperature are commonly used to control adsorption behavior in RP chromatography. Several studies have been performed to compare the effect of a modulation of the mobile phase composition to that of the temperature.[65–67] It is found that the temperature is a less effective operating parameter for reducing retention compared to the solvent strength. Strong temperature effects are evidenced only for ionizable compounds; therefore, manipulating of the temperature is a useful tool mainly in ion-exchange and ion-pair chromatography.[67] For neutral components, a change of 4°C–5°C in temperature corresponds to approximately 1% changes in composition of the mobile phase.[65,66] But the temperature range of interest is usually limited from the ambient temperature up to 50°C–60°C since classical bonded silica stationary phases are thermally unstable and operating at higher temperature requires robust, thermally stable, but expensive new generation columns. The use of higher temperature can also cause decomposition of temperature-labile analytes.

However, combining solvent strength and temperature can have a powerful effect on the separation in RP chromatography in terms of selectivity and efficiency. The acceptance of importance of temperature is based on well-known effects of the decrease of viscosity of the mobile phase and the increase in analyte diffusivity. The first effect leads to lowering backpressure over the column, allowing increased flow rates. This gives improvement of the process productivity, which is directly related to the speed of the separation process. The consequence of increased diffusivity is faster mass-transport kinetics, which is reported to result in improvement of column efficiency.

The relationship between retention factor (k) and column temperature (T) in RPLC is often described by the van't Hoff equation:

$$\ln k = -\frac{\Delta H^0}{RT} + \frac{\Delta S^0}{R} + \ln \phi$$

where

ΔH^0 and ΔS^0 are retention enthalpy and entropy
R is gas constant
ϕ is phase ratio

If the retention of polar compounds in HILIC is through partitioning between the mostly organic mobile phase and the water-rich liquid layer on the stationary phase surface as proposed by Alpert,[6] the van't Hoff equation should apply to HILIC. These propositions are fulfilled just when ϕ = constant.

There are studies, where the relationship between retention factor (k) and water content in the mobile phase explains partitioning and/or adsorptive retention of analytes.[68] Besides, slopes of the relation graph of analytes are higher for analytes having more polar interactive sites, and typically negative enthalpy values are higher for more polar phases and more polar analytes.

In ion-exchange HPLC, enthalpy changes generally are large and dominate the retention interaction. Enthalpy changes can be either positive or negative (exothermic or endothermic reactions), and in ion exchange, generally cause significant changes in retention. Thus, separation at higher temperature in ion-exchange chromatography is quite commonly found in the literature reports.

In the RP-HPLC, retention interactions are usually exothermic with relatively small changes in enthalpy. This usually manifests itself in only modest decreases of retention time as temperature increases in contrast to the changes in ion-exchange chromatography.[69–71]

When elevated temperature is discussed in chromatographic circles, it is focused on reduction of system backpressure followed by various operational benefits. Lower pressure causes less stress on the hardware (valves, seals, etc.) and also enables the use of higher flow rates, if desired. This becomes important when some of the modern columns packed with smaller sized stationary phase particles are used that have a higher backpressure than the more commonly used 5 μm material.

Column temperature can tamper analyte transferring enthalpy from mobile to stationary phases, especially when ionic interactions are involved.[66] Dong and Huang studied basic compounds, having negative slopes of van't Hoff curves.[72] This fact means that retention times behave non-normally, increasing with raising of temperature. The authors observed simultaneous increase in column plate number, thus raising column temperature improved the resolution of the tested compounds.

We want to present a study of the relationship between the column temperature and retention of both mildronate and its related impurities. There is a number of publications dealing with studies of column temperature effects in HILIC,[3,4,20,73,74] but no information was found for the compounds as described in Table 14.1.

This chapter presents an investigation on the relationship between chromatographic retention and column temperature effect on various stationary phases (silica and zwitterionic sulfobetaine, columns 3, 5, 6 in Table 14.2).

Temperature effects were determined for three mobile phases: (1) 10% water with 0.1% formic acid/90% acetonitrile; (2) 10% water with 0.1% formic acid/90% methanol; and (3) 15% 5 mM ammonium formate (pH 5)/85% acetonitrile. In this study, the temperature effect on retention was investigated by varying column temperature from 30°C to 55°C, and the retention data for the test compounds were used to construct van't Hoff plots for the three columns mentioned above. Figures 14.19 through 14.21 show the retention behavior of compounds 1–7 as a function of temperature ($\log k$ vs. $1/T$).

With a silica column and acetonitrile mobile phases, the retention increased as the temperature was increased which indicates either a positive enthalpy for interaction of the analytes with the stationary phase and/or substantial entropy contributions. This temperature effect is opposite to what is typically observed under RP conditions in HPLC, i.e., decreasing retention with increasing temperature. This result provides

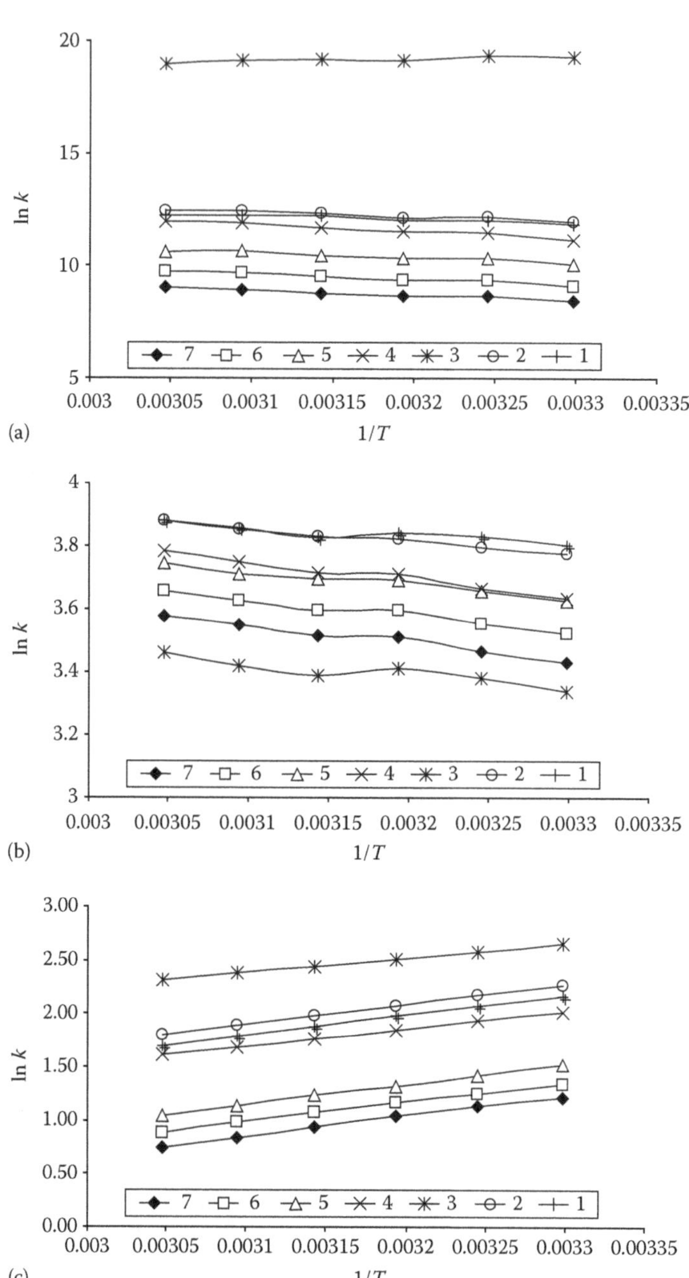

FIGURE 14.19 Retention as function of temperature. Mobile phase: 90% ACN. Stationary phases: (a) Atlantis HILIC Silica, (b) Waters Spherisorb Silica, and (c) ZIC HILIC.

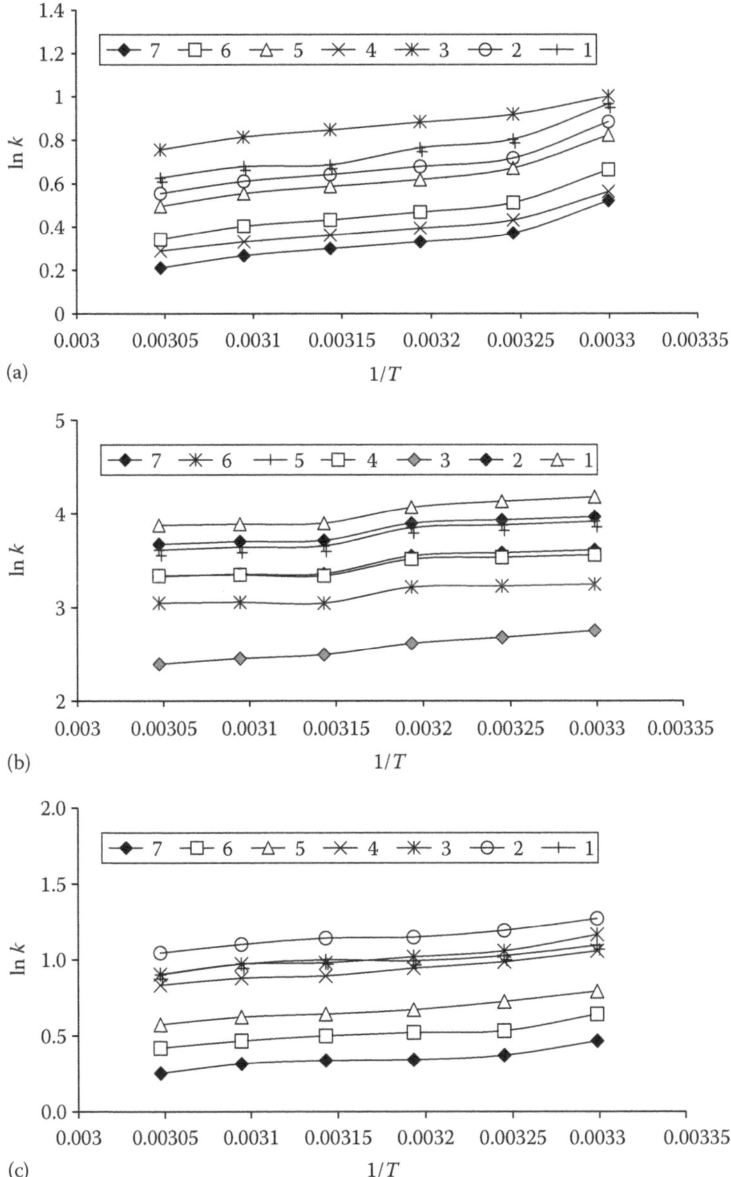

FIGURE 14.20 Retention as function of temperature. Mobile phase: 90% MeOH. Stationary phases: (a) Atlantis HILIC Silica, (b) Waters Spherisorb Silica, and (c) ZIC HILIC.

other routes for improving selectivity via temperature control. Raising temperature can increase retention, thus improving *R* values. Also, higher temperatures result in lower viscosity and faster mass transfer which can decrease peak widths to provide another means for improving resolution depending on the extent of retention and diffusion in the stationary phase.[75]

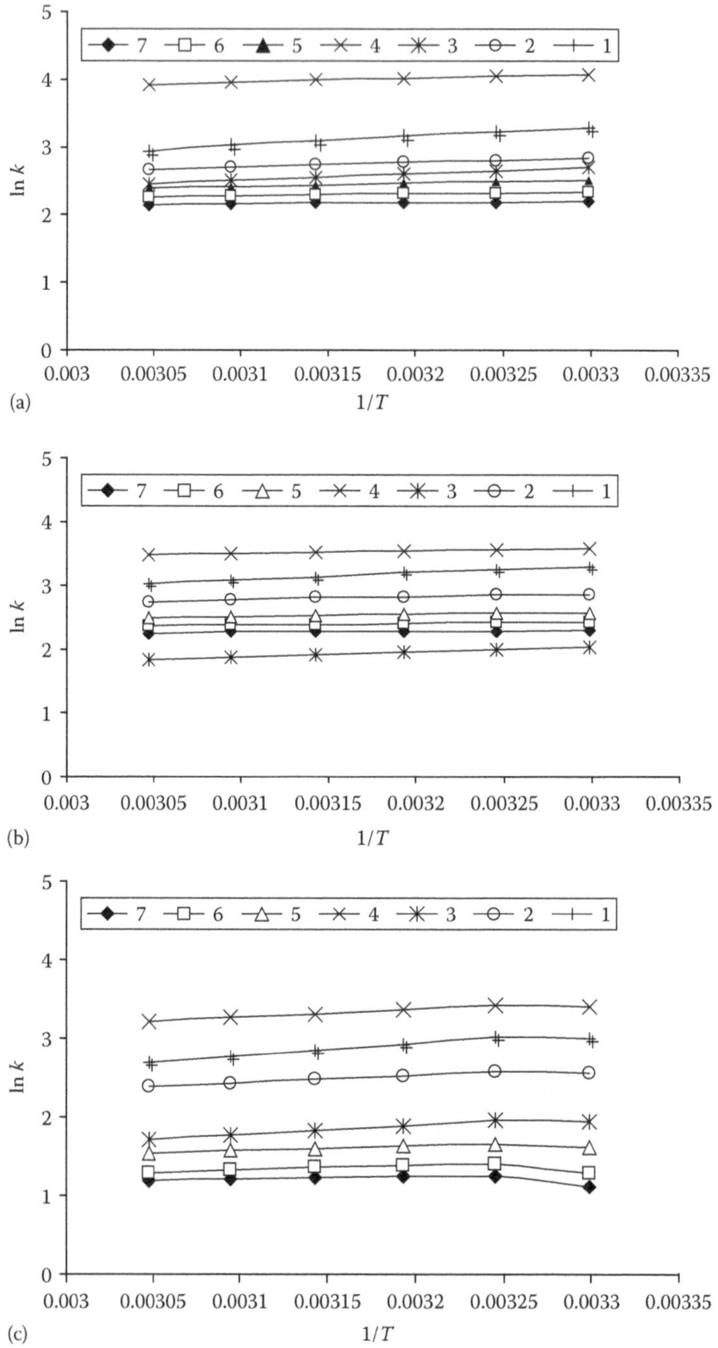

FIGURE 14.21 Retention as function of temperature. Mobile phase: 15% 5 mM ammonium formate (pH 5), 85% ACN. Stationary phases: (a) Atlantis HILIC Silica, (b) Waters Spherisorb Silica, and (c) ZIC HILIC.

Linear relationship between k and temperature was observed in the temperature range of 30°C–55°C for the test compounds on all columns with R^2 ranging from 0.950 to 0.998. This is similar to what is usually observed in RPLC,[76,77] even though nonlinear van't Hoff plots have also been reported and attributed to the temperature dependency of both enthalpy (ΔH^0) and phase ratio (ϕ) in RPLC.[78,79] Generally, when nonlinear van't Hoff plots are observed, it is assumed that the enthalpy and entropy change with temperature. This study did not find any nonlinear behavior of HILIC sorbents.

References to temperature in HPLC may refer to the temperature of the column packing, the mobile phase, or the column-heating device. The temperature inside the column is of primary interest, but difficult to be measured without physically affecting the column. A simpler way to determine the column temperature is to use a flow-through thermocouple to measure the temperature of the mobile phase entering and/or exiting the column. An average of the inlet and outlet mobile-phase temperature will give a reasonable value for the average column temperature and information about the temperature gradient along the column. The column oven temperature generally will differ from the temperature inside the column, but as long as the column oven temperature is constant, this measurement will provide a suitable reference value for most applications.

MeOH is a protic solvent as an organic component in the mobile phase. When MeOH was employed as the organic part of mobile phase instead of ACN, the retention of seven test compounds exhibited disorderly changes with increasing temperature, as shown in Figure 14.20a through c. Some of them displayed increasing retention with elevated temperature, others showed the opposite. The different temperature effects are the result of a multiple retention mechanisms. In addition to hydrophilic and ion-exchange interactions, solvent–analyte and solvent–stationary-phase hydrogen bonding are also involved.[19]

All these results show that temperature plays a more complex role on the retention of analytes in MeOH than in ACN due to the multiplicity and complexity of the retention mechanisms.

With temperature increase, the trend of retention increase was observed at pH 2.1; further the retention values of the analytes significantly decrease of retention was achieved at pH 5.0. These results reflect a transitional retention mechanism with the pH change. As the buffer pH varies from 2.1 to 5.0, the ionization of the silanols on the silica surface gradually increases and the analytes are kept in a protonated state,[80] offering more opportunity for cation exchange. The major mechanism governing the retention is transformed from hydrophilic to ion-exchange interaction. When ion exchange becomes the dominant mechanism, the pK_a of the analyte plays an important role in affecting the retention and selectivity. According to the previous work,[81,82] an increase in temperature can cause a reduction of protonated species. Thus, the retention of the test compounds at higher temperature would decrease as observed in Figures 14.19 and 14.21 due to the weakened ion-exchange contribution.

The impact of temperature on retention for various analytes could be different. As discussed above, at pH 5, because ion-exchange interaction is stronger, the pK_a of the bases becomes an important parameter governing retention. As proposed in previous literature,[81,82] the pK_a of analytes decreases with increasing temperature, and the smaller the pK_a of the base is, the more it decreases. Since the pK_a value of compound 1 (pK_a 9.80) is smaller than that of compound 2 (pK_a 6.9), a greater reduction of the protonated species must be obtained for compound 1. This implies that the ion-exchange contribution to the retention factor is smaller.

It has been found that temperature plays a different role when the type and content of the organic modifier or pH in the mobile phase are changed. For the seven test compounds examined in an organic-rich environment, temperature response varied with the type of the organic solvent.

Column temperature is an important parameter that affects the retention of polar compounds in HILIC. The effect of column temperature on the retention of hydrazine-based compounds was investigated on silica and zwitterionic sulfobetaine columns in the temperature range of 30°C–55°C with different organic solvents and different pHs. At an increased temperature on silica column, the retention factor depends (increases or decreases) on the organic solvent (ACN, MeOH) and pH. But on zwitterionic sulfobetaine sorbent, the retention factor changes significantly depending on the type of organic solvents and pH.

14.7 APPLICATION RESULTS AND METHOD VALIDATION

As a result of this systematic investigation of different columns and mobile phases, the HILIC method using a zwitterionic sulfobetaine stationary phase was practically used to determine six impurities in the hydrazine-based pharmaceutical compound mildronate. The validation was performed with ZIC-HILIC column with mobile phase 85% ACN/15% 5 mM ammonium formate (pH 5.0) with LC-MS-MS according to the ICH Guidelines.[83–86] The validation parameters included specificity, accuracy, linearity, precision, limit of detection (LOD), and limit of quantitation (LOQ). The method specificity has been demonstrated by fair separation of all the known and potential degradation products. At the start of the study, an assumption was made that any impurity may be present at concentrations not more that 0.1% of assay concentration (the preliminary quality specification). The concentration of the main compound (mildronate) 1.0 mg/mL was used during the validation.

Validation results for the impurities method are summarized in Table 14.3. The method was tested on real samples of mildronate, produced in JSC "Grindeks."

TABLE 14.3
Summary of Results of Validation

Compound	Precision, RSD% (n = 6)	Mean Accuracy, %	R^2	LOD, % of Assay	LOQ, % of Assay
1	9.4	96.9–104.7	0.991	0.003	0.01
2	3.3	105.7–114.0	0.997	0.0003	0.001
3	7.7	111.2–116.4	0.998	0.0005	0.002
5	2.7	102.1–114.4	0.992	0.0004	0.001
6	2.1	104.2–109.6	0.998	0.0001	0.0004
7	2.7	103.1–108.4	0.998	0.00006	0.0002

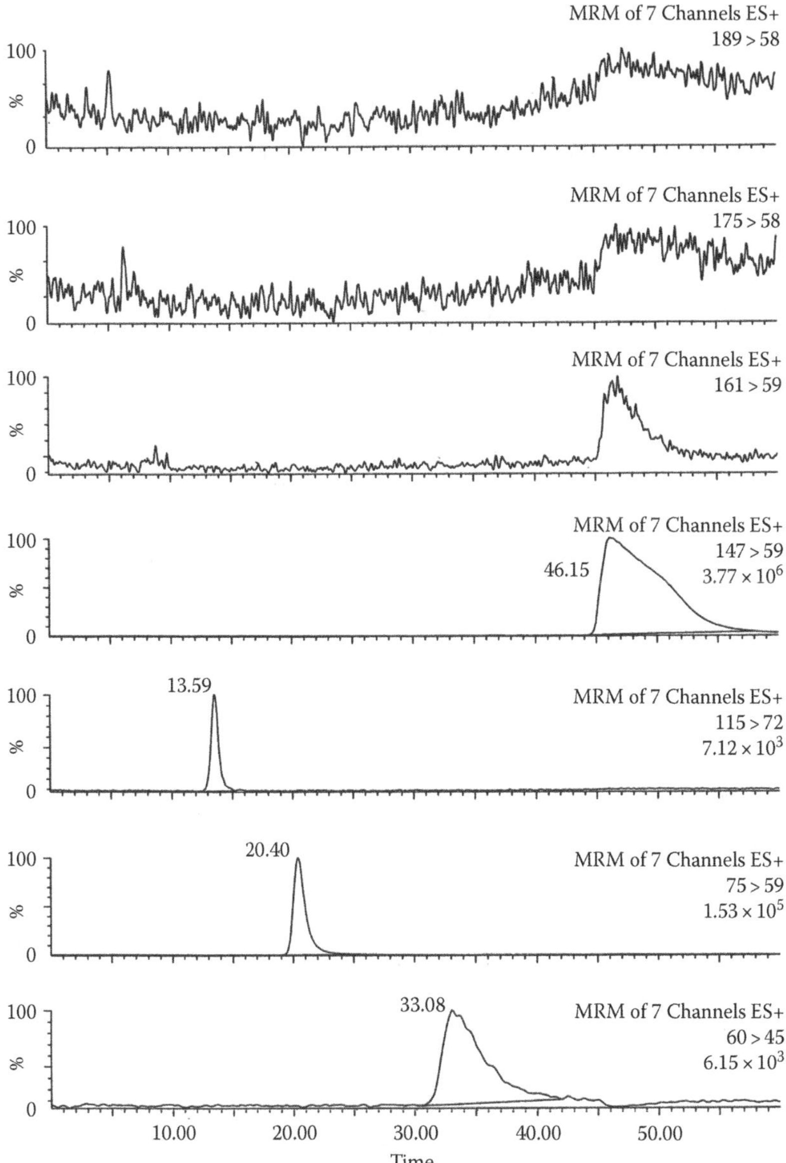

FIGURE 14.22 MRM chromatogram of technical batch 2374-06 of mildronate. Column: ZIC-HILIC; Concentration of mildronate 1 mg/mL. Mobile phase: acetonitrile–5 mM ammonium formate pH 5.0 (85/15, v/v). Specific transitions of compounds 1–7 (from bottom to top).

The validated method was used to analyze technical batch (Figure 14.22) and commercial batch (Figure 14.23). Figure 14.22 shows that technical product contains visible quantities of related substances, but Figure 14.23 shows only traces of them. Comparative results of analysis are presented in Table 14.4. The developed method is suitable to control the quality of commercial mildronate API.

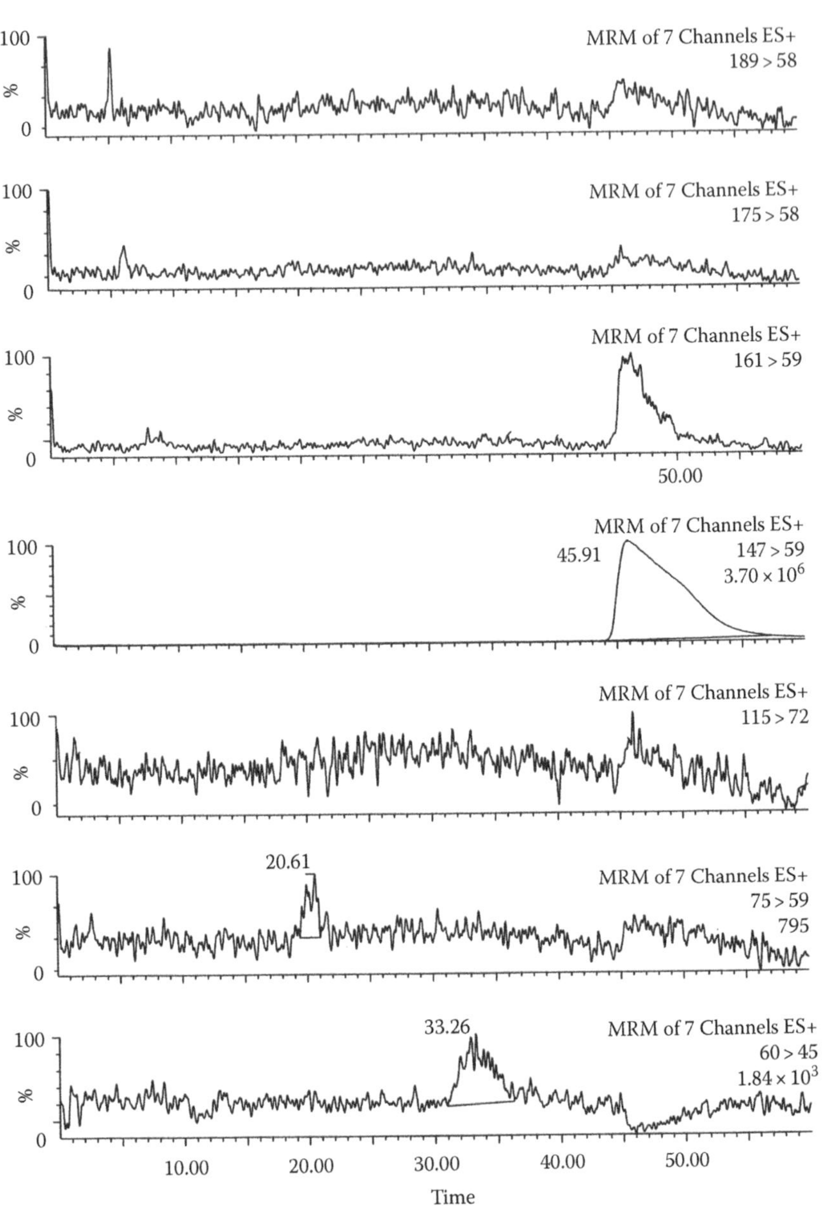

FIGURE 14.23 MRM chromatogram of production batch 990208 of mildronate. Column: ZIC-HILIC; Concentration of mildronate 1 mg/mL. Mobile phase: acetonitrile–5 mM ammonium formate pH 5.0 (85/15, v/v). Specific transition of compound 1–7 (from bottom to top).

TABLE 14.4
Quantitative Results of Analyses of Technical and Commercial Mildronate, in Percent

Compound	Technical, Batch 2374-06	Commercial, Batch 990208
1	0.20	0.01
2	0.35	Traces (~0.0003)
3	0.02	Not found
5	Not found	Not found
6	Not found	Not found
7	Not found	Not found

14.8 SUMMARY

HILIC is a valuable alternative to RP and ion-pair chromatography. It has been shown that HILIC separations of quaternary hydrazine derivatives are possible using both specialized HILIC type and common silica columns. Specialized columns show weaker retention and are more suitable for separations at high concentrations of acetonitrile. Variation of polar stationary phase is used as a tool to change the selectivity of separation for the test analytes. It was shown that the variation of acetonitrile concentration is effective to adjust retention but does not significantly influence the selectivity.

The pH and buffer concentration are of primary importance for adjusting retention of cations and cationic esters, but has little effect on zwitterions. The retention of cations and cationic esters follow regularities expected for a dual (partition—ion exchange) mechanism. Retention data of zwitterionic compounds suggest that the role of ion exchange is less significant for this group of substances. HILIC is an efficient tool in method development for extremely polar and ionic hydrazine derivatives. ZIC-HILIC column was tested even more, the method validation having been performed for quantification of related substances. The validated method is suitable for purity analysis of mildronate drug substance.

ACKNOWLEDGMENTS

The successful completion of this work owes a great deal to cooperation provided by Imants Davidsons and Dr. Osvalds Pugovics, Dr. Helena Kazoka, JSC "Grindeks," University of Latvia and Latvian Organic Institute, and we have pleasure in expressing our gratitude to them.

REFERENCES

1. Strege, M.A., *Anal. Chem.*, 70 (1998), 2439.
2. Olsen, B.A., *J. Chromatogr. A*, 913 (2001), 113.
3. Guo, Y. and Gaiki, S., *J. Chromatogr. A*, 1074 (2005), 71.
4. Guo, Y., *J. Liq. Chromatogr. Relat. Technol.*, 28 (2005), 497.
5. Chrums, S.A., *J. Chromatogr. A*, 720 (1996), 75.

6. Alpert, A.J., *J. Chromatogr.*, 499 (1990), 177.
7. Rückert, H. and Samuelson, O., *Sven. Kem. Tidskr.*, 66 (1954), 337.
8. Samuelson, O. and Sjöström, E., *Sven. Kem. Tidskr.*, 64 (1952), 305.
9. Hemström, P. and Irgum, K., *J. Sep. Sci.*, 29 (2006), 1784.
10. Strege, M.A., *Am. Pharm. Rev.*, 2 (1999), 53.
11. Strege, M.A., Stevenson, S., and Lawrence, S.M., *Anal. Chem.*, 72 (2000), 4629.
12. Garbis, S.D., Melse-Boonstra, A., West, C.E., and vanBreemen, R.B., *Anal. Chem.*, 73 (2001), 5358.
13. Tolstikov, V.V. and Fiehn, O., *Anal. Biochem.*, 301 (2002), 298.
14. Person, M., Hazotte, A., Elfakir, C., and Lafosse, M., *J. Chromatogr. A*, 1081 (2005), 174.
15. Schlichtherle-Cerney, H., Affloter, M., and Cerny, C., *Anal. Chem.*, 75 (2003), 2349.
16. Baughman, T.M., Wright, W.L., and Hutton, K.A., *J. Chromatogr. B*, 852 (2007), 505.
17. Peru, M., Kuchta, L., Headley, V., and Cessna, J., *J. Chromatogr. A*, 1107 (2006), 152.
18. Xuan, Y., Scheuermann, B., Meda, R., Hayen, H., Wiren, N., and Weber, G., *J. Chromatogr. A*, 1136 (2006), 73.
19. Li, R.P. and Huang, J.X., *J. Chromatogr. A*, 1041 (2004), 163.
20. Liu, M., Chen, E., Ji, R., and Semin, D., *J. Chromatogr. A*, 1188 (2008), 255.
21. Alpert, A., *J. Anal. Chem.*, 80 (2008), 62.
22. Boersema, P.J., Mohammed, S., and Heck, A.J.R., *Anal. Bioanal. Chem.*, 391 (2008), 151.
23. Vikingsson, S., Kronstrand, R., and Josefsson, M., *J. Chromatogr. A*, 1187 (2008), 46.
24. Yoshida, T. *Anal. Chem.*, 69 (1997), 3038.
25. McCalley, D.V., *J. Chromatogr. A*, 1171 (2007), 46.
26. Gustavsson, S., Samskog, J., Markides, K., and Langstrom, K., *J. Chromatogr. A*, 937 (2001), 41.
27. Dunphy, R. and Burinsky, D.J., *J. Pharm. Biomed. Anal.*, 31 (2003), 905.
28. Guo-ying, C., Run-feng, L., Xin, H., and Chun-hua, S., *Chin. Pharm. J.*, 40 (2005), 864.
29. Run-feng, L., Guo-ying, C., Xin, H., and Jun-ren, Z., *Chin. J. Pharm. Anal.*, 26 (2006), 358.
30. Ivanovskaya, E.A., Anisimova, L.S., Belikhmaer, Ya.A., Koshelskaya, O.A., and Sokolova, A.A., *Pharm. Chem. J.*, 29 (No. 3) (1995), 219.
31. Yun-Feng, L., Xin, H., and Kai-Shun, B., *J. Chromatogr. B*, 852 (2007), 35.
32. Sahartova, O., Shatz, V., and Kalvins, I., *J. Pharm. Biomed. Anal.*, 11 (1993), 1045.
33. Hmelnickis, J., Pugovics, O., Kazoka, H., Viksna, A., Susinskis, I., and Kokums, K. *J. Pharm. Biomed. Anal.*, 48 (2008), 69.
34. Liu, M., Ostovic, J., Chen, E.X., and Cauchon, N., *J. Chromatogr. A*, 1216 (2009), 2362.
35. Jagota, N.K., Chetram, A.J., and Nair, J.B., *J. Pharm. Biomed. Anal.*, 16 (1998), 1083.
36. Carlin, A., Gregory, N., and Simmons, J., *J. Pharm. Biomed. Anal.*, 17 (1998), 885.
37. Gyllenhaal O., Groenberg, I., and Vessman, J., *J. Chromatogr.*, 511 (1990), 303.
38. Selim, S. and Warner, C.R., *J. Chromatogr.*, 166 (1978), 507.
39. Timbrell, J.A., Wright, J.M., and Smith, C.M., *J. Chromatogr.*, 138 (1977), 165.
40. Seifart, H.I., Gent, W.I., Parkin, D.P. et al., *J. Chromatogr. B*, 674 (1995), 269.
41. Abdou, H.M., Medwick, T., and Bailey, L.C., *Anal. Chim. Acta*, 93 (1977), 221.
42. Kirchherr, H., *J. Chromatogr.*, 617 (1993), 157.
43. *A Practical Guide to HILIC*. SeQuant AB, Umea, Sweden, April 2006.
44. Snyder, L.R., *J. Chromatogr.*, 255 (1983), 3.
45. Scott, R.P.W., *Adv. Chromatogr.*, 20 (1982), 169.
46. Vervoort, R.J.M., Ruyter, E., Debets, A.J.J. et al., *J. Chromatogr. A*, 931 (2001), 67.
47. Grumbach, E.S., Wagrowski-Diehl, D.M., Mazzeo, J.R. et al., Application notes Hydrophilic Interaction Chromatography Using Silica Columns for the Retention of Polar Analytes and Enhanced ESI-MS Sensitivity, Waters Corporation, www.waters.com

48. Castells, R.C., Castells, C.B., and Castillo, M.A., *J. Chromatogr. A*, 775 (1997), 73.
49. Zhou, T. and Lucy, C.A., *J. Chromatogr. A*, 1123 (2008), 8.
50. Jandera, P., *J. Sep. Sci.*, 31 (2008), 1421.
51. McKeown, A.P., Euerby, M.R., Lomax, H. et al., *J. Sep. Sci.*, 24 (2001), 835.
52. Hu, W., Takeuchi, T., and Haraguchi, H., *Anal. Chem.*, 65(17) (1993), 2204.
53. Hu, W., *Langmuir*, 15(21) (1999), 7168.
54. Hu, W. and Haddad, P.R., *Trends Anal. Chem.*, 17(2) (1998), 73.
55. Okada, T. and Patil, J.M., *Langmuir*, 14(21) (1998), 6241.
56. Cook, H.A., Hu, W., Fritz, J.S. et al., *Anal. Chem.*, 73 (2001), 3022.
57. Cook, H.A., Diconoski, G., and Haddad, P.R., *J. Chromatogr. A*, 997(1–2) (2003), 13.
58. Risley, D.S. and Pack, B.W., *LC/GC North America*, 24(8) (2006), 12.
59. Jonsson, T. and Appelblad, P., *LCGC*, 17(Suppl.) (2004), 72.
60. Jiang, W., Fischer, G., Girmay, Y. et al., *J. Chromatogr. A*, 1127 (2006), 82.
61. Guo, D.C., Mant, C.T., Taneja, A.K. et al., *J. Chromatogr.*, 359 (1986), 499.
62. Pack, B.W. and Risley, D.S., *J. Chromatogr. A*, 1073 (2005), 269.
63. Wang, X., Li, W., and Rasmussen, H., *J. Chromatogr. A*, 1083 (2005), 58.
64. Normal Phase Chromatography: TSKgel Amide-80, *Tosoh Bioscience 2008 Laboratory Products Catalogue*, www.separations.us.tosohbioscience.com
65. Chen, M.H., *J. Chromatogr. A*, 788 (1991), 51.
66. Dolan, J.W., *J. Chromatogr. A*, 965 (2002), 195.
67. Kiriden, W., Poole, C.F., and Koziol, W.W., *J. Chromatogr. A*, 1060 (2004), 177.
68. Wu, J.Y., Bicker, W., and Lindner, W., *J. Sep. Sci.*, 31 (2008), 1492.
69. Melander, W.R., Stoveke, J., and Horvath, Cs., *J. Chromatogr.*, 185 (1979), 111.
70. Melander, W.R., Nahum, A., and Horvath, Cs., *J. Chromatogr.*, 185 (1979), 129.
71. Nahum, A. and Horvath, Cs., *J. Chromatogr.*, 203 (1981), 53.
72. Dong, L. and Huang, J., *Chromatographia*, 65 (2007), 519.
73. Guo, Y. and Huang, A., *J. Pharm. Biomed. Anal.*, 31 (2003), 1191.
74. Tanaka, H., Zhou, X., and Masayoshi, O., *J. Chromatogr. A*, 987 (2003), 119.
75. Pesek, J., Matyska, T., Fisher, M., and Sana, R., *J. Chromatogr. A*, 1204 (2008), 48.
76. Philipsen, A., Claessens, A., Lind, H., Klumperman, B., and German, L., *J. Chromatogr. A*, 790 (1997), 101.
77. Kayillo, S., Dennis, R., Wormell, P., and Shalliker, A., *J. Chromatogr. A*, 967 (2002), 173.
78. Cole, L. and Dorsey, J., *Anal. Chem.*, 64 (1992), 1317.
79. Chester, T. and Coym, J., *J. Chromatogr. A*, 1003 (2003), 101.
80. Nawrocki, J., Rigney, R.P., McCormick, A., and Carr, W., *J. Chromatogr. A*, 657 (1993), 229.
81. Buckenmaier, S.M.C., McCalley, D.V., and Euerby, M.R., *J. Chromatogr. A*, 1060 (2004), 117.
82. Buckenmaier, S.M.C., McCalley, D.V., and Euerby, M.R., *J. Chromatogr. A*, 1026 (2004), 251.
83. ICH Guideline Q2A: Text on Validation of Analytical Procedures: Terms and definitions, *International Conference on Harmonization*, Red. Reg. (60 FR 11260), March 1, 1995.
84. ICH Guideline Q3A; Impurities in New Drug Substances, *International Conference on Harmonization*, Fed. Reg. (68 FR 6924), February 11, 2003.
85. ICH Guideline Q2B: Validation of Analytical Procedures: Methodology, *International Conference on Harmonization*, Fed. Reg. (62 FR 27463), May 19, 1997.
86. ICH Guideline Q3B: Impurities in New Drug Products, *International Conference on Harmonization*, Fed. Reg. (62 FR 27454), May 19, 1997.

15 HILIC Retention Behavior and Method Development for Highly Polar Basic Compounds Used in Pharmaceutical Synthesis

Minhui Ma and Min Liu

CONTENTS

15.1 INTRODUCTION

A large proportion of pharmaceuticals are basic or alkaline organic compounds. The analysis of basic compounds has traditionally presented significant challenges to reversed-phase (RP) HPLC due to severe peak tailing caused by secondary interactions between charged basic analytes and active residual silanols on RP stationary phases.[1] The peak-tailing problem has been largely addressed in the 1990s with the development of the new generation of RP-HPLC columns based on high-purity Type B silica.[2,3] RP-HPLC methods with high efficiency and good peak shape can now be readily developed for basic compounds. The exception is a group of basic compounds that are highly polar and consequently poorly retained on RP columns even with 100% aqueous mobile phase. In pharmaceutical analysis, many substances, such as starting materials, reaction reagents, intermediates, process impurities, or drug substance may fall into this group.

Hydrophilic interaction chromatography (HILIC) is an alternative technique for the analysis of polar compounds.[4] It uses a polar stationary phase (such as silica or polar stationary phases covalently bonded on silica) in conjunction with a mobile phase containing a higher percentage of organic solvent (often acetonitrile). In HILIC, the elution order is analogous to nonaqueous normal-phase LC. It can be regarded as a variant of normal-phase LC. Early applications of HILIC started in the 1970s and focused mainly on the analysis of sugars and oligosaccharides.[5–8] HILIC applications have since expanded to include peptides,[4,9–12] proteomics,[13] and small polar compounds in biological fluids,[14–16] natural products,[16,17] pharmaceuticals,[18–30] and others. Examples of pharmaceutical applications included the analyses of starting materials with their impurities,[18–20] intermediates or drug substances with their reaction reagent residues,[21] process impurities,[22–26] and counterions.[27–29] Since 2003 a number of review articles on both the fundamentals and the applications of HILIC have been published.[11–16,31–37] A large number of new HILIC columns have also been introduced by column manufacturers in recent years.[38]

The analysis of basic compounds by mixed HILIC and cation exchange mode was first reported by Jane in 1975,[39] in which the separation of a wide range of basic drugs of forensic interest was achieved on an unmodified silica column using methanol–aqueous ammonia/ammonium nitrate as mobile phase. Since then, similar methods using silica columns and methanol–aqueous buffer at neutral or alkaline pH as mobile phase have been utilized as an alternative to RP-HPLC for the analysis of basic pharmaceutical compounds.[30,40–43] More recent applications of HILIC methods in pharmaceutical analysis use not only silica but also other polar stationary phases, and the mobile phases are mostly acetonitrile and aqueous buffer at acidic or neutral pH.[44,45]

In this chapter, we present two case studies of systematic HILIC method development for the analysis of highly polar basic compounds. The first method was developed for purity and stability monitoring of a pharmaceutical starting material 4-(aminomethyl)pyridine (4-AMP)[19] and the second for simultaneous determination of genotoxic hydrazine and 1,1-dimethyl hydrazine residues in a pharmaceutical intermediate.[21] Although the retention of polar compounds in HILIC has been proposed to mainly involve a partition mechanism between the largely organic mobile phase and a water-enriched layer partially immobilized on the surface of the stationary phase,[4,8,46] the exact retention mechanism for HILIC is still open to debate.[11,31,32,34,36] As a result, method development guidelines for HILIC are generally unavailable. Therefore, a systematic approach to HILIC method development was carried out, which involved the evaluation and optimization of stationary phase, mobile phase strength, type of organic modifier, buffer type, buffer pH, ionic strength, and column temperature. Also, many highly polar basic compounds contain nitrogen but lack UV chromophore. In the second case study, alcohol is exploited to replace acetonitrile as an organic modifier in the mobile phase, which enables the use of chemiluminescent nitrogen detection (CLND). Both methods have been validated and used to support pharmaceutical synthesis.

15.2 HILIC METHOD FOR PURITY AND STABILITY MONITORING OF 4-AMP

4-AMP is a commonly used starting material for the synthesis of active pharmaceutical ingredient (API). A stability-indicating analytical method is required to monitor its purity and stability. The purity method originally used for lot release by suppliers was a gas chromatography method with flame ionization detection (GC-FID). But the suitability of this GC-FID method for stability monitoring had not been properly evaluated or challenged until discoloration and solidification were noticed after the starting material was stored for an extended period of time at room temperature. The GC-FID method failed to provide any meaningful information for the investigation. Therefore, the degradation products were suspected to be nonvolatile. The separation effort using RP-HPLC was not successful because 4-AMP and its related compounds are highly polar and poorly retained. An ion-pair HPLC method developed in our lab was able to detect the degradation products, but was not compatible with mass detector for the identification of the unknown degradation products. For this reason, a MS compatible HILIC method was developed.

15.2.1 Method Development

15.2.1.1 HILIC-Column Screening

As in RP-LC, different column chemistries are available for HILIC, which provide different selectivity. An extensive number of HILIC columns were screened. Table 15.1 shows the physical characteristics of the HILIC columns used in this study. For the column screening, a suitable sample mixture as shown in Table 15.2 was prepared by spiking potential impurities 4-methylpyridine (4-MP), 2-(aminomethyl)pyridine (2-AMP), and 3-(aminomethyl)pyridine (3-AMP) at 1% level (5 μg/mL) into an aged 4-AMP sample that contained degradation products (Degradant-1 and Degradant-2). All standards and samples were prepared in acetonitrile/water (90/10, v/v). The concentration for the main component 4-AMP was 0.5 mg/mL. All experiments for this study were carried out on an Agilent 1100 series HPLC system with UV detection at 254 nm. Toluene was used as a void volume marker.

Column-screening experiments were performed under the same conditions and the chromatograms are shown in Figures 15.1 and 15.2 for the seven bonded polar stationary phases and the five unmodified silica stationary phases, respectively. Under the screening conditions used, only two bonded polar phases (TSK-Gel Amide-80 and ZIC-HILIC) showed complete separation of all potential impurities and degradation products from the main peak (4-AMP) while satisfactory separation were obtained from all five unmodified silica columns. Those results can partially explain why many published separations of strongly basic small molecules in HILIC mode were achieved on unmodified silica phases.[14,23,24,30,39–43,45,47]

TABLE 15.1
Physical Characteristics of the HILIC Columns Used in This Study

Column Name	Functionality	Dimension (mm)	Particle Size (μm)	Surface Area (m²/g)	Pore Size (Å)
TSK-Gel Amide-80	Carbamoyl	250 × 4.6	5	350	80
ZIC-HILIC	Sulfoalkylbetaine zwitterionic	150 × 4.6	5	140	200
Develosil 100 Diol-5	Dihydroxypropane	250 × 4.6	5	350	100
Polyhydroxyethyl A	Poly(2-hydroxyethyl aspartamide)	100 × 4.6	3	N/A	100
Luna NH_2	3-Aminopropyl	150 × 4.6	3	400	100
Luna CN	Propylcyano	150 × 4.6	3	400	100
Hypersil Gold PFP	Pentafluorophenylpropyl	150 × 4.6	3	220	170
Luna Silica (2)	Bare silica	150 × 4.6	3	400	100
Kromasil Silica	Bare silica	150 × 4.6	3.5	340	100
Atlantis HILIC	Bare silica	150 × 4.6	3	330	100
Alltima HP HILIC	Bare silica	150 × 4.6	3	230	120
HyPurity Silica	Bare silica	150 × 4.6	3	200	190

Source: Adapted from Liu, M. et al., *J. Chromatogr. A*, 1188, 255, 2008. With permission from Elsevier.

It is interesting to note that similar separation profiles were observed with all unmodified silica phases as well as the two bonded polar phases (TSK-Gel Amide-80 and ZIC-HILIC) except for the poor separation between the two degradants. The retention times increased in the order of analyte hydrophilicity: 4-MP, 2/3/4-AMP, and Degradant-1/-2, as predicted from the log D values calculated using the ACD software as shown in Table 15.2. The much shorter retention of 2-AMP compared with the regioisomers 3- and 4-AMP was somewhat surprising because similar log D values should result in similar retention if liquid–liquid partition is the only retention mechanism. Such difference may imply the significant contribution of specific surface interaction to their retention. The separations on all five unmodified (Type B) silica columns are very comparable. The increase in retention time correlated well with the increase in specific surface area. High comparability among Type B silica columns in the HILIC mode has been reported recently for the separation of a mixture of acidic, neutral, and strongly basic compounds.[45] Judging from the slightly improved peak shape and separation between the two degradants, the Atlantis HILIC column was selected for further method development.

15.2.1.2 Mobile-Phase Components

Similar to RP-HPLC method development, mobile-phase components in the HILIC mode can also significantly influence analyte retention and selectivity. In this study, mobile phase strength, pH, and ionic strength are evaluated.

TABLE 15.2
Structure and Physical Properties of 4-AMP and Its Related Compounds

Compound	1	2	3	4	5	6
	4-MP (synthetic impurity)	2-AMP (isomeric impurity)	3-AMP (isomeric impurity)	4-AMP (main compound)	Degradant-1 (degradant)	Degradant-2 (degradant)
Molecular weight	93.13	108.14	108.14	108.14	305.38	305.38
Structure	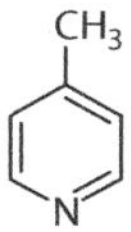	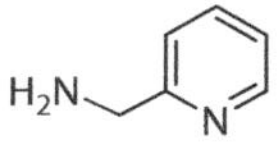	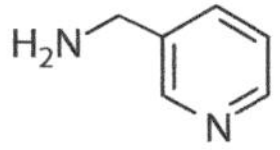	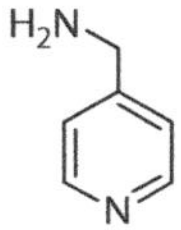	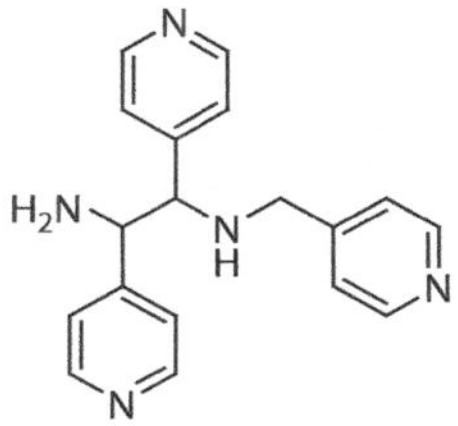	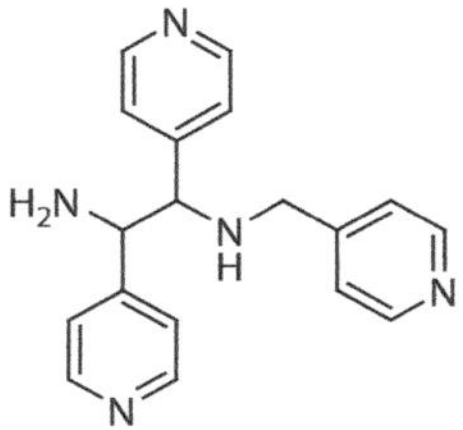
pK_a	5.94	8.70 2.18	8.34 2.27	7.81 2.91	5.65 5.38 5.32 3.71 0.23	5.65 5.38 5.32 3.71 0.23
Log D at pH 3.0	−1.18	−3.56	−3.56	−3.70	−4.76	−4.76

Source: Adapted from Liu, M. et al., *J. Chromatogr. A*, 1188, 255, 2008. With permission from Elsevier.

Note: The pK_a (i.e., ${}^{w}_{w}pK_a$) and log D values predicted by ACD software version 8.0 are aqueous based and thus can be used as a reference to polarity and ionization only. Compounds 5 and 6 are diastereomers which were trimerized from 4-AMP.

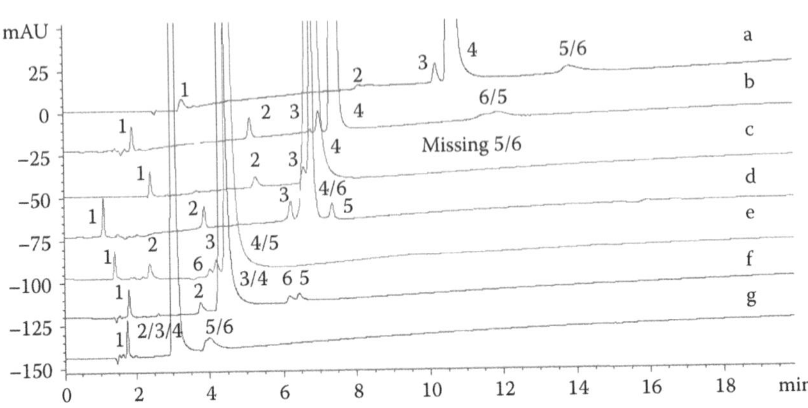

FIGURE 15.1 Separation of 4-AMP and its related compounds on different HILIC stationary phases. Column temperature is at 30°C. Mobile phases A and B are acetonitrile and 50 mM ammonium formate (pH 3.0), respectively. Gradient is from 10% B to 50% B in 20 min. Flow rate is 1.5 mL/min and detection is at 254 nm. Compounds 1–6 are 4-MP, 2-AMP, 3-AMP, 4-AMP, Degradant-1, and Degradant-2, respectively. Columns: (a) TSK-Gel Amide-80, (b) SeQuant ZIC-HILIC, (c) Developsil 100 Diol-5, (d) PolyHydroxyEthyl A, (e) Phenomenex Luna-NH_2, (f) Phenomenex Luna CN, (g) Hypersil Gold PFP. (Reproduced from Liu, M. et al., *J. Chromatogr. A*, 1188, 255, 2008. With permission from Elsevier.)

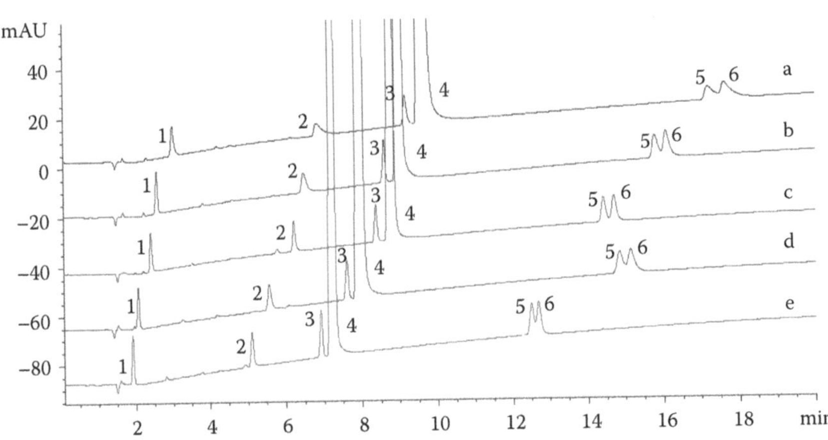

FIGURE 15.2 Separation of 4-AMP and its related compounds on bare silica HILIC columns from different manufactures. Chromatographic conditions are as in Figure 15.1. Compounds 1–6 are 4-MP, 2-AMP, 3-AMP, 4-AMP, Degradant-1, and Degradant-2, respectively. Columns: (a) Phenomenex Luna Silica(2), (b) Phenomenex Kromasil Silica, (c) Waters Atlantis HILIC, (d) Alltech Alltima HP HILIC, (e) ThermoElectron HyPurity Silica. (Reproduced from Liu, M. et al., *J. Chromatogr. A*, 1188, 255, 2008. With permission from Elsevier.)

15.2.1.2.1 Mobile-Phase Strength

The mobile-phase strength and the gradient program used for column screening as shown in Figure 15.2c are acceptable for routine pharmaceutical analysis in terms of peak separation and overall run time. To help understand the retention mechanisms of polar basic compounds on unmodified silica in HILIC mode, mobile-phase strength was evaluated over a range of 5%–40% water (95%–60% acetonitrile) with isocratic elution. The effect of the stronger solvent concentration (volume fraction of water) in the mobile phase on the retention factors (k') showed the typical HILIC mode behavior (Figure 15.3A).

As mentioned in the introduction, the retention mechanisms in HILIC mode are still not well understood. In RP-LC, retentions are believed to follow a partition mechanism and can be described by a semi-empirical equation:[3,31,34,48,49]

$$\log k' = \log k'_w - S\varphi \tag{15.1}$$

where

k'_w is theoretically equal to the retention factor in pure weaker solvent of the mobile phase

φ is the concentration (volume fraction) of the stronger solvent in mobile phase

S is the slope of log k' versus φ when fitted to a linear regression model

When Equation 15.1 is applied to the HILIC mode, the stronger solvent is water.

On the other hand, the retention of polar analytes in nonaqueous normal phase systems containing a nonpolar and a polar solvents can be described by the mechanism of localized adsorption through displacement of the adsorbed polar solvent (the Snyder–Soczewinski model).[3,31,34,50]

$$\log k' = \log k'_B - \left(\frac{A_S}{n_B}\right) \log N_B \quad (\text{or } \log k' = \log k'_B - S' \log \varphi) \tag{15.2}$$

where

N_B is the mole fraction of the polar (or stronger) solvent B in a binary organic mobile phase

k'_B is the retention factor if pure polar solvent B is used as mobile phase

A_S and n_B are the cross-sectional areas occupied by the solute molecule (analyte) on surface and the polar solvent B, respectively

The ratio (A_S/n_B) is the number of solvent B molecules necessary to displace one adsorbed molecule of the analyte. This equation can be written in terms of volume fraction (φ) as well.

Figure 15.3B and C show the log–linear (i.e., log k' versus volume fraction of water) and the log–log (i.e., log k' versus logarithm of water volume fraction) plots corresponding to the partition and the adsorption models, respectively. Due to the difficulties in determining very low retention factors with accuracy, 4-MP was not included in the plots. For 4-AMP and its regioisomers 2-AMP and 3-AMP, good linear relationships were observed in the log–log plot ($R^2 > 0.998$), indicating that adsorption contributes to their retention. However, linear relationships were not observed for the two degradants in either of the two plots over the range of water concentrations evaluated in this study. If we assume that liquid–liquid partition is indeed one of the retention mechanisms involved in the HILIC mode and that it can be described by Equation 15.1, the observed results for the two degradants suggest that neither partition nor adsorption dominates their retention processes over the entire range of

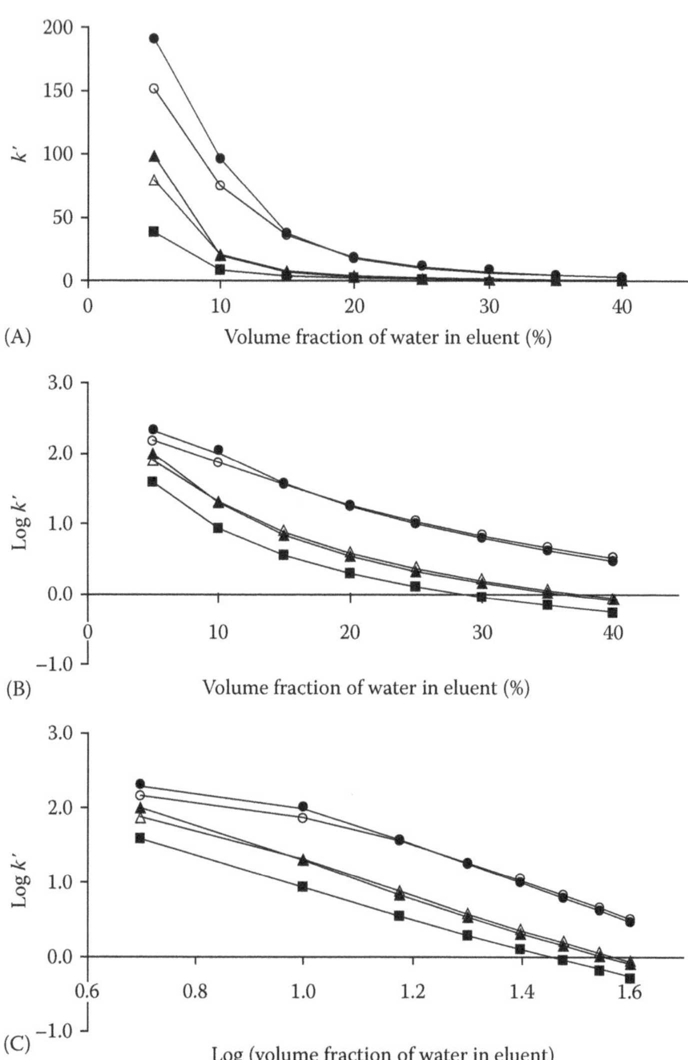

FIGURE 15.3 (A) Plot of retention factor (k') versus the volume fraction of water in eluent. (B) Plot of log k' versus the water volume fraction in eluent. (C) Plot of log k' versus logarithm of the water volume fraction in eluent. An Atlantis HILIC column is used. Column temperature is at 30°C. Isocratic runs with the mobile-phase compositions of acetonitrile/50 mM ammonium formate (pH 3.0) from 60/40 to 95/5. Flow rate is 1.5 mL/min and detection is at 254 nm. 4-MP (—◆—), 2-AMP (—■—), 3-AMP (—▲—), 4-AMP (—△—), Degradant-1 (—○—), Degradant-2 (—●—). (Reproduced from Liu, M. et al., *J. Chromatogr. A*, 1188, 255, 2008. With permission from Elsevier.)

water concentrations evaluated in this study. Such nonlinear relationships for both the log–linear and the log–log plots have been previously reported.[31,51] This is not surprising considering the complexity and multiplicity of the retention mechanisms that may be involved in the HILIC mode. As a result, multiple linear regressions based on quantitative structure–retention relationship models have been recently applied to predict HILIC retentions.[37,51] On the other hand, a closer examination of the results from current study and those published by others[31,44,51,52] suggests that strongly basic compounds or weakly basic compounds but under low buffer concentrations are more likely to follow a linear log–log relationship, probably due to stronger ionic interaction with HILIC stationary phases.

15.2.1.2.2 Mobile-Phase pH

For basic and acidic compounds, their retention depends on both the mobile-phase pH and their $\mathrm{p}K_a$ values because the mobile-phase pH controls the ionization of the basic and acidic compounds as well as the silanols on silica surface. In this study, mobile phase A was pure acetonitrile and mobile phase B was 50 mM ammonium formate, which was adjusted to the desired pH using formic acid. Because the mobile-phase pH and $\mathrm{p}K_a$ values of acidic or basic compounds are strongly influenced by the organic content in the mobile phase, there is a need to distinguish the following terms[53–56]: aqueous-buffer pH (${}^{w}_{w}\mathrm{pH}$) and mobile-phase buffer pH (${}^{s}_{w}\mathrm{pH}$ and ${}^{s}_{s}\mathrm{pH}$) after mixing with organic solvent. There are two different pH scales (${}^{s}_{w}\mathrm{pH}$ and ${}^{s}_{s}\mathrm{pH}$) for mobile phase because pH electrode may be calibrated in water (W) or in the same aqueous–organic solvent (S). The ${}^{s}_{s}\mathrm{pH}$ has the ordinary physical meaning and is directly related to the ionized fraction of an analyte in the mobile phase, but it is more difficult to measure than ${}^{s}_{w}\mathrm{pH}$. To distinguish the variation of $\mathrm{p}K_a$, the same notations used for pH should be applied, i.e., ${}^{w}_{w}\mathrm{p}K_a$, ${}^{s}_{w}\mathrm{p}K_a$, and ${}^{s}_{s}\mathrm{p}K_a$.[53]

In this study, gradient elution was used and the mobile-phase ${}^{s}_{w}\mathrm{pH}$ values were not measured. The effect of the measured aqueous buffer ${}^{w}_{w}\mathrm{pH}$ (in the range of 2.5–5.0) on analyte retention and separation is shown in Figure 15.4. When ${}^{w}_{w}\mathrm{pH}$ was raised, the retention times of 4-AMP and its regioisomers, 2-AMP and 3-AMP, increased, whereas the retention times of 4-MP and the two degradants decreased. To have a meaningful discussion about the pH effect, we need to make an estimate of the more meaningful ${}^{s}_{s}\mathrm{pH}$ and ${}^{s}_{s}\mathrm{p}K_a$ values based on relevant information available in the literature. In a recent study, McCalley measured ${}^{s}_{w}\mathrm{pH}$ of 0.1 M ammonium formate buffer in 85% acetonitrile over the aqueous buffer ${}^{w}_{w}\mathrm{pH}$ range of 3.0–4.5.[45] He observed significant differences between ${}^{w}_{w}\mathrm{pH}$ and ${}^{s}_{w}\mathrm{pH}$: ${}^{w}_{w}\mathrm{pH}$ 3.0 (${}^{s}_{w}\mathrm{pH} \sim 5.2$), ${}^{w}_{w}\mathrm{pH}$ 3.5 (${}^{s}_{w}\mathrm{pH} \sim 5.7$), and ${}^{w}_{w}\mathrm{pH}$ 4.5 (${}^{s}_{w}\mathrm{pH} \sim 6.5$). Under specific mobile-phase composition and temperature, the ${}^{s}_{w}\mathrm{pH}$ value is related to ${}^{s}_{s}\mathrm{pH}$ by a constant value (δ), $\delta = {}^{s}_{w}\mathrm{pH} - {}^{s}_{s}\mathrm{pH}$. It has been reported that at 30°C the δ values are −1.709, −0.923, −0.580, and −0.388 for 90%, 80%, 70%, and 60% (v/v) acetonitrile, respectively.[55] As shown in Figure 15.2c, 4-MP, the 2-/3-/4-AMP isomers, and the two degradants eluted at about 90%, 80%–75%, and 65%–60% acetonitrile of the gradient, respectively. Therefore, a significant difference between ${}^{w}_{w}\mathrm{pH}$ and ${}^{s}_{s}\mathrm{pH}$ is expected especially for early eluting peaks. On the other hand, it has been reported that the ${}^{s}_{s}\mathrm{p}K_a$ value of pyridinium first decreased gradually from the original aqueous ${}^{w}_{w}\mathrm{p}K_a$ 5.23 to reach a minimum value of 4.5 at 60% acetonitrile and then increased to a level close to the original aqueous

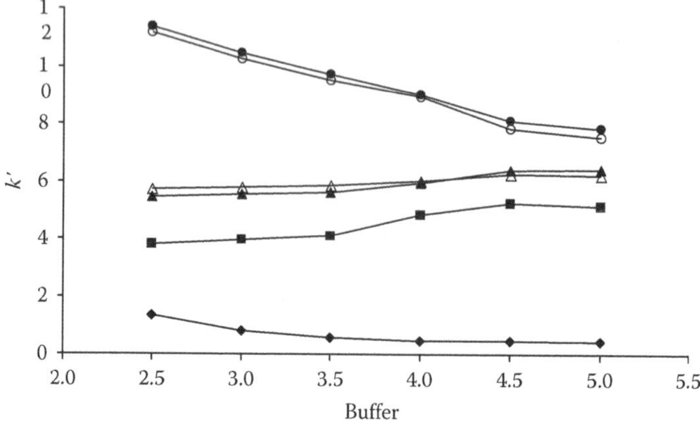

FIGURE 15.4 Effect of pH on the retention of 4-AMP and its related compounds on an Atlantis HILIC column. Column temperature is 30°C. Mobile phase A is ACN while mobile phase B is 50 mM ammonium formate with a pH ranging from 2.5 to 5.0. The gradient is from 10% B to 50% B in 20 min. Flow rate is 1.5 mL/min and UV detection at 254 nm. 4-MP (—◆—), 2-AMP (—■—), 3-AMP (—▲—), 4-AMP (—△—), Degradant-1 (—○—), Degradant-2 (—●—). (Reproduced from Liu, M. et al., *J. Chromatogr. A*, 1188, 255, 2008. With permission from Elsevier.)

${}^{w}_{w}\text{p}K_a$ value at about 90% acetonitrile.[53,54] Therefore, the differences between ${}^{w}_{w}\text{p}K_a$ and ${}^{s}_{s}\text{p}K_a$ are expected to be minimal to moderate for the analytes.

Based on the above information, the insensitivity of the 2-/3-/4-AMP retention times to mobile-phase pH may be explained by the fact that they are stronger bases with ${}^{s}_{s}\text{p}K_a$ values still significantly larger than the mobile-phase ${}^{s}_{s}\text{pH}$ over the range studied (${}^{w}_{w}\text{pH}$ 2.5–5.0). The small increase in retention over the pH range may be due to the increase in ionization of those more acidic silanol groups because a large increase in (bulk) silanol ionization on the Atlantis Silica stationary phase occurs only above ${}^{w}_{w}\text{pH}$ 9.0 (or ${}^{s}_{w}\text{pH}$ ~ 8.0) as reported by McCalley.[45] On the other hand, for the observed decrease in retention times of 4-MP and the two degradants, one may assume that they are weaker bases with ${}^{s}_{s}\text{p}K_a$ values close to the mobile-phase ${}^{s}_{s}\text{pH}$ over the range studied (${}^{w}_{w}\text{pH}$ 2.5–5.0). Thus, an increase in the mobile-phase pH reduced their ionization, which resulted in shorter retention. The retention factor of 4-MP reached a minimum at ${}^{w}_{w}\text{pH}$ 4.0 or higher while the retention factors of the two degradants continued to decrease. This observation may be explained by the significant difference in the mobile-phase ${}^{s}_{s}\text{pH}$ experienced between 4-MP and the two degradants during gradient elution. At ${}^{w}_{w}\text{pH}$ 4.0 or higher, it is likely that almost all 4-MP exist in the neutral form in a mobile phase of ~90% acetonitrile and with a ${}^{s}_{s}\text{pH}$ value much higher than ${}^{w}_{w}\text{pH}$.

15.2.1.2.3 Ionic Strength of Mobile Phase

For ionizable compounds, buffer concentration or ionic strength may significantly influence their retention, if the retention mechanism involves ionic interaction. Figure 15.5 shows the effect of ammonium formate buffer concentration at pH 3.0.

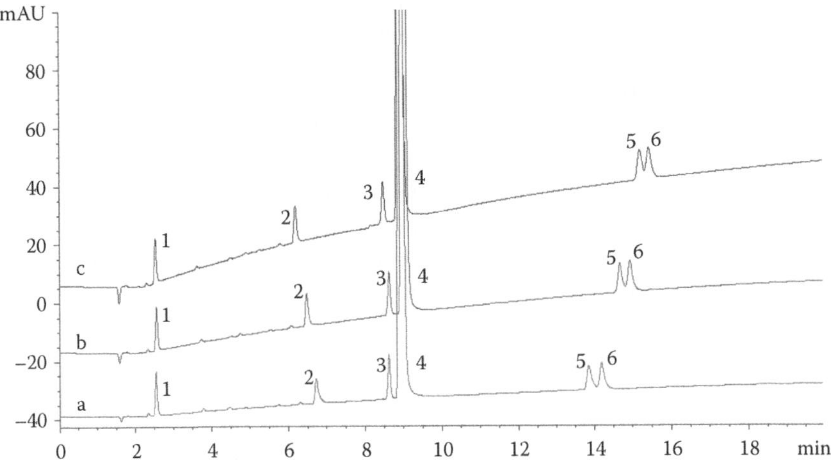

FIGURE 15.5 Effect of buffer ionic strength on the separation of 4-AMP and its related compounds on Atlantis HILIC column. Column temperature is 30°C. Mobile phase A is ACN and mobile phase B is 10 mM (a), 50 mM (b), and 100 mM (c) ammonium formate pH 3.0, respectively. Gradient is from 10% B to 50% B in 20 min. Flow rate is 1.5 mL/min and UV detection at 254 nm. Compounds 1–6 are 4-MP, 2-AMP, 3-AMP, 4-AMP, Degradant-1, and Degradant-2, respectively. (Reproduced from Liu, M. et al., *J. Chromatogr. A*, 1188, 255, 2008. With permission from Elsevier.)

As buffer concentration varied from 10 to 100 mM, different retention behaviors were observed for the analytes. No change in retention time was noticed for 4-MP over the buffer concentration range, which may suggest that the molecule exists mostly in the neutral form. This is consistent with what has been discussed in the previous section. For 2-AMP and 3-AMP, their retention times were reduced with the increase in buffer concentration. This is the typical behavior of ion-exchange retention mechanism, which has been widely reported and well understood.[3,24,40,45] The retention time change for 2-AMP was larger than 3-AMP. Such retention time change appeared to be minimal for 4-AMP. These results suggested that the contribution of ionic interaction to their retention decreased in the order: 2-AMP > 3-AMP > 4-AMP, which correlated to their basicity (Table 15.2). The retention behavior of the two degradants was different. Their retention times increased with the increase in buffer concentration, which has been referred to as a salting-out effect in the literature.[57] Similar results were obtained in the isocratic mode (data not shown). This unusual behavior has been previously reported by Guo and Gaiki for the retention of salicylic acid, aspirin, and cytosine on four different columns in the HILIC mode.[44] They hypothesized that the buffer is preferentially dissolved in the water-rich liquid layer on the stationary phase, which would result in an increase in volume or hydrophilicity of the liquid layer and lead to stronger retention of the analytes by the partition mechanism.

When buffer concentration was reduced, it was noticed that the tailing of 2-AMP peak increased significantly, indicating a secondary retention mechanism. The tailing factors at 100, 50, and 10 mM were 1.7, 1.9, and 2.2, respectively. After taking into consideration of baseline drift, peak tailing, and selectivity, we selected the 50 mM buffer concentration for the final method.

15.2.1.3 Column Temperature

In liquid chromatography, column temperature has been increasingly recognized as one of the important parameters for optimizing chromatographic method to increase efficiency and to reduce analysis time. Column temperature can also influence analyte retention and selectivity because the distribution equilibrium constant (K_{eq}) of an analyte between the mobile phase and the stationary phase is temperature dependent. For isocratic retention, the dependence of analyte retention factor, k', on column temperature, T, is well known:

$$\ln k' = \frac{-\Delta H^{\circ}}{RT} + \frac{\Delta S^{\circ}}{R} + \ln \beta \tag{15.3}$$

where

ΔH° and ΔS° are the enthalpy and entropy changes, respectively, associated with the transfer of the analyte from the mobile phase to the stationary phase
R is the gas constant
β is the phase ratio of the column

The van't Hoff plot of ln k' versus $1/T$ is linear if the phase ratio, ΔH° and ΔS° do not change significantly with temperature and if the retention mechanism remains the same. The slope of the plot will give the enthalpy change of the transfer.

In RP-HPLC, the effect of temperature on retention and selectivity has been widely studied.[58–61] For nonionizable analytes, temperature generally has a minor effect on retention and selectivity. It has been shown that on silica-based C18 column, a temperature increase of 5°C has roughly the same effect on retention as a 1% increase in acetonitrile concentration.[62] For ionizable analytes, strong temperature effects on retention and selectivity are evidenced at mobile-phase pH close to analyte pK_a.[63–66] In this case, the apparent enthalpy change obtained from the van't Hoff plot is not only determined by the enthalpies of transfer for the neutral and the ionized species, but also by the enthalpies of the secondary equilibrium processes (e.g., protonations for both analyte and buffer, ion pairing), as well as the ratio of the conjugate base and acid forms of the buffering agent. As a consequence, the van't Hoff plots of the retention factors may not yield straight lines depending on the relative magnitude of the pertinent enthalpy changes. Even a negative slope was predicted by the theoretical model for certain combinations of enthalpy changes.[63,67] Under the limit conditions when the analyte is present only as the neutral or the ionized form, the theoretical model predicted linear van't Hoff plots.[63] Recent studies have taken the effect of organic solvents on buffer pK_a and pH into the consideration.[67–69]

In the HILIC mode, studies have been performed on temperature effect.[32,43,44,47,70,71] Like in RP-LC, temperature can be a significant variable that assists in optimizing chromatographic conditions. Such studies may also help understand retention mechanism. Similar to the findings in RP-LC, linear and nonlinear van't Hoff plots with

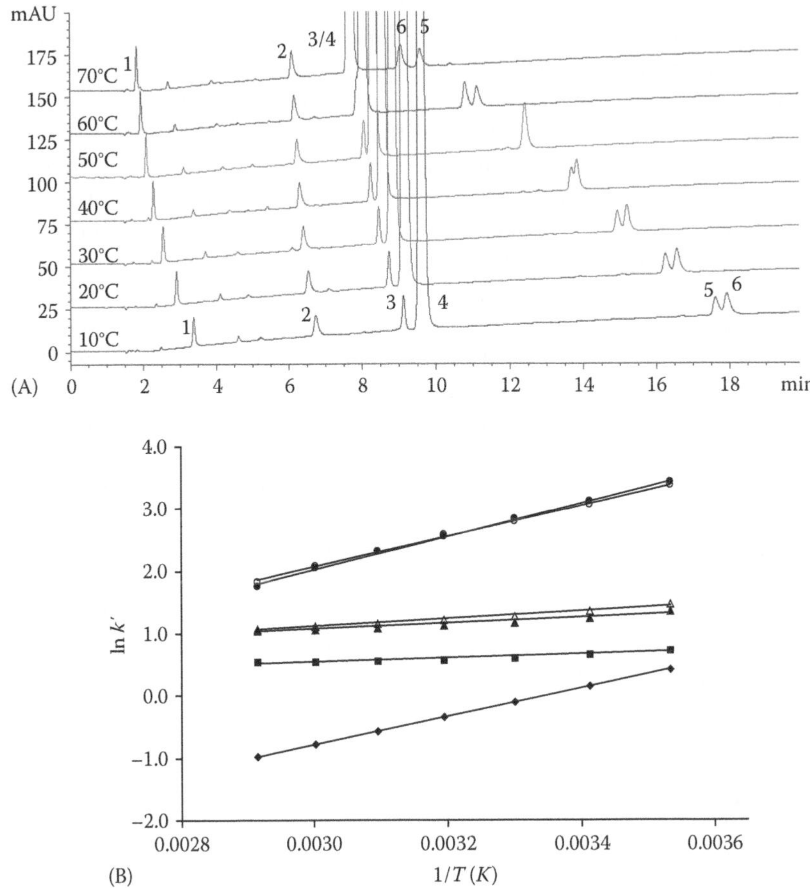

FIGURE 15.6 (A) Effect of column temperature on the separation of 4-AMP and its related compounds on Atlantis HILIC column. Column temperature is in the range of 10°C–70°C. Mobile phases A and B are ACN and 50 mM ammonium formate pH 3.0, respectively. Gradient is from 10% B to 50% B in 20 min. Flow rate is 1.5 mL/min and UV detection is at 254 nm. (B) Plot of the logarithm of retention factor (ln k') versus the reciprocal of column temperature ($1/T$) under isocratic conditions. Mobile phase is 80/20 (v/v) ACN/50 mM ammonium formate pH 3.0. Other conditions are as in (A). 4-MP (—◆—), 2-AMP (—■—), 3-AMP (—▲—), 4-AMP (—△—), Degradant-1 (—○—), Degradant-2 (—●—). (Modified from Liu, M. et al., *J. Chromatogr. A*, 1188, 255, 2008. With permission from Elsevier.)

both positive and negative slopes have been observed in the HILIC mode.[32,44,43,47,70,71] In this study, the temperature effect was evaluated over the range from 10°C to 70°C. Figure 15.6A shows the temperature effect on the retention and separation of 4-AMP and its related compounds using the Atlantis HILIC column with a gradient elution. The most noticeable changes in the chromatograms over the temperature range are the drastic decrease in the retentions of 4-MP and the two degradants as well as the switch of elution order between the two degradants. In addition, the resolution between 3- and 4-AMP is totally lost at temperatures above 50°C.

In order to construct the van't Hoff plots as shown in Figure 15.6B, the experiment was also performed under isocratic conditions. For all analytes, the slopes (or the apparent enthalpy changes) are negative, demonstrating that the overall chromatographic process is an exothermic process. As expected, 4-MP and the two degradants have much larger slopes than 2-, 3-, and 4-AMP. This observation is consistent with the fact that the mobile-phase ${}^{s}_{s}pH$ is close to the ${}^{s}_{s}pK_a$ values of 4-MP and the two degradants but still significantly lower than those of 2-, 3-, and 4-AMP as discussed in previous section. It has been reported that amine compounds have relatively large enthalpies of ionization while carboxylic acids (e.g., acetic acid) and inorganic acid (e.g., H_3PO_4 and $H_2PO_4^-$) have negligible enthalpies of ionization.[63,67,68,72] Therefore, all the analytes in this study are expected to have decreased ${}^{s}_{s}pK_a$ values at elevated temperatures while the ${}^{s}_{s}pK_a$ value (and the mobile-phase ${}^{s}_{s}pH$) for the formate buffer remains almost the same over the temperature range studied. As a result, the ratio of the neutral to the cationic forms of 4-MP as well as the two degradants will be increased at elevated temperatures, leading to reduced retention. On the other hand, since the ${}^{s}_{s}pK_a$ values of 2-, 3-, and 4-AMP are much higher than the mobile-phase ${}^{s}_{s}pH$, such significant changes in retention is not expected. Since all the plots are linear, the apparent enthalpy changes are estimated from the slopes of the van't Hoff plot, which are 19, 2.4, 4.1, 5.1, 20, and 22 kJ/mol for 4-, 2-, 3-, 4-AMP, Degradant-1, and Degradant-2, respectively. Considering the overall separation profile, column temperature 30°C was a good compromise and was selected for the final method.

15.2.2 Method Validation

Based on the evaluation of all the above chromatographic parameters, final method conditions are summarized as follows:

Column:	Water Atlantis HILIC, 3 µm, 150 × 4.6 mm
Column temperature:	30°C
Mobile phase A:	100% acetonitrile
Mobile phase B:	50 mM ammonium formate (pH 3.0)
Gradient:	10% B–50% B in 20 min
Flow rate:	1.5 mL/min
UV detection:	254 nm
Sample diluent:	Acetonitrile/water (90/10, v/v)
Sample concentration:	0.5 mg/mL 4-AMP
System suitability sample:	Aged 4-AMP sample spiked with 1% 4-MP, 2-, and 3-AMP

This method has been fully validated for lot release and stability testing of 4-AMP, which included specificity, accuracy, precision, linearity, limit of detection (LOD) and limit of quantitation (LOQ), and robustness. The method validation results are summarized in Table 15.3.

TABLE 15.3
Validation Summary for the Determination of 4-AMP Assay and Related Impurities

Experiment	Results				
Specificity of the method with respect to potential impurities of synthesis and degradation	High degree of specificity to 4-AMP demonstrated by absence of interference from all known potential impurities and degradation products. Peak purity verified by photo diode array and mass spectrometry.				
Accuracy and precision for 4-AMP assay	Acceptable recovery and precision demonstrated at 80%, 100%, and 120% of the nominal assay concentration:				
	Level	80%	100%	120%	
	%Mean recovery	100.5	99.9	100.3	
	%RSD ($n = 3$):	0.64	0.88	0.60	
Accuracy and precision for impurities and degradants	Acceptable recovery and precision demonstrated for all impurities at 0.05%, 0.5%, and 0.1% level:				
		Level	0.05%	0.5%	1.0%
	4-MP	%Mean recovery	91.1	101.3	99.1
		%RSD ($n = 3$)	5.7	1.3	0.1
	2-AMP	%Mean recovery	94.0	98.9	101.1
		%RSD ($n = 3$)	6.7	0.7	1.1
	3-AMP	%Mean recovery	105.8	98.7	100.5
		%RSD ($n = 3$)	4.1	2.0	1.1
	Degradant-1	%Mean recovery	102.0	100.6	100.6
		%RSD ($n = 3$)	3.3	0.9	0.8
	Degradant-2	%Mean recovery	102.2	101.7	100.6
		%RSD ($n = 3$)	4.0	0.7	0.7
Linearity	Acceptable linearity of 4-AMP demonstrated over a range of 0.02%–120% of nominal with an R value of 1.0000. Acceptable linearity demonstrated for each impurity and degradant over the range of LOQ to 1% with R above 0.9997.				
LOD/LOQ	Acceptable LOD/LOQ demonstrated for each impurity and degradant: LOD = 0.02% (S/N ≥ 3), LOQ = 0.05% (S/N ≥ 10).				
Robustness	Acceptable robustness demonstrated with respect to mobile phase ratios, buffer pH, column lot, flow rate, column temperature, and UV wavelength.				

15.3 HILIC-CLND METHOD FOR HYDRAZINE RESIDUES IN AN INTERMEDIATE

Hydrazine and methyl-substituted hydrazines are commonly used as reducing agents in pharmaceutical synthesis. Hydrazine is also a degradation product in some pharmaceutical preparations such as hydralazine and isoniazid.[73] Since hydrazine and substituted hydrazine are genotoxic and potentially carcinogenic,[74] their residues in synthetic intermediate or API must be monitored and controlled. Due to the lack of UV chromophore in all hydrazine-related molecules, most analytical procedures involve derivatization followed by GC,[73,75] HPLC,[76] or TLC.[77] To avoid derivatization, ion chromatography with electrochemical detector has been used for direct analysis of hydrazine residue in a pharmaceutical intermediate.[78] Here, we present a HILIC method using a CLND[79–82] for direct analysis of hydrazine and 1,1-dimethylhydrazine in a pharmaceutical intermediate. The method development was focused on the use of alcohol rather than acetonitrile as a weak eluent in the mobile phase with systematic evaluation of method parameters. To aid the exploration of retention mechanism, methylhydrazine and 1,2-dimethylhydrazine were also included in the study because they provide additional variability in analyte hydrophilicity (Table 15.4).

15.3.1 Method Development

15.3.1.1 Alcohol as a Weak Eluent

As mentioned in the introduction, methanol as an organic modifier in the mobile phase has been used mostly with aqueous buffers at neutral or alkaline pH for separation of basic drugs on silica columns.[30,39–43] Under such mobile-phase conditions, the silanol groups of the silica stationary phase are at least partially ionized. The retention of basic compounds involves ion exchange although other retention mechanisms (such as interaction with surface silanol and siloxane groups) also play a role, depending on pH, methanol concentration, and the type of buffer in the mobile phase.[40,42,43,57,83–85] At acidic pH, the retention of basic compounds on silica is significantly reduced

TABLE 15.4
Structure and Physical Properties of Hydrazines

Compound	1	2	3	4
	1,2-Dimethylhydrazine	1,1-Dimethylhydrazine	Methylhydrazine	Hydrazine
MW	60.10	60.10	46.07	32.05
Structure	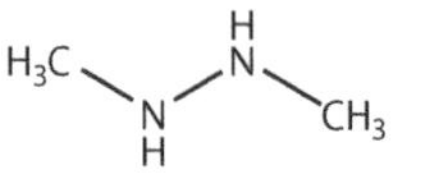 $H_3C{-}NH{-}NH{-}CH_3$	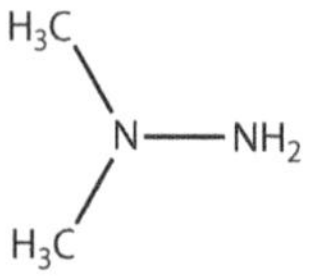 $(H_3C)_2N{-}NH_2$	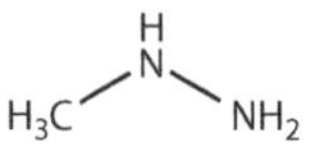 $H_3C{-}NH{-}NH_2$	$H_2N{-}NH_2$
${}^{w}_{w}pK_a$	7.52	7.21	7.87	8.07

Source: ${}^{w}_{w}pK_a$ values are Adapted from Hinman, R.L., *J. Org. Chem.*, 23, 1587, 1958.

due to the decrease in the degree of silanol ionization.[40,42,57,84] With the reduction of the electrostatic interaction, the eluotropic strength of a mobile phase constituted of methanol and aqueous buffer becomes too strong to be useful.[19,24] Therefore, in the recent HILIC applications of silica columns, the less polar solvent acetonitrile is the most commonly used organic modifier.[14,19,24,45,86–88] Acetonitrile has much weaker hydrogen-bonding capability and will not compete as strongly as methanol with analytes for hydrogen-bond interaction with the stationary phase. Nevertheless, the use of methanol as a weak eluent has been shown to offer different selectivity.[24,32,47,89] Another advantage of using methanol or other short-chain alcohols as a weak eluent is the enhanced buffer solubility and in some cases sample solubility as well.[30] The use of alcohol as a weak eluent also makes it possible to use a nitrogen-specific detector CLND.

To increase the retention of basic compounds in acidic mobile phase with alcohol as a weak eluent, it requires that the stationary phase possesses some electrostatic interaction under acidic conditions. Based on our column screening, the sulfoalkylbetaine bonded zwitterionic stationary phase, ZIC-HILIC, provides adequate retention of hydrazines. Because of the concurrent presence of a quaternary ammonium group and a sulfonate group in close proximity of 1:1 ratio, the ZIC-HILIC column does not show strong ion exchange interaction. But the zwitterionic stationary phase maintains a certain level of unique electrostatic interaction with both anionic and cationic analytes and is capable of simultaneously separating anions and cations.[29,90] This ZIC-HILIC stationary phase also has a slight electrostatic contribution from the sulfonic group and this low negative excess charge is independent of pH.[31,90] A recent study has shown decent retention and separation of opioids and their glucuronides on the ZIC-HILIC column with a mobile phase containing 80/20 (v/v) methanol/ammonium formate (1 mM, pH 4).[89] Figure 15.7 shows the retention of hydrazines on a ZIC-HILIC column with different alcohols as a weak eluent. The retention times of hydrazines increased drastically as the less polar ethanol or isopropyl alcohol was used. Methanol is a much stronger elution solvent because of its strong hydrogen-bond interactions with both stationary phase and the analytes. This result is consistent with the findings in the literature.[24,30] In this study, ethanol was selected to be the weak eluent because it provided adequate retention and separation of hydrazines with acceptable peak shape.

15.3.1.2 Combined ZIC and HILIC Modes

Since the introduction of zwitterion ion chromatography (ZIC) (also called electrostatic ion chromatography or EIC) by Hu et al. in 1993,[91] the retention mechanism of ZIC has been a subject of continuous debate.[92–96] Until the introduction of the covalently bonded zwitterionic stationary phases (more specifically the silica-based ZIC-HILIC phase) by Irgum and colleagues,[90,97–99] ZIC applications were mostly performed on zwitterionic stationary phases that were formed by physical adsorption of zwitterionic surfactants onto a C18 stationary phase. Almost all studies on ZIC-retention mechanism were focused on the separation of inorganic anions using such dynamically coated zwitterionic stationary phases under fully or highly aqueous mobile-phase conditions.[91–96] The introduction of sulfoalkylbetaine-bonded zwitterionic stationary phase ZIC-HILIC eliminated the need for the hydrophobic

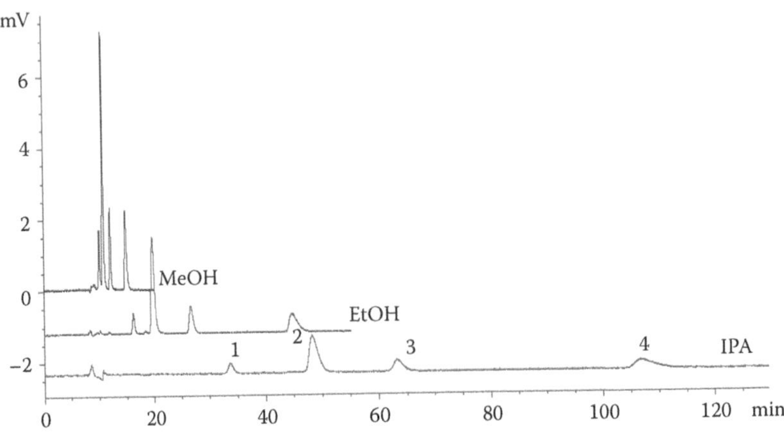

FIGURE 15.7 Effect of type of alcohol on the retention of hydrazines on a ZIC-HILIC column. For each separation, isocratic run was performed in the mobile phase of TFA/water/alcohol (0.1/10/90, v/v/v), column temperature at 30°C, flow rate at 0.4 mL/min with a splitter and CLND. The CLND system was set at 1050°C combustion furnace, 50 mL/min argon, 280 mL/min oxygen, 75 mL/min makeup (argon), 30 mL/min ozone, 5°C cooler, gain ×1, 750 V on PMT. The analyte concentrations were about 30–70 μg/mL in water/ethanol (20/80, v/v). Acetonitrile at 0.1% (v/v) level in ethanol was used as a void time marker. Injection volume was 10 μL. 1: 1,2-dimethylhydrazine, 2: 1,1-dimethylhydrazine, 3: methylhydrazine, 4: hydrazine. (Reproduced from Liu, M. et al., *J. Chromatogr. A*, 1216, 2362, 2009. With permission from Elsevier.)

C18 substrate and also made it possible to use high organic content in the mobile phase. Consequently, the ZIC-HILIC phase has found more applications in the HILIC mode with its unique combined zwitterionic and HILIC interactions. The retention mechanism on ZIC-HILIC stationary phase is complex and very limited mechanistic studies have been performed so far.[31] Most applications of ZIC-HILIC used an acetonitrile–water mobile phase with organic buffers such as ammonium acetate or format. Nevertheless, typical HILIC retention behavior of hydrazines was also observed on the ZIC-HILIC column using an alcohol–water mobile phase under acidic conditions. Hydrazines were eluted in the order of increased hydrophilicity (i.e., 1,2-dimethylhydrazine, 1,1-dimethylhydrazine, methylhydrazine, and hydrazine) and their retentions decreased with the increase in mobile-phase strength (water content). Because the retention mechanism on ZIC-HILIC column is not well understood, in addition to the effect of mobile-phase strength other critical method parameters such as type of acid additive, ionic strength, buffer type, and column temperature were systematically evaluated during the hydrazine method development.[21]

15.3.1.3 Type of Acid Additive

The effects of three different acids, TFA, formic acid, and acetic acid, and their concentrations in the mobile phases on the retention and separation of hydrazines are shown in Figure 15.8 and Table 15.5, respectively. In all cases, the elution orders of the hydrazines were the same, but their retentions were strongly influenced by the acid type. The shortest retention and best peak shape were observed with the

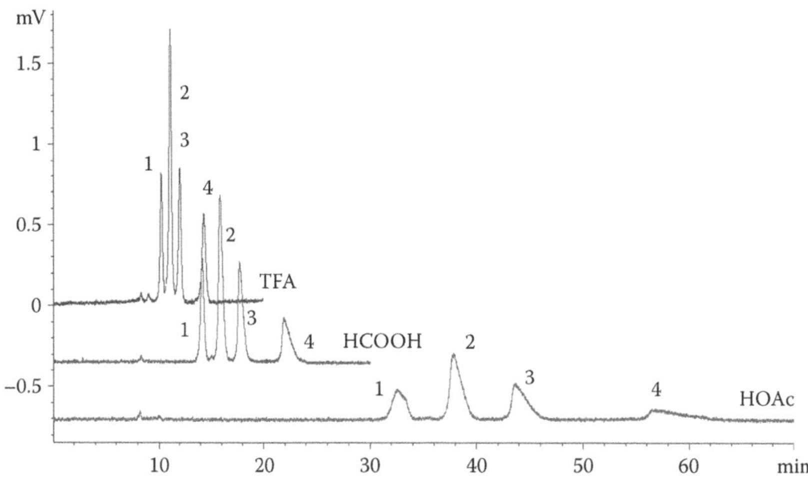

FIGURE 15.8 Effect of type of acid additives on the retention of hydrazines on a ZIC-HILIC column. For each separation, isocratic elution was performed with acid/water/ethyl alcohol (0.1/30/70, v/v/v) as mobile phase, column temperature at 30°C, flow rate at 0.4 mL/min with a splitter and CLND. 1: 1,2-dimethylhydrazine, 2: 1,1-dimethylhydrazine, 3: methylhydrazine, 4: hydrazine. Other conditions as in Figure 15.7. (Reproduced from Liu, M. et al., *J. Chromatogr. A*, 1216, 2362, 2009. With permission from Elsevier.)

strongest acid, TFA. Similar effect had been reported when peptides were analyzed on the TSK-Gel Amide-80 column using the three different acids.[10] In addition, as shown in Table 15.5, minimal change in retention was observed with the change of TFA concentrations, while the most significant change in retention was noticed with the increase in acetic acid concentration. These observations may be interpreted by the differences in acidity and ion-pairing capability among the three different acids.

The pH values shown in Table 15.5 were measured before mixing with the organic modifier. Considering the relatively high aqueous ${}^{w}_{w}pK_a$ values of 1,2-dimethylhydrazine (7.52), 1,1-dimethylhydrazine (7.21), methylhydrazine (7.87), and hydrazine (8.07),[100] all hydrazines are expected to be protonated in the ethanol–water mobile phase with TFA (${}^{w}_{w}pK_a$ 0.52), and likely with formic acid (${}^{w}_{w}pK_a$ 3.75) as well, even though the ${}^{s}_{s}pK_a$ values of TFA and formic acid in 80% ethanol are expected to be higher than their ${}^{w}_{w}pK_a$. With acetic acid (${}^{w}_{w}pK_a$ of 4.75), however, its ${}^{s}_{s}pK_a$ in 80% ethanol might increase to a level close to the ${}^{s}_{s}pK_a$ values of the hydrazines, resulting in only partial protonation of the hydrazines. It has been reported that in 80/20 (v/v) methanol/water the ${}^{s}_{s}pK_a$ values of formic acid and acetic acid are increased to about 5.2 and 6.5, respectively, while the ${}^{s}_{s}pK_a$ values of protonated neutral bases such as ammonium and anilinium are about 0.7 units lower than ${}^{w}_{w}pK_a$.[56]

In ZIC, the retention of ionic analytes on sulfoalkylbetaine stationary phase is strongly affected by mobile-phase composition.[92–96,98] Without electrolyte in the mobile phase, the positive and negative charges of one sulfoalkylbetaine zwitterion are likely paired with the negative and positive charges, respectively, of adjacent zwitterions. A small amount of salt in the mobile phase may breakup the self-association and results in an abrupt change in surface morphology. The retention of

TABLE 15.5
Effect of Acid Modifier on the Retention Time (Minutes) of Hydrazines

	TFA			
Analyte	**0.05% (pH 2.26)**	**0.1% (pH 1.93)**	**0.15% (pH 1.77)**	**0.2% (pH 1.65)**
Hydrazine	21.86	20.74	20.45	20.03
Methylhydrazine	15.33	15.15	15.24	14.98
1,1-Dimethylhydrazine	13.20	13.07	13.17	12.98
1,2- Dimethylhydrazine	11.61	11.50	11.59	11.42
	HCOOH			
	0.1% (pH 2.65)	**0.25% (pH 2.44)**	**0.5% (pH 2.29)**	**0.75% (pH 2.19)**
Hydrazine	39.78	36.18	36.03	35.44
Methylhydrazine	27.36	25.01	24.96	24.58
1,1-Dimethylhydrazine	22.18	20.36	20.24	19.88
1,2-Dimethylhydrazine	18.71	17.29	17.20	16.91
	HOAc			
	0.1% (pH 3.24)	**0.25% (pH 3.04)**	**0.5% (pH 2.89)**	**0.75% (pH 2.79)**
Hydrazine	119.16	79.40	63.96	56.47
Methylhydrazine	76.23	53.94	43.55	38.62
1,1-Dimethylhydrazine	59.73	42.72	34.47	30.56
1,2- Dimethylhydrazine	48.35	34.34	28.49	25.57

Source: Adapted from Liu, M. et al., *J. Chromatogr. A*, 1216, 2362, 2009. With permission from Elsevier.
Notes: (1) The pH (i.e., ${}^{w}_{w}$pH) was measured before mixing with organic solvent. Retention is in minutes.
(2) Chromatographic conditions: the mobile phase was acid/water/ethanol (various/20/80, v/v/v). Other chromatographic conditions were the same as those described in Figure 15.8.

ionic analytes is strongly influenced by the eluent anion present in the mobile phase and its ion-pairing capability with the quaternary ammonium group of the sulfoalkylbetaine stationary phase. In general, strong ion-pairing eluent (or chaotropic ions in the Hofmeister series) such as ClO_4^- will impart negative charge to the stationary phase, resulting in decreased retention of analyte anions or increased retention of analyte cations. In addition, analyte cations especially monovalent inorganic cations are much weakly retained due to their weak interaction with the sulfonate group.[92–96] Considering the pK_a values of the three different acids and their ion-pairing capability, formic acid and acetic acid are less likely than TFA to form strong ion pairs with either the quaternary ammonium group of the stationary phase or the positively charged hydrazines in the mobile phase. Stronger ion pair in the mobile phase would lead to weaker electrostatic interaction with the sulfoalkylbetaine stationary phase as well as decreased hydrophilic partition into the water-rich layer on the stationary phase, resulting in shorter retention. However, this cannot explain the drastic difference in retention between formic acid and acetic acid as additive because the two acids are not expected to have a big difference in ion-pairing capability.[101,102] Other factors might play important roles. For example, with acetic acid, it is possible to have strong HILIC mode retention of the neutral hydrazines, which may partition into the immobilized water-rich layer without experiencing electrostatic repulsion from the positively charged quaternary ammonium group of the stationary phase.

15.3.1.4 Type of Buffer and Ionic Strength

Under acidic conditions, the retention of protonated hydrazines on the ZIC-HILIC stationary phase involves electrostatic interaction. Therefore, the effects of buffer type and buffer concentration are expected to be significant. In ZIC, the effect of salt type and concentration on the retention behavior of inorganic anions has been extensively studied in 100% aqueous mobile phase.[90,92–99] The information available for the ZIC-HILIC phase in the HILIC mode is limited to the effect of ammonium acetate buffer on the retention of acidic compounds[44,103] and inorganic anions and cations.[29] The use of alcohol as the weak eluent in this study made it possible to evaluate five different types of commonly used buffers, ammonium formate, ammonium acetate, ammonium phosphate, sodium phosphate, and triethylamine phosphate, in a ethanol–aqueous (80/20) mobile phase.

The effect of five different buffers on the retention of the hydrazines is summarized in Table 15.6. Because CLND requires nitrogen-free eluent and is not compatible with high salt concentration in the mobile phase, a refractive index (RI) detector was used for this experiment. Since the HCl salts of hydrazine and 1,2-dimethylhydrazine were used and RI is a universal detector, the chloride peak was also observed because both cation and anion can be retained on the ZIC-HILIC column. The results in Table 15.6 showed that among the three ammonium salts, the retention time increased for hydrazines and decreased for chloride ion in the following order: formate < acetate < phosphate ($H_2PO_4^-$). This retention order is difficult to explain based on the relative strength of the interactions between the buffer anions and the quaternary ammonium group of the sulfoalkylbetaine stationary phase because the interactions with formate and acetate are expected

TABLE 15.6
Effect of Buffer Type on the Retention Time (Minutes)

Name	Ammonium Formate	Ammonium Acetate	Ammonium Phosphate	Sodium Phosphate	Triethylamine Phosphate
Hydrazine	26.00	27.87	44.94	37.39	>60
Methylhydrazine	20.02	21.53	33.03	29.42	46.75
1,1-Dimethylhydrazine	18.05	19.27	27.39	25.01	32.91
1,2-Dimethylhydrazine	14.96	16.27	22.23	20.85	29.18
Chloride	14.58	14.04	10.57	12.99	8.71

Source: Adapted from Liu, M. et al., *J. Chromatogr. A*, 1216, 2362, 2009. With permission from Elsevier.

Note: Chromatographic conditions: the mobile phase was 50 mM buffer pH 4/ethanol (20/80, v/v); other chromatographic conditions were the same as those described in Figure 15.9.

to be stronger than phosphate ($H_2PO_4^-$) according to the Hofmeister series.[101,102] We again hypothesize that the observed order of retention is related to the effect in the mobile phase where the interaction of protonated hydrazines with formate or acetate is stronger than with phosphate, resulting in weaker electrostatic interactions with the stationary phase. The results in Table 15.6 also showed the effect of buffer cations on the retention of hydrazines and chloride. Shorter retention of hydrazines was observed with sodium phosphate while the longest retention was observed with triethylammonium phosphate ($Na^+ < NH_4^+ < TEA^+$). This elution order is consistent with the affinity sequence observed for conventional strong cation exchanger.[104] TEA^+ is a weaker eluent because it contains larger and more hydrophobic groups, which results in a more diffuse positive charge and weaker interaction with the sulfonate group.[96]

The effect of buffer concentration/ionic strength on the retention of hydrazines is shown in Figure 15.9. The retention of the positively charged hydrazines decreased when ammonium formate buffer concentration was increased from 5 to 90 mM, whereas the retention of the negatively charged chloride ion increased but quickly reached a plateau at about 30 mM ammonium formate. Similar results have been reported for the effect of ammonium acetate buffer concentration on the retention of inorganic cations on the ZIC-HILIC column by Risley and Pack.[29] They explained the effect by a two-part competition mechanism, i.e., increasing ammonium concentration in the mobile phase reduces the interaction of analyte cations with the sulfonate group of the stationary phase and also decreases the opportunity for analyte cations to partition into the enriched aqueous layer in the HILIC mode. However, such competition for interaction with the sulfonate group of the ZIC-HILIC phase is not as strong as the effect of salt concentration on a conventional strong cation exchanger. In conventional cation-exchange chromatography, a plot of the logarithm of the analyte retention factor (log k') versus the logarithm of buffer cation concentration is linear with a slope of −1 for singly charged analyte and monovalent buffer cation.[104,105] Figure 15.9B shows the plot of log k' versus log (buffer concentration) in the range of 5–30 mM. Within this range, linear relationships were observed for all hydrazines and for chloride. However, the observed slopes (about −0.3 for hydrazines and +0.1 for chloride) are much smaller than −1 or +1. Such decreased slopes have been reported for inorganic anions and are believed to be typical in ZIC mode because both electrostatic attraction and repulsion are present.[95,106–108] In ZIC-HILIC mode, the effect of buffer concentration on HILIC interaction may also play a role.

The retention behavior of chloride ion at low ammonium formate buffer concentrations (5–30 mM) is similar to the increased retention of iodide in an eluent with less chaotropic anions such as SO_4^{2-} in the ZIC mode as reported by Cook et al.[95] Formate is less chaotropic than chloride in the Hofmeister series.[95,101,102] Guo and Gaiki observed an increase in the retention of salicylic acid on three different types of HILIC columns (TSK-Gel Amide-80, HILIC Silica, and ZIC-HILIC) when ammonium acetate concentration in mobile phase was increased from 5 to 20 mM.[44] They hypothesized that the increased retention was due to weakened electrostatic repulsion at higher salt concentrations. Similar interpretations have been made by Takegawa et al.[103] and Alpert.[109]

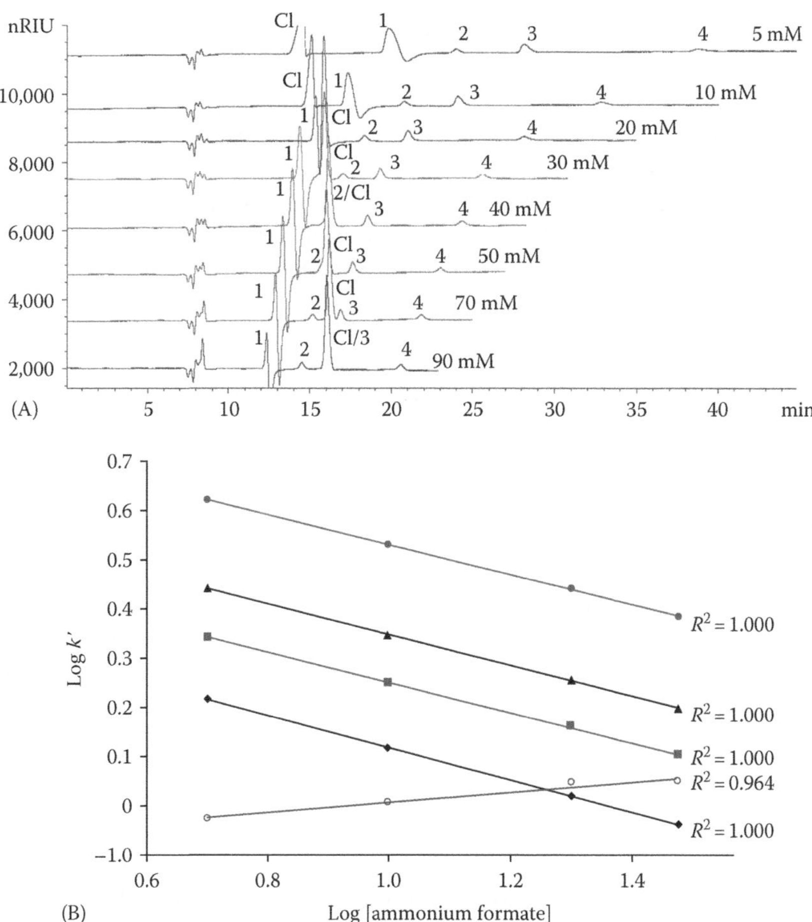

FIGURE 15.9 (A) The effect of ionic strength on the retention of hydrazines on a ZIC-HILIC column. Column temperature was 30°C. Isocratic runs with the mobile phase of 5–90 mM ammonium formate buffer pH 3.0/ethanol (20/80, v/v). (The buffer concentrations refer to the concentration before mixing with organic solvent.) Flow rate was 0.4 mL/min with RI detection. The analyte concentrations were about 0.8–1.2 mg/mL in water/ethanol (20/80, v/v). The injection volume was 1 μL. 0.5% Toluene (v/v) in ethanol was used as a void volume marker. Chloride is from the HCl salts of hydrazine and 1,2-dimethylhydrazine. 1: 1,2-dimethylhydrazine, 2: 1,1-dimethylhydrazine, 3: methylhydrazine, 4: hydrazine. (B) Plot of logarithm of retention factor (k') against logarithm of buffer concentration (5–30 mM), hydrazine (—•—), methyl hydrazine (—▲—), 1,1-dimethyl hydrazine (—▪—), 1,2-dimethyl hydrazine (—◆—), chloride (—○—). (Reproduced from Liu, M. et al., *J. Chromatogr. A*, 1216, 2362, 2009. With permission from Elsevier.)

15.3.2 Method Validation

Based on method development, the final method conditions are summarized as follows:

Column:	ZIC-HILIC, 5 μm, 250 × 4.6 mm
Column temperature:	30°C
Mobile phase:	0.1/30/70 (v/v) TFA/water/ethanol
Flow rate:	0.4 mL/min (0.2 mL/min to CLND by post-column flow splitter)
Detector parameters:	CLND, 1050°C combustion furnace, 50 mL/min argon, 280 mL/min oxygen, 75 mL/min makeup (argon), 30 mL/min ozone, 5°C cooler, gain ×1, 750 V on PMT
Injection volume:	20 μL
Sample diluent:	DMSO/ethanol (30/70, v/v)
Sample concentration:	10 mg/mL intermediate
Standard:	2 μg/mL each of 1,1-hydrazine and hydrazine
System suitability:	10 mg/mL intermediate spiked with 0.02% 1,1-hydrazine and hydrazine

Column temperature was evaluated during method development over the range of 10°C–60°C. Only a slight decrease in retention times was observed. This is understandable because no significant temperature effect is expected for ionizable compounds if mobile phase pH is not close to their pK_a values. Column temperature 30°C was selected for the final method. The intermediate is hydrophobic and very soluble in DMSO. In HILIC mode, the sample diluent DMSO/ethanol (30/70 v/v) is a weaker solvent compared with the mobile phase.

The optimized method has been validated for its performance and suitability for simultaneous analysis of trace amounts of hydrazine and 1,1-dimethylhydrazine in

TABLE 15.7
Method Validation Summary for the Determination of 1,1-Dimethylhydrazine and Hydrazine Residues

Experiment	Results				
Specificity	No interference from the intermediate and other potential impurities				
Accuracy	Acceptable recovery demonstrated for 1,1-dimethylhydrazine and hydrazine at 0.02% and 0.04% level spiked into 10 mg/mL intermediate:				
		1,1-Hydrazine		Hydrazine	
	Spiking level	0.02%	0.04%	0.02%	0.04%
	%Mean recovery ($n = 2$)	101	103	84	103
Linearity	Acceptable linearity demonstrated for 1,1-dimethylhydrazine and hydrazine in the range of 0.01%–0.1% with coefficients of determination 0.9998 and 0.9996, respectively				
Repeatability	Acceptable injection repeatability of 1,1-dimethylhydrazine and hydrazine at 0.02% level: 3.9% and 5.0%, respectively				
LOD/LOQ	Acceptable signal-to-noise ratio for reporting limit at 0.02% for 1,1-dimethylhydrazine (S/N = 28) and hydrazine (S/N = 9)				

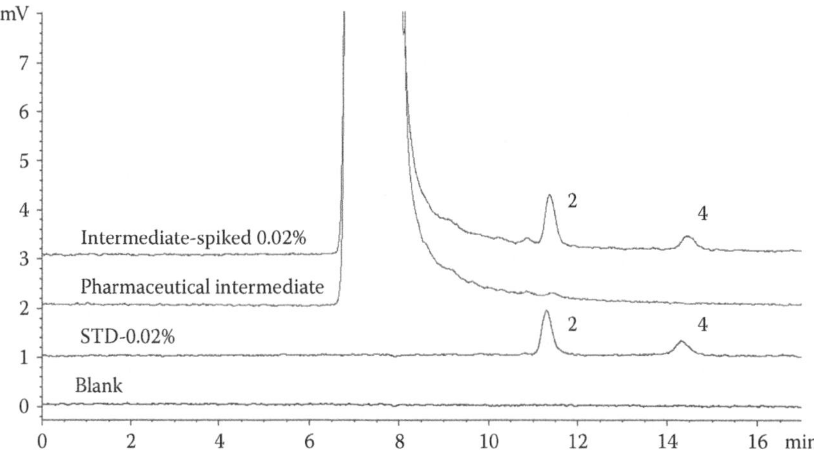

FIGURE 15.10 Simultaneous analysis of 1,1-dimethylhydrazine and hydrazine in a pharmaceutical intermediate on a ZIC-HILIC column. Mobile phase was TFA/water/ethanol (0.1/30/70, v/v/v). Flow rate was 0.4 mL/min with a splitter and CLND. Injection volume was 20 μL. Pharmaceutical intermediate sample contained 10 mg/mL of the intermediate in DMSO/ethanol (30/70, v/v). The spiked intermediate sample consisted of 10 mg/mL intermediate, 2 μg/mL 1,1-dimethylhydrazine and 2 μg/mL hydrazine in DMSO/ethanol (30/70, v/v). The standard (STD) solution had 2 μg/mL each of 1,1-dimethylhydrazine and hydrazine in water/ethanol (30/70, v/v). 2: 1,1-dimethylhydrazine, 4: hydrazine. Other conditions are as in Figure 15.7. (Reproduced from Liu, M. et al., *J. Chromatogr. A*, 1216, 2362, 2009. With permission from Elsevier.)

a pharmaceutical intermediate to support process development for an early phase program. The method validation results are summarized in Table 15.7 and typical chromatograms are shown in Figure 15.10. The method sensitivity is adequate for early-phase process development. But for late-phase process development and drug-substance release testing, higher sensitivity is needed to meet the requirement of the recently implemented EMEA guideline on genotoxic impurities. A separate method based on ion chromatography with electrochemical detection has been developed in our laboratory for this purpose.[110]

REFERENCES

1. McCalley, D.V., *J. Sep. Sci.*, 2003, 26, 187.
2. McCalley, D.V., Liquid chromatographic separations of basic compounds. In *Advances in Chromatography*, ed. E. Grushka and N. Grinberg. CRC Press, Boca Raton, FL, 2008, pp. 305–350.
3. Snyder, L.R., Kirkland, J.J., and Giajch, J.L., *Practical HPLC Method Development*, 2nd. Wiley-Interscience, New York, 1997.
4. Alpert, A.J., *J. Chromatogr.*, 1990, 499, 177.
5. Havlicek, J. and Samuelson, O., *Anal. Chem.*, 1975, 47, 1854.
6. Linden, J.C. and Lawhead, C.L., *J. Chromatogr.,* 1975, 105, 125.
7. Palmer, J.K., *Anal. Lett.*, 1975, 8, 215.
8. Churms, S.C., *J. Chromatogr.*, 1996, 720, 75.

9. Zhu, B.Y., Mant, C.T., and Hodges, R.S., *J. Chromatogr.*, 1991, 548, 13.
10. Yoshida, T., *Anal. Chem.*, 1997, 69, 3038.
11. Yoshida, T., *J. Biochem. Biophys. Methods*, 2004, 60, 265.
12. Mant, C.T. and Hodges, R.S., *J. Sep. Sci.*, 2008, 31, 2754.
13. Boersema, P.J., Mohammed, S., and Heck, A.J.R., *Anal. Bioanal. Chem.*, 2008, 391, 151.
14. Naidong, W., *J. Chromatogr. B*, 2003, 796, 209.
15. Iwasaki, Y., Ishii, Y., Ito, R. et al., *J. Liq. Chromatogr. Relat. Technol.*, 2007, 30, 2117.
16. Nguyen, H.P. and Schug, K.A., *J. Sep. Sci.*, 2008, 31, 1465.
17. Strege, M.A., *Anal. Chem.*, 1998, 70, 2439.
18. Gavin, P.F., Olsen, B.A., Wirth, D.D., and Lorenz, K.T., *J. Pharm. Biomed. Anal.*, 2006, 41, 1251.
19. Liu, M., Chen, E.X., Ji, R., and Semin, D., *J. Chromatogr. A*, 2008, 1188, 255.
20. Strege, M.A., Durant, C., Boettinger, J., and Fogarty, M., *LCGC North Am.*, 2008, 26, 632.
21. Liu, M., Ostovic, J., Chen, E.X., and Cauchon, N., *J. Chromatogr. A*, 2009, 1216, 2362.
22. Olsen, B.A., *J. Chromatogr. A*, 2001, 913, 113.
23. McKeown, A.P., Euerby, M.R., Lomax, H. et al., *J. Sep. Sci.*, 2001, 24, 835.
24. Li, P. and Huang, J., *J. Chromatogr. A*, 2004, 1041, 163.
25. Wang, X., Li, W., and Rasmussen, H.T., *J. Chromatogr. A*, 2005, 1083, 58.
26. Hmelnickis, J., Pugovičs, O., Kažoka, H. et al., *J. Pharm. Biomed. Anal.,* 2008, 48, 649.
27. Guo, Y. and Huang, A., *J. Pharm. Biomed. Anal.*, 2003, 31, 1191.
28. McClintic, C., Remick, D.M., Peterson, J.A., and Risley, D.S., *J. Liq. Chromatogr. Relat. Technol.*, 2003, 26, 3093.
29. Risley, D.S. and Pack, B.W., *LCGC N. Am.*, 2006, 24, 776.
30. Ali, M.S., Ghori, M., Rafiuddin, S. et al., *J. Pharm. Biomed. Anal.,* 2007, 43, 158.
31. Hemström, P. and Irgum, K., *J. Sep. Sci.*, 2006, 29, 1784.
32. Hao, Z., Xiao, B., and Weng, N., *J. Sep. Sci.*, 2008, 31, 1449.
33. Ikegami, T., Tomomatsu, K., Takubo, H. et al., *J. Chromatogr. A*, 2008, 1184, 474.
34. Jandera, P., *J. Sep. Sci.*, 2008, 31, 1421.
35. Dejaegher, B., Mangelings, D., and Heyden, Y.V., *J. Sep. Sci.*, 2008, 31, 1438.
36. Wang, C., Jiang, C., and Armstrong, D.W., *J. Sep. Sci.*, 2008, 31, 1980.
37. Jinno, K., Quiming, N.S., Denola, N.L. et al., *Anal. Bioanal. Chem.,* 2009, 393, 137.
38. Majors, R.E. *LCGC North Am.*, 2008, 26, 238.
39. Jane, I., *J. Chromatogr.*, 1975, 111, 227.
40. Bidlingmeyer, B.A., Del Rios, J.K., and Korpi, J., *Anal. Chem.*, 1982, 54, 442.
41. Richardson, H. and Bidlingmeyer, B.A., *J. Pharm. Sci.*, 1984, 73, 1480.
42. Schmid, R.W. and Wolf, Ch., *Chromatographia*, 1987, 24, 713.
43. Bidlingmeyer, B.A. and Henderson, J., *J. Chromatogr. A*, 2004, 1060, 187.
44. Guo, Y. and Gaiki, S., *J. Chromatogr. A*, 2005, 1074, 71.
45. McCalley, D.V., *J. Chromatogr. A*, 2007, 1171, 46.
46. McCalley, D.V. and Neue, U.D., *J. Chromatogr. A*, 2008, 1192, 225.
47. Dong, L. and Huang, J., *Chromatographia*, 2007, 65, 519.
48. Snyder, L.R., Dolan, J.W., and Gant, J.R., *J. Chromatogr.*, 1979, 165, 3.
49. Jandera, P., Churáčk, J., and Svoboda, L., *J. Chromatogr.*, 1979, 174, 35.
50. Snyder, L.R. and Poppe, H., *J. Chromatogr.*, 1980, 184, 363.
51. Jin, G., Guo, Z., Zhang, F., Xue, X., Jin, Y., and Liang, X., *Talanta*, 2008, 76, 522.
52. Yoshida, T., *J. Chromatogr. A*, 1998, 811, 61.
53. Rosés, M. and Bosch, E., *J. Chromatogr. A*, 2002, 982, 1.
54. Espinosa, S., Bosch, E., and Rosés, M., *Anal. Chem.*, 2000, 72, 5193.
55. Gagliardi, L.G., Castells, C.B., Rafols, C., Rosés, M., and Bosch, E., *Anal. Chem.*, 2007, 79, 3180.
56. Bosch, E., Bou, P., Allemann, H., and Rosés, M., *Anal. Chem.*, 1996, 68, 3651.
57. Cox, G.B. and Stout, R.W., *J. Chromatogr. A*, 1987, 384, 315.

58. Greibrokk, T. and Anderson, T., *J. Chromatogr. A*, 2003, 1000, 743.
59. Dolan, J.W., *J. Chromatogr. A*, 2002, 965, 195.
60. Heinisch, S. and Rocca, J.-L., *J. Chromatogr. A*, 2009, 1216, 642.
61. Rosés, M., Subirats, X., and Bosch, E., *J. Chromatogr. A*, 2009, 1216, 1756.
62. Chen, M.H. and Horváth, C., *J. Chromatogr. A*, 1997, 788, 51.
63. Melander, W.R., Stoveken, J., and Horváth, C., *J. Chromatogr.*, 1979, 185, 111.
64. Zhu, P.L., Dolan, J.W., Snyder, L.R. et al., *J. Chromatogr. A*, 1996, 756, 51.
65. Li, J., *Anal. Chim. Acta*, 1998, 369, 21.
66. McCalley, D.V., *J. Chromatogr. A*, 2000, 902, 311.
67. Castells, C.B., Gagliardi, L.G., Ràfols, C., Rosés, M., and Bosch, E., *J. Chromatogr. A*, 2004, 1042, 23.
68. Gagliardi, L.G., Castells, C.B., Ràfols, C., Rosés, M., and Bosch, E., *J. Chromatogr. A*, 2005, 1077, 159.
69. Heinisch, S., Puy, G., Barrioulet, M.-P., and Rocca, J.-L., *J. Chromatogr. A*, 2006, 1118, 234.
70. Hao, Z., Lu, C.Y., Xiao, B., Weng, N. et al., *J. Chromatogr. A*, 2007, 1147, 165.
71. Guo, Y., Srinivasan, S., and Gaiki, S., *Chromatographia*, 2007, 66, 223.
72. Castells, C.B., Ràfols, C., Rosés, M., and Bosch, E., *J. Chromatogr. A*, 2003, 1002, 41.
73. Matsui, F., Robertson, D.L., and Lovering, E.G., *J. Pharm. Sci.*, 1983, 12, 948.
74. Hydrazine/Hydrazine sulfate (CASRN 302-01-2) in EPA's online Integrated Risk Information System (IRIS). http://www.epa.gov/iris/subst/0352.html (accessed June 25, 2009).
75. Carlin, A., Gregory, N., and Simmons, J., *J. Pharm. Biomed. Anal.*, 1998, 17, 885.
76. Monograph: Hydralazine Hydrochloride. In *United States Pharmacopeia*. United States Pharmacopeial Convention, Rockville, MD, 2007, USP 30, p. 2284.
77. Monograph: Povidone. In *United States Pharmacopeia*. United States Pharmacopeial Convention, Rockville, MD, 2007, USP 30, p. 2994.
78. Jagota, N.K., Chetram, A.J., and Nair, J.B., *J. Pharm. Biomed. Anal.*, 1998, 16, 1083.
79. Fujinari, E.M. and Courthaudon, L.O., *J. Chromatogr. A*, 1992, 592, 209.
80. Yan, X., *J. Chromatogr. A*, 1999, 842, 267.
81. Bhattachar, S.N., Wesley, J.A., and Seadeek, C., *J. Pharm. Biomed. Anal.*, 2006, 41, 152.
82. Yan, B., Zhao, J., Leopold, K., Zhang, B., and Jiang, G., *Anal. Chem.*, 2007, 79, 718.
83. Flanagan, R.J. and Jane, I., *J. Chromatogr. A*, 1985, 323, 173.
84. Lingeman, H., van Munster, H.A., Beynen, J.H. et al., *J. Chromatogr. A*, 1986, 352, 261.
85. Law, B., *J. Chromatogr. A*, 1987, 407, 1.
86. Naidong, W., Lee, J.W., Jiang, X. et al., *J. Chromatogr. A*, 1999, 735, 255.
87. Naidong, W., Shou, W., Chen, Y.-L., and Jiang, X., *J. Chromatogr. A*, 2001, 754, 387.
88. Grumbach, E.S., Wagrowski-Diehl, D.M., Mazzeo, J.R. et al., *LCGC North Am.*, 2004, 22, 1010.
89. Vikingsson, S., Kronstrand, R., and Josefsson, M., *J. Chromatogr. A*, 2008, 1187, 46.
90. Jiang, W. and Irgum, K., *Anal. Chem.*, 2002, 74, 4682.
91. Hu, W., Takeuchi, T., and Haraguchi, H., *Anal. Chem.*, 1993, 65, 2204.
92. Hu, W., Haddad, P.R., Hasebe, K. et al., *Anal. Chem.*, 1999, 71, 1617.
93. Okada, T. and Patil, J.M., *Langmuir*, 1998, 14, 6241.
94. Patil, J.M. and Okada, T., *Anal. Commun.*, 1999, 36, 9.
95. Cook, H.A., Hu, W., Fritz, J.S., and Haddad, P.R., *Anal. Chem.*, 2001, 73, 3022.
96. Fritz, J.S., *J. Chromatogr. A*, 2005, 1085, 8.
97. Jiang, W. and Irgum, K., *Anal. Chem.*, 1999, 71, 333.
98. Viklund, C., Sjögren, A., Irgum, K., and Nes, I., *Anal. Chem.*, 2001, 73, 444.
99. Jiang, W. and Irgum, K., *Anal. Chem.*, 2001, 73, 1993.
100. Hinman, R.L., *J. Org. Chem.*, 1958, 23, 1587.
101. Roberts, J.M., Diaz, A.R., Fortin, D.T. et al., *Anal. Chem.*, 2002, 74, 4927.

102. Dai, J., Mendonsa, S.D., Bowser, M.T., Lucy, C.A., and Carr, P.W., *J. Chromatogr. A*, 2005, 1069, 225.
103. Takegawa, Y., Deguchi, K., Ito, H. et al., *J. Sep. Sci.*, 2006, 29, 2533.
104. Ståhlberg, J., *J. Chromatogr. A*, 1999, 855, 3.
105. Rocklin, R.D., Pohl, C.A., and Schibler, J.A., *J. Chromatogr. A*, 1987, 411, 107.
106. Hu, W., Haddad, P.R., Tanakar, K., and Hasebe, K., *Analyst*, 2000, 125, 241.
107. Hu, W., Hasebe, K., Tanakar, K., and Haddad, P.R., *J. Chromatogr. A*, 1999, 805, 161.
108. Jiang, W., Fischer, G., Girmay, Y., and Irgum, K., *J. Chromatogr. A*, 2006, 1127, 82.
109. Alpert, A.J., *Anal. Chem.*, 2008, 80, 62.
110. Lai, S., Wang, S., and Ostovic, J., unpublished data.

16 Retention of Polar Acidic and Basic Drugs by HILIC

Mohammed Shahid Ali and
Aamer Roshanali Khatri

CONTENTS

16.1 INTRODUCTION

Drug analysis has evolved to cater to the requirements of drug discovery and pharmaceutical quality control. High-performance liquid chromatography (HPLC) is the most widely used separation technique for qualitative and quantitative determination of various chemicals in pharmaceutical industry. Prior to the 1970s, few reliable HPLC methods were available to the laboratory scientist. Till then, hydrophilic stationary phase and hydrophobic mobile phase were used as normal phase (NP) chromatography. Discovery of reversed-phase (RP) liquid chromatography in the late 1970s allowed improved separation for a wide variety of compounds. Since a substantial number of drugs in pharmaceutical research and discovery are ionic and polar hydrophilic, RP-HPLC methods may be difficult for highly polar compounds because of inadequate retention due to their highly polar nature. Achieving retention and separation of polar compounds is an ongoing challenge for chromatographers.[1–3] Such compounds need highly or 100% aqueous mobile phases but this can lead to wetting of stationary phases (e.g., C_{18}).[4,5] Attempts to reduce the risk of dewetting of C_{18} phases by means of polar embedding or polar end capping is not successful since these stationary phases provide lower retention than standard C_{18} phases.[2] Lack of retention for hydrophilic compounds in RP-HPLC is largely due to solovophilic factors, where polar functional groups have an ability to form favorable dipolar bonds with the solvent to become solvated. Since nonpolar stationary phase cannot offer similar bonding, the solutes stay in solution and elute in void volume.[6] In these cases, ion-exchange chromatography is the obvious choice where retention is accomplished by coulombic interaction mechanism. Another approach is the use of ion-pairing chromatography (IPC) where the ion-pair reagents in mobile phase mediate the retention of analytes on RP columns. Ion-pair reagents are soap-like ionic molecules having a charge opposite to that of molecule of interest, as well as a hydrophobic region to interact with stationary phase. The counter ion combines with the ions of the eluents, becoming ion pairs in the stationary phase, which results in retention. Several IPC methods have been reported for polar compounds but its use is limited due to expensive ion-pair reagents and its incompatibility with electrospray ionization mass spectrometry (ESI-MS).[7] To retain compounds that are highly hydrophilic and uncharged, derivatization techniques are used where one or more of the functional groups are converted into hydrophobic groups by a chemical reaction.[8] Derivatization may be limited to single-component analysis and not suitable for multicomponents where one or a few of the solutes may not be derivatized to give detectable molecules under similar conditions. Also, derivatization may not be suitable for closely eluting related compounds and the procedure is time consuming.

Hydrophilic-interaction chromatography (HILIC) has been used for the analysis of small polar and basic compounds. A high-organic, low-aqueous mobile phase is used to retain analytes with increasing orders of hydrophilicity. HILIC has emerged as a viable option to analyze polar hydrophilic compounds. Although the first application of HILIC is associated with the analysis of sugars about 30 years ago, HILIC was later recognized as a subset of chromatography. In 1990, Andrew Alpert first described the term HILIC, also called reverse reversed-phase or aqueous NP chromatography to distinguish it from the NP chromatography.[9] According to Alpert, retention is proportional to the polarity of the solute and inversely proportion to the polarity of the mobile phase.[9] HILIC has been proved to be a useful technique for the separation of polar and basic compounds because of its complementary selectivity to that of RP-HPLC.

Therefore, the HILIC approach was applied to separate neutral, acidic, and basic pharmaceutical compounds in different formulations. Methods were validated as per ICH Guidelines. The retention mechanism was studied and the effects of chromatographic parameters were evaluated.

16.2 ANALYSIS OF PHARMACEUTICAL INGREDIENTS

We have developed a few analytical methods suitable for drug substance like pseudoephedrine hydrochloride (PSH), diphenhydramine hydrochloride (DPH), and dextromethorphan hydrobromide (DXH) in cough–cold liquid formulation;[10] metformin hydrochloride (MFH) and related compounds cyanoguanidine (CGD) and melamine (MLN) in tablets[11]; brimonidine tartrate (BT),[12] sodium cromoglycate (SCG) in ophthalmic solutions[13]; and brimonidine tartrate (BT), maleic acid (MAC), and timolol maleate (TIM) in ophthalmic solutions. The unique selectivity on silica column caused due to hydrophilic interaction, ion exchange, and RP retention[9,14–17] facilitates the analysis of simple single-component polar analytes to multicomponent analytes of varying nature. The detailed chromatographic conditions for these compounds are described in Table 16.1. The composition of the selected formulations for detailed study is described herein.

16.2.1 Basic Amines in Cough–Cold Formulations

The test formulation was syrup with active ingredients as PSH (30.0 mg), DPH (12.5 mg), and DXH (15.0 mg) in each 5 mL. Other active ingredients were menthol and sodium citrate with excipients as saccharin sodium, flavor, and sodium benzoate (as preservative). Standard and test solutions for injection in LC system were prepared in diluent (water and methanol, 50:50, v/v) to obtain final concentrations of analytes as PSH: 300 µg/mL, DPH: 125 µg/mL, and DXH: 150 µg/mL.

16.2.2 MFH and Related Compounds in Tablets

Tablet formulations containing 500, 850, and 1000 mg of MFH were evaluated. The inactive components present in the formulation were excipients like hypromellose, magnesium stearate, Povidone K 30, polyethylene glycol, talc, and croscarmellose sodium. Solutions of MFH were prepared in the diluent (acetonitrile and water, 50:50, v/v) to provide a concentration of 150 µg/mL for injection in LC system.

16.2.3 BT in Ophthalmic Solution

Brimo ophthalmic solution manufactured by Jamjoom Pharmaceuticals Company Limited, Jeddah, Saudi Arabia, was used to evaluate determination of BT. The formulation comprised of BT 2.0 mg/mL as active ingredient. Other ingredients included citric acid monohydrate, polyvinyl alcohol, sodium chloride, sodium citrate, water, and benzalkonium chloride (as preservative). Standard and test solutions were prepared in diluent (acetonitrile and water, 92:8, v/v) to contain BT 100 µg/mL for injection in LC system.

16.2.4 SCG in Ophthalmic Solution

Test samples were Optidrin ophthalmic solution manufactured by Jamjoom Pharmaceuticals Company Limited with composition as SCG (20 mg), benzalkonium chloride as preservative (0.1 mg) per mL. Other components were polysorbate 80, disodium edetate, sodium chloride, hypromellose (hydroxypropyl methylcellulose), and water. Standard and test solutions were prepared in diluent (acetonitrile and water, 86:14, v/v) to contain 200 µg of SCG per mL for injection in LC system.

TABLE 16.1
Summary of Chromatographic Conditions

Analyte	Buffer	Mobile Phase	Column	Flow Rate (mL/min)	Injection Volume (μL)	Detection
PSH, DPH, and DXH	6.0 g ammonium acetate and 10 mL triethylamine per liter water, pH adjusted to 5.2 with orthophosphoric acid 85% (OPA)	Methanol + buffer (95:5, v/v), apparent pH of mobile phase ≈7.2.	Supelcosil™ LC–Si, 25 cm × 4.6 mm, packed with 5 μm silica particles (Supelco, Switzerland; Part No. 58295)	1.2	10	254 nm for PSH and DPH, changed to 280 nm at 9.0 min for DXH
MFH, CGD, and MLN	25 mM NaH_2PO_4, pH adjusted to 3.0 with OPA	Acetonitrile + buffer (84:16, v/v)	Atlantis® HILIC–Si, 25 cm × 4.6 mm, packed with 5 μm silica particles (Waters, Ireland; Part No. 18600203)	2.0	10	218 nm
BT	5.0 g ammonium acetate per liter water	Acetonitrile + buffer (92:8, v/v), pH adjusted to 7.1 with glacial acetic acid.	Supelcosil™ LC–Si, 25 cm × 4.6 mm, packed with 5 μm silica particles (Supelco, Buchs, Switzerland; Part No. 58295)	2.0	10	254 nm
SCG	30 mM ammonium acetate, pH adjusted to 3.0 with glacial acetic acid	Acetonitrile + buffer (86:14, v/v), apparent pH of mobile phase ≈5.6.	Atlantis® HILIC–Si, 25 cm × 4.6 mm, packed with 5 μm silica particles (Waters, Ireland; Part No. 18600203)	2.0	10	326 nm
BT, MAC, and TIM	5 g ammonium acetate per liter water	Acetonitrile + buffer (88:12, v/v), (no pH adjusted)	Atlantis® HILIC–Si, 25 cm × 4.6 mm, packed with 5 μm silica particles (Waters, Ireland; Part No. 18600203)	2.0	10	280 nm

16.3 RETENTION MECHANISM

HILIC is a variation of NP chromatography.[18] The retention mechanism is multimodal on silica[14] and is believed to be governed by hydrophilic interaction (partitioning). Superimposed electrostatic interaction between analytes and silica (ion exchange) functions as another mode of retention and is controlled by buffer (e.g., ammonium acetate). Besides, hydrophilic interaction may also include adsorption of analytes by silica gel.

HILIC uses polar stationary phases such as underivatized silica,[1,19–23] or silica packings modified with various functionalities, such as aminopropyl,[24–26] polyhydroxyethyl aspartamide,[9] cyclodextrin,[27] 3-cyanopropyl,[28] 2,3-dihydroxypropyl,[29,30] amide,[31,32] poly(succinimide).[33] Besides, a wide variety of silica packings with functionalities like polyamine, carboxymethyl, polyaspartic acid, poly(2-sulfoethyl aspartamide), polymeric sulfoalkylbetaine zwitterionic are also available. In addition, polar packings with functionalities like 2,3-dihydroxypropyl on methacrylic copolymer support, amine on poly(vinyl alcohol) gel support, polyamine on PVA copolymer support, mono-, di-, and triamine bonded by a propyl spacer on zirconia support, sulfonic acid on styrene/divinyl benzene support are used for the analysis of carbohydrates.[6] We used underivatized silica for the development of HILIC methods.

In HILIC, the stationary phase is polar material such as silica and the mobile phase is highly organic (e.g., acetonitrile or methanol or a combination of one or more organic solvents) with a small amount of aqueous solvent and counter ion (e.g., ammonium acetate or sodium dihydrogen phosphate) where compounds elute in order of increasing hydrophilicity.[9] Nonpolar compounds elute from the column first, while polar solutes show strong interaction with silanol groups on the silica surface that results in enhanced retention. This polar selectivity of silica is a helpful tool in achieving the separation of polar amine compounds from weak acids, e.g., cough–cold formulation matrix.

Mobile-phase condition optimized for analysis of a basic analyte was found unsuitable for an acidic or another basic analyte, and vice versa in RP-HPLC due to resolution problem and peak shape. Resolution between PSH, DPH, and DXH peaks could not be achieved with different mobile phases having combination of phosphate buffer, ammonium acetate buffer with methanol, or acetonitrile in conjunction with different stationary phases such as octadecylsilane, octylsilane, and nitrile groups chemically bonded to porous silica particles. Cough–cold formulations usually require mixtures of acidic and basic compounds for their effectiveness and therefore are difficult to separate by RP-HPLC using RP columns (e.g., C_{18} column). Due to above limitations of RP-HPLC in dealing with such complex formulations, interferences from other matrix components like saccharin sodium, sodium benzoate, citric acid, flavor, and sodium citrate present in formulation were also observed. Therefore, separation of active components from matrix components is a key for developing a rugged and robust analytical method. These interferences in RP-HPLC are due to ionization constant of active components having basic pK_a (PSH, DPH, and DXH) and other matrix components having acidic pK_a like saccharin sodium ($pK_a = 1.8$), citric acid ($pK_a = 3.15$, 4.77, and 6.40), and sodium benzoate ($pK_a = 4.2$). However, the pK_a profiles of active ingredients and matrix components were proved to be very useful in present HILIC procedure employing slightly basic mobile phase because of their ionization characteristics.[10] This resulted in unique selectivity of these compounds in HILIC mode and unambiguous separation of PSH, DPH, and DXH were achieved with elution of matrix components in dead volume region. Like RP-HPLC and NP-HPLC, selectivity and separation of mixtures in HILIC technique too, depends on the mobile phase pH, buffer salts, aqueous or organic solvents, sample-solution media type, and the components themselves. Critical parameters were studied and evaluated and described as follows.

16.3.1 Effect of Organic Solvent

Under HILIC conditions, solvent strength from weakest to strongest is tetrahydrofuran (THF) < acetone < acetonitrile < isopropanol < ethanol < methanol < water, where water is the strongest eluting solvent. The retention time (RT) of the described compounds was found to be increased with the increase in organic solvent concentration in the mobile phase. Therefore, it is necessary to consider the proportion of different organic solvents while optimizing mobile phase.

16.3.1.1 Choice of Organic Solvent

Selection of organic solvent depends on the nature of molecule. For example, methanol was used for the analysis of basic amines (PSH, DPH, and DXH) since acetonitrile provided much higher retention and resulted broad peak shape and tailing. While the analysis of MFH, BT, and SCG required acetonitrile to be used as the organic solvent. Figure 16.1 shows impact of acetonitrile and methanol on the

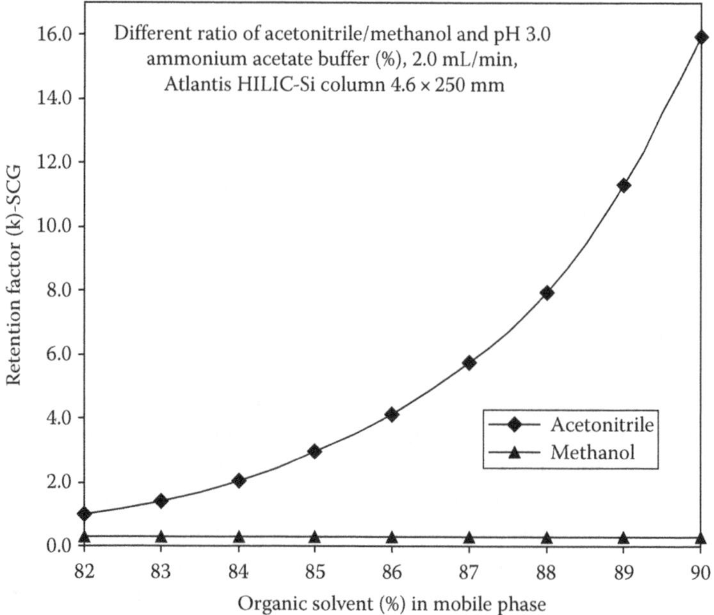

FIGURE 16.1 Effect of acetonitrile and methanol on the retention of SCG. (Reproduced from Ali, M.S. et al., *J. Sep. Sci.*, 31, 1645, 2008. With permission from Wiley-VCH Verlag Gmbh & Co. KGaA.)

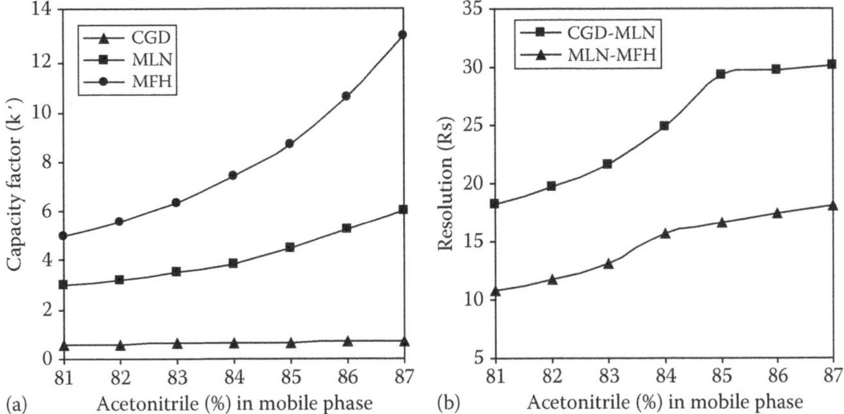

FIGURE 16.2 Effect of acetonitrile on (a) retention and (b) resolution of MFH, MLN, and CGD. (From Ali, M.S. et al., *Chromatographia*, 67, 517, 2007. With kind permission from Springer Science + Business Media.)

retention factor of SCG, where SCG is found to be unretained with methanol as organic solvent. In general, adequate retention was achieved using the appropriate organic solvent concentration for all subject compounds irrespective of basic and acidic nature of the drug substances. The impact of organic solvent change may be different for compounds in a given mixture. For example, related compounds CGD and MLN are analyzed together with MFH. The effect of acetonitrile is shown in Figure 16.2, where insignificant change in retention factor (k) of CGD was observed as compared to MLN and that of MFH. This, in turn, affects the resolution between CGD and MLN, and that between MLN and MFH. In order to maintain an enriched water layer around the silica particles inside the HILIC column, 5% water is minimum requirement in the mobile phase. Among the described methods, least organic concentration was that for simultaneous determination of MFH, MLN, and CGD where 84% acetonitrile was used while highest organic concentration was employed for the simultaneous determination of PSH, DPH, and DXH which required 95% methanol as organic modifier in the mobile phase.

16.3.1.2 Use of Alternate Solvent

Many HILIC procedures use acetonitrile as part of mobile phase. Acetonitrile is a by-product during the manufacture of acrylonitrile. Acrylonitrile is co-polymerized with butadiene and styrene to make ABS, a plastic used in cars etc, and the market for it collapsed in 2008–2009. This has resulted in the shortage of acetonitrile and its price went very high, and affected the analytical activities worldwide. One of the most feasible options is to redevelop the analytical method using another solvent or combination of different solvents depending on the solvent strengths, e.g., analysis of sucrose in iron–sucrose injection as described in USP 2009. Chromatographic conditions for this analysis is to use a mixture of acetonitrile and water (79:21, v/v) at a flow rate of 2.0 mL/min as mobile phase, a 4 mm × 25 cm column that contains packing USP L8 (aminopropylsilane), and refractive index detector. This is a HILIC

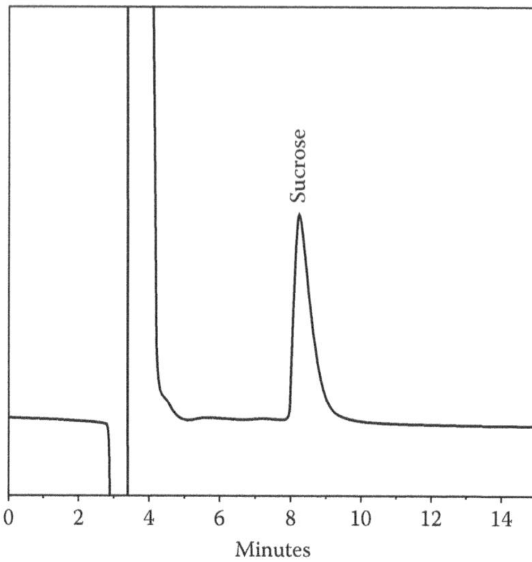

FIGURE 16.3 Chromatogram of iron–sucrose solution showing separation of sucrose.

method since it uses of 79% organic solvent on a polar amino column and the analyte being hydrophilic sucrose. When the chromatograms are recorded for standard and test solution under prescribed conditions, typical RT for sucrose is about 8 min. In order to replace acetonitrile from the mobile phase, we need a mobile phase that possesses similar strength as described in the USP method. Mobile phase was, therefore, optimized using THF, methanol, and water with the consideration of column back pressure increase due to the use of highly viscous solvents. Mobile phase consisting of THF, methanol, and water (60:35:5, v/v/v) at a flow rate of 1.5 mL/min using a Spherisorb 5 μm NH_2 4.6 mm × 25 cm column (Waters, USA) provided an RT of about 8 min upon injection of 10 μL standard and test solution (Figure 16.3). The correlation coefficient (R^2) was found to be 0.999 for the standard solutions of 13, 16, 18, 21, and 23 mg of sucrose per mL. Injection volume 20 μL provided distorted peak shapes for solutions with higher concentration of sucrose (>21 mg/mL). Acetonitrile is considered a stronger solvent than THF under HILIC conditions. Mobile-phase mixture containing THF will require another solvent which is stronger than acetonitrile to bring the cumulative solvent strength similar to that described in USP 2008 for sucrose analysis in iron–sucrose injection. For other HILIC methods having mobile phases with significant proportion of acetonitrile, similar experiments can be used to redevelop alternative methods for long-term use.

16.3.2 Mobile-Phase pH and Ion Exchange

In HILIC, more hydrophilic compounds are retained longer on a silica column whereas hydrophobic compounds elute out of a silica column first. It is important to control mobile-phase pH since ionic interactions vary with pH. It is expected that PSH, DPH, and DXH would also elute out of a silica column in the same order based on their hydrophilicity or hydrophobicity. It must be noted that PSH is highly soluble followed by DPH which is freely soluble and DXH is sparingly soluble in water. The elution pattern found was not based on their solubilities as expected. PSH eluted first followed by DPH and DXH. The retention behavior of these compounds was found to be influenced by the organic modifier. With every increase in the percentage volume of organic modifier, here methanol, the individual RT of these compounds increased indicating involvement of hydrophilic interactions in the separation mechanism. Another factor that influences the separation in this case is the ion-exchange interaction. This interaction is very strong and influences largely the separations achieved. All the compounds in the present example are ionizable compounds and the ionized species that carried the maximum charge (ionization degree) elutes out first. Mobile-phase pH selection is a technique that works well for ionizable compounds, because the retention characteristics of ionizable compounds are a function of mobile-phase pH.[4,34] In HILIC, selection of an appropriate pH of the mobile phase is very critical while achieving separation of ionizable compounds. Therefore, it is important to understand the process of ionization with varying pH. It is useful to have an idea of the pK_a values of the sample components, because changes in mobile pH have the largest effect on analyte retention when the pH is near the pK_a of the analyte. Changes in pH during a separation attempt involving ion-exchange phenomena can make a great difference in the chromatogram. PSH, DPH, and DXH are salts of weak bases and get completely ionized when dissolved in water. They exist as conjugate acids of respective weak bases and halide counter ions namely Cl^-, Cl^-, and Br^-. PSH is a salt of the weak base pseudoephedrine and strong acid HCl; DPH is a salt of the weak base diphenhydramine and strong acid HCL; and DXH is a salt of the weak base dextromethorphan and strong acid HBr. The pH of the aqueous solution in which they exist is the most influential factor for their ionization.

Percentage of ionization at a given pH can be calculated by[35]

$$\%\ \text{Ionization} = \frac{100}{[1 + \text{antilog}(\text{pH} - \text{p}K_w + \text{p}K_b)]} \tag{16.1}$$

where

pH is the value at which % ionization is calculated
pK_w is the ionic product of water (14.00 at 25°C)
pK_b is the basicity or dissociation constant

Conjugate acid–base pairs are linked by the expression

$$\text{p}K_w = \text{p}K_a + \text{p}K_b \tag{16.2}$$

Hence, at 25°C, Equation 16.1 can be written as

$$\%\ \text{Ionization} = \frac{100}{[1 + \text{antilog}(\text{pH} - \text{p}K_a)]} \tag{16.3}$$

We can calculate the degree of ionization for a given compound with its known pK_a value by substituting the value of antilog, which can be obtained by a scientific calculator. For example, % ionization for DXH at pH 7.3, can be calculated as follows:

$$\begin{aligned}
\%\,\text{Ionization (DXH at pH 7.3)} &= \frac{100}{[1+\text{antilog}(\text{pH}-\text{p}K_a)]} \\
&= \frac{100}{[1+\text{antilog}(7.3-8.3)]} \\
&= \frac{100}{[1+\text{antilog}(-1.0)]} \\
&= 90.91\%
\end{aligned}$$

It is well known that basic compounds are completely ionized at pH value 2 units below their pK_a value and are completely unionized at pH values 2 units higher than pK_a values. The degree of ionization for PSH, DPH, and DXH at different mobile-phase pH values is shown in Table 16.2. If the pH value of the mobile phase is lower than the pK_a value of the compounds being analyzed, they will exhibit an elution pattern based on the ionization behaviors. The positively charged species undergo cation exchange with negatively charged silanol groups.[15] Basic compounds like PSH, DPH, and DXH will exist as cations and undergo ion-exchange interactions with the anionic silanol groups originating from the silica of the stationary phase. PSH is the most charged ion among the analytes and hence shows the least RT, eluting after matrix peaks. This is an indicator of high ion-exchange

TABLE 16.2
Ionization of Compounds at Different Mobile-Phase pH

Buffer		PSH (pK_a 9.5)		DPH (pK_a 9.1)		DXH (pK_a 8.3)	
pH	**MP pH[a]**	**pH–pK_a**	**% Ionization**	**pH–pK_a**	**% Ionization**	**pH–pK_a**	**% Ionization**
5.5	7.5	−2.0	99.01	−1.6	97.55	−0.8	86.32
5.3	7.3	−2.2	99.37	−1.8	98.44	−1.0	90.91
5.2	7.2	−2.3	99.50	−1.9	98.76	−1.1	92.64
5.0	7.0	−2.5	99.68	−2.1	99.21	−1.3	95.23
4.6	6.7	−2.8	99.84	−2.4	99.60	−1.6	97.55
4.2	6.3	−3.2	99.94	−2.8	99.84	−2.0	99.01
4.0	6.1	−3.4	99.96	−3.0	99.90	−2.2	99.37

Source: Ali, M.S. et al., *J. Pharm. Biomed. Anal.*, 43, 158, 2007. With permission from Elsevier.

[a] Measured pH of mobile phase consisting of methanol and buffer (6.0 g ammonium acetate and 10 mL triethylamine per liter, at specified pH) in the ratio 95:5, v/v.

activity of the cation of PSH and the residual silanol groups of the surface of silica column. Another important factor in understanding the separation mechanism is the presence of buffer salts, which create the counter ion, present in the mobile phase. The counter ions promote ion-exchange mechanism responsible for lowering the retention of analytes. Use of appropriate buffers in mobile phase with pH at least 2 units apart of the pK_a values of the compounds, here lower than compound pK_a because of limitation of use of high mobile-phase pH in conjunction with the used silica column, can help in shifting the equilibrium of ionization by suppression of the ionization process. The counter ions in the mobile phase compete with the ions of compounds and thus influence the retention behavior through the ion-exchange interactions and lowering of the RTs. The increasing RT of PSH, DPH, and DXH, can therefore, be attributed to the electrostatic interaction with the silanol of silica and counterions.

Effect of pH of the retention of these compounds is shown in Figure 16.4, which is a direct result of the positively charged ions interacting with the silanol groups influenced by varying degree of ionization. The study of this retention behavior

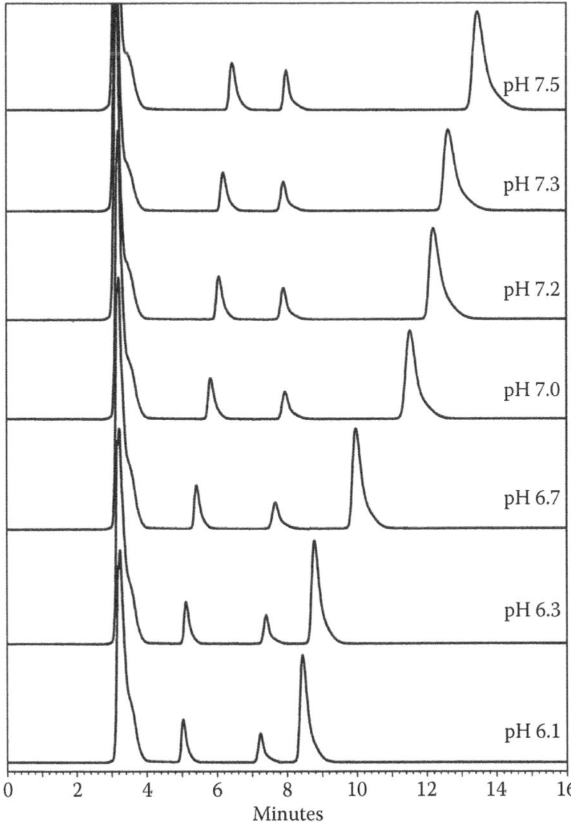

FIGURE 16.4 Overlaid chromatograms of test solution at different mobile-phase pH showing their retention behaviors. (From Ali, M.S. et al., *J. Pharm. Biomed. Anal.*, 43, 158, 2007. With permission from Elsevier.)

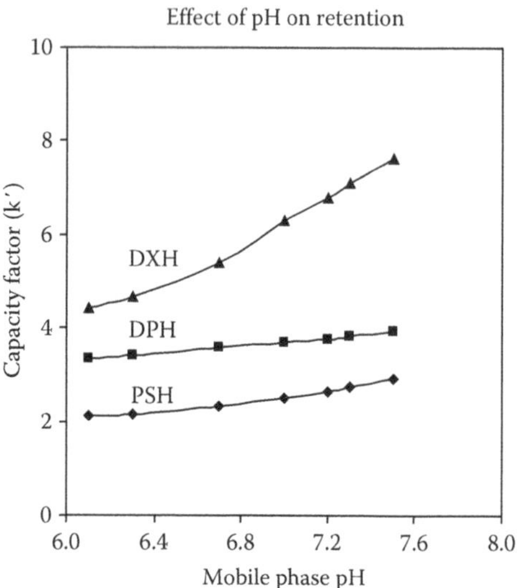

FIGURE 16.5 Plot of mobile-phase pH vs. capacity factor of PSH, DPH, and DXH. (From Ali, M.S. et al., *J. Pharm. Biomed. Anal.*, 43, 158, 2007. With permission from Elsevier.)

was carried out with a mobile phase comprising of 6.0 g/L ammonium acetate and 10 mL/L triethylamine.[10] The pH range studied using this buffer was 4.0–5.5. The final mobile phase comprised of an organic modifier methanol and the above buffer in volume ratio of 95:5. The pH of the final mobile phase was as per the studied pH values as mentioned in Table 16.2. Change in the pH of mobile phase caused DXH RTs to vary more significantly as compared to PSH and DPH in Figure 16.5. Lower pH values of 6.1 in the study led to higher ionization of PSH (99.96%), DPH (99.90%), and DXH (99.37%); thus, resulting in lower resolution and RTs. At the pH values of 6.1 and lower, the ionization of the compounds will lead to 100% ionization and more of ion exchange in the column leading to lower RTs. The peaks are not well resolved, some times co-eluting and thus making it difficult and unsuitable for quantitative determinations. However, at pH values higher than 7.5, the ionization of DXH is least. Due to low ionization, DXH shows higher RT and same is the case with PSH and DPH as seen in Figure 16.5. Mobile phases of values higher than 7.5 resulted in longer retention and peak broadening and hence making them unsuitable for quantitative applications. Operating on silica columns at pH values of 7.5 and above is not suggested due to loss of silica, creation of voids leading to changes in RT, resolution loss, and peak symmetry. The pH of the mobile phase is a very critical parameter and needs to be optimized for any application based on the desired separation and the ionization degree of compounds to provide the necessary chromatographic behaviors. The optimum chromatographic conditions for the separation of PSH, DPH, and DXH are as summarized in Table 16.1. This was achieved at mobile phase pH 7.2 (buffer pH 5.2) that provided optimal selectivity with fair resolution and retention. The compounds at this pH show adequate retention and resolution with acceptable peak shapes, because of their ionization degree (PSH-99.50%, DPH-98.76%, and DXH-92.64%) suitable for optimum cation exchange interactions.

An increasing trend in the individual retention of PSH, DPH, or DXH is seen in the studied mobile phase pH range 6.1–7.5. At lower mobile phase pH, e.g., pH 6.1, the positive charge densities of the analytes is highest. As the mobile-phase pH increases (toward pH 7.5), the %ionization decreases, thereby causing weakening of electrostatic repulsive force and increase in retention.

BT is positively charged and shows decrease of RT as pH of mobile phase is increased between the ranges of 6.5 and 7.5.[12] The pK_a value of BT is 7.4 and at the pH values between 6.5 and 7.5, it exists as a cation with decreasing charge density which leads to lower electrostatic attraction as mixed-mode effect and hence lower RTs are observed.

Based on the separation mechanism explained above the behavior of the compounds like MFH, CGD, and MLN can also be described. The charge density of these compounds will vary depending on the ionization state and it is highly influenced and controlled by the mobile phase pH. This tool is very helpful in controlling the desired RT and achieving the optimum separation goal. Molecules having lesser degree of ionization (positive) have been shown to possess higher capacity factor.[10] The pK_a value of MFH is reported to be 2.8 and 11.5.[36] MFH possess a charge of +2 at pH values below 2.8 and a charge of +1 between pH values of 2.8 and 11.5. Due to this varying charge exhibited by MFH, the RT of MFH is explainable due to the increase in electrostatic attraction of MFH with the silanol groups in a mixed-mode effect over the pH range 3.7–5.3 (Figure 16.6) due to increase in negatively charged silanols of silica surface, although slight decrease in its charge density shall occur. MFH will exist as an un-ionized species at pH

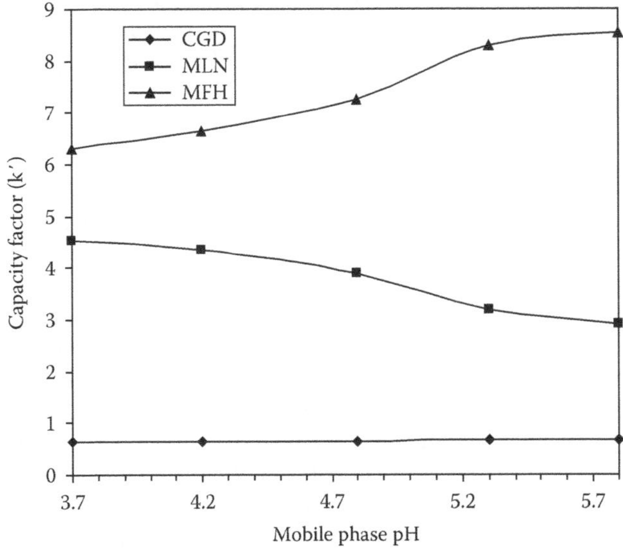

FIGURE 16.6 Effect of mobile-phase pH on the retention of MFH, CGD, and MLN. (From Ali, M.S. et al., *Chromatographia*, 67, 517, 2007. With kind permission from Springer Science + Business Media.)

values above 5.8 and hence RT of MFH is not influenced significantly beyond this pH value since charge density (degree of ionization) will not change significantly and extent of ionization of MFH will approach zero as mobile-phase pH increases. Experimental conditions of pH value beyond 6.0 are not recommended due to the limitation of the silica column as described earlier.[37] The pK_a value of MLN is reported to be 5.1.[38] Over the experimental range of pH from 3.7 to 5.3, MLN becomes less ionized. On loss of its charge, MLN seems to be less influenced by the negative charge carried by the silanol groups of silica column. Figure 16.6 shows the decrease in RT of MLN over the pH range used in experimentation. It must be noted for the present separation method that the ionization degree influences the retention for MFH while the retention of MLN is dependent on the ionization state. CGD found to be the least impacted by the changes in the pH of the mobile phase which shows that its ionization is unaffected by changes in pH in the range 3.7–5.3. Higher selectivity of the method is mainly due to the hydrophilic interactions exhibited by the different polarity of the analytes. Hydrophilicity of the compounds namely, MFH, MLN, and CGD dictates the elution pattern in this example of separation.

Another example that is worth looking at to understand the mechanism is that of SCG, an anion with pK_a value of 1.9. Over the pH range of the experimentation, increase of negative charge leading to complete ionization is observed followed by the decrease in capacity factor in the pH range of 4.2–5.9.[13] The limitation of the operating pH values for the column used in the experiments apply as before and hence maximum pH value studied is 5.9. The lowering of capacity factor for SCG with increase in mobile phase pH from 4.2 to 5.9 is characterized by the increase in the negative charge of surface silanol groups in the silica column leading to strengthening of electrostatic repulsion between SCG and silanols. For the retention behavior of anionic species dependent on degree of their ionization, the impact on retention by change in mobile phase pH is therefore opposite to that of cationic species.

16.3.3 Ionic Strength

Correct choice of buffer is crucial for the maintenance of mobile phase pH because buffers themselves are ionic and may take part in the ion-exchange process. It is important to add a buffer in the mobile phase to control the ionization states of both the analyte and the stationary phase, which affects the acid–base equilibrium between analyte and stationary phase. Decreased retention was observed for PSH, DPH, and DXH with increase in ionic strength (ammonium acetate or triethylamine) in the mobile phase. Since the main mechanism in HILIC is believed to be the partitioning of the analytes into the aqueous phase that forms a stagnant layer on the stationary-phase surface, the decrease in RT might be best understood by a shift in equilibrium concentrations. As the NH_4^+ concentration increases preferentially in the aqueous layer, there is less opportunity for the analyte counter ions to partition into the aqueous layer.[39] Thus, the cations are swept through the column (mainly in the organic layer) with less interaction with the negatively charged silanols of column and the aqueous phase. Decreased resolution, capacity factor, and peak tailing were observed with buffer strength increase in the mobile phase. Increase in the ratio of buffer in mobile phase (e.g., methanol and buffer 90:10, instead of 95:5, v/v) showed decreased retention and resolution. This retention behavior seen above is indicative of HILIC because of opposite chromatographic effect than that found in RP-HPLC.[10] Therefore, the decrease in retention of PSH, DPH, and DXH with increased buffer concentration is attributed to the increase in the electrostatic repulsion of these compounds due to the presence of NH_4^+ ions.

Similar behavior was observed for BT where increased ammonium acetate strength resulted in decreased retention of BT. As buffer strength increases, the electrostatic attractive interactions arising from negatively charged silanols and cationic analyte are weakened, which causes a decrease in the retention.

Buffer plays an important role in peak shape and retention of ionic compounds. Ammonium acetate and ammonium formate buffers usually are advantageous for extension of a HILIC method to ESI-MS analysis because of their volatility. These buffers provide good solubilities in organic solvents like methanol and acetonitrile at high proportions required in HILIC. Precipitation due to buffer solubility in the mobile phase is generally not a problem with acetate buffers. At low wavelength, these buffers may interfere with analysis and decrease the sensitivity of the method by providing noisy baseline. Hence, use of these buffers at low wavelength (less than 225 nm) should be limited. At wavelengths above 225 nm, acetate buffer can be used conveniently and safely. Because of the above factor, sodium dihydrogen phosphate buffer, which provides low absorption, was selected for the analysis of MFH, CGD, and MLN at 218 nm (Figure 16.7). Significant increase in retention of MFH was observed compared to MLN and CGD with increase in buffer strength (Figure 16.8) at the optimized mobile phase pH. These compounds shall be influenced by the increase in ionic strength depending on their ionization degree. Significant increase in the retention of MFH in buffer strength range 20–40 mM phosphate buffer was observed because of availability of negatively charged buffer ions in the stagnant aqueous layer of silica column.

It has been previously established that ionic concentration has drastic effect on the retention and peak shape of ions.[34] Effect of ionic strength on the retention of SCG was studied using 20, 25, 30, 35, and 40 mM solutions of ammonium acetate.[13] The pH of each solution was adjusted to pH 3.0 with glacial acetic acid. Mobile phase comprising acetonitrile and each buffers (86:14, v/v) under similar chromatographic condition were employed and retention of SCG was noted. Figure 16.9 shows increased retention with the increase in ammonium acetate buffer strength (mM). As buffer strength (NH_4^+) increases, repulsive electrostatic interactions which are arising from negatively charged silanols and anionic cromoglicate are weakened. This accounts for the increased retention of SCG. Buffer strength of 30 mM was considered the optimal choice (k value ≈ 4.2).

16.3.4 Stationary Phase

A wide variety of HILIC stationary phases are available involving underivatized silica and derivatized silica.[6] We used bare silica columns from Waters (Atlantis HILIC silica) and Supelco (LC-Si) for the method development. One of the most

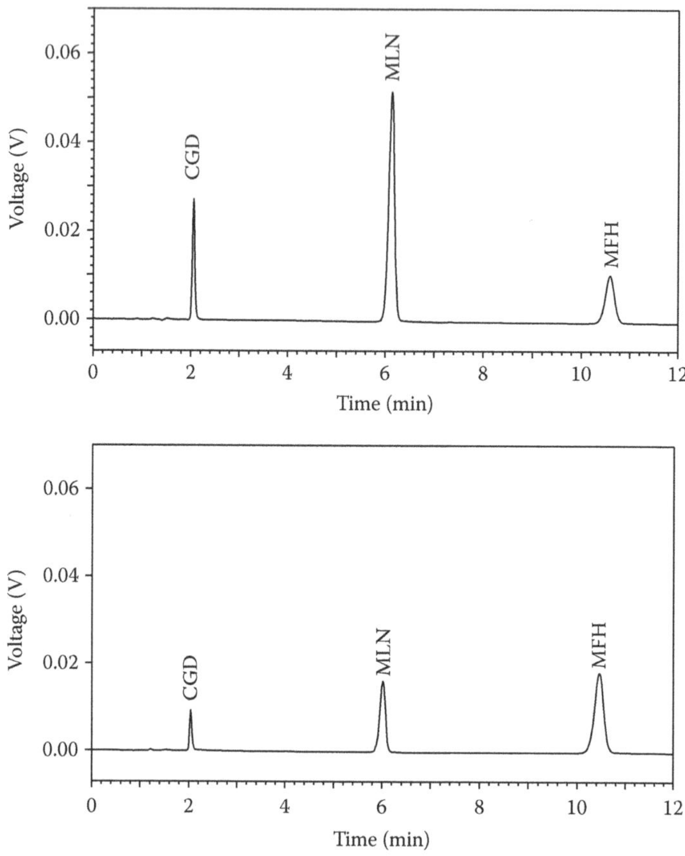

FIGURE 16.7 Chromatogram showing separated peaks of CGD, MLN, and MFH at different wavelengths (top) 218 nm (bottom) 232 nm. (From Ali, M.S. et al., *Chromatographia*, 67, 517, 2007. With kind permission from Springer Science + Business Media.)

important advantages of using bare silica columns is the low column back pressure exhibited by using the high viscous organic solvents like methanol and THF, which facilitates operating chromatography at high flow rates. Another property of bare silica packing is the absence of ligands that may detach from the packings and appear as spurious peaks in the MS-spectra. Silica columns from different manufacturers can provide a slight/significant difference in retention behavior and resolution for the particular analytes.[6] This impact can be due to the presence of metal ions which affects the acidity of the silanol groups, or the purity of silica due to differences in manufacturing procedures. Specific silica columns for HILIC applications are commercially available which are designed to provide better peak shape, adequate retention, and resolution. Presence of water shall cause formation of negatively charged silanol groups which in turn attracts cationic analytes (e.g., BT, MFH) for ion-exchange mechanism. However, the same silanol group shall

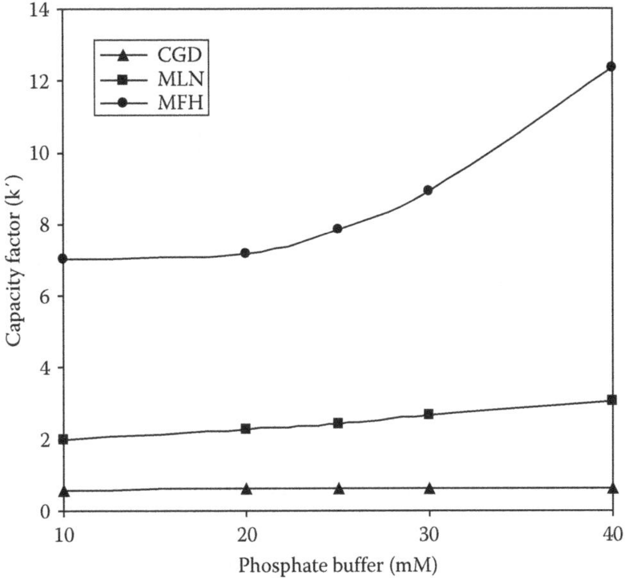

FIGURE 16.8 Effect of ionic strength on the retention of MFH, CGD, and MLN. (From Ali, M.S. et al., *Chromatographia*, 67, 517, 2007. With kind permission from Springer Science + Business Media.)

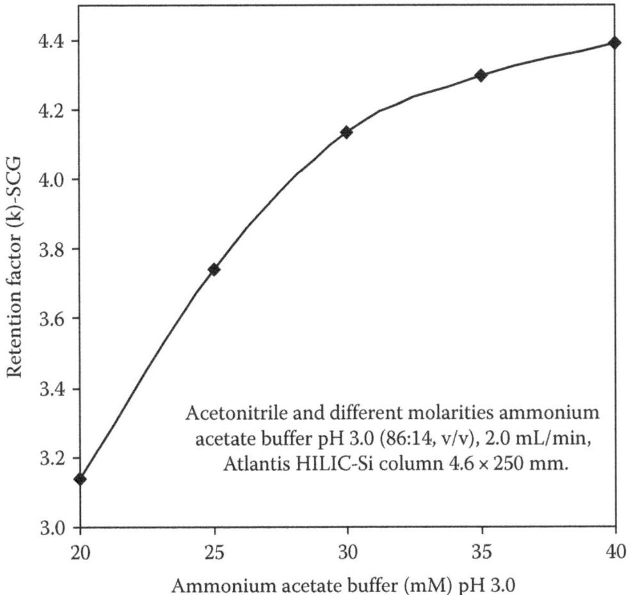

FIGURE 16.9 Effect of ionic strength on the retention of SCG. (Reproduced from Ali, M.S. et al., *J. Sep. Sci.*, 31, 1645, 2008. With permission from Wiley-VCH Verlag Gmbh & Co. KGaA.)

exhibit electrostatic repulsion with anionic analytes (e.g., SGC) and ion exchange shall not take place. The use of electrolytic buffer is required to control the electrostatic interactions due to mixed-mode separation of cationic and anionic analytes on silica columns.

To promote ion exchange for anionic analytes, silica modified with triazole has been used. Positively charged triazole is suitable to retain acidic compounds via anion exchange. The impact of pH and ionic strength on retention behavior of cationic and anionic analytes shall be opposite to that of bare silica column packing. Therefore, selection of column should take consideration of the analyte type(s), and separation goal. For example, separation of BT (cationic analyte), TIM (cationic analyte), and MAC (anionic analyte) (as impurity of TIM) was achieved on Atlantis HILIC-Si column 4.6 × 250 mm using acetonitrile and buffer (ammonium acetate, 5 g/L) in the ratio 88:12, v/v, at a flow rate of 2.0 mL/min with detection wavelength at 280 nm (Figure 16.10).

Nonpolar mobile phase comprising of solvents like hexane, toluene, etc. are used in NP LC with silica columns. Commercially available silica columns, e.g., Suplecosil™ LC-Si are shipped in nonpolar solvents like hexane and ethyl acetate (98:2). To use such columns in HILIC separation, they need to be flushed with comparatively less polar solvents like isopropyl alcohol followed by acetonitrile and methanol to avoid immiscibility. Subsequently, the column can be hydrated with 5% water. Columns that are designated for HILIC mode need not be subjected to this conversion process as they are shipped in solvents like acetonitrile and little amount of water. The operating pH range of commercially available columns also

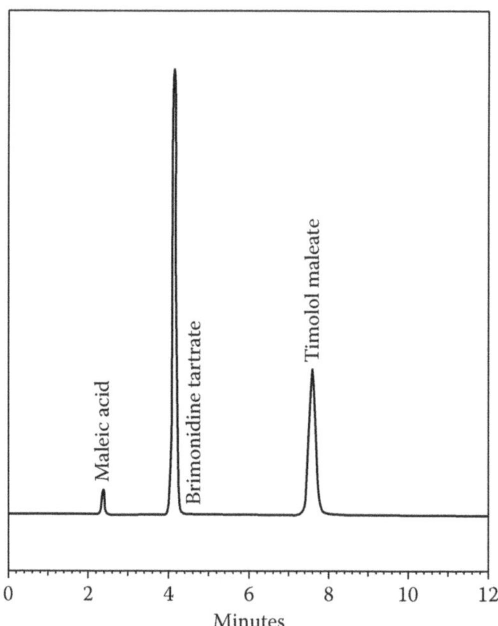

FIGURE 16.10 Chromatogram showing separated peaks of maleic acid, brimonidine tartrate, and timolol maleate.

vary depending on the column material, e.g., Suplecosil™ LC-Si, HALO HILIC column (2–8), Agilent Zorbax HILIC plus column (1–8), Atlantis HILIC-Silica column (2–6) as recommended by column manufacturers.

16.3.5 Diluent and Injection Volume

Solvents used for the sample preparation influence the peak shape and sensitivity of determination in chromatographic analysis.[34] Solvents that resemble the mobile phase composition are known to provide excellent results. The mobile phase used in the analysis of SCG is acetonitrile and 30 mM ammonium acetate pH 3.0 in volume ratio of 86:14. Similar sample diluent composition applies to other methods too. Normally, injection volumes of 5–50 μL provides sharp peak shapes and enhanced sensitivity as long as the diluent composition is maintained similar to mobile phase. Various combinations of diluents were used to examine the impact of diluent on peak shape. Sharp and symmetrical peak shapes were obtained on injection volume of 5 μL with sample diluents having acetonitrile and water 70:30, 60:40, 50:50, 40:60, 30:70 (v/v). Peak distortion was seen with these diluents following injection volume higher than 10 μL. It is important to note that water alone as diluent provided peak distortion in all the described methods. Diluent comprising of methanol and water (50:50, v/v) for PSH, DPH, and DXH determination with injection volume of 5–30 μL showed good peak shapes for all compounds as shown in Figure 16.11.

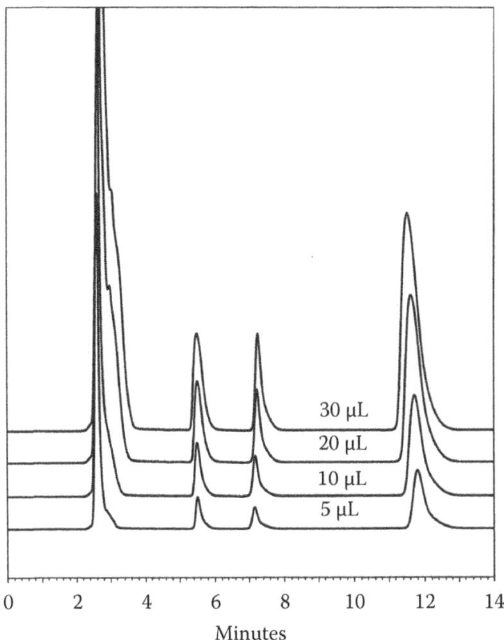

FIGURE 16.11 Overlaid chromatograms of test solution following injection volume of 5, 10, 20, and 30 μL. (From Ali, M.S. et al., *J. Pharm. Biomed. Anal.*, 43, 158, 2007. With permission from Elsevier.)

16.3.6 Validation of Methods

Validation of all the methods developed in author's lab was done as per the requirements of ICH Guidelines.[40] Parameters such as specificity, accuracy, precision range, robustness, and system suitability were examined and found to be acceptable for the intended analyses.

Specificity is defined as the ability of the analytical procedure to assess unequivocally the analyte in the presence of component like impurities, degradants, matrix, etc. that may be present in the analytical samples. Forced degradation, as part of specificity evaluation, is necessary while developing stability-indicating assay methods. Specificity is the most challenging parameter to achieve for a stability-indicating method. Blank matrix solution should not provide interfering peak(s) and impurity peak(s) should be separated from the main analyte peaks.

Degradations of PSH, DPH, and DXH were simulated by subjecting the solutions to acidic hydrolysis (HCl), alkaline hydrolysis (NaOH), oxidation (H_2O_2), photolytic (UV/sunlight), and heat as recommended by the ICH Guidelines. The purity of analyte peaks were evaluated using photodiode array detector and found to be homogeneous. Figure 16.12 shows the chromatogram of oxidation study with maximum degradation. Two impurities namely CGD and MLN were separated from the main MFH peak with fair resolution (Figure 16.7) and could be quantified using the same method. Chromatograms of the stressed samples revealed complete separation of

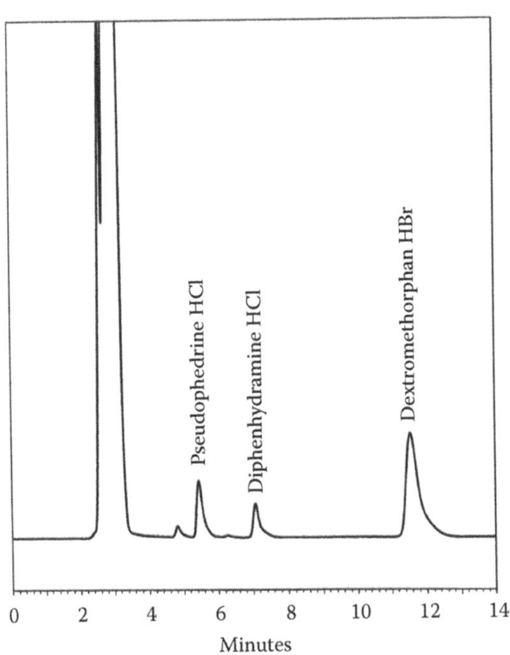

FIGURE 16.12 Chromatogram of test solution subjected to forced degradation by oxidation (H_2O_2) showing degradation peaks. (From Ali, M.S. et al., *J. Pharm. Biomed. Anal.*, 43, 158, 2007. With permission from Elsevier.)

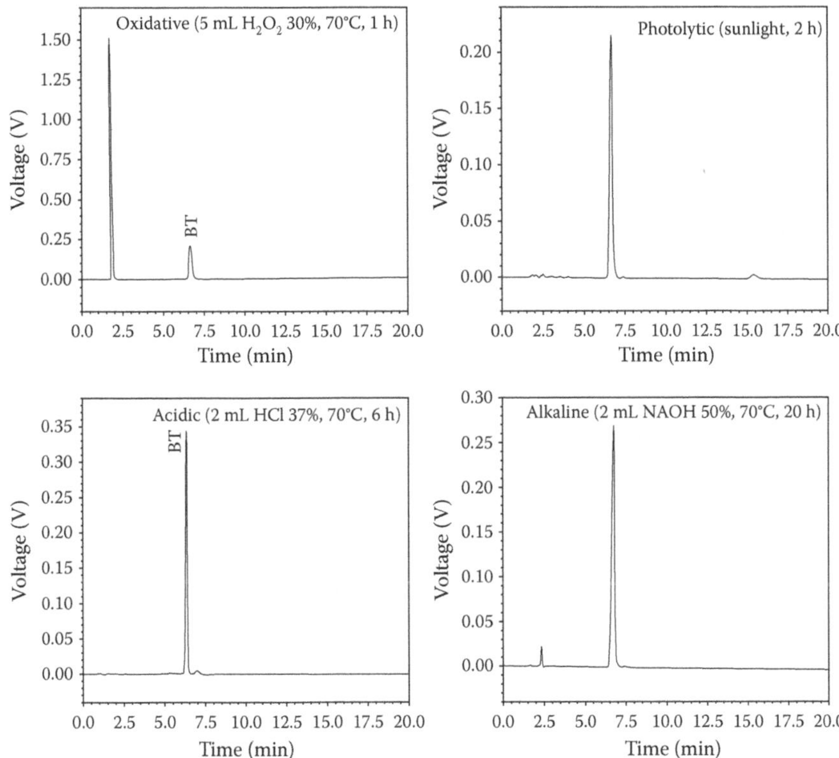

FIGURE 16.13 Chromatograms of BT obtained after various stress conditions. (From Ali, M.S. et al., *Chromatographia*, 70, 539, 2009. With kind permission from Springer Science + Business Media.)

secondary peak(s) from the parent peak BT (Figure 16.13) with peak purity of 100%. Similarly, complete resolution of SCG from degradation compounds was observed in stress studies (Figure 16.14). The HILIC method used to quantify SCG could not retain its two impurities namely SCG impurity 1 (Cromolyn diethyl ester) and SCG impurity 2 (1,3-bis(2-acetyl-3-hydroxyphenoxy)-2-propanol) because of their non-polar and hydrophobic nature. These impurities eluted in dead volume and did not interfere with SCG quantification. Method-validation data are presented in Tables 16.3 and 16.4. The data suggests that the methods are specific, accurate, precise, and robust for the intended analyses.

16.4 ADVANTAGES

By utilization of the unique selectivity of HILIC, an analyst is now able to separate a wide variety of difficult to retain polar compounds. High organic mobile phase is ideal for efficient desolvation and compound ionization in ESI-MS. It can provide

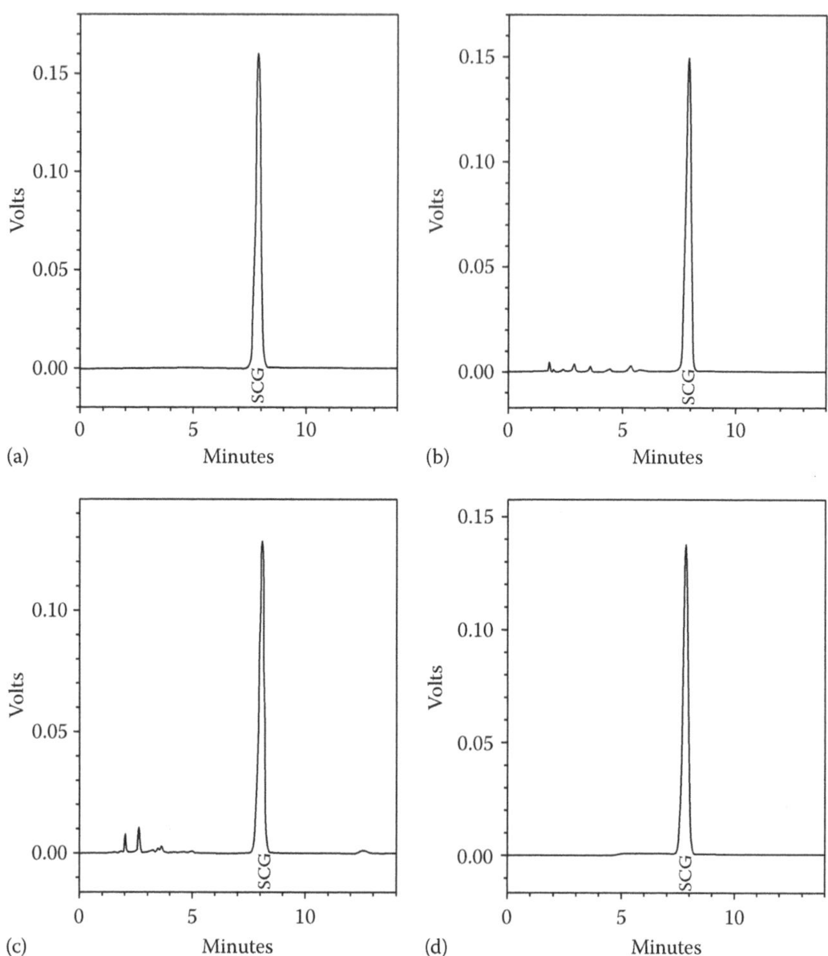

FIGURE 16.14 Chromatograms of SCG obtained due to various stress conditions. (a) Standard (no stress), (b) Oxidative (5% H_2O_2, 5 mL, 80°C, 5 h), (c) Photolytic-254 nm (24 h), and (d) Basic (0.1 N NaOH, 5 mL, RT, 2 h). (Reproduced from Ali, M.S. et al., *J. Sep. Sci.*, 31, 1645, 2008. With permission from Wiley-VCH Verlag Gmbh & Co. KGaA.)

enhanced sensitivity and lower limits of detection in ESI-MS. The methodology has the potential to replace many RP-HPLC methods by providing superior peak shapes and good resolution. HILIC can be operated at high flow rate due to less viscous acetonitrile used in the mobile phase, which provided lower column back pressure and could be useful in high-throughput analysis.

TABLE 16.3
Summary of Method-Validation Data

		Results and Observations			
		Determination of PSH, DPH, and DXH			
Validation Parameters	**Data Element**	**PSH**	**DPH**	**DXH**	**Determination of BT**
Specificity	Peak purity	Spectrally pure peaks with degraded solutions	Spectrally pure peaks with degraded solutions	Spectrally pure peaks with degraded solutions	Spectrally pure BT peak with degraded solutions
	Resolution	>2.0 between secondary peak(s) and main peaks	>2.0 between secondary peak(s) and main peaks	>2.0 between secondary peak(s) and main peaks	>2.0 between secondary peak(s) and BT peak
Linearity	R^2	0.99974 (50%–200%)	0.99965 (50%–200%)	0.99985 (50%–200%)	0.99998 (20%–200%)
Accuracy	Recovery	99.0%–101.6% ($n = 5$) (50%–150%)	99.4%–100.6% ($n = 5$) (50%–150%)	100.1%–101.8% ($n = 5$) (50%–150%)	99.3%–101.2% ($n = 8$) (20%–200%)
	Average	100.7% ($n = 5$)	100.1% ($n = 5$)	100.8% ($n = 5$)	100.3% ($n = 8$)
	RSD	0.99% ($n = 5$)	0.75% ($n = 5$)	0.61% ($n = 5$)	0.31% ($n = 8$)
Instrumental precision	RSD	0.56% ($n = 6$)	0.66% ($n = 6$)	0.59% ($n = 6$)	0.22% ($n = 6$)
Repeatability	RSD	<1.0% ($n = 6$)	<1.0% ($n = 6$)	<1.0% ($n = 6$)	<0.53% ($n = 5$)
Intermediated precision	RSD	0.66% ($n = 20$)	0.64% ($n = 20$)	0.91% ($n = 20$)	0.62% ($n = 12$)
	Recovery	100.5% ($n = 20$)	100.2% ($n = 20$)	101.1% ($n = 20$)	100.63% ($n = 12$)
Robustness	T	1.90–1.93	1.92–1.95	1.95–1.96	1.28–1.33
	N	>4300	>8000	>4500	>7500
	k	>2.4	>3.4	>6.3	>3.0
	Rs	N/A	5.0–5.4	9.2–9.8	N/A
	Peak area (RSD)	0.25% ($n = 6$)	0.33% ($n = 6$)	0.18% ($n = 6$)	<0.4% ($n = 6$)
Limit of detection (LOD)	S/N ratio	8.5 (0.75 μg/mL)	3.3 (0.3125 μg/mL)	15.7 (0.375 μg/mL)	3.2 (0.005 μg/mL)
Limit of quantitation (LOQ)	S/N ratio	16.5 (1.5 μg/mL)	10.1 (0.625 μg/mL)	32.7 (0.750 μg/mL)	12.7 (0.02 μg/mL)
	RSD (at LOQ)	3.3% ($n = 6$)	2.8% ($n = 6$)	3.5% ($n = 6$)	1.9% ($n = 6$)
Solution stability (after 24 h)	Assay (%)	No significant change	No significant change	No significant change	No significant change
	Secondary peaks	Not found	Not found	Not found	Not found

TABLE 16.4
Summary of Method-Validation Data

		Results and Observations			
		Determination of MFH, CGD, and MLN			
Validation Parameters	**Data Element**	**MFH (Assay)**	**CGD (as Impurity)**	**MLN (as Impurity)**	**Assay of SCG**
Specificity	Peak purity	Spectrally pure peaks with degraded solutions	Spectrally pure peaks with degraded solutions.	Spectrally pure peaks with degraded solutions.	Spectrally pure SCG peak with degraded solutions.
	Resolution	>2.0 between secondary, MFH, MLN, and CGD peak(s)	>2.0 between secondary, MFH, MLN, and CGD peak(s)	>2.0 between secondary, MFH, MLN, and CGD peak(s)	>2.0 between secondary peak(s) and SCG peak
Linearity (25%–200%)	R^2	0.9999	0.99985	0.99985	0.99999
Accuracy	Recovery	99.8%–101.0% ($n = 8$) (20%–200%)	96.4%–102.3% ($n = 4$) (50%–200%)	96.7%–101.9% ($n = 4$) (50%–200%)	99.8%–100.7% ($n = 7$) (25%–200%)
	Average	100.27% ($n = 8$)	99.7% ($n = 12$)	99.6% ($n = 12$)	100.4% ($n = 7$)
	RSD	0.56% ($n = 8$)	1.66% ($n = 12$)	1.79% ($n = 12$)	0.31% ($n = 7$)
Instrumental precision	RSD	0.65%	1.76%	1.58%	0.17% ($n = 6$)
Repeatability	RSD	<1.35% (50%–200%)	<1.35% (50%–200%)	<1.35% (50%–200%)	<0.2% ($n = 3$) (50%–150%)
Intermediate precision	RSD	0.71% ($n = 12$)	1.58% ($n = 12$)	1.50% ($n = 12$)	0.64% ($n = 12$)
	Recovery	100.3% ($n = 12$)	100.2% ($n = 12$)	100.4% ($n = 12$)	100.12% ($n = 12$)
Robustness	T	1.05–1.12	1.11–1.13	1.03–1.06	0.90–1.12
	N	>15,000	>6,500	>12,000	>3,500
	k	>7.0	>0.6	>3.5	>2.0
	Rs	N/A	22.0–28.0 (CGD–MLN)	14.1–17.6 (MLN–MFH)	N/A
	Peak area (RSD)	0.14% ($n = 6$)	0.55% ($n = 6$)	0.51% ($n = 6$)	<1.0% ($n = 6$)
Limit of detection (LOD)	S/N ratio	2.9 (100 ng/mL)	2.8 (5.0 ng/mL)	3.8 (25 ng/mL)	2.9 (25 ng/mL)
Limit of quantitation (LOQ)	S/N ratio	10.4 (350 ng/mL)	10.7 (25 ng/mL)	11.2 (75 ng/mL)	12.3 (100 ng/mL)
	RSD (at LOQ)	1.5% ($n = 6$)	1.9% ($n = 6$)	2.0% ($n = 6$)	2.9% ($n = 6$)
Solution stability (after 24 h)	Assay (%)	No significant change	No significant change	No significant change	No significant change
	Secondary peaks	Not found	Not found	Not found	Not found

16.5 CONCLUSION

The described HILIC methods are suitable for the analysis of drug substances, such as PSH, DPH, DXH, MFH, CGD, MLN, and BT in various pharmaceutical products. The methods are superior to those GC or RPLC methods described previously. The use of silica column has been proved to be beneficial to retain polar compounds. Unique selectivity is resulted due to mixed-mode retention mechanisms. Various polar acidic and basic analytes can be separated using HILIC. Diluent has a significant impact on analyte peak shape and should be kept similar to the mobile-phase composition if possible. Yet, water as a solvent proved to be another useful entity because of its affinity to silica and creating an enriched layer for retention of polar analytes. It is believed that many analytical problems posed by small and polar compounds can be solved by HILIC.

REFERENCES

1. Olsen, B. A. 2001. Hydrophilic interaction chromatography using amino and silica columns for the determination of polar pharmaceuticals and impurities. *J. Chromatogr. A* 913:113–122.
2. Layne, J. 2002. Characterization and comparison of the chromatographic performance of conventional, polar-embedded, and polar-endcapped reversed-phase liquid chromatography stationary phases. *J. Chromatogr. A* 957:149–164.
3. Hanai, T. 2003. Separation of polar compounds using carbon columns. *J. Chromatogr. A* 989:183–196.
4. Neue, U. D., Grumbach, E. S., Mazzeo, J. R., Tran, K., Wagrowski-Diehl, D. M. 2003. Method development in reversed-phase chromatography, Bioanalytical Separations, In *Handbook of Analytical Separations*, ed. I. D. Wilson. Elsevier Science B.V., New York, pp. 185–214.
5. Reid, T. S., Henry, R. A. 1999. *Am. Lab.* 31:24–28.
6. Hemstroem, P., Irgum, K. 2006. Review. Hydrophilic interaction chromatography. *J. Sep. Sci.* 29:1784–1821.
7. Gustavsson, S. A., Samskog, J., Markides, K., Langstrom, B. 2001. Studies of signal suppression in liquid chromatography-electrospray ionization mass spectrometry using volatile ion-pairing reagents. *J. Chromatogr. A* 937:41–47.
8. Lunn, G., Hellwig, L. C. 1998. *Handbook of Derivatization Reactions for HPLC*. Wiley & Sons, Inc., New York.
9. Alpert, A. J. 1990. Hydrophilic interaction chromatography (HILIC): A new method for separation of peptides, nucleic acids and other polar solutes. *J. Chromatogr.* 499:177–196.
10. Ali, M. S., Ghori, M., Rafiuddin, S., Khatri, A. R. 2007. A new hydrophilic interaction liquid chromatographic (HILIC) procedure for the simultaneous determination of pseudoephedrine hydrochloride (PSH), diphenhydramine hydrochloride (DPH) and dextromethorphan hydrobromide (DXH) in cough-cold formulations. *J. Pharm. Biomed. Anal.* 43:158–167.
11. Ali, M. S., Rafiuddin, S., Ghori, M., Khatri, A. R. 2007. Simultaneous determination of metformin hydrochloride, cyanoguanidine and melamine in tablets by mixed-mode hydrophilic interaction liquid chromatography. *Chromatographia* 67:517–525.
12. Ali, M. S., Khatri, A. R., Munir, M. I., Ghori, M. 2009. A stability-indicating assay of brimonidine tartrate in ophthalmic solution and stress testing by hydrophilic interaction chromatography (HILIC). *Chromatographia* 70:539–544.

13. Ali, M. S., Al-Jawi, D. A., Al-Hetari, Y., Ghori, M., Khatri, A. R. 2008. Stability-indicating assay of sodium cromoglicate in ophthalmic solution using mixed-mode hydrophilic interaction chromatography. *J. Sep. Sci.* 31:1645–1650.
14. Naidong, W. 2003. Bioanalytical liquid chromatography tandem mass spectrometry methods on underivatized silica columns with aqueous/organic mobile phases. *J. Chromatogr. B* 796:209–224.
15. Bidlingmeyer, B. A., Del Rios, J. K., Korpi, J. 1982. Separation of organic amine compounds on silica gel with reversed-phase eluents. *Anal. Chem.* 54:442–447.
16. Cox, G. B., Stout, R. W. 1987. Study of the retention mechanism for basic compounds on silica under 'pseudo-reversed-phase' conditions. *J. Chromatogr.* 384:315–336.
17. Strege, M. A., Stevenson, S., Lawrence, S. M. 2000. Mixed mode anion-cation exchange/hydrophilic interaction liquid chromatography-electrospray mass spectrometry as an alternative to reversed phase for small molecule drug discovery. *Anal. Chem.* 72:4629–4633.
18. Rabel, F. M., Caputo, A. G., Butts, E. T. 1976. Separation of carbohydrates on a new polar bonded phase material. *J. Chromatogr.* 126:731–740.
19. Naidong, W., Shou, W., Chenn, Y. L., Jiang, X. 2001. Novel liquid chromatographic-tandem mass spectrometric methods using silica columns and aqueous-organic mobile phases for quantitative analysis of polar ionic analytes in biological fluids. *J. Chromatogr. B* 754:387–399.
20. Naidong, W., Shou, W. Z., Addison, T., Maleki, S., Jiang, X. 2002. Liquid chromatography-tandem mass spectrometric bioanalysis using normal-phase columns with aqueous-organic mobile phases—A novel approach of eliminating evaporation and reconstitution steps in 96-well SPE. *Rapid Commun. Mass Spectrom.* 16:1965–1975.
21. Grumbach, E. S., Diehl, D. M., McCabe, D. R., Mazzeo, J. R., Neue, U. D. 2003. Hydrophilic interaction chromatography for enhanced ESI-MS sensitivity and retention of polar basic analytes. *LCGC N. Am.* Suppl. S:53–54.
22. Grumbach, E. S., Diehl, D. M., Mazzeo, J. R. 2004. A sensitive ESI-MS HILIC method for the analysis of acetylocholine and choline. *LCGC N. Am.* Suppl. S:74–75.
23. Su, J., Hirji, R., Zhang, L., He, C., Selvaraj, G., Wu, R. 2006. Evaluation of the stress-inducible production of choline oxidase in transgenic rice as a strategy for producing the stress-protectant glycine betaine. *J. Exp. Bot.* 57:1129–1135.
24. Neue, U. D. 1997. *HPLC Columns Theory, Technology and Practice*. John Wiley & Sons, Hoboken, NJ, pp. 217–223.
25. Tolstikov, V. V., Fiehn, O. 2002. Analysis of highly polar compounds of plant origin: Combination of hydrophilic interaction chromatography and electrospray ion trap mass spectrometry. *Anal. Biochem.* 301:298–307.
26. Beumer, J. H., Joseph, E., Egorin, M. J., Covey, J. M., Eiseman, J. L. 2006. Quantitative determination of zebularine (NSC 309132), a DNA methyltransferase inhibitor, and three metabolites in murine plasma by high-performance liquid chromatography coupled with on-line radioactivity detection. *J. Chromatogr. B Analyt. Technol. Biomed. Life Sci.* 831:147–155.
27. Strege, M. A. 1998. Hydrophilic interaction chromatography-electrospray mass spectrometry analysis of polar compounds for natural product drug discovery. *Anal. Chem.* 70:2439–2445.
28. McClintic, C., Remick, D. M., Peterson, J. A., Risley, D. S. 2003. Novel method for the determination of piperazine in pharmaceutical drug substances using hydrophilic interaction chromatography and evaporative light scattering detection. *J. Liq. Chromatogr. Relat. Technol.* 26:3093–3104.
29. Tanaka, H., Zhou, X., Masayoshi, O. 2003. Characterization of a novel diol column for high-performance liquid chromatography. *J. Chromatogr. A* 987:119–125.

30. Uutela, P., Reinila, R., Piepponen, P., Ketola, R. A., Kostiainen, R. 2005. Analysis of acetylcholine and choline in microdialysis samples by liquid chromatography/tandem mass spectrometry. *Rapid Commun. Mass Spectrom.* 19:2950–2956.
31. Higley, T. J., Yoshida, T. 2003. Separation of peptides by hydrophilic interaction chromatography using TSKgel Amide-80. *LCGC N. Am.*, June: Suppl. S:20–21.
32. Takahashi, N. 1996. Three-dimensional mapping of N-linked oligosaccharides using anion-exchange, hydrophobic and hydrophilic interaction modes of high-performance liquid chromatography. *J. Chromatogr. A* 720:217–225.
33. Alpert, A. J., Shukla, M., Shukla, A. K., Zieske, L. R. et al. 1994. Hydrophilic-interaction chromatography of complex carbohydrates. *J. Chromatogr. A* 676:191–202.
34. Neue, U. D., Pheobe, C. H., Tran, K., Cheng, Y. F., Lu, Z. 2001. Dependence of reversed-phase retention of ionizable analytes on pH, concentration of organic solvent and silanol activity. *J. Chromatogr. A* 925:49–67.
35. Florence, A. T., Attwood, D. 2006. *Physicochemical Principles of Pharmacy*, 4th edn., Pharmaceutical Press, London, U.K., pp. 75–82.
36. Van de Merbel, N. C., Wilkens, G., Fowless, S., Oosterhuis, B., Jonkman, J. H. G. 1998. LC phases improve, but not all assays do: Metformin bioanalysis revisited. *Chromatographia* 47:542–546.
37. Column Care and Use Instructions—Atlantis™ Columns, Waters Corporation, Milford, MA, p. 8.
38. Hirt, R. C., Schmitt, R. G. 1958. Ultraviolet absorption spectra of derivatives of symmetric triazine—II: Oxo-triazines and their acyclic analogs. *Spectrochim. Acta* 12:127–138.
39. Hu, W., Tao, H., Haraguchi, H. 1994. Evaluation of sulfobetaine-type zwitterionic stationary phases for ion chromatographic separation using water as a mobile phase. *Anal. Chem.* 66:2514–2520.
40. ICH Harmonized Tripartite Guideline, *Validation of Analytical Procedures: Text and Methodology, Q2(R1)*, November 2005, ICH Steering Committee, ICH Secretariat, Geneva, Switzerland.

17 Retention and Selectivity of Polar Stationary Phases for Hydrophilic Interaction Chromatography

Yong Guo

CONTENTS

17.1 INTRODUCTION

Small polar compounds, although representing only a small portion of therapeutic drugs, are often very challenging to pharmaceutical analysis. Lack of sufficient retention on reversed-phase columns makes it difficult to analyze polar compounds by reversed-phase liquid chromatography (RPLC). Hydrophilic interaction chromatography (HILIC) provides a viable alternative to ion-pairing or normal phase methods, which are typically used for the analysis of polar and/or ionizable compounds. Recent reviews on HILIC reveal increasing popularity of HILIC in pharmaceutical analysis of small polar drugs (Hemstrom and Irgum 2006, Dejaegher and Heyden 2010, Jian et al. 2010). The term "hydrophilic interaction chromatography" (HILIC) was first coined by Andrew Alpert in 1990 (Alpert 1990); however, similar separation technique had been used for sugar analysis much earlier (Linden and Lawhead 1975, Palmer 1975). Alpert also postulated the retention mechanism for HILIC as hydrophilic partitioning of polar solutes between the organic solvent-rich mobile phase and a water-rich liquid layer immobilized on the surface of polar stationary phases (Alpert 1990). McCalley and Neue (2008) recently demonstrated possible presence of a water layer on porous silica surface by measuring the elution volume of benzene using a mobile phase containing high levels of acetonitrile. More published studies seem to support hydrophilic partitioning as the major retention mechanism for HILIC, but there are mounting evidences that other interactions (e.g., adsorption, electrostatic, hydrogen-bonding, dipole–dipole interaction) also play very important roles in HILIC separation (Olsen 2001, Li and Huang 2004, Guo and Gaiki 2005, Hao et al. 2008).

Most HILIC separation was performed on the polar stationary phases that were typically used in normal-phase chromatography (e.g., amino, cyano, and silica phases) in the early days. The normal-phase columns had to be first converted into the reversed-phase conditions and then used for HILIC separation. With increasing popularity of HILIC, more and more polar phases with diverse functional groups have been designed and developed for HILIC in recent years (Dejaegher et al. 2008, Ikegami et al. 2008, Jandera 2008). The availability of various polar phases provides method development chemists with the opportunity to find an appropriate phase for the desired separation. At the same time, it also presents a big challenge to select the optimal phase during method development in a systematic manner in a short time. As in RPLC, the choice of a stationary phase should be based on a thorough understanding of the retention and selectivity characteristics of the available stationary phases. This chapter focuses on the retention and selectivity of various polar phases commercially available.

17.2 POLAR STATIONARY PHASES FOR HILIC

Underivatized silica is a popular phase for HILIC application, particularly in bioanalytical fields (Wen 2003, Jian et al. 2010). Irreversible adsorption of solutes is not as problematic in HILIC as in normal-phase chromatography due to the presence of a significant level of water in the mobile phase. The silica columns for normal-phase chromatography have been used for HILIC separation (Olsen 2001, Hao et al. 2007); however, the columns need to be washed extensively to remove normal-phase solvents. Significant difference in retention has been reported among the silica columns from different manufactures (Olsen 2001, Hao et al. 2008). A new superficially porous silica column has been shown to have greater resistance to overloading for ionized basic compounds than the reversed-phase columns (McCalley 2008). There are more and more silica columns on the market that are developed exclusively for HILIC application, such as Atlantis HILIC silica column. The silica columns promoted for HILIC application are typically packed and stored in aqueous organic solvents instead of normal-phase solvents.

In addition to the silica phase, a wide variety of polar functional groups have been employed as stationary phases in HILIC (Hemstrom and Irgum 2006, Ikegami et al. 2008, Jandera 2008). Most bonded phases used for HILIC can be classified into three major categories based on the charge characteristics of the functional groups, namely, neutral, cationic, and zwitterionic groups as listed in Table 17.1. The polar phases in Table 17.1 are presented by the conventional names of the functional groups.

TABLE 17.1
Polar Stationary Phases Commonly Used in HILIC

Phase Type	Phase Name	Functional Group
Neutral	Amide	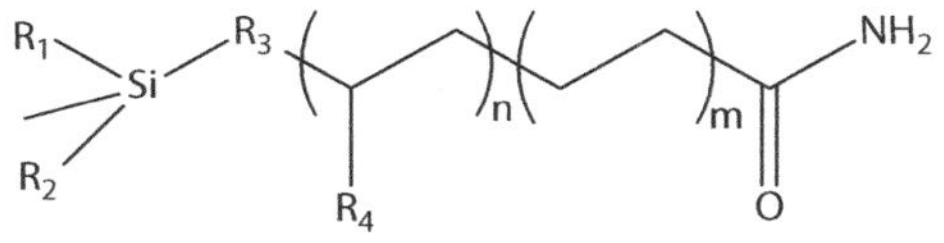
	Aspartamide	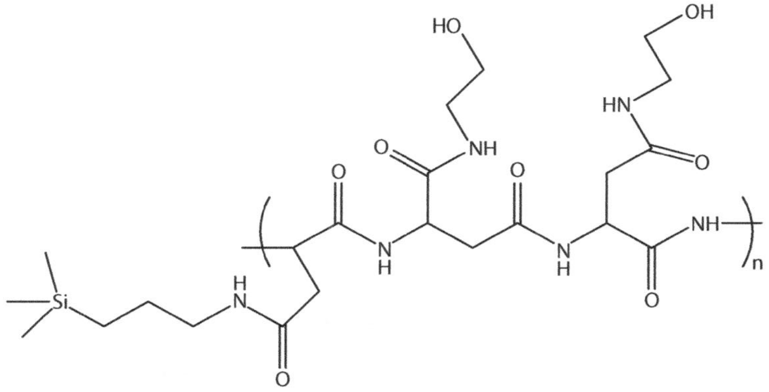
	Diol	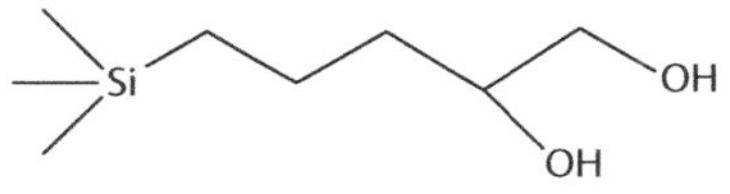

(continued)

TABLE 17.1 (continued)
Polar Stationary Phases Commonly Used in HILIC

Phase Type	Phase Name	Functional Group
Cationic	Amino	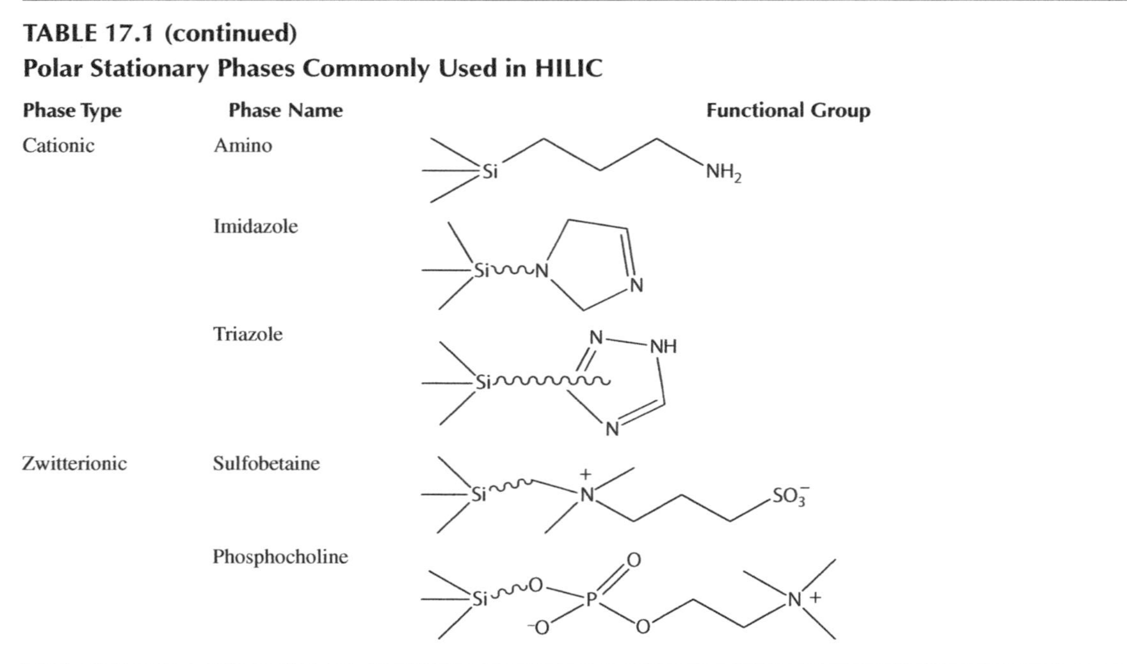
	Imidazole	
	Triazole	
Zwitterionic	Sulfobetaine	
	Phosphocholine	

The neutral group includes many popular HILIC phases, such as amide, diol, and cross-linked diol phases. The amide phase was originally developed for carbohydrate analysis under "aqueous normal phase" condition in 1988, but has become one of the most popular phases for HILIC. The functional group on the amide phase is carbamoyl moiety positioned at the distal end of an aliphatic chain attached to the silica surface. The diol phase is the same as the one used in normal-phase chromatography. Some diol columns are packed in reversed-phase solvents, but some are packed in normal-phase solvents, which need extensive washing to convert to aqueous conditions. The cross-linked diol phase (structure not shown) is covalently bonded to the silica surface and the diol groups are highly cross-linked through ether linkages (Expolore Luna HILIC 2007). Similar to the cross-linked diol phase, both polyhydroxy and polyvinyl alcohol (PVA) phases contain a layer of polyhydroxylated and PVA polymer bound to the silica surface, respectively, and have many hydroxyl groups (–OH) present on the polymer surface, which render these phases polar. The aspartamide phase in this group is worth special mention since it was the first stationary phase specifically developed for HILIC separation. The aminopropylated silica was first coated with a layer of polysuccinimide, which was then treated with ethanolamine to generate the final stationary phase. It is not as commonly used as other phases possibly due to relatively lower efficiency (Hemstrom and Irgum 2006).

The second category includes the stationary phases that can be positively charged in the pH range typically used for HILIC separation (pH 3–8). The amino, imidazole, and triazole phases are bonded to the silica surface through an aliphatic linker. The amino phase has been used for HILIC separation even before the term HILIC was invented, and still remains very popular for various applications. Many amino columns are prepared for normal-phase separation, and have significant differences in retention for polar solutes when run in HILIC mode (Guo and Huang 2003). The amino phase is also not very stable over a long period of time. Polyamine phases containing secondary and tertiary amine groups are more stable, but have different selectivity than the conventional amino phase. The cationic group also includes the imidazole and triazole phases specially designed for HILIC application. The imidazole phase is attached to the silica surface through the secondary amine moiety on the imidazole ring; however, the point of attachment on the triazole phase is unknown. The positive charges on these phases can have direct impact on retention and selectivity for charged solutes through electrostatic interactions.

The third group is the zwitterionic phases specially designed for HILIC. Two representative zwitterionic phases are sulfobetaine and phosphocholine phases, which differ in the placement of charged groups (Jiang 2003, Jiang et al. 2006). The quaternary amine group bearing positive charge is at the proximal end of the sulfobetaine phase, but at the distal end of the phosphocholine phase. The sulfobetaine phase has been gaining popularity in HILIC rapidly in recent years; however, the phosphocholine phase has not been commercialized at the time of writing.

17.3 SELECTIVITY OF VARIOUS POLAR STATIONARY PHASES IN HILIC

In RPLC, stationary phase chemistry (e.g., C_{18} vs. phenyl phase) has been widely used to change selectivity and achieve desired separation (Snyder et al. 1997, Kazakevich and Lobrutto 2007). Although a variety of functional groups have been used as the stationary phase (Table 17.1), there is a lack of critical understanding of the selectivity of various polar phases in HILIC. With the availability of more and more HILIC phases, this has become particularly important since the knowledge about various HILIC phases is useful to select appropriate columns during method development.

The selectivity of the polar phases presented in Table 17.1 has been experimentally investigated under the same chromatographic conditions. A group of nucleosides and nucleic acid bases were used as the model compounds for this investigation (Guo and Gaiki 2005). The model compounds were eluted isocratically using a mobile phase of acetonitrile and ammonium acetate solution (85/15, v/v). The selectivity of the amide, aspartamide, sulfobetaine, and silica phases is well illustrated by the chromatograms shown in Figure 17.1. Major selectivity difference is observed among the selected phases. For example, adenosine and uridine are not separated on the amide phase, but almost baseline resolved on the aspartamide phase. The selectivity of the sulfobetaine phases is similar to that of the aspartamide phase. In comparison, the silica phase has a very different selectivity for the model compounds than the amide, aspartamide and sulfobetaine phase.

Figure 17.2 shows the separation of the model compounds on the diol, cross-linked diol, polyhydroxy, and PVA phases, which all have –OH groups on the packing surface. These phases share relatively similar selectivity for the model compounds. The only difference is that some phases (e.g., cross-linked diol and polyhydroxy phase) have better selectivity for cytidine and guanosine. In addition, it is interesting to note that the silica phase (Figure 17.1) seems to be very different from the phases shown in Figure 17.2. This indicates that the silica phase may have some specific interactions with the model compounds.

The separation of the model compounds on the cationic phases (e.g., amino, imidazole, and triazole phase) is shown in Figure 17.3. Different selectivity of the three phases is well depicted by the elution patterns. For examples, adenosine and uridine barely resolved on the amino phase, but well separated on the imidazole and triazole phases with opposite elution order. It is worth noting that the model compounds are not charged at the mobile-phase pH used in this study. The difference in selectivity most likely results from specific interactions (e.g., hydrogen bonding, dipole interaction) between the stationary phases and solutes.

The selectivity of the various polar phases has also been evaluated using a group of organic acids, which include acetylsalicylic acid and some of its metabolites (Guo et al. 2007). Figures 17.4 through 17.6 show the separation of the acidic compounds on various polar phases under the same chromatographic conditions. As shown in Figure 17.4, the amide and aspartamide phases share similar selectivity for the acidic compounds, but is different than the sulfobetaine phase. For example, hippuric acid and α-hydroxyhippuric acid have opposite elution order on the sulfobetaine phase compared to the amide and aspartamide phases. In addition, the silica phase also shows very different selectivity than the other phases, particularly for gentisic acid.

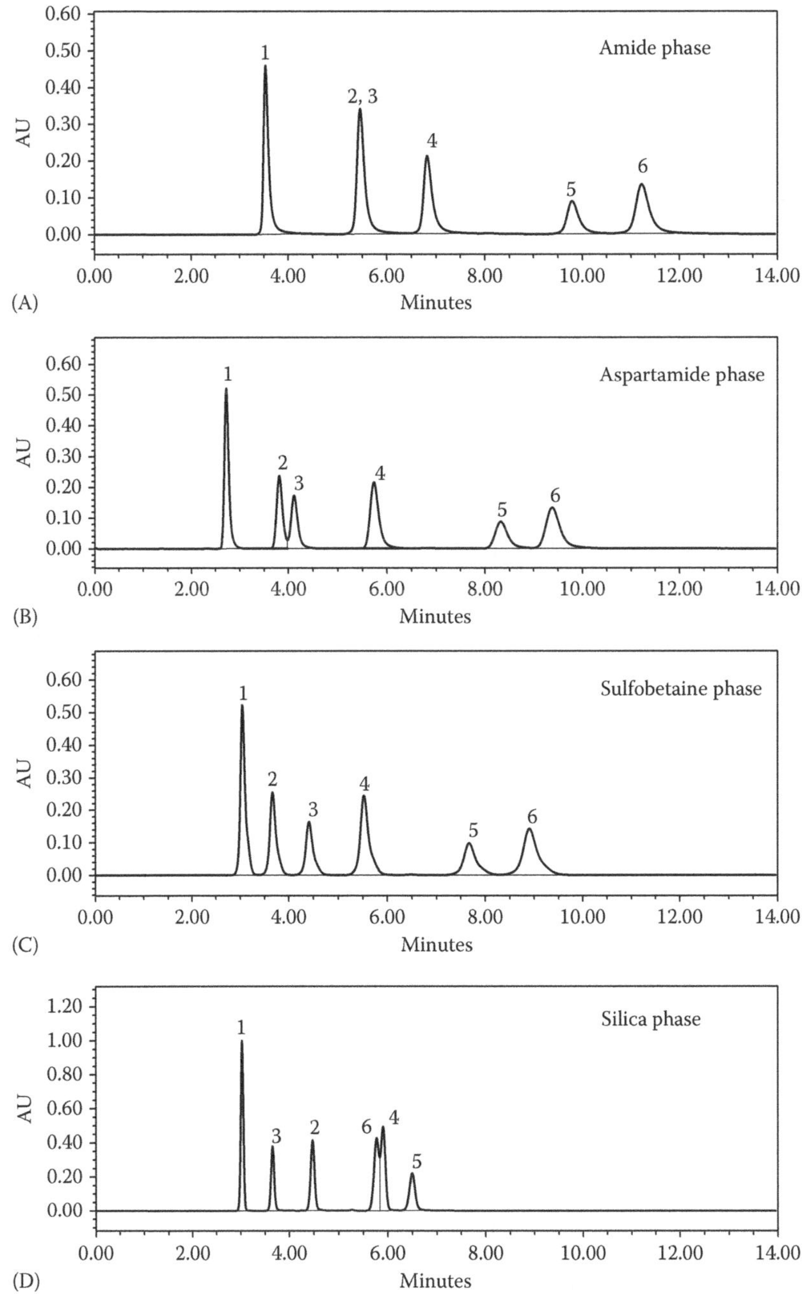

FIGURE 17.1 Separation of neutral model compounds on (A) amide, (B) aspartamide, (C) sulfobetaine, and (D) silica phase. Mobile phase, acetonitrile/water (85/15, v/v) containing 10 mM ammonium acetate. Column dimension, 250 mm × 4.6 mm ID, 5 µm particle size. Column temperature, 30°C. Flow rate, 1.5 mL/min. UV detection at 248 nm. Peak label: (1) uracil, (2) adenosine, (3) uridine, (4) cytosine, (5) cytidine, and (6) guanosine.

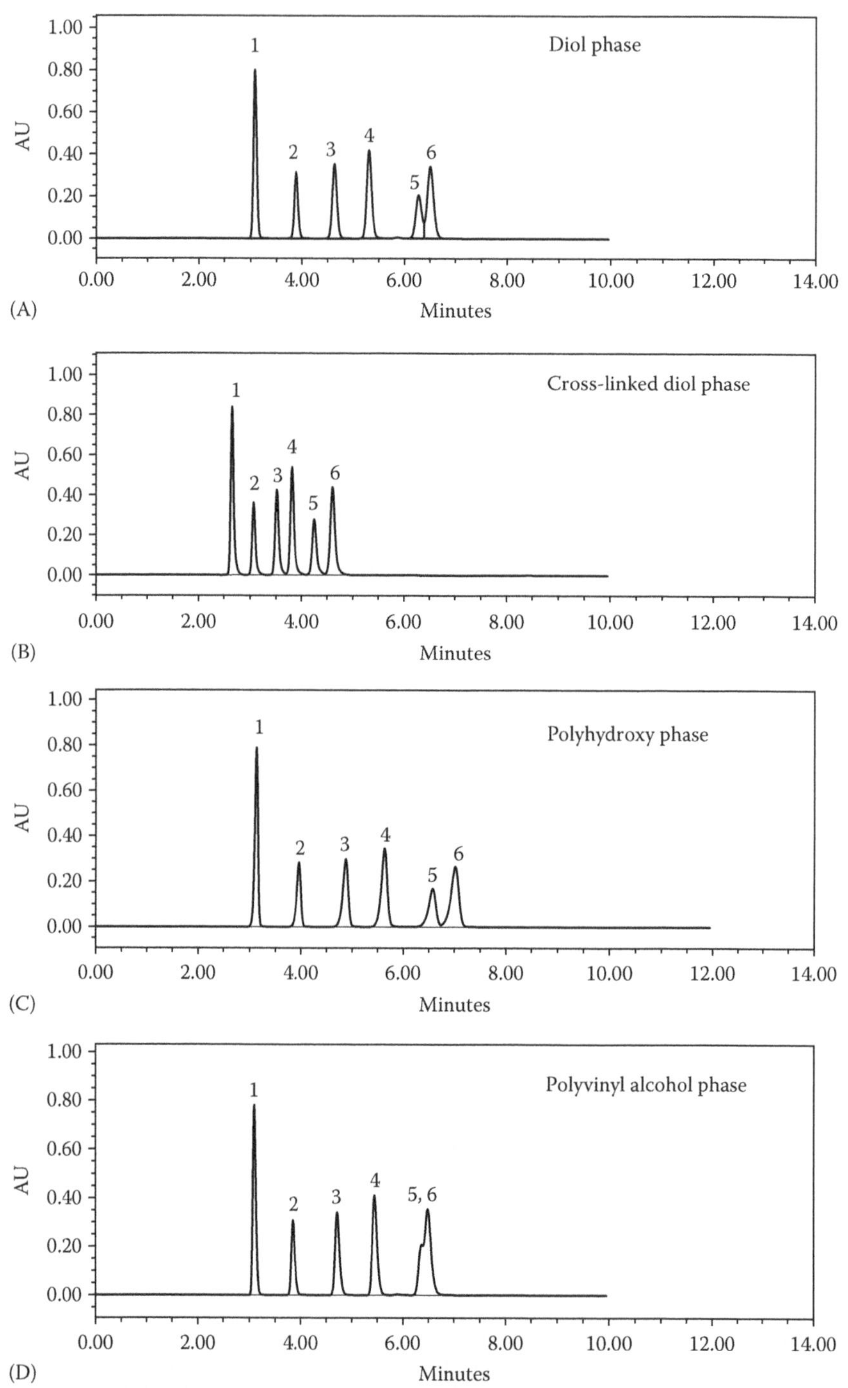

FIGURE 17.2 Separation of neutral model compounds on (A) diol, (B) cross-linked diol, (C) polyhydroxy, and (D) PVA phase. Other conditions are the same as Figure 17.1.

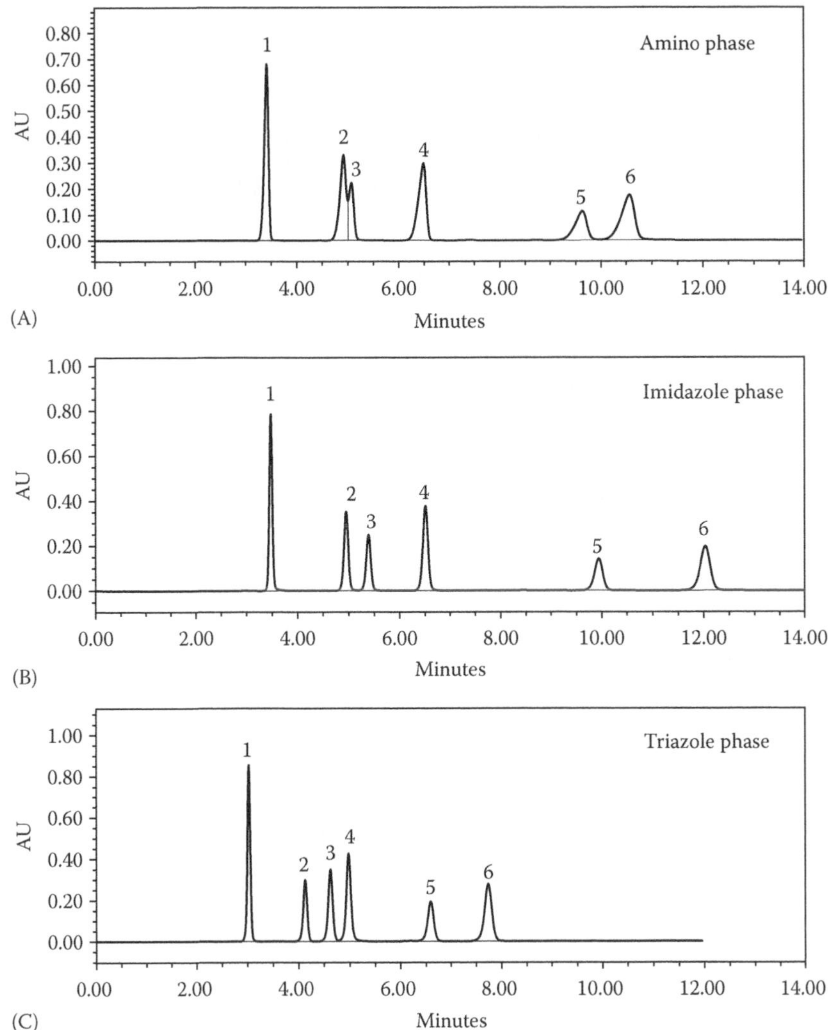

FIGURE 17.3 Separation of neutral model compounds on (A) amino, (B) imidazole, and (C) triazole phase. Other conditions are the same as Figure 17.1.

Figure 17.5 shows the separation of the acidic compounds on the diol, cross-linked diol, polyhydroxy, and PVA phases. The diol, polyhydroxy, and PVA phases display similar selectivity for the acidic compounds with a small difference in the resolution of hippuric acid and α-hydroxyhippuric acid. In contrast, the cross-linked diol phase seems to be much less selective for the acids with two pairs co-eluting, possibly due to insufficient retention.

The separation of the acids on the amino, imidazole and triazole phase is shown in Figure 17.6. The elution order of the acids on the amino phase bears some

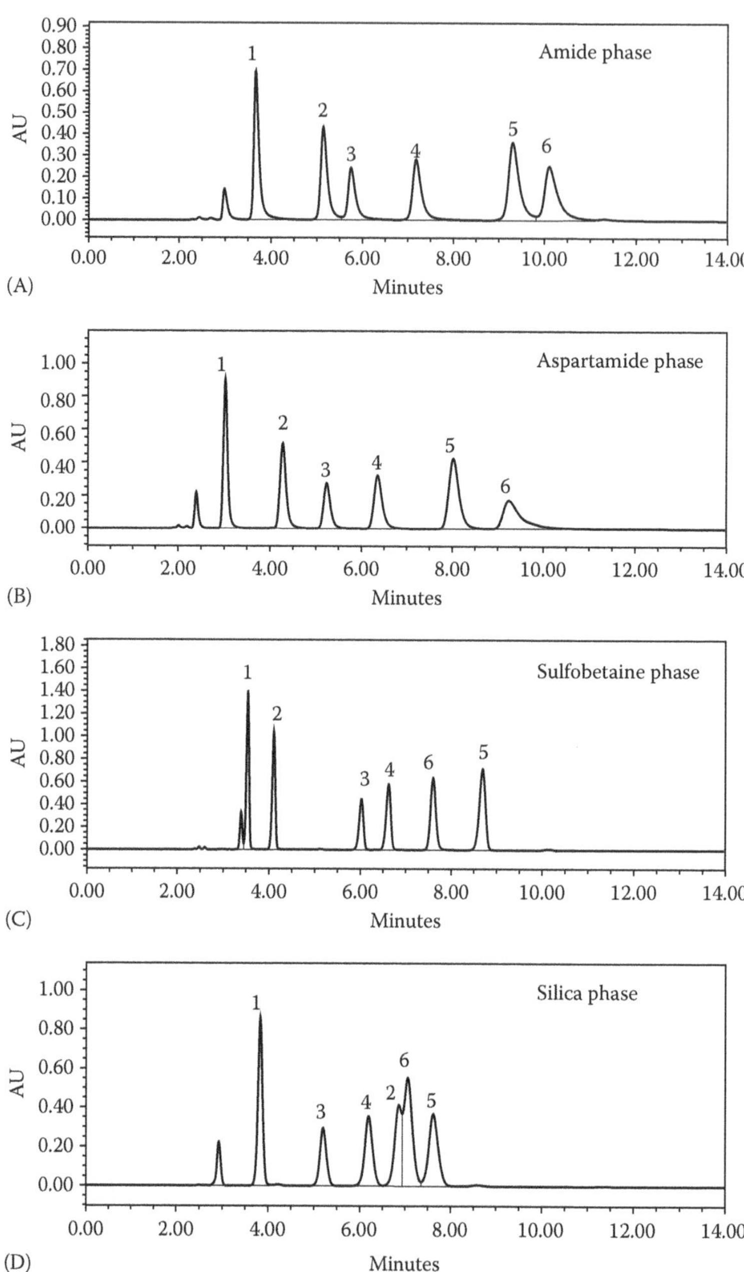

FIGURE 17.4 Separation of acidic model compounds on (A) amide, (B) aspartamide, (C) silica, and (D) sulfobetaine phase. Mobile phase, acetonitrile/water (85/15, v/v) containing 20 mM ammonium acetate. Column dimension, 250 mm × 4.6 mm ID, 5 μm particle size. Column temperature, 30°C. Flow rate, 1.0 mL/min. UV detection at 228 nm. Peak label: (1) salicylic acid, (2) gentisic acid, (3) acetylsalicylic acid, (4) salicyluric acid, (5) hippuric acid, and (6) α-hydroxyhippuric acid.

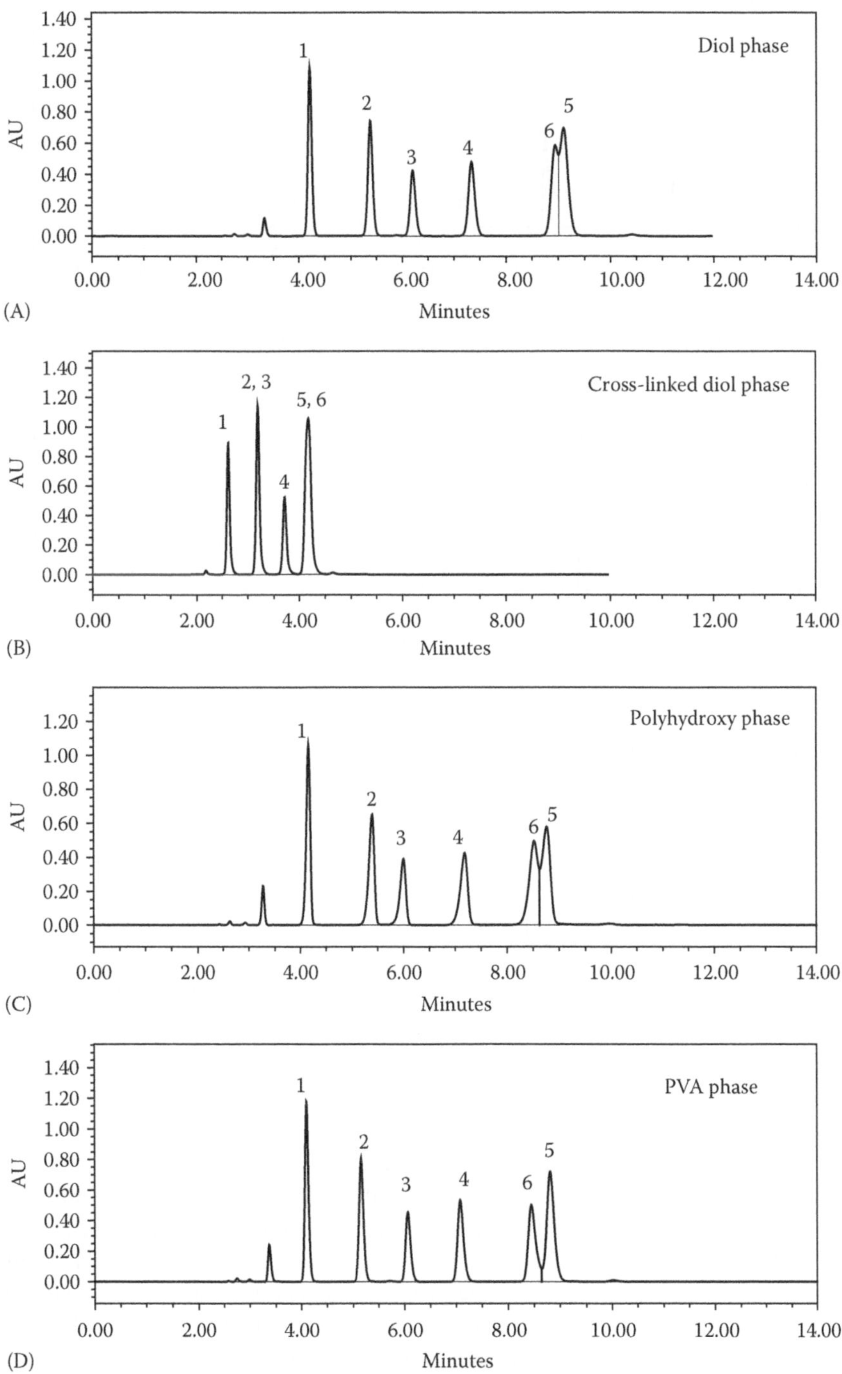

FIGURE 17.5 Separation of acidic model compounds on (A) diol, (B) cross-linked diol, (C) polyhydroxy, and (D) PVA phase. Other conditions are the same as Figure 17.4.

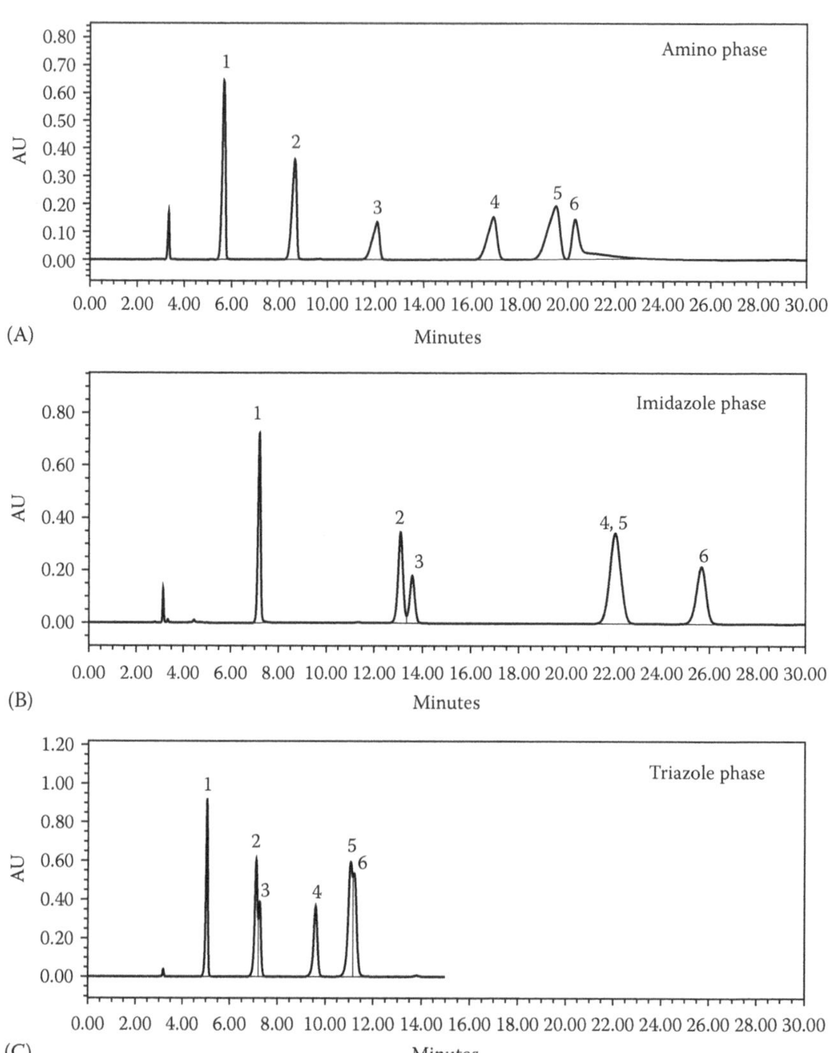

FIGURE 17.6 Separation of acidic model compounds on (A) amino, (B) imidazole, and (C) triazole phase. Other conditions are the same as Figure 17.4.

resemblance to that on the amide and aspartamide phases as shown in Figure 17.4, but α-hydroxyhippuric acid has a significant tailing. The selectivity of the imidazole and triazole phases is very different from the amino phase, for example, gentisic acid and acetylsalicylic acid are well resolved on the amino phase, but only partially resolved on the imidazole and triazole phases. It is also interesting to compare the selectivity of the imidazole and triazole phases. On the imidazole phase, salicyluric acid co-elutes with hippuric acid, but is separated from α-hydroxyhippuric acid. In comparison, salicyluric acid is separated from hippuric acid, but nearly co-elutes with α-hydroxyhippuric acid on the triazole phase. It is important to point out that the acids were negatively charged and the amino and imidazole groups were positively charged in the mobile phase (pH ~ 6); however, it is not clear whether the triazole phase was charged or not. The difference in selectivity among the cationic phases is at least partially related to the electrostatic interactions between the negatively charged solutes and positively charged functional groups.

17.4 RETENTIVITY OF VARIOUS POLAR STATIONARY PHASES IN HILIC

Stronger retention for polar compounds is the primary reason for the increasing popularity of HILIC. In addition to selectivity, retentivity is another factor important to achieve desired resolution according to Equation 17.1 (Snyder et al. 1997):

$$R_s = \frac{1}{4} \cdot \sqrt{N} \cdot \frac{\alpha - 1}{\alpha} \cdot \frac{k'}{1 + k'} \tag{17.1}$$

where

N is the number of theoretical plates
α and k' are the separation and retention factors, respectively

Since HILIC separation is based on hydrophilic partitioning of the solutes between the mobile phase and a water-rich liquid layer immobilized on the surface of packing materials, weak retentivity of the stationary phases can limit the content of the organic solvents in the mobile phase to a narrow range, thus limiting selectivity.

In RPLC, toluene is usually used as the reference compound to evaluate relative retention of different stationary phases. However, there has not been a compound widely accepted as the reference compound to evaluate the retention of the stationary phases in HILIC. Cytosine was used as the reference compound to compare the relative retentivity of various polar phases listed in Table 17.1. The capacity factors of cytosine on the selected polar phases, in the order of increasing retention, are graphically presented in Figure 17.7. It is worth noting that hydrophilic partitioning was likely the primary retention mechanism since cytosine is uncharged under the experimental conditions. The bar chart in Figure 17.7 clearly demonstrates different retentivity of various polar phases in HILIC. The cross-linked diol phase has the least retention for cytosine. The amide and aspartamide phases, on the other hand, have stronger retention. Interestingly, most phases have relatively similar retention for cytosine, such as PVA, triazole, silica, diol, amino sulfobetaine, polyhydroxy, and imidazole phases. The diversity of the chemical structure in these phases does not seem to make a significant difference in the retentivity. This may be a reflection of hydrophilic partitioning as the underlining retention mechanism for these stationary phases as opposed to strong specific interactions.

Salicyluric acid was also used as the reference compound to evaluate the relative retention of the polar phases. As shown in Figure 17.7, there are some similarities, and also noticeable differences in the relative retention of the polar phases when salicyluric acid, instead of cystosine, is used as the reference compound.

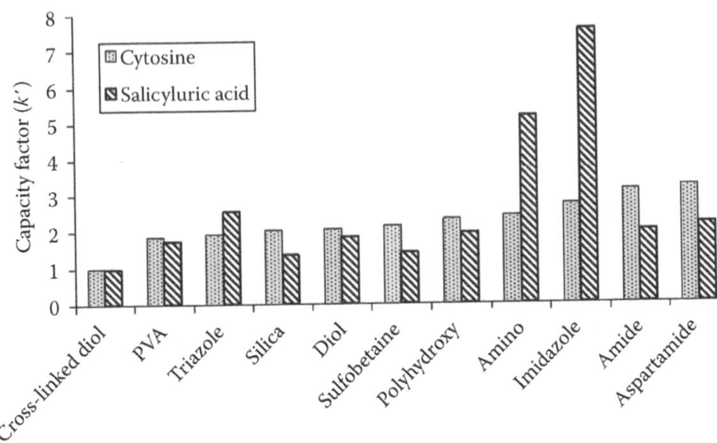

FIGURE 17.7 Capacity factors of cytosine and salicyluric acid on various HILIC phases. Conditions are the same as Figure 17.1 for cytosine and Figure 17.4 for salicyluric acid.

The cross-linked diol phase is the least retentive phase for both cytosine and salicyluric acid. The other neutral phases (e.g., the diol, polyhydroxy, and PVA phases) also show similar retentivity for salicyluric acid. However, the cationic phases, the amino and imidazole phase in particular, have much stronger retention for the acid, which can be attributed to the electrostatic attraction between the negatively charged acid and positively charged functional groups under the experimental conditions. In contrast, the silica, amide, and aspartamide phases seem to have reduced retention for the acid compared to the other phases. All these phases have negative charges on the packing surface, silica from deprotonated silanol groups, amide from the residual silanol groups, and aspartamide from residual negative charges from C-terminus. It is possible that the reduced retention of the acid on these phases is related to electrostatic repulsion of the negatively charged acid. However, there is not sufficient evidence to indicate that the electrostatic repulsion is the only factor contributing to the decrease in retention.

The above examples demonstrate that the electrostatic interactions between the charged stationary phases and solutes are very important to the retention and selectivity in HILIC. The cationic phases (e.g., amino phase) are particularly subjective to the electrostatic interactions. The retention can be strengthened for the negatively charged compounds, but weakened for the positively charged compounds on the cationic phases. The negative charges found in the silica, amide, and aspartamide phases can also induce electrostatic interactions with charged solutions. However, the electrostatic interactions on these phases are not overwhelmingly strong, in comparison to the cationic phase. On the other hand, some neutral phases (e.g., diol, cross-linked diol, polyhydroxy, and PVC phases) are less vulnerable to the electrostatic interactions, thus have relatively similar retentivity regardless of the charge state of the solutes. In RPLC, it is rather straightforward to determine the relative retentivity of the stationary phase based on the hydrophobicity of the functional groups (Snyder et al. 1997, Kazakevich and Lobrutto 2007).

For example, the C_{18} phase has stronger retention than the C_8 or phenyl phase. In HILIC, however, it is difficult to link the relative retentivity to the functional group. Empirical comparison can provide useful guidance for column selection in HILIC method development.

17.5 FACTORS AFFECTING RETENTION AND SELECTIVITY IN HILIC

In addition to stationary phase chemistry, other chromatographic parameters, such as organic solvents, mobile-phase pH, salt concentration, and column temperature can also have direct impact on the retention and selectivity in HILIC. The mobile phase for HILIC separation typically contains organic solvents (e.g., acetonitrile, ethanol, and THF) at high levels (above 60%, v/v), which is needed to effect hydrophilic interaction (Guo and Gaiki 2005, Hemstrom and Irgum 2006). The effect of organic content in the mobile phase on retention is depicted in Figure 17.8, which shows the curves of log k' vs. acetonitrile content in the mobile phase for acetylsalicylic acid and cytosine on four different stationary phases (i.e., amino, amide, silica, and sulfobetaine phases). The retention of the model compounds typically increases with the acetonitrile content in a nonlinear fashion in the range of 65%–95% (v/v). At lower acetonitrile content (<85%), log k' increases almost linearly with the acetonitrile content in the mobile phase for both the neutral and acidic compounds on the selected phases, except acetylsalicylic acid on the amino phase. However, the retention increases much faster at the acetonitrile content above 85%, indicating that the retention is very sensitive to small changes in the organic content. Interestingly, the retention of acetylsalicylic acid on the amino phase remains almost unchanged in the acetonitrile range of 65%–80%, but increases as the acetonitrile content goes above 80%. This observation is likely related to the electrostatic interaction between the negatively charged acid and positively charged amino phase. The dominant retention mechanism might be the electrostatic attraction at low acetonitrile levels. As the acetonitrile content increases, the retention might be more controlled by hydrophilic partitioning, thus more influenced by the acetonitrile content in the mobile phase.

Mobile-phase pH is another chromatographic parameter critical to the retention of ionizable compounds in HILIC, which is subjected to the influence of the mobile-phase pH through ionization. Typically, charged solutes are more retained than uncharged ones in HILIC (Guo and Gaiki 2005, Hemstrom and Irgum 2006). The plots in Figure 17.9 show the variation of the retention time of acetylsalicylic acid and cytosine with the mobile-phase pH in the range of 3.3–6.5 on the amino, amide, silica, and sulfobetaine phases. For acetylsalicylic acid, the retention time is relatively unchanged between pH 6.5 and 4.8. As the mobile-phase pH decreases from 4.8 to 3.3, there is a gradual decrease in retention time on the amide, silica, and sulfobetaine phase, but a large drop in retention time at pH 3.3 on the amino phase. Acetylsalicylic acid has a pK_a ~ 3.5 and is predominantly charged at pH 6.5 and 4.8, but is partially uncharged at pH close to its pK_a through protonation. The uncharged acid has less retention, resulting in a gradual decrease in retention time at lower mobile-phase pH on the amide, silica, and sulfobetaine phases. On the

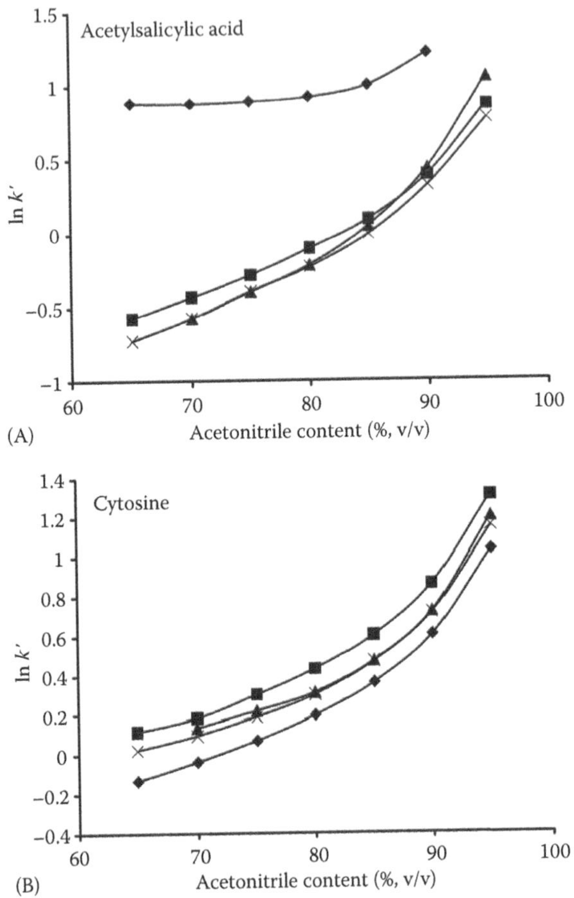

FIGURE 17.8 The plots of log k' vs. acetonitrile content (%) for (A) acetylsalicylic acid and (B) cytosine on (♦) amino, (■) amide, (▲) silica, and (×) sulfobetaine phase. Column temperature 30°C. The mobile phase contains 5 mM ammonium acetate. (From Guo, Y. and Gaiki, S., *J. Chromatogr. A*, 1074, 71, 2005. With permission from Elsevier.)

other hand, the stationary phase can also change its charge states at different pH, thus introducing electrostatic interaction with the charged solutes. The amino phase is positively charged in the pH range from 3.3 to 6.5, which can induce the electrostatic attraction with the negatively charged acid and influence the retention. When acetylsalicylic acid becomes partially charged at pH 3.3, the electrostatic attraction is significantly reduced, thus resulting in a big drop in retention time, as shown in Figure 17.9. In comparison, cytosine has a very different behavior at different mobile-phase pHs. The retention time of cytosine only varies slightly on the silica, amide, and sulfobetaine phases in the pH range 3.3–6.8. On the amino phase, however, the retention time of cytosine decreases slightly from pH 6.5 to 4.8, and drops significantly below pH 4.8. Cytosine has two pK_a values: pK_{a_1} ~ 4.6 and pK_{a_2} ~ 12.2 in water. At the mobile-phase pH below pK_{a_1}, cytosine becomes positively charged,

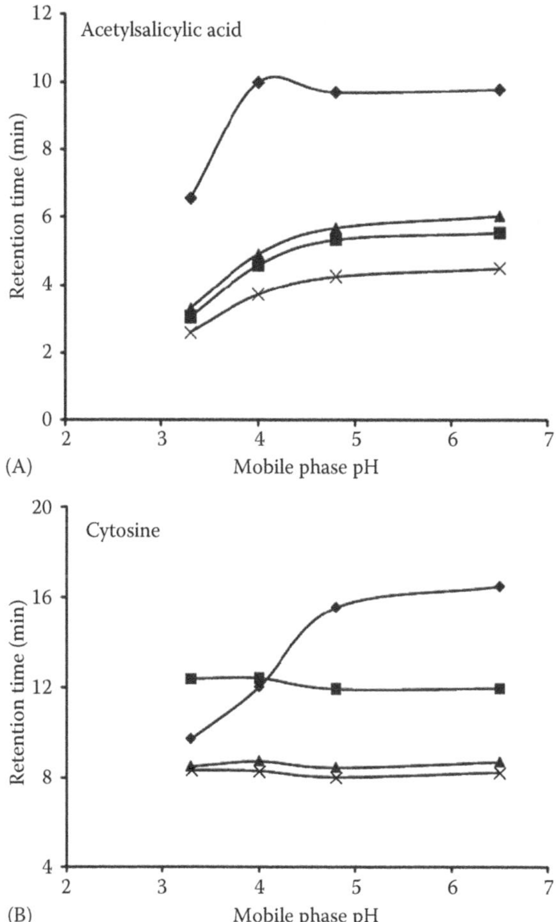

FIGURE 17.9 The effect of mobile-phase pH on the retention of (A) acetylsalicylic acid and (B) cytosine on (♦) amino, (■) amide, (▲) silica, and (×) sulfobetaine phase. Column temperature 30°C. Mobile phase, acetonitrile/water (90/10, v/v) containing 10 mM ammonium formate. Mobile-phase pH is the pH values of ammonium acetate solutions.

inducing electrostatic repulsion from the positively charged amino phase, and leading to reduced retention time below pH 4.8.

The electrostatic interactions between the charged solutes and stationary phases can be modulated by adding buffer salts to the mobile phase. The salt ions can reduce the electrostatic interaction between the charged solutes and stationary phase. Therefore, the salt concentration in the mobile phase can have a direct impact on the retention in HILIC. Figure 17.10 shows the effect of salt concentration on the capacity factor of salicyluric acid and cytosine on the amino, amide, silica, and sulfobetaine phases. On the amino phase, the capacity factor of salicyluric acid decreases drastically as the salt concentration increases from 5 to 40 mM. This is a direct result of reduced electrostatic attraction between

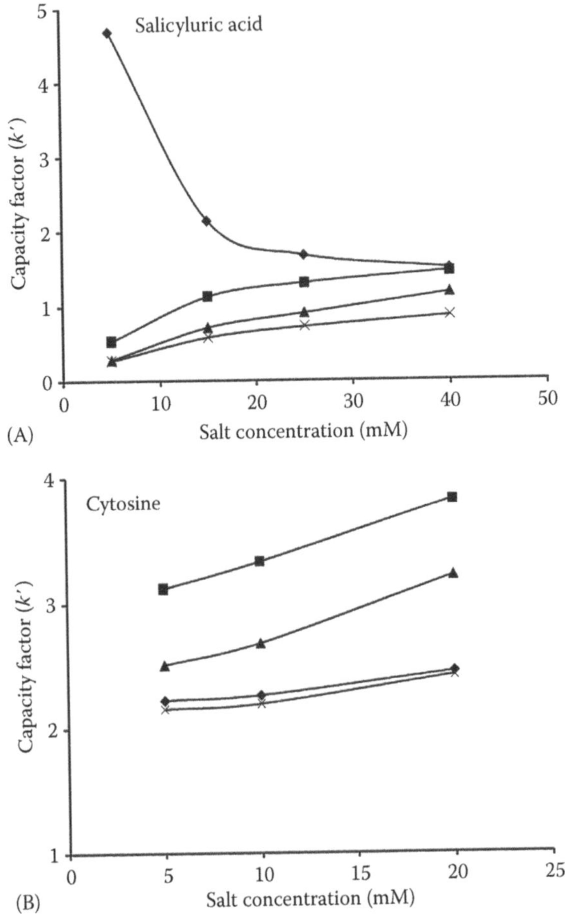

FIGURE 17.10 The effect of salt concentration on the retention of salicyluric acid and cytosine on (♦) amino, (■) amide, (▲) silica, and (×) sulfobetaine phase. Column temperature 30°C. (A) The mobile phase for salicyluric acid is acetonitrile/water (80/20, v/v) containing 5–40 mM ammonium acetate. (B) The mobile phase for cytosine is acetonitrile/water (85/15) containing 5–20 mM ammonium acetate.

the negatively charged acid and positively charged amino groups. In contrast, a small, but significant increase in the retention of salicyluric acid is observed on the silica, amide, and sulfobetaine phases as the salt concentration increases, especially from 5 to 15 mM. This could be due to reduced electrostatic repulsion between the negatively charged acid and the negative charges on the silica and amide phases. Furthermore, the increase in retention with the salt concentration on the sulfobetaine phase indicates that there is not any significant electrostatic attraction between the negatively charged acid and positive quaternary amine moiety. Additionally, a similar trend is also observed for cytosine on all the four phases, as shown in Figure 17.10. The salt concentration seems to have more effect on the amide and silica phases than on the amino and sulfobetaine phases.

The electrostatic interaction cannot explain the effect of salt concentration on the retention for cytosine, which is not charged under the experimental conditions. The increase in retention may be related to some changes of the stationary phase induced by the salt ions. It is possible that higher salt concentrations might increase the polarity of the stagnant liquid layer on the stationary phase; however, the exact mechanism for increased retention is not clear.

In addition to retention, selectivity can also be affected by the salt concentration. Figure 17.11 shows the separation of the acidic model compounds on the silica phase with 10 and 20 mM ammonium acetate in the mobile phase. As expected, the retention of the acids increases at 20 mM salt concentration. More interestingly, there is also a change in the selectivity for acetylsalicylic acid, salicyluric acid, and α-hippuric acid. The three acids elute close to each other at 10 mM ammonium acetate, but are well separated at 20 mM. This is another piece of evidence that salt concentration might induce changes in the stationary phase, thus leading to selectivity change.

Column temperature is another parameter that can alter retention and selectivity in RPLC, and can have similar effect in HILIC. Increasing temperature generally

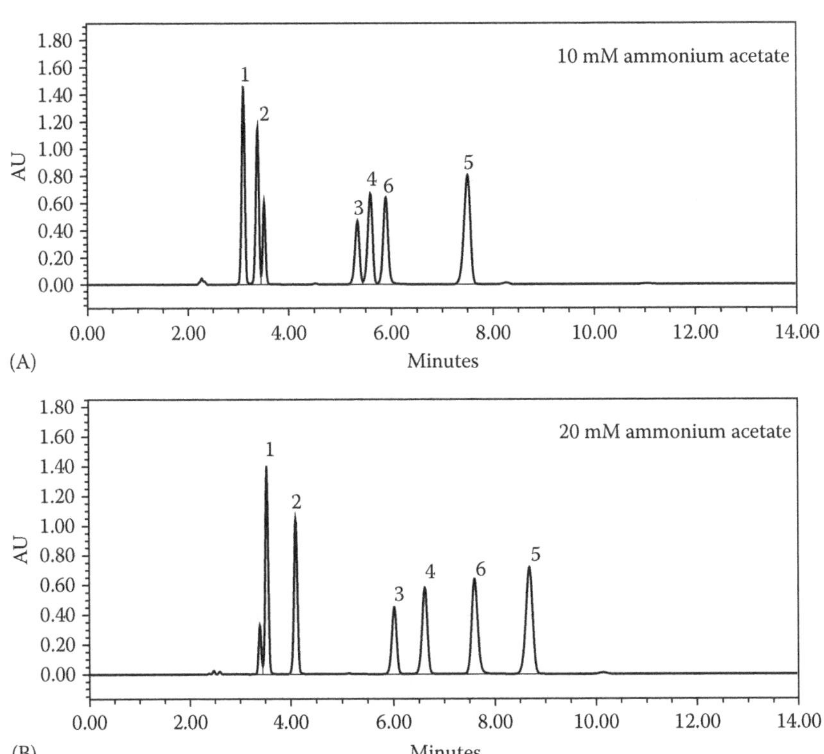

FIGURE 17.11 Separation of the acidic model compounds on the silica phase. Mobile phase, acetonitrile/water (85/15, v/v) containing (A) 10 and (B) 20 mM ammonium acetate. Column dimension, 250 mm × 4.6 mm ID, 5 μm particle size. Column temperature, 30°C. Flow rate, 1.0 mL/min. UV detection at 228 nm. Peak label: (1) salicylic acid, (2) gentisic acid, (3) acetylsalicylic acid, (4) salicyluric acid, (5) hippuric acid, and (6) α-hydroxyhippuric acid.

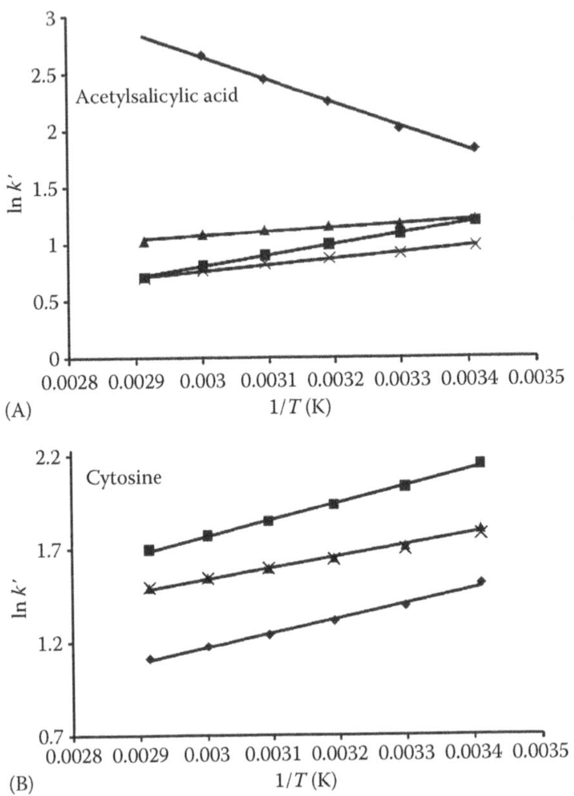

FIGURE 17.12 Effect of column temperature on the retention of (A) acetylsalicylic acid and (B) cytosine on (♦) amino, (■) amide, (▲) silica, and (×) sulfobetaine phase. Mobile phase, acetonitrile/water (90/10, v/v) containing 10 mM ammonium acetate. (From Guo, Y. and Gaiki, S., *J. Chromtogra. A*, 1074, 71, 2005. With permission of Elsevier.)

leads to reduced retention, and can often improve peak shape due to faster diffusion at higher temperatures. Figure 17.12 shows the van't Hoff plots for acetylsalicylic acid and cytosine on the amino, amide, silica, and sulfobetaine phases. A general trend of decreasing retention at higher temperature is observed on the amide, sulfobetaine and silica phases, and the amide phase seems to be more sensitive to the temperature change than the other phases judging from the slope of the van't Hoff plots. On the amino phase, however, an increase in retention is observed for acetylsalicylic acid with increasing temperatures. Similar retention increase at higher temperatures has also been reported for basic compounds on the silica phase (Hao et al. 2007). The amino phase is known to have significant electrostatic interactions with the negatively charged acid at low salt concentrations. The non-classic van't Hoff behavior may be associated with the electrostatic interactions.

The retention behavior of the acidic compound at different temperatures has also been found to change with the salt concentration in the mobile phase. Figure 17.13

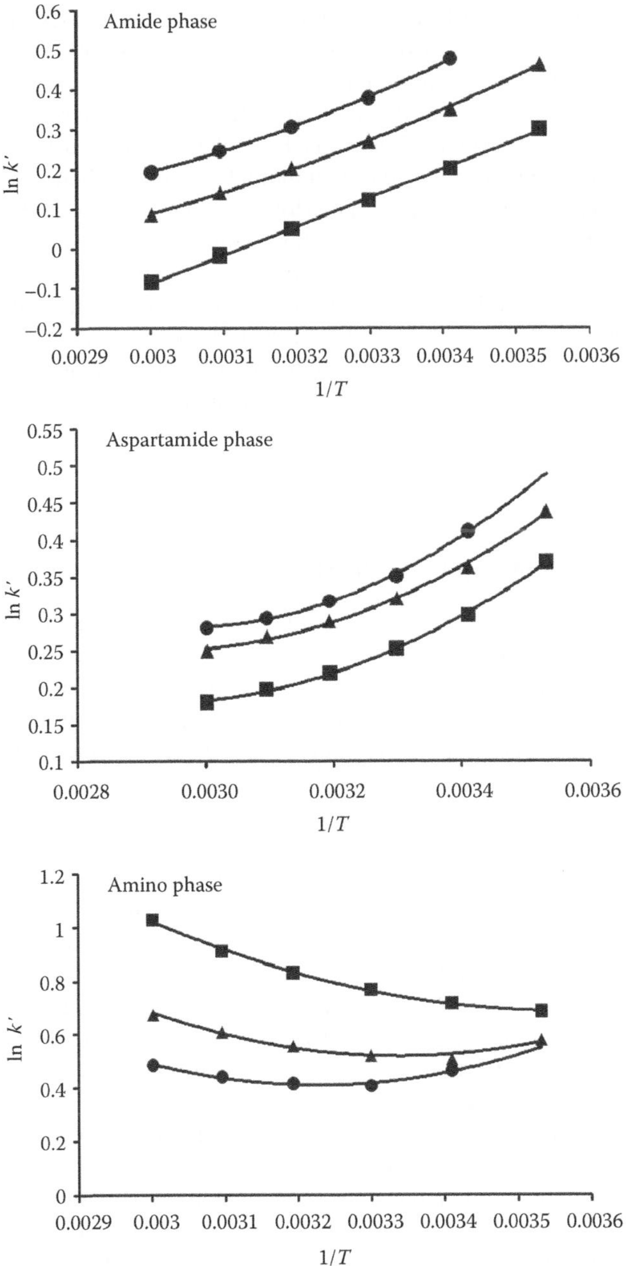

FIGURE 17.13 Effect of salt concentration on van't Hoff plots on amide, aspartamide, and amino phase. Column temperature, 30°C. Mobile phase, acetonitrile/water (80/20, v/v) containing (•) 40 (▲) 25, and (■) 15 mM ammonium acetate.

shows the van't Hoff plots of salicyluric acid on the amino, amide, and aspartamide phases with 15, 25, and 40 mM ammonium acetate in the mobile phase. On the amide phase, a linear van't Hoff plot is observed for salicyluric acid at 15 mM, but a small curvature in the van't Hoff plots appears at 25 and 40 mM ammonium acetate. In comparison, the curvature in the van't Hoff plots is more pronounced on the aspartamide phase even at 15 mM, but the curvature remains the same at higher salt concentrations. The deviation of the van't Hoff plots from linearity indicates that the retention might be under the control of multiple forces in addition to hydrophilic interaction. On the amino phase, the van't Hoff plot for salicyluric acid also indicates an increase in retention at higher temperatures similar to acetylsalicylic acid, but a small curvature is observed at low temperatures even at 15 mM ammonium acetate. At higher salt concentration (25 and 40 mM), the curvature in the van't Hoff plots is more pronounced. The van't Hoff plot at 40 mM curves upward in the low temperature range, indicating that the acid returns to "normal" temperature behavior possibly due to sufficient suppression of the electrostatic interaction between the acid and amino phase at high salt concentration.

17.6 SEPARATION OF POSITIONAL ISOMERS IN HILIC

The separation of positional isomers is of great interest to analytical chemists since many synthetic impurities and metabolites of drugs are positional isomers, and is also very challenging due to the similarity of the positional isomers not only in structure, but also in chromatographic properties. Reversed-phase and normal-phase chromatography have been applied to the separation of position isomers with certain success (Kazakevich and Lobrutto 2007). However, HILIC has not been well established as a viable technique for the separation of positional isomers.

Caffeine metabolites include some mono-methylated, dimethylated xanthines, and dimethylated uric acids. Caffeine and its metabolites (including 3-mono-methylated xanthines, 3-dimethylated xanthines, and 3-dimethylated uric acids) were separated on the amide and silica phases. As shown in Figure 17.14, an interesting feature of HILIC separation is that the positional isomers eluted as clusters are based on the degree of methylation, i.e., trimethylated xanthine (caffeine) eluting first, then dimethylated xanthines, followed by mono-methylated xanthines. Dimethylated uric acids are strongly retained in HILIC and a gradient was needed to elute them in a reasonable time. The amide phase shows nearly baseline separation of all the positional isomers. However, only dimethylated uric acid isomers were resolved on the silica phase. The isomers of mono- and dimethylated xanthines essentially co-eluted under the same conditions. Comparing to reversed-phase separation, HILIC separation has totally different selectivity in terms of eluting order, particularly for dimethylated uric acids (Safranow and Machoy 2005).

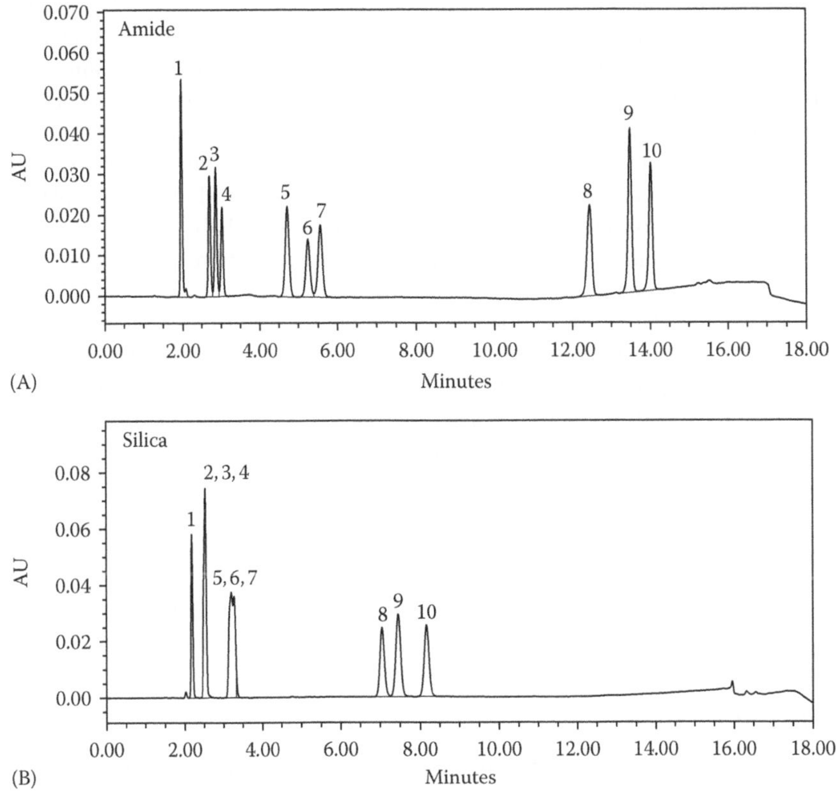

FIGURE 17.14 Separation of positional isomers of mono-methylated, dimethylated xanthines, and dimethylated uric acids on (A) amide and (B) silica phase. Column dimension, 150 mm × 4.6 mm ID, 3 μm particle size. Column temperature, 30°C. Flow rate, 1.2 mL/min. Mobile phase, acetonitrile/water (90/10, v/v) containing 10 mM ammonium acetate. UV detection at 240 nm. Peak labels: (1) caffeine, (2) 1,3-dimethylxanthine, (3) 1,7-dimethylxanthine, (4) 3,7-dimethylxanthine, (5) 1-methylxanthine, (6) 3-methylxanthine, (7) 7-methylxanthine, (8) 1,3-dimethyluric acid, (9) 1,7-dimethyluric acid, (10) 3,7-dimethyluric acid.

17.7 CONCLUSIONS

More and more polar phases with diverse chemistry become available for HILIC applications. Most polar phases have been shown to have very different selectivity and retentivity for polar compounds. The cross-linked diol phase has been found to have the weakest retentivity, and the amide and aspartamide phases have stronger retentivity than other phases. Most polar phases show very different selectivity for the polar compounds, but the diol, polyhydroxy, and PVA phases seem to have rather similar selectivity. The retention and selectivity of some phases with ionizable functional groups (e.g., the amino and silica phase) is highly subjective to electrostatic interactions with charged solutes. It is important to point out that there has been no correlation found between the structure of the functional groups and the retentivity or selectivity. In addition, the retentivity and selectivity are also highly influenced by the chromatographic parameters, such as organic solvent, mobile-phase pH, salt concentration, and column temperature. The retention of the polar solutes is very sensitive to the change in the content of organic solvent in the mobile phase, especially when the organic solvent content is above 85% (v/v). Mobile-phase pH can influence the retention and selectivity of polar solutes through the ionization of both the ionizable compounds and stationary phases. Comparing to reversed-phase separation, the salt concentration seems to have a more significant effect on the retention and selectivity in HILIC. In addition to attenuating the electrostatic interaction, the salt concentration seems to be related to the retention of even neutral compounds. However, the mechanism by which the salt exerts its effect is not fully understood. The temperature study on the amide and aspartamide phases indicates that the retention on the polar phases might be under multiple mechanisms depending on the salt concentration.

ACKNOWLEDGMENT

The author would like to recognize the contribution of Sheetal Gaiki, Emma Huang, and other colleagues for many of the findings reported in this chapter.

REFERENCES

Alpert, A.J. 1990. Hydrophilic-interaction chromatography for the separation of peptides, nucleic acids and other polar compounds. *J. Chromatogr.*, 499: 177–196.

Dejaegher, B. and Heyden, Y.V. 2010. HILIC methods in pharmaceutical analysis. *J. Sep. Sci.*, 33: 698–715.

Dejaegher, B., Mangelings, D., and Heyden, Y.V. 2008. Method development for HILIC assays. *J. Sep. Sci.*, 31: 1438–1448.

Guo, Y. and Gaiki, S. 2005. Retention behavior of small polar compounds on polar stationary phases in hydrophilic interaction chromatography. *J. Chromatogr. A*, 1074: 71–80.

Guo, Y. and Huang, A.H. 2003. A HILIC method for the analysis of tromethamine as the counter ion in an investigational pharmaceutical salt. *J. Pharm. Biomed. Anal.*, 31: 1191–1201.

Guo, Y., Srinivasan, S., and Gaiki, S. 2007. Investigating the effect of chromatographic conditions on retention of organic acids in hydrophilic interaction chromatography using a design of experiment. *Chromatographia*, 66: 223–229.

Hao, Z., Lu, Y., Xiao, B., and Weng, N. 2007. Separation of amino acids, peptides and corresponding Amadori compounds on a silica column at elevated temperature. *J. Chromatogr. A*, 1147: 165–171.

Hao, Z., Xiao, B., and Weng, N. 2008. Impact of column temperature and mobile phase components on selectivity of hydrophilic interaction chromatography (HILIC). *J. Sep. Sci.*, 31: 1449–1464.

Hemstrom, P. and Irgum, K. 2006. Hydrophilic interaction chromatography. *J. Sep. Sci.*, 29: 1784–1821.

Ikegami, T., Tomomatsu, K., Takubo, H., Horie, K., and Tanaka, N. 2008. Separation efficiencies in hydrophilic interaction chromatography. *J. Chromatogr. A*, 1184: 474–503.

Jandera, P. 2008. Stationary phases for hydrophilic interaction chromatography, their characterization and implementation into multidimensional chromatography concepts. *J. Sep. Sci.*, 31: 1421–1437.

Jian, W., Edom, R.W., Xu, Y., and Weng, N. 2010. Recent advances in application of hydrophilic interaction chromatography for quantitative bioanalysis. *J. Sep. Sci.*, 33: 681–697.

Jiang, W. 2003. *Zwitterionic Separation Materials for Liquid Chromatography and Capillary Electrophoresis*, PhD Thesis, Umea, Sweden: Umea University.

Jiang, W., Fischer, G., Girmay, Y., and Irgum, K. 2006. Zwitterionic stationary phase with covalently bonded phosphorylcholine type polymer grafts and its applicability to separation of peptides in the hydrophilic interaction chromatography mode. *J. Chromatogr. A*, 1127: 82–91.

Kazakevich, Y. and Lobrutto, R. Eds. 2007. *HPLC for Pharmaceutical Scientists*. Hoboken, NJ: John Wiley & Sons, Inc.

Li, R.P. and Huang, J.X. 2004. Chromatographic behavior of epirubicin and its analogues on high-purity silica in hydrophilic interaction chromatography. *J. Chromatogr. A*, 1041: 163–169.

Linden, J.C. and Lawhead, C.L. 1975. Liquid chromatography of saccharides. *J. Chromatogr.*, 105: 125–133.

McCalley, D.V. 2008. Evaluation of the properties of a superficially porous silica stationary phase in hydrophilic interaction chromatography. *J. Chromatogr. A*, 1193: 85–91.

McCalley, D.V. and Neue, U.D. 2008. Estimation of the extent of the water-rich layer associated with the silica surface in hydrophilic interaction chromatography. *J. Chromatogr. A*, 1192: 225–229.

Olsen, B.A. 2001. Hydrophilic interaction chromatography using amino and silica columns for determination of polar pharmaceuticals and impurities. *J. Chromatogr. A*, 913: 113–122.

Palmer, J.K. 1975. A versatile system for sugar analysis via liquid chromatography. *Anal. Lett.*, 8: 215–224.

Safranow, K. and Machoy, Z. 2005. Simultaneous determination of 16 purine derivatives in urinary calculi by gradient reversed-phase high-performance liquid chromatography with UV detection. *J. Chromatogr. B*, 819: 229–235.

Snyder, L.R., Kirkland, J.J., and Glajch, J.L. 1997. *Practical HPLC Method Development*, 2nd Ed. Hoboken, NJ: John Wiley & Sons, Inc.

Wen, N. 2003. Bioanalytical liquid chromatography tandem mass spectrometry methods on underivatized silica column with aqueous/organic mobile phases. *J. Chromatogr. B*, 796: 209–224.

Expolore Luna HILIC, www.phenomenex.com, Phenomenex, 2007.

18 HILIC-MS/MS for the Determination of Polar Bioactive Substances: Representative Applications in the Fields of Pharmacokinetics and Metabolic Profiling

Yannis L. Loukas and Yannis Dotsikas

CONTENTS

18.1 INTRODUCTION

HILIC or hydrophilic interaction liquid chromatography (originally called hydrophilic interaction chromatography) is a liquid chromatographic technique for the separation of polar and hydrophilic compounds. The expression "aqueous normal phase" is another term that sometimes is used for this technique. Thus, in HILIC, we should have a column with a hydrophilic stationary phase and a solvent (eluent) composed of water, buffer, and a high concentration of a water-miscible organic solvent like acetonitrile or methanol. Typically, in HILIC applications, the mobile phase consists of acetonitrile at a concentration between 50% and 95% in an aqueous buffer such as ammonium formate or ammonium acetate. These buffers are soluble in organic solvents and are volatile, compatible with mass spectrometry (MS) or evaporative light scattering (ELSD) detectors that are used mostly in HILIC applications.

In HILIC, compounds elute in the reverse order to that of reversed-phase liquid chromatography (RPLC). A compound that elutes first in an RPLC column should have the highest retention in HILIC, and vice versa. Hydrophilic compounds are problematic to separate in RPLC, such as, acids, bases, ions, sugars, and other charged and neutral compounds, could have a better and easier separation in HILIC. The applications of HILIC have rapidly increased in the last few years and have gained increasing popularity, especially in the separation of small organic, polar, and hydrophilic molecules in the area of pharmaceutics (pharmacokinetics, bioequivalence studies), metabolomics, biomarkers discovery, pesticides, etc.

Mass spectrometry is one of the most popular detectors of HPLC methods today. Electrospray ionization (ESI) sensitivity in an HILIC-MS method could increase dramatically (10–100 times) compared to that in an RPLC-MS method due to the higher content of organic solvent used in the mobile phase, which lowers surface tension, thereby simplifying droplet formation during the ESI process and significantly improving the formation of ions in the gas phase. Moreover, when solid-phase extraction (SPE) is used during sample preparation, the analytes that are eluted with solvents compatible with HILIC columns would allow direct injections into MS without any evaporation and reconstitution, simplifying the whole procedure.

18.2 APPLICATION OF HILIC-MS/MS IN THE FIELD OF PHARMACOKINETICS–BIOEQUIVALENCE STUDIES

Pharmaceutical drug discovery and development has been extremely benefited from the use of high-performance liquid chromatography (HPLC) coupled with triple quadrupole detectors (MS/MS). A search of the keywords LC-MS/MS in scientific databases produces more than 20,000 records showing the applicability of this technique. Most of these studies include two main parts: (1) sample preparation and (2) the development and validation of the LC-MS method. HILIC offers advantages to both parts. During sample preparation, the most commonly used procedures are liquid–liquid extraction (LLE), solid-phase extraction (SPE), and protein precipitation (PP). Despite the wide application of LLE in plasma sample preparation, this procedure is hampered by time-consuming steps, along with decreased sensitivity in some methods. The direct injection of LLE overcomes both drawbacks and offers an alternative for plasma samples analysis. Moreover, a typical SPE protocol involves labor-intensive and time-consuming steps of column conditioning, sample loading, washing, elution, evaporation, and final reconstitution. It has been estimated that about half of the sample preparation time is consumed during sample evaporation and reconstitution.[1] However, the last two steps might be eliminated for compounds with certain physicochemical properties by the replacement of the usually employed RPLC methods with HILIC. As far as the PP procedure is concerned, successful combinations with HILIC-MS/MS have been previously reported for the quantification of several drug candidates in biological matrices.[2–4] However, all the reported methods involved mixing plasma samples with the crash solvent (most often acetonitrile), followed by the time-consuming steps of centrifugation, transfer of the supernatant to new vials or plates, as well as in some cases, evaporation and reconstitution. The applicability of an HILIC procedure could be summarized in the following examples.

18.3 BIOEQUIVALENCE STUDY OF CARVEDILOL

In the pharmacokinetic study of carvedilol, using cisapride as an internal standard, HILIC-MS/MS with an Atlantis HILIC Silica Column was applied.[5,6] The multiple reaction monitoring (MRM) mode was employed for the quantification: m/z 407.2 → 99.9 for carvedilol and m/z 466.1 → 183.8 for cisapride as shown in Figure 18.1. As the mobile phase was a mixture of acetonitrile–ammonium formate (50 mM, pH 4.5) (90:10, v/v), the higher organic content resulted in sensitivity improvement via the enhancement of ionization efficiency. Because of the higher sensitivity of the HILIC–MS/MS method compared to that of RPLC-MS/MS, the plasma sample volume (50 μL) used in this study was smaller than that (200 μL) in RPLC-MS/MS[6] with the same lower limit of quantitation (LLOQ) (0.1 ng/mL). The described method was successfully applied to the bioequivalence study of orally administered carvedilol after a single oral dose of carvedilol (25 mg tablet).

18.4 PHARMACOKINETICS STUDIES OF DONEPEZIL, LORATADINE, AND CETIRIZINE

Another example describes the behavior of three pharmaceutical molecules of medium polarity—Donepezil (DNP), Loratadine (LOR), and Cetirizine (CTZ) as shown in Figure 18.2, and the superior applicability of HILIC-MS/MS for quantitation during their bioequivalence studies.[7] The MRM was performed at m/z 380.6 → 91.2 for DNP, 389.0 → 201.2 for CTZ, and 383.3 → 337.2 for LOR. DNP, CTZ, and LOR solutions in acrylonitrile (ACN) were injected in triplicate onto C18 and silica YMC analytical columns under various mobile-phase compositions. In particular, the mobile phase consisted of an ACN/formic acid 10 mM mixture in ratios ranging from 95% ACN to 90% aqueous buffer (v/v). The ACN percentage of the mobile-phase composition using either a C18 or a YMC HILIC silica column resulted in the following results: at lower ACN concentrations (20%–40%), the C18 column revealed much higher retention than the silica column for CTZ and LOR, while similar results were obtained for DNP as shown in Figure 18.3. The retention

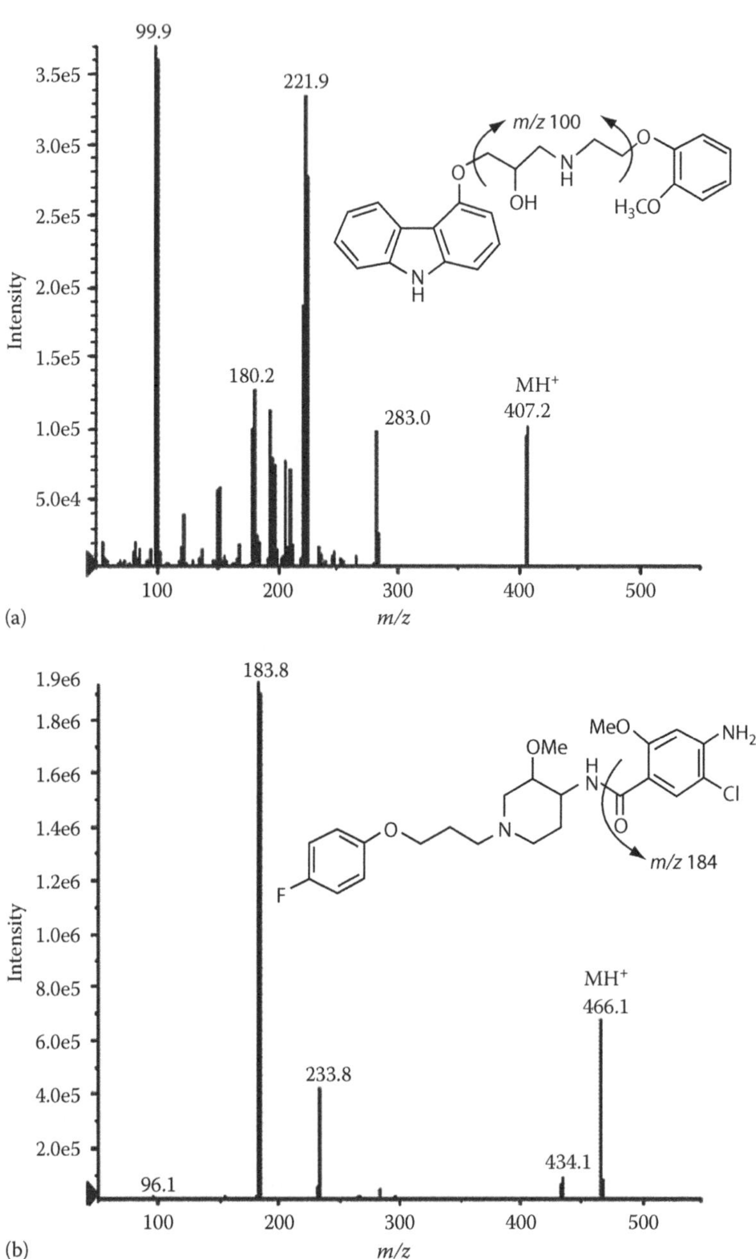

FIGURE 18.1 Product ion mass spectra of (a) carvedilol and (b) cisapride (IS). (From Jeong, D.W. et al., *J. Pharm. Biomed. Anal.*, 44, 547, 2007. With permission from Elsevier.)

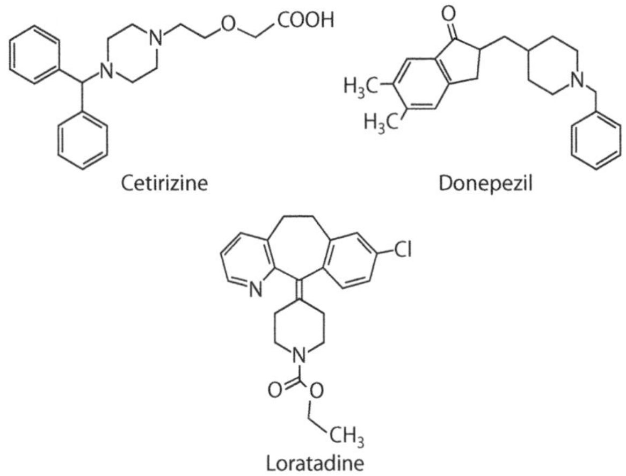

Cetirizine

Donepezil

Loratadine

FIGURE 18.2 Chemical structures of donepezil, loratadine, and cetirizine. (From Apostolou, C. et al., *Biomed. Chromatogr.*, 22, 1393, 2008. With permission from John Wiley and Sons.)

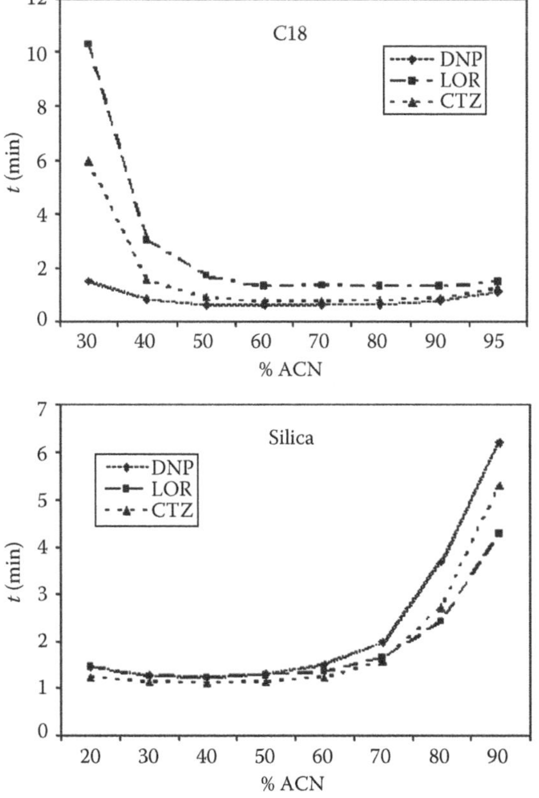

FIGURE 18.3 Retention time vs. ACN concentration in the mobile phase for C18 and silica columns. (From Apostolou, C. et al., *Biomed. Chromatogr.*, 22, 1393, 2008. With permission from John Wiley and Sons.)

mechanism under these conditions favors the employment of the C18 column as silica only presents a pseudo-reversed-phase profile.[8] However, the lower sensitivity and incompatibility with ESI-MS discourage the employment of such chromatographic conditions. At medium ACN concentrations (50%–70%), silica presented a slightly higher retention than C18 for all three compounds. A relatively lower retention at the silica column under these conditions was due to the mixed modes of separation and the complicated retention mechanisms. However, when the ACN concentration was higher than 70%, all three compounds were retained much more efficiently with silica than with the C18 column. The retention mechanism with the silica column was mainly hydrophilic interactions, while C18 column retention was mainly the interactions of the analytes with the remaining uncapped silanol bases of the packing material. The retention times from C18 were similar to those produced from silica with a medium ACN concentration, however, the increased sensitivity favored the application of HILIC conditions of the LC-MS/MS analysis for all three compounds. In particular, ACN concentrations higher than 80% in the mobile phase were proven to be suitable for efficient retention under HILIC mechanism. As a result, in all HILIC methods developed, the mobile phase consisted of at least 80%–85% ACN, hydrophilic interactions being the main mechanism governing the retention of all compounds. HILIC-MS/MS was superior compared

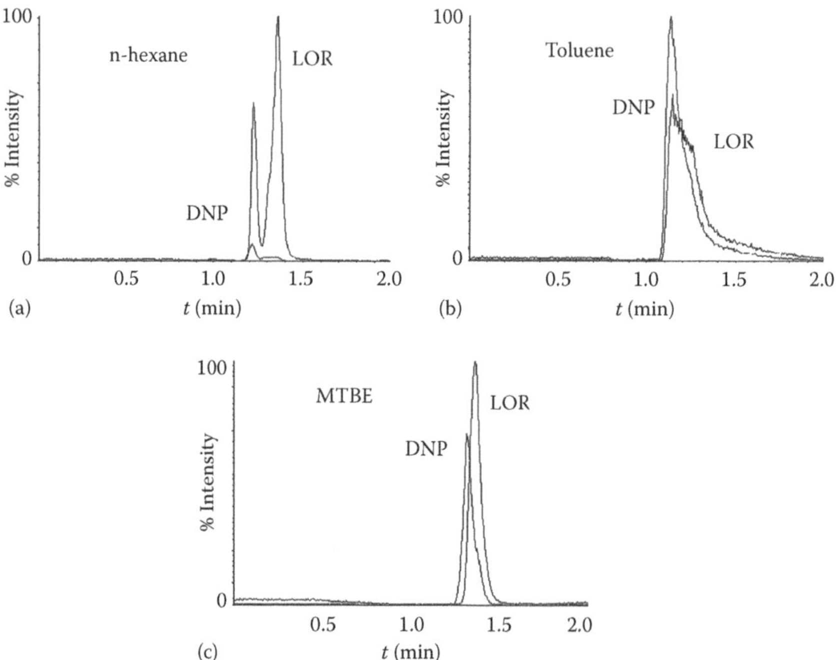

FIGURE 18.4 LLE organic solvent optimization for direct injection on the silica column. (From Apostolou, C. et al., *Biomed. Chromatogr.*, 22, 1393, 2008. With permission from John Wiley and Sons.)

to RPLC during the LC-MS/MS development phase of the above three molecules. Furthermore, the results of direct injections to MS/MS after LLE, SPE, and PP are shown in Figures 18.4 through 18.7. As a result, HILIC is proven to be a highly attractive alternative and complementary to the RPLC technique for pharmaceuticals of medium to high polarity.

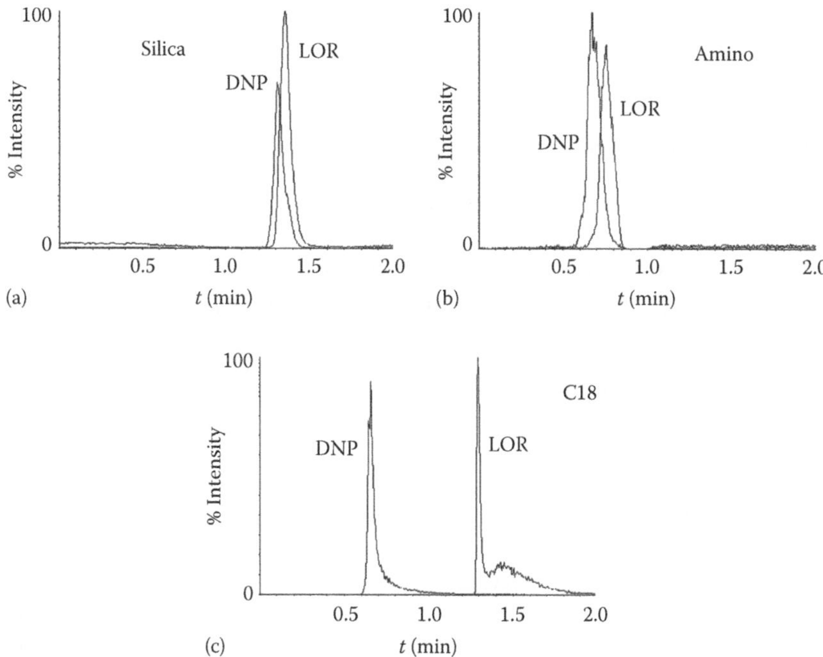

FIGURE 18.5 Column selection for direct injection of LLE with methyl butyl ether (MTBE) as the organic solvent. (From Apostolou, C. et al., *Biomed. Chromatogr.*, 22, 1393, 2008. With permission from John Wiley and Sons.)

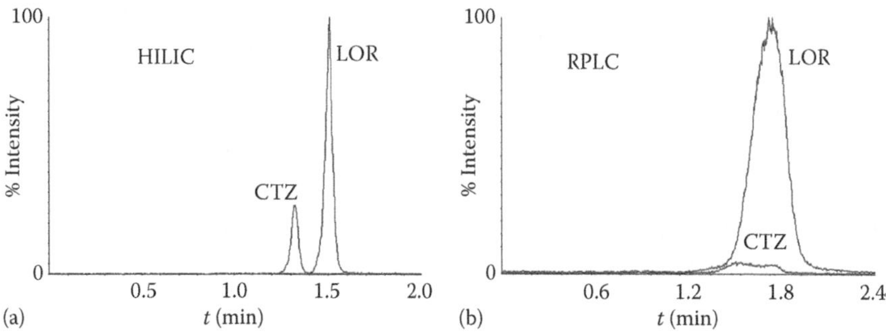

FIGURE 18.6 HILIC and RPLC chromatograms comparison for CTZ determination employing direct injection of SPE. (From Apostolou, C. et al., *Biomed. Chromatogr.*, 22, 1393, 2008. With permission from John Wiley and Sons.)

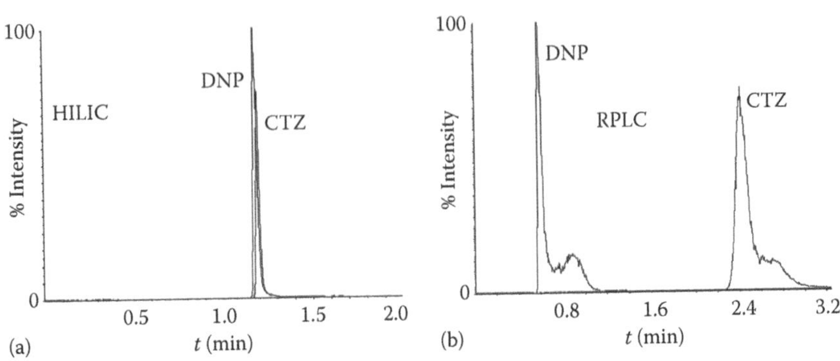

FIGURE 18.7 HILIC and RPLC chromatograms comparison for CTZ determination employing direct injection of PP or PPT. (From Apostolou, C. et al., *Biomed. Chromatogr.*, 22, 1393, 2008. With permission from John Wiley and Sons.)

18.5 DETERMINATION OF GABAPENTIN IN HUMAN PLASMA USING HILIC WITH TANDEM MASS SPECTROMETRY

Gabapentin, 2-(aminomethyl-1-cyclohexyl)acetic acid, was originally developed as a structural analogue of the inhibitory neurotransmitter–aminobutyric acid (GABA) to reduce spinal reflex during the treatment of spasticity and was later found to have anticonvulsant activity in various seizure models.[9] Gabapentin is an amino acid and lacks a chromophore that would allow trace level analysis by absorption spectrophotometry. Moreover, HPLC with UV[10] or fluorescence[11] detection, capillary electrophoresis with UV/fluorescence detection,[12] and gas chromatography (GC) with flame ionization[13] and mass spectrometry (MS) detection[14] have been used, but these methods require derivatization and protein precipitation. The described method presents the successful application of HILIC-MS/MS for the determination of polar gabapentin using metformin as an internal standard. The zwitterionic characteristic of gabapentin makes it extremely difficult to be retained in RP-HPLC[15] and to be extracted from biological samples. The HILIC method with a silica column and a low-aqueous–high-organic mobile phase is used for the retention of polar gabapentin and metformin. By increasing the content of water, a stronger elution solvent in the HILIC mode reduced the retention times of gabapentin and metformin on the silica column. The higher sensitivity of HILIC-MS/MS (ca 20pg on-column) was achieved using a smaller sample volume (i.e., 10 μL of human plasma) in comparison to those (ca 200–2000pg on-column) of RPLC-MS/MS with 100–200 μL of human plasma [15]. The described method was used successfully for a bioequivalence study of orally administered gabapentin in healthy volunteers.

18.6 PHARMACOKINETIC STUDY OF THE PEPTIDE DRUG TASPOGLUTIDE

Another alternative to the SPE is the online cleanup, where the solid-phase extraction is performed online by using two different types of columns operating simultaneously through a divert 10-port valve[16,17] as shown in Figure 18.8. The first column is the trapping column (TC) for cleaning the biological sample and retaining the

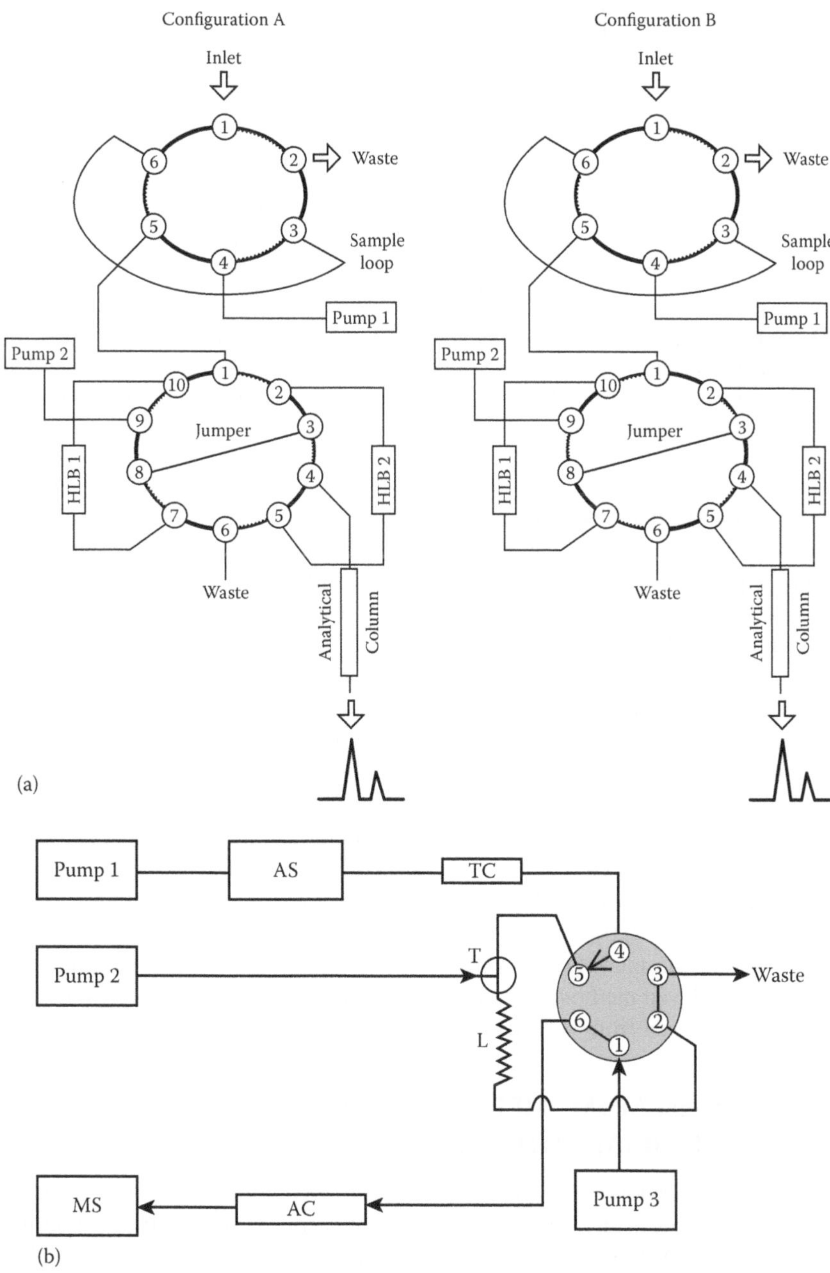

FIGURE 18.8 Scheme of column switching LC-MS/MS system. (a) Two trapping columns and one analytical column. (From Heinig, K. and Wirz, T., *Anal. Chem.*, 81, 3705, 2009. With permission from American Chemical Society.) (b) One trapping column and one analytical column. (From Kousoulos, C. et al., *Talanta*, 30, 360, 2007. With permission from Elsevier.)

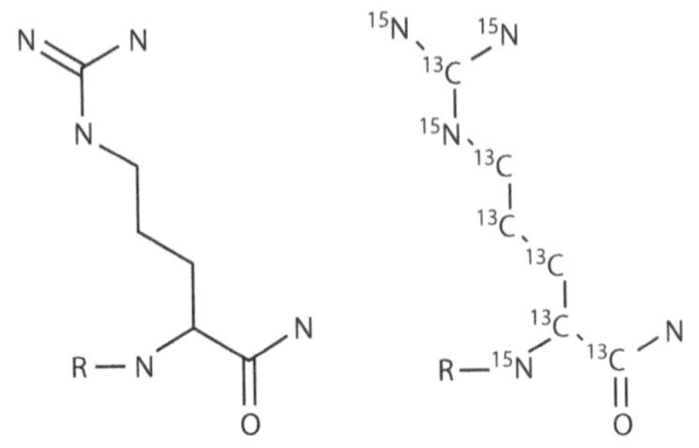

Taspoglutide Internal standard

FIGURE 18.9 Structures of taspoglutide (drug) and internal standard (IS), R = His-Aib-Glu-Gly-Thr-Phe-Thr-Ser-Asp-Val-Ser-Ser-Tyr-Leu-Glu-Gly-Gln-Ala-Ala-Lys-Glu-Phe-Ile-Ala-Trp-Leu-Val-Lys-Aib. (From Kousoulos, C. et al., *Talanta*, 30, 360, 2007. With permission from Elsevier.)

analytes of interest, while the second one should be the analytical column (AC) for the separation of analytes. During the cleaning phase of the procedure, the valve directs the mobile phase to waste, usually with a high flow rate, while during the analytical phase, the valve diverts the flow to the detector. A very impressive and novel application of an HILIC column as a TC was used in the study of the peptide drug, taspoglutide, MW 3339.7 g/mol, and [$^{13}C_6$,$^{15}N_4$-taspoglutide] (IS), MW 3347.7 g/mol, synthesized at F. Hoffmann-La Roche as shown in Figure 18.9. The TC was a 50 mm × 2.1 mm Atlantis HILIC Si, 3 μm with precolumn filter (Waters). The AC consisted of two Zorbax Poroshell 300-SB C18, 75 mm × 2 mm, 5 μm columns (Agilent) in series. The HILIC TC was an excellent choice for online SPE followed by RP chromatography because (1) large volumes of highly organic sample solutions could be directly applied without prior dilution and (2) the polar analyte and the IS were selectively enriched, while the lipophilic constituents were rinsed off. According to the authors, the described method performed reliably with respectable LLOQs of 10.0 and 50.0 pg/mL in human and animal plasma, respectively. The low required human plasma volume of only 250 μL compared to the 1.5 mL plasma in the previously employed method provided the possibility for repeat analyses or reduced blood volumes taken from volunteers or patients.

18.7 HILIC-ESI-MS/MS METHOD FOR THE QUANTITATION OF POLAR METABOLITES OF ACRYLAMIDE IN HUMAN URINE

The carcinogen acrylamide (AA) is formed during food processing.[18] AA is metabolized to mercapturic acids, which are excreted with urine with the pathway described in Figure 18.10. A HILIC-MS/MS method using a zwitterionic stationary phase (ZIC-HILIC) was developed and validated to quantitate the mercapturic acids of AA (AAMA) and glycidamide (GAMA), and AAMA-sulfoxide in human urine. In contrast to RPLC, the application of ZIC-HILIC resulted in the efficient retention and separation of these highly polar compounds. The analysis was based again on online cleanup with the use of a Stability BS-C17 AC (5 μm, 3 mm × 33 mm, Ammerbuch,

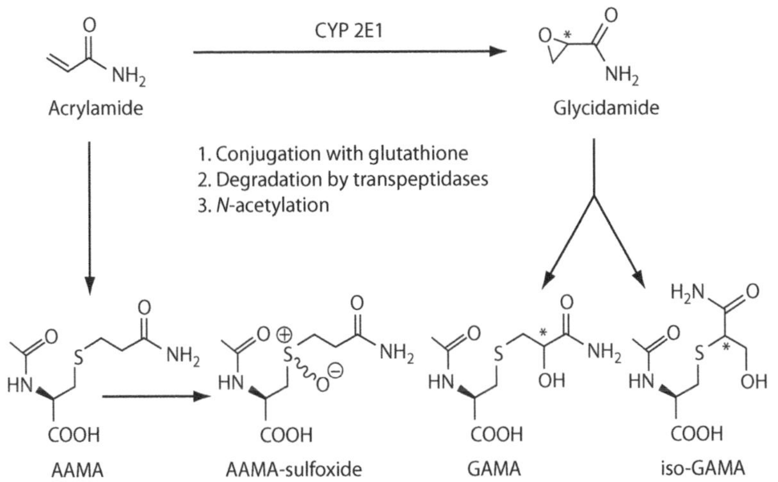

FIGURE 18.10 Biotransformation of acrylamide in the human body. (From Kopp, E.K. et al., *J. Agric. Food Chem.*, 56, 9828, 2008. With permission from American Chemical Society.)

Germany) and the AC (ZIC-HILIC, 3.5 μm, 2.1 mm × 150 mm, SeQuant AB, Umeå, Sweden) with a mobile phase consisting of a 14% ammonium acetate buffer and 86% acetonitrile.

Urine samples were obtained from healthy human subjects (three female and three male subjects, body weights between 52 and 75 kg, and ages between 23 and 28 years). Urine samples from the subjects were collected over a predefined time frame of 72 h in intervals of 8 h. All subjects were nonsmokers and did not drink alcoholic beverages 72 h before and during the study. Because of the baseline separation of the isobaric AAMA-sulfoxide and GAMA, which share a fragmentation of *m/z* 249.2 to *m/z* 120.0, the interferences between these metabolites could be avoided (Figure 18.11). Peaks representing AAMA-sulfoxide may easily be misinterpreted as GAMA if the peaks were not clearly separated. An inefficient separation of GAMA and AAMA-sulfoxide may explain some of the high GAMA concentrations reported in human urine samples.[19] Glycidamide, the precursor of GAMA, is the DNA-reactive AA-metabolite supposedly responsible for tumor induction after AA administration in rodents.[20] Therefore, an overestimation of GAMA excretion due to interference with AAMA-sulfoxide in human urine may result in an overestimation of the potential risk of health effects due to AA exposures for humans.

18.8 HILIC-MS/MS TECHNIQUE FOR SENSITIVE MONITORING OF THE CHANGES OF URINARY ESTROGEN CONJUGATES

Estrogens play important roles in the development and maintenance of secondary sexual characteristics—pregnancy and long-bone maturation.[21] They also have been found to be associated with the development of breast cancer. Conjugation to sulfate and glucuronide is one of the major estrogen metabolism pathways, and these conjugates are excreted mainly through urine. Free estrogens in urine usually occur

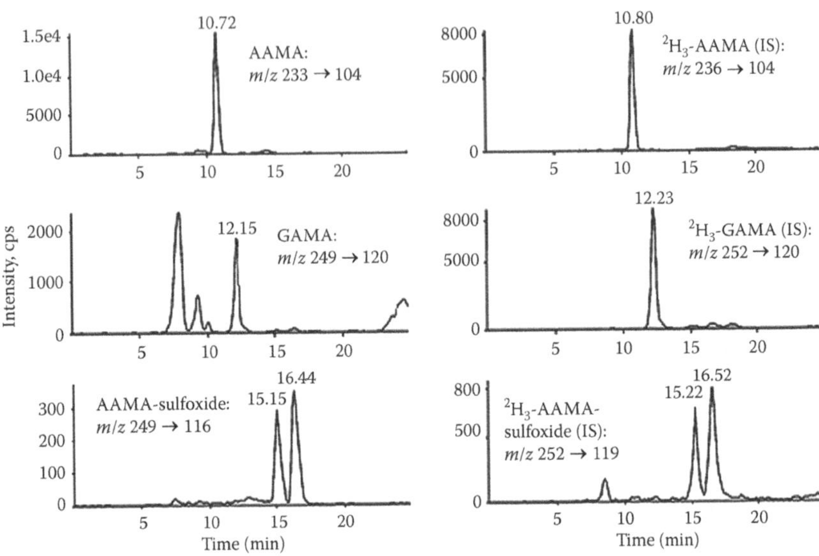

FIGURE 18.11 HILIC-ESI-MS/MS in the multiple reaction monitoring mode (MRM). Chromatogram showing the mass traces for the mercapturic acids of acrylamide excreted in the urine of a nonexposed human subject. The corresponding mass traces for the internal standards are shown on the right-hand side. (From Kopp, E.K. et al., *J. Agric. Food Chem.*, 56, 9828, 2008. With permission from American Chemical Society.)

at extremely low levels and are not detectable. Estrogen metabolism varies significantly from person to person suggesting that individual profiles of estrogens and their metabolites may provide information on variations in estrogen metabolism and cancer risk. The accurate measurement of these estrogen conjugates may further the study of the roles they play in estrogen-related physiological processes.

This study demonstrates the superiority of an HILIC-MS/MS method for the determination of seven estrogens in human urine. Their structures appear in Figure 18.12. HILIC separation of the analytes was performed on a TSK-Gel Amide-80 (2.0 mm × 150 mm, 5 μm, 80 Å; Tosoh Bioscience, Montgomeryville, PA) with a guard cartridge (2.0 mm × 10 mm) at room temperature. Isocratic elution was used and the mobile phase was acetonitrile/aqueous ammonium acetate (5 mM, pH 6.80) (85/15, v/v). The eluent from the LC column was directly transferred into the ion source of the mass spectrometer without postcolumn splitting. The advantages of the HILIC-MS/MS compared to that of RPLC-MS/MS, concerning the limits of quantitation, appear in Table 18.1.

18.9 COMBINATION OF HILIC-MS AND RPLC-MS FOR PROFILING POLAR URINE METABOLITES

Complex biological samples, such as urine, contain a very large number of endogenous metabolites reflecting the metabolic state of an organism (nicotine, creatinine, tryptophane, lidocaine, β-hydroxyethyl theophylline, caffeine, salicylic

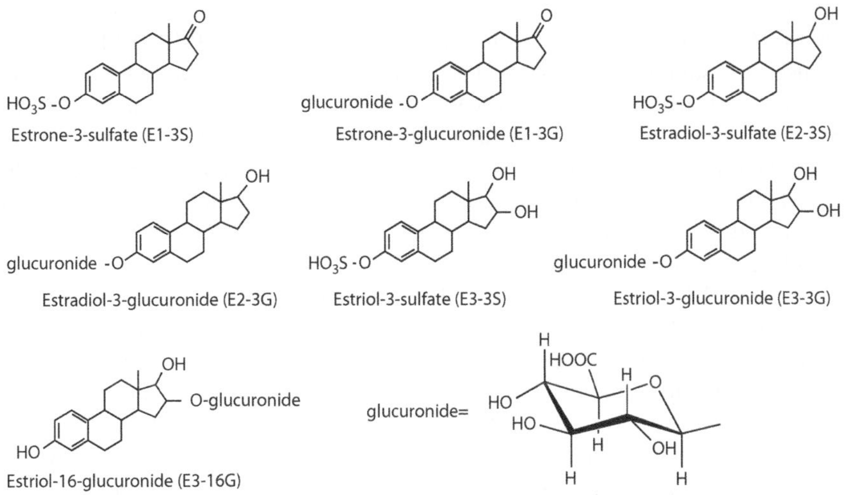

FIGURE 18.12 Structures of the seven estrogen conjugates and of glucuronide. (From Qin, F. et al., *Anal. Chem.*, 80, 3404, 2008. With permission from American Chemical Society.)

TABLE 18.1
Comparison of the Limits of Quantification or Detection (LOQ or LOD) Obtained Using the HILIC-MS/MS with Those Obtained Using the RPLC-MS/MS Methods

Analytes	HILIC-MS/MS[a] (ng/mL)	RPLC-MS/MS[b] (ng/mL)[22]	RPLC-MS/MS[c] (ng/mL)[23]	RPLC-MS/MS[d] (ng/mL)[24]
E1-3S	0.002	0.02	0.2	30
E2-3S	0.005	0.03	0.2	
E3-3S	0.02	0.052	0.2	20
E1-3G	0.02	0.035		
E2-3G	0.02	0.07		
E3-16G	0.2	0.24		
E3-3G	1.0	0.6		

Source: Qin, F. et al., *Anal. Chem.*, 80, 3404, 2008. With permission from American Chemical Society.
[a] LOQ is based on 1.0 mL of urine sample.
[b] LOQ is based on 5.0 mL of urine sample.
[c] LOQ is based on 0.1 mL of urine sample.
[d] LOD is based on 3.0 mL of urine sample.

acid, *p*-nitrophenol and cholic acid, hydroxyproline and uric acid).[25] Metabolite patterns can provide a comprehensive signature of the physiological state of an organism as well as insights into specific biochemical processes. Although the metabolites excreted in urine are commonly highly polar, the samples are generally analyzed using reversed-phase liquid chromatography mass spectrometry

(RP-LC/MS). In this published work, a method for detecting highly polar metabolites by HILIC-ESI-MS is described as a complement to RP-LC/ESI-MS. SPE was used for sample preparation. The adsorbent was activated and conditioned first with 1 mL methanol and then with 1 mL ammonium acetate buffer (10 mM, pH 4). An aliquot of 0.5 mL of rat urine was loaded onto the SPE column (30 mg, Waters Oasis HLB). In the washing step, a 0.5 mL ammonium acetate buffer (10 mM, pH 4) was used and 0.5 mL methanol was used for elution. Both the wash fraction and the eluate were collected separately and filtered through syringe filters before injection.

For the HILIC analysis, a ZIC®-HILIC column (3.5 µm, 2.1 mm × 100 mm) from SeQuant AB (Umeå, Sweden) was used together with a C4 precolumn (5 µm, 2.1 mm × 10 mm) from Thermo. Mobile phase A consisted of 5 mM ammonium acetate (adjusted to pH 4 with formic acid), while mobile phase B consisted of acetonitrile and 0.025% formic acid. The SPE sorbent is a hydrophilic–lipophilic based (HLB) copolymer that is more suitable for polar compounds. It would be advantageous to use the same stationary phase both for the sample preparation and for the analytical column. (When the wash fraction was injected and RPLC was used, the amount of nonretained compounds was higher compared to what was obtained after analyzing the eluate.) By using these two complementary setups with a ZIC-HILIC column for analysis of the wash fraction and a C18 column for the analysis of the eluate, the number of metabolites to be detected could be increased. As shown in Table 18.2, some of the substances examined retained on the RP C18 column and others on the HILIC column.

TABLE 18.2
Retention Times for 11 Compounds Using HILIC and RP-HPLC

	Retention Time[a]	
Compound	**HILIC**	**RP-HPLC**
Hydroxyproline ($C_5H_9NO_3$)	12.7	1.9
Uric acid ($C_5H_4N_4O_3$)	10.5	2.06
Nicotine ($C_{10}H_{14}N_2$)	14.6	2.6
Creatinine ($C_4H_7N_3O$)	11.5	2.2
Tryptophane ($C_{11}H_{12}N_2O_2$)	13.4	4.3
Lidocaine ($C_{14}H_{22}N_2O$)	11.2	9.3
β-Hydroxyethyl theophylline ($C_9H_{12}N_4O_3$)	3.4	5.4
Caffeine ($C_8H_{10}N_4O_2$)	3.3	7.5
Salicylic acid ($C_7H_6O_3$)	2.7	7.5
p-Nitrophenol ($C_6H_5NO_3$)	3.2	11.7
Cholic acid ($C_{24}H_{40}O_5$)	3.3	13.5

Source: Idborga, H. et al., *J. Chromatogr. B*, 828, 9, 2008. With permission from Elsevier.

[a] The void volume corresponding time (t_0) was calculated to be 2.3 min for ZIC®-HILIC and 2.1 min for RPLC that might differ slightly from experimental results.

18.10 QUANTIFICATION OF METHYLMALONIC ACID AND HOMOCYSTEIN IN SERUM AND URINE WITH HILIC-MS

Homocysteine (Hcy) and methylmalonic acid (MMA) are the most specific markers for inborn errors of methionine and propionic acid metabolism.[26–28] The different methods that have been developed for measuring Hcy and MMA in serum, plasma, urine, and cerebrospinal fluid include GC-MS, LC-MS/MS, and capillary electrophoresis. The main obstacles to overcome are related to the low physiological concentrations of MMA in human serum and the fact that both markers are hydrophilic, nonvolatile compounds. Finally, MMA is a structural isomer of succinic acid (SA) that may cause interference. Concentrations of SA in the serum are usually considerably higher than MMA. Current methods require extraction and derivatization steps to yield MMA derivatives that are compatible either with GC-MS or RP-HPLC/MS techniques.

Recently, all the above problems have been overcome by using HILIC-MS and specifically by using the ZIC-HILIC column. A 50 mm, 4.6 mm, 5 µm, baseline separation was achieved within 3 min. The analysis was performed with the following conditions: (1) column temperature: 30°C, (2) mobile phase: acetonitrile/ammonium acetate (100 mM, pH 6.8); 75/25 (v/v), and (3) flow-rate: 1.0 mL/min with a split: 100 µL/min to MS. The ZIC®-HILIC column is indeed a suitable tool for separation of Hcy, MMA, and SA. Combined with MS detection, a physiological relevant concentration can easily be quantified with the possibility of processing up to 20 samples per hour.

In conclusion, the combination of polar stationary phases and aqueous/organic mobile phases could significantly enhance HILIC-MS/MS method sensitivity. The low column backpressure due to high-organic mobile phases has made higher flow rates feasible. Furthermore, a direct injection of the organic solvent extracts from the application of LLE, SPE, and PPT onto the hydrophilic columns is compatible with HILIC-MS/MS systems. All the above make HILIC a useful partner or an alternative to RP LC-MS/MS, especially for the analysis of polar compounds.

REFERENCES

1. Zhou W, Zhou S, Pelzer M, Liu C, Jiang X, and Weng N. 2004. Impact of high flow rate on sensitivity and resolution in quantitative bioanalytical LC-MS/MS using monolithic and HILIC columns. *Proceedings of the 52nd ASMS Conference on Mass Spectrometry and Allied Topics*, May 23–27, Nashville, Tennessee.
2. Li W, Li Y, Francisco DT, and Naidong W. 2005. Hydrophilic interaction liquid chromatographic tandem mass spectrometric determination of atenolol in human plasma. *Biomedical Chromatography* **19**: 385–393.
3. Shou WZ, Bu H, Addison T, and Jiang X. 2002. Development and validation of a liquid chromatography/tandem mass spectrometry (LC/MS/MS) method for the determination of ribavirin in human plasma and serum. *Journal of Pharmaceutical and Biomedical Analysis* **29**: 83–94.
4. Brown SD, White CA, and Bartlett MG. 2002. Hydrophilic interaction liquid chromatography/electrospray mass spectrometry determination of acyclovir in pregnant rat plasma and tissues. *Rapid Communications in Mass Spectrometry* **16**: 1871–1876.

5. Jeong DW, Kima YH, Ji HY, Youn YS, Lee KC, and Lee HS. 2007. Analysis of carvedilol in human plasma using hydrophilic interaction liquid chromatography with tandem mass spectrometry. *Journal of Pharmaceutical and Biomedical Analysis* **44**: 547–552.
6. Borges NCC, Mendes GD, Silva DO, Rezende VM, Barrientos-Astigarraga RE, and Nucci GD. 2005. Quantification of carvedilol in human plasma by high-performance liquid chromatography coupled to electrospray tandem mass spectrometry: Application to bioequivalence study. *Journal of Chromatography B* **822**: 253–262.
7. Apostolou C, Kousoulos C, Dotsikas Y, and Loukas YL. 2008. Comparison of hydrophilic interaction and reversed-phase liquid chromatography coupled with tandem mass spectrometric detection for the determination of three pharmaceuticals in human plasma. *Biomedical Chromatography* **22**: 1393–1402.
8. Cox GB and Stout RW. 1987. Study of the retention mechanism for basic compounds on silica under 'pseudo-reversed-phase' conditions. *Journal of Chromatography* **384**: 315.
9. Ji HY, Jeong DW, Kim YH, Kim HH, Yoon YS, Lee KC, and Lee HS. 2006. Determination of gabapentin in human plasma using HILIC-MS/MS. *Rapid Communications in Mass Spectrometry* **20**: 2127–2132.
10. Zhu Z and Neirinck L. 2002. High-performance liquid chromatographic method for the determination of gabapentin in human plasma. *Journal of Chromatography B* **779**: 307–312.
11. Chollet DF, Goumaz L, Juliano C, and Anderegg G. 2000. Fast isocratic high-performance liquid chromatographic assay method for the simultaneous determination of gabapentin and vigabatrin in human serum. *Journal of Chromatography B* **746**: 311–314.
12. Garcia LL, Shihabi ZK, and Oles K. 1995. Determination of gabapentin in serum by capillary electrophoresis. *Journal of Chromatography B* **669**: 157–162.
13. Hooper WD, Kavanagh MC, and Dickinson RG. 1990. Determination of gabapentin in plasma and urine by capillary column gas chromatography. *Journal of Chromatography B* **529**: 167–174.
14. Gambelunghe C, Mariucci G, Tantucci M, and Ambrosini MV. 2005. Gas chromatography–tandem mass spectrometry analysis of gabapentin in serum. *Biomedical Chromatography* **19**: 63–67.
15. Ifa DR, Falci M, Moraes ME, Bezerra FA, Moraes MO, and de Nucci G. 2001. Gabapentin quantification in human plasma by high-performance liquid chromatography coupled to electrospray tandem mass spectrometry. Application to bioequivalence study. *Journal of Mass Spectrometry* **36**: 188–194.
16. Heinig K and Wirz T. 2009. Determination of taspoglutide in human and animal plasma using liquid chromatography–tandem mass spectrometry with orthogonal column-switching. *Analytical Chemistry* **81**: 3705–3713.
17. Kousoulos C, Dotsikas Y, and Loukas YL. 2007. Turbulent flow and ternary column-switching on-line clean-up system for high-throughput quantification of risperidone and its main metabolite in plasma by LC-MS/MS Application to a bioequivalence study. *Talanta* **30**: 360–367.
18. Kopp EK, Sieber M, Kellert M, and Dekant W. 2008. Rapid and sensitive HILIC-ESI-MS/MS quantitation of polar metabolites of acrylamide in human urine using column switching with an online trap column. *Journal of Agricultural and Food Chemistry* **56**: 9828–9834.
19. Bjellaas T, Janak K, Lundanes E, Kronberg L, and Becher G. 2005. Determination and quantification of urinary metabolites after dietary exposure to acrylamide. *Xenobiotica* **35**: 1003–1018
20. Manjanatha MG, Aidoo A, Shelton SD, Bishop ME, McDaniel LP, Lyn-Cook LE, and Doerge DR. 2006. Genotoxicity of acrylamide and its metabolite glycidamide administered in drinking water to male and female Big Blue mice. *Environmental and Molecular Mutagenesis* **47**: 6–17

21. Qin F, Zhao Y, Sawyer B, and Li X. 2008. Hydrophilic interaction liquid chromatography–tandem mass spectrometry determination of estrogen conjugates in human urine. *Analytical Chemistry* **80**: 3404–3411.
22. D'Ascenzo G, Di Corcia A, Gentili A, Mancini R, Mastropasqua R, Nazzari M, and Samperi R. 2003. Fate of natural estrogen conjugates in municipal sewage transport and treatment facilities. *The Science of the Total Environment* **302**: 199–209.
23. Zhang HW and Henion J. 1999. Quantitative and qualitative determination of estrogen sulfates in human urine by liquid chromatography/tandem mass spectrometry using 96-well technology. *Analytical Chemistry* **71**: 3955–3964.
24. Volmer DA and Hui JPM. 1997. Rapid determination of corticosteroids in urine by combined solid phase microextraction/liquid chromatography/mass spectrometry. *Rapid Communications in Mass Spectrometry* **11**: 1926–1933.
25. Idborga H, Zamania L, Edlunda P, Schuppe-Koistinenb I, and Jacobsson S. 2008. Metabolic fingerprinting of rat urine by LC/MS: Part 1. Analysis by hydrophilic interaction liquid chromatography–electrospray ionization mass spectrometry. *Journal of Chromatography B* **828**: 9–13.
26. Lakso H, Appelblad P, and Schneede J. 2008. Quantification of methylmalonic acid in human plasma with hydrophilic interaction liquid chromatography separation and mass spectrometric detection. *Clinical Chemistry* **54**: 2028–2035.
27. Rinaldo P, Hahn S, and Matern D. 2006. Inborn errors of amino acid, organic acid and fatty acid metabolism. In: *Tietz Textbook of Clinical Chemistry and Molecular Diagnostics 5th Edition*. Burtis CA, Ashwood ER, and Burns DE, Eds. Elsevier Sanders Company, St. Louis, MO, pp. 2207–2247.
28. Watkins D and Rosenblatt DS. 2001. Cobalamin and inborn errors of cobalamin absorption and metabolism. *The Endocrinologist* **11**: 98–104.

19 Method Development and Analysis of Mono- and Diphosphorylated Nucleotides by HILIC HPLC-ESI-MS

Samuel H. Yang, Hien P. Nguyen, and Kevin A. Schug

CONTENTS

19.1 INTRODUCTION

Reversed-phase high-performance liquid chromatography (RP-HPLC) has been widely used as the standard for the high-throughput analysis of soluble polar compounds. In RP-HPLC, separation of polar compounds is achieved giving an elution order based on increasing hydrophobicity. Although a powerful technique, RP-HPLC is not ideal for highly hydrophilic compounds as they are poorly retained and are often eluted at the dead volume; this behavior subjects components to coelution with other matrix components and limits quantitative analysis. Compounds that are poorly retained in RP-HPLC can potentially be evaluated using normal-phase liquid chromatography (NP-HPLC), where a nonpolar mobile phase and a polar stationary phase are employed. However, poor solubility of hydrophilic compounds in nonpolar solvents severely limits the application of NP-HPLC to these analytes. Also, when coupling NP-HPLC to a detection technique such as electrospray ionization–mass spectrometry (ESI-MS), additional problems arise. For instance, nonpolar solvents cause a drastic loss in sensitivity due to their poor ionization efficiency[1]; some nonpolar solvents can also potentially damage plastic (e.g., PEEK) tubing commonly used with HPLC-ESI-MS instruments.

Hydrophilic interaction chromatography (HILIC) is a mode of chromatography that conveniently overcomes the limitations of both RP-HPLC and NP-HPLC for analysis of hydrophilic compounds.[2–5] HILIC can separate polar analytes while using a mobile phase system ideally suited for ESI-MS, specifically a hydroorganic solvent mixture with high polar organic solvent content.[1] HILIC mobile phases differ from RP-HPLC in elution strength. Generally, HILIC uses a lower aqueous and higher organic content in order to achieve retention of polar analytes on stationary phases with polar interaction sites. In this mode, water plays the role of the stronger eluting solvent, and thus, the lower aqueous content facilitates enhanced interaction between the analyte and the polar stationary phase, and enables retention.[6] With the development of HILIC, a wide range of hydrophilic compounds, which had previously been difficult to separate, can now be rapidly analyzed by HPLC-ESI-MS with sufficient capacity and resolving power.[7–15]

A large amount of interest currently focuses on the development of analytical techniques to quantitate and characterize biomolecules.[16–21] Nucleotides are a class of biomolecules that hold an utmost importance in the living cell and play a critical role in cell proliferation and metabolism regulation. For example, cyclic adenosine monophosphate (cAMP) is a key regulator of many protein kinases, such as protein kinase A (PKA), which performs signal transfer initiated by specific hormones in a wide variety of cellular functions (e.g., in glycogen and lipid metabolism).[22–26] Even more so, single nucleotides act as a form of energy storage in the form of adenosine triphosphate (ATP) where cleavage of the gamma phosphate gives adenosine diphosphate (ADP) and releases energy that drives enzymatic reactions.[27–30] Nucleotides can also act as regulators of enzymes based on their degree of phosphorylation, such as in the case of small guanosine triphosphates (GTPases) in the G-protein signaling pathway. Binding of these GTPases to guanosine diphosphate (GDP)/guanosine triphosphate (GTP) causes inactivation/activation of another regulation mechanism for the subsequent enzyme downstream in the signaling cascade.[31–34]

Despite such significance of nucleotides in cellular functions, a method involving facile sample preparation with rapid separation and detection of phosphorylated nucleotides by HILIC-ESI-MS has had minimal progress when compared to other classes of biomolecules. Separation of the nitrogen bases of nucleosides has been reported and is tenable.[35] The same success cannot be claimed, however, when considering the nucleotide molecule consisting of all three of its major components, specifically, a nitrogen base, a 5-membered sugar ring, and an attached phosphate tail. Chromatographic methods generally involve the addition of a hydrophobic ion-pairing reagent into the mobile phase that interacts with the nucleotides resulting in retention on a standard RP column. One such common additive is *N,N*-dimethylhexylamine (DMHA).[36–38] Although appreciable retention can be obtained from this method, it also has some limitations. The overwhelming interference from the ion-pairing reagent in the mobile phase causes a dramatic loss of sensitivity as well as a high background signal when coupled to ESI-MS detection. The interaction of the ion-pairing reagent with the nucleotide, which can access a variety of protonation states, can be unpredictable leading to sporadic results when sampling a wide range of nucleotides and nucleotide derivatives. Additionally, the nature of these ion-pairing reagents causes heavy contamination of the capillary lines in the HPLC-MS system as well as long column equilibration times. Although some work has successfully applied HILIC for the quantification of a single nucleotide,[39] there is still a lack of a method to rapidly perform separations on a wide range of nucleotides.

HILIC is a logical choice in building a method for the analysis of highly hydrophilic nucleotides. Keeping the themes of rapid and reproducible analysis in mind, we have focused on the development of a set of optimal conditions for the separation and quantification of a mixture of mono- and diphosphorylated nucleotides. A comparative study was also conducted between multiple types of commercial HILIC columns and their ability to resolve various nucleotides within the mixture. The data allow insights to be drawn into some of the interactions between the nucleotides and the various HILIC stationary phases that are available to the scientific community.

19.2 EXPERIMENTAL SECTION

19.2.1 Chemicals and Materials

LC-MS grade water and acetonitrile were supplied by Burdick and Jackson (Muskeegon, MI) and formic acid was obtained from J.T. Baker (Phillipsburg, NJ). Ammonium formate was obtained from Acros Organics (Morris Plains, NJ). Six of the nucleotides, adenosine 5′-diphosphate (ADP), guanosine 5′-monophosphate (GMP), guanosine 5′-diphosphate (GDP), thymine 5′-monophosphate (TMP), cytosine 5′-monophosphate (CMP), and uridine 5′-monophosphate (UMP), were purchased from Sigma-Aldrich (Milwaukee, WI). The other two nucleotides, adenosine 5′-monophosphate (AMP) and uridine 5′-diphosphate (UMP), were purchased from Fluka Analytical (Buchs, Switzerland). Table 19.1 shows a list of the eight nucleotides, including their structures, monoisotopic masses, and calculated log P values. Six different HILIC columns were tested—Phenomenex (Torrance, CA) Luna HILIC (10 cm × 2.0 mm, 3 μm particle diameter (d_p), 180 Å pore size); Tosoh Bioscience (Kyoto, Japan) TSK-Gel Amide-80 (10 cm × 2.0 mm, 3 μm d_p, 80 Å pore size); Varian (Palo Alto, CA) Polaris 3 Amide (10 cm × 2.0 mm, 3 μm d_p, 200 Å pore size); Varian Pursuit XRs 3 Diol (10 cm × 2.0 mm, 3 μm d_p, 100 Å pore size); Varian Polaris 3 Amide-C18 (10 cm × 2.0 mm, 3 μm d_p, 200 Å pore size); and Thermo-Fisher Scientific, Inc. (West Palm Beach, FL) BETASIL Cyano (10 cm × 1 mm, 5 μm d_p, pore size not available). In-depth comparisons of column chemistries have been reserved for the discussion section.

TABLE 19.1
List of Nucleotides Used in This Study

Compound Number	Compound Name	Compound Structure	Mass[a] (amu)	Log D (pH 7.0)[b]
1	Adenosine 5′-monophosphate (AMP)	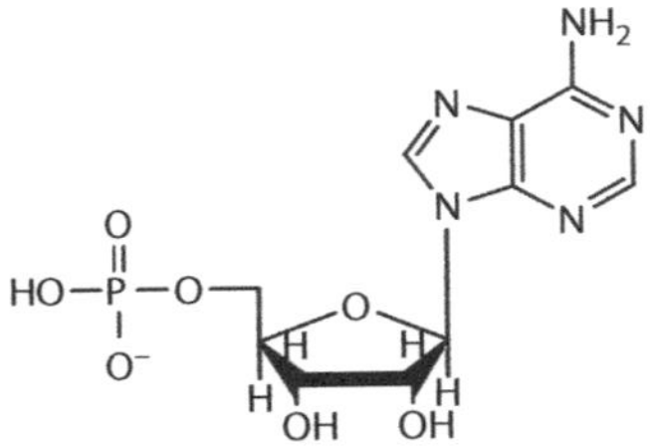	346.06	–4.28
2	Adenosine 5′-diphosphate (ADP)	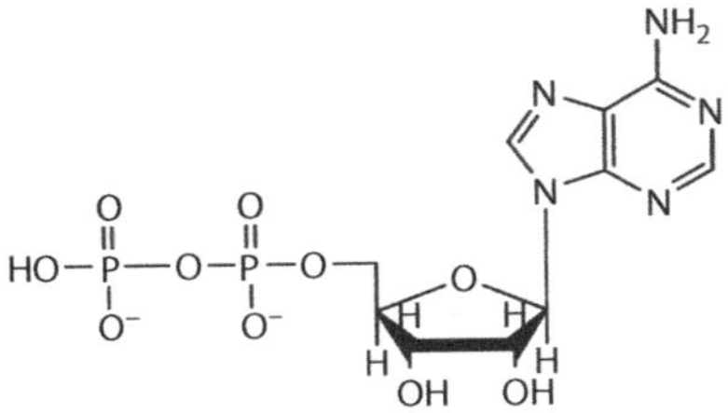	425.01	–7.71
3	Guanosine 5′-monophosphate (GMP)	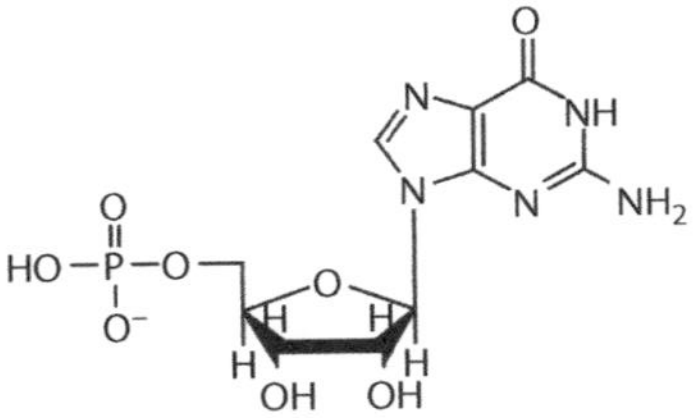	362.05	–4.49

4	Guanosine 5′-diphosphate (GDP)	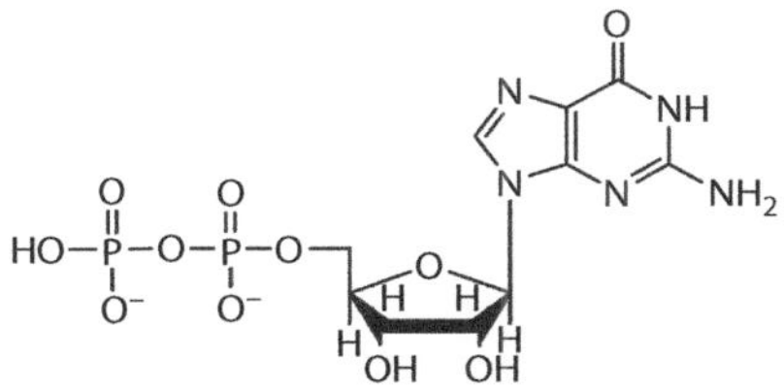	441.01	–7.92
5	Uridine 5′-monophosphate (UMP)	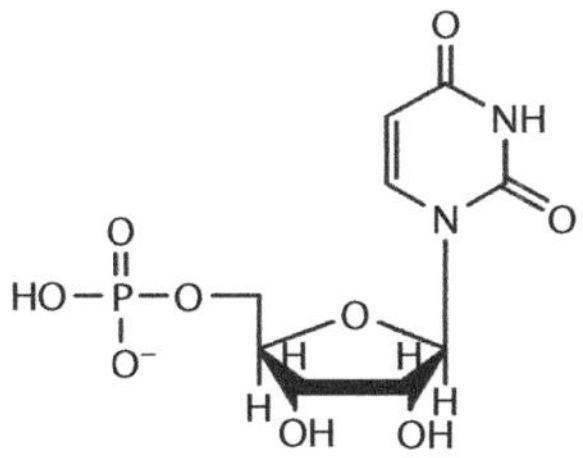	323.03	–5.67
6	Uridine 5′-diphosphate (UDP)	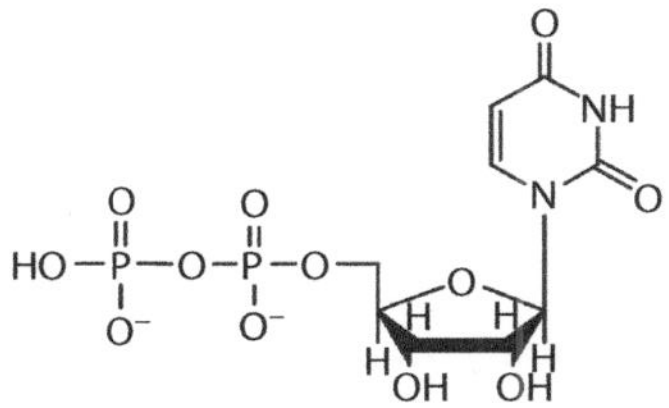	401.99	–9.09

(*continued*)

TABLE 19.1 (continued)
List of Nucleotides Used in This Study

Compound Number	Compound Name	Compound Structure	Mass[a] (amu)	Log D (pH 7.0)[b]
7	Thymine 5′-monophosphate (TMP)	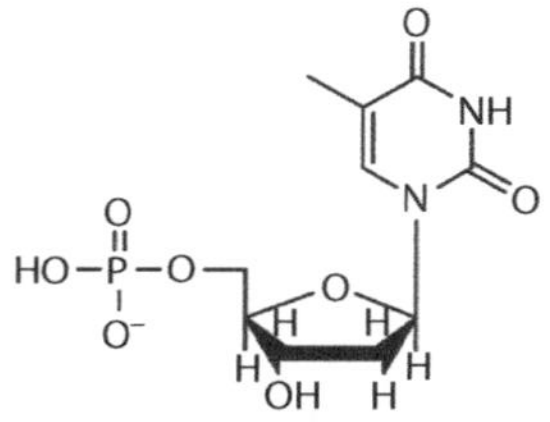	321.05	–5.67
8	Cytidine 5′-monophosphate (CMP)	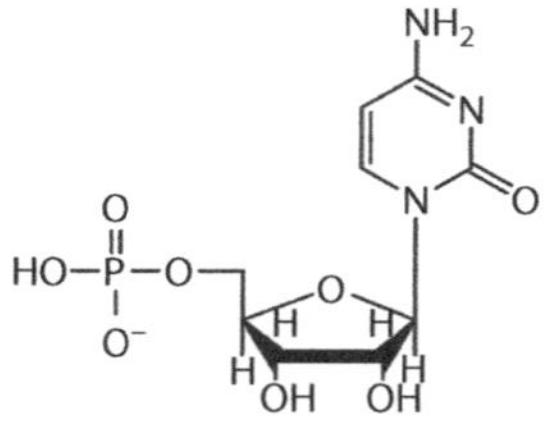	322.04	–5.63

[a] Monoisotopic.
[b] Log P calculated with ACD/labs log P calculator for compound in its neutral unionized form.

Samples were made by first preparing 10 mM stock solutions of each nucleotide in water followed by dilution to make tested experimental concentrations at 50, 100, and 200 μM. Samples were premixed in a composition of water and ACN (30:70) prior to injection onto the column in order to emulate starting mobile-phase compositions. All samples were prepared at room temperature. Water/ACN compositions in excess of 80% organic content were not possible due to precipitation of the nucleotides at a high organic content. A constant flow rate of 200 μL/min was used throughout all chromatographic separations for 2 mm I.D. columns and a flow rate of 50 μL/min was used for the BETASIL cyano column with 1 mm I.D. Mobile phase A consisted of water with 20 mM ammonium formate and 0.05% formic acid (v/v) (pH 3.98), while mobile phase B was composed of a mixture of water/ACN (10:90) with 20 mM ammonium formate and 0.05% formic acid (v/v). The best separation (on the Tosoh Amide-80 column) was achieved using a gradient program consisting of a 5 min 80% ACN isocratic period in the beginning, then an acetonitrile gradient of 80% ACN to 50% ACN in a 15 min period, and finally followed by a final 5 min isocratic period at 50% ACN. Optimal gradient profiles were also obtained for the Phenomenex Luna HILIC and amide-C18 columns in which only a single change was made where the gradient programs are 80% ACN to 50% ACN in 25 min, and 87.5% ACN to 50% ACN in 15 min, respectively. Capacity factors (k′) were calculated based on the equation $(t_R - t_0)/t_0$ where t_R was the retention time of the sample analyte and t_0 was the estimated dead time. Injection volumes of 15 μL were made for all sample solutions.

19.2.2 Instrumentation

All measurements were performed on a Shimadzu LCMS 2010 (Shimadzu Scientific Instruments, Inc., Columbia, MD), which included a dual high-pressure-mixing LC-20AD pump system and an SIL-20A HT autosampler coupled to a quadrupole mass analyzer, equipped with a conventional ESI source. Electrospray ionization was performed in both the positive and the negative modes with a spray capillary voltage of 4.5 and −3.5 kV, respectively. Scan speeds used over the entire runtime were set at 250 amu/s at a range of 320–450 (*m/z*). A positive-/negative-ionization mode switching time interval of 1 Hz was used. Other ion source parameters are as follows: nebulizing gas, 1.2 L/min; drying gas flow rate, 6000 kPA; curved desolvation line temperature, 250°C; and CDL heat block temperature, 250°C. The detector voltage was set to 1.60 kV. Data analysis was performed using LCM Solutions (version 3.4) software.

19.2.3 Physiochemical Parameters

Calculations of octanol–water partition coefficients (log P), pH-dependent octanol–water partition coefficients (log D), and pK_a values were performed for each nucleotide using Advanced Chemistry Development, Inc. (ACD/Labs) log P, log D, and pK_a calculators (version 9), respectively. Values presented for log P (Table 19.1) and log D (Table 19.2) are expected to contain approximately 10% relative uncertainty. The calculated uncertainties in pK_a values range from 0.1 to 0.5 pK_a units according to the output of the program.

TABLE 19.2
Comparison of Log D Values, Charge States (z), and % Species of Charge State z Present at pHs 4.7, 7.0, and 10.0

	pH 4.7			pH 7.0			pH 10.0		
	Log D	z	**% Species**	**Log D**	z	**% Species**	**Log D**	z	**% Species**
AMP	−3.0	−1	89.5	−4.3	−2	81.6	−4.7	−2	99.8
ADP	−7.2	−1	90.8	−7.7	−2	56.7	−8.4	−3	99.8
GMP	−3.2	−1	97.5	−4.5	−2	81.2	−5.6	−2	85.8
GDP	−7.7	−2	99.0	−7.9	−3	55.9	−9.3	−4	86.0
UMP	4.4	−1	97.7	−5.7	−2	78.3	−7.0	−2	98.0
UDP	−8.8	−2	99.2	−9.1	−3	54.5	−10.7	−4	98.1
TMP	−4.4	−1	97.7	−5.7	−2	81.7	−6.7	−2	85.0
CMP	−4.3	−1	94.2	−5.6	−2	81.6	−6.1	−2	99.0

19.2.4 Method Validation

Sample preparation followed the same protocol used in method development. For the establishment of calibration curves, 10 different concentrations of nucleotide mixtures were tested in triplicate with a range of 0–350 μM. Multiple curve fitting approaches were investigated to find the best fit to the experimental data. A set of seven replicate measurements at 50 μM was used for the calculation of the limit of detection for each nucleotide.

19.3 RESULTS AND DISCUSSION

19.3.1 Optimal pH

Method development for the separation of nucleotides with HILIC begins with the determination of an optimal pH for the system. Manipulation of the hydrogen ion activity directly affects analyte selectivity and retention; significant changes in these chromatographic characteristics can result with only small changes in pH. For HILIC HPLC-ESI-MS of nucleotides, control over the pH of the system holds even greater importance because of the existence of different protonation states of each nucleotide at different pHs. Depending on the type of nucleotide, whether it is (a) purine- or pyrimidine-based, or (b) mono-, di-, or triphosphate, the pH causes populations of different protonation states in varying percentages. Knowing that this phenomenon occurs, it becomes crucial to find an optimal pH that isolates all nucleotides in the mixture with a dominant protonation state. The presence of a mixture of different protonation states would obscure chromatographic peaks for each nucleotide possibly leading to severe peak broadening, multiple and inconsistent retention times, less-than-optimal ionization efficiency, or other unpredictable results. Thus, an appreciation of the pK_a and log D values for each nucleotide can be helpful when selecting a suitable pH that accommodates all the nucleotides.

A representation of the calculations obtained for the nucleotide, UMP (**5**), is shown in Figure 19.1A, where the percent compositions of protonation states are

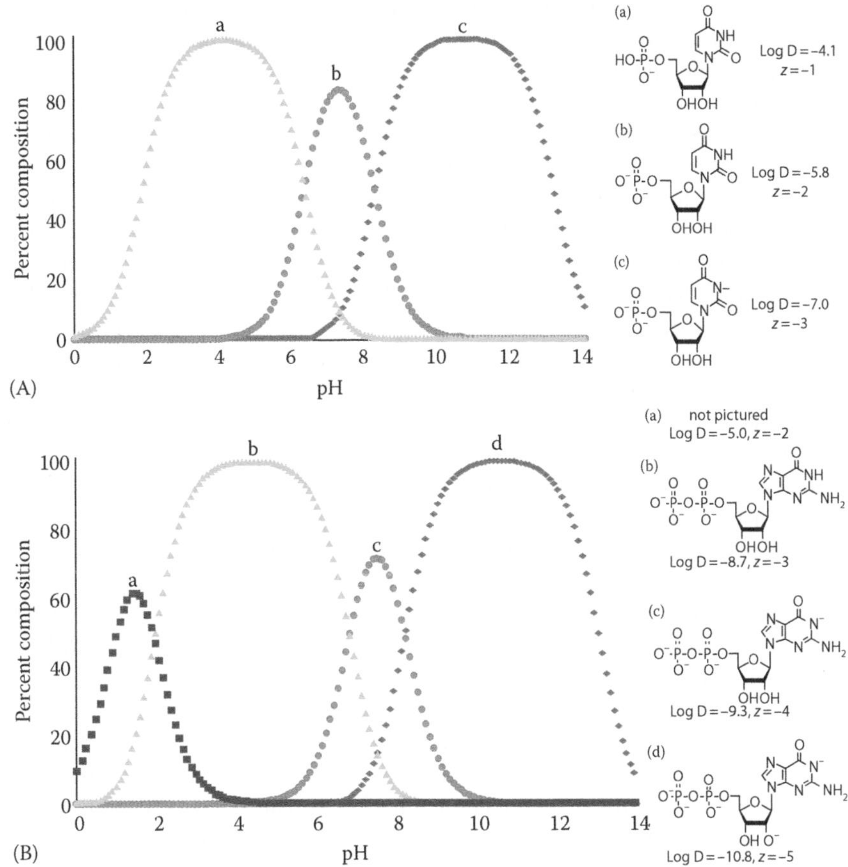

FIGURE 19.1 Percent composition of predicted protonation states for (A) UDP (**5**) and (B) UDP (**6**) as a function of pH. z denotes charge (protonation) state.

plotted against the pH. From these data, it is clear that two pH settings are the most feasible to be experimentally tested; specifically, pH 4.1 and pH 11.1 are predicted to place over 90% of the species population in a single protonation state (−1 and −3, respectively). In comparison, for UDP (**6**) (Figure 19.1B), the most optimal pHs for separation are predicted to be 4.5 and 10.6 ($z = -3$ and −5, respectively). The different nucleotides each have their own optimal pH values based on similar analyses. After a comparison between all eight nucleotides, it was concluded that the most suitable pHs that satisfy the optimal ranges for all nucleotides, and where all exist predominantly in a single ion form, were 4.7 and 10.0, as shown in Table 19.2. An operating pH of 10.0 was ruled out due to the increase in the solubility of the silica-based stationary phases at this high pH.[1] Thus, acidic mobile phases were investigated to optimize the separations.

It is important to note that control (and measurement) of the pH during HILIC separations is not straightforward. While it is relatively simple to accurately set the pH of the aqueous mobile phase, to perform the same pH control on the organic mobile phase is impractical with a standard pH meter. The large presence of acetonitrile, a weak base with a low dielectric constant, in the organic mobile phase reduces charge separation and the ionizability of compounds.[40,41] This phenomenon results in a pH shift from the desired pH of 4.7 to higher values, when a large proportion of acetonitrile is present. Adding more acid to the organic mobile phase until the pH reaches 4.7 is not a viable option as it causes more complications (i.e., a variable ionic strength) when using a gradient profile in the separation scheme, and it cannot be reliably measured. Due to the change in organic composition during the gradient chromatographic elutions, the exact pH cannot be determined during the run. Still, it is reasonable that most acidic species will be affected in a similar fashion when the dielectric constant of the medium is changed. Percentages of formic acid (0.01%, 0.05%, 0.1%, and 0.5%) were varied to find the optimal pH (data not included) at which the best separation and the strongest ESI-MS signal intensities were recorded. The addition of 0.05% (v/v) formic acid with 20 mM ammonium formate buffer solution, equating to a pH of 3.96 in the aqueous mobile-phase reservoir, was found to be optimal; an equal amount of acid was added to the organic mobile-phase mixture to ensure consistency of the acid content throughout the experimental testing.

While operating under acidic conditions, logic would normally dictate for the analyte to be preferentially detected in the positive-ionization mode. This generally accepted idea is a misconception, as it is not always the case, and the HILIC separation of nucleotides represents one such exception. According to Boyd, the ESI-MS analysis of amino acids can follow a "wrong-way-round" electrospray ionization pattern in that the amino acids can still be detected in the negative-ionization mode despite working under acidic conditions.[42] Conversely, amino acid detection can also be achieved in the positive-ionization mode while operating under basic conditions.[42,43] In this study, the detection of nucleotides under acidic conditions was generally most sensitive with the positive-ionization mode. However, the negative-ionization mode shows appreciable signal intensity for the nucleotides as well. For certain nucleotides, the signal intensity in the negative-ionization mode equaled or even surpassed that of the positive-ionization mode. Figure 19.2 shows extracted ion chromatograms of GMP (**3**) and UDP (**6**), retained on the amide column, along with their respective mass spectra in both the positive- and negative-ionization modes. Both positive-mode spectra exhibit $[M + H]^+$ signals, with matching $[M - H]^-$ signals in the negative mode; the differences lie within their signal strengths. While the GMP $[M + H]^+$ signal was nearly an order of magnitude greater than its corresponding $[M - H]^-$ signal, the UDP $[M + H]^+$ and $[M - H]^-$ signals were within the same order of magnitude.

The rest of the nucleotides that were analyzed also exhibited either of the two patterns between the positive- and negative-ionization modes. Interestingly, UMP (**5**), UDP (**6**), and TMP (**7**) are the three nucleotides that follow the latter trend that were mentioned, previously. An explanation for such behavior can be found in the structure of their nearly identical nucleoside bases, which are unique from other nucleoside bases in that they contain two carbonyl groups attached to the pyrimidine ring as opposed to incorporation of at least one amine group onto the pyrimidine ring. The presence of an amine group facilitates sensitive detection in the positive-ionization mode. It is also important to note the presence of an $[UDP + NH_4]^+$ adduct

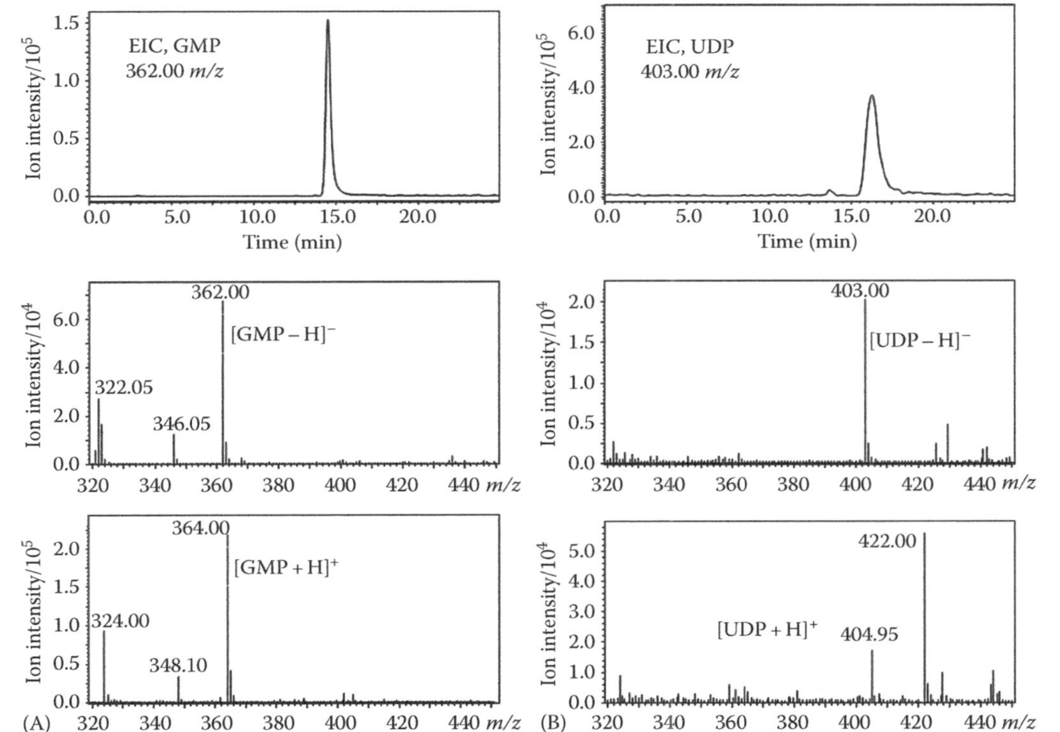

FIGURE 19.2 (A) Extracted ion chromatogram of GMP (**3**) on Tosoh TSK-Gel Amide-80 column with gradient separation. Mass spectra of the peak in the negative- and positive-ionization modes show signals for $[GMP - H]^-$ and $[GMP + H]^+$, respectively. (B) Extracted ion chromatogram of UDP (**6**) under the same conditions. Mass spectra of the peak in the negative- and positive-ionization modes show signals for $[UDP - H]^-$ and $[UDP + H]^+$, respectively.

in the mass spectra, which is unique to UDP as it is not seen on any other of the analyzed nucleotides. The presence of this ammonium adduct must be taken into consideration as it represents a significant fraction of the UDP ions and will affect the calculated sensitivity values in method validation. Therefore, due to the apparent viability of both the positive- and negative-ionization modes in the detection of the nucleotides, both modes were scanned during each chromatographic run. In addition, both modes were considered and individually analyzed in further method validation calculations.

19.3.2 Column Chemistries

A total of five types of HILIC columns were tested in developing a method for the separation of nucleotides. The columns were obtained from four different manufacturers allowing a comparison between some of the different types of columns that are currently commercially available. The types of stationary phases include polar functional groups such as amide- (TSK-Gel Amide and Polaris Amide-C18) or dihydroxy-bonded (diol) variants (Luna HILIC and XRs Pursuit 3 Diol). Polaris Aminopropyl and BETASIL Cyano columns were also studied. Theoretically, the amide- and amino-functionalized stationary phases should be the most suited to perform separations on acidic analytes. Hydrogen bonding and electrostatic forces increase the interaction between the amide/amino groups with the negatively charged phosphate of the nucleotides yielding significant retention. The two bridged diol columns and the cyano column can also interact with the negatively charged analytes through hydrogen bonding. The Polaris Amide-C18 has a polar embedded amide group underneath a C18 chain that could allow for both hydrogen bonding and hydrophobic interactions to contribute to analyte retention.

19.3.3 Gradient Method Development

Multiple chromatographic parameters were optimized leading to the best separations from each type of column. Although buffer pH has already been considered as previously mentioned, mobile phase compositions and gradients as well as ionic strengths were extensively tested. Isocratic experiments were first performed on each of the columns at different mobile-phase compositions in order to give a rough estimate of retention times and selectivity on each of the columns. From these data, the optimal gradient profiles were designed. A range of different buffer ionic strengths (10, 20, 30, 40, and 100 mM ammonium formate) were tested (data not included) to find the optimal buffer concentration, which was determined to be 20 mM ammonium formate. Other buffer systems were also tested (i.e., ammonium carbonate, ammonium bicarbonate, and ammonium acetate), but did not provide acceptable performance (data not included).

The best separations, which were most effective on the Tosoh TSK-Gel Amide-80 column, were achieved using a gradient program consisting of a 5 min isocratic period in the beginning, followed by a gradient of 80% ACN to 50% ACN in 15 min, and then a 5 min isocratic period at 50% ACN. Separation was achieved on the Phenomenex Luna HILIC using a similar gradient program as that of the Tosoh amide, except with a gradient of 80% ACN to 50% ACN in 25 min. Other gradients profiles were tested with slightly different modifications to the gradient for optimization of retention on other columns. For the amide-C18 column, the optimal conditions for retention required a gradient profile that started at 87.5% ACN to 50% ACN in 15 min following the same program as before with the 5 min isocratic periods before and after the gradient. The other three columns (aminopropyl, diol, and cyano) were also tested with different gradient profiles containing different variations, but the retention and separation achieved from these columns were not as successful as those for the first three columns described earlier.

19.3.4 Solubility Issues

Ideally, an analyte sample should be premixed in the initial mobile-phase composition of the separation to reduce the equilibration of the sample upon injection onto the column. Preparation of the nucleotide samples in the initial mobile-phase compositions (such as 20:80 water/ACN and 12.5:87.5 water/ACN) was not possible due to the precipitation of the nucleotides at a high organic content. Being negatively charged and thus highly polar, the nucleotides have low solubility in organic solvents such as acetonitrile. The precipitation of solids not only obscures the chromatographic results, but also causes clogging of the electrospray needle and leads to exceedingly high back pressures, which requires immediate maintenance to resolve. As a result, a composition of 30:70 water/ACN was determined to be the highest amount of organic solvent that can be premixed with nucleotides without causing precipitation. Some higher contents of acetonitrile would initially dissolve the nucleotides (such as a 25:75 water/ACN composition), but precipitation would still occur after the passage of time within the autosampler vial as well as after the elution of the nucleotides from the column, again causing the electrospray needle to become blocked.

19.3.5 Separation Results

Figure 19.3 shows the extracted ion chromatograms of each nucleotide (**1–8**) obtained following separation with the Tosoh Amide-80 column. Separation was observed for all eight of the nucleotides with distinct retention times except for compounds **1** and **5**. The purine diphosphates **2** and **4** exhibit broadened peaks and low signal intensities, a recurring theme in this investigation. A further discussion on these results with the purine diphosphates is given in the following text. The extracted ion chromatogram for CMP (**8**) at *m/z* 322 contains two different peaks at different retention times due to the proximity of its molecular mass to that of TMP (**7**). The first peak corresponds to the +1 isotope of TMP (**7**) whose molecular mass is exactly one proton less than CMP (**8**), whereas the second peak corresponds to CMP (**8**). All assigned signals for the respective extracted ion chromatograms were assigned by mass as well as by analysis of single component mixtures.

Extracted ion chromatograms of each nucleotide (**1–8**) collected from separation on the Phenomenex Luna HILIC column are shown in Figure 19.4. The Luna HILIC column also provided appreciable retention and selectivity for many of the

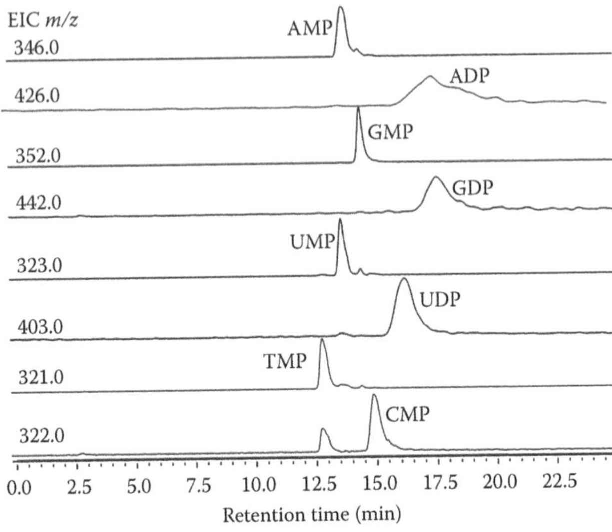

FIGURE 19.3 Extracted ion chromatograms of AMP (**1**), ADP (**2**), GMP (**3**), GDP (**4**), UMP (**5**), UDP (**6**), TMP (**7**), CMP (**8**) obtained by separation on a Tosoh TSK-Gel Amide-80 column with a gradient program.

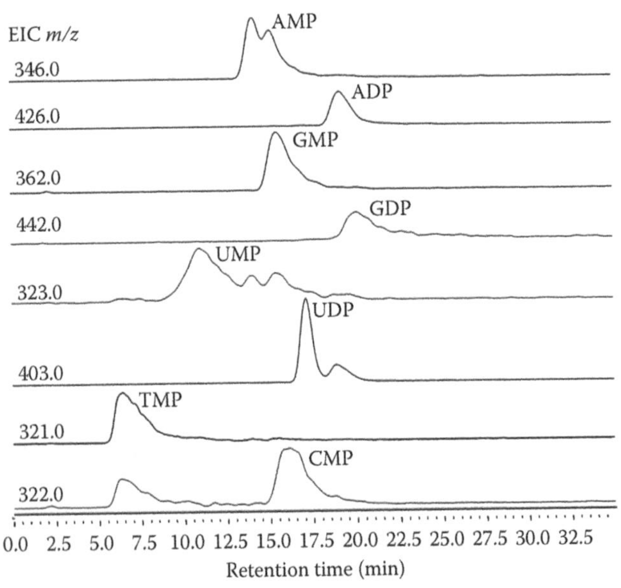

FIGURE 19.4 Extracted ion chromatograms of AMP (**1**), ADP (**2**), GMP (**3**), GDP (**4**), UMP (**5**), UDP (**6**), TMP (**7**), CMP (**8**) obtained by separation on a Phenomenex Luna HILIC column with a gradient program.

compounds, comparable to that of the Tosoh Amide-80 column. The peaks do however, suffer in shape and quality exhibiting a high degree of peak broadening and peak tailing in all extracted ion chromatograms. Asymmetrical peaks are typically indicative of multiple interaction modes between the stationary phase and the analyte.[44–46] The same trend with a diminished sensitivity of the purine diphosphates (**2, 4**) is also observed as well as the same dual peaks are seen from the chromatogram of CMP (**8**), which can again be attributed to the +1 isotope of TMP (**7**), which has the same mass-to-charge ratio as CMP (**8**).

The Varian Polaris 3 Amide-C18 column also provided some separation of the nucleotides, although it was significantly less pronounced than that for the Tosoh amide or the Phenomenex Luna HILIC. The majority of the monophosphates (**1, 5, 7,** and **8**) eluted near the dead volume peak giving capacity factors less than 1. However, the diphosphates (**2, 4,** and **6**) and GMP (**3**) were minimally retained by the unique stationary phase yielding capacity factors within a range of 1.0–1.2. Whereas the other two columns were able to successfully resolve the nucleotides, the amide-C18 was only able to separate the nucleotides into two major peaks with overall lower retention compared to the Tosoh Amide-80 and Phenomenex Luna HILIC columns. Once again, the purine diphosphates (**2** and **4**) showed a loss in sensitivity with broadened peaks.

Table 19.3 displays calculated capacity factors (k′) of each nucleotide on each of the three columns that exhibited retention. From the results of this investigation, it is clear that the Tosoh amide was the most effective for the analysis of the mixture of nucleotides and gave the best retention, resolution, and quality of peak shapes out of the three columns. The Varian XRs 3 Diol and the Thermo BETASIL Cyano columns did not provide appreciable retention of the nucleotide analytes, and thus their results were not reported in the comparison. The results from the Varian aminopropyl column were unique in that elution of the nucleotides was not observed, which leads to the speculation that they may still be retained on the column.

An explanation of the results can be made based on the different column stationary phases and their ability to separate the nucleotide mixture. As mentioned previously,

TABLE 19.3
Comparison of Capacity Factors (k′) of Each Nucleotide with Different HILIC Columns

	AMP (1)	ADP (2)	GMP (3)	GDP (4)	UMP (5)	UDP (6)	TMP (7)	CMP (8)
Tosoh TSK-Gel Amide-80 (t_0 = 1.0 min)	12.5	16.0	13.7	16.9	12.4	15.4	10.8	14.4
Phenomenex Luna HILIC (t_0 = 1.25 min)	10.4	14.3	11.3	15.0	7.3	12.7	4.4	11.8
Varian Polaris 3 Amide-C18 (t_0 = 1.23 min)	0.5	1.2	1.1	N.D.	0.3	1.1	0.3	0.4

the amide-and amino-functionalized stationary phases should theoretically perform the best in the separation of these negatively charged analytes. This conjecture has been verified by the results obtained with the Tosoh Amide-80 column as it indeed performed the most effective separation. One obvious question that arises is why the Varian aminopropyl column did not have as much success as the Tosoh amide column. The gradient elutions of the aminopropyl column, as well as the washings (data not shown), did not contain ion signals corresponding to the nucleotide masses and thus indicated that the nucleotides may not have been eluted off the column under the typical HILIC mobile-phase conditions used in this study. Perhaps the acidic mobile-phase conditions caused strong binding interactions between the amino group and the nucleotide phosphate groups and thus prevented the elution of the nucleotides. Retention of nucleotides on the Polaris aminopropyl column has been reported,[39] but the basic mobile-phase conditions (pH 9.45) approach the solubility limit of the silica-based stationary phase and may cause damage to the column over prolonged usage.[1] Interestingly, the Varian Polaris 3 amide-C18 column was still capable of achieving minimal retention of the nucleotides despite the presence of the large, hydrophobic C18 chain attached to the amide nitrogen, which would intuitively hinder the hydrophilic interaction between the amide group and the nucleotides. The Thermo cyano and Varian diol columns did not provide appreciable retention of the nucleotides under the experimental conditions investigated. Both types of columns have also been reported to be well suited for the separation of basic compounds,[47] so the findings for these columns are not surprising, except to note that the Luna HILIC (also, a diol stationary phase) did indeed provide adequate retention. Without an in-depth knowledge of the exact stationary-phase chemistry, no further explanation of this disparity can currently be made.

Throughout the analysis, there was an obvious distinction between the monophosphates and the diphosphates. The monophosphates eluted before the diphosphates on a consistent basis, which was expected. The presence of two phosphate groups increases the hydrophilicity of the diphosphate nucleotides relative to the monophosphates, and increases retention. The calculated values for log D (see Figure 19.1 and Table 19.2) support this assertion. Within the subgroups of mono- and diphosphates, the nucleotides containing purine bases (**1–4**) were retained more than the pyrimidine nucleotides (**5–8**). The N-7 nitrogen in the imidazole ring of purine bases can act as an additional polar group accentuating the interaction between the polar stationary phase and the nucleotide. The imidazole nitrogen is also suspect in causing the pattern of low sensitivity and broadened peaks seen throughout all analyses. Protonation on this imidazole nitrogen is possible and could provide a positive charge that neutralizes the negative charge from the phosphate group, leading to poor ionization efficiency.[48,49] This deleterious effect is even more pronounced in the case of guanine-based compounds **2** and **4**, because of an additional protonation site on N-1 nitrogen, which is not present in adenosine nucleotides **1** and **3**. Although the purine nucleotides consistently had lower signal intensities than the pyrimidine nucleotides, the diphosphate purines suffered a much greater loss in sensitivity and quality of peak shape than their monophosphate counterparts due to the adverse effects caused by both the second phosphate and the purine base. The same negative effects were not seen in the pyrimidine diphosphate, UDP (**6**); it exhibited both sharp peaks and high ion-signal intensity.

Among the pyrimidines, TMP (**7**) had the least retention because of the presence of a methyl group on the C-5 carbon of the nucleoside base as well as the absence of a hydroxyl group at the C-2 carbon on the ribose sugar ring. TMP (**7**) is the only tested nucleotide with a deoxyribose sugar because it naturally exists solely in a deoxy-form, one of the clear distinctions between deoxyribonucleic acids (DNA) and ribonucleic acids (RNA). It is also important to note that UMP (**5**) and CMP (**8**), which differ between the attachment of a carbonyl group and an amino group, respectively, on the C-4 carbon of the nucleoside base, switch elution orders when separated on the Tosoh Amide-80 column and the Phenomenex Luna HILIC column. The amide group retains like-amine groups present on the CMP (**8**) nucleoside base more than the carbonyl group on the UMP (**5**).

19.3.6 Method Validation

The establishment of calibration curves for each nucleotide in both the positive- and negative-ionization mode was achieved by use of seven chosen concentration points within a range of 0–350 μM. A fitting of the resultant plots was then performed and lead to the discovery that not all generated calibration curves were uniformly linear, the details of which are reported in Table 19.4. Some nucleotide curves were best fit to a polynomial (x^2) regression or even a segmented, linear regression containing two distinct linear trends within the same calibration curve. It is also interesting to note that this nonlinear fit did not carry through between both positive- and negative-ionization modes. An analysis of the calibration curves lead to the finding that nucleotide curves that were nonlinear in one specific ionization mode would be linear in the opposite ionization mode with the exception of ADP (**2**), which produced nonlinear calibration curves in both modes. Figure 19.5 presents the notable calibration curves of nucleotides that illustrate these results. In the positive-ionization mode, CMP (**8**) clearly fit a square-polynomial with an R^2-value of 0.999, whereas when the analysis was performed in the negative-ionization mode, a linear regression was the best fit to the curve with an R^2-value of 0.991. Conversely, in the case of UDP (**6**), the opposite was observed where the linear fit was best achieved on the calibration curve generated from the positive-ionization mode, while the curve from the negative-ionization mode was best fit by a square polynomial regression. Both regression fits of UDP calibration curves had correlation coefficients better than $R^2 = 0.988$. Further questions arise from the two calibration curves obtained from the method validation of ADP (**2**), which not only displayed a square-polynomial fit for its curve from the negative-ionization mode, but also produced a segmented, linear fit for the calibration curve from the positive mode as seen from Figure 19.5. There are clearly two separate linear functions in the ADP positive-mode calibration curve where the first line retains linearity from 0 to 150 μM, and the second line begins at 150 μM and maintains linearity to 300 μM. Currently, the cause of these nonlinear regressions is difficult to surmise. It can only be speculated that some mechanism occurs during the chromatographic elution that reduces the signal detection of certain nucleotides at low concentrations up to a certain threshold, after which the signal intensities of the nucleotide rise sharply. It is possible that poor peak shape for some of the compounds could also account for this odd behavior. Fortunately, with the exception of ADP (**2**), the issue of nonlinearity can be

TABLE 19.4
Selected Method Validation Parameters for Positive- and Negative-Mode Ionization of Eight Nucleotides

		Positive Mode			Negative Mode		
	$t_R \pm$ SD (min)	Best Fit	Range (R^2)	LOD (ng)	Best Fit	Range (R^2)	LOD (ng)
AMP	14.4 ± 0.1	Linear	20–300 µM (0.991)	96	Linear	10–300 µM (0.991)	74
ADP	17.80 ± 0.06	Linear, segmented	10–150 µM (0.994); 150–300 µM (0.987)	116	Polynomial (x^2)	10–300 µM (0.998)	~60
GMP	15.45 ± 0.08	Linear	10–300 µM (0.994)	92	Linear	10–300 µM (0.980)	74
GDP	17.92 ± 0.06	Linear	20–300 µM (0.970)	162	Polynomial (x^2)	10–300 µM (0.972)	~70
UMP	14.3 ± 0.1	Polynomial (x^2)	10–300 µM (0.999)	~50	Linear	10–300 µM (0.982)	82
UDP	16.73 ± 0.06	Linear	20–300 µM (0.988)	127	Polynomial (x^2)	10–300 µM (0.998)	~60
TMP	13.0 ± 0.2	Linear	10–300 µM (0.993)	71	Linear	10–300 µM (0.961)	110
CMP	16.04 ± 0.08	Polynomial (x^2)	10–300 µM (0.999)	~50	Linear	10–300 µM (0.991)	103

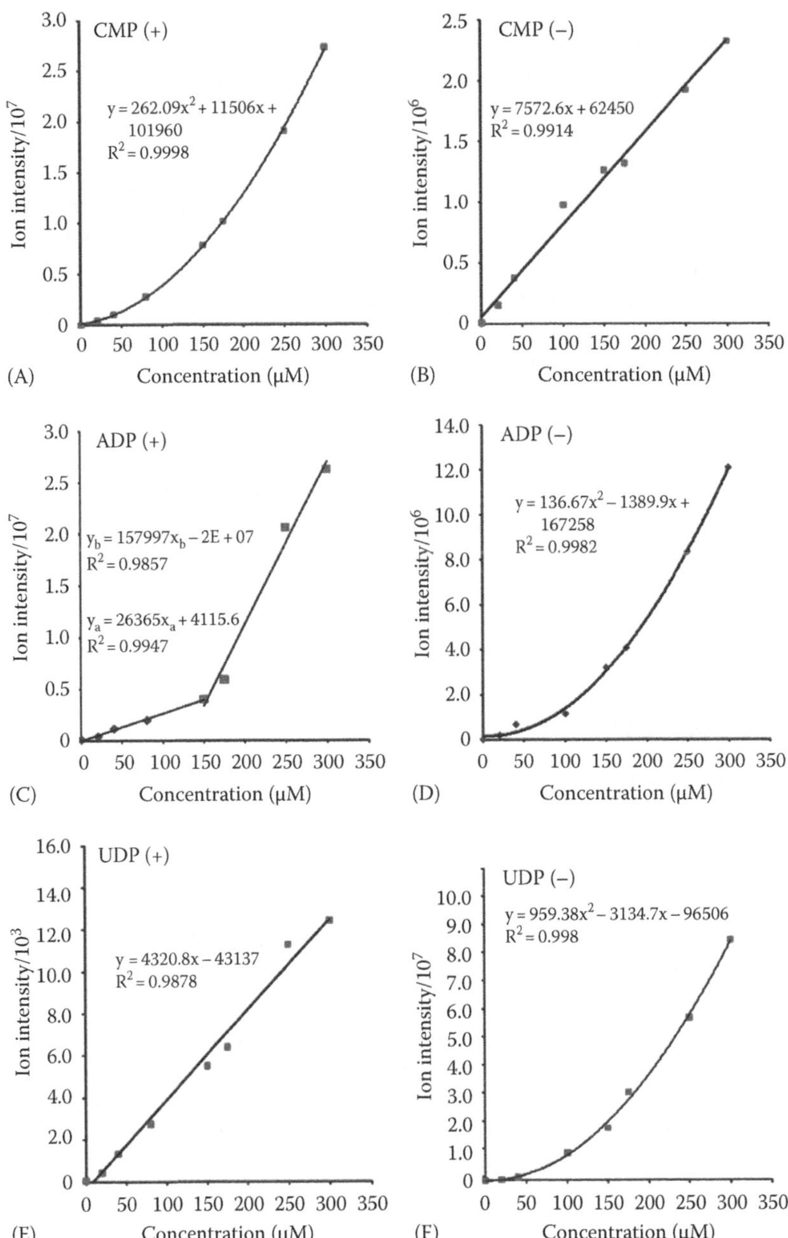

FIGURE 19.5 Calibration curves obtained for CMP (**8**) performed in the (A) positive-ionization mode and (B) negative-ionization mode; for ADP (**2**) performed in the (C) positive-ionization mode and (D) negative-ionization mode; and for UDP (**6**) performed in the (E) positive-ionization mode and (F) negative-ionization mode. All analyses were performed with the optimized method on the Tosoh Amide-80 column.

circumvented with an analysis of the nucleotide in the opposite ionization mode, which does retain linearity within the reported range. The calibration curves maintained correlation to their best fit regression primarily within a range of 10–300 μM with some exceptions (i.e., the positive-ionization mode of ADP (**2**), GDP (**4**), and UDP (**6**)). At a concentration of 350 μM (data not shown), all nucleotide calibration curves deviated from their best fit and were thus excluded from the calibration curves.

The data indicate good reproducibility of the developed method with regard to retention times (<2% RSD). The limit of detection (LOD) determinations for each nucleotide in each ionization mode indicate moderate levels of sensitivity ranging in the detection of 48–162 ng of analyte in a 15 μL injection volume. The detection limits of the nonlinear plots were only roughly determined by successive decreases of analyte concentration until the nucleotide mass spectral signal was unable to be detected. In calculating the LOD for uridine 5′-diphosphate (UDP) in the positive-ionization mode, both the signal intensities from $[UDP + H]^+$ and the $[UDP + NH_4]^+$ ions were taken into account by plotting the summation of both signal intensities against concentration after which the LOD was then calculated as mentioned, previously. The division of the total signal intensity between the $[UDP + H]^+$ ion and the $[UDP + NH_4]^+$ adduct leads to an expected decrease in sensitivity for this specific nucleotide. According to the table, the best ionization mode for optimal LOD is analyte dependent. AMP, ADP, GMP, GDP, and UDP show the lowest LOD in the negative-ionization mode, whereas UMP, TMP, and CMP are detected with the lowest LOD in the positive-ionization mode.

19.4 CONCLUDING REMARKS

From this investigation, an HILIC-ESI-MS method was developed for the separation of eight different nucleotides. In addition, six different commercial columns were tested. It is clear that the Tosoh Amide-80 column is most suited for achieving the separation of nucleotides, whereas the Phenomenex Luna HILIC column showed the capability in doing so as well. With the use of ESI-MS detection, overlapping signals from unresolved analytes can be deconvoluted for nucleotide detection, however, poor peak shapes in some cases, arising from multiple interactions modes between the analyte and the stationary phase, were difficult to avoid.

The optimization of pH was crucial in the development of the method to control the protonation states of the different nucleotides. The described method for the nucleotides is expected to lead to a further development of rapid enzyme kinetic assays. Quantification of nucleotide enzyme substrates or enzyme products can now be achieved using a more simplified HPLC-ESI-MS protocol that does not require the use of ion-pairing reagents. However, the investigation of additional stationary-phase chemistries (e.g., zwitterionic) may be necessary to achieve the best possible analysis conditions.

ACKNOWLEDGMENTS

The authors wish to acknowledge support from the University of Texas at Arlington, Shimadzu Scientific Instruments, Inc., and Varian, Inc. in performing this work.

REFERENCES

1. Nguyen, H. P.; Schug, K. A. The advantages of ESI-MS detection in conjunction with HILIC mode separations: Fundamentals and applications. *J. Sep. Sci.* 2008, *31*, 1465–1480.
2. Shou, W. Z.; Weng, N. Simple means to alleviate sensitivity loss by trifluoroacetic acid (TFA) mobile phases in the hydrophilic interaction chromatography–electrospray tandem mass spectrometric (HILIC–ESI/MS/MS) bioanalysis of basic compounds. *J. Chromatogr. B* 2005, *825*, 186–192.
3. Li, W.; Li, Y.; Li, A. C.; Zhou, S.; Weng, N. Simultaneous determination of norethindrone and ethinyl estradiol in human plasma by high performance liquid chromatography with tandem mass spectrometry—Experiences on developing a highly selective method using derivatization reagent for enhancing sensitivity. *J. Chromatogr. B* 2005, *825*, 223–232.
4. Song, Q.; Junga, H.; Li, A. C. et al. Automated 96-well solid phase extraction and hydrophilic interaction liquid chromatography–tandem mass spectrometric method for the analysis of cetirizine (ZYRTEC®) in human plasma—With emphasis on method ruggedness. *J. Chromatogr. B* 2005, *814*, 105–114.
5. Hao, Z.; Xiao, B.; Weng, N. Impact of column temperature and mobile phase components on selectivity of hydrophilic interaction chromatography (HILIC). *J. Sep. Sci.* 2008, *31*, 1449–1464.
6. Alpert, A. J. Hydrophilic-interaction chromatography for the separation of peptides, nucleic acids and other polar compounds. *J. Chromatogr. A* 1990, *499*, 177–196.
7. Gianotti, V.; Chiuminatto, U.; Mazzucco, E.; Gosetti, F.; Bottaro, M.; Frascarolo, P.; Gennaro, M. C. A new hydrophilic interaction liquid chromatography tandem mass spectrometry method for the simultaneous determination of seven biogenic amines in cheese. *J. Chromatogr. A* 2008, *1185*, 296–300.
8. Wang, Y.; Lu, X.; Xu, G. Development of a comprehensive two-dimensional hydrophilic interaction chromatography/quadrupole time-of-flight mass spectrometry system and its application in separation and identification of saponins from *Quillaja saponaria. J. Chromatogr. A* 2008, *1181*, 51–59.
9. Zhang, H.; Guo, Z.; Zhang, F.; Xu, Q.; Liang, X. HILIC for separation of co-eluted flavonoids under RP-HPLC mode. *J. Sep. Sci.* 2008, *31*, 1623–1627.
10. Zhang, H.; Wei, Z. G.; Jiatao, L. et al. Purification of flavonoids and triterpene saponins from the licorice extract using preparative HPLC under RP and HILIC mode. *J. Sep. Sci.* 2009, *32*, 526–535.
11. Appelblad, P.; Jonsson, T.; Jiang, W.; Irgum, K. Fast hydrophilic interaction liquid chromatographic separations on bonded zwitterionic stationary phase. *J. Sep. Sci.* 2008, *31*, 1529–1536.
12. Dernovics, M.; Lobinski, R. Speciation analysis of selenium metabolites in yeast-based food supplements by ICPMS assisted hydrophilic interaction HPLC hybrid linear ion trap/orbitrap MS. *Anal. Chem.* 2008, *80*, 3975–3984.
13. Qin, F.; Zhao, Y.-Y.; Sawyer, M. B.; Li, X.-F. Hydrophilic interaction liquid chromatography–tandem mass spectrometry determination of estrogen conjugates in human urine. *Anal. Chem.* 2008, *80*, 3404–3411.
14. Jiang, Z.; Smith, N. W.; Ferguson, P. D.; Taylor, M. R. Hydrophilic interaction chromatography using methacrylate-based monolithic capillary column for the separation of polar analytes. *Anal. Chem.* 2007, *79*, 1243–1250.
15. Wang, C.; Jiang, C.; Armstrong, D. W. Considerations on HILIC and polar organic solvent-based separations: Use of cyclodextrin and macrocyclic glycopetide stationary phases. *J. Sep. Sci.* 2008, *31*, 1980–1990.
16. Park, S. K.; Liao, L.; Kim, J. Y.; Yates, J. R. A computational approach to correct arginine-to-proline conversion in quantitative proteomics. *Nat. Methods* 2009, *6*, 184–185.

17. Ebhardt, H. A.; Xu, Z.; Fung, A. W.; Fahlman, R. P. Quantification of the post-translational addition of amino acids to proteins by MALDI-TOF mass spectrometry. *Anal. Chem.* 2009, *81*, 1937–1943.
18. Zuberovic, A.; Wetterhall, M.; Hanrieder, J.; Bergquist, J. CE MALDI-TOF/TOF MS for multiplexed quantification of proteins in human ventricular cerebrospinal fluid. *Electrophoresis* 2009, *30*, 1836–1843.
19. Xia, B.; Feasley, C. L.; Sachdev, G. P.; Smith, D. F.; Cummings, R. D. Glycan reductive isotope labeling for quantitative glycomics. *Anal. Biochem.* 2009, *387*, 162–170.
20. Ding, L.; Cheng, W.; Wang, X.; Ding, S.; Ju, H. Carbohydrate monolayer strategy for electrochemical assay of cell surface carbohydrate. *J. Am. Chem. Soc.* 2008, *130*, 7224–7225.
21. Boersema, P. J.; Mohammed, S.; Heck, A. J. R. Hydrophilic interaction liquid chromatography (HILIC) in proteomics. *Anal. Bioanal. Chem.* 2008, *391*, 151–159.
22. Ren, Y. L.; Garges, S.; Adhya, S.; Krakow, J. S. Cooperative DNA binding of heterologous proteins: Evidence for contact between the cyclic AMP receptor protein and RNA polymerase. *Proc. Natl. Acad. Sci. U.S.A.* 1988, *85*, 4138–4142.
23. Ling, E.; Danilov, Y. N.; Cohen, C. M. Modulation of red cell band 4.1 function by cAMP-dependent kinase and protein kinase C phosphorylation. *J. Biol. Chem.* 1988, *263*, 2209–2216.
24. Yamano, S.; Tanaka, K.; Matsumoto, K.; Toh-e, A. Mutant regulatory subunit of 3′,5′-cAMP-dependent protein kinase of yeast *Saccharomyces cerevisiae. Mol. Gen. Genet.* 1987, *210*, 413–418.
25. Bubis, J.; Neitzel, J. J.; Saraswat, L. D.; Taylor, S. S. A point mutation abolishes binding of cAMP to site A in the regulatory subunit of cAMP-dependent protein kinase. *J. Biol. Chem.* 1988, *263*, 9668–9673.
26. Corbin, J. D.; Cobb, C. E.; Beebe, S. J. et al. Mechanism and function of cAMP- and cGMP-dependent protein kinases. *Adv. Second Messenger Phosphoprotein Res.* 1988, *21*, 75–86.
27. Swierczynski, J.; Klimek, J.; Zelewski, L. Correlation between the malate dependent progesterone and citrate biosynthesis in the mitochondrial fraction of human term placenta. The stimulatory effect of ADP and ATP. *J. Steroid Biochem. Mol. Biol.* 1986, *24*, 591–595.
28. Wittenberg, B. A.; Wittenberg, J. B. Myoglobin-mediated oxygen delivery to mitochondria of isolated cardiac myocytes. *Proc. Natl. Acad. Sci. U.S.A.* 1987, *84*, 7503–7507.
29. Kun, E. Kinetics of ATP-dependent Mg^{2+} flux in mitochondria. *Biochemistry* 1976, *15*, 2328–2336.
30. Sarkar, N. K. The effect of ions and ATP on myosin and actomyosin. *Enzymologia* 1950, *14*, 237–245.
31. Geyer, M.; Schweins, T.; Herrmann, C.; Prisner, T.; Wittinghofer, A.; Kalbitzer, H. R. Conformational transitions in p21ras and in its complexes with the effector protein Raf-RBD and the GTPase activating protein GAP. *Biochemistry* 1996, *35*, 10308–10320.
32. Valencia, A.; Chardin, P.; Wittinghofer, A.; Sander, C. The ras protein family: Evolutionary tree and role of conserved amino acids. *Biochemistry* 1991, *30*, 4637–4648.
33. Klebe, C.; Bischoff, F. R.; Ponstingl, H.; Wittinghofer, A. Interaction of the nuclear GTP-binding protein Ran with its regulatory proteins RCC1 and RanGAP1. *Biochemistry* 1995, *34*, 639–647.
34. Riese, M. J.; Wittinghofer, A.; Barbieri, J. T. ADP ribosylation of Arg41 of Rap by ExoS inhibits the ability of Rap to interact with its guanine nucleotide exchange factor, C3G. *Biochemistry* 2001, *40*, 3289–3294.
35. Guo, Y.; Gaiki, S. Retention behavior of small polar compounds on polar stationary phases in hydrophilic interaction chromatography. *J. Chromatogr. A* 2005, *1074*, 71–80.

36. Tuytten R.; Lemiere, F.; Van Dongen, W.; Esmans. E. L.; Slegers, H. Short capillary ion-pair high-performance liquid chromatography coupled to electrospray (tandem) mass spectrometry for the simultaneous analysis of nucleoside mono-, di- and triphosphates. *Rapid Commun. Mass Spectrom.* 2002, *16*, 1205–1215.
37. Xing, J.; Adrienne, A. A.; Zhao, T. N. Liquid chromatographic analysis of nucleosides and their mono-, di- and triphosphates using porous graphitic carbon stationary phase coupled with electrospray mass spectrometry. *Rapid Commun. Mass Spectrom.* 2004, *18*, 1599–1606.
38. Cordell, R. L.; Hill, S. J.; Ortori, C. A.; Barrett, D. A. Quantitative profiling of nucleotides and related phosphate-containing metabolites in cultured mammalian cells by liquid chromatography tandem electrospray mass spectrometry. *J. Chromatogr. B* 2008, *871*, 115–124.
39. Vincenzo, P.; Giuliano, C.; Zhang, R. et al. HILIC LC-MS for the determination of 2′-C-methyl-cytidine-triphosphate in rat liver. *J. Sep. Sci.* 2009, *32*, 1275–1283.
40. Subirats, X.; Bosch, E.; Rosés, M. Retention of ionisable compounds on high-performance liquid chromatography XVIII: pH variation in mobile phases containing formic acid, piperazine, tris, boric acid or carbonate as buffering systems and acetonitrile as organic modifier. *J. Chromatogr. A* 2009, *1216*, 2491–2498.
41. Kostiainen, R.; Kauppila, T. J. Effect of eluent on the ionization process in liquid chromatography–mass spectrometry. *J. Chromatogr. A* 2009, *1216*, 685–699.
42. Mansoori, B. A.; Volmer, D. A.; Boyd, R. K. “Wrong-way-round” electrospray ionization of amino acids. *Rapid Commun. Mass Spectrom.* 1997, *11*, 1120–1130.
43. Zhou, S.; Cook, K. D. Protonation in electrospray mass spectrometry: Wrong-way-round or right-way-round? *J. Am. Soc. Mass Spectrom.* 2000, *11*, 961–966.
44. De Schutter, J. A.; De Moerloose, P. Polar contributions of the stationary phase to the reversed-phase ion-pair high-performance liquid chromatographic separation of quaternary ammonium drugs. *J. Chromatogr. A* 1988, *437*, 83–95.
45. Sýkora, D.; Tesarová, E.; Popl, M. Interactions of basic compounds in reversed-phase high-performance liquid chromatography influence of sorbent character, mobile phase composition, and pH on retention of basic compounds. *J. Chromatogr. A* 1997, *758*, 37–51.
46. Gilar, M.; Yu, Y.-Q.; Ahn, J.; Fournier, J.; Gebler, J. C. Mixed-mode chromatography for fractionation of peptides, phosphopeptides, and sialylated glycopeptides. *J. Chromatogr. A* 2008, *1191*, 162–170.
47. Liu, M.; Chen, E. X.; Ji, R.; Semin, D. Stability-indicating hydrophilic interaction liquid chromatography method for highly polar and basic compounds. *J. Chromatogr. A* 2008, *1188*, 255–263.
48. Enke, C. G. A predictive model for matrix and analyte effects in electrospray ionization of singly-charged ionic analytes. *Anal. Chem.* 1997, *69*, 4885–4893.
49. Sherman, C. L.; Brodbelt, J. S. An equilibrium partitioning model for predicting response to host-guest complexation in electrospray ionization mass spectrometry. *Anal. Chem.* 2003, *75*, 1828–1836.

20 Hydrophilic Interaction Liquid Chromatography–Based Enrichment Protocol Coupled to Mass Spectrometry for Glycoproteome Analysis

Cosima Damiana Calvano

CONTENTS

20.1 INTRODUCTION

Hydrophilic interaction liquid chromatography (HILIC) was first introduced in the 1970s[1] for carbohydrates analysis and subsequently defined by Alpert in 1990[2] as a variant of normal-phase liquid chromatography (NPLC). However, HILIC offers a series of advantages compared to NPLC. In the latter, stationary phase is polar and the mobile phase consists of apolar solvents, resulting in increased retention with increased polarity of the analyzed samples and/or stationary phase and/or decreased polarity of the mobile phase.[3,4]

Nevertheless, the nonpolar solvents used as mobile phases are often dangerous for environment, expensive, and quite toxic and not completely suitable to solubilize polar and hydrophilic compounds. In HILIC, a hydrophilic stationary phase and a partly aqueous organic solvent mobile phase are used. Similarly to NPLC, the retention increases with increased polarity of the analyzed compounds and/or stationary phase and/or decreased polarity of the mobile phase. However, contrary to NPLC, HILIC uses aqueous-organic solvents with a high organic-solvent fraction as mobile phase, which demonstrate an increased solubility for the polar and hydrophilic compounds. For this reason, HILIC applications have seen a considerable increase of interest[5–10] for the analysis of many categories of polar compounds, charged as well as uncharged, in complex sample mixtures.[11]

Another reason for the increase in popularity over NPLC is the great compatibility of the buffer conditions used in the liquid chromatography–electrospray ionization–mass spectrometry (LC–ESI–MS) technique.

The interfacing with ESI–MS is often a problem with NPLC, since totally organic, nonpolar eluents do not enable an efficient ionization contrary to HILIC where the presence of water as strongly eluting solvent favors the ESI process.[12,13]

The high organic content of the buffers gives HILIC two added advantages: high sensitivity in ESI–MS[12,14,15] and faster separations due to the low viscosity of HILIC eluents.[16] In fact, higher portions of organic modifier within the elution solvents could result in better electrospray sensitivity due to the decreasing of surface tension and solvation energies for polar compounds. Further, solvents with higher viscosities such as those used in reverse phase tend to be less volatile and have higher surface tension. These solvents will originate larger and less efficiently charged droplets at the capillary tip and desolvatation of the electrosprayed eluent droplets will require higher interface drying gas temperatures. Then, the higher volatility of the organic modifier permits a more efficient desolvation of the electrosprayed droplet. As a result, HILIC works best for solutes that are challenging in reverse phase and can be used as "orthogonal" separation to reverse phase, which allows a multidimensional separation of complex samples[17–19] saving time and preventing sample losses.[20]

This approach can be useful to catch and separate solutes that have retention on both RP and HILIC-type stationary phases such as peptides or proteins, as shown by the increased number of HILIC application in the proteomics field.[21] In particular, due to the selectivity toward polar group, HILIC has shown to be a useful tool for the enrichment of posttranslational modifications (PTMs) such as glycosylation,[22] *N*-acetylation,[18] and phosphorylation[23] in proteomics applications.

In this chapter, an introduction to the most diffuse PTM, i.e., glycosylation, will be effort together with a summary of HILIC mechanism, materials, and applications for polar compounds. Moreover, a focus on the use of HILIC in glycoproteomics for the enrichment of PTMs at both peptide and protein levels will be presented. It will be described how SPE/HILIC scheme is among the preferred combinations when the aim is to enrich glycocompounds. The coupling of this enrichment strategy to different spectrometric techniques such as matrix-assisted laser desorption ionization (MALDI) and ESI will be illustrated for the analysis of complex sample mixtures such as biological fluids. A comparison with some other chromatographic materials for the enrichment yield will be shortly illustrated.

20.2 PROTEIN MODIFICATIONS: GLYCOSYLATION

The characterization of the different PTMs is receiving much attention in proteomics research since they have a significant effect on protein function.[24] Glycosylation is one of the most abundant PTMs of proteins[25,26] known to be a highly complex and dynamic process associated with membrane-bound and secreted proteins. The glycosylation of proteins is involved in many different biological roles including signaling, transcription, apoptosis, differentiation, adhesion, and infiltration, oncogenic transformation, and metastasis.[27–30]

Alterations of protein glycosylation sites, alteration in the behavior of peptide backbone due to glycan composition, size and shape, branching and heterogeneity of proteins on the cell surface, different protein folding, and the assembly of protein complexes in body fluids are implicated in several inflammatory diseases, cancers, and congenital disorders.[31–41]

Glycoproteins generally exist as populations of glycosylated variants, also known as glycoforms, of a single homogeneous peptide at one glycosylation site bringing more than 100 alternative glycans with variable composition, linkage, branching, and anomericity. This glycan heterogeneity contributes to the structural complexity and diversity of glycoproteins. There are five known types of protein glycosylations[42–44]: *N*-Glycosylation, *O*-Glycosylation, *C*-Mannosylation, phosphoglycosylation, and glypiation. In *O*-glycosylation oligosaccharide chains are covalently attached to serine or threonine residues, in *C*-glycosylation to tryptophan residues, in phosphoglycosylation to serine residues, in glypiation to proline residues, and in *N*-glycosylation to asparagine residues. However, *N*-glycosylation is characterized in greatest detail. In this case, glycosylation occurs only at the sequence Asn-X-Ser/Thr, where X can be any amino acid except proline. Figure 20.1 shows some typical structures of *N*-linked and *O*-linked oligosaccharide chains. A suitable method to verify whether a candidate glycoprotein is *N*-glycosylated is to treat the protein with enzymes called *N*-glycanase that cleave the glycan residues and examine the product by SDS-PAGE.

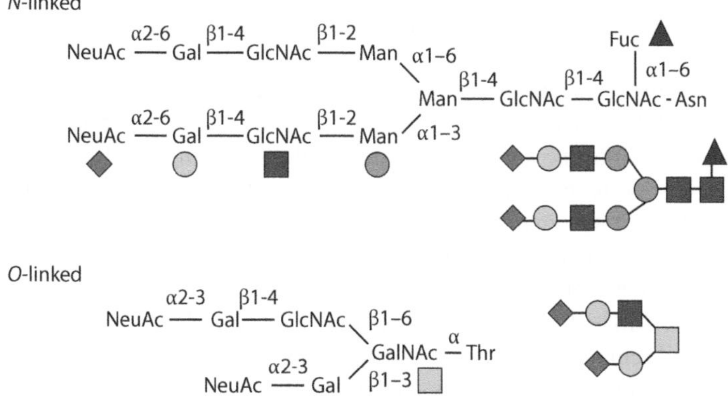

FIGURE 20.1 Some typical structures of *N*-linked and *O*-linked oligosaccharide chains found attached to proteins.

If a decrease in molecular weight is observed for the product of the *N*-glycanase reaction in comparison to the undigested glycoprotein, then the removal of *N*-linked carbohydrate has occurred and the protein was glycosylated.

N-linked glycans attached to glycoproteins fall into three major classes, i.e., high-mannose, complex, and hybrid, all of which have a common core structure consisting of two *N*-acetylglucosamine (GlcNAc) residues, three glucose (Glc), and nine mannose (Man) monosaccharides. In high-mannose type, several mannose residues ranging from 5 to 9 are attached to the core structure. Complex-type *N*-glycans may contain branching structures originating from the GlcNAc residues known as antennae. Each core mannose can have several galactose, GlcNAc, sialic acid, and fucose residues attached to the core structure giving rise to more than two antennae. Hybrid-type glycans contain structural elements of both high-mannose- and complex-type glycans. The full characterization of all glycoproteins in a complex sample is a very difficult task involving the determination of glycan size, branching points, and linkages of the carbohydrate monomers for each glycosylation site.

20.3 TROUBLES IN GLYCOPROTEOME ANALYSIS

The characterization of posttranslationally modified proteins can be approached in two ways. In fact, it is possible to investigate the modification while still attached to the target protein or following its release from the target protein. Each of the strategies for different types of glycosylation uses different experimental protocols.

However, in all cases, variable composition, heterogeneity, linkage, branching, and anomericity of the constituent sugars contribute to the structural complexity of glycoprotein glycans. As a consequence, a large number of techniques are often necessary to fully determine the structure of a glycan or a glycoprotein. These techniques may include ^{1}H and ^{13}C-nuclear magnetic resonance (NMR) spectroscopy for the determination of the structure and configuration of the carbohydrates,[45,46] electrophoresis,[47] gas chromatography[48] (GC), and liquid chromatography (LC),[49,50] possibly combined with mass spectrometry (MS).[51–53] However, the study of protein glycosylation had limited applications in biomedical and glycobiological fields since each of these strategies has its challenges inherent to the technique itself and to the complexity of the modifications, the sensitivity with which it can be analyzed, the relatively extensive and time-consuming analytical procedures employed.

As far as NMR is concerned, it is less sensitive than other techniques and requires rather homogeneous glycoprotein samples, resulting in a few suitable techniques for most proteomic studies, where the amount of sample is a limiting factor. Electrophoresis has shown to be unable to generate a quantitative or even qualitative separation for glycoprotein and GC is often associated with very time-consuming sample preparation protocols including derivatization.

Indeed, LC–MS technique overcomes most of these problems even if three major drawbacks still remain.[54] First, glycosylations can be highly labile, overall on threonine or serine residues, complicating analyses and reproducibility. Second, glycosylated peptides have poor ionization efficiencies with respect to unmodified peptides. Third, glycan modifications are extremely heterogeneous complicating the analysis, as a single peptide may be modified by several forms of glycans with different structures and masses, resulting in a signal dilution of glycopeptide ions overall in the case of complex protein mixtures.

Although many complications, recent advances in sample preparation and MS have demonstrated the ability to isolate and identify both the peptide backbone and glycan. It has been demonstrated that it is always advantageous and often necessary to enrich for glycopeptides prior to MS analysis. For this reason, there is growing interest in methods that can selectively isolate glycopeptides from proteolytic digests of complex protein mixtures for subsequent analysis by LC–MS.[55–59]

20.4 ENRICHMENT METHODS FOR GLYCOPROTEOME

Several recent studies highlight that a particular modification in proteins or peptides could be used as target to develop enrichment strategies using affinity chromatography methods.[60–62] For example, in glycoproteins, the glycan moiety can be used to select the modified proteins/peptides in complex peptide mixtures using glycan-specific structural features (e.g., affinity enrichment by antibodies and lectins[63–70]) or glycan-specific chemical properties (e.g., derivatization methods[71–75]). Among affinity enrichment techniques, increasing interest has grown toward lectin-based chromatography used extensively to purify glycoproteins, glycopeptides and oligosaccharides from different biological samples. Lectins are usually plant glycoproteins that specifically bind to certain oligosaccharide structures on glycoconjugates. Different types of lectins are easily immobilized on solid supports showing a variety of selectivities and binding specificities to different soluble carbohydrates or to carbohydrate moieties,[67,68] which are a part of glycoproteins or glycolipids. However, each lectin species isolates only a subset of glycan species making difficult an analysis of a complex unknown sample. Besides the specific lectin-affinity enrichment methods, approaches based on physical and chemical features of glycopeptides and glycoproteins are most valuable. First, the diffuse characteristic that tryptic glycopeptides arising from a complex peptide/glycopeptide mixture have higher masses than nonglycosylated permits a significant enrichment by size exclusion chromatography.[76–79] Recently, a chemical derivatization followed by the immobilization using periodate oxidation and hydrazide chemistry[58] has been used to selectively and efficiently isolate, identify, and quantify *N*-linked glycopeptides. However, in this workflow sugars are destructed and removed from glycan moieties causing a loss of structural information even if a relative quantitation of glycopeptides can be attempted by the inclusion of stable isotope labeling.[80] The problem frequently met in these chemistry-based methods relates to the large amounts of starting material and to the generation of side products. It has recently demonstrated that complex sample fractionation, purification, and preconcentration can be achieved by using "custom-made" miniaturized solid-phase extraction (SPE)[81,82] made with a variety of chromatographic functionalities and selectivity. These kinds of SPE columns, cheap, easy to prepare, simple to use usually require very little sample material. For example, SPE columns using graphite powder show a good recovery of very hydrophilic peptides, such as glycopeptides[71]; SPE columns using titanium dioxide microparticles show a good recovery of sialic acid modified peptides.[83] HILIC has mainly been used for the analysis of small, polar molecules such as carbohydrates, oligonucleotides, and amino acids, and more recently for the separation of peptides and glycopeptides.[84–86] It has established to be a useful method for purification of glycosylated peptides due to the retention determined by the hydrophilic character and size of the glycan moiety. HILIC could reduce the complexity of peptide/glycopeptide mixtures allowing depletion of hydrophobic peptides, and retention of hydrophilic glycopeptides.

20.5 HILIC: MECHANISM AND MATERIAL

As stated in the introduction, HILIC can be considered a variant of normal-phase chromatography, where the retention mechanism is supposed to be partitioning of the analyte between a water-enriched layer covering the hydrophilic stationary phase and a relatively hydrophobic bulk eluent, typically composed of 5%–40% water in acetonitrile.[2] Gregor et al.[87] described the existence of a water-enriched layer on the surface of hydrophilic stationary phases, while Rückert and Samuelson[88] proposed an absorption of nonelectrolytes in ion exchange resins by means of a stagnant water layer.

Several studies demonstrated that analytes are retained on the hydrophilic stationary phase by hydrogen bonding, ionic interactions, and dipole–dipole interactions. Yoshida[89] proposed an adsorptive retention mechanism for peptides in HILIC explaining that elution patterns were similar to the normal-phase separation and suggested hydrogen bonding as the principal interaction mode. Guo and Gaiki[13,90] supposed a partitioning mechanism since they found two experimental evidences that could not be explained otherwise: (a) no strong specific interaction between the solutes and four different HILIC stationary phases and (b) the increase in retention time with salt concentration. However, it is always more diffuse and recognized the definition introduced by Alpert[2] who stated that the term HILIC should be used in the case of (a) strongly eluting solvent consisting of water and (b) retention mechanism based on partitioning.

Recently, several stationary phases have been specifically introduced for HILIC approaches each of them displaying different retention characteristics, separation selectivity, and requiring distinct buffer constitutions for each application.[91] The use of buffered eluents is highly recommended to reduce electrostatic interactions between charged analytes and charged silanol groups or other charged species on the stationary phase. The preferred buffer salts are ammonium salts of formate or acetate due to their suitable buffering range, high solubility in acetonitrile rich eluents, and compatibility with MS techniques.

It is important to highlight that the type of the stationary phase together with the composition of the mobile phase can be plainly tuned to solve specific separation problems in the mixed-mode HILIC. The most diffuse phases used include neat silica (Atlantis) containing functional groups such as siloxanes, derivatized silica carrying polymeric coatings like poly(2-hydroxyethyl aspartamide) (PolyHydroxyethyl A), cation exchanger polysulfoethyl A,[92] weak cation exchanger Polycat A,[93] weak anion exchanger PolyWAX,[94,95] TSK-Gel amide 80,[98,99] and zwitterionic (ZIC)-HILIC.[22] Figure 20.2 illustrates some structures of typical HILIC phases.

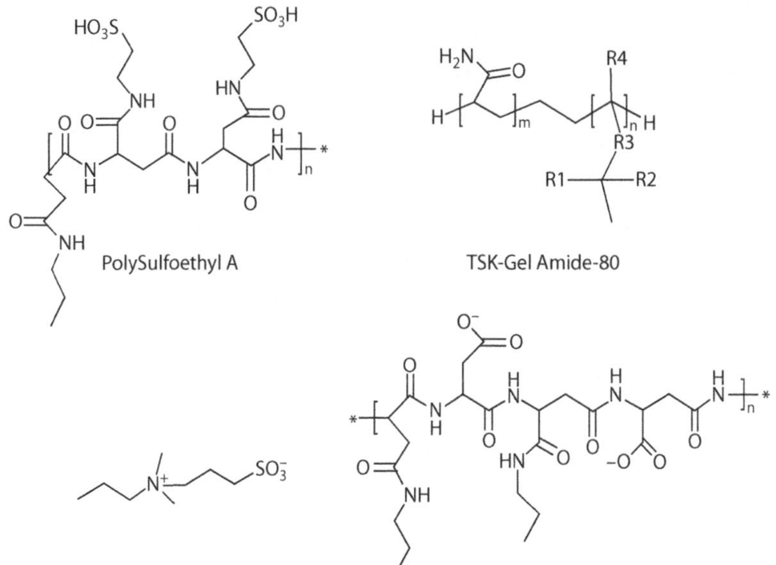

PolySulfoethyl A

TSK-Gel Amide-80

ZIC-HILIC

PolyCAT A

FIGURE 20.2 Chemical structures of the functional groups in common HILIC stationary phases. (Reproduced from Boersema, P.J. et al., *Anal. Bioanal. Chem.*, 391, 151, 2008. With permission.)

The stationary phase of silica(-based) columns contains negative charge at basic pH due to deprotonation of silanol groups, then the HILIC retention is increased when pH is below pK_a of basic compounds. In such cases, negatively charged surface electrostatically repels negatively charged compounds like sialic acids requiring an acidic pH and/or ion pairing reagent to avoid unwanted interactions between basic compounds and deprotonated silanol groups. In the case of amine and aminopropyl phases an opposite effect is observed, where at pH below 9, positively charged amino groups act as weak anion-exchangers requiring a basic pH and/or ion pairing reagent. It is easy to understand as a pH shift could have a great effect on the ionic interactions by reducing or introducing charged groups on both analyte and stationary phase. Moreover, another critical factor when changing the mobile phase is that too much increased ionic strength results in a reduction of ionic interactions between analyte and stationary phase, leaving the hydrogen bonding as the major retentive effect. The presence of ions in the mobile phase will also weaken ionic interactions since they can act as counterions for both charged analytes and groups of the stationary phase and will decrease the ionization efficiency of ESI. Further disadvantages of this kind of bare columns are the possibility to observe phenomena of dissolution and hydrolysis that limited stability in aqueous solvents and phenomena of irreversible adsorption of compounds onto stationary phases due to their highly polar character (silica columns) and to high reactive stationary phases (aminopropyl columns). More recently, zwitterionic sulfoalkylbetaine stationary phases[71] have been introduced for HILIC separations of small organic ionic compounds, inorganic salts and proteins.[98,99] These charged stationary phases exhibit permanent negative and positive charges given by strongly acidic sulfonic acid groups and strongly basic quaternary ammonium groups separated by a short alkyl spacer acting as a weak-cation exchanger.[5]

The chromatographic properties of zwitterionic materials significantly differ from other HILIC phases. The 1:1 molar ratio of the two oppositely charged groups assures a very low net negative surface charge on the bonded layer, only little affected by pH, and ascribed to large distance of the sulfonic groups from the silica gel surface. Therefore, for zwitterionic stationary phase, both ion-exchange and electrostatic interactions are considered weaker than typical ion exchangers and bonded amino phases, or "bare" silica, amide and other bonded phases, respectively. In fact, in this case, the residual silanol groups are efficiently protected by self-association of oppositely charged functional groups. As schematized in Figure 20.3, the sulfoalkylbetaine-bonded phases strongly adsorb water by hydrogen bonding and the bulk layer of water becomes part of the stationary phase. This layer plays an important role in determining the retention mechanism, in part explicated by partitioning of the analytes between the bulk mobile phase and the adsorbed water layer, in part by polar interactions (such as hydrogen-bonding and dipole–dipole interactions) in the stationary phase, and in smaller extent by weak electrostatic interactions between analytes carrying either positive or negative charges. In any case, the retention increases with increasing sample polarity. Different trades are commercially available for zwitterionic columns such as ZIC–HILIC (on silica gel support) and ZIC–pHILIC (on polymer support). Takegawa et al.[100] demonstrated the potential of ZIC–HILIC by the separation of 2-aminopyridine-labeled *N*-glycans of human serum IgG.

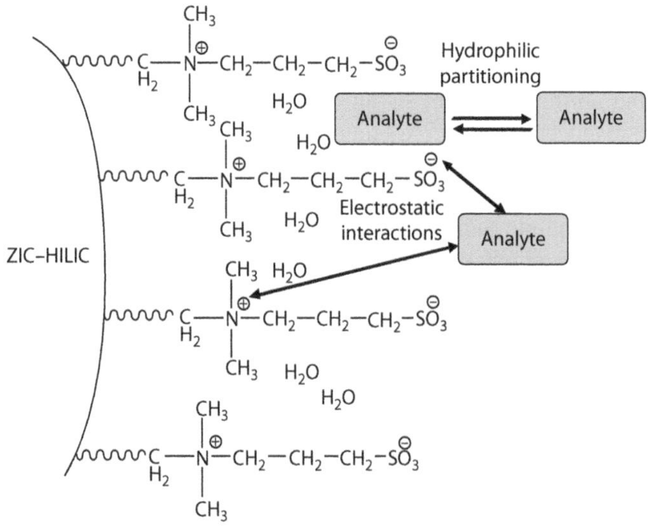

FIGURE 20.3 Mechanism of HILIC separation including partition and adsorption driven by hydrophilic and electrostatic interactions on a sulfoalkylbetaine-bonded phase.

20.6 HILIC: FROM GLYCOMICS TO GLYCOPROTEOMICS

The emergence of glycoproteomics has extended the applicability of HILIC from only being used for carbohydrate analysis, via peptides, to small molecules and even whole proteins. Many applications of HILIC–ESI–MS for neutral oligosaccharides have been described in literature such as the analysis of plant sugar residues,[101] the analysis of reduced *O*-glycans,[102,103] of fluorescently labeled *N*-glycans,[104–109] and the stereoisomers (anomers) of the reducing end carbohydrate ring of *N*-glycans.[107,108] A particular force of this technique in the field of oligosaccharides analyses is the tolerance of the chromatographic media to modifications: since the retention is mainly established by the hydrogen bonding of the multiple glycan moiety, polar substituents as sulfates or various hydrophobic tags remain all compatible with HILIC separation of oligosaccharides. This strength explains also why the separations of monosaccharides (mannose, fucose, galactose, glucuronic acid, *N*-acetylglucosamine, *N*-acetylneuraminic acid) present in *N*-linked glycoproteins[110] are also determined with HILIC. Many other applications are found in pharmaceutical field such as the determination of mannitol, gemcitabine, and sodium cation in various formulations[111] or fine particle dose of lactose in a dry inhalation product,[112] in agricultural field such as the determination of spectinomycin and lincomycin in run-off waters from lands assigned to crop cultivation,[113] or in biological field such as the determination of 2-aminopyridine derivatized *N*-glycans and *N*-glycopeptides of human serum immunoglobulin[114] and of α-1-acid glycoprotein.[100]

As seen in the overview above, the more recent applications of hydrophilic separation materials referred to protein separations. At the present, the separation of proteins in their native and overall of protein or peptide posttranslational modified state remains a very challenging task in LC difficult to run in a routine fashion. The translation of this technique from small polar compounds to proteomics has been quite simple because no significant differences between the chromatographic behavior of small molecules and large macromolecules as proteins are observed if HILIC interactions are dominating. In theory, the large size of proteins means small diffusion coefficients and lower number of interactions with the stationary phase per unit of time. Decreased diffusion that is on the base of mass transfer mechanism in chromatography, results in decreased *efficiency* in the separation of biological macromolecules compared to small molecules in the same experimental conditions. Protein separations therefore rely on *selectivity*. However, the chromatographic behavior of oligosaccharides and glycopeptides can be similar if hydrophilic interactions dominate on other interactions correlated to analyte size such as in size exclusion chromatography. This phenomenon has been observed in the separation of oligosaccharides and glycopeptides varying the acetonitrile concentration in the mobile phase.[87] In fact, at acetonitrile concentrations of up to 40% many silica-based and polymeric stationary phases show conventional gel filtration behavior for oligosaccharides in size exclusion chromatography. If the acetonitrile concentrations is up to or higher than 50%, an inversion of the elution sequence is observed and decreasing polarity of the mobile phase results in increased retention. At these acetonitrile concentrations, the hydrophilic interaction of oligosaccharides with the stationary phase is favored compared to the gel filtration effects.

This observation was confirmed for *N*-glycopeptides: the separation of a tryptic digest of horseradish peroxidase in 50% acetonitrile resulted in the separation of glycoforms. In this case, it was noted that glycoforms with disaccharides attached to the *N*-glycosylation site eluted earlier than the glycoforms with a heptasaccharide *N*-glycan structure, following HILIC behavior rather than size exclusion chromatography. The same considerations can be carried out for HILIC applied as SPE for oligosaccharides and glycopeptides in the complex mixtures containing salts, nonglycosylated peptides and proteins. Recently, more global proteomics approaches were reported that targeted protein glycosites using SPE–HILIC to enrich glycopeptides.[22,115–118]

20.7 GLYCOPROTEOMICS BY HILIC AND MASS SPECTROMETRY

MS has proved to be the election technique for glycan analysis due to its high sensitivity, selectivity, and possibility to automation. First applications in this area were run out using GC–MS still used today for the detection of monosaccharide composition.[119,120] Laborious protocols for oligosaccharides derivatization, scarce volatility of bigger molecules determined a significant expansion of soft ionization techniques such as fast atom bombardment, electrospray ionization, and matrix-assisted laser desorption/ionization in glycosylation analysis. Additionally, since ESI and MALDI methods typically do not fragment the precursor ions and can hyphenate with LC online and offline, respectively, they are well suited for analyzing complex mixtures. MALDI and ESI MS analyses of the recovered glycopeptides are very sensitive and rapid means to generate a glycan profile for individual glycosylation sites in proteins.[121] Almost all kinds of *N*-linked, *O*-linked, and GPI-anchored peptides have been purified and characterized by MALDI or ESI tandem MS.[22,98,115] Even better results can be achieved by using techniques allowing to achieve high mass accuracy in the MS and MS/MS mode such high-resolution mass spectrometers like quadrupole time of flight (Q-TOF), ion trap Fourier transform ion cyclotron resonance, or ion trap Orbitrap instruments. A more straightforward and detailed assignment of glycan species and structure elucidation[122] can be reached thanks to these techniques.[123] The ion-trap-type instruments offer MS capabilities, sometimes useful for detailed analysis of glycans.

There are two strategies of characterizing glycans using MS. In the first strategy, glycans are cleaved from the glycoprotein chemically or enzymatically and subjected to purification before mass spectral analysis.[124,125] This approach reveals scarce results when more than one glycosylation site is present on the glycoprotein, because it does not permit to correlate glycan composition with the different attachment sites. The second approach characterizes glycopeptides after subjecting the glycoprotein to proteolysis. In this strategy, called also "glycosylation site-specific analysis," the glycan composition can be correlated to the attachment site on glycopeptide, due to the presence of the peptide portion on each glycopeptides. In this kind of studies, *N*-linked glycans have received more attention compared with *O*-linked glycans, probably because the former contain a consensus sequence easily recognized for enzymatic treatments. However, glycopeptides analysis can also be applied to *O*-linked glycopeptides. In this chapter, the protocol proposed is more suitable for

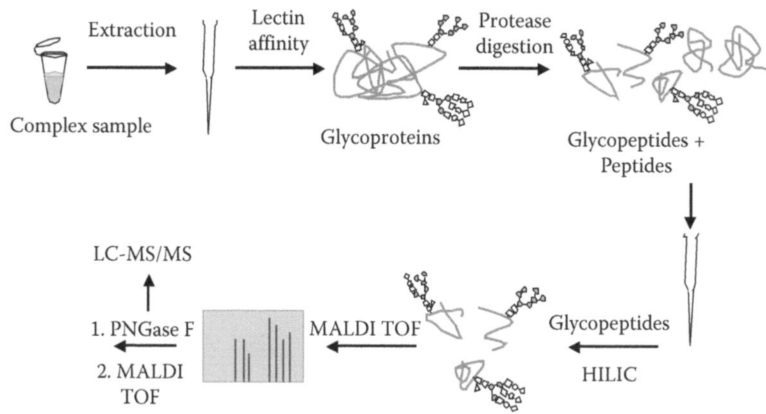

FIGURE 20.4 Strategy showing glyco-capture procedure for the identification of glycoproteins.

N-linked glycopeptides analysis. A basic approach involves the use of different forms of proteolytic digestion, combined with the use of affinity enrichment material in microcolumns and sensitive MS detection and characterization. The digestion protocol can be carried out on complex mixtures composed of proteins and glycoproteins or can be performed on simpler sample mixtures where a previous purification at proteins level has been carried out. Then a variant to a basic approach consists of a selective capture of glycoproteins by means of gel electrophoresis or affinity chromatography (Figure 20.4).

For a protein-specific analysis, to isolate a broad range of glycoproteins, columns having immobilized lectins are employed. Some examples of lectins include concanavalin A (ConA), wheat germ agglutinin and jacalin that can be used alone in a single affinity column or mixed in different columns in series or combined to get better coverage.[126,127]

After protein enrichment, it can be possible to isolate glycoproteins by SDS-PAGE and perform an in-gel digestion or to perform directly an in-solution digestion on the enriched fraction. Peptides and glycopeptides are obtained and a further enrichment for glycosylated peptides can be made by using ZIC–HILIC microcolumns. The relative abundance of glycopeptides was significantly increased following enrichment by HILIC. It was observed for a standard protein as fetuin that using reverse phase SPE for purification generated a spectrum dominated by nonglycosylated peptides while, in comparison, SPE with HILIC resulted in a spectrum full of glycosylated peptides. Moreover, for different proteins a different mass spectrum profile was obtained after HILIC, but, in all cases, the higher mass region was enriched by glycosylated peptides. An example is illustrated in Figure 20.5 where MALDI TOF spectra of the enriched portion of different protein digests (fetuin, α-glycoprotein, and RNase B) are shown.

Small diversions from this protocol have been reported and include lectin-mediated affinity capture at the peptide level[116] and the further separation of glycopeptides by strong cation exchange.[115] Figure 20.6 shows as example a comparison between spectra resulting from the enriched fraction of α-1-glycoprotein tryptic

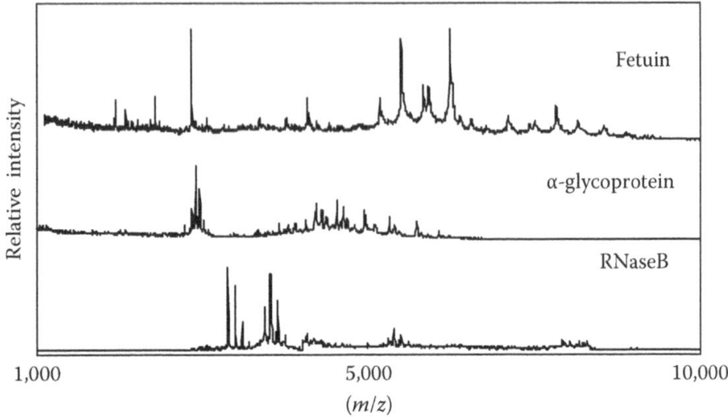

FIGURE 20.5 MALDI TOF mass spectra obtained after HILIC purification on different protein tryptic digests.

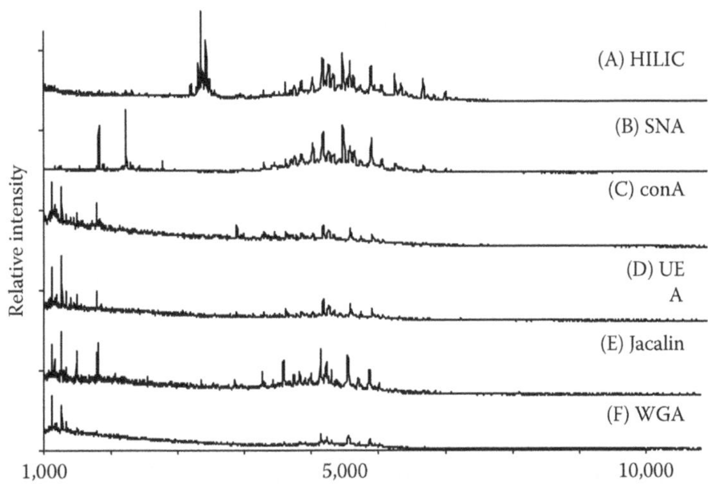

FIGURE 20.6 MALDI TOF MS spectra obtained from an α-1-glycoprotein digest followed by (A) HILIC purification, (B) *Sambucus nigra* lectin (SNA) purification, (C) Concanavalin (ConA) purification, (D) *Urex europaeus* (UE) purification, (E) Jacalin purification, and (F) wheat germ agglutinin (WGA) purification.

digest by HILIC and different lectins microcolumns. As seen, similar results are obtained, even if HILIC has a broader versatility also on different proteins.

The recovered glycopeptides can be directly analyzed using either MALDI MS/MS or LC–ESI–MS/MS or can be subject to an enzymatic deglycosylated prior to MS analysis.

The common enzyme used for deglycosylation is *N*-glycosidase F (Glycopeptidase F, GPase F) that removes *N*-linked oligosaccharides from the polypeptide backbone. The cleavage occurs between the innermost residue of the oligosaccharide and the asparagine residue to which the oligosaccharide is linked. In this enzymatic reaction,

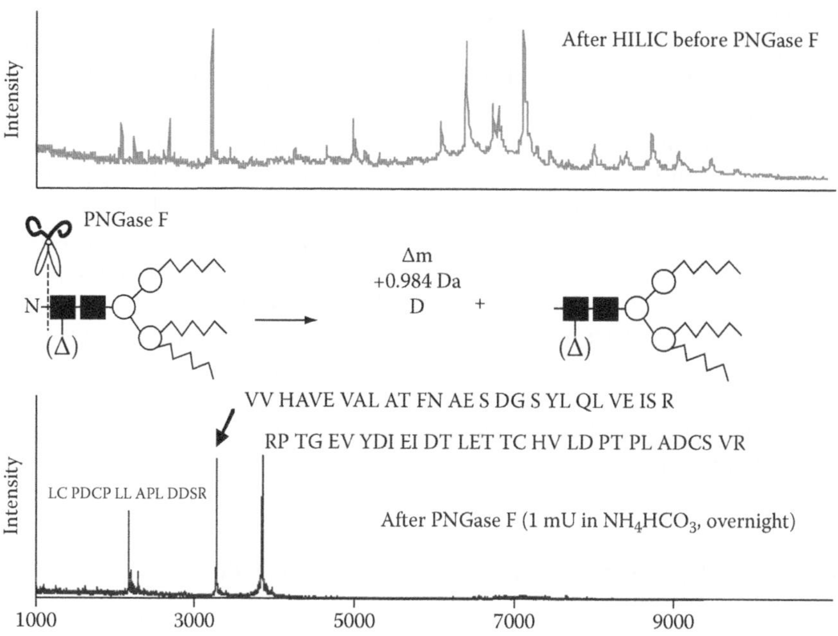

FIGURE 20.7 MALDI TOF MS spectra obtained before and after PNGase F digestion on fetuin protein tryptic digest.

the asparagine is converted to an aspartate residue with a concomitant mass increase of 0.98 Da. Figure 20.7 shows the results obtained after PNGase F digestion on fetuin protein digest. One drawback of this approach is that it does not distinguish between enzymatic asparagine-to-aspartate conversion and other causes of deamidation, in vivo or in vitro. For this reason, in the effort to unambiguous localize glycosylation sites it is opportune tagging the site using ^{18}O isotope labeling during PGNase digestion[115,116] or to perform enzymatic deglycosylation using endo-δ-*N*-acetylglucosaminidases, which cleave the glycosidic bond between two proximal GlcNAc residues in the chitobiose core leaving one GlcNAc residue attached to the peptide.

If this approach is followed in conjunction with mass spectrometric analysis of glycopeptides,[126] it is possible to observe in the mass spectrum an oxonium ion of GlcNAc at *m/z* 204.08 a marker indicating the presence of an *N*-glycan attachment site in a peptide.

The oxonium ion helps in identifying by a simple full scan analysis whether or not the peaks in the mass spectrum correspond to glycopeptides. After verification, the glycopeptides should be characterized in the peptide portions and in the glycan portions and the assignment should be validated with MS/MS data.

If a PNGase F digestion has been performed instead, a 0.98 Da mass shift is observed in the spectrum. However, when glycopeptides are subjected to collision-induced dissociation (CID), other marker ions at *m/z* 163 (Hex^+), *m/z* 204 ($HexNAc^+$), *m/z* 366 ($Hex\text{-}HexNAc^+$), and *m/z* 292 ($NeuNAc^+$) can be observed.[128,129] These ions can be used as diagnostic ions to verify the presence of glycopeptides in the mass spectrum. When the peptide is unknown, a rapid way to identify marker ions is to look for a pair of peaks that are distant 120 or 266 Da. This pair of peaks corresponds to [Peptide + Glc-NAc + H]$^+$ and [Peptide + C_2H_2NHAc]$^+$ or [Peptide + GlcNAc + Fucose + H]$^+$ and [Peptide + C_2H_2NHAc]$^+$, respectively.

After determining which ions correspond to glycopeptides in the mass spectrum, it is necessary to proceed with the compositional assignment of the peptide portion usually separately from the glycan portion recurring to MS/MS data analysis. Usually MS/MS data obtained by the CID of glycopeptides are dominated by the glycosidic cleavage of the sugar moieties and y- and b-type ions from the peptide portion. One way to recognize the peptide is by identifying the glycan composition from MS/MS experiments first and then deducting this mass from the precursor ions so that the remaining mass can be used to obtain the peptide mass.

Finally, the glycopeptide compositions can be characterized using databases such as GlycoPep DB (http://hexose.chem.ku.edu)[130–132] or GlycoMod (http://us.expasy.org/tools/glycomod/).[133,134] To use these tools, the mass spectrum is converted into a peak list and searched against these databases for possible matches.

20.8 APPLICATION TO BIOMARKER DISCOVERY

In a survey of the SWISS-PROT database results that more than 50% of proteins are glycosylated. Biological fluids are a rich source of glycoproteins and are very informative since *N*-glycosylated proteins arise from tissues and organs constituting useful biomarkers of their status. In fact, diseased tissues are expected to release disease-specific proteins and the detection and characterization of such disease-specific biomarkers is of great importance. Then glycan heterogeneity and biological variations as well as glycan-related differences between healthy and diseased individuals can be discovered in a biological fluid such as serum as a diagnostic and preventing tool. The main problem in glycoproteins/peptides analysis in serum is their very low concentrations. In fact, 99% of serum protein content is constituted by albumin, transferrins, immunoglobulins, and complement factors represent, while the remaining 1% is represented by lower abundance circulatory proteins, shed proteins, and apoptotic and necrotic cells.[134] It is necessary to isolate glycoproteins from the most abundant background serum proteins prior to protein and peptide analysis by MS.

The proposed strategy[135] here uses different immobilized lectins for glycoproteins extraction and HILIC resin for the enrichment of glycopeptides prior to MALDI MS and LC–ESI–MS/MS. HILIC material was packed into GELoader tips (Figure 20.8) in microcolumns previously for improving sensitivity and sequence coverage for proteins. Enzymatic release of *N*-linked glycans with peptide *N*-glycosidase F was used for the determination of glycosylation sites. A protocol for HILIC preparation[136] can be easily reproduced for serum samples:

1. ZIC–HILIC chromatographic media, (ZIC–HILIC, silica 10 μm, 200 Å, Sequant AB, Umeå, Sweden), suspended in acetonitrile or methanol.
2. C8 StageTips made from either GELoader tips or 10 μL disposable syringe tips.
3. Disposable 1 mL syringe, fitted to the GELoader with a cut down 200 μL tip.

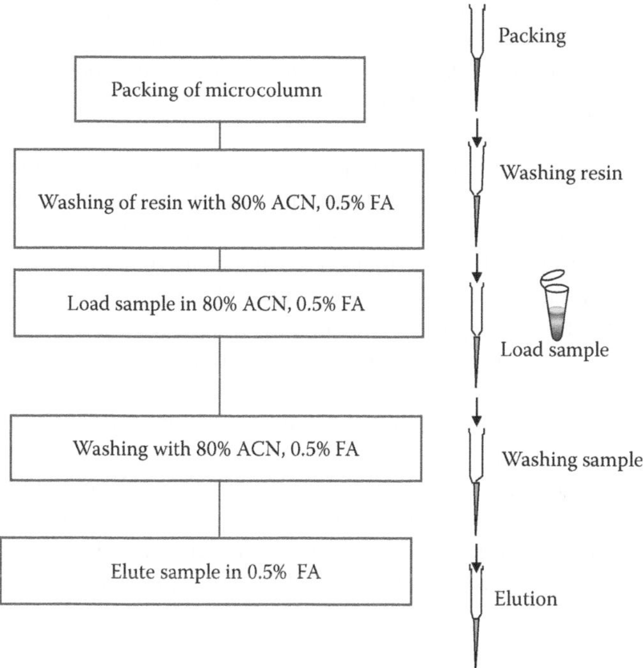

FIGURE 20.8 Illustration of all steps for HILIC SPE preparation.

4. HILIC wash: 80% acetonitrile, 19.5% water, 0.5% formic acid—can be stored at 4°C for up to 1 week.
5. HILIC elute: 0.5% aqueous formic acid—can be stored at 4°C for up to 1 week.
6. 1–20 pmol of the digested sample is made up in 10–20 μL of HILIC wash solution.
7. Prepare the HILIC microcolumns. Vortex the ZIC–HILIC bead–containing solution and deposite a few microlitres of the resulting slurry into 10 μL of HILIC wash solution that was loaded into a StageTip. The HILIC beads are packed on top of the C8 plug by applying gentle air pressure with the plastic syringe. The length of the column is dependent on the amount of peptides you wish to analyze, with a 3–5 mm column sufficient for up to 20 pmol.
8. Clean the microcolumn with 15 μL of HILIC elution solution, flushing the solution through with gentle pressure from the syringe.
9. Condition the column by flushing with 30 μL of HILIC wash solution.
10. Load the sample containing glycopeptides onto the column using the syringe.
11. Wash unbound material from the column with 20–40 μL of HILIC wash solution.
12. Elute the glycosylated peptides from the HILIC material with 7–15 μL of HILIC elute solution—*retain eluate*.

13. Elute these glycopeptides with relatively hydrophobic peptides from the plug with 3 µL of HILIC—wash solution—*pool eluate*.
14. Dry the glycopeptides down in a vacuum centrifuge and store until required for mass spectrometric analysis.

A comparison between HILIC and lectin-based techniques on glycopeptides mixtures is also assessed. Four different sample preparation protocols (Figure 20.9) were tested in terms of glycoprotein identification efficiency, *N*-linked glycan site determination, and reproducibility. The four glycopeptide enrichment strategies provided a comparable number of identified glycoproteins (Figure 20.10). The main difference between the four strategies is in the number of identified glycopeptides. If very complex samples are analyzed without a prepurification (Strategies A and B), a lower number of glycosylation sites are recovered. The presence of very abundant nonglycosylated serum proteins and their derived tryptic peptides exceeds the capacity of the HILIC resin (Strategy B, Figure 20.9). In contrast, more informative results were obtained when introducing an enrichment step at the glycoprotein level as well as at the glycopeptide level. Many of the glycopeptides and *N*-glycosylation sites were determined by both lectin (Strategy C, Figure 20.9) and HILIC (Strategy D, Figure 20.9) and only few peptides derived from nonglycosylated human serum albumin were detected.

These studies highlight that HILIC works very well on simple protein mixtures or even single proteins, but an unambiguous characterization of peptide glycan structures at the proteome level such as in serum is still rather challenging. HILIC then can be used as an effective and relatively simple tool for the targeted analysis of

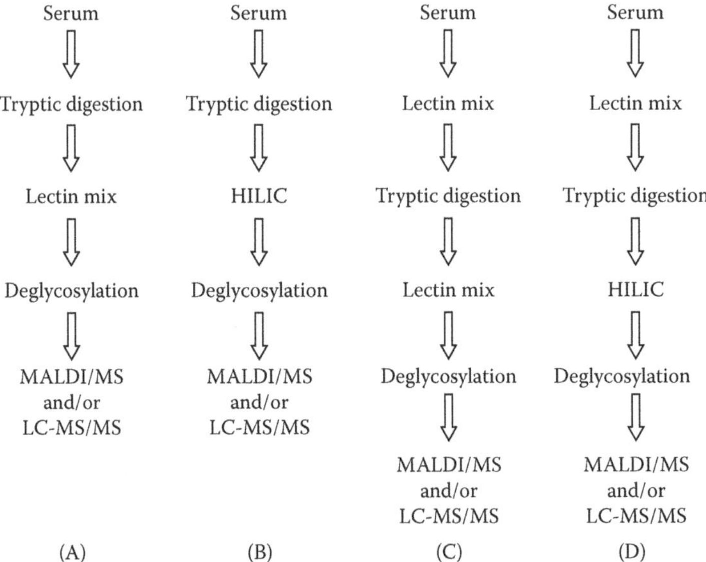

FIGURE 20.9 Purification methods including all preparative steps. The strategies are indicated as (A) lectin only, (B) HILIC only, (C) lectin-lectin, and (D) lectin-HILIC. (Taken from Calvano, C.D. et al., *J. Proteomics*, 71, 304, 2008. With permission from Elsevier.)

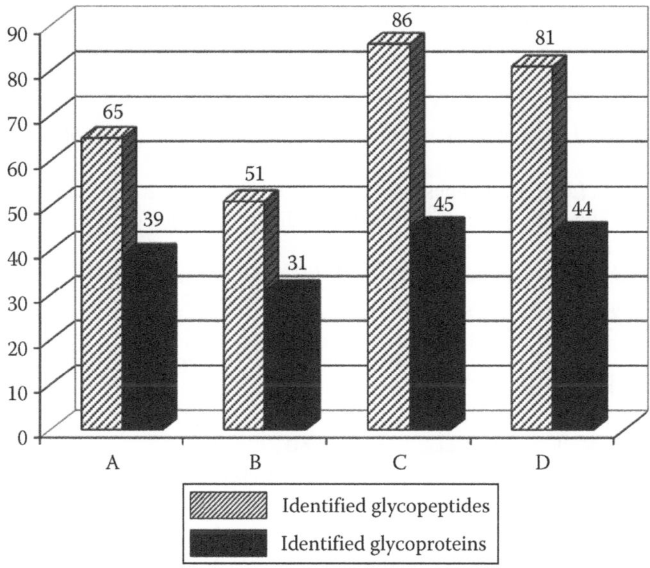

FIGURE 20.10 Graphical comparison of the results obtained from the employed methods for *N*-glycosylation analysis as described in Figure 20.9. (Taken from Calvano, C.D. et al., *J. Proteomics*, 71, 304, 2008. With permission from Elsevier.)

protein glycosylation, but further studies are necessary to assess if HILIC might also have an important role to play in not only glycosite elucidation but also in the compositional analysis.

REFERENCES

1. Linden, J.C., Lawhead, C.L., *J. Chromatogr. A*, 1975, 105, 125.
2. Alpert, A.J., *J. Chromatogr. A*, 1990, 499, 177.
3. Rizzi, A., Retention and selectivity, in: Katz, E., Eksteen, R., Schoenmakers, P., Miller, N. (Eds.), *Handbook of HPLC*, Marcel Dekker, New York, 1998, p. 1.
4. Caude, M., Jardy, A., Normal-phase liquid chromatography, in: Katz, E., Eksteen, R., Schoenmakers, P., Miller, N. (Eds.), *Handbook of HPLC*, Marcel Dekker, New York, 1998, p. 325.
5. Hemström, P., Irgum, K. *J Sep. Sci.*, 2006, 29, 1784.
6. Strege, M.A., Stevenson, S., Lawrence, S.M. *Anal. Chem.*, 2000, 72, 4629.
7. Risley, D.S., Strege, M.A., *Anal. Chem.*, 2000, 72, 1736.
8. Sarg, B., Koutzamani, E., Helliger, W., Rundquist, I., Lindner, H.H., *J. Biol. Chem.*, 2002, 277, 39195.
9. Sarg, B., Helliger, W., Hoertnagl, B., Puschendorf, B., Lindner, H., *Arch. Biochem. Biophys.*, 1999, 372, 333.
10. Lindner, H., Sarg, B., Grunicke, H., Helliger, W., *J. Cancer Res. Clin. Oncol.*, 1999, 125, 182.
11. Xu, R.N., Fan, L., Rieser, M.J., El-Shourbagy, T.A., *J. Pharm. Biomed. Anal.*, 2007, 44, 342.
12. Naidong, W., *J. Chromatogr. B*, 2003, 796, 209.
13. Guo, Y., Gaiki, S., *J. Chromatogr. A*, 2005, 1074, 71.
14. Grumbach, E.S., Wagrowski-Diehl, D.M., Mazzeo, J.R., Alden, B., Iraneta, P.C., *LCGC N. Am.*, 2004, 22, 1010.

15. Shou, W.Z., Naidong, W., *J. Chromatogr. B*, 2005, 825, 186.
16. Shou, W., Chen, Y., Eerkes, A., Tang, Y. et al., *Rapid Commun. Mass Spectrom.*, 2002, 16, 1613.
17. Gilar, M., Olivova, P., Daly, A.E., Gebler, J.C., *Anal Chem.*, 2005, 77, 6426.
18. Boersema, P.J., Divecha, N., Heck, A.J.R., Mohammed, S., *J. Proteome Res.*, 2007, 6, 937.
19. Wang, X.D., Li, W.Y., Rasmussen, H.T., *J. Chromatogr. A*, 2005, 1083, 58.
20. Weng, N., Shou, W., Addison, T., Maleki, S., Jiang, X., *Rapid Commun. Mass Spectrom.*, 2002, 16, 1965.
21. Boersema, P.J., Mohammed, S., Heck, A.J.R., *Anal. Bioanal. Chem.*, 2008, 391, 151.
22. Hagglund, P., Bunkenborg, J., Elortza, F., Jensen, O.N., Roepstorff, P., *J. Proteome Res.*, 2004, 3, 556.
23. McNulty, D.E., Annan, R.S., in: *55th ASMS Conference on Mass Spectrometry*, Indianapolis, IN, June 3–7, 2007.
24. Jensen, O.N., *Nat. Rev. Mol. Cell. Biol.*, 2006, 7, 391.
25. Bertozzi, C.R., Kiessling, L.L., *Science*, 2001, 291, 2357.
26. Rudd, P.M., Elliott, T., Cresswell, P., Wilson, I.A., Dwek, R.A., *Science*, 2001, 291, 2370.
27. Gewinner, C., Hart, G., Zachara, N., Cole, R., Beisenherz-Huss, C., Groner, B., *J. Biol. Chem.*, 2004, 279, 3563.
28. Alikhani, Z., Alikhani, M., Boyd, C.M., Nagao, K., Trackman, P.C., Graves, D.T., *J. Biol. Chem.*, 2005, 280, 12087.
29. Phan, U.T., Waldron, T.T., Springer, T.A., *Nat. Immunol.*, 2006, 7, 883.
30. Partridge, E.A., Le Roy, C., Di Guglielmo, G.M., Pawling, J., Cheung, P., Granovsky, M., Nabi, I.R., Wrana, J.L., Dennis, J.W., *Science*, 2004, 306, 120.
31. Durand, G., Seta, N., *Clin. Chem.*, 2000, 46, 795.
32. Bakry, N., Kamata, Y., Simpson, L.L., *J. Pharmacol. Exp. Ther.*, 1991, 258, 830.
33. Becker, J.W., Reeke, G.N., Jr., Wang, J.L., Cunningham, B.A., Edelman, G.M., *J. Biol. Chem.*, 1975, 250, 1513.
34. Yang, Z., Hancock, W.S., *J. Chromatogr. A*, 2004, 1053, 79.
35. Turner, G.A., *Clin. Chim. Acta*, 1992, 208, 149.
36. Hellwage, J., Kuhn, S., Zipfel, P.F., *Biochem. J.*, 1997, 326, 321.
37. Junnikkala, S., Jokiranta, T.S., Friese, M.A., Jarva, H., Zipfel, P.F., Meri, S., *J. Immunol.*, 2000, 164, 6075.
38. Yang, Z., Harris, L.E., Palmer-Toy, D.E., Hancock, W.S., *Clin. Chem.*, 2006, 52, 1897.
39. Zhao, J., Patwa, T., Qiu, W., Shedden, K., Hinderer, R., Misek, D., Anderson, M., Simeone, D., Lubman, D., *J. Proteome Res.*, 2007, 6, 1864.
40. Donate, F., Juarez, J.C., Guan, X., Shipulina, N.V., Plunkett, M.L., Tel-Tsur, Z., Shaw, D.E., Morgan, W.T., Mazar, A.P., *Cancer Res.*, 2004, 64, 5812.
41. Olsson, A.-K., Larsson, H., Dixelius, J., Johansson, I., Lee, C., Oellig, C., Bjork, I., Claesson-Welsh, L., *Cancer Res.*, 2004, 64, 599.
42. Aebersold, R., *J. Proteome Res.*, 2005, 4, 1104.
43. Kannagi, R., Izawa, M., Koike, T., Miyazaki, K., Kimura, N., *Cancer Sci.*, 2004, 95, 377.
44. Varki, A. (Eds.), *Essentials of Glycobiology*, Cold Spring Harbour Press, Cold Spring Harbour, NY, 1999.
45. Voisin, S., Houliston, S.R., Kelly, J., Brisson, J.R., Watson, D., Bardy, S.L., Jarrell, K.F., Logan, S.M., *J. Biol. Chem.*, 2005, 280, 16586.
46. Yamaguchi, Y., Takizawa, T., Kato, K., Arata Y., Shimada I., *J. Biomol. NMR*, 2000, 18, 357.
47. Zhuang, Z., Starkey, J.A., Mechref, Y., Novotny, M.V., Jacobson, S.C., *Anal. Chem.*, 2007, 79, 7170.
48. Priem, B., Gitti, R., Bush, C.A., Gross, K.C., *Plant Physiol.*, 1993, 102, 445.
49. Maslen, S., Sadowski, P., Adam, A., Lilley, K., Stephens E., *Anal. Chem.*, 2006, 78, 8491.

50. Hashii, N., Kawasaki, N., Itoh, S., Hyuga, M., Kawanishi, T., Hayakawa, T., *Proteomics*, 2005, 5, 4665.
51. Siluka, D., Kima, H.S., Colea, T., Wainera, I.W., *J. Pharm. Biomed. Anal.*, 2008, 48, 960.
52. Liu, Z., Mutlib, A.E., Wang, J., Talaat, R.E., *Rapid Commun. Mass Spectrom.*, 2008, 22, 3434.
53. Jauregui, O., Sierra, A.Y., Carrasco, P., Gratacos, E., Hegardt, F.G., Casals, N., *Anal. Chim. Acta*, 2007, 599, 1.
54. Raman, R., Raguram, S., Venkataraman, G., Paulson, J.C., Sasisekharan, R., *Nat. Methods*, 2005, 2, 817.
55. Lewandrowski, U., Moebius, J., Walter, U., Sickmann, A., *Mol. Cell Proteomics*, 2006, 5, 226.
56. Wada, Y., Tajiri, M., Yoshida, S., *Anal. Chem.*, 2004, 76, 6560.
57. Tajiri, M., Yoshida, S., Wada, Y., *Glycobiology*, 2005, 15, 1332.
58. Zhang, H., Li, X.J., Martin, D.B., Aebersold, R., *Nat. Biotechnol.*, 2003, 21, 660.
59. Ghosh, D., Krokhin, O., Antonovici, M., Ens, W., Standing, K.G., Beavis, R.C., Wilkins, J.A., *J. Proteome Res.*, 2004, 3, 841.
60. Kakehi, K., Kinoshita, M., in: N.G. Karlsson and N.H. Packer (Eds.), *Glycomics: Methods and Protocols, Methods in Molecular Biology*, Springer, 2009, vol. 534, 1.
61. Tomana, M., Niedermeier, W., Mestecky, J., Schrohenloher, R.E., Porch, S., *Anal. Biochem.*, 1976, 72, 389.
62. Kubota, K., Sato, Y., Suzuki, Y., Goto-Inoue, N., Toda, T., Suzuki, M., Hisanaga, S., Suzuki, A., Endo, T., *Anal. Chem.*, 2008, 80, 3693.
63. Nawarak, J., Phutrakul, S., Chen, S.T., *J. Proteome Res.*, 2004, 3, 383.
64. Uematsu, R., Furukawa, J., Nakagawa, H., Shinohara, Y., Deguchi, K., Monde, K., Nishimura, S., *Mol. Cell Proteomics*, 2005, 4, 1977.
65. Qiu, R., Regnier, F.E., *Anal. Chem.*, 2005, 77, 2802.
66. Gabius, H.J. et al., *Biochim. Biophys. Acta*, 2002, 1572, 165.
67. Wang, Y., Wu, S.L., Hancock, W.S., *Biotechnol. Prog.*, 2006, 22, 873.
68. Drake, R.R., Schwegler, E.E., Malik, G., Diaz, J., Block, T., Mehta, A., Semmes, O., *Mol. Cell Proteomics*, 2006, 5.10, 1957.
69. Wang, L., Li, F., Sun, W., Wu, S., Wang, X., Zhang, L., Zheng, D., Wang, J., Gao, Y., *Mol. Cell Proteomics*, 2006, 5.3, 560.
70. Monzo, A., Bonn, G.K., Guttman, A., *Trends Anal. Chem.*, 2007, 26, 423.
71. Larsen, M.R., Cordwell, S.J., Roepstorff, P., *Proteomics*, 2002, 2, 1277.
72. Gerard, C., *Methods Enzymol.*, 1990, 182, 529.
73. Brittain, S.M., Ficarro, S.B., Brock, A., Peters, E.C., *Nat. Biotechnol.*, 2005, 23, 463.
74. Mirzaei, H., Regnier, F., *Anal. Chem.*, 2005, 77, 2386.
75. Zhang, Q., Tang, N., Brock, J.W.C., Mottaz, H.M., Ames, J.M., Baynes, J.W., Smith, R.D., Metz, T.O., *J. Proteome Res.*, 2007, 6, 2323.
76. Nice, E.C., Rothacker, J., Weinstock, J., Lim, L., Catimel, B., *J.Chromatogr. A*, 2007, 1168, 190.
77. Madera, M., Mechref, Y., Novotny, M.V., *Anal. Chem.*, 2005, 77, 4081.
78. Qiu, R., Regnier, F.E., *Anal. Chem.*, 2005, 77, 7225.
79. Thaysen-Andersen, M., Hojrup, P., *Am. Biotechnol. Lab.*, 2006, 24, 14.
80. Kaji, H., Saito, H., Yamauchi, Y., Shinkawa, T., Taoka, M., Hirabayashi, J., Kasai, K., Takahashi, N., Isobe, T., *Nat. Biotechnol.*, 2003, 21, 667.
81. Callesen, A.K., Mohammed, S., Bunkenborg, J., Kruse, T.A., Cold, S., Mogensen, O., Christensen, R., Vach, W., Jørgensen, P.E., Jensen, O.N., *Rapid Commun. Mass Spectrom.*, 2005, 19, 1578.
82. Aresta, A., Calvano, C.D., Palmisano, F., Zambonin, C.G., Monaco, A., Tommasi, S., Pilato, B., Paradiso, A., *J. Pharm. Biomed. Anal.*, 2008, 46, 157.

83. Larsen, M.R., Jensen, S.S., Jakobsen, L.A., Heegaard, N.H., *Mol. Cell Proteomics*, 2007, 6, 1778.
84. Churms, S.C., *J. Chromatogr. A*, 1996, 720, 75.
85. Kieliszewski, M.J., Oneill, M., Leykam, J., Orlando, R., *J. Biol. Chem.*, 1995, 270, 2541.
86. Zhang, J., Wang, D.I., *J. Chromatogr. B*, 1998, 712, 73.
87. Gregor, H.P., Collins, F.C., Pope, M., *J. Colloid Sci.*, 1951, 6, 304.
88. Rückert, H., Samuelson, O., *Sven. Kem. Tidskr.*, 1954, 66, 337.
89. Yoshida, T., *J. Biochem. Biophys. Methods*, 2004, 60, 265.
90. Guo, Y., Huang, A.H., *J. Pharm. Biomed. Anal.*, 2003, 31, 1191.
91. Kind, T., Tolstikov, V., Fiehn, O., Weiss, R.H., *Anal. Biochem.*, 2007, 363, 185.
92. Zhu, B.-Y., Mant, C.T., Hodges, R.S., *J. Chromatogr. A*, 1991, 548, 13.
93. Lindner, H., Sarg, B., Meraner, C., Helliger, W., *J. Chromatogr. A*, 1996, 743,137.
94. Ytterberg, J.A., Ogorzalek-Loo, R.R., Boontheung, P., Loo, J.A., in: *55th ASMS Conference on Mass Spectrometry*, Indianapolis, IN, June 3–7, 2007.
95. Alpert, A.J., *Anal. Chem.*, 2008, 80, 62.
96. Yoshida, T., *Anal. Chem.*, 1997, 69, 3038.
97. Tomiya, N., Awaya, J., Kurono, M., Endo, S., Arata, Y., Takahashi, N., *Anal. Biochem.*, 1988, 171, 73–90.
98. Omaetxebarria, M.J., Hagglund, P., Elortza, F., Hooper, N.M., Arizmendi, J.M., Jensen, O.N., *Anal Chem.*, 2006,78, 3335.
99. Callesen, A.K., Vach, W., Jørgensen, P.E., Cold, S., Tan, Q., Depont Christensen, R., Mogensen, O., Kruse, T.A., Jensen, O.N., Madsen, J.S., *J. Proteome Res.*, 2008, 4, 1419.
100. Takegawa, Y., Deguchi, K., Ito, H., Keira, T., Nakagawa, H., Nishimura, S., *J. Sep. Sci.*, 2006, 29, 2533.
101. Alpert, A.J., Shukla, M., Shukla, A.K., Zieske, L.R., Yuen, S.W., Ferguson, M.A., Mehlert, A., Pauly, M., Orlando, R., *J. Chromatogr. A*, 1994, 676, 191.
102. Thomsson, K.A., Karlsson, N.G., Hansson, G.C., *J. Chromatogr. A*, 1999, 854, 131.
103. Thomsson, K.A., Karlsson, H., Hansson, G.C., *Anal. Chem.*, 2000, 72, 4543.
104. Charlwood, J., Birrell, H., Bouvier, E.S., Langridge, J., Camilleri, P., *Anal. Chem.*, 2000, 72, 1469.
105. Saba, J.A., Shen, X., Jamieson, J.C., Perreault, H., *J. Mass Spectrom.*, 2001, 36, 563.
106. Saba, J.A., Kunkel, J.P., Jan, D.C., Ens, W.E., Standing, K.G., Butler, M., Jamieson, J.C., Perreault, H., *Anal. Biochem.*, 2002, 305, 16.
107. Wuhrer, M., Koeleman, C.A.M., Deelder, A.M., Hokke, C.H., *Anal. Chem.*, 2004, 76, 833.
108. Wuhrer, M., Koeleman, C.A.M., Hokke, C.H., Deelder, A.M., *Int. J. Mass Spectrom.*, 2004, 232, 51.
109. Geyer, H., Wuhrer, M., Resemann, A., Geyer, R., *J. Biol. Chem.*, 2005, 280, 40731.
110. Karlsson, G., Winge, S., Sandberg, H., *J. Chromatogr. A*, 2005, 1092, 246.
111. Risley, D.S., Yang, W.Q., Peterson, J.A., *J. Sep. Sci.*, 2006, 29, 256.
112. Beilmann, B., Langguth, P., Häusler, H., Grass, P., *J. Chromatogr. A*, 2006, 1107, 204.
113. Peru, K.M., Kuchta, S.L., Headley, J.V., Cessna, A.J., *J. Chromatogr. A*, 2006, 1107, 152.
114. Takegawa, Y., Deguchi, K., Keira, T., Ito, H., Nakagawa, H., Nishimura, S.-I., *J. Chromatogr. A*, 2006, 1113, 177.
115. Hagglund, P., Matthiesen, R., Elortza, F., Hojrup, P., Roepstorff, P., Jensen, O.N., Bunkenborg, J., *J. Proteome Res.*, 2007, 6, 3021.
116. Kaji, H., Yamauchi, Y., Takahashi, N., Isobe, T., *Nat. Protoc.*, 2006, 1, 3019.
117. Thaysen-Andersen, M., Thogersen, I.B., Nielsen, H.J., Lademann, U., Brunner, N., Enghild, J.J., Hojrup, P., *Mol. Cell Proteomics*, 2007, 6, 638.
118. Kaji, H., Kamiie, J., Kawakami, H., Kido, K., Yamauchi, Y., Shinkawa, T., Taoka, M., Takahashi, N., Isobe, T., *Mol. Cell Proteomics*, 2007, 6, 2100.
119. DeHoffman, E., Stroobant, V., *Oligosaccharides in Mass Spectrometry: Principles and Applications*, 2nd edn., John Wiley & Sons Ltd., West Sussex, U.K., 2001, pp. 292–320.

120. Montilla, A., van de Lagemaat, J., Olano, A., del Castillo, M.D., *Chromatographia*, 2006, 63, 453.
121. Mortz, E., Sareneva, T., Julkunen, I., Roepstorff, P., *J. Mass Spectrom.*, 1996, 31, 1109.
122. Harvey, D.J., *Proteomics*, 2005, 5, 1774.
123. Harvey, D.J., *Expert Rev. Proteomics*, 2005, 2, 87.
124. Mechref, Y., Novotny, M.V., *Chem. Rev.*, 2002, 102, 321.
125. Zaia, J., *Mass Spectrom. Rev.*, 2004, 23, 161.
126. Wuhrer, M., Catalina, M.I., Deelder, A.M., Hokke, C.H., *J. Chromatogr. B*, 2007, 849, 115.
127. Yamamoto, K., Tsuji T., Osawa, T., *Mol. Biotechnol.*, 1995, 3, 25.
128. Conboy, J.J., Henion, J.D., *J. Am. Soc. Mass Spectrom.*, 1992, 3, 804.
129. Huddleston, M.J., Bean M.F., Carr, S.A., *Anal. Chem.*, 1993, 65, 877.
130. Dalpathado, D.S., Irungu, J., Go, E.P., Butnev, V.Y., Norton, K., Bousfield, G.R., Desaire, H., *Biochemistry*, 2006, 45, 8665.
131. Go, E.P., Rebecchi, K.R., Dalpathado, D.S., Bandu, M.L., Zhang, Y., Desaire, H., *Anal. Chem.*, 2007, 79, 1708.
132. Bykova, N.V., Rampitsch, C., Krokhin, O., Standing, K.G., Ens, W., *Anal. Chem.*, 2006, 78, 1093.
133. Larsen, M.R., Hojrup, P., Roepstorff, P., *Mol. Cell Proteomics*, 2005, 4, 107.
134. Roberts, G., Keyser, J.W., Baum, M., *Br. J. Surg.*, 1975, 62, 816.
135. Calvano, C.D., Zambonin, C.G., Jensen, O.N., *J. Proteomics*, 2008, 71, 304.
136. Selby, D.S., Larsen, M.R., Calvano, C.D., Jensen, O.N., *Glycosylation. Methods in Molecular Biology: Functional Proteomics*, 2008, 484, 263.

21 Development and Application of Methods for Separation of Carbohydrates by Hydrophilic Interaction Liquid Chromatography

Goran Karlsson

CONTENTS

21.1 INTRODUCTION

21.1.1 Carbohydrates

Carbohydrates (sugars or saccharides) are a large group of organic polar compounds that contain aldehydes or ketones, or their derivatives, which have many hydroxyl groups in their chemical structure.[1] Monosaccharides are the simplest type of carbohydrates, which can assume cyclic forms (hemiketals or hemiacetals) and which can also react with other carbohydrates to form di-, oligo-, and polysaccharides. Carbohydrates are the most abundant class of biomolecules, having diverse functions such as the storage of energy (e.g., starch in plants and glycogen in animals) or the formation of structural components (e.g., cellulose in plants and chitin and cartilage in animals). Additionally, carbohydrates and their derivatives play major roles in biochemical systems, as in the vital parts of proteins (including enzymes, antibodies, and hormones), and are thus crucial for the functioning of physiological processes such as the immune system, fertilization, pathogenesis, blood clotting, and development. The inappropriate glycosylation of proteins, for example, can lead to a reduced biological activity, decreased half-life in circulation, or unwanted immunogenicity.

21.1.2 Techniques for Separation and Detection of Carbohydrates

For more than 30 years, carbohydrates have been separated by aqueous normal-phase liquid chromatography (NPLC), using amino-propyl columns with about 75% acetonitrile and 25% water for isocratic elution.[2] Thus, these methods can be included in the liquid chromatography group now denoted HILIC (hydrophilic interaction liquid chromatography), which is a variant of NPLC. In contrast to conventional NPLC, HILIC uses water for elution in a mostly organic mobile phase. NPLC, including HILIC, in contrast to reversed-phase liquid chromatography (RPLC), which is the most used liquid chromatography method, provides good retention of highly polar compounds, both charged and uncharged, including carbohydrates. The elution order in HILIC is roughly reversed when compared with that of RPLC,[3] and, thus, HILIC gives strong retention of compounds that are weakly or not at all bound in RPLC, which allows them to be separated by this technique. HILIC is used primarily for the separation of very polar compounds with a negative value of log octanol–water partition coefficient (log P).

The direct detection of carbohydrates by UV absorbance or fluorescence is usually not possible, and, during the 1970s and 1980s, the universal refractive index (RI) detector was often used. Later, this rather insensitive detector was replaced by evaporative light scattering detection (ELSD), mass spectrometry (MS), and other detection techniques. Gas chromatography (GC) of trimethylsilyl (TMS) derivatized monosaccharides was introduced more than 50 years ago[4] and later was successfully combined with mass spectrometry (GC-MS),[5] which was a major step in determining the structure and concentration of carbohydrates. Nuclear magnetic resonance (NMR) is a powerful technique for determining carbohydrate structure and provides complimentary information to the MS data. Another approach for carbohydrate detection uses derivatization of the reducing end of the mono-, di-, or oligosaccharide, where a hydrophobic chromophore[6] or fluorophore[7] is attached, which is followed with RPLC for the separation and detection by absorbance or fluorescence, respectively. Derivatization of carbohydrates for fluorescence detection is especially useful for detecting carbohydrate derivates with a very high sensitivity.

Ligand-exchange chromatography (LEC) has been successfully used for the separation of monosaccharides.[8] With this method, hydroxyl groups of carbohydrates form a complex with multivalent metal ions (such as Pb^{2+}), and separation follows on a cation-exchange chromatography column. Additionally, ordinary anion-exchange chromatography can be used, for example, for sialic acid[9] and glucuronic acid[10] containing carbohydrates. Boronate easily forms esters with *cis*-diols in carbohydrates, which was used by Bourne more than 40 years ago to enhance the separation of carbohydrates by paper chromatography.[11] Boronate affinity chromatography, with boronate covalently coupled to the gel matrix, was later, for example, developed for the separation between hemoglobin and a glycosylated form of hemoglobin.[12] Fluorophore-assisted carbohydrate electrophoresis, where a charged fluorophore is covalently bound to the carbohydrate, is used for the separation of simple and complex carbohydrates.[13] Other electrophoretic techniques for the separation of carbohydrates, such as capillary electrophoresis (CE)[14] and capillary electrochromatography (CEC),[15] which achieve high efficiency and resolution, are powerful alternatives to liquid chromatography methods for analyzing complex carbohydrates derived from biological samples. Many other techniques have been used for the separation of different types of simple and complex carbohydrates, including affinity chromatography using columns coupled with a suitable lectin (e.g., concavalin A, which has a high affinity for mannose-containing carbohydrates),[16] and supercritical fluid chromatography.[17]

High-performance anion-exchange chromatography (HPAEC) is a high-resolution technique for analyzing carbohydrates that uses a strongly alkaline mobile phase (pH about 13), where hydroxyl groups of the carbohydrates are ionized and negatively charged and thus can be bound to an anion-exchange chromatography column, and the elution is with a buffer of increasing ionic strength. For example, a sodium acetate gradient may be used for the elution of complex carbohydrates, or in an isocratic mode can be used to separate mono- and disaccharides. HPAEC is often combined with pulsed amperometric detection (PAD) for the very sensitive detection of carbohydrates.[18] This technique is now widely used and considered to have excellent separating and detection capabilities for analyzing carbohydrates. Besides the high resolution and selectivity of HPAEC-PAD chromatography, this system has an advantage, compared to fluorescence-labeling techniques, for example, in that derivatization of the carbohydrates is not required for PAD detection. Nevertheless, PAD requires relatively more maintenance (particularly for the gold membrane in the working electrode) to avoid baseline drift and other distortions of the chromatogram. A comprehensive description and evaluation of all useful methods and techniques for separation and analysis of carbohydrates is beyond the scope of this book. For review articles that cover some of the methods and techniques, see Cataldi et al. (HPAEC-PAD),[19] Campa et al. (CE),[20] Klampfl (CEC-MS),[21] and Zaia (MS).[22]

21.1.3 HILIC in the Separation of Carbohydrates

HILIC was defined by Alpert in 1990[3] as a variant of normal-phase chromatography, where a hydrophilic stationary phase is used in combination with a mostly organic aqueous mobile phase, often acetonitrile (sometimes propanol or acetone), and elution is usually performed by increasing the polarity of the mobile phase by increasing the water concentration. Elution by increasing the ionic strength can also be used, especially when running ion-exchange chromatography (IEX) columns in HILIC mode (i.e., using about 60%–90% organics), according to Alpert.[3] Aqueous NPLC, now defined as HILIC, has been used since 1975 for separating simple and complex carbohydrates.[2] In those days, RI detectors were often used in analytical chromatography, so that gradient elution was not possible. In addition to acetonitrile and water, a volatile buffer, such as ammonium acetate, which is compatible with an electrospray ionization (ESI) interface MS, is often included in the mobile phase to avoid electrostatic interaction with ionized silanols in the silica solid phase. The high concentration of acetonitrile facilitates the detection by ESI MS, a technique which is widely used today, and which is used for numerous purposes due to high specificity and sensitivity. HILIC can be readily combined with an ESI interface in a liquid chromatography-mass spectrometry (LC-MS) system. This is in contrast to HPAEC systems, where some kind of post-column ion-suppression device is required before the ESI interface,[23] and in contrast to ordinary NPLC that uses 100% organics, which does not facilitate ionization.

In HILIC, the retention depends on the distribution of the analytes between a partially immobilized aqueous layer, bound to the gel matrix surface, and the mobile phase. In addition, hydrogen bonding and electrostatic interactions between the analyte and the stationary phase are also believed to be more important than comparable mixed-mode retention mechanisms in other liquid chromatography techniques, for example, RPLC.[3,24,25] This is in contrast to conventional NPLC, where mixtures of organics, without water, are used in the mobile phase, and where the separation mainly depends on surface adsorption. A higher number of monosaccharides in more complex carbohydrates usually give an increased retention in HILIC, for example, the disaccharide sucrose elutes after its monosaccharides glucose and fructose (Figure 21.1). Fucose, lacking one hydroxyl group, is thus equivalent to 6-deoxy-L-galactose and has a lower retention compared to corresponding monosaccharides, for instance, galactose and glucose (Figure 21.2), which both have an intact 6′ hydroxyl group. In addition, aminosugars (e.g., *N*-acetyl-D-glucosamine) have a stronger retention on HILIC columns, compared to corresponding nonaminosugars.[24]

Using water in HILIC mobile phases is advantageous because it increases the solubility of polar compounds, and the water content does not need to be completely controlled, as is the case in ordinary NPLC. An increased interest in HILIC has been

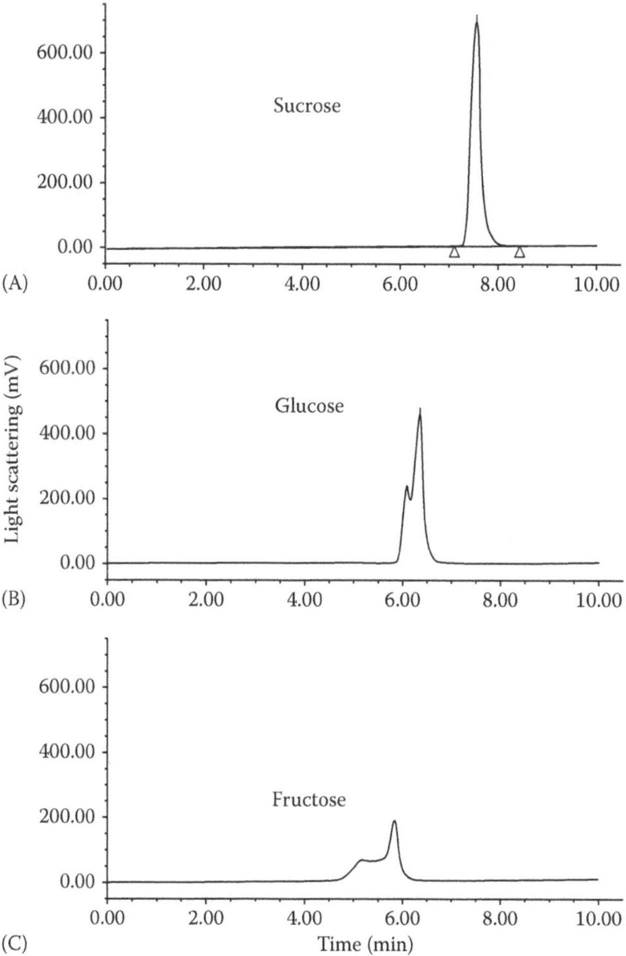

FIGURE 21.1 Hydrophilic interaction chromatography of (A) sucrose, (B) glucose, and (C) fructose. A PolyHydroxyethyl A column was used and the elution was isocratically performed with 25% water and 75% acetonitrile, at ambient temperature (about 22°C). The reducing sugars glucose and fructose show typical mutarotation double-peaks. ELS was used for detection. (Reproduced from Karlsson, G. et al., *J. Chromatogr. Sci.*, 42, 361, 2004, Fig. 1. With permission from Preston Publications, a Division of Preston Industries, Inc.)

seen for more than a decade due to the desire to analyze more complex biological samples, such as in studies of proteomics and metabolomics. During this period, several liquid chromatography column manufacturers have introduced new HILIC columns with different types of ligands (poly-succinimide-derived, amide, diol, and ion-exchange ligands, including the zwitterionic type, etc.) as well as the bare silica columns, suitable for carbohydrate analysis. See the work by Hemström and Irgum for a comprehensive review of HILIC.[25]

In the following sections, method development and applications of HILIC methods for separating and analyzing carbohydrates are discussed.

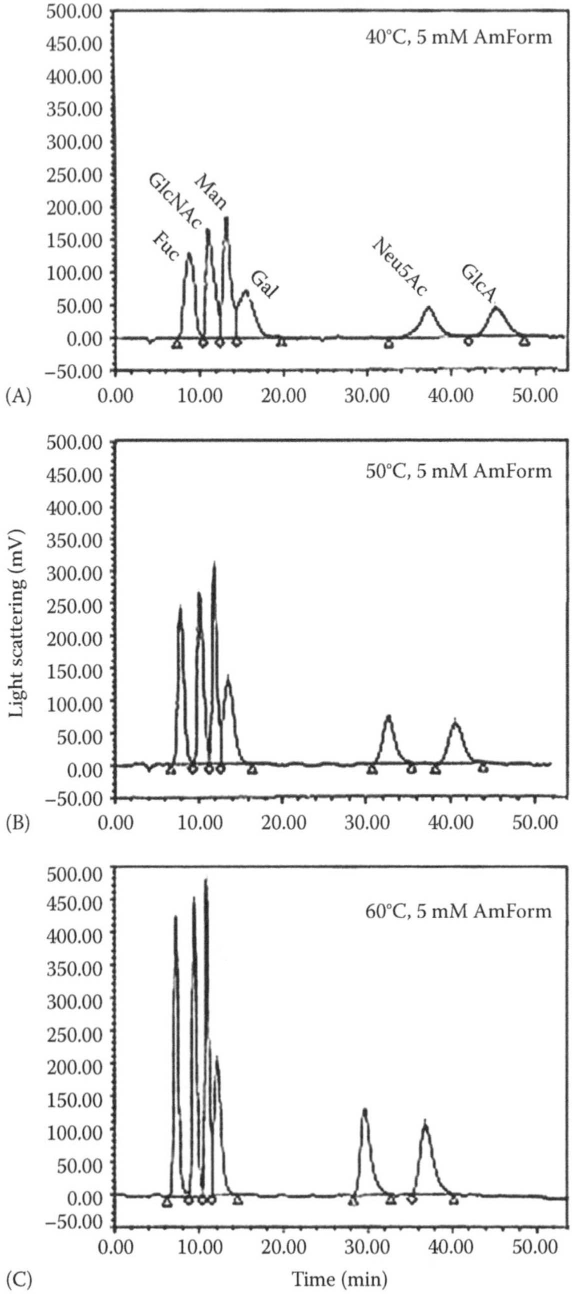

FIGURE 21.2 Optimization of the separation of monosaccharides by HILIC. The influence of temperature and ammonium formate (AmForm, pH 5.5) concentration on the separation is shown in A–E, and the selectivity results are given in F. A TSK-Gel Amide-80 column, and 82% acetonitrile, in water, was used in all experiments: (A) 40°C, 5 mM AmForm; (B) 50°C, 5 mM AmForm; (C) 60°C, 5 mM AmForm;

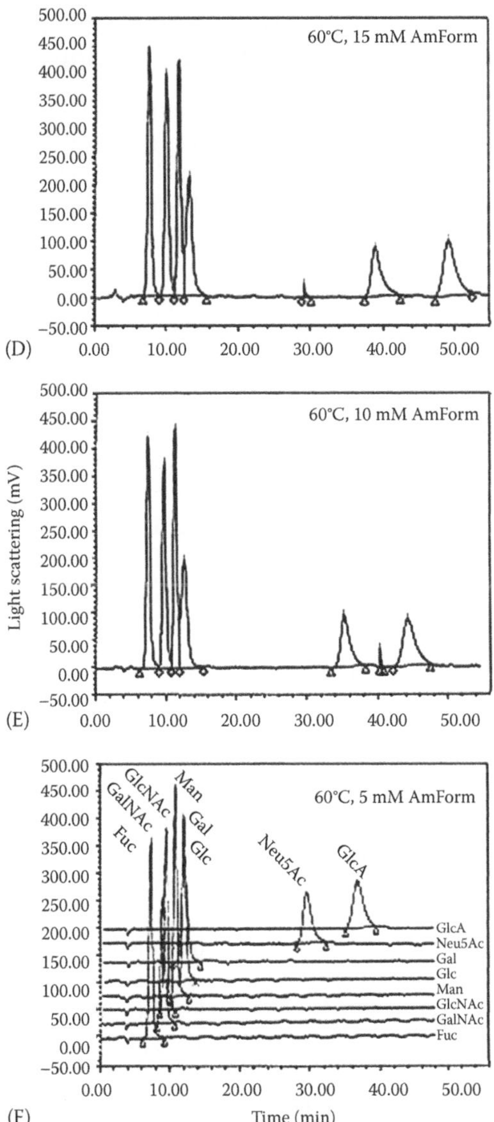

FIGURE 21.2 (continued) (D) 60°C, 15 mM AmForm; (E) 60°C, 10 mM AmForm; and (F) 60°C, 5 mM AmForm. Mixtures of all six monosaccharides [L-fucose (Fuc), D-mannose (Man), D-galactose (Gal), *N*-acetyl-D-glucosamine (GlcNAc), *N*-acetylneuraminic acid (Neu5Ac, the most common sialic acid), and D-glucuronic acid (GlcA)] were injected in A–E. In F, which shows the overlay of eight chromatograms, the same six monosaccharides and *N*-acetyl-D-galactosamine (GalNAc) and D-glucose (Glc) were analyzed separately. Increasing the temperature gave an increased efficiency and resolution, while increasing the ammonium formate concentration gave increased retention for the carboxylic acid-containing monosaccharides (*N*-acetylneuraminic acid and D-glucuronic acid). The conditions in C and F represent the optimized method. ELS was used for detection. (Reproduced from Karlsson, G. et al., *J. Chromatogr. A*, 1092, 246, 2005. With permission from Elsevier.)

21.2 DEVELOPMENT AND OPTIMIZATION OF HILIC METHODS FOR SEPARATING AND ANALYZING CARBOHYDRATES

In the development of HILIC methods for separating and analyzing carbohydrates and other compounds, many variables should be considered, including column type, mobile-phase composition, temperature, flow rate, sample preparation, sample concentration and volume, and mode of detection. Different approaches and aims of the analysis can be used, depending on whether the analysis is for quantification, structure determination, impurity determination, or obtaining a chromatographic profile of a mixture of complex carbohydrates. For example, if the aim is to quantify one specific carbohydrate, then a baseline resolved peak (Rs > 1.5) of this compound is preferred, but if the aim is to obtain a chromatographic profile of a mixture of very complex carbohydrates, the baseline resolution of all peaks is usually not possible. In addition, the mobile-phase composition must be compatible with the detector, which is important for MS detection and characterization. Recent review articles that cover the method development of HILIC methods include the works by Ikegami et al.,[26] Dejaegher et al.,[27] and Hao et al.[28] The excellent book by Snyder et al.[29] is also recommended for its coverage of general high-performance liquid chromatography (HPLC) method development.

21.2.1 Use of Different Stationary Phases

Amino-propyl was the first HILIC stationary phase and was used much during the 1970s and 1980s. In the past decade, as several new stationary phases for HILIC have been commercially available, amino-propyl columns are less frequently used. This can be partially explained by the fact that amino-propyl columns are unstable and that a reactive Schiff's base can be formed by the amino group. Schiff's bases can bind to aldehydes, including reducing carbohydrates, which may change the functionality of the HILIC column, though this has not been shown to occur for reducing carbohydrates in the HILIC conditions. In 1990, Alpert defined HILIC and introduced a new column material of the succinimide type (polyhydroxyethyl aspartamide silica, PolyHydroxyethyl A, from PolyLC), for separating various polar compounds.[3] A few years later, a similar succinimide-based column (PolyGlycoplex, from PolyLC), intended mainly for analyzing carbohydrates, including oligosaccharides, was described by the same author.[24] Bare silica columns, adapted for HILIC,[30] have also been commercialized in the last decade (e.g., Atlantis, manufactured by Merck). This column type gives a low retention, compared to the other types of HILIC columns, and thus needs a higher concentration of organics in the mobile phase to bind and separate carbohydrates. Diol columns, which readily give hydrogen bonding, are more stable than bare silica and have been mostly used for protein separation, though this column type has also been occasionally used for HILIC of carbohydrates. Cyano-propyl columns, to some extent, have been used in HILIC; however, these columns give a low retention, due to the inability to form hydrogen bonds. They also suffer from low stability in aqueous mobile phases.[25]

The amide column is more stable than several other HILIC column types and has been frequently used to separate several polar substances, including simple and complex carbohydrates.[31–35] Reports suggest that amide columns have a higher efficiency than that of PolyHydroxyethyl A columns.[34] HILIC columns with zwitterionic (sulfoalkylbetaine) ligands were introduced by Irgum and have been used to separate different types of carbohydrates.[25,36,37] Columns with zwitterionic groups are very stable and have been previously used for conventional cation-exchange chromatography. When the organic (acetonitrile) concentration exceeds about 50%, the hydrophilic forces mainly dominate, in comparison to electrostatic interactions, when using an ion-exchange column in HILIC mode (i.e., with about 60%–90% organics in the mobile phase). The ionic interactions that are present at more than 50% acetonitrile will be less prominent and will contribute to the mixed-mode retention. The amide (e.g., the Amide-80 column, from Tosoh) and zwitterionic (e.g., the ZIC HILIC column, from SeQuant/Merck) types of HILIC columns are now among the most popular stationary phases for HILIC separations. One difference between the two is that a higher flow rate can be used for the amide-type column. User-made monolithic HILIC columns, with polyacrylamide-coated silica, for achieving a high flow rate and high resolution, have been used for separating galactose, sucrose, and lactose.[38] Bare silica monolithic columns are available from Merck and can be used for HILIC. Commercially available cyclodextrin HILIC columns have been used for separating mono-, di-, and oligosaccharides.[39] Cyclodextrins, which are composed of five or more 1-4 linked α-D-glucopyranoside units, are relatively hydrophobic in the interior of their toroid structure, which is used for RPLC separations. The outside, however, is more hydrophilic and is used for HILIC separations. A column length of 15 cm, with an inner diameter (ID) of 4.6 mm, is generally recommended, but shorter columns, with an ID of 2.1 mm, are more suitable for MS detection, because a lower flow rate can be used and that it gives less dilution of the sample. The particle size of 3 μm, which increases the resolution compared to the more traditionally 5 μm particles, and a pore size of about 100 Å (10 nm) is often used, except for larger molecules where a larger pore size is recommended. For example, in the case of hyaluronic acid with a high molecular mass (up to 5 million Da), 4000 Å pore size has been used.[10] A 1.7 μm particle size bare silica HILIC column, Acquity (Waters), is available for use in ultrahigh-performance liquid chromatography (UPLC). Generally, silica-based columns tolerate a pH of 2–7, while polymeric-based HILIC columns may tolerate a pH up to 10 (or higher). See Table 21.1 for information about some different HILIC column types and manufacturers.

21.2.2 Mobile-Phase Composition

The first choice for an organic for HILIC is acetonitrile, which has excellent chromatographic properties, at least partly because of its low viscosity and therefore high diffusion rate and efficiency, so that the resolution will be superior to that of other organics (such as propanol) that are used in HILIC. The eluotropic strength of solvents used in HILIC is water > methanol > ethanol > propanol > acetonitrile > acetone > tetrahydrofuran (THF). Alcohols (e.g., methanol) more easily form hydrogen bonds with analytes and also with water than acetonitrile. This will affect the partially immobilized aqueous layer that is bound to the gel matrix surface, which will lower

TABLE 21.1
Applications for HILIC Separation of Carbohydrates

Application/Analyte	Column Brand Name and Manufacturer	Support and Functionality	Mobile Phase[a]	Detection	Reference
Mono-, di-, and oligosaccharides	Cyclobond III and I, Advanced Separation Technologies	Silica, α- and β-cyclodextrin	92%–70% acetonitrile	UV and RI	39 Appl. 3.1
Highly polar compounds from plant samples (including mono- and oligosaccharides)	Amide-80, Tosoh	Silica, amide	85%–45% acetonitrile and 1.0–3.6 mM ammonium acetate (pH 5.5) dual gradient	ESI MS-MS	34 Appl. 3.2
Monosaccharides and oligosaccharides	Amide-80, Tosoh	Silica, amide	Isocratic 80% or 60% acetonitrile and 80%–60% acetonitrile gradient	Fluorescence of post-column derivatization with benzamidine	33
Monosaccharides in *N*-linked oligosaccharides	Amide-80, Tosoh	Silica, amide	82% acetonitrile, 5 mM ammonium formate, pH 5.5	ELSD	32
Mono-, di-, and trisaccharides	User-made	Monolithic silica, modified with polyacrylamide	80% acetonitrile, 13 mM ammonium acetate	ESI MS/MRM	38
Mono- and disaccharides	Luna Amino, Phenomenex	Silica, amino	80% acetonitrile, 40 μM cesium acetate	MS/MRM	40
Highly polar compounds from plants (mostly mono- and disaccharides)	ZIC HILIC, SeQuant/Merck	Silica, sulfoalkylbetaine zwitterionic	90%–10% acetonitrile and 0.5–4.5 mM ammonium acetate dual gradient, in 0.1% formic acid, pH 4	MS-MS	41

Glucuroconjugate of propofol	Atlantis, Merck	Bare silica	87% acetonitrile, 12 mM ammonium acetate, pH 5	ESI MS-MS	30
Sucrose in a protein formulation	PolyHydroxyethyl A, PolyLC	Silica/poly-succinimide, polyhydroxyethyl aspartamide	75% acetonitrile	ELSD	42
Oligosaccharides, complex carbohydrates	PolyGlycoplex, PolyLC	Silica/poly-succinimide derived	10 mM TEAP, 80% methanol, pH 4.4	MS and UV 200/230 nm	24
N-linked glycans and *N*-glycopeptides	ZIC-HILIC, SeQuant/Merck	Silica, sulfoalkylbetaine zwitterionic	77%–63% acetonitrile, in 5 mM ammonium acetate	ESI MS and UV 215 nm, of peptides and 2-AP glycans	36 Appl. 3.3
Oligosaccharides	Polymer-NH_2, Astec	Polymer, amino	70%–5% acetonitrile, 2.9%–4.9% acetic acid/0.9%–2.9% TEA, dual gradient, in 1% THF	Fluorescence of 2-AB derivatives	35 Appl. 3.4
Glucosinolates	ZIC-HILIC, SeQuant/Merck	Silica, polymeric sulfoalkylbetaine zwitterionic	70% acetonitrile, 15 mM ammonium formate, pH 4.5	UV (235 nm)	37 Appl. 3.5
Glucosaminoglycans	Amide-80 capillary column, Tosoh	Polymeric, amide	82%–46% acetonitrile and 10–30 mM formic acid, pH 4.4, dual gradient	MS-MS	43 Appl. 3.6
N-linked oligosaccharides from fetuin	Amide-80, Tosoh	Silica, amide	64%–52% acetonitrile, 5 or 25 mM ammonium acetate, pH 5.5	Fluorescence of 2-AB derivatives	31
Oligosaccharides	GlycoSep N, ProZyme	Silica, amide	75%–58% acetonitrile and 12–29 mM ammonium acetate, pH 4.4, dual gradient	MS and fluorescence of 2-aminoacridone derivatives	44

[a] Water or aqueous buffer comprises the remaining part of the mobile phases.

the polar retention of carbohydrates to the column. Changing from an aprotic organic solvent (e.g., acetonitrile) to a protic organic solvent (e.g., propanol) can sometimes change the selectivity. However, the efficiency in the chromatographic system, when using propanol or other alcohols, will unfortunately decrease, compared to that when using acetonitrile, and, usually, it is best to stick to acetonitrile. An alternative, though seldom used, is to use a ternary mobile phase, for example, 70% acetonitrile/15% methanol/15% water, in ammonium acetate buffer, to change the selectivity without losing much efficiency. Water is sometimes completely replaced with the very polar organic methanol, by using 75% acetonitrile/25% methanol, for example. However, water is usually included in the HILIC mobile phases.

To obtain good chromatographic results, the water content in an HILIC mobile phase should be no less than 5% (or 3%). For an initial investigative analysis, a scouting gradient, for example, with 85%–50% acetonitrile, in water (i.e., 15%–50% water) with a buffer component, may be used for most HILIC columns, in a volume that corresponds to about 25 column volumes. If all peaks elute in a minor part of the gradient, it would usually be best to then try an isocratic run, with a concentration of acetonitrile corresponding to about the middle part where the peaks eluted in the gradient. If the peaks have a more spread-out elution, in the larger part of the gradient, then a gradient should be used. A more time-consuming alternative would be to use a series of isocratic runs with increasing organic concentrations, where it would be best to start with a very strong mobile phase, for example, 50% acetonitrile in a buffered aqueous solution, and then stepwise (initially 10% steps, followed by 5% steps) increase the acetonitrile concentration as necessary, usually to about 60%–80% for most carbohydrates. For an optimized isocratic system, the retention factor k (formerly called the capacity factor k') is ideally 2–10 (1–20 may also be acceptable, especially for more complex samples) for the separated compounds. If early or late eluters (peaks with very low or high k values, i.e., outside the 1–20 range) appear in the chromatogram, a gradient would likely be necessary for the separation. For determining a suitable mobile-phase composition for an isocratic run, as judged from the scouting gradient, compensating adjustments should be made for the gradient delay volume (also called the dwell volume) and the column dead volume. If a gradient is chosen, the gradient range can be decreased (e.g., to 80%–60% acetonitrile), provided that all peaks appear in that range. A number of gradient variations, with different concentration steepness, gradient times, and acetonitrile concentration ranges should be tested to obtain the desired resolution, peak shape, and analysis time. If a small difference occurs in the polarity between the carbohydrates, for example, if just a number of monosaccharides need to be separated, the first choice would be to try an isocratic separation. In contrast, if mixtures of carbohydrates that contain components of substantially different size and/or differences in the number of carboxyl- or other charged groups, i.e., a large difference in total polarity, are to be separated, then a gradient is usually necessary. Using a gradient will avoid peak broadening for peaks with long retention times and will accomplish peak compression, giving high, narrow peaks with high efficiency and improved detection sensitivity over the whole gradient. After an 80%–60% acetonitrile linear gradient, in water, the column should generally be equilibrated in about 5–10 column volumes with the initial conditions (80% acetonitrile). See Figure 21.3 for a comparison between isocratic and gradient elutions of carbohydrates.

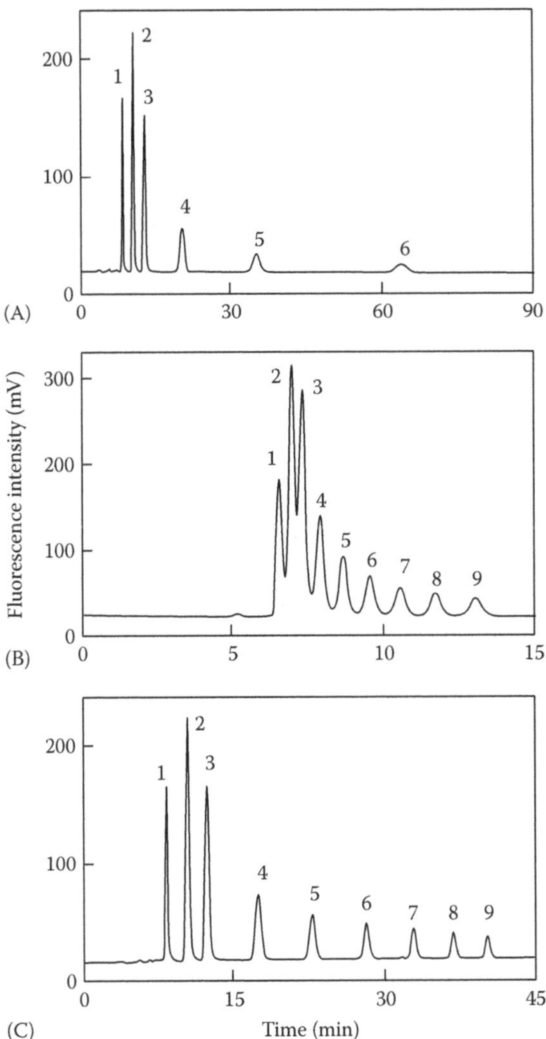

FIGURE 21.3 Simultaneous separation of three monosaccharides and six oligosaccharides. A TSK-Gel Amide-80 column was used at 80°C together with post-column labeling by benzamidine followed by fluorescence detection. (A) Isocratic elution with acetonitrile:water (80:20); (B) isocratic elution with acetonitrile:water (60:40); and (C) linear gradient, 80%–60% acetonitrile, in water. A mixture of (1) D-ribose, (2) D-fructose, (3) D-glucose and the laminarioligosaccharides (4) laminaribose, (5) laminaritriose, (6) laminaritetraose, (7) laminaripentaose, (8) laminarihexaose, and (9) laminariheptaose was separated. By using the gradient, narrow, high, and well-resolved peaks were obtained for each of the separated carbohydrates. (Reproduced from Kakita, H. et al., *J. Chromatogr. A*, 961, 77, 2002. With permission from Elsevier.)

As an alternative to ordinary linear gradients, a segmented gradient may also be considered, which can be successfully used in cases where an adjusted (more shallow) gradient does not give the desired resolution. If, for example, several crowded low-resolved peaks appear in the middle of a 80%–60% acetonitrile gradient, in addition to some well-separated peaks before and after the crowded part, it might help to first run an 80% to about 70% acetonitrile linear gradient, as before, followed by a very shallow gradient part, or an isocratic plateau, to achieve an improved resolution in the crowded part, and finally a continuation of the linear gradient, down to 60% acetonitrile. Furthermore, the flow rate can also, if required, be optimized (0.5–1.0 and about 0.2 mL/min are usually recommended for 4.6 and 2.1 mm ID columns, respectively, see column information from the respective manufacturers). In addition to a high concentration of acetonitrile and some water, a buffer (5–20 mM, up to 200 mM), primarily formic acid, acetic acid, ammonium formate, ammonium acetate, ammonia, or ammonium carbonate, depending on the desired pH, is recommended for addition to the mobile phases. Buffers are recommended by many column manufacturers to avoid electrostatic interaction with charged free silanols in the silica stationary phase. The pH of the mobile phase can influence the retention of ionizable carbohydrates in HILIC. The charged state of the ionizable carbohydrate depends on whether the pH is below or above the pK_a, which in turn affects the hydrophilicity and thereby the retention of the analyte. Thus, the pH should be changed if the desire is to change the selectivity due to the charge of ionizable analytes. In addition, when the pH increases, for example, from 3 to 6, more silanols in the stationary phase are ionized, and an increased retention of positively charged compounds is usually obtained. Even if only neutral carbohydrates are analyzed, ionized contaminants might be present in low concentrations in the sample. Hence, buffers in the mobile phase are beneficial for reproducibility and performance in HILIC. The buffer capacity of the mobile phase should also be of concern; it is recommended to set the pH to ±1.0 pH unit from the pK_a, preferably as close as possible to the pK_a of the used buffer, to obtain an acceptable buffer capacity. This is not strictly followed, however, and trifluoroacetic acid, for example, has a pK_a of 0.3, but is often used as a "buffer" (or "additive" in this case) at pH 2 in RPLC, even though the buffer capacity is very low at that pH. The pH should be adjusted, if necessary, before the organics are added to the mobile phase.

The buffer component (e.g., ammonium acetate), in addition to the buffering effect, can influence retention and selectivity in HILIC in several ways. An increased salt concentration can increase the polarity of the partly immobilized aqueous layer on the surface of the column material, thereby increasing the retention of polar compounds in an HILIC system, which supports the partitioning theory of retention in HILIC.[3,24,25] On the contrary, for charged column HILIC materials (e.g., NH_2 columns at neutral or acidic pH), an increased salt concentration will decrease electrostatic interactions and thereby decrease the retention of negatively charged carbohydrates, giving the elution due to an ion-exchange effect. Elution in a simultaneously dual mode, where an increasing water gradient gives the main elution, and an increasing buffer concentration (e.g., ammonium acetate) is used in the gradient, has been used for separating various carbohydrates (Table 21.1). By adding a salt gradient, an improved selectivity and resolution can sometimes be obtained, especially for charged HILIC columns, for example, amino columns in combination with negatively charged carbohydrates.[35] This elution mode has been called HILIC-IEX (or HILIC weak anion-exchange [WAX] chromatography). In HILIC-IEX, the elution is sometimes used in a segmented mode; first, the acetonitrile concentration is decreased to a suitable level in the gradient, which elutes uncharged carbohydrates; and then the ion strength is increased to elute negatively charged carbohydrates which are more strongly bound to the column.

For amide columns, an increased concentration of ammonium acetate or ammonium formate (10–25 mM instead of 5 mM) increases the retention of ionized mono- and oligosaccharides that contain carboxylic groups, possibly due to increased binding of these carbohydrates to enriched salts in the partly immobilized aqueous layer on the column gel matrix surface (Figures 21.2 and 21.4). Inorganic buffer salts (e.g., phosphates) should generally be avoided due to their low solubility in HILIC mobile phases. Nevertheless, phosphates have the advantage of a high detection sensitivity at low UV wavelengths (about 210 nm), and the solubility of phosphates in acetonitrile is higher at acidic pH than at alkaline pH, due to a less charged state. Buffer salts that have been used in HILIC include ammonium acetate, ammonium formate, cesium acetate, triethylammonium phosphate (TEAP), and sodium perchlorate (see Table 21.1).

21.2.3 Temperature and Other Variables

In HILIC, column oven temperature of about 30°C–60°C (occasionally up to 100°C) is often used. However, the possibility of temperature-dependent degradation of the carbohydrates should be considered, and elevated temperatures should be avoided for labile compounds. A higher temperature is often preferred because it decreases the viscosity of the mobile phase and increases the diffusion and the chromatographic efficiency, giving a high resolution. Mutarotation, i.e., the specific rotation of cyclic carbohydrates as they reach equilibrium between their α- and β-anomeric forms of reducing sugars, can give typical disturbing double-peaks in HILIC. An elevated temperature and a more alkaline pH,[45] as well as organic amines[46] (e.g., triethylamine),[24] accelerate the rate of the anomer interconversion sufficiently to reach a dynamic equilibrium of the anomers that avoids the double-peak appearance. In Figure 21.1, typical anomeric mutarotation double-peaks are shown for the reducing monosaccharides glucose and fructose, but not for the non-reducing disaccharide, sucrose, in an HILIC system run at ambient temperature (about 22°C), with a PolyHydroxyethyl A column. Later, an amide column was used at temperatures up to 60°C, with a similar mobile phase, to avoid the mutarotation double-peaks (Figure 21.2).

In HILIC, as in most other chromatographic techniques, the retention time is usually decreased by elevated temperatures, with other conditions being unchanged, which can be compensated by increasing the percentage of organics. Figure 21.2 shows an optimized separation of monosaccharides, where different temperatures and concentrations of ammonium formate were evaluated on an amide column.

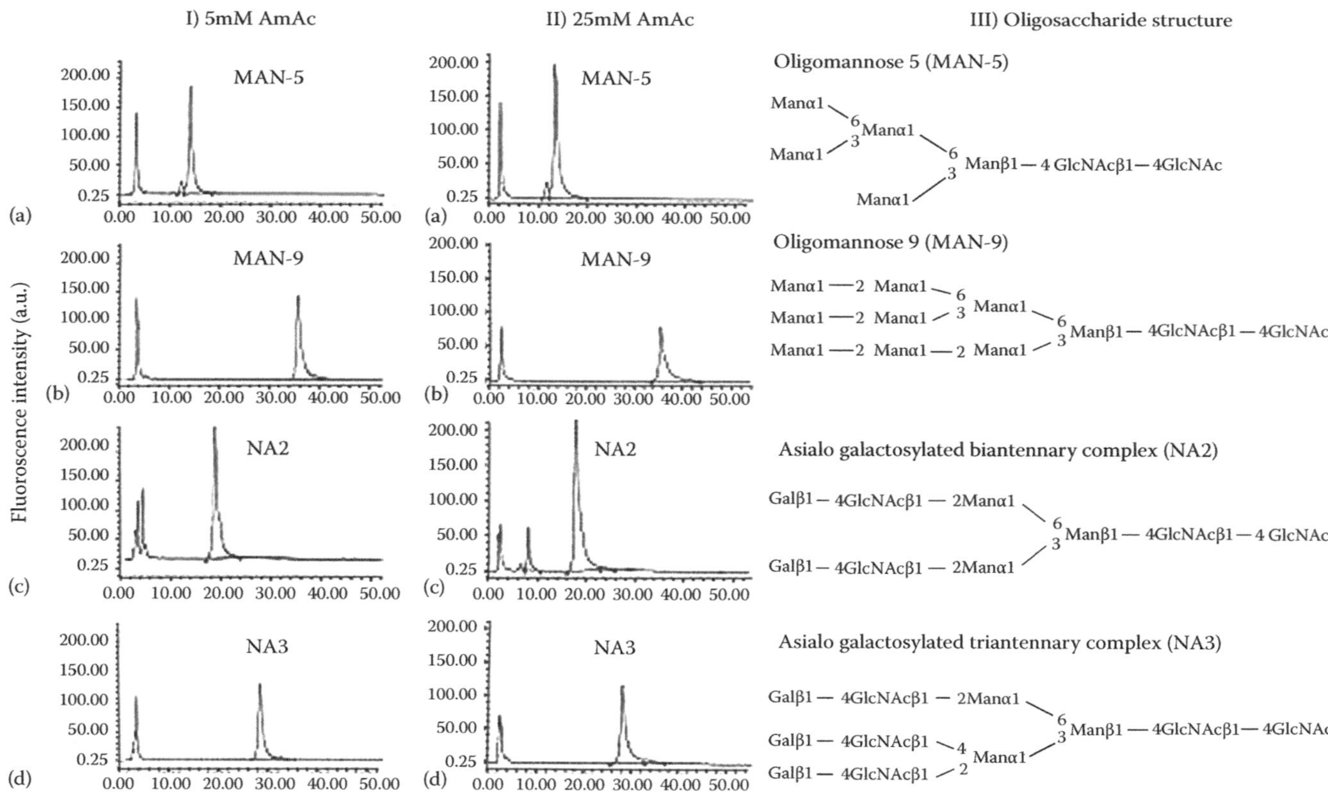
I) 5mM AmAc
II) 25mM AmAc
III) Oligosaccharide structure
Fluoroscence intensity (a.u.)
(a)
(b)
(c)
(d)
MAN-5
MAN-9
NA2
NA3
200.00
150.00
100.00
50.00
0.25
0.00 10.00 20.00 30.00 40.00 50.00
Oligomannose 5 (MAN-5)
Manα1 —6 Manα1 —3 Manα1 —6 Manβ1 — 4 GlcNAcβ1 — 4GlcNAc
Manα1 —3
Oligomannose 9 (MAN-9)
Manα1 —2 Manα1 —6 Manα1
Manα1 —2 Manα1 —3 Manα1 —6 Manβ1 — 4GlcNAcβ1 — 4GlcNAc
Manα1 —2 Manα1 —2 Manα1 —3
Asialo galactosylated biantennary complex (NA2)
Galβ1 — 4GlcNAcβ1 — 2Manα1 —6 Manβ1 — 4GlcNAcβ1 — 4 GlcNAc
Galβ1 — 4GlcNAcβ1 — 2Manα1 —3
Asialo galactosylated triantennary complex (NA3)
Galβ1 — 4GlcNAcβ1 — 2Manα1 —6 Manβ1 — 4GlcNAcβ1 — 4GlcNAc
Galβ1 — 4GlcNAcβ1 —4 Manα1 —3
Galβ1 — 4GlcNAcβ1 —2

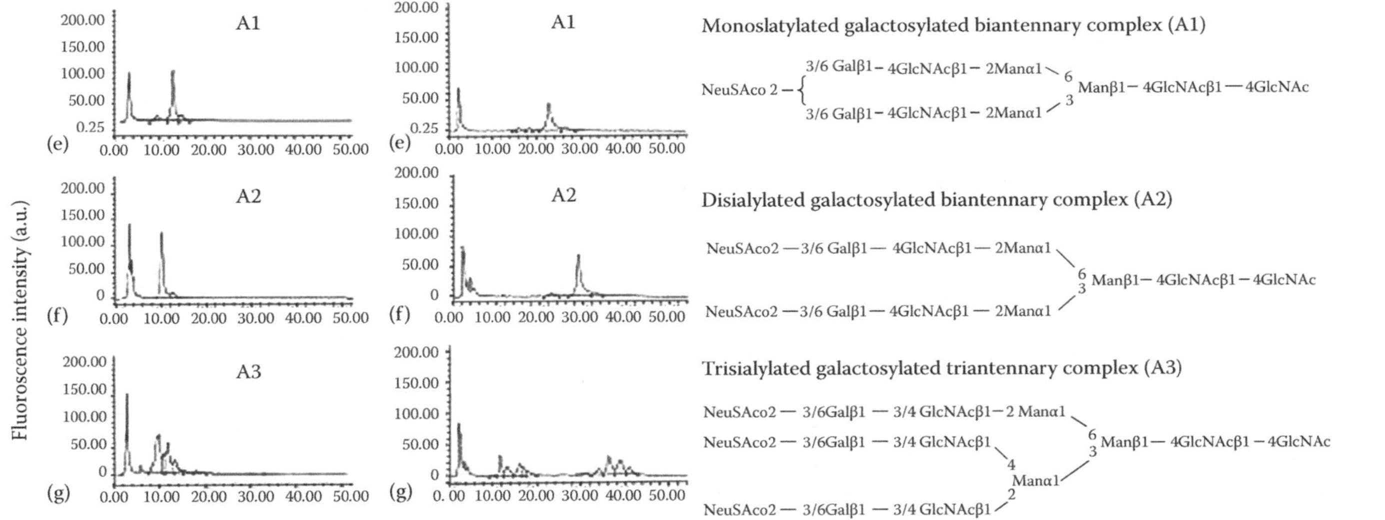

FIGURE 21.4 Influence of ammonium acetate (AmAc) concentration on the retention of 2-aminobenzamide (2-AB)-labeled oligosaccharides. A mixture of seven 2-AB-labeled glycan standards was separated by the Amide-80 column at different concentrations of ammonium acetate, using a segmented acetonitrile gradient (64%–52%) in ammonium acetate (pH 5.5) and water for elution at 45°C. (I) 5 mM ammonium acetate; (II) 25 mM ammonium acetate; (III) structures of the seven, *N*-linked glycan standards purchased from ProZyme; abbreviated MAN-5, MAN-9, NA2, NA3, A1, A2, and A3. *N*-acetyl-D-glucosamine (GlcNAc), D-mannose (Man), D-galactose (Gal), and *N*-acetylneuraminic acid (Neu5Ac) are indicated in the structures. The higher concentration of ammonium acetate (25 mM) increased the retention for *N*-acetylneuraminic acid-containing oligosaccharides. (Reproduced from Karlsson, G. et al., *J. Chromatogr. Sci.*, 46(1), 68, 2008, Figs. 1,3,4. With permission from Preston Publications, a Division of Preston Industries, Inc.)

Autosampler- and needle-wash solutions should be used, as they are for many HPLC systems, that contain 80% acetonitrile, 20% water, and no buffer, to avoid possible disturbances from very small amounts of washing solution that contain a high percentage of water that could otherwise give peak broadening.

21.2.4 Sample Preparation Strategy

Ideally, the injected sample should have the same percentage of organics as in the initial condition of the HILIC analysis, and never less than 50% organics, to avoid peak broadening and decreased chromatographic resolution. However, when injecting very small sample volumes, a higher water concentration is tolerated, which can be useful because carbohydrates, especially glycans, often have low solubilities in high concentrations of organic solvents. Usually, samples with about 70% acetonitrile are sufficient for avoiding peak broadening and precipitation. In addition, the injected sample should not contain particles or interfering substances and should be injected in the lowest possible volume, preferably 5–25 and 0.5–5 µL for 4.6 and 2.1 mm ID columns, respectively. For samples containing a lot of salts or proteins, precipitation will usually occur when the acetonitrile is added. The precipitate should then be removed by centrifugation (e.g., 10,000 × g for 10 min), and the clear supernatant can be injected on the HILIC system. The recovery of carbohydrates may decrease when the relative volume of the pellet increases, since pellets often contain approximately 50% liquid, and, in addition, some binding forces may occur between the carbohydrates and the precipitate. To remove particles from the HILIC samples, a recommended approach would be to use centrifugation (see above) or filtration with a syringe filter (based on poly-tetrafluoroethene or similar chemical-inert materials) having a 0.45 µm pore size. The sample concentration depends on the loading capacity of the selected column, and on the complexity and solubility of sample components. The sample solubility should be tested in the mobile phase prior to injection.

The removal of large amounts of proteins, lipids, and other nonpolar substances, which may disturb the chromatographic performance or the detection, in a sample, can easily be achieved by RPLC solid-phase extraction (SPE) cartridges (often with 100 mg dry gel per cartridge), usually with C-18 groups bound to the gel matrix. By combining RPLC with an HILIC analysis, a so-called orthogonal separation, where the two separation mechanisms are totally independent of each other, is achieved, which is analytically advantageous. The dry gel SPE cartridges are (i) first, activated with methanol (or sometimes acetonitrile), for wetting and cleaning of the gel; (ii) conditioned with, for example, ammonium acetate buffer (or water) to remove the methanol and equilibrate the cartridge; (iii) loaded with sample; and finally (iv) washed with buffer (or water). During steps (iii and iv), all polar substances, including carbohydrates, are eluted. These should be collected and concentrated, for example, by centrifugal evaporation, and the dry carbohydrates can be dissolved in a small volume of water, followed by an addition of acetonitrile to a final concentration of about 70%, before being injected on the HILIC column. If the nonpolar substances are desired, they can easily be eluted from the SPE cartridge by increasing the organic concentration. Cartridges of porous graphitic carbon are useful for binding and separating underivatized oligosaccharides, by RPLC, from fluorescent dyes and other interfering compounds.

Another approach is to collect fractions from a previous RPLC analysis and directly inject these on an HILIC system, or more conveniently, use an automatic column-switching system to perform the procedure. However, it is often necessary to concentrate the sample and to add more organics before the sample is injected on the HILIC system. Sample preparation with ion-exchange SPE cartridges can also be used before performing analytical HILIC, and HILIC SPE cartridges are also commercially available. In addition, liquid–liquid extraction and other extraction techniques are used for sample preparation. For the specific cleavage of *N*-linked glycans from glycoproteins, peptide *N*-glycosidase F (PNGase F), or hydrazine, which also cleaves O-linked carbohydrates, can be used, followed by an RPLC SPE step, for example, to separate glycans and protein. See the book by David and Moldoveanu for a comprehensive review of sample preparation techniques and strategies in chromatography.[47]

21.2.5 Detection of Carbohydrates

The detectors for analytical liquid chromatography can be divided into universal detectors, which respond to all compounds independent of their physicochemical properties, and selective detectors, which respond to just some of these properties. Although direct detection of carbohydrates by UV absorbance or fluorescence is usually not possible, some carbohydrates that have a high number of carboxylic groups, for example, hyaluronic acid, can be detected by absorbance at 210 nm, using a buffer and salt of low absorbance (e.g., sodium phosphate buffer and sodium sulfate in a salt gradient) to avoid interference.[10]

The RI detector, which is a common universal detector, has been used for a long time for carbohydrate analyses. However, RI detectors suffer from disadvantages such as low sensitivity, noncompatibility with gradient elution, and being affected by small fluctuations in temperature and pressure. In the past two decades, ELSD has been used as a semiuniversal HPLC detector, as it is compatible with gradient elution and is widely used for detecting nonvolatile compounds such as carbohydrates.[42,48,49] However, ELSD is not sensitive enough for detecting low concentrations of carbohydrates. ELSD is based on the light scattering detection of nonvolatile compounds after nebulization and evaporation of the mobile phase.

The derivatization of reducing carbohydrates with chromophoric reagents, for example, *p*-aminobenzoic ethyl ester,[6] followed by absorbance detection was a great improvement for detecting carbohydrates. Derivatization with fluorescent groups (e.g., 2-aminobenzamide (2-AB), 2-anthranilic acid (2-AA), benzamidine, or 2-aminoacridone) is often performed to detect glycans with very high sensitivity.[50] However, disadvantages of the derivatization procedure include the possibility of carbohydrate degradation, fluctuating yield of derivates, and toxicity of the reagents. Figure 21.5 shows a schematic fluorescence-labeling procedure, using 2-AB.[51] The labeling works effectively for reducing carbohydrates, and the procedure can

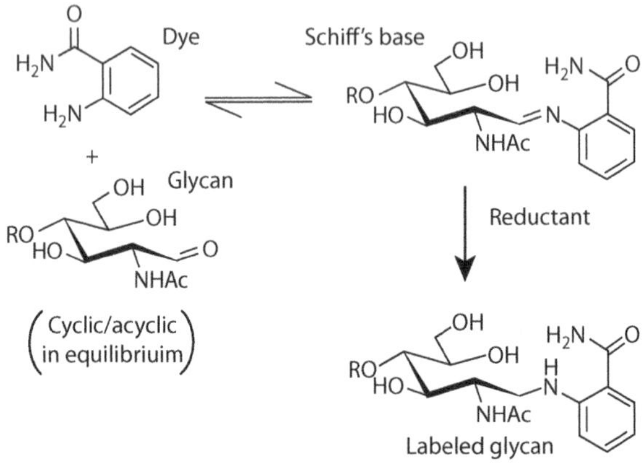

FIGURE 21.5 Labeling of carbohydrates by 2-aminobenzamide (2-AB). The Schiff's base formation requires a glycan with a free reducing sugar, which, in equilibrium, is between the ring closed (cyclic) and ring open (acyclic) forms. The primary amino group of the dye performs a nucleophilic attack on the carbonyl carbon of the acyclic reducing sugar residue to form a partially stable Schiff's base. The Schiff's base imine group is then chemically reduced by cyanoborohydride to give a stable, labeled glycan. The fluorescent-labeled glycan has a distinct fluorescence with maximum excitation and emission at 330 and 420 nm wavelengths, respectively. (Reproduced from Signal 2-AB labeling kit, Prod. No. GKK-404. Product information from ProZyme, San Leandro, CA. With permission from ProZyme.)

become more complicated for nonreducing carbohydrates, possibly involving the post-column hydrolysis of complex carbohydrates and labeling of the generated reducing sugars. After labeling, a clean-up step is usually necessary to remove most of the unreacted dye, to avoid a strong fluorescent interfering signal. See the review by Anumula for a comprehensive description and evaluation of fluorescence labeling in the detection of carbohydrates.[50]

NMR is a powerful and complicated technique for detecting and characterizing carbohydrates. NMR requires expertise and intricate equipment and operates on the principle of detecting which nuclei distort a magnetic field owing to their property of spin. NMR provides complimentary information to MS in obtaining comprehensive structural data, and NMR and MS can be used in parallel with liquid chromatography. For a review of LC-NMR-MS, refer to Corcoran and Spraul.[52]

MS is close to being a universal LC detector, and different MS systems, including triple quadrupoles, ion traps, and quadrupoles coupled to time-of-flight instruments, are increasingly used for analysis and identification of carbohydrates.[26,34,40,41,43] The principle of MS is that the analytes are first ionized and then a mass analyzer differentiates the ions according to their mass-to-charge ratio (*m/z*), with the ion beam being measured by a detector. The ESI interface, between the LC column and the MS inlet, is commonly used and fully compatible with HILIC mobile phases with a high percentage of organics, water, and low concentrations of volatile buffers. ESI MS is usually performed in the positive mode, generating positively charged carbohydrates by the addition of a proton $[M + H]^+$ (or another cation such as sodium ion $[M + Na]^+$), or may be performed in the negative mode, for carbohydrates containing anionic groups, such as deprotonized carboxylic acid groups $[M - H]^-$. In simple MS, where one MS analyzer is used, only *m/z* is detected and the analytes with same *m/z* cannot be discriminated. In tandem mass spectrometry (MS-MS), two MS analyzers operate in discrete zones and ions can be fragmented by energy transfers (usually by collisions with gas molecules) in the region between the zones. MS-MS is a powerful technique for determining structure and for quantifying carbohydrates and other compounds. Sequential enzymatic digestion, with exoglucosidases that have a narrow specificity, in combination with LC-MS, is also a useful tool for determining complex carbohydrate structures. One drawback in using glycosidases (or other digesting enzymes) is in the risk of incomplete cleavage, due to the steric hindrance or suboptimal incubation conditions. Multiple reaction monitoring (MRM), also called single reaction monitoring, is an MS-MS technique, where both analyzers are set to monitor specified pairs of parent (precursor) and product ions, which is useful for quantifying carbohydrates.[38,40] MRM is usually performed with an internal standard that has exactly the same structure as the analyte but is labeled with a few atoms of a stable heavy isotope (e.g.,^{2}H,^{13}C,^{15}N) to increase accuracy and precision. Short descriptions of a few HILIC-MS/MRM methods are given in Table 21.1. For a review of the analysis of carbohydrates by MS, see Zaia.[22]

Thus, MS is often the preferred HILIC detection technique, because of its high accuracy, precision, and sensitivity, and the peaks from contaminating compounds can be directly identified. Fluorescence labeling and detection is very sensitive and the method can be used for low concentrations of reducing carbohydrates. ELSD is also convenient if high concentrations of carbohydrates are to be measured. All of these detection techniques involve some optimization, for example, MS may need adjustments of the cone voltage, capillary voltage, scan time, and other settings, and the fluorescence labeling and clean-up procedures may also need some optimization. ELSD should be optimized by testing different temperatures and gas flows.

21.2.6 Short Outline for the Development of an HILIC Method

This outline has been successfully used to develop HILIC methods for several years at the author's laboratory. The following discussion should be considered as general input for developing an HILIC method for separating and analyzing carbohydrates. In developing an HILIC method, many aspects should be considered. For example, what is the aim of the work? Is it for quantifying one compound or many compounds, for determining the structure of a complex carbohydrate, or for obtaining the chromatographic profile of a mixture of glycans? How complex is the sample and what interfering substances are known to be present? Is a suitable sample preparation method available? Which detection technique should be used: an ESI MS (or ESI MS-MS) system on-line to the LC or fluorescence labeling and detection?

The chemical characteristics of the analyte (structure, hydrophilicity, ionizable groups, stability, solubility in possible mobile phases, etc.) should be considered, and published analytical methods should be investigated with regard to the analyte (or similar substances), including sample preparation procedures and detection techniques. In developing an HILIC method, especially when little work has been done in the area, the best approach is to begin with a column with a documented good selectivity and the resolution necessary for the type of separation being developed, together with an 85%–50% acetonitrile scouting gradient, in water and 10 mM ammonium acetate (or ammonium formate), at a column temperature of about 40°C, with a high but suitable flow rate for the selected column. The organic percentage for an isocratic separation, or possibly gradient, should be optimized for the analysis. If the desired resolution, selectivity, peak shape, detection response, or analysis time is not obtained, additional variables should be optimized, for example, by testing increased temperatures, changes in pH, buffer type and concentration, and slightly altered flow rates. In addition, sample preparation procedures and detection settings may need to be optimized. Further optimization for very complex or difficult samples can be derived by testing segmented and dual gradients, other types of HILIC columns, or other organics. In the next section, some applications for the HILIC analysis of carbohydrates are described, as valuable inputs to developing a new method.

21.3 HILIC APPLICATIONS FOR THE SEPARATION OF CARBOHYDRATES

In this section, a number of HILIC applications for the separation and analysis of different types of carbohydrates, using different detection techniques and columns, are described, which may be used to identify suitable initial conditions in developing a method. Some HILIC analyses of several classes of carbohydrates are briefly described in Table 21.1. HILIC of different types of carbohydrates is covered in several reviews, including the separation of low-molecular-weight carbohydrates in food and beverages,[53] marine samples,[54] glycoalkaloids in potatoes and tomatoes,[55] glucosaminoglycans (GAGs),[56] and for the assessment of protein glycosylation.[57]

21.3.1 HILIC Separation of Mono-, Di-, and Oligosaccharides by Cyclodextrin Columns in Combination with UV and RI Detection

Cyclodextrins, which are made of starch, are a family of cyclic oligosaccharides. The most used cyclodextrins are α-, β-, and γ-cyclodextrin, which are composed of six, seven, and eight monosaccharides, respectively. Cyclodextrin has been used in liquid chromatography to obtain the specific separation of chiral substances, and cyclodextrin columns are often used for various RPLC applications and for the HILIC separation of carbohydrates. Figure 21.6 shows a chromatogram of the HILIC analysis of a mixture of mono-, di-, and oligosaccharides, with high resolution of the separated components in the analyzed mixture of the commercially obtained carbohydrates. This example is from Armstrong et al.,[39] where both α- and β-cyclodextrin columns, Cyclobond III and I (250 × 4.6 mm ID, 5 μm particle size, from Advanced Separation Technologies), respectively, were tested with isocratic and gradient elution, using acetonitrile–water or acetone–water, together with UV or RI for detection. The retention was related to the size and the number of hydroxyl groups of the

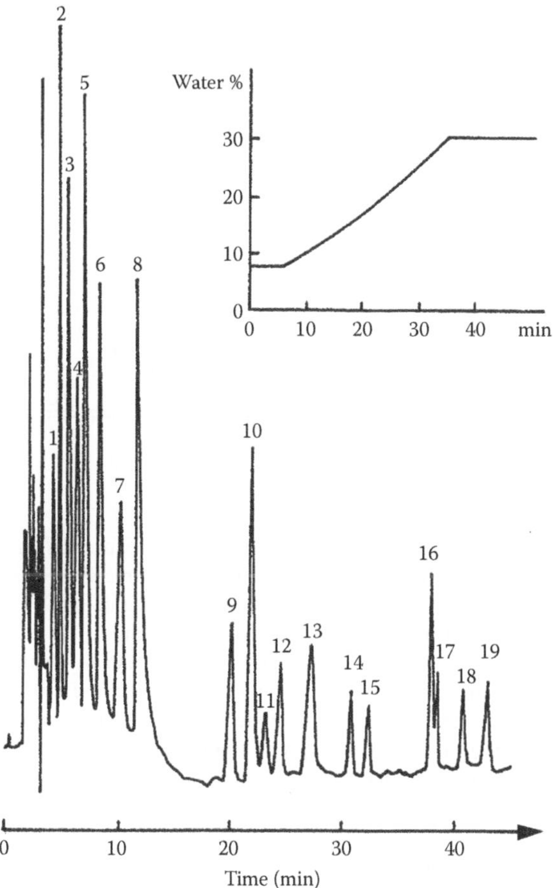

FIGURE 21.6 The gradient separation of carbohydrates on a β-cyclodextrin column (Cyclobond I), used at 22°C. An acetonitrile–water gradient, flow rate 1.0 mL/min, was used, and the change in the mobile-phase composition is shown in the figure inset: (1) Phenyl-β-D-glucopyranoside, (2) 2-deoxy-D-ribose, (3) ribose, (4) xylose, (5) talose, (6) sorbose, (7) glucose, (8) sorbitol, (9) sucrose, (10) turanose, (11) maltose, (12) lactose, (13) melibiose, (14) melezitose, (15) maltotriose, (16) stachyose, (17) α-cyclodextrin, (18) β-cyclodextrin, and (19) γ-cyclodextrin. (Reproduced from Armstrong, D.W. and Jin, H.L., *J. Chromatogr.*, 462, 219, 1989. With permission from Elsevier.)

carbohydrate. The efficiency and stability of the cyclodextrin columns were found to be superior to amino-alkyl columns.

21.3.2 HILIC Separation by an Amide-80 Column and MS Analysis of Carbohydrates from Plants

In this example, from Tolstikov and Fiehn,[34] an Amide-80 column (250 × 2 mm ID, 5 μm particle size, from Tosoh) at ambient temperature, was used to separate

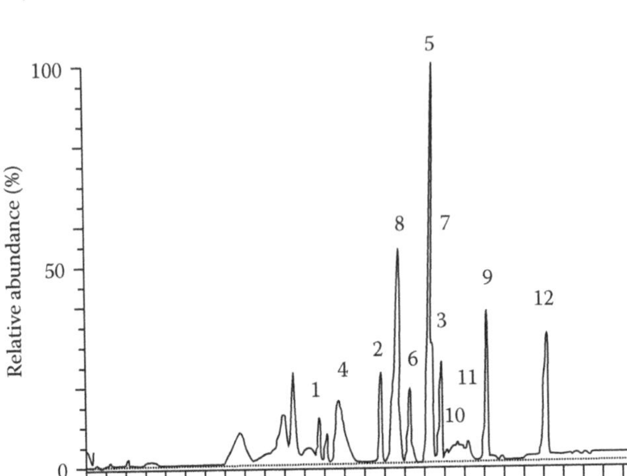

FIGURE 21.7 HILIC-MS base peak chromatograms of a mixture of standards. A TSK-Gel Amide-80 column was used with a dual gradient (10–60 min), 85%–45% acetonitrile and 1.0–3.6 mM ammonium acetate (pH 5.5), in water, at a flow rate of 0.15 mL/min. A 10 µL sample was injected. Identified peaks: (1) *N*-acetyl-D-glucosamine, (2) sucrose, (3) D(+) raffinose, (4) L-methionine, (5) *N*-methyl-L-deoxynojirimycin, (6) L-alanyl-L-alanine, (7) 1,4-dideoxy-1,4-imino-D-arabinitol, (8) uridine-5-diphosphoglucose, (9) stachyose, (10) glucosaminic acid, (11) 2-amino-2-deoxy-D-glucose, and (12) maltoheptaose. An electrospray interface coupled to an ion trap mass spectrometer, with stepwise fragmentation, was used for identification. (Reproduced from Tolstikov, V.V. and Fiehn, O., *Anal Biochem*. 301, 298, 2002. With permission from Elsevier.)

plant-related substances (oligosaccharides, glycosides, amino sugars, amino acids, and sugar nucleotides) from *Cucurbita maxima* leaves (Figure 21.7). Samples from phloem exudates from the petioles of the herb were analyzed. The goal of the study was metabolomic, i.e., quantification of every metabolite in a biological system. HILIC is very useful for metabolomics since the metabolites are generally very polar. For identification, ESI (both in positive and in negative modes, using continuous polarity switching) MS-MS was used on-line with the HILIC system. Quantification of the analytes was performed using external standards. The Amide-80 column gave a superior resolution compared to another tested HILIC column, a PolyHydroxyethyl A column (PolyLC).

21.3.3 HILIC Separation by a ZIC-HILIC Column and MS Analysis of Tryptic Glycopeptides from Human Immunoglobulin

Most proteins are glycosylated, and the linked carbohydrates are important in various biological functions. Thus, glycosylation must be analyzed and evaluated, for example, when producing recombinant glycoproteins for pharmaceutical use. The *N*-linked glycans from glycoproteins have complex structures and, because of the many possible isomers, require sophisticated analyses. Usually, glycopeptides are analyzed by RPLC, the most often applied liquid chromatography technique, but sometimes this does not give the desired separation, especially for small and hydrophilic peptides, which have low or no retention in RPLC. On the contrary, these hydrophilic peptides, regardless of being glycosylated, usually have a strong retention in HILIC systems, so that a combination of RPLC and HILIC gives a broader sequence coverage of the protein, compared to the analysis with RPLC alone.

In the example from Takegawa et al.,[36] a zwitterionic type of column, ZIC-HILIC (150 × 2.1 mm ID, 3.5 μm particle size, from SeQuant/Merck) was used at 40°C and combined with MS analysis of the tryptic glycopeptides from human immunoglobulin G (IgG) (Figure 21.8). Trypsination was performed according to a standard procedure, and the samples were injected in 20 μL 80% acetonitrile. Elution was performed by a gradient (0–120 min), 77%–63% acetonitrile, in water and 5 mM ammonium acetate, at a flow rate of 0.2 mL/min. *N*-glycans and tryptic peptides were analyzed after 2-aminopyridine derivatization (PA). The HILIC system was run on-line with an ESI MS system for detection and identification. The ZIC HILIC column gave a high resolution and a high selectivity for glycopeptides, including the isomeric separation of different glycans on the same peptide.

21.3.4 Analysis of 2-Aminobenzoic Acid-Derivatized Oligosaccharides by an NH_2 Column and Fluorescence Detection

In this application, from Anumula and Dhume,[35] 2-aminobenzoic acid (2-AA)-derivatized oligosaccharides from the well-characterized standard glycoprotein bovine fetuin were analyzed. An amino polymeric column (Polymer-NH_2, 250 × 4.6 mm ID, 5 μm particle size, from Astec) was used at 50°C and a dual gradient (0–80 min), with decreasing organics (70%–5% acetonitrile), and increasing ionic strength (2.9%–4.9% acetic acid and 0.9%–2.9% triethylamine; TEA), in water and 1% THF, at a flow rate of 1.0 mL/min, was used for elution. The sample preparation of the glycans was performed by enzymatic cleavage by PNGase F and hydrazionolysis, and detection of the 2-AA derivates was by fluorescence. By combining elution in HILIC mode (increasing water) and IEX mode (increasing ion strength), sometimes called HILIC-IEX (also known as NP-IEX or NP-HPAEC), a very high resolution can be obtained. The same sample was also analyzed by the reference high-resolution technique, HPAEC-PAD, and the results were similar with regard to resolution and selectivity. The HILIC-IEX method, for separation of fluorescence-labeled glycans, gave even more defined peaks than did the HPAEC-PAD method. In both methods, oligosaccharides containing zero to four sialic acids were eluting in separate regions, with consecutively increased retention times, as shown for the HILIC-IEX method in Figure 21.9.

21.3.5 Separation of Glucosinolates by a ZIC-HILIC Column and Detection by Absorbance

Glucosinolates, which are synthesized in plants, are a group of glucosides with structures that include a sulfonated oxime, a thio-linked glycone, and an additional

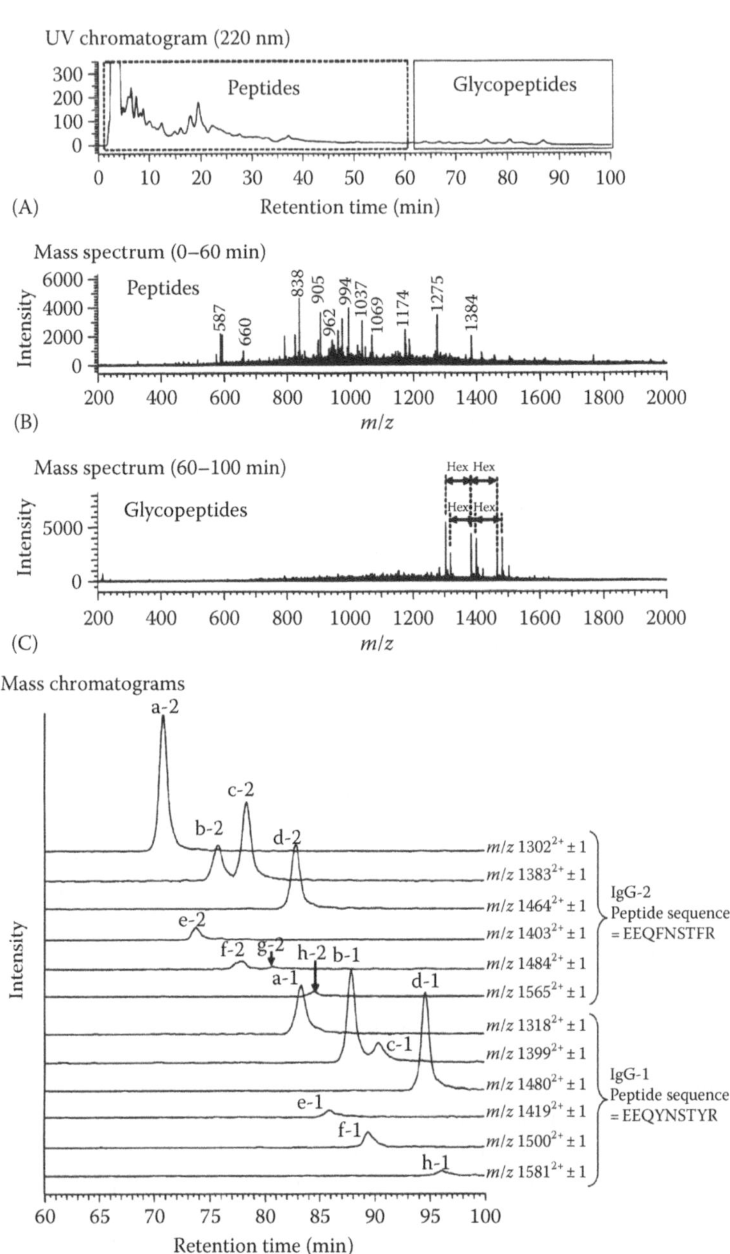

FIGURE 21.8 ZIC-HILIC separation and mass spectrometry analysis of tryptic glycopeptides from human immunoglobulin G (IgG). (A) UV (220nm) chromatogram, (B) accumulated mass spectrum, 0–60min, (C) accumulated mass spectrum, 60–100min, and (D) mass chromatograms of molecular ions of major *N*-glycopeptides of IgG-1. The table shows structure, exact mass values, and observed integer mass values of PA-labeled *N*-glycans and major *N*-glycopeptides from human IgG. (Reproduced from Takegawa, Y. et al., *J. Chromatogr. A*, 1113, 177, 2006. With permission from Elsevier.)

Summary of structures, annotations, exact mass values, and observed integer mass values of PA N-glycans and major N-glycopeptides from human serum IgG

Structures (n = 1–3)	PA N-glycans (R1 = PA)	IgG-1 N-glycopeptides (R2 = EEQYNSTYR)	IgG-2 N-glycopeptides (R3 = EEQFNSTER)
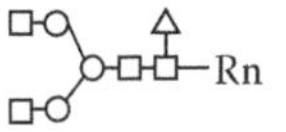 –Rn	a E.m. 1540.6 Obs. 1541 (z = 1)	a-1 E.m. 2633.0 Obs. 1318 (z = 2)	a-2 E.m. 2601.1 Obs. 1302 (z = 2)
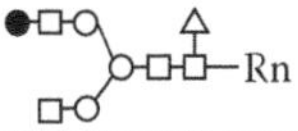 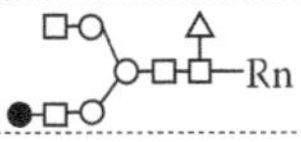 –Rn –Rn	b/c E.m. 1702.7 Obs. 1703 (z = 1)	b-1/c-1 E.m. 2795.1 Obs. 1399 (z = 2)	b-2/c-2 E.m. 2763.1 Obs. 1383 (z = 2)
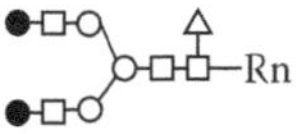 –Rn	d E.m. 1864.7 Obs. 933 (z = 2), 1865 (z = 1)	d-1 E.m. 2957.1 Obs. 1480 (z = 2)	d-2 E.m. 2925.2 Obs. 1464 (z = 2)
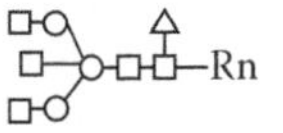 –Rn	e E.m. 1743.7 Obs. 873 (z = 2), 1744 (z = 1)	e-1 E.m. 2836.1 Obs. 1419 (z = 2)	e-1 E.m. 2804.1 Obs. 1403 (z = 2)
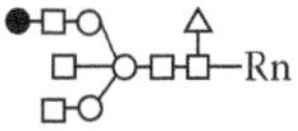 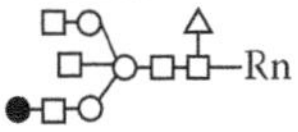 –Rn –Rn	f/g E.m. 1905.7 Obs. 954 (z = 2), 1906 (z = 1)	f-1 E.m. 2998.2 Obs. 1500 (z = 2)	f-2/g-2 E.m. 2966.2 Obs. 1484 (z = 2)
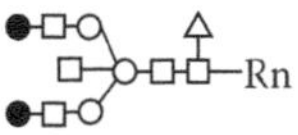 –Rn	h E.m. 2067.8 Obs. 1035 (z = 2)	h-1 E.m. 3160.2 Obs. 1581 (z = 2)	h-2 E.m. 3128.2 Obs. 1565 (z = 2)

(○) Mannose (Man); (●) Galactose (Gal); (△) Fucose (Fuc); (□) N-acetyl-glucosamine (GlcNAc); PA: 2-aminopyridine; E.m.: exact mass; Obs.: Observed mass; z: number of charges.

FIGURE 21.8 (continued)

side-group R. Glucosinolates are precursors of the cancer-protective isothiocyanates. Wade et al.[37] used a zwitterionic ZIC-HILIC column (150 × 4.6 mm ID, 5 µm particle size, from SeQuant/Merck) at 25°C with elution by 70% acetonitrile, in water and 15 mM ammonium formate, pH 4.5, flow rate 0.5 mL/min, to separate glucosinolates (Figure 21.10). Detection was performed by absorbance at 235 nm. The results from the ZIC-HILIC column were more reproducible than those from a PolyHydroxyethyl A column.

21.3.6 Separation of GAGs by an Amide-80 Column Combined with MS Analysis

GAGs are composed of long, unbranched polysaccharides of a repeating disaccharide unit and are the important components in connective tissue, including connective joint tissues. The analysis of GAGs is important to research in several fields, including studies of osteoarthritis.

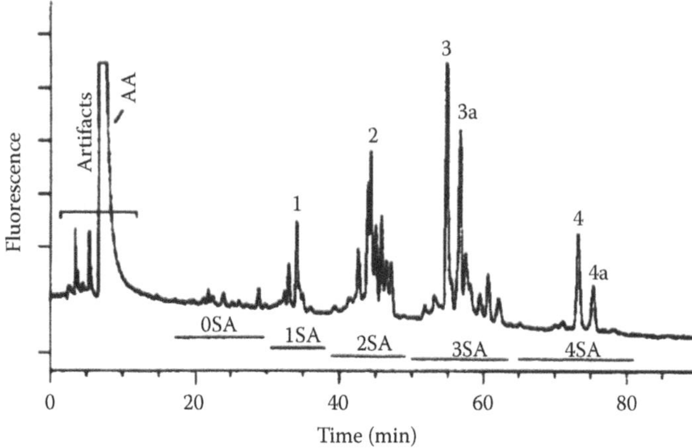

FIGURE 21.9 Fetuin oligosaccharide map of 2-aminobenzoic acid (2-AB)-labeled oligosaccharides separated by HILIC on an amino polymeric column. Bars indicate the regions where glycans with zero to four sialic acids (SA) elute. (Reproduced from Anumula, K.R. and Dhume, S.T., *Glycobiology*, 8, 685, 1998. With permission from Oxford University Press.)

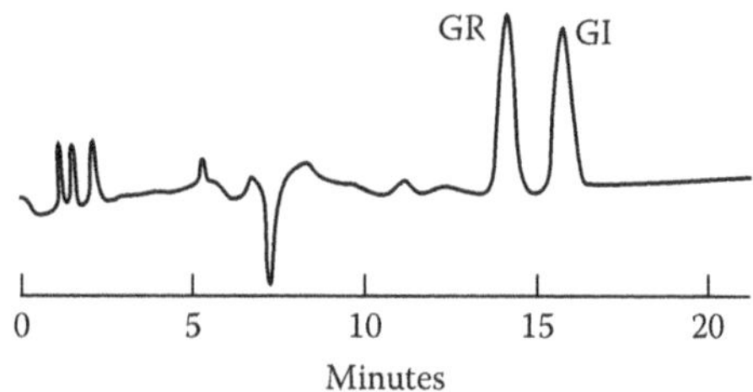

FIGURE 21.10 The separation of glucosinolates by a ZIC-HILIC column using 70% acetonitrile in water for elution. GR and GI denote glucoraphanin (4-(methylsulfonyl) butyl glucosinolate) and glucoiberin (3-(methylsulfonyl) butyl glucosinolate), respectively. Detection by absorbance at 235 nm. (Reproduced from Wade, K.L. et al., *J. Chromatogr. A*, 1154, 469, 2007. With permission from Elsevier.)

Hitchcock et al.[43] used an Amide-80 capillary column (150 × 0.25 mm ID, 5 μm, from Tosoh) together with a dual gradient (0–50 min), 82%–46% acetonitrile in water and 10–30 mM formic acid, pH 4.4, at a flow rate of 0.1 mL/min for elution (Figure 21.11). MS-MS was used for detection and identification. GAGs were extracted from joint tissues and partly depolamirized, using chondroitinase. The oligosaccharide products were labeled with differently stable isotopes, using (d_0 or d_4) 2-anthranilic acid before the HILIC MS-MS analysis. By this method, the structures of different regions of chondroitin sulfate from human and bovine joint tissues were determined.

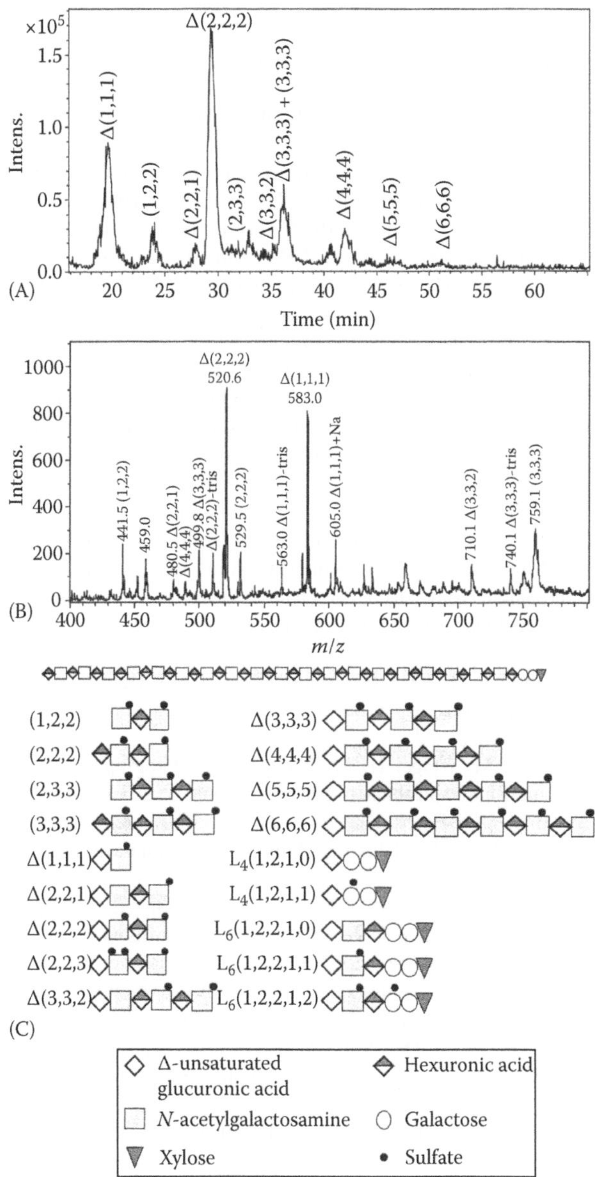

FIGURE 21.11 The separation of glycosaminoglycans (GAGs) by an Amide-80 capillary column, using a negative acetonitrile gradient, in water, and a positive formic acid gradient, combined with MS detection. (A) Base peak chromatogram of chondroitin lyase depolamirized chondroitin sulfate and dermatan sulfate from d_4-2-AA-juvenile bovine cartilage mixed with d_0-2-AA chondroitin sulfate internal oligosaccharide standard. GAG oligosaccharide chains ranging from disaccharide to dodecasaccharide elute at 15–55 min. (B) Average mass spectrum of all eluted oligosaccharides in the sample mixture. (C) Oligosaccharide structures found in connective tissue samples. (Reproduced from Hitchcock, A.M. et al., *Proteomics*, 8, 1384, 2008. With permission from Wiley-VCH Verlag Gmbh & Co. KGaA.)

REFERENCES

1. Campbell MK, Farrell SO. *Biochemistry*, 6th edn. Thomson Brooks & Cole, Belmont, CA, 2009.
2. Linden JC, Lawhead CL. 1975. Liquid chromatography of saccharides. *J. Chromatogr.* 105:125–133.
3. Alpert AJ. 1990. Hydrophilic-interaction chromatography for the separation of peptides, nucleic acids and other polar compounds. *J. Chromatogr.* 499:177–196.
4. Langer SH, Pantages P, Wender I. 1958. Gas–liquid chromatographic separation of phenols as trimethylsilyl ethers. *Chem. Ind.* (*Lond.*) 1664–1665.
5. Hansson GC, Karlsson H. 1993. Gas chromatography and gas chromatography-mass spectrometry of glycoprotein oligosaccharides. *Methods Mol. Biol.* 14:47–54.
6. Kwon H, Kim J. 1993. Determination of monosaccharides in glycoproteins by reverse-phase high-performance liquid chromatography. *Anal. Biochem.* 215:243–252.
7. Ambrosius M, Kleesiek K, Götting C. 2008. Quantitative determination of the glycosaminoglycan *delta*-disaccharide composition of serum, platelets and granulocytes by reversed-phase high-performance liquid chromatography. *J Chromatogr. A* 1201:54–60.
8. Gey MH, Unger KK. 1996. A strategy for chromatographic and structural analysis of monosaccharide species from glycoproteins. *Fresenius J. Anal. Chem.* 356:488–94.
9. Nakagawa H, Kawamura Y, Kato K, Shimada I, Arata Y, Takahashi N. 1995. Identification of neutral and sialyl N-linked oligosaccharide structures from human serum glycoproteins using three kinds of high-performance liquid chromatography. *Anal. Biochem.* 226:130–138.
10. Karlsson G, Bergman R. 2003. Determination of the distribution of molecular masses of sodium hyaluronate by high-performance anion-exchange chromatography. *J. Chromatogr. A* 986:67–72.
11. Bourne EJ, Lees EM, Weigel H. 1963. Paper chromatography of carbohydrates and related compounds in the presence of benzeneboronic acid. *J. Chromatogr.* 11:253–257.
12. Herold DA, Boyd JC, Bruns DE, Emerson JC, Burns KG, Bray RE et al. 1983. Measurement of glycosylated hemoglobins using boronate affinity chromatography. *Ann. Clin. Lab. Sci.* 13:482–8.
13. Buzzega D, Maccari F, Volpi N. 2008. Fluorophore-assisted carbohydrate electrophoresis for the determination of molecular mass of heparins and low-molecular-weight (LMW) heparins. *Electrophoresis* 29:4192–4202.
14. Soga T, Heiger DN. 1998. Simultaneous determination of monosaccharides in glycoproteins by capillary electrophoresis. *Anal. Biochem.* 261:73–78.
15. Liu CY, Chen TH, Misra TK. 2007. A macrocyclic polyamine as an anion receptor in the capillary electrochromatographic separation of carbohydrates. *J. Chromatogr. A* 1154:407–415.
16. Santori F, Hubble J. 2003. Isocratic separation of monosaccharides using immobilized Concanavalin A. *J. Chromatogr. A* 1003:123–6.
17. Herbreteau B, Lafosse M, Morin-Allory L, Dreux M. 1990. Analysis of sugars by supercritical fluid chromatography using polar packed columns and light-scattering detection. *J. Chromatogr.* 505:299–305.
18. Panagiotopoulos C, Sempéré R, Lafont R, Kerhervé P. 2001. Sub-ambient temperature effects on the separation of monosaccharides by high-performance anion-exchange chromatography with pulsed amperometric detection. Application to marine chemistry. *J. Chromatogr. A* 920:13–22.
19. Cataldi TR, Campa C, De Benedetto GE. 2000. Carbohydrate analysis by high-performance anion-exchange chromatography with pulsed amperometric detection: The potential is still growing. *Fresenius J. Anal. Chem.* 368:739–758.

20. Campa C, Coslovi A, Flamigni A, Rossi M. 2006. Overview on advances in capillary electrophoresis-mass spectrometry of carbohydrates: A tabulated review. *Electrophoresis* 27:2027–2050.
21. Klampfl CW. 2004. Review coupling of capillary electrochromatography to mass spectrometry. *J. Chromatogr. A* 1044:131–144.
22. Zaia J. 2008. Mass spectrometry and the emerging field of glycomics. *Chem. Biol.* 15:881–892.
23. Schols HA, Mutter M, Voragen AG, Niessen WM, van der Hoeven RA, van der Greef J et al. 1994. The use of combined high-performance anion-exchange chromatography-thermospray mass spectrometry in the structural analysis of pectic oligosaccharides. *Carbohydr. Res.* 261:335–342.
24. Alpert AJ, Shukla M, Shukla AK, Zieske LR, Yuen SW, Ferguson MA et al. 1994. Hydrophilic-interaction chromatography of complex carbohydrates. *J. Chromatogr. A* 676:191–202.
25. Hemström P, Irgum K. 2006. Hydrophilic interaction chromatography. *J. Sep. Sci.* 29:1784–1821.
26. Ikegami T, Tomomatsu K, Takubo H, Horie K, Tanaka N. 2008. Separation efficiencies in hydrophilic interaction chromatography. *J. Chromatogr. A* 1184:474–503.
27. Dejaegher B, Mangelings D, Vander Heyden Y. 2008. Method development for HILIC assays. *J. Sep. Sci.* 31:1438–1448.
28. Hao Z, Xiao B, Weng N. 2008. Impact of column temperature and mobile phase components on selectivity of hydrophilic interaction chromatography (HILIC). *J. Sep. Sci.* 31:1449–1464.
29. Snyder LR, Barker J, Ando DJ, Kirkland JJ, Glajch JL. 2002. *Practical HPLC Method Development: With Mass Spectrometry*, 2nd revised edn. Wiley Interscience, New York.
30. Cohen S, Lhuillier F, Mouloua Y, Vignal B, Favetta P, Guitton J. 2007. Quantitative measurement of propofol and in main glucuroconjugate metabolites in human plasma using solid phase extraction-liquid chromatography-tandem mass spectrometry. *J. Chromatogr. B* 854:165–172.
31. Karlsson G, Swerup E, Sandberg H. 2008. Combination of two hydrophilic interaction chromatography methods that facilitates identification of 2-aminobenzamide-labeled oligosaccharides. *J. Chromatogr. Sci.* 46:68–73.
32. Karlsson G, Winge S, Sandberg H. 2005. Separation of monosaccharides by hydrophilic interaction chromatography with evaporative light scattering detection. *J. Chromatogr. A* 1092:246–249.
33. Kakita H, Kamishima H, Komiya K, Kato Y. 2002. Simultaneous analysis of monosaccharides and oligosaccharides by high-performance liquid chromatography with post-column fluorescence derivatization. *J. Chromatogr. A* 961:77–82.
34. Tolstikov VV, Fiehn O. 2002. Analysis of highly polar compounds of plant origin: Combination of hydrophilic interaction chromatography and electrospray ion trap mass spectrometry. *Anal. Biochem.* 301:298–307.
35. Anumula KR, Dhume ST. 1998. High resolution and high sensitivity methods for oligosaccharide mapping and characterization by normal-phase high performance liquid chromatography following derivatization with highly fluorescent anthranilic acid. *Glycobiology* 8:685–694.
36. Takegawa Y, Deguchi K, Keira T, Ito H, Nakagawa H, Nishimura S. 2006. Separation of isomeric 2-aminopyridine derivatized N-glycans and N-glycopeptides of human serum immunoglobulin G by using a zwitterionic type of hydrophilic-interaction chromatography. *J. Chromatogr. A* 1113:177–181.
37. Wade KL, Garrard IJ, Fahey JW. 2007. Improved hydrophilic interaction chromatography method for the identification and quantification of glucosinolates. *J. Chromatogr. A* 1154:469–472.

38. Ikegami T, Horie K, Saad N, Hosoya K, Fiehn O, Tanaka N. 2008. Highly efficient analysis of underivatized carbohydrates using monolithic-silica-based capillary hydrophilic interaction (HILIC) HPLC. *Anal. Bioanal. Chem.* 391:2533–2542.
39. Armstrong DW, Jin HL. 1989. Evaluation of the liquid chromatographic separation of monosaccharides, disaccharides, trisaccharides, tetrasaccharides, deoxysaccharides and sugar alcohols with stable cyclodextrin bonded phase columns. *J. Chromatogr.* 462:219–232.
40. Rogatsky E, Jayatillake H, Goswami G, Tomuta V, Stein D. 2005. Sensitive LC MS quantitative analysis of carbohydrates by Cs^+ attachment. *J. Am. Soc. Mass Spectrom.* 16:1805–1811.
41. Antonio C, Larson T, Gilday A, Graham I, Bergström E, Thomas-Oates J. 2008. Hydrophilic interaction chromatography/electrospray mass spectrometry analysis of carbohydrate-related metabolites from *Arabidopsis thaliana* leaf tissue. *Rapid Commun. Mass Spectrom.* 22:1399–1407.
42. Karlsson G, Hinz AC, Winge S. 2004. Determination of the stabilizer sucrose in a plasma-derived antithrombin process solution by hydrophilic interaction chromatography with evaporative light-scattering detection. *J. Chromatogr. Sci.* 42:361–365.
43. Hitchcock AM, Yates KE, Costello CE, Zaia J. 2008. Comparative glycomics of connective tissue glycosaminoglycans. *Proteomics* 8:1384–1397.
44. Charlwood J, Birrell H, Bouvier ES, Langridge J, Camilleri P. 2000. Analysis of oligosaccharides by microbore high-performance liquid chromatography. *Anal. Chem.* 72:1469–1474.
45. Eliasson AC (ed.). 2006. *Carbohydrates in Food*, 2nd edn. Taylor & Francis, London, U.K.
46. Brons C, Olieman C. 1983. Study of the high-performance liquid chromatographic separation of reducing sugars, applied to the determination of lactose in milk. *J. Chromatogr.* 259:79–86.
47. David V, Moldoveanu SC (eds). 2002. Sample preparation in chromatography. *Journal of Chromatography Library*, Vol. 65. Elsevier Science, Amsterdam, the Netherlands.
48. Morin-Allory L, Herbreteau B. 1992. High-performance liquid chromatography and supercritical fluid chromatography of monosaccharides and polyols using light-scattering detection: Chemometric studies of the retentions. *J. Chromatogr.* 590:203–213.
49. Wei Y, Ding M. 2000. Analysis of carbohydrates in drinks by high-performance liquid chromatography with a dynamically modified amino column and evaporative light scattering detection. *J. Chromatogr. A* 904:113–117.
50. Anumula KR. 2006. Advances in fluorescence derivatization methods for high-performance liquid chromatographic analysis of glycoprotein carbohydrates. *Anal. Biochem.* 350:1–23.
51. Signal 2-AB labeling kit, Prod. No. GKK-404. Product information from ProZyme, San Leandro, CA.
52. Corcoran O, Spraul M. 2003. LC-NMR-MS in drug discovery. *Drug Discov. Today* 8:624–631.
53. Montero CM, Dodero MC, Sanchez DA, Barroso CG. 2004. Analysis of low molecular weight carbohydrates in food and beverages: A review. *Chromatographia* 59:15–30.
54. Panagiotopoulos C, Sempéré R. 2005. Analytical methods for the determination of sugars in marine samples: A historical perspective and future directions. *Limnol. Oceanogr. Methods* 3:419–454.
55. Friedman M. 2004. Analysis of biologically active compounds in potatoes (*Solanum tuberosum*), tomatoes (*Lycopersicon esculentum*), and jimson weed (*Datura stramonium*) seeds. *J. Chromatogr. A* 1054:143–155.
56. Sasisekharan R, Raman R, Prabhakar V. 2006. Glycomics approach to structure-function relationships of glycosaminoglycans. *Annu. Rev. Biomed. Eng.* 8:181–231.
57. Brooks SA. 2006. Protein glycosylation in diverse cell systems: Implications for modification and analysis of recombinant proteins. *Expert Rev. Proteomics* 3:345–359.

22 Hydrophilic Interaction Liquid Chromatography in the Characterization of Glycoproteins

Joanne E. Nettleship

CONTENTS

22.1 INTRODUCTION TO GLYCOPROTEINS

Glycosylation is one of the most common posttranslational modifications of proteins—it is estimated that over 50% of human proteins are glycosylated.[1] Glycoproteins have great importance in biological systems with the glycan moiety acting as a key to the correct folding and action of the protein.[2] The characterization of glycoproteins is important in the diagnosis and treatment of human diseases. For biopharmaceutical industry, consistent generation of the same glycoforms in different batches of recombinant protein is essential to retain the drug activity; thus glycoprotein analysis plays an important role in the production process. In addition, characterization of glycoproteins from patients with a specific disease can lead to the discovery of new biomarkers or targets for drug development.

This review discusses the importance of glycoproteins in biological systems in Section 22.1.1 along with a review of glycan classes in Section 22.1.2. Some of the challenges faced in characterizing glycoproteins and the use of mass spectrometry (MS) for this analysis are given in Section 22.1.3. Various enrichment and separation techniques for both glycoproteins and glycopeptides are discussed in Section 22.2. Finally, some examples in the use of hydrophilic interaction liquid chromatography (HILIC) for glycoprotein analysis are given in Section 22.3.

22.1.1 Importance of Glycosylation in Biological Systems and How Variation Can Lead to Disease

Carbohydrates have been shown to be involved in many biological processes such as inter- and intracellular signaling, immune functions, cell adhesion and division, viral replication, parasitic infections, and protein regulation.[3–7] It is the complexity of the glycan chains (discussed in Section 22.1.2) that enables glycoproteins to encode a large amount of information for specific molecular recognition. Glycosylation also determines protein folding, stability, and kinetics.[2] For example, Limjindaporn et al. have observed the interaction of the glycosylated envelope protein of dengue virus with chaperones in the human host, namely, immunoglobulin heavy chain–binding protein, calnexin, and calreticulin.[8] These interactions were shown to be important in virion production since decreasing expression of the chaperones reduced the production of infectious virions.[8] In addition, Rudd et al. comment in their review that almost all the key proteins involved in immune response are glycoproteins.[5] For example, the folding and assembly of both peptide-loaded major histocompatibility complex antigens and the T cell receptor complex involve the use of specific glycoforms.[5]

Glycosylation is found in all eukaryotes and has also been observed in bacteria and archaea.[9,10] For instance, the intestinal bacterium *Bacteroides fragilis* uses an *O*-glycosylation system for the incorporation of exogenous fucose into proteins in order to successfully colonize the mammalian gut.[11] In addition, *O*-glycosylation of pilin has been shown to occur in *Neisseria gonorrhoeae* whereas glycosylation-deficient mutants show the attenuated virulence.[12,13]

Glycosylation has been implicated in many disease states, such as cancer, atherosclerosis and rheumatoid arthritis, inflammatory diseases, and congenital disorders.[14–19]

In neurodegenerative diseases, for example, Creutzfeldt–Jakob disease, Alzheimer's disease and Parkinson's disease, glycosylation modification of vital enzymes has been shown to be important.[18] A key enzyme, acetylcholinesterase has been shown to take altered glycosylation in the postmortem brain and cerebrospinal fluid of Alzheimer's patients.[20] In addition, there is altered glycosylation of a related enzyme, butylcholinesterase, in cerebrospinal fluid.[20] Unfortunately, the sensitivity is too low to use as a biomarker for diagnosing Alzheimer's disease from other types of dementia using these alterations.[21]

Glycosylation has also been implicated in the transmissibility of influenza A viruses. The ability of the virus to adapt to a human host is governed by the binding specificity of the viral surface hemagglutinin to long α2–6 sialylated glycan receptors in the tissues of human upper respiratory tract.[22] Srinivasan et al. showed that a single amino acid mutation in the viral hemagglutinin led to a change in the binding specificity from α2–6 sialylated to mixed α2–3/α2–6 sialylated glycans, which resulted in inefficient transmission of the virus.[22] The subtle difference in glycan binding changes the human-to-human transmissibility of the virus and could act as an indication of the probability of an influenza pandemic.

In addition to having a role in disease, glycosylation is being used as a target for vaccine development, for example, in AIDS[23,24] and in cancer.[25] For instance, Ni et al. are examining the possibility of a carbohydrate-based HIV-1 vaccine by synthesizing oligomannose-containing glycoconjugates, which resemble a cluster on HIV-1 gp120.[24] In the field of cancer immunotherapy, several carbohydrate antigens are vaccination targets.[25] Slovin et al. have used Thomsen–Friedenreich (TF) self-antigen, a core disaccharide of *O*-glycosylated complex glycoproteins expressed on malignant cells, in clinical trials as a target for specific delivery of a synthetic TF cluster-keyhole limpet hemocyanin conjugate vaccine in the treatment of prostate cancer.[26]

22.1.2 Types of Glycosylation and Glycan Patterns

The glycans are attached to glycoproteins through specific amino acids, namely, asparagine of *N*-glycosylation, and serine or threonine of *O*-glycosylation. There are some rare examples of other glycan linkages such as covalent *O*-glycosidic linkages with hydroxylysine in collagens[27] and through lysine residues as seen in diabetes.[28]

N-linked glycans mostly contain an *N*-acetylglucosamine (GlcNAc) joined via an amide bond to the asparagine residues of a protein (Figure 22.1), whereas, most commonly (e.g., mucins), *O*-glycosylation occurs via an *N*-acetylgalactosamine (GalNAc) moiety linked to the hydroxyl group of either serine or threonine (Figure 22.2). There are other, rare linkages such as *O*-GlcNAc in cytosolic and nuclear proteins,[29] fucose (Fuc)-serine linkages in epidermal growth factor domains,[30] and mannose (Man)-serine linkages found in proteins from *Saccharomyces cerevisiae*[31] and *Pichia pastoris*[32] as well as glycoproteins found in the brain.[33]

N-linked glycans consist of a core unit, which becomes elaborated with variable antennae regions (Figure 22.1) that are formed by addition or deletion of sugar units to a common precursor. There are three subgroups of *N*-glycans classified according to the type and location of the sugar residues added to the core. Glycans trimmed by α-glucosidases and α-mannosidases without any additions contain only mannose

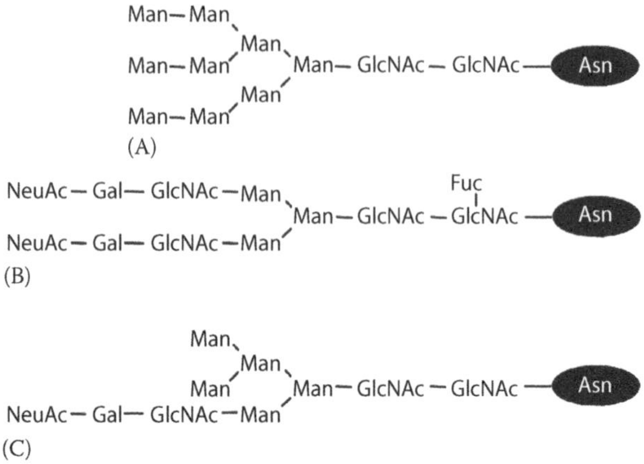

Man—Man
Man
Man—Man Man—GlcNAc—GlcNAc—Asn
Man
Man—Man
(A)

NeuAc—Gal—GlcNAc—Man Fuc
Man—GlcNAc—GlcNAc—Asn
NeuAc—Gal—GlcNAc—Man
(B)

Man
Man
Man Man—GlcNAc—GlcNAc—Asn
NeuAc—Gal—GlcNAc—Man
(C)

FIGURE 22.1 Examples of the three classes of *N*-glycans showing (A) high mannose, (B) complex, and (C) hybrid-type glycans.

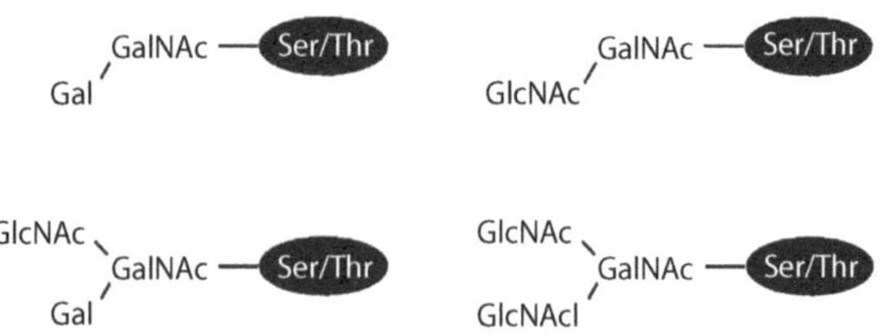

FIGURE 22.2 Four of the common *O*-glycan core structures.

attached to the core structure with the composition of $Man_{5-9}GlcNAc_2$. These glycans are commonly known as "high-mannose type" (Figure 22.1A): Glycans containing *N*-acetyllactosamine (LacNAc composed of Gal followed by GlcNAc) in their antennal region are classified as "complex type" (Figure 22.1B). These complex glycans have a variable number of antennae formed by addition of GlcNAc to the α-mannose residues of the core structure. In mammalian proteins, the antennae are often composed of tandem repeats of LacNAc formed by sequential addition of GlcNAc and Gal residues. The third class of glycans is the "hybrid type," which contains both a mannose and a LacNAc attached to the glycan core (Figure 22.1C). Complex-type glycans are usually capped by an α-linked sialyl or fucosyl group—the backbone sugars are generally β-linked. In addition to these subclasses of *N*-glycosylation patterns, a GlcNAc residue can be attached to the 4-position of the β-mannose resulting in a bisecting residue. A fucose (Fuc) moiety can also be added to the core GlcNAc residue. The presence or absence of this core-fucosylation further adds to the diversity of glycan structures (Figure 22.1B).

The structures of *O*-linked glycans are even more diverse than *N*-linked glycans[34] with sizes ranging from single monosaccharides to large polysaccharides similar in size to the complex-type *N*-glycans. Large *O*-glycans can be divided into two domains, the core and the antennae. There are four mucin-type *O*-glycan core structures, which are the most widespread in mammalian glycoproteins (Figure 22.2) though these are by no means the limit of *O*-glycan core structure diversity.[35] This structural variation continues with elongation of the core structures with a variety of sugar residues such as Gal, Fuc, GlcNAc.

In addition to the size and structure of the glycan, glycoproteins show variation in the occupancy of glycosylation sites. A potential glycosylation site in a protein may be unoccupied due to steric limitations at the site, possibly as a result of another glycosylation site being in close proximity. Moreover, more than one glycan structure may exist at a glycosylation site within a glycoprotein population. This variation in the structure of the attached glycan gives rise to protein glycoforms.

22.1.3 Challenges in Glycosylation Analysis by MS

Compared to the analysis of non-glycosylated proteins, glycoprotein analysis faces a greater challenge due to the complexity and heterogeneity of the oligosaccharide chains. This challenge is highlighted by the fact that over 50% of proteins in the Swiss-Prot database are predicted to be glycosylated, though only around 10% of these glycoproteins have been characterized.[1] Several steps are usually needed for the complete structural analysis of a glycoprotein: identification of the protein and the analysis of protein amino acid backbone, determination of the glycosylation sites and the occupancy of these sites, and analysis of the glycan structure and quantitation of each glycan attached to a particular glycosylation site.

N-glycans can be released from the protein backbone by treatment with endoglycosidases such as peptide-*N*-glycosidase F (PNGase F), which removes the glycan and converts the asparagine residue to an aspartic acid,[36] or a combination of endo-β-*N*-acetylglucosaminidases that leave one GlcNAc residue at the glycosylation site.[37] The situation is more challenging with *O*-glycosylation sites as there are no universal enzymes for the removal of all the *O*-glycans. However, the glycans can be removed by alkaline β-elimination[38] or by base-catalyzed β-elimination in the presence of ethylamine.[39]

22.1.3.1 Mass Spectrometry

MS in combination with a separation technique is one of the most powerful tools for characterization of glycoproteins because it provides structural information about the glycosylation with low sample consumption.[40] Although MS responses of glycoproteins and glycopeptides are normally weaker than non-glycosylated analytes,[35,41] MS is the principle analysis technique used in glycoproteomics. There are two main mass spectrometric ionization methods used in the characterization of glycoproteins: matrix-assisted laser desorption/ionization (MALDI) and electrospray ionization (ESI). Dalpathado and Desaire have recently reviewed the use of MS for glycosylated proteins,[42] and hence only a brief discussion is included in this chapter.

22.1.3.2 MALDI-MS

MALDI was developed for the analysis of large molecular weight molecules and can be used for the analysis of proteins as well as oligosaccharides. The method is very sensitive, and samples can be analyzed at femtomol to picomol levels. MALDI can tolerate a wide variety of sample compositions including most biological buffers.[40] For analysis by MALDI, the sample is mixed with a matrix such as 2,5-dihydroxybenzoic acid, which absorbs photon energy at an irradiation laser wavelength (often a nitrogen laser at 337 nm). Ionization occurs during laser pulsing in which the matrix transfers the absorbed energy to the analyte. This results in predominantly singly charged analyte ions, which are subsequently directed toward the mass analyzer.

MALDI-TOF (MALDI time-of-flight) analysis of intact glycoproteins generally yields information on average carbohydrate content[43] due to the glycan heterogeneity, although useful spectra have been obtained.[44–46] Isotopic resolution of glycoproteins can be achieved for proteins with a molecular mass under about 10 kDa.[40] The correct choice of matrix is essential to gain the best result possible for intact glycoproteins; for instance, larger glycoproteins give enhanced signals when sinapinic acid is used as the matrix.[47]

The characterization of glycoproteins often involves digestion of the protein using a protease such as trypsin and analysis of the resultant glycopeptides. These glycopeptides can be analyzed using MALDI-MS in a number of ways: the intact glycopeptide can be measured to facilitate the determination of the existence of a glycan, the peptide backbone can be fragmented giving information on the peptide sequence and glycosylation site, and the oligosaccharide can be fragmented to reveal the glycan structure. In some cases, multidimensional chromatography is needed prior to the analysis of glycopeptides from a digestion mixture containing several glycoproteins, for example, cellular extracts, in order to reduce sample complexity.

22.1.3.3 ESI-MS

The other ionization method for MS, which is used extensively in protein analysis, is ESI. This is a soft ionization technique, which normally generates multiple charged ion species for the mass measurement of higher molecular weight proteins. Usually, ESI produces little or no fragmentation in the ionization source. In ESI, the sample is ionized through a metal capillary at a high voltage. The electric field generates a mist of charged droplets as the sample emerges from the capillary. The solvent is gradually evaporated with the aid of heat and gas flow, thus the droplet size is constantly decreased until naked sample ions are generated, which are then directed into the analyzer for mass measurement.

As with MALDI, analysis of intact glycoproteins via ESI-MS results in an average molecular weight. Glycoforms become more difficult to be completely resolved when protein mass and/or glycosylation sites increase, although smaller glycoproteins have been fully analyzed by this method.[48–50] The characterization of glycopeptides can be performed using ESI-MS with or without fragmentation to yield information on the peptide backbone as well as the attached glycans. The ESI-MS is usually coupled with a chromatography system allowing for separation of peptides immediately before analysis.

22.2 CHARACTERIZATION OF GLYCOPROTEINS

22.2.1 Glycoprotein Enrichment from Complex Solutions

The heterogeneity of oligosaccharide chains in glycoproteins leads to difficulties in resolving the structural differences of glycoproteins and in particular of glycoforms of the same protein. The challenge is increased when the glycoprotein is present in cellular extracts or fluids, as the protein of interest is often available in minute

TABLE 22.1
Examples of the Separation of Glycoproteins from Complex Solutions

Enrichment Technique	Glycoprotein(s)	Source of Glycoprotein	Reference
Concanavalin A lectin affinity chromatography	Urine glycoproteins	Human urine	51
Mixed lectin affinity chromatography	Glycoproteome analysis	Human serum	65
Concanavalin A magnetic beads	RNase B (spiked into sample)	Human serum	63
Capillary zone electrophoresis	Trisialotransferrin	Human serum	70
Capillary isoelectric focusing	Alpha-1-acid glycoprotein	Human serum	135
Capillary gel electrophoresis	Influenza A virus glycoproteins	Mammalian cell culture	136
Micellar electrokinetic chromatography	Antithrombin	Human plasma	137
2D SDS-PAGE	Liver cancer glycoproteins	Human serum	138
Hydrophilic interaction chromatography	IgE-FcεRIα	Mammalian cell culture	See Section 22.3.1
Phenylboranate chromatography	Erythropoietin	Mammalian cell culture	84

quantities.[51,52] Conventional chromatographic techniques for protein separation such as ion exchange, size exclusion, and reversed-phase chromatography can be used for glycoprotein analysis.[53] Some of the common separation techniques specific for glycoproteins are discussed in this section, with Table 22.1 giving some examples.

22.2.1.1 Lectin Affinity Chromatography

The use of lectin affinity chromatography for glycoprotein separation was first described by Donnelly and Goldstein in 1970[54] and remains one of the most widely used methods for the purification of glycoproteins. There are several recent reviews on this topic[55–57] and only a brief description is included here. The sugar chains of a glycoprotein are bound to lectin, which is immobilized on a matrix; unbound material is then washed away prior to the elution of the retained glycoprotein. Numerous plant lectins, in immobilized format, are commercially available, which can be used for glycan binding. However, a majority of the lectins are selective for a certain sugar type. A mixture of glycopeptides can be purified using a lectin with broad specificity such as wheat germ agglutinin or concanavalin A,[58] or a specific lectin can be used for targeted purification of glycopeptides. For example, lentil lectin only binds high mannose glycans with core fucosylation.[59] Broad specificity is of particular use in the initial steps of a purification protocol or in glycoproteomics where the aim is to identify all the glycoproteins from a given species.[60,61] The choice of lectins with specific sugar binding affinity has to be made through screening exercises where the most appropriate immobilized lectin for isolation of a particular glycoform can be discovered.[62] This method of lectin glycan profiling can be miniaturized and has proved useful in determining the types of oligosaccharide attached to a protein.[60] In addition to miniaturization, lectins are also attached to magnetic beads to facilitate automation of such protocols.[63]

Lectin affinity chromatography may be used in series to increase the specificity of a purification protocol. For example, Qiu and Regnier[55] coupled a concanavalin A column to a *Sambucus nigra* agglutinin column to study the degree of sialylation in human serum glycoproteins. In addition, different lectins may be mixed together to achieve the specificity required.[64] For example, Yang and Hancock used a mixture of the agarose-bound lectins concanavalin A, wheat germ agglutinin and Jacalin lectin, to study the glycoproteome from human serum.[65]

Although a selected lectin can be specific to a certain type of glycan, lectin affinity chromatography often lacks the desired specificity for targeted glycans because other oligosaccharides may interact weakly with the matrix and are co-purified with the glycoprotein of interest.[66] Therefore, only limited information on structure of the glycan chain can be obtained using lectin affinity chromatography. Despite this disadvantage, lectin chromatography is of great importance in the enrichment of glycoproteins from complex solutions and is often the first technique employed in a series of experiments for the analysis of glycoproteins (Section 22.2.3).

22.2.1.2 Capillary Electrophoresis

High-performance capillary electrophoresis (CE) is a useful technique for the separation of glycoforms of a particular protein due to its high resolving power. Several modes of separation can be used for different applications, such as capillary zone electrophoresis (CZE), capillary isoelectric focusing (CIEF), capillary gel electrophoresis (CGE), and micellar electrokinetic chromatography (MEKC), with some examples given in Table 22.1. In general, CE consists of a fused silica or glass capillary that is often modified with a coating such as a covalently bonded polymer.[67] The glycoforms are separated using either a positive or negative voltage depending on the coating and background electrolyte. For instance, a negative voltage would be used for glycoprotein separation with a fused silica capillary containing a polybrene coating with 20% MeOH, 1 M acetic acid as the background electrolyte.[68] CE is also used in a microchip format, thus making it amenable for the separation of very low abundance glycoproteins and offering a shorter analysis time.[69]

CE has been used extensively to monitor glycoform populations in the production of recombinant glycoproteins for therapeutic use.[67,70] CE can also be used as a diagnostic tool.[71] In addition to the separation of glycoforms, CE has been used to separate glycoproteins from complex sample solutions.[72,73] For example, CE has been applied to the separation of transferrin from albumin-depleted human serum samples obtained either from healthy patients or those with congenital disorder of glycosylation.[74]

22.2.1.3 Other Separation Techniques

In addition to lectin affinity chromatography and CE, other methods that are often employed for glycoprotein enrichment, including 2D SDS-PAGE, HILIC and phenylboronate chromatography, are discussed here.

Two-dimensional (2D) electrophoresis is often used to separate proteins for analysis by in-gel digestion followed by MS.[75,76] When applied to glycoproteins, two problems with this technique must be taken into consideration: glycoproteins bind less SDS than non-glycosylated proteins due to their carbohydrates and therefore have a different mobility within the gel.[77] Furthermore, glycoproteins tend to resolve into more than one spot because of the differences in molecular weight and/or isoelectric point caused by various glycoforms.[35,77] The identification of glycoproteins from an SDS gel image is aided by glycoprotein-specific stains such as periodic acid-Schiff and Pro-Q Emerald® by Molecular Probes.[78,79]

Although HILIC can be used for the separation and concentration of intact glycoproteins (see Section 22.3.1),[80,81] it is more commonly associated with glycopeptide separations and will be discussed extensively in Section 22.2.2.1.

Glycoproteins can be enriched using an agarose-based resin with a phenylboronate ligand such as the *m*-aminophenylboronic acid beads commercialized by Clontech.[82] This ligand binds to the *cis*-diol groups of glycans such as mannose by forming a 5-membered ring (Figure 22.3A). The protein can be released by lowering the pH or using an elution buffer containing Tris or sorbitol.[83] For example, this technique has been used by Zanette et al. for the purification of erythropoietin from mammalian cell cultures.[84]

22.2.2 Glycopeptide Enrichment

Glycopeptide separations face similar challenges as those for glycoprotein as both contain oligosaccharide chains with great heterogeneity. One of the main issues with glycopeptide identification is the low abundance of these glycopeptides—only about 2%–5% of the peptides produced from a glycoprotein are glycosylated.[85,86] In addition, glycopeptides normally produce lower MS response compared with the non-glycosylated peptides.[35,41] The situation becomes much more difficult when a complex sample such as human serum is digested and analyzed using a shotgun proteomics approach—99% of the total protein mass of human serum is made up of only 22 proteins.[87] Thus glycopeptide information is masked by the heavy presence of abundant non-glycosylated peptides.

Several methods of glycopeptide enrichment are described in the following text. These methods can be used either for shotgun proteomics or as the second step in a tandem approach to glycoprotein analysis.

22.2.2.1 Hydrophilic Interaction Liquid Chromatography

HILIC is a useful tool for the separation of very polar or hydrophilic compounds. The stationary phase is hydrophilic and may be charged depending on the type of ligands used as well as mobile-phase pH values. The aqueous component of the mobile phase is an integral part of the matrix as a water-enriched liquid layer is established within the stationary phase (Figure 22.4). The mobile phase for HILIC typically contains water and acetonitrile, and is therefore amenable to mass-spectrometric analysis. Separation by HILIC is achieved by peptides partitioning into the water-enriched stationary phase through both hydrogen bonding and dipole–dipole interactions.[88] For charged HILIC stationary phases such as zwitterionic interaction chromatography

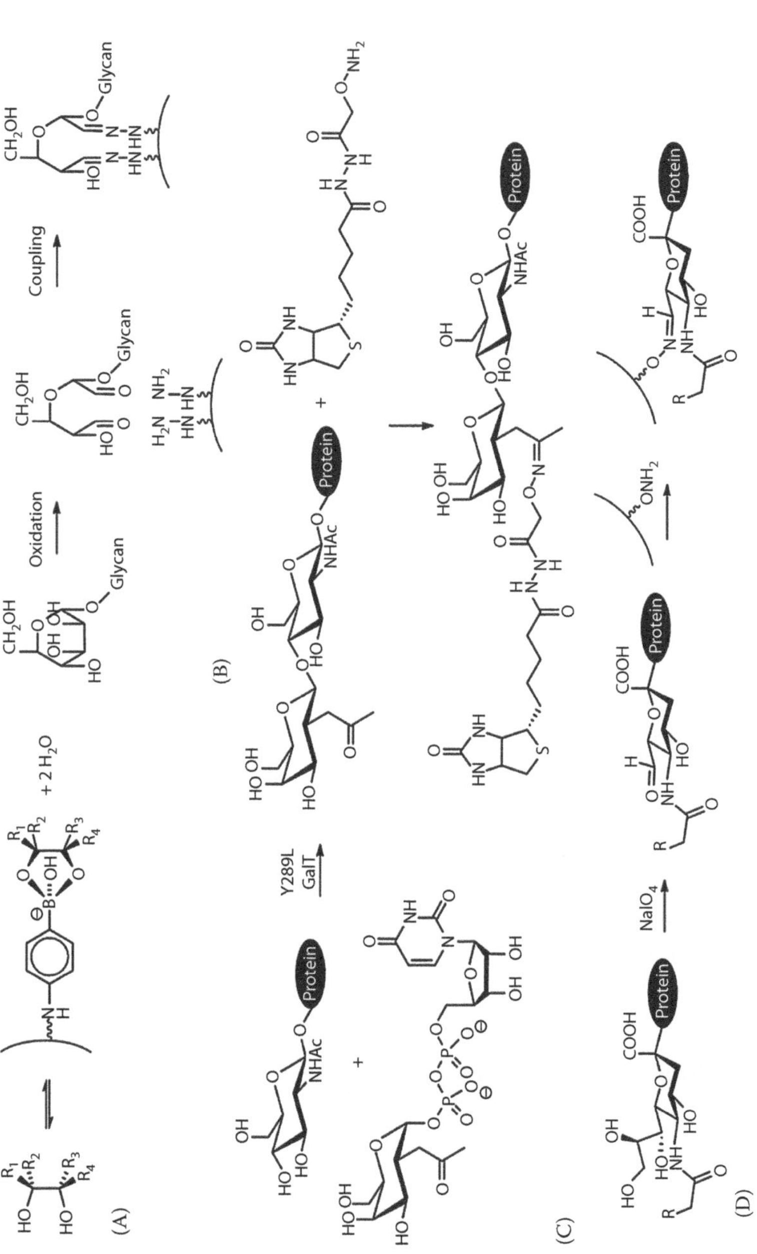

FIGURE 22.3 Chemical modifications of glycoproteins and peptides. (A) Reversible interaction of 1,2-*cis*-diol sugars with phenylboronate beads. (B) Oxidation of a carbohydrate to an aldehyde followed by coupling to hydrazine resin. (C) Scheme for the detection of *O*-GlcNAc glycosylated proteins where the final product binds to streptavidin-HRP. (D) Selective oxidation of sialic acid–containing glycoproteins and oxime bond coupling to a solid support (R = H or OH).

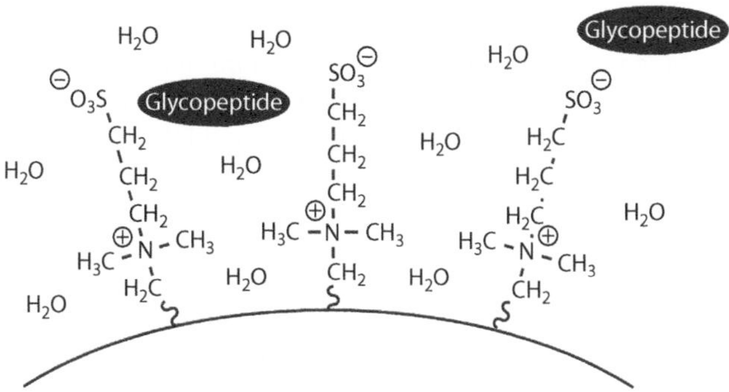

FIGURE 22.4 The structure of ZIC-HILIC resin showing the water-enriched liquid layer and interactions with glycopeptides by both hydrophilic partitioning (left-hand side) and electrostatic interactions (right-hand side).

(ZIC)-HILIC, the retention of analytes is influenced by the electrostatic interactions imposed by the matrix on the water layer (Figure 22.4). Because the stationary phase is zwitterionic, salts or buffers are not needed in the mobile phase since the electrostatic forces of the stationary phase are counterbalanced by the proximity of an ion of the opposite charge (Figure 22.4).[89]

Glycopeptides are more hydrophilic than non-glycosylated peptides due to their sugar chains; therefore, glycopeptides elute later than non-glycosylated peptides during HILIC separation. As a result, HILIC is becoming a powerful technique in the enrichment of glycopeptides in a sample. The retention of a glycopeptide on a HILIC column depends on the hydrophilicity and size of the glycan chain but not on the glycan type.[90] However, some glycoforms can be resolved, such as the separation of differentially sialylated glycopeptides from interferon-γ where the number of sialyl groups correlated to the retention time of the glycopeptides on the HILIC column.[91]

In general, ZIC-HILIC has proved to be more useful than standard HILIC for the separation of glycopeptides as the zwitterionic nature of ZIC-HILIC eliminates the use of buffers in the mobile phase and is therefore easily interfaced with MS.[92] This technique is now widely used in glycoproteomics[37,93–97] to significantly increase the relative abundance of glycopeptides over non-glycosylated peptides.[98] Hagglund et al. first used this method in glycoproteomics to reduce the complexity in a glycoprotein digest sample by removal of non-glycosylated peptides. Upon the separation of glycopeptides using the HILIC method, the analysis of the *N*-glycosylation sites was successfully accomplished after the removal of the complex sugars to leave one GlcNAc residue.[37] Totally, 62 glycosylation sites were identified using HILIC enrichment from a digest of the glycoproteins in human plasma.[37] Kaji et al. analyzed the diversity of glycoproteins in *Caenorhabditis elegans* using an initial lectin capture of glycoproteins followed by enrichment of glycopeptides by HILIC.[95] In these experiments, 829 glycoproteins were identified, which contained 1465 *N*-glycosylation sites.[95]

For recombinant glycoproteins or simple protein mixtures, HILIC may be used for the enrichment[36,99] or for the separation of glycopeptides followed by the analysis of the product with or without glycan removal.[100] For example, offline ZIC-HILIC enrichment of glycopeptides allows for the fast analysis of glycosylation site occupancy of recombinant glycoproteins by ESI-MS.[36] Full mass spectrometric analysis of peptides with complex glycans has been performed on small glycopeptides from a nonspecific enzymatic digestion (e.g., pronase). Using HILIC, these glycopeptides were separated according to the size of the glycan chain.[99,100] Information about the glycans was obtained from MS/MS data, and the identification of the peptide backbone and the glycan attachment site were made based on MS^3 spectra.[100] This method was used to identify two glycans with different structures that were attached alternatively to one glycosylation site of *Dolichos biflorus* lectin.[100]

As an alternative to the HILIC resins and columns described above, glycopeptides can be enriched based on hydrophilic interaction with carbohydrate gel matrices such as cellulose or sepharose.[101,102] This method has been used to analyze the *N*-glycosylation sites of apolipoprotein B-100 showing that 17 out of the 19 potential glycosylation sites were occupied.[102] It has also been used for the differential analysis of site-specific glycans from cellular fibronectin.[102]

22.2.2.2 Lectin Affinity Chromatography

Similar to glycoprotein enrichment, glycopeptides can be purified using lectin affinity chromatography. This can be done either using a broad specificity lectin such as the widely used concanavalin A resin to capture *N*-glycosylated peptides,[58,103,104] or using sequential or multi-lectin columns to capture all glycopeptides including ones with *O*-linked glycans,[61,105] or using a specific lectin to capture a particular group of glycopeptides.[106,107] Lectin affinity chromatography is often used in the "glyco-catch" method where the lectin column is used in an initial enrichment of the glycoprotein and then the same column is used again for the enrichment of the glycopeptides after glycoprotein digestion.[61] This method has been used to identify soluble glycopeptides from a *C. elegans* lysate.[108]

22.2.2.3 Hydrazine Chemistry

Both glycoproteins and glycopeptides can be trapped using hydrazine chemistry. The vicinal hydroxyl groups of carbohydrate moiety of *N*-linked glycopeptides are oxidized to aldehydes with periodate first. These aldehyde groups are then covalently bound to hydrazine resin (Figure 22.3B).[109] After the glycoprotein is captured using hydrazine resin, it is then digested in situ using trypsin to remove non-glycosylated peptides. The peptide portion of *N*-linked glycopeptides is then released by PNGase F treatment, leaving the oligosaccharides bound to the resin. A scheme for the modification of glycopeptides by hydrazine chemistry is shown in Figure 22.3B. One of the main advantages of this method is its ability to immobilize glycoproteins, digest in situ, and then remove all unwanted peptides before elution of the glycopeptides, thus abolishing the need for both glycoprotein and glycopeptide enrichment. Sun et al. have used hydrazine resin to capture the glycopeptides from a tryptic digest of the microsomal fraction of an ovarian cancer cell line. The approach yielded a 19- to 45-fold enrichment of glycopeptides in the sample.[85] This method has also been used for the identification of glycoproteins in the plasma membrane from HeLa cell lysates,[110] and for the analysis of glycoproteins in human serum[109] and in plasma.[111] In these cases, the glycoprotein was immobilized on the hydrazine matrix before digestion and then the glycopeptides were released by PNGase F treatment.

22.2.2.4 Ion-Pairing Normal-Phase Chromatography

Although standard normal phase chromatography is not as widely used as HILIC chromatography due to reproducibility problems, the separation of glycosylated and non-glycosylated peptides by this technique can be improved by the addition of inorganic monovalent ions such as NaCl, LiCl, or NaOH.[112] This ion-pairing interaction increases the hydrophobicity difference between peptides with glycans and the ones without glycans, and consequently results in a more efficient separation than with a normal-phase column used under standard conditions. The improvement has been exemplified by the analysis of glycopeptides from breast cancer cells.[112]

22.2.2.5 Others

In addition to the methods discussed above, glycopeptides may also be enriched from complex peptide solutions by techniques such as size-exclusion chromatography (SEC), graphitized carbon and CE. The logic behind SEC is based on the fact that glycopeptides tend to have a higher mass than non-glycosylated peptides,[113] whereas the use of graphitized carbon is mainly because of the high hydrophobicity of the materials for retaining small glycopeptides.[114]

CZE is a useful technique for the enrichment of glycopeptides because it can be highly selective. By carefully choosing the conditions for electrophoretic and electrokinetic countermigration, this technique can be used to enrich small and highly hydrophilic glycopeptides, which otherwise are difficult to analyze using a standard reversed-phase HPLC approach. These glycopeptides normally elute near the void volume in HPLC separation.[73] Bindila et al. used CZE in the analysis of glycopeptides and amino acids from human urine of a patient with *N*-acetylhexosaminiclase deficiency.[115] CZE showed good separation efficiency and resolution according to the degree of sialylation and the type of amino acids allowing structural identification of single components of the complex sample.[115]

Most methods described above do not alter the glycan structure. However, highly specific enrichment of glycopeptides can be achieved using chemical modifications.[116] For example, glycopeptides with glycans containing *cis*-diol can be isolated using phenylboranic acid by the method discussed for glycoprotein enrichment in Section 22.2.1.3. The scheme shown in Figure 22.3A illustrates this approach.[63,117] This method is particularly useful because it binds both *N*- and *O*-glycopeptides.

O-GlcNAc-containing peptides can be analyzed using modification of the glycan with galactosyltransferase, which introduces a ketone-labeled galactose to the peptide. This ketone group is then biotinylated using a Schiff-base reaction (Figure 22.3C); thus glycopeptides can be purified using streptavidin affinity chromatography.[118] This method has been used to identify glycoproteins from the mammalian brain[119] and HeLa cell lysates.[120]

Sialylated glycopeptides can be enriched by selective oxidation of the terminal sialic acid residue with periodate followed by ligation of the newly formed aldehyde group with a polymer substrate such as aminooxy-functionalized polyacrylamide (Figure 22.3D). After washing to remove non-sialylated peptides, the sialylated glycopeptides are eluted with trifluoroacetic acid, which cleaves the bond between the sialic acid and adjacent galactose residue.[121] The authors exemplified the method by enriching glycopeptides from human α-fetoprotein, bovine pancreas fibrinogen and human erythropoietin.[121]

22.2.3 Tandem Methods

For many glycoproteomic applications, the protein samples are rather complex.[35,86] Hence more than one enrichment method is needed in practice. Generally, the tandem enrichment method consists of initial glycoprotein selection followed by tryptic digestion and subsequent glycopeptide enrichment, although in some cases more than two methods are used in series. For example, Kaji et al. used an initial lectin capture of glycoproteins followed by tryptic digestion and purification of the glycopeptides by both lectin affinity chromatography and HILIC in their analysis of *C. elegans* glycoproteins.[95]

Two techniques that are widely used in the most common tandem methods are lectin affinity chromatography (for glycoprotein capture) coupled to an additional step of enrichment of the glycopeptides either using lectin or HILIC. If the same lectin affinity column is used for both glycoprotein and glycopeptide enrichment, the technique is known as the glyco-catch method, which is described briefly in Section 22.2.2.2.[61]

The protocol used by Kaji et al. combined lectin affinity chromatography and HILIC for the identification of glycoproteins and their glycosylation sites from mouse liver.[94] The purity of the glycopeptides was increased to 70%–100% after the HILIC step. Totally, around 2500 *N*-glycosylation sites in 1200 proteins were identified in the work.[94] The analysis was finished within a week, highlighting the high-throughput nature of the technique. A similar tandem strategy was also described by Bunkenborg et al. in which two lectin steps of enrichment for both glycoproteins and glycopeptides were employed for glycoproteomic analysis of human body fluids.[122] The method was exemplified using the model proteins α-1-acid glycoprotein, fetuin, ovalbumin, and ribonuclease B.[122]

Recently, Calvano et al. assessed these tandem techniques for the characterization of serum glycoproteins and found 86 *N*-glycosylation sites in 45 proteins by the glyco-catch method and 81 *N*-glycosylation sites in 44 proteins by lectin affinity followed by HILIC.[80] The methods were shown to be complementary as different glycopeptides were isolated using the differing tandem approaches. An overlap of 63 *N*-glycosylation sites from 38 proteins were demonstrated in the report.[80]

In addition to those using lectin as the initial step, any of the methods described here can be used in series to yield a tandem method for glyco-enrichment. For example, more than one type of CE can be used in tandem.[123] It should be noted, however, that compatibility between the various techniques and the mass spectrometer needs to be taken into consideration.

22.3 EXAMPLES OF GLYCOPROTEIN CHARACTERIZATION USING HILIC

The examples in this chapter all use human IgE-FcεRIα as the test glycoprotein. IgE-FcεRIα is the α-chain ectodomain of the high-affinity immunoglobulin-E (IgE) receptor FcεRI, which is found on the surface of effector cells within the immune system. IgE-FcεRIα is associated with allergic response, anaphylaxis and antiparasitic immunity.[124,125] Two glycosylation variants have been used: the wild type, denoted wt-IgE-FcεRIα, which contains six *N*-glycosylation sites (N21, N42, N50, N74, N135, and N166) and one *O*-glycosylation site (T142), and a triple mutant referred to as mut_3-IgE-FcεRIα in this chapter. Based on the work of Garman et al.,[126] the triple mutant lacks glycosylation sites at residues N74, N135, and T142, all of which are variably occupied (see Section 22.3.3).

The glycans attached to IgE-FcεRIα were simplified either by expressing the protein in human embryonic kidney (HEK) 293T cells in the presence of kifunensine, which inhibits *N*-glycosylation processing leaving sugars in the form Man_{5-9} $GlcNAc_2$, or by producing the protein in ricin-resistant HEK 293S cells, which have no GnTI activity giving predominantly $Man_5GlcNAc_2$ glycans.[127,128]

22.3.1 Glycoprotein Enrichment

In the proof-of-principle example described below, ZIC-HILIC was used to enrich the recombinant glycoprotein mut_3-IgE-FcεRIα from a mixture containing three other non-glycosylated proteins. The purified protein has about 1 mg/mL concentration with about 25% of the protein being glycoprotein (Figure 22.5, lane 1). After the glycoproteins in the sample were bound to the ZIC-HILIC resin and the resin washed; the eluted glycoprotein was estimated to have approximately 70% purity (Figure 22.5, lane 2). The method is of particular use in recombinant protein production when protein yield is low; therefore, enrichment of glycoproteins is needed prior to analysis by SDS-PAGE or MS. It is to be noted, however, that most media used for mammalian cell culturing contains other glycosylated proteins, especially from components of fetal calf serum. These proteins are also enriched using this method along with the glycoprotein of interest. The method can be extended to enrich glycoproteins present in cellular extracts or fluids.[80,81] The product after glycoprotein enrichment can then be used for further analysis, for example, determination of glycosylation site occupancy (Section 22.3.3) and characterization of glycan structure (Section 22.3.4).

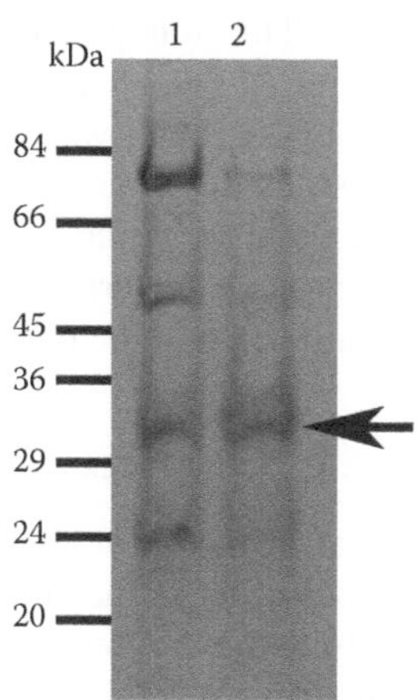

FIGURE 22.5 SDS-PAGE gel showing enrichment of mut_3-IgE-FcεRIα (lane 2) from a mixture also containing three non-glycosylated proteins (lane 1). Enrichment is from ~25% to 70% measured using densitrometry of the bands on the gel. The position of mut_3-IgE-FcεRIα is shown by the arrow on the right-hand side.

22.3.2 Characterization of Glycoforms

Although HILIC is not commonly used for the analysis of glycoforms, it can be useful when studying proteins with low degree of glycan heterogeneity. In the example described here, mut_3-IgE-FcεRIα was grown in HEK 293T cells in the presence of kifunensine[128] producing only $Man_{5-9}GlcNAc_2$ sugars.[127]

TABLE 22.2
Glycoform Separation of the Tryptic Peptide Containing N166 from mut$_3$-IgE-FcεRIα Grown in HEK 293T Cells in the Presence of Kifunensine

Retention Time (Minutes)	Glycoform	Expected *m/z*	Observed *m/z*
11.59	$Man_9GlcNAc_2$	1956.35	1957.31
11.48	$Man_8GlcNAc_2$	1875.33	1873.18
11.41	$Man_7GlcNAc_2$	1794.30	1793.60

Oligosaccharide structures are given along with the expected and observed *m/z* values for the $[M + 2H]^{2+}$ ions observed.

After tryptic digest of the protein, the resulting glycopeptides were separated using a ZIC-HILIC column with isocratic elution of 55% acetonitrile, 45% water, 10 mM ammonium acetate buffer pH 8.0, and analyzed by ESI-MS. $[M + 2H]^{2+}$ ions detected for the tryptic glycopeptide containing N166 are given in Table 22.2. The difference in retention time between various glycoforms was not as great as the results for sialylated glycopeptides by Zhang and Wang[91] and by Takegawa et al.[129] This was probably caused by the greater hydrophobicity introduced by the large glycan-containing peptide (17 amino acids, average hydrophilicity[130] is equal to −0.2 with values below 0 indicating a hydrophobic peptide). Glycoform analysis by this method is particularly useful in tracking batch-to-batch glycoform variation in the production of a recombinant glycoprotein. In the current example, glycoform characterization was used to analyze the glycan heterogeneity prior to crystallization of the protein.[126]

22.3.3 Characterization of Glycosylation Site Occupancy

Several methods use HILIC for *N*-glycosylation site occupancy analysis. These methods differ in the labeling of the glycosylation site prior to peptide analysis. If the glycoprotein is purified and the sequence is known, the glycosylation site can be readily determined by simply identifying the peptides whose molecular weight sees a 1 Da increase when compared to the theoretical calculation. The mass change in peptide molecular weight is brought about by PNGase F treatment, which converts the asparagine residue to an aspartic acid.[36] HILIC glycopeptide enrichment can also be undertaken offline in a batch mode to enrich all glycopeptides prior to PNGase F treatment. Table 22.3 shows the results from glycosylation site occupancy analysis for recombinant wt-IgE-FcεRIα produced in mammalian cells. The data indicates that four of the *N*-glycosylation sites are always occupied, with two being variably occupied and no sites being unoccupied (Figure 22.6). The results verify mutation studies.[126,131] The *O*-glycosylation site occupancy cannot be investigated by this method. However, the identification of relevant non-glycosylated peptide implies that the *O*-glycosylation site could be either not occupied at all or only variably occupied

TABLE 22.3
Results from Glycosylation Site Analysis of wt-IgE-FcεRIα

Glycosylation Site	Occupancy
N21	Fully occupied
N42	Fully occupied
N50	Fully occupied
N74	Variably occupied
N135	Variably occupied
N166	Fully occupied
T142	Either variably or not occupied

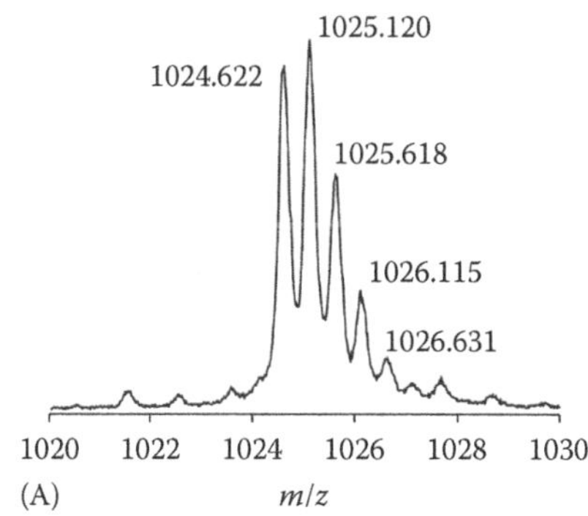

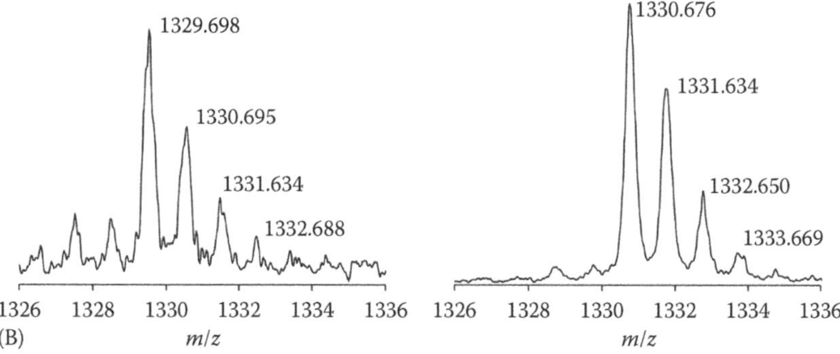

FIGURE 22.6 MS data for (A) the $[M + 2H]^{2+}$ ion of the tryptic peptide containing N166, which is fully occupied, and (B) the $[M + H]^{+}$ ions for the peptide containing N74, which is variably occupied (+0 Da and +1 Da) generated by digestion with chymotrypsin.

(Table 22.3). Mutation studies have shown the *O*-linked glycan was not essential for cell growth.[126,131]

If the glycoprotein is in a complex mixture and the sequence is unknown, unambiguous assignment of the glycosylation sites is performed by either stable isotope labeling with ^{18}O,[95] or by leaving one GlcNAc residue attached to the site using a combination of endo-β-*N*-acetylglucosaminidases instead of PNGase F.[37]

22.3.4 Full Characterization of Glycoproteins

For complete characterization of a glycoprotein, the protein backbone, the occupancy of the glycosylation sites and the glycan structures need to be thoroughly analyzed. In the case of a purified recombinant glycoprotein containing only *N*-linked glycans, the protein backbone can be analyzed simply by the treatment of the sample with PNGase F followed by MS analysis of the intact protein (Figure 22.7A). It is to be noted that the expected mass should be 1 Da less for

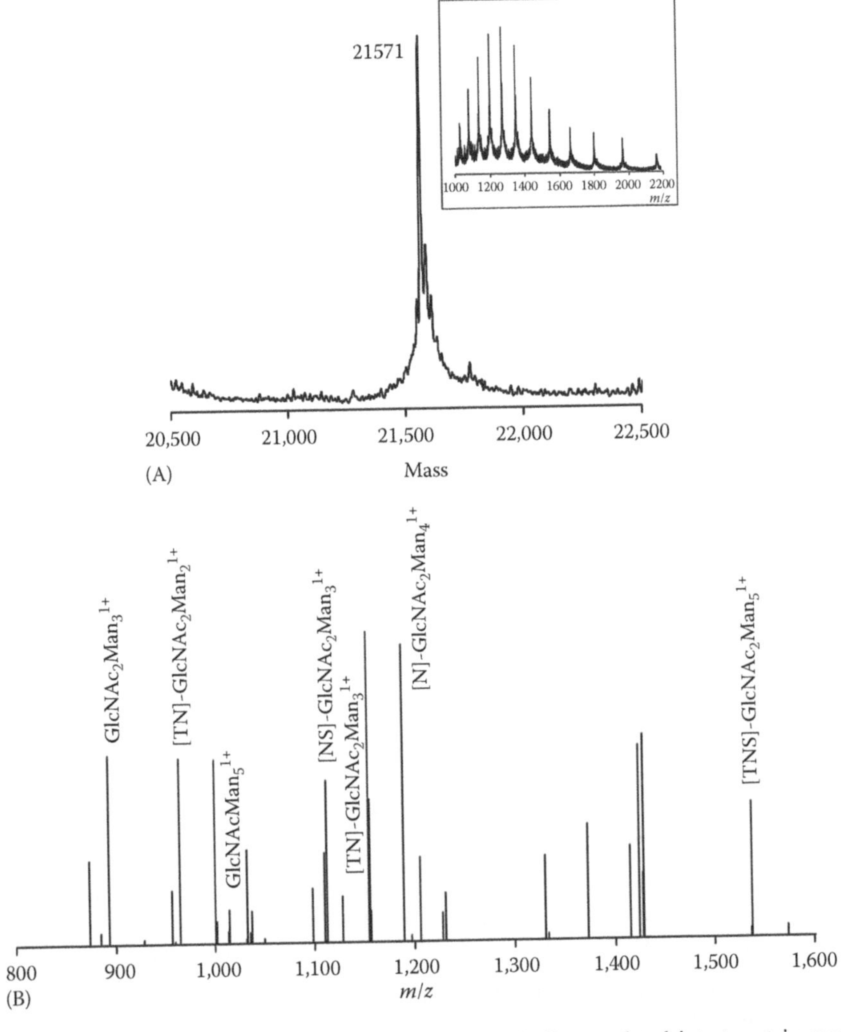

FIGURE 22.7 Characterization of mut$_3$-IgE-FcεRIα. (A) Deconvoluted intact protein mass spectrum with combined raw data shown in the inset. (B) Fragment ion spectrum (MS/MS) of the pronase-generated glycopeptide TNS of mut$_3$-IgE-FcεRIα with a Man$_5$GlcNAc$_2$ glycan moiety attached. MS/MS data obtained from the doubly protonated precursor ion at *m/z* = 769.3.

each fully occupied *N*-glycosylation site because of the conversion of asparagine to aspartic acid upon PNGase F treatment. If identification of the glycoprotein is needed, for example, the protein has been enriched from a complex sample, the protein is digested with trypsin and subjected to MS/MS analysis followed by the identification of the peptides using a database searching software such as Mascot.[132] Because the removal of *O*-linked sugars presents some difficulty (Section 22.1.3), tryptic digestion is recommended for proteins containing these glycans. Digestion is followed by analysis of the non-glycosylated peptides in order to identify the protein.

In general, glycosylation site occupancy and analysis of glycan structure are performed separately. However, both analyses can be efficiently combined using the method of Wuhrer et al.[100] This combined method analyzes glycopeptides directly to obtain information on both glycan structure and glycosylation site. This method can be used for both *N*- and *O*-linked glycans. Such information is lost when the glycans are removed before analysis. After the identification of the protein backbone (if needed) using either intact protein MS or analysis of tryptic peptides, the glycoprotein is subjected to digestion with pronase. Nonspecific proteolysis with pronase results in small peptides of two to eight amino acids, which are amenable to detection by MS.[133] The glycopeptides are separated using normal phase chromatography or HILIC before analysis by multistage MS fragmentations. Dodds et al. have improved the method by immobilization of pronase to eliminate auto-proteolytic fragments from the final peptide sample.[134]

An example is shown here where mut$_3$-IgE-FcεRIα with Man$_5$GlcNAc$_2$ glycans was digested with pronase followed by HILIC and mass spectrometric analysis (Figure 22.7B). The HILIC column effectively separated the pronase-generated peptides using isocratic elution in 55% acetonitrile, 45% water, 10 mM ammonium acetate buffer pH 8.0. The glycopeptides elute in the retention time window from 13.7 to 18.0 min. Glycopeptides were readily identified by screening for neutral loss of terminal monosaccharide units—a neutral loss of 81 is indicative of a doubly charged precursor ion containing a terminal hexose. Figure 22.7B shows the MS/MS spectrum for the doubly charged ion at $m/z = 769.3$, which corresponds to the tripeptide TNS containing the glycosylation site N50. The identification of the peaks in the spectrum in Figure 22.7B was facilitated by prior knowledge of the glycan pattern,[127] which was of particular use as the MS/MS data contained ions from fragmentation of both glycosidic linkages and peptide bonds. The data obtained, however, did confirm that only Man$_5$GlcNAc$_2$ glycan was attached to mut$_3$-IgE-FcεRIα as expected due to the protein being produced from HEK 293S cells.[127,128] As indicated by Wuhrer et al., this technique gives most information when used in combination with other glycan analysis methods including conventional glycan release followed by MS of the oligosaccharides.[100]

Although full characterization of glycoproteins by this method is complex, it has been successfully applied to model glycoproteins such as ribonuclease B,[100,134] horse radish peroxidise,[99,100] κ-casein,[134] and chicken egg albumin.[134] In addition, it is also used to characterize a previously uncharacterized protein, *D. biflorus* lectin.[100]

22.4 SUMMARY

In this chapter, the challenges associated with glycoprotein characterization are shown to be associated with the diversity and heterogeneity of the glycan moieties attached to a protein. Various general methods for the isolation and separation of glycoproteins and glycopeptides are presented along with their benefits and limitations. In many cases, glycoprotein characterization is seen to require more than one step of enrichment in order to gain the maximum amount of information from each sample. These tandem methods usually consist of a step of isolation of the glycoprotein from solution followed by digestion and enrichment or separation of the resulting glycopeptides.

The use of HILIC in glycoprotein characterization was highlighted by four worked examples using the human-secreted protein IgE-FcεRIα. In the first experiment, the intact glycoprotein was enriched from a mixture also containing non-glycosylated proteins using a batch method. Several glycoforms attached to an *N*-glycosylation site were separated using column chromatography where the elution volume was related to the size of the glycan. The *N*-glycosylation sites were characterized for occupancy using a simple batch method for the enrichment of glycopeptides before PNGase F treatment and MS analysis. In the final example, both the occupancy of a glycosylation site and the composition of the attached glycan were characterized using digestion with pronase followed by HILIC and MS/MS analysis.

ACKNOWLEDGMENTS

Within the Oxford Protein Production Facility, I wish to thank Nahid Rahman for production and purification of mut_3-IgE-FcεRIα and Ray Owens for critical reading of this manuscript. I would also like to thank Jens Loebermann and Rebecca Beavil at King's College, London, for the gift of wt-IgE-FcεRIα.

REFERENCES

1. Apweiler, R., Hermjakob, H., and Sharon, N., On the frequency of protein glycosylation, as deduced from analysis of the SWISS-PROT database, *Biochim. Biophys. Acta* 1473 (1), 4–8, 1999.
2. Mitra, N., Sinha, S., Ramya, T. N., and Surolia, A., N-linked oligosaccharides as outfitters for glycoprotein folding, form and function, *Trends Biochem. Sci.* 31 (3), 156–63, 2006.
3. Helenius, A. and Aebi, M., Intracellular functions of N-linked glycans, *Science* 291 (5512), 2364–2369, 2001.
4. Huang, Y., Huang, J. H., Xie, Q. J., and Yao, S. Z., Carbohydrate-protein interactions, *Prog. Chem.* 20 (6), 942–950, 2008.
5. Rudd, P. M., Elliott, T., Cresswell, P., Wilson, I. A., and Dwek, R. A., Glycosylation and the immune system, *Science* 291 (5512), 2370–2376, 2001.
6. Rabinovich, G. A. and Toscano, M. A., Turning 'sweet' on immunity: Galectin-glycan interactions in immune tolerance and inflammation, *Nat. Rev. Immunol.* 9 (5), 338–352, 2009.
7. Hann, S. R., Role of post-translational modifications in regulating c-Myc proteolysis, transcriptional activity and biological function, *Semin. Cancer Biol.* 16 (4), 288–302, 2006.

8. Limjindaporn, T., Wongwiwat, W., Noisakran, S., Srisawat, C., Netsawang, J., Puttikhunt, C., Kasinrerk, W., Avirutnan, P., Thiemmeca, S., Sriburi, R., Sittisombut, N., Malasit, P., and Yenchitsomanus, P. T., Interaction of dengue virus envelope protein with endoplasmic reticulum-resident chaperones facilitates dengue virus production, *Biochem. Biophys. Res. Commun.* 379 (2), 196–200, 2009.
9. Abu-Qarn, M., Eichler, J., and Sharon, N., Not just for Eukarya anymore: Protein glycosylation in Bacteria and Archaea, *Curr. Opin. Struct. Biol.* 18 (5), 544–550, 2008.
10. Upreti, R. K., Kumar, M., and Shankar, V., Bacterial glycoproteins: Functions, biosynthesis and applications, *Proteomics* 3 (4), 363–379, 2003.
11. Fletcher, C. M., Coyne, M. J., Villa, O. F., Chatzidaki-Livanis, M., and Comstock, L. E., A general O-glycosylation system important to the physiology of a major human intestinal symbiont, *Cell* 137 (2), 321–331, 2009.
12. Szymanski, C. M. and Wren, B. W., Protein glycosylation in bacterial mucosal pathogens, *Nat. Rev. Microbiol.* 3 (3), 225–237, 2005.
13. Aas, F. E., Vik, A., Vedde, J., Koomey, M., and Egge-Jacobsen, W., *Neisseria gonorrhoeae* O-linked pilin glycosylation: Functional analyses define both the biosynthetic pathway and glycan structure, *Mol. Microbiol.* 65 (3), 607–624, 2007.
14. Daniels, M. A., Hogquist, K. A., and Jameson, S. C., Sweet 'n' sour: The impact of differential glycosylation on T cell responses, *Nat. Immunol.* 3 (10), 903–910, 2002.
15. Dwek, M. V., Ross, H. A., and Leathem, A. J., Proteome and glycosylation mapping identifies post-translational modifications associated with aggressive breast cancer, *Proteomics* 1 (6), 756–762, 2001.
16. Freeze, H. H., Update and perspectives on congenital disorders of glycosylation, *Glycobiology* 11 (12), 129R-143R, 2001.
17. Marquardt, T. and Denecke, J., Congenital disorders of glycosylation: Review of their molecular bases, clinical presentations and specific therapies, *Eur. J. Pediatr.* 162 (6), 359–379, 2003.
18. Hwang, H., Zhang, J., Chung, K. A., Leverenz, J. B., Zabetian, C. P., Peskind, E. R., Jankovic, J., Su, Z., Hancock, A. M., Pan, C., Montine, T. J., Pan, S., Nutt, J., Albin, R., Gearing, M., Beyer, R. P., Shi, M., and Zhang, J., Glycoproteomics in neurodegenerative diseases, *Mass Spectrom. Rev.* 8, 8, 2009.
19. Saldova, R., Royle, L., Radcliffe, C. M., Abd Hamid, U. M., Evans, R., Arnold, J. N., Banks, R. E., Hutson, R., Harvey, D. J., Antrobus, R., Petrescu, S. M., Dwek, R. A., and Rudd, P. M., Ovarian cancer is associated with changes in glycosylation in both acute-phase proteins and IgG, *Glycobiology* 17 (12), 1344–1356, 2007.
20. Saez-Valero, J. and Small, D. H., Acetylcholinesterase and butyrylcholinesterase glycoforms are biomarkers of Alzheimer's disease, *J. Alzheimers Dis.* 3 (3), 323–328, 2001.
21. Saez-Valero, J., Fodero, L. R., Sjogren, M., Andreasen, N., Amici, S., Gallai, V., Vanderstichele, H., Vanmechelen, E., Parnetti, L., Blennow, K., and Small, D. H., Glycosylation of acetylcholinesterase and butyrylcholinesterase changes as a function of the duration of Alzheimer's disease, *J. Neurosci. Res.* 72 (4), 520–526, 2003.
22. Srinivasan, A., Viswanathan, K., Raman, R., Chandrasekaran, A., Raguram, S., Tumpey, T. M., Sasisekharan, V., and Sasisekharan, R., Quantitative biochemical rationale for differences in transmissibility of 1918 pandemic influenza A viruses, *Proc. Natl. Acad. Sci. U.S.A.* 105 (8), 2800–2805, 2008.
23. Wang, L. X., Toward oligosaccharide- and glycopeptide-based HIV vaccines, *Curr. Opin. Drug Discov. Dev.* 9 (2), 194–206, 2006.
24. Ni, J., Song, H., Wang, Y., Stamatos, N. M., and Wang, L. X., Toward a carbohydrate-based HIV-1 vaccine: Synthesis and immunological studies of oligomannose-containing glycoconjugates, *Bioconjug. Chem.* 17 (2), 493–500, 2006.
25. Slovin, S. F., Keding, S. J., and Ragupathi, G., Carbohydrate vaccines as immunotherapy for cancer, *Immunol. Cell Biol.* 83 (4), 418–428, 2005.

26. Slovin, S. F., Ragupathi, G., Musselli, C., Fernandez, C., Diani, M., Verbel, D., Danishefsky, S., Livingston, P., and Scher, H. I., Thomsen-Friedenreich (TF) antigen as a target for prostate cancer vaccine: Clinical trial results with TF cluster (c)-KLH plus QS21 conjugate vaccine in patients with biochemically relapsed prostate cancer, *Cancer Immunol. Immunother.* 54 (7), 694–702, 2005.
27. Hart, G. W., Glycosylation, *Curr. Opin. Cell. Biol.* 4 (6), 1017–1023, 1992.
28. Rajeswari, P., Natarajan, R., Nadler, J. L., Kumar, D., and Kalra, V. K., Glucose induces lipid peroxidation and inactivation of membrane-associated ion-transport enzymes in human erythrocytes in vivo and in vitro, *J. Cell. Physiol.* 149 (1), 100–109, 1991.
29. Wells, L., Vosseller, K., and Hart, G. W., Glycosylation of nucleocytoplasmic proteins: Signal transduction and O-GlcNAc, *Science* 291 (5512), 2376–2378, 2001.
30. Harris, R. J. and Spellman, M. W., O-linked fucose and other post-translational modifications unique to EGF modules, *Glycobiology* 3 (3), 219–224, 1993.
31. Mormeneo, S., Zueco, J., Iranzo, M., and Sentandreu, R., O-linked mannose composition of secreted invertase of *Saccharomyces cerevisiae*, *FEMS Microbiol. Lett.* 48 (3), 271–274, 1989.
32. Duman, J. G., Miele, R. G., Liang, H., Grella, D. K., Sim, K. L., Castellino, F. J., and Bretthauer, R. K., O-Mannosylation of *Pichia pastoris* cellular and recombinant proteins, *Biotechnol. Appl. Biochem.* 28 (Pt 1), 39–45, 1998.
33. Kogelberg, H., Chai, W., Feizi, T., and Lawson, A. M., NMR studies of mannitol-terminating oligosaccharides derived by reductive alkaline hydrolysis from brain glycoproteins, *Carbohydr. Res.* 331 (4), 393–401, 2001.
34. Hounsell, E. F., Davies, M. J., and Renouf, D. V., O-linked protein glycosylation structure and function, *Glycoconj. J.* 13 (1), 19–26, 1996.
35. Geyer, H. and Geyer, R., Strategies for analysis of glycoprotein glycosylation, *Biochim. Biophys. Acta* 1764 (12), 1853–1869, 2006.
36. Nettleship, J. E., Aplin, R., Aricescu, A. R., Evans, E. J., Davis, S. J., Crispin, M., and Owens, R. J., Analysis of variable N-glycosylation site occupancy in glycoproteins by liquid chromatography electrospray ionization mass spectrometry, *Anal. Biochem.* 361 (1), 149–151, 2007.
37. Hagglund, P., Bunkenborg, J., Elortza, F., Jensen, O. N., and Roepstorff, P., A new strategy for identification of N-glycosylated proteins and unambiguous assignment of their glycosylation sites using HILIC enrichment and partial deglycosylation, *J. Proteome Res.* 3 (3), 556–566, 2004.
38. Greis, K. D., Hayes, B. K., Comer, F. I., Kirk, M., Barnes, S., Lowary, T. L., and Hart, G. W., Selective detection and site-analysis of O-GlcNAc-modified glycopeptides by beta-elimination and tandem electrospray mass spectrometry, *Anal. Biochem.* 234 (1), 38–49, 1996.
39. Hanisch, F. G., Jovanovic, M., and Peter-Katalinic, J., Glycoprotein identification and localization of O-glycosylation sites by mass spectrometric analysis of deglycosylated/alkylaminylated peptide fragments, *Anal. Biochem.* 290 (1), 47–59, 2001.
40. Morelle, W. and Michalski, J. C., The mass spectrometric analysis of glycoproteins and their glycan structures, *Curr. Anal. Chem.* 1 (1), 29–57, 2005.
41. Annesley, T. M., Ion suppression in mass spectrometry, *Clin. Chem.* 49 (7), 1041–1044, 2003.
42. Dalpathado, D. S. and Desaire, H., Glycopeptide analysis by mass spectrometry, *Analyst* 133 (6), 731–738, 2008.
43. Sei, K., Nakano, M., Kinoshita, M., Masuko, T., and Kakehi, K., Collection of alpha1-acid glycoprotein molecular species by capillary electrophoresis and the analysis of their molecular masses and carbohydrate chains. Basic studies on the analysis of glycoprotein glycoforms, *J. Chromatogr. A* 958 (1–2), 273–281, 2002.

44. Gimenez, E., Benavente, F., Barbosa, J., and Sanz-Nebot, V., Towards a reliable molecular mass determination of intact glycoproteins by matrix-assisted laser desorption/ionization time-of-flight mass spectrometry, *Rapid Commun. Mass Spectrom.* 21 (16), 2555–2563, 2007.
45. Zaia, J., Boynton, R., Heinegard, D., and Barry, F., Posttranslational modifications to human bone sialoprotein determined by mass spectrometry, *Biochemistry* 40 (43), 12983–12991, 2001.
46. Ongay, S., Puerta, A., Diez-Masa, J. C., Bergquist, J., and de Frutos, M., CIEF and MALDI-TOF-MS methods for analyzing forms of the glycoprotein VEGF 165, *Electrophoresis* 30 (7), 1198–1205, 2009.
47. Beavis, R. C. and Chait, B. T., Cinnamic acid derivatives as matrices for ultraviolet laser desorption mass spectrometry of proteins, *Rapid Commun. Mass Spectrom.* 3 (12), 432–435, 1989.
48. Hui, J. P., White, T. C., and Thibault, P., Identification of glycan structure and glycosylation sites in cellobiohydrolase II and endoglucanases I and II from *Trichoderma reesei*, *Glycobiology* 12 (12), 837–849, 2002.
49. Gong, B., Cukan, M., Fisher, R., Li, H., Stadheim, T. A., and Gerngross, T., Characterization of N-linked glycosylation on recombinant glycoproteins produced in *Pichia pastoris* using ESI-MS and MALDI-TOF, *Methods Mol. Biol.* 534, 213–223, 2009.
50. Karas, M., Bahr, U., and Dulcks, T., Nano-electrospray ionization mass spectrometry: Addressing analytical problems beyond routine, *Fresenius J. Anal. Chem.* 366 (6–7), 669–676, 2000.
51. Wang, L., Li, F., Sun, W., Wu, S., Wang, X., Zhang, L., Zheng, D., Wang, J., and Gao, Y., Concanavalin A-captured glycoproteins in healthy human urine, *Mol. Cell Proteomics* 5 (3), 560–562, 2006.
52. White, K. Y., Rodemich, L., Nyalwidhe, J. O., Comunale, M. A., Clements, M. A., Lance, R. S., Schellhammer, P. F., Mehta, A. S., Semmes, O. J., and Drake, R. R., Glycomic characterization of prostate-specific antigen and prostatic acid phosphatase in prostate cancer and benign disease seminal plasma fluids, *J. Proteome Res.* 7, 7, 2009.
53. Bonner, P. L. R., *Protein Purification*, Taylor & Francis, New York, 2007.
54. Donnelly, E. H. and Goldstein, I. J., Glutaraldehyde-insolubilized concanavalin A: An adsorbent for the specific isolation of polysaccharides and glycoproteins, *Biochem. J.* 118 (4), 679–680, 1970.
55. Qiu, R. and Regnier, F. E., Use of multidimensional lectin affinity chromatography in differential glycoproteomics, *Anal. Chem.* 77 (9), 2802–2809, 2005.
56. Monzo, A., Bonn, G. K., and Guttman, A., Boronic acid-lectin affinity chromatography. 1. Simultaneous glycoprotein binding with selective or combined elution, *Anal. Bioanal. Chem.* 389 (7–8), 2097–2102, 2007.
57. Mechref, Y. and Novotny, M. V., Structural investigations of glycoconjugates at high sensitivity, *Chem. Rev.* 102 (2), 321–369, 2002.
58. Bunkenborg, J., Pilch, B. J., Podtelejnikov, A. V., and Wisniewski, J. R., Screening for N-glycosylated proteins by liquid chromatography mass spectrometry, *Proteomics* 4 (2), 454–465, 2004.
59. Chen, H., Xu, X., Lin, H. H., Chen, S. H., Forsman, A., Aasa-Chapman, M., and Jones, I. M., Mapping the immune response to the outer domain of a human immunodeficiency virus-1 clade C gp120, *J. Gen. Virol.* 89 (Pt 10), 2597–2604, 2008.
60. Hirabayashi, J., Concept, strategy and realization of lectin-based glycan profiling, *J. Biochem.* 144 (2), 139–147, 2008.
61. Hirabayashi, J. and Kasai, K., Separation technologies for glycomics, *J. Chromatogr. B Anal. Technol. Biomed. Life Sci.* 771 (1–2), 67–87, 2002.

62. Wiener, M. C. and van Hoek, A. N., A lectin screening method for membrane glycoproteins: Application to the human CHIP28 water channel (AQP-1), *Anal. Biochem.* 241 (2), 267–268, 1996.
63. Sparbier, K., Koch, S., Kessler, I., Wenzel, T., and Kostrzewa, M., Selective isolation of glycoproteins and glycopeptides for MALDI-TOF MS detection supported by magnetic particles, *J. Biomol. Tech.* 16 (4), 407–413, 2005.
64. Wang, Y., Wu, S. L., and Hancock, W. S., Approaches to the study of N-linked glycoproteins in human plasma using lectin affinity chromatography and nano-HPLC coupled to electrospray linear ion trap—Fourier transform mass spectrometry, *Glycobiology* 16 (6), 514–523, 2006.
65. Yang, Z. and Hancock, W. S., Approach to the comprehensive analysis of glycoproteins isolated from human serum using a multi-lectin affinity column, *J. Chromatogr. A* 1053 (1–2), 79–88, 2004.
66. Mega, T., Oku, H., and Hase, S., Characterization of carbohydrate-binding specificity of concanavalin A by competitive binding of pyridylamino sugar chains, *J. Biochem.* 111 (3), 396–400, 1992.
67. Kamoda, S. and Kakehi, K., Evaluation of glycosylation for quality assurance of antibody pharmaceuticals by capillary electrophoresis, *Electrophoresis* 29 (17), 3595–3604, 2008.
68. Balaguer, E. and Neususs, C., Glycoprotein characterization combining intact protein and glycan analysis by capillary electrophoresis-electrospray ionization-mass spectrometry, *Anal. Chem.* 78 (15), 5384–5393, 2006.
69. Dolnik, V. and Liu, S., Applications of capillary electrophoresis on microchip, *J. Sep. Sci.* 28 (15), 1994–2009, 2005.
70. Ramdani, B., Nuyens, V., Codden, T., Perpete, G., Colicis, J., Lenaerts, A., Henry, J. P., and Legros, F. J., Analyte comigrating with trisialotransferrin during capillary zone electrophoresis of sera from patients with cancer, *Clin. Chem.* 49 (11), 1854–1864, 2003.
71. Mischak, H., Coon, J. J., Novak, J., Weissinger, E. M., Schanstra, J. P., and Dominiczak, A. F., Capillary electrophoresis-mass spectrometry as a powerful tool in biomarker discovery and clinical diagnosis: An update of recent developments, *Mass Spectrom. Rev.* 30, 30, 2008.
72. Mechref, Y. and Novotny, M. V., Glycomic analysis by capillary electrophoresis-mass spectrometry, *Mass Spectrom. Rev.* 30, 30, 2008.
73. Amon, S., Zamfir, A. D., and Rizzi, A., Glycosylation analysis of glycoproteins and proteoglycans using capillary electrophoresis-mass spectrometry strategies, *Electrophoresis* 29 (12), 2485–2507, 2008.
74. Sanz-Nebot, V., Balaguer, E., Benavente, F., Neususs, C., and Barbosa, J., Characterization of transferrin glycoforms in human serum by CE-UV and CE-ESI-MS, *Electrophoresis* 28 (12), 1949–1957, 2007.
75. Issaq, H. and Veenstra, T., Two-dimensional polyacrylamide gel electrophoresis (2D-PAGE): Advances and perspectives, *Biotechniques* 44 (5), 697–698, 700, 2008.
76. Carrette, O., Burkhard, P. R., Sanchez, J. C., and Hochstrasser, D. F., State-of-the-art two-dimensional gel electrophoresis: A key tool of proteomics research, *Nat. Protoc.* 1 (2), 812–823, 2006.
77. Loster, K. and Kannicht, C., 2-dimensional electrophoresis: Detection of glycosylation and influence on spot pattern, *Methods Mol. Biol.* 446, 199–214, 2008.
78. Miller, I., Crawford, J., and Gianazza, E., Protein stains for proteomic applications: Which, when, why?, *Proteomics* 6 (20), 5385–5408, 2006.
79. Wu, J., Lenchik, N. J., Pabst, M. J., Solomon, S. S., Shull, J., and Gerling, I. C., Functional characterization of two-dimensional gel-separated proteins using sequential staining, *Electrophoresis* 26 (1), 225–237, 2005.

80. Calvano, C. D., Zambonin, C. G., and Jensen, O. N., Assessment of lectin and HILIC based enrichment protocols for characterization of serum glycoproteins by mass spectrometry, *J. Proteomics* 71 (3), 304–317, 2008.
81. Schneider, U., Protein concentration by hydrophilic interaction chromatography combined with solid phase extraction, *Methods Mol. Biol.* 424, 63–9, 2008.
82. Clontech, Glycoprotein enrichment and detection, *Clontechniques* 23 (3), 6–8, 2008.
83. Brena, B. M., Batista-Viera, F., Ryden, L., and Porath, J., Selective adsorption of immunoglobulins and glucosylated proteins on phenylboronate-agarose, *J. Chromatogr.* 604 (1), 109–115, 1992.
84. Zanette, D., Soffientini, A., Sottani, C., and Sarubbi, E., Evaluation of phenylboronate agarose for industrial-scale purification of erythropoietin from mammalian cell cultures, *J. Biotechnol.* 101 (3), 275–287, 2003.
85. Sun, B., Ranish, J. A., Utleg, A. G., White, J. T., Yan, X., Lin, B., and Hood, L., Shotgun glycopeptide capture approach coupled with mass spectrometry for comprehensive glycoproteomics, *Mol. Cell Proteomics* 6 (1), 141–149, 2007.
86. Liu, X., Ma, L., and Li, J. J., Recent developments in the enrichment of glycopeptides for glycoproteomics, *Anal. Lett.* 41 (2), 268–277, 2008.
87. Issaq, H. J., Xiao, Z., and Veenstra, T. D., Serum and plasma proteomics, *Chem. Rev.* 107 (8), 3601–3620, 2007.
88. Alpert, A. J., Hydrophilic-interaction chromatography for the separation of peptides, nucleic acids and other polar compounds, *J. Chromatogr.* 499, 177–196, 1990.
89. Viklund, C., Sjogren, A., Irgum, K., and Nes, I., Chromatographic interactions between proteins and sulfoalkylbetaine-based zwitterionic copolymers in fully aqueous low-salt buffers, *Anal. Chem.* 73 (3), 444–452, 2001.
90. Wuhrer, M., Catalina, M. I., Deelder, A. M., and Hokke, C. H., Glycoproteomics based on tandem mass spectrometry of glycopeptides, *J. Chromatogr. B Anal. Technol. Biomed. Life Sci.* 849 (1–2), 115–128, 2007.
91. Zhang, J. and Wang, D. I., Quantitative analysis and process monitoring of site-specific glycosylation microheterogeneity in recombinant human interferon-gamma from Chinese hamster ovary cell culture by hydrophilic interaction chromatography, *J. Chromatogr. B Biomed. Sci. Appl.* 712 (1–2), 73–82, 1998.
92. SeQuant, *A Practical Guide to HILIC*, Umea, Sweden, 2008.
93. Hagglund, P., Matthiesen, R., Elortza, F., Hojrup, P., Roepstorff, P., Jensen, O. N., and Bunkenborg, J., An enzymatic deglycosylation scheme enabling identification of core fucosylated N-glycans and O-glycosylation site mapping of human plasma proteins, *J. Proteome Res.* 6 (8), 3021–3031, 2007.
94. Kaji, H., Yamauchi, Y., Takahashi, N., and Isobe, T., Mass spectrometric identification of N-linked glycopeptides using lectin-mediated affinity capture and glycosylation site-specific stable isotope tagging, *Nat. Protoc.* 1 (6), 3019–3027, 2006.
95. Kaji, H., Kamiie, J., Kawakami, H., Kido, K., Yamauchi, Y., Shinkawa, T., Taoka, M., Takahashi, N., and Isobe, T., Proteomics reveals N-linked glycoprotein diversity in *Caenorhabditis elegans* and suggests an atypical translocation mechanism for integral membrane proteins, *Mol. Cell Proteomics* 6 (12), 2100–2109, 2007.
96. Thaysen-Andersen, M., Thogersen, I. B., Nielsen, H. J., Lademann, U., Brunner, N., Enghild, J. J., and Hojrup, P., Rapid and individual-specific glycoprofiling of the low abundance N-glycosylated protein tissue inhibitor of metalloproteinases-1, *Mol. Cell Proteomics* 6 (4), 638–647, 2007.
97. Omaetxebarria, M. J., Hagglund, P., Elortza, F., Hooper, N. M., Arizmendi, J. M., and Jensen, O. N., Isolation and characterization of glycosylphosphatidylinositol-anchored peptides by hydrophilic interaction chromatography and MALDI tandem mass spectrometry, *Anal. Chem.* 78 (10), 3335–3341, 2006.

98. Boersema, P. J., Mohammed, S., and Heck, A. J., Hydrophilic interaction liquid chromatography (HILIC) in proteomics, *Anal. Bioanal. Chem.* 391 (1), 151–159, 2008.
99. Yu, Y. Q., Fournier, J., Gilar, M., and Gebler, J. C., Identification of N-linked glycosylation sites using glycoprotein digestion with pronase prior to MALDI tandem time-of-flight mass spectrometry, *Anal. Chem.* 79 (4), 1731–1738, 2007.
100. Wuhrer, M., Koeleman, C. A., Hokke, C. H., and Deelder, A. M., Protein glycosylation analyzed by normal-phase nano-liquid chromatography—mass spectrometry of glycopeptides, *Anal. Chem.* 77 (3), 886–894, 2005.
101. Wada, Y., Tajiri, M., and Yoshida, S., Hydrophilic affinity isolation and MALDI multiple-stage tandem mass spectrometry of glycopeptides for glycoproteomics, *Anal. Chem.* 76 (22), 6560–6565, 2004.
102. Tajiri, M., Yoshida, S., and Wada, Y., Differential analysis of site-specific glycans on plasma and cellular fibronectins: Application of a hydrophilic affinity method for glycopeptide enrichment, *Glycobiology* 15 (12), 1332–1340, 2005.
103. Demelbauer, U. M., Zehl, M., Plematl, A., Allmaier, G., and Rizzi, A., Determination of glycopeptide structures by multistage mass spectrometry with low-energy collision-induced dissociation: Comparison of electrospray ionization quadrupole ion trap and matrix-assisted laser desorption/ionization quadrupole ion trap reflectron time-of-flight approaches, *Rapid Commun. Mass Spectrom.* 18 (14), 1575–1582, 2004.
104. Uematsu, R., Furukawa, J., Nakagawa, H., Shinohara, Y., Deguchi, K., Monde, K., and Nishimura, S., High throughput quantitative glycomics and glycoform-focused proteomics of murine dermis and epidermis, *Mol. Cell Proteomics* 4 (12), 1977–1989, 2005.
105. Yang, Z. and Hancock, W. S., Monitoring glycosylation pattern changes of glycoproteins using multi-lectin affinity chromatography, *J. Chromatogr. A* 1070 (1–2), 57–64, 2005.
106. Schwientek, T., Mandel, U., Roth, U., Muller, S., and Hanisch, F. G., A serial lectin approach to the mucin-type O-glycoproteome of Drosophila melanogaster S2 cells, *Proteomics* 7 (18), 3264–3277, 2007.
107. Xiong, L. and Regnier, F. E., Use of a lectin affinity selector in the search for unusual glycosylation in proteomics, *J. Chromatogr. B Anal. Technol. Biomed. Life Sci.* 782 (1–2), 405–418, 2002.
108. Hirabayashi, J., Hayama, K., Kaji, H., Isobe, T., and Kasai, K., Affinity capturing and gene assignment of soluble glycoproteins produced by the nematode *Caenorhabditis elegans*, *J. Biochem.* 132 (1), 103–114, 2002.
109. Zhang, H., Li, X. J., Martin, D. B., and Aebersold, R., Identification and quantification of N-linked glycoproteins using hydrazide chemistry, stable isotope labeling and mass spectrometry, *Nat. Biotechnol.* 21 (6), 660–666, 2003.
110. McDonald, C. A., Yang, J. Y., Marathe, V., Yen, T. Y., and Macher, B. A., Combining results from lectin affinity chromatography and glycocapture approaches substantially improves the coverage of the glycoproteome, *Mol. Cell Proteomics* 8 (2), 287–301, 2009.
111. Liu, T., Qian, W. J., Gritsenko, M. A., Camp, D. G., II, Monroe, M. E., Moore, R. J., and Smith, R. D., Human plasma N-glycoproteome analysis by immunoaffinity subtraction, hydrazide chemistry, and mass spectrometry, *J. Proteome Res.* 4 (6), 2070–2080, 2005.
112. Ding, W., Hill, J. J., and Kelly, J., Selective enrichment of glycopeptides from glycoprotein digests using ion-pairing normal-phase liquid chromatography, *Anal. Chem.* 79 (23), 8891–8899, 2007.
113. Alvarez-Manilla, G., Atwood, J., III, Guo, Y., Warren, N. L., Orlando, R., and Pierce, M., Tools for glycoproteomic analysis: Size exclusion chromatography facilitates identification of tryptic glycopeptides with N-linked glycosylation sites, *J. Proteome Res.* 5 (3), 701–708, 2006.

114. Larsen, M. R., Hojrup, P., and Roepstorff, P., Characterization of gel-separated glycoproteins using two-step proteolytic digestion combined with sequential microcolumns and mass spectrometry, *Mol. Cell Proteomics* 4 (2), 107–119, 2005.
115. Bindila, L., Peter-Katalinic, J., and Zamfir, A., Sheathless reverse-polarity capillary electrophoresis-electrospray-mass spectrometry for analysis of underivatized glycoconjugates, *Electrophoresis* 26 (7–8), 1488–1499, 2005.
116. Bond, M. R. and Kohler, J. J., Chemical methods for glycoprotein discovery, *Curr. Opin. Chem. Biol.* 11 (1), 52–58, 2007.
117. Xu, Y. W., Wu, Z. X., Zhang, L. J., Lu, H. J., Yang, P. Y., Webley, P. A., and Zhao, D. Y., Highly specific enrichment of glycopeptides using boronic acid-functionalized mesoporous silica, *Anal. Chem.* 81 (1), 503–508, 2009.
118. Khidekel, N., Arndt, S., Lamarre-Vincent, N., Lippert, A., Poulin-Kerstien, K. G., Ramakrishnan, B., Qasba, P. K., and Hsieh-Wilson, L. C., A chemoenzymatic approach toward the rapid and sensitive detection of O-GlcNAc posttranslational modifications, *J. Am. Chem. Soc.* 125 (52), 16162–16163, 2003.
119. Khidekel, N., Ficarro, S. B., Peters, E. C., and Hsieh-Wilson, L. C., Exploring the O-GlcNAc proteome: Direct identification of O-GlcNAc-modified proteins from the brain, *Proc. Natl. Acad. Sci. U.S.A.* 101 (36), 13132–13137, 2004.
120. Tai, H. C., Khidekel, N., Ficarro, S. B., Peters, E. C., and Hsieh-Wilson, L. C., Parallel identification of O-GlcNAc-modified proteins from cell lysates, *J. Am. Chem. Soc.* 126 (34), 10500–10501, 2004.
121. Kurogochi, M., Amano, M., Fumoto, M., Takimoto, A., Kondo, H., and Nishimura, S., Reverse glycoblotting allows rapid-enrichment glycoproteomics of biopharmaceuticals and disease-related biomarkers, *Angew. Chem. Int. Ed. Engl.* 46 (46), 8808–8813, 2007.
122. Bunkenborg, J., Hagglund, P., and Jensen, O. N., Modification-specific proteomic analysis of glycoproteins in human body fluids by mass spectrometry, in *Proteomics of Human Body Fluids*, ed. Thongboonkerd, V., Humana Press, Totowa, NJ, 2007, pp. 107–128.
123. Kasicka, V., Recent developments in CE and CEC of peptides, *Electrophoresis* 29 (1), 179–206, 2008.
124. Gould, H. J. and Sutton, B. J., IgE in allergy and asthma today, *Nat. Rev. Immunol.* 8 (3), 205–217, 2008.
125. Kraft, S., Rana, S., Jouvin, M. H., and Kinet, J. P., The role of the FcepsilonRI beta-chain in allergic diseases, *Int. Arch. Allergy Immunol.* 135 (1), 62–72, 2004.
126. Garman, S. C., Wurzburg, B. A., Tarchevskaya, S. S., Kinet, J. P., and Jardetzky, T. S., Structure of the Fc fragment of human IgE bound to its high-affinity receptor Fc epsilonRI alpha, *Nature* 406 (6793), 259–266, 2000.
127. Chang, V. T., Crispin, M., Aricescu, A. R., Harvey, D. J., Nettleship, J. E., Fennelly, J. A., Yu, C., Boles, K. S., Evans, E. J., Stuart, D. I., Dwek, R. A., Jones, E. Y., Owens, R. J., and Davis, S. J., Glycoprotein structural genomics: Solving the glycosylation problem, *Structure* 15 (3), 267–273, 2007.
128. Nettleship, J. E., Rahman-Huq, N., and Owens, R. J., The production of glycoproteins by transient expression in Mammalian cells, *Methods Mol. Biol.* 498, 245–263, 2009.
129. Takegawa, Y., Deguchi, K., Ito, H., Keira, T., Nakagawa, H., and Nishimura, S., Simple separation of isomeric sialylated N-glycopeptides by a zwitterionic type of hydrophilic interaction chromatography, *J. Sep. Sci.* 29 (16), 2533–2540, 2006.
130. Hopp, T. P. and Woods, K. R., Prediction of protein antigenic determinants from amino acid sequences, *Proc. Natl. Acad. Sci. U.S.A.* 78 (6), 3824–3828, 1981.
131. Letourneur, O., Sechi, S., Willette-Brown, J., Robertson, M. W., and Kinet, J. P., Glycosylation of human truncated Fc epsilon RI alpha chain is necessary for efficient folding in the endoplasmic reticulum, *J. Biol. Chem.* 270 (14), 8249–8256, 1995.

132. Perkins, D. N., Pappin, D. J., Creasy, D. M., and Cottrell, J. S., Probability-based protein identification by searching sequence databases using mass spectrometry data, *Electrophoresis* 20 (18), 3551–3567, 1999.
133. An, H. J., Peavy, T. R., Hedrick, J. L., and Lebrilla, C. B., Determination of N-glycosylation sites and site heterogeneity in glycoproteins, *Anal. Chem.* 75 (20), 5628–5637, 2003.
134. Dodds, E. D., Seipert, R. R., Clowers, B. H., German, J. B., and Lebrilla, C. B., Analytical performance of immobilized pronase for glycopeptide footprinting and implications for surpassing reductionist glycoproteomics, *J. Proteome Res.* 8 (2), 502–512, 2009.
135. Lacunza, I., Kremmer, T., Diez-Masa, J. C., Sanz, J., and de Frutos, M., Comparison of alpha-1-acid glycoprotein isoforms from healthy and cancer patients by capilliary IEF, *Electrophoresis* 28 (23), 4447–4451, 2007.
136. Schwarzer, J., Rapp, E., and Reichl, U., N-glycan analysis by CGE-LIF: profiling influenza A virus hemagglutinin N-glycosylation during vaccine production, *Electrophoresis* 29 (20), 4203–4214, 2008.
137. Romisch, J., Donges, R., Stauss, H., Inthorn, D., Muhlbayer, D., Jochum, M., and Hoffmann, J. N., Quantification of antithrombin isoform proportions in plasma samples of healthy subjects, sepsis patients, and in antithrombin concentrates, *Pathophysiol. Haemost. Thromb.* 32 (3), 143–150, 2002.
138. Block, T. M., Camunale, M. A., Lowman, M., Steel, L. F., Romano, P. R., Fimmel, C., Tennant, B. C., London, W. T., Evans, A. A., Blumberg, B. S., Dwek, R. A., Mattu, T. S., and Mehta, A. S., Use of targeted glycoproteins that correlate with liver cancer in woodchucks and humans, *Proc. Natl. Acad. Sci. U.S.A.* 102 (3), 779–784, 2005.

23 Analysis of Protein Glycosylation and Phosphorylation Using HILIC-MS

Morten Thaysen-Andersen, Kasper Engholm-Keller, and Peter Roepstorff

CONTENTS

23.1 INTRODUCTION

Glycosylation and phosphorylation represent the majority of posttranslational protein modifications and it is widely accepted that they are substantially involved in numerous essential cellular processes such as inter- and intracellular signaling, metabolism, protein synthesis and degradation, and cell survival. For this reason, functional glycomics, glycoproteomics, and phosphoproteomics are rapidly growing research areas. One of the most prominent challenges associated with these disciplines is the substoichiometric presence of these modifications, resulting from substantial heterogeneity of glycosylation as well as frequently a low degree of phosphorylation of a given site. Fractionation or enrichment of the modified proteins is consequently essential to alleviate this problem. However, the significant hydrophilicities associated with these biomolecules limit the use of traditional purification techniques such as reversed-phase liquid chromatography (RPLC) setups. In contrast, hydrophilic interaction chromatography (HILIC), which features the opposite characteristics of RPLC, is an attractive technique for the analysis of hydrophilic compounds and applications usually involve detection by mass spectrometry (MS) to benefit from its high sensitivity, accuracy, and resolution as well as its high throughput potential.

In this chapter, we describe various HILIC applications for detailed studies of glycans, glycopeptides, and phosphopeptides. The applications range from sample preparations using HILIC in solid-phase extraction (SPE) formats to chromatographic separation using analytical and capillary-scale HILIC columns with both off- and online detection. The advantages of combining HILIC strategies with downstream MS detection are emphasized.

23.2 BACKGROUND

Following synthesis, proteins are often modified by covalent attachment of various chemical groups. These posttranslational modifications (PTMs) can potentially alter the physicochemical properties of the original proteins and thereby modulate their activities. In particular, glycosylation and phosphorylation, which occur frequently on proteins, are known to be involved in a multitude of cellular processes.

Protein glycosylation is the attachment of carbohydrate moieties (glycans) to the polypeptide backbone and includes two major glycosylation types although several other less abundant types are known. In *N*-linked glycosylation, the glycans are linked to asparagine residues in the restricted sequence Asn-Xaa-Thr/Ser/Cys, where Xaa is any residue except for proline. In contrast, no consensus sequence is known for *O*-linked glycans, which are linked to the polypeptide chain through serine and threonine residues. Often substantial heterogeneity arises from the linkage of numerous glycan structures to a given glycosylation site. It is estimated that more than half of all proteins are glycosylated[1] and the biological roles of glycosylation are extremely diverse, spanning the spectrum from conformational stability and protection against degradation to molecular and cellular recognition in development, growth, and cellular communication.[2]

Protein phosphorylation is the covalent coupling of a phosphate group to an amino acid side chain in a polypeptide backbone and is catalyzed by protein kinases, which accounts for ~1.7% of the genes in the human genome.[3] It has been estimated that up to 30% of the proteins in eukaryotes are phosphorylated.[4,5] In mammalian cells, phosphorylation occur on the amino acids serine, threonine, and tyrosine, while the basic amino acids histidine, arginine, and lysine are also prone to phosphorylation in prokaryotes. The relative abundances of phosphoserine, phosphothreonine, and phosphotyrosine, have been estimated to be approximately 90%, 10%, and 0.05%, respectively.[6] Reversible phosphorylation of proteins is an essential regulatory mechanism, which can alter catalytic activity, stability, and interaction with other biomolecules and is thus involved in most cellular processes such as inter-/intracellular signaling, metabolism, protein synthesis and degradation, and apoptosis.[7]

Considering the significant involvement of protein glycosylation and phosphorylation in molecular/cellular processes and their abundances, it is no surprise that these two PTMs have been intensely investigated. Although, understanding of the role of the modifications is beginning to emerge on the individual protein level, the complete involvement of the PTMs is far from known. One of the main challenges in deciphering the PTM code arises from the fact that the modifications are produced by enzymatic processing (non-template driven), meaning that the level and type of modification of a given protein can vary depending on cell-type and the physiological condition of the cell. The heterogeneity and substoichiometry of the PTMs is another obstacle limiting the characterization, which makes sample enrichment essential before PTM proteins analysis. Furthermore, the hydrophilicity associated with the peptides bearing phosphorylations or glycosylations is posing yet another challenge, as most traditional techniques such as reversed-phase liquid chromatography are designed for analyzing molecules based on hydrophobicity. Together these challenges call for sensitive, robust and reproducible techniques that allow separation and enrichment of peptides containing hydrophilic PTMs.

HILIC in conjunction with MS represents such a technique. The reversed retention characteristics compared to RPLC,[8] makes HILIC ideal for the analysis of polar and hydrophilic compounds such as glycosylated and phosphorylated species. An additional benefit is that hydrophilic analytes have relatively high solubility in the polar aqueous/organic mobile phases used in HILIC. Furthermore, the use of high concentration of organic solvent, mainly acetonitrile (ACN) and volatile buffer systems is ideal for downstream MS detection in an on- or off-line configuration. HILIC has experienced a sudden increase in popularity,[9] which has been promoted by the need to analyze hydrophilic analytes in mixtures and the increased use of LC-MS. This, in turn, has been induced by some advances on the MS instrumental side e.g., development of fast and accurate mass spectrometers with high resolving power (e.g., Orbitrap) and the introduction of new fragmentation techniques (e.g., electron transfer dissociation [ETD]), which has been shown to be very useful for fragmentation of peptides containing labile PTMs such as phosphorylation and glycosylation. As a consequence, the HILIC-MS combination represents an attractive approach to study hydrophilic and labile PTMs, and a number of new applications have been published over the last 5–6 years. These include applications for studying protein glycosylation and phophorylation spanning from sample enrichment and cleanup during sample preparation to traditional chromatographic setups in which modified peptides are separated using HILIC in an off- or online combination with MS.

23.3 ANALYSIS OF GLYCANS AND GLYCOPEPTIDES USING HILIC

23.3.1 HILIC SPE for Sample Preparation of Glycans and Glycopeptides

The combination of a relative poor ionization efficiency of glycopeptides with respect to the unmodified peptides and the "dilution" of signals resulting from the distribution of multiple glycoforms at the linkage sites, makes glycopeptide enrichment essential prior to MS detection. Among various other sample preparation techniques i.e., lectins,[10,11] graphitized carbon,[12,13] titanium dioxide (TiO_2) (for negatively charged compounds),[14] hydrazide chemistry[15] and boronic acids,[16] which are described in detail in recent reviews.[17,18] HILIC SPE has been used increasingly to enrich for glycopeptides. The rationale is that the hydrophilic contribution from the glycan moiety is often sufficient to generate a rather unique overall hydrophilicity among the glycopeptides, which can be used as a parameter to separate them from the less hydrophilic non-glycosylated peptides on polar stationary phases using hydrophilic interactions. The hydrophilicity of glycans is primarily a result of the large number of hydroxy groups, and, as the hydrophilic contribution increases with glycan size (and charge), applications have mainly been presented for *N*-linked glycopeptides, since *N*-glycans tend to be larger than *O*-linked glycans. For traditional glycomic approaches, HILIC SPE also represents an ideal technique for clean-up (desalting) of released glycans prior to MS.

The HILIC enrichment has been performed utilizing a variety of stationary phases and SPE formats on samples ranging from purified glycoproteins to extremely complex matrix, such as serum and other body fluids. Examples of various commercial and noncommercial SPE formats are shown in Figure 23.1.

When choosing a format, it is important to consider the type of stationary phase, column/membrane compatibility, and loading capacity/speed as well as mobile phase. Ideally, it is preferred that the non-glycosylated peptides appear in the flow-through, while the glycopeptides are selectively retained on the column and can be eluted separately. As salts and detergents are eluted in the flow-through/wash fractions, desalting of the sample is simultaneously performed. Desalting is important, as salts interfere with the MS detection if present. Solvents used for HILIC SPE purifications are usually similar to the ones used in regular HILIC separation (see Section 23.3.2 HILIC separation of glycans and glycopeptides), but vary slightly depending on the specific experiment and the downstream detection methods.

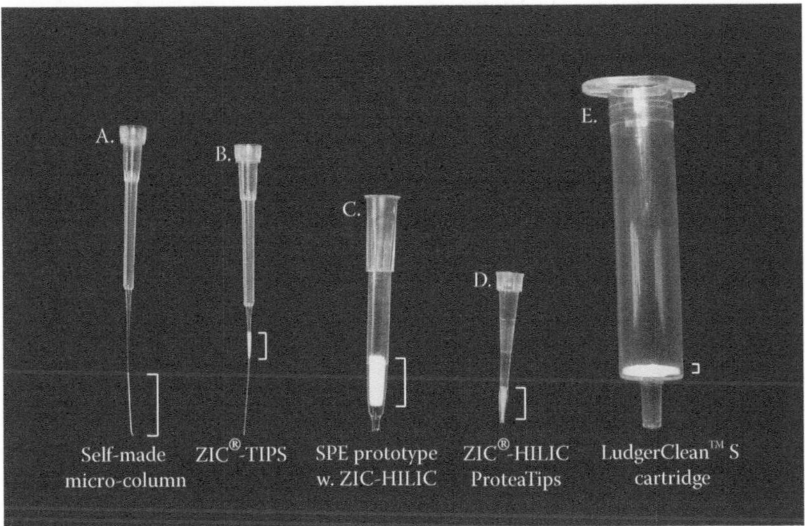

FIGURE 23.1 Examples of HILIC SPE formats. (A) Self-made micro-column packed in GELoader tips (Eppendorf, Germany). See Thaysen-Andersen et al.[20] for details on column preparation and handling. Various HILIC SPE materials and column volumes can be used, which are the major advantages of this format when optimizing the purifications for individual experiments. (B) The format of the commercial available ZIC®-TIPS (Merck Sequant, Umeå, Sweden) is similar to the self-made micro-columns; however, the zwitter-ionic HILIC SPE material (ZIC-HILIC) forms a column higher in the tip. (C) This SPE prototype from Merck Sequant (discontinued format) has a higher column volume and consequently a higher capacity. (D) ZIC®-HILIC ProteaTips (Protea Biosciences, Morgantown, WV) is a variant of the ZIC®-TIPS format using the same stationary phase. The tip is packed in standard P10 pipette tips and specific washing/loading and elution solvents can be purchased together with the columns. (E) LudgerClean™ (Ludger, Oxfordshire, U.K.) is a commercial purification cartridge containing a hydrophilic binding membrane. The size of the membrane enables high capacity (up to 20 μg glycan).

Therefore, it is recommended to consult the listed references for details. Following column equilibration, load and wash HILIC SPE in a suitable mobile phase of high organic content, a one-step elution is normally performed by switching to an aqueous mobile phase either with or without a low concentration of organic solvent (typically 0%–30% ACN). Thus, water is the stronger eluting solvent, and a small amount of weak acid (e.g., 0.5%–2% formic acid) is often included to generate protonated species for MS. The one-step elution is carried out since fractionation of the retained compounds often is not required. The use of MS-friendly solvents allows subsequent analysis of the enriched glycopeptide/glycan fractions by electrospray ionization MS (ESI-MS) or matrix-assisted laser desorption/ionization MS (MALDI-MS).

23.3.1.1 Selective Enrichment of Glycopeptides from Purified/Semi-Purified Glycoprotein

One very useful HILIC SPE application is the enrichment of glycopeptides from relatively simple mixtures i.e., tryptic digests of purified or semi-purified glycoproteins.

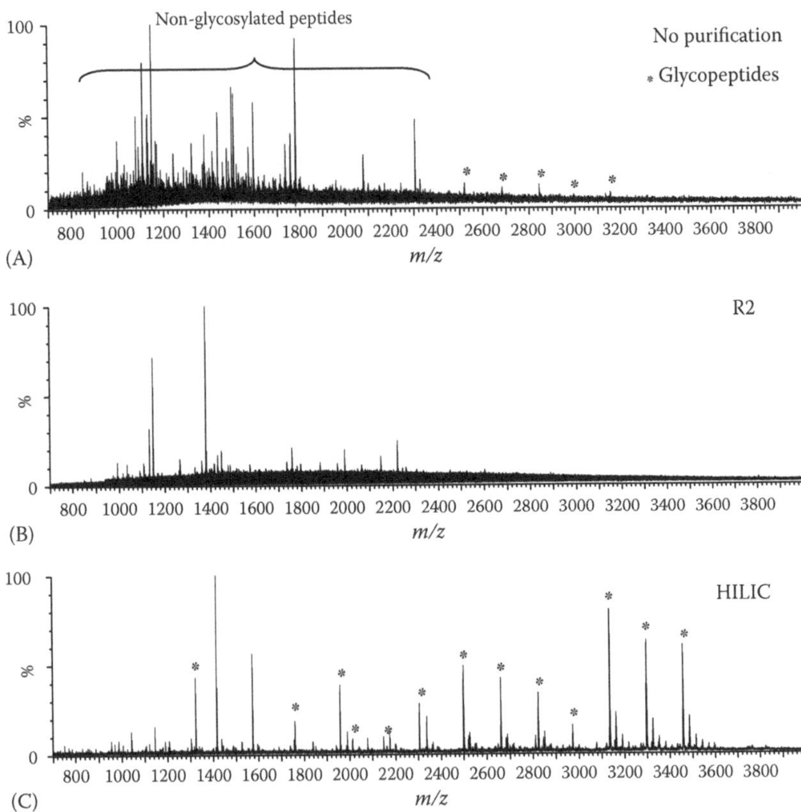

FIGURE 23.2 MALDI-Q-TOF MS of a peptide mixture generated from purified recombinant plasminogen-activator inhibitor 1 (PAI-1). Three different sample preparations are compared: (A) the peptide mixture spotted directly on the target as dried droplet with no purification, (B) R2 (hydrophobic) purification in micro-column format, and (C) ZIC-HILIC micro-column purification. The glycopeptides (marked with $*$) were barely observable without purification and completely absent in the R2 purification. In contrast, the glycopeptides were highly enriched in the HILIC SPE enabling easy characterization of the glyco-conjugates based on their mass and on the subsequent fragmentation pattern using MS/MS. Recombinant PAI-1 was kindly provided by Professor Peter Andreasen, University of Aarhus, Denmark.

Here, HILIC has been shown to be capable of selectively retaining the glycopeptides. The resulting depletion of non-glycosylated peptides is essential when MALDI-MS or infusion ESI-MS is used, where no further analyte separation is performed prior to MS. This is exemplified in Figure 23.2, where a peptide mixture obtained by tryptic digestion of the *N*-glycosylated protein plasminogen-activator inhibitor 1 (PAI-1) was prepared using three different approaches, all with MALDI-quadrupole time-of-flight (Q-TOF) MS (and MS/MS) as detection.

When the sample was spotted directly on the target without any purification, the signals for the non-glycosylated peptides were dominating, and the glycopeptides were barely detectable due to suppression effects and signal dilution. Upon desalting using the conventional hydrophobic resin (R2), the signals for the glycopeptides were completely absent, indicating that the hydrophilic glycopeptides either were non-retained on the hydrophobic stationary phase or were fully suppressed by the non-glycosylated peptides. Upon HILIC purification on micro-columns, in contrast, the signals for the glycopeptides were clearly present in the high mass region as well as some fragmentation products in the lower mass region, whereas most of the non-glycosylated peptides were depleted. This demonstrates the importance of glycopeptide enrichment. The enrichment was performed using self-made micro-columns in GELoader tips packed with zwitter-ionic material (ZIC-HILIC, Sequant/Merck, Uppsala, Sweden) (see Figure 23.1A), but other HILIC phases and formats are expected to give similar results.

The method for rapid and sensitive site-specific glycoprofiling of *N*-glycosylated proteins is illustrated with some examples shown in Figure 23.3. Salivary glycoproteins were separated by SDS-PAGE and their glycoprofiles were determined using MALDI-TOF MS and MS/MS following HILIC SPE enrichment of the glycopeptides.

Specifically, salivary tissue inhibitor of metalloproteinases-1 (TIMP-1), which has two occupied *N*-glycosylation sites, was found to be extremely heterogeneously glycosylated and it was possible to map 54 glycopeptide forms to the two sites. The glycosylation pattern of extra parotid glycoprotein (EP-GP) containing a single occupied *N*-linked site was also characterized to be heterogeneous. Both profiles were rich in glycans containing terminal fucose residues, which is a known feature of saliva glycoproteins.[19] Recently, the sensitivity of the approach was illustrated by the glycoprofiling of low abundant glycoproteins in plasma[20] and the strategy could be used to investigate the potential of glycosylated proteins as biomarkers.[21] The same approach using ZIC-HILIC SPE has also been used for the selective purification and enrichment of glycophosphatidylinosito (GPI)-anchored peptides.[22]

Other formats than the HILIC GELoader tips have been proven useful for purification. For example, Yu et al. presented an approach where a 96-well micro-elution plate was packed with 5 mg HILIC stationary phase i.e., aminopropyl silica per well.[23] In this setup, both released *N*-linked glycans and pronase-generated glycopeptides were shown to be selectively enriched and desalted using a mobile phase consisting of ACN and aqueous ammonium citrate (10 mM).[24] In another study, hydrophilic affinity separation of oligosaccharides and glycopeptides was obtained by partitioning with cellulose or Sepharose in a microcentrifuge tube in a batch mode.[25,26] Using a mobile phase consisting of 1-butanol/ethanol/water (4:1:1, v/v) impressive separation was achieved.

23.3.1.2 Enrichment and Desalting of Released Glycans

Enzymatic or chemical release of *N*- and *O*-linked glycans from glycoprotein are commonly used approaches in glycomics to determine the global glycan profile. Due to lower mass and the hydrophilic nature, the released glycans are easily purified from the proteins. Following isolation using high molecular mass cutoff filters/membranes or graphitized carbon columns, the glycans are often derivatized using a variety of methods e.g., fluorescence labeling[27] or permethylation[28] in order to enhance detection and identification of the glycans. The introduction of salts and detergents in these derivatization steps calls for thorough desalting of the sample

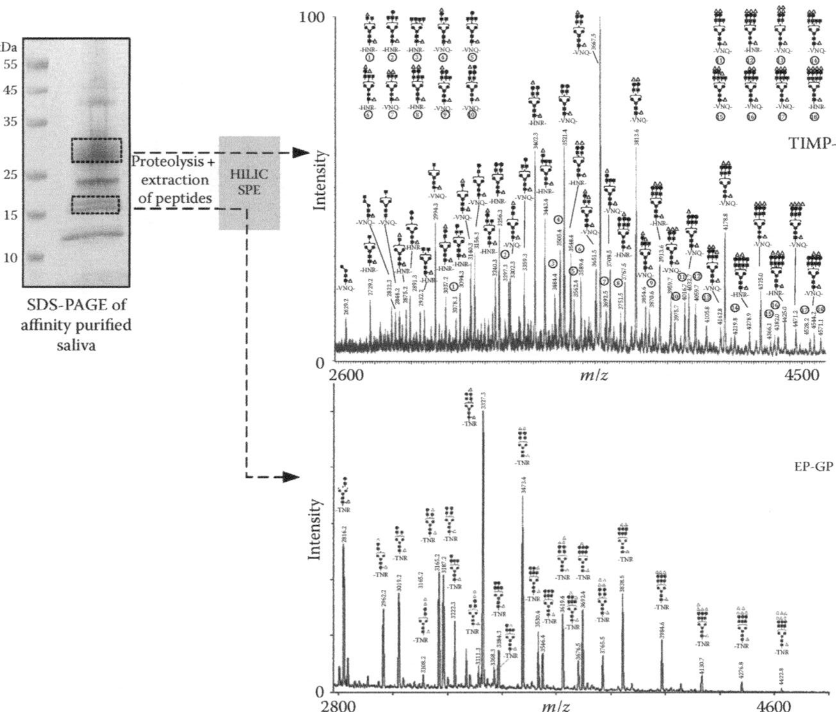

FIGURE 23.3 Rapid and sensitive site-specific glycoprofiling of affinity purified salivary *N*-glycoproteins separated on SDS-PAGE using MALDI-MS. The stained gel bands were excised, digested with trypsin, and the extracted peptide mixtures were purified using ZIC-HILIC SPE as previously described.[20] The signals for glycopeptides could be assigned based on MS and MS/MS data. The upper gel band (and mass spectrum) was found to contain tissue inhibitor of metalloproteinases-1 (TIMP-1) and the lower band contained extra parotid glycoprotein (EP-GP). Monosaccharide code: HexNAc: filled square, Hex: filled circle, Fucose: open triangle.

prior to MS. Together with graphitized carbon purification, HILIC SPE represents a widely used method for glycan cleanup and, as presented next, desalting of glycans can be performed without loss of quantitative information.

23.3.1.3 Quantitative Desalting of Glycopeptides and Glycans Using HILIC SPE

HILIC, in contrast to lectins, is considered to be a non-biased matrix that do not select for certain structural classes or subclasses of the glycome/glycoproteome. However, all sample handling, including purification, can potentially generate a bias by changing the composition of the species in a given mixture resulting in loss of quantitative information. In a recent study, a number of SPE materials including these with HILIC character were investigated for bias when glycopeptides and released (free) glycans of ribonuclease B (RNase B) were desalted.[29] Here, a fixed amount of sample (glycans/glycopeptides) was applied to HILIC columns of varying volume and the glycoprofile of the retained fraction was subsequently determined using MALDI-TOF-MS. The obtained profiles were then evaluated against a reference profile obtained by fluorescence detection of 2-aminobenzamide (2-AB) labeled glycans after separation by HPLC. For ZIC-HILIC SPE desalting of both glycans and glycopeptides, biases were introduced, when the column capacities were exceeded due to the higher ratios of sample amount versus column volume (>50 fmol/nL), as shown in Figure 23.4A,B. This was a result of competitive binding where the normally abundant Man 5–6 glycopeptides/glycans were outcompeted by the more hydrophilic Man 7–9 glycopeptides/glycans. However, using columns of sufficient capacities for the applied amount of glycan and glycopeptides (50 fmol/nL), the purifications were essentially non-biased.

The same trend was observed for a variety of other HILIC stationary phases i.e., PolyHydroxyEthyl A, TSK-Gel amide 80, PolySulfoEthyl A, and LudgerClean S, as shown in Figure 23.4C. This observation stresses the need for sufficient column capacity when performing HILIC SPE purification. For enrichment from complex mixtures, however, it should be noted that columns with capacities far greater than needed may increase the binding of non-glycosylated peptides (Thaysen-Andersen, unpublished data). Thus, the column capacity should be optimized to the analyte amount and the sample complexity.

23.3.1.4 Enrichment of Glycopeptides from Complex Mixture Using HILIC SPE

Self-made ZIC-HILIC SPE micro-columns have also been used to enrich glycopeptides from complex mixtures. This approach has mostly been used to determine *N*-glycosylation sites of glycoproteins and a general workflow, based on the available literature, is presented in Figure 23.5.

In 2004, Hagglund et al. enriched glycopeptides from lectin-purified plasma using HILIC SPE. Following the removal of the glycan moiety except for the innermost *N*-acetylglucosamine (GlcNAc) residue using endo-β-*N*-acetylglucosaminidase, 62 glycosylation sites were unambiguously identified from 37 glycoproteins using LC-ESI-MS/MS.[30] The unambiguous assignment was elicited by the unique mass

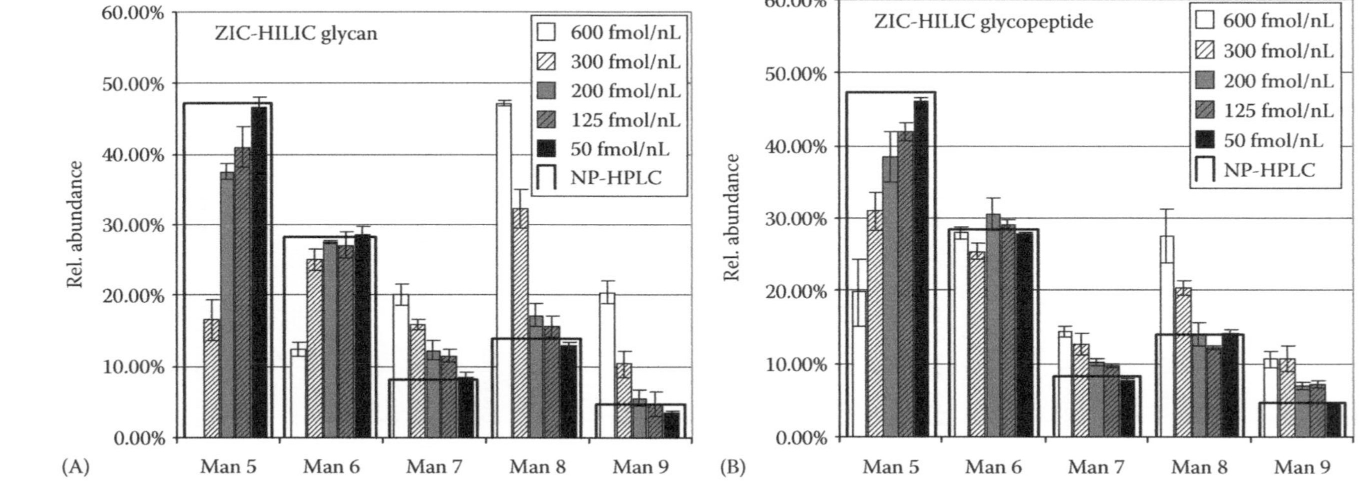

FIGURE 23.4 Test for bias of ZIC-HILIC SPE self-made micro-columns when used for desalting of bovine pancreatic RNase B (A) glycans and (B) glycopeptides. Fixed sample amounts were applied to columns of different volumes (see insert for sample amount/column volume ratios and color/pattern coding of the bars) and the retained fractions were glycoprofiled using MALDI-TOF MS. These profiles were compared to a reference HPLC profile generated using fluorescence labeled (2-AB) *N*-glycan (black outline). The correct glycoprofile as evaluated by the reference could be obtained for both the glycans and glycopeptides MS quantitation using rather low sample amount/column volume ratios (50 fmol/nL). Significant biases as illustrated by an overrepresentation of more hydrophilic glycoforms (Man 7–9) were introduced when this ratio was exceeded (>50 fmol/nL). Thus, columns of sufficient capacities were needed to avoid loss of quantitative information.

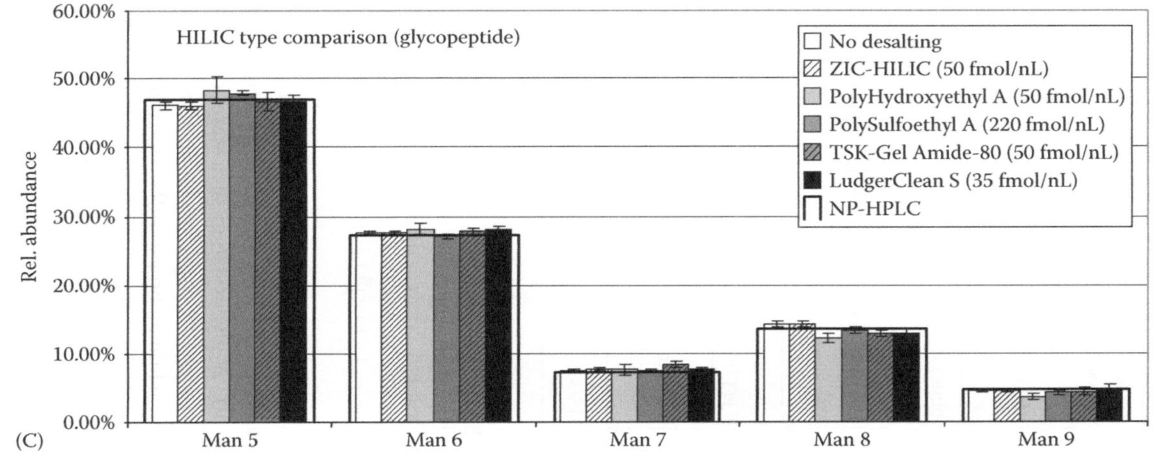

FIGURE 23.4 (continued) (C) The same tendency to introduce a bias has been shown for other HILIC SPE materials[29]; however, the correct glycoprofile could be obtained for all the materials tested when columns of sufficient capacities were used. The ratios used for obtaining non-biased profiles are listed for the individual HILIC types.

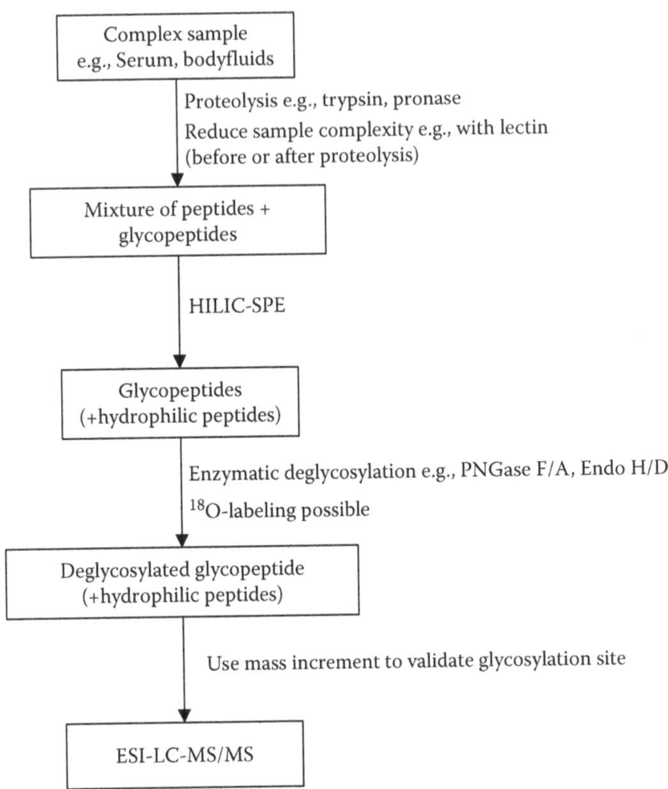

FIGURE 23.5 General workflow for the identification of *N*-linked glycosylation sites from complex samples. The enrichment of glycopeptides from the peptide mixture is essential and can be performed using HILICSPE. Following partial or complete deglycosylation using endoglycosidase H/D or PNGase F/A deglycosylated peptides can be identified in an ESI-LC-MS/MS setup. The identification is based on the mass increment associated with the deglycosylation and can be combined with O^{18} labeling.

increment of the (glyco)peptide moiety of 203 and 349 Da corresponding to the mass of the innermost GlcNAc without and with an attached fucose residue, respectively. Using the similar strategy, 103 *N*-glycosylation sites were identified from Cohn IV fraction of human plasma.[31] Lately, 63 *N*-glycosylation sites were mapped out from 32 glycoproteins obtained from human breast milk using complete enzymatic deglycosylation with *N*-glycosidase F and the mass increment of 0.98 Da as being the positive identifier of a glycosylation site in the LC-ESI-MS.[32] The major pitfall of using the asparagine-to-aspartate conversion is that the approach does not distinguish between deglycosylation-induced conversion and other causes of deamidation, either in vivo or in vitro. Deglycosylation in heavy isotopic water ($H_2{}^{18}O$) has been performed as the +2.98 Da mass shift is more easily recognized,[33] but the drawback here is that the ^{18}O also can be incorporated in the C-terminus of tryptic peptides in case of residual tryptic activity, thereby complicating the identification and increasing the false positive rate. In another study utilizing ZIC-HILIC SPE, glycopeptides were enriched from an extremely complex mixture of tryptic peptides obtained from serum with or without previous lectin purification.[34] Following enzymatic deglycosylation, the *N*-glycosylation sites were identified using both MALDI-MS and LC-ESI-MS/MS. In this study, 86 (with lectin purification) and 81 (without lectin purification) sites were identified. A limited overlap between the identified sites led to the conclusion that these two approaches were complementary.

23.3.2 HILIC Separation of Glycans and Glycopeptides

Although the separation efficiency of HILIC may not be as good as that of RP-HPLC separations (wider peaks and consequently lower resolution), HILIC is still an efficient method for retention and separation of polar compounds, such as glycans and glycopeptides. Applications include separations of different glycoforms of partially purified glycan/glycopeptide fractions with off- or online MS detection as well as more crude separations of non-glycosylated species from glyco-conjugates in prefractionation approaches. In contrast to RP-HPLC, which represents a rather homogeneous type of chromatography, the retention mechanism of HILIC varies depending on the functional groups of the stationary phase and the analyte, as well as the nature of the mobile phase. Thus, the stationary and mobile phases are essential parameters to consider for optimal analyte retention and separation. As reviewed recently, a selection of HILIC stationary phases are available and will not be covered in details here.[35] For the mobile phase, solvents with a high content of organic component (40%–97%) in water are generally used. Dissolving glycans and glycopeptides in high organic mobile phases could be difficult, and as a consequence, an 80% organic solvent water mixture is typically used in the mobile phase as starting condition. ACN is by far the most popular organic modifier that hydrates the stationary phase and generates reproducible results. At least 3% water should be present. Suitable buffers include ammonium salts of acetate and formate due to their volatibility and excellent solubility in organic solvent. These salts have been shown to minimize the electrostatic interactions (repulsions or attractions) between charged stationary phases and analytes (i.e., sialic acids).[36] Formate and other weak acids are often used as mobile phase additives for adjusting pH when using online HILIC-MS detection.

23.3.2.1 HILIC Separation of Glycans with Off-Line MS Detection

Separation and profiling of released glycans using HILIC (alternatively called NP-HPLC by other authors in literature) coupled with a fluorescence detector has been the standard technique for decades together with high-performance anion-exchange chromatography with pulsed amperometric detection.[37–39] Often analytical-scale HILIC columns are used and detection limits around 50–100 fmol for 2-AB labeled glycans have been reported.[40] In a typical setup, separation of glycans is performed on a TSK-Gel amide 80 column (TOSOH, Japan) in a mixture of high ACN amount and 50 mM ammonia formate in water, pH 4.4, as the eluent. Gradient elution is normally performed by slowly decreasing the concentration of the organic solvent, and a 120–180 min run is frequently used. Although this approach is limited in resolution, sensitivity, and throughput, it benefits from easy quantitation of the separated glycans by measuring the peak area of the eluting analytes. Another advantage is that

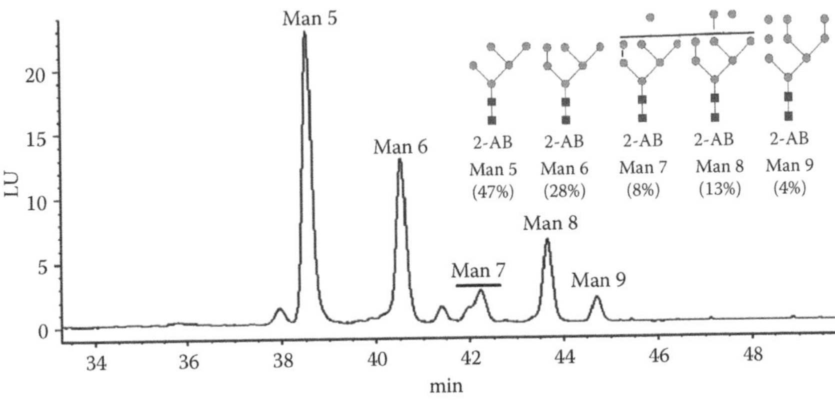

FIGURE 23.6 HILIC separation (traditionally named as NP-HPLC in the literature for this type of application) of released and 2-AB labeled RNase B *N*-glycans using a TSK-Gel amide 80 column (2 × 150 mm, 5 μm). The five glycoforms (Man 5–9), which are illustrated, could easily be separated using gradient elution. Isomeric glycan structures were observed for Man 7, which eluted as two peaks. Based on fluorescence detection, the relative abundances of the glycoforms could be determined using the peak area of the eluting analytes. Starting mobile phase of 80% ACN in aqueous 50 mM ammonium formate, pH 4.4, was used and a gradient of decreasing ACN content was introduced (not shown).

not only the size but also the structure of the glycans affect the retention time often providing information of isomeric structures, that cannot be distinguished in MS due to identical *m/z*. The identification of the eluting compounds is usually performed by matching retention times to an existing library and to a reference dextran ladder or simply by collection of the eluted fractions with subsequent MS detection. The five glycoforms of RNase B (not counting isomeric structures) could easily be separated using a TSK-Gel amide 80 HILIC column and a relatively short gradient of 60 min (Figure 23.6). The retention pattern illustrates longer retention times for the larger glycans in HILIC mode due to increased hydrophilicity.

More complex mixtures can be separated by extending the gradient or including an additional separation step. For example, HILIC fractions can be applied to a second chromatographic dimension with online MS detection. This was recently performed in a study where HILIC-separated glycans were collected and applied to RP-nano-LC-ESI-MS.[41] In another variant, capillary-scale HILIC separation of glycans has been combined with automatic spotting on a MALDI target plate with subsequent automated MS acquisition.[42,43]

23.3.2.2 HILIC Separation of Glycopeptides with Off-Line MS Detection

Although not widely used, glycopeptides can be separated by HILIC and detected by their UV absorbance. Since glycans have almost no UV absorbance, the detection is based on the absorbance of the peptide moiety. The eluted fractions are then collected and characterized using MS in an off-line setup. Figure 23.7 shows an example of this approach, where an RNase B tryptic digest was separated on a TSK-Gel amide 80 column. The non-glycosylated peptides of the mixture eluted early in the

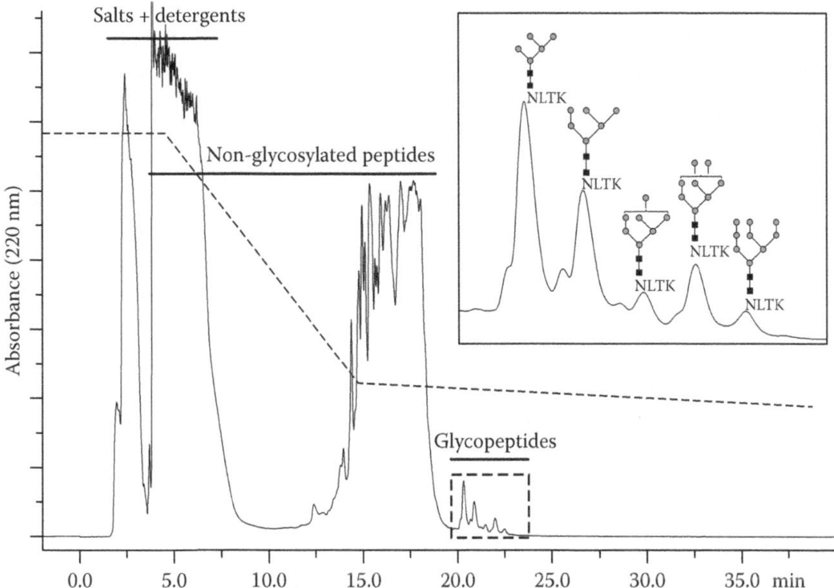

FIGURE 23.7 HILIC separation of a peptide mixture generated by trypsin digestion of bovine pancreatic RNase B. A TSK-Gel amide 80 column (2 × 150 mm, 5 μm) was used and the gradient of decreasing ACN concentration is indicated with a dotted line. Salts, detergents, and very hydrophobic peptides eluted in flow-through or very early in the gradient, whereas the majority of the non-glycosylated peptides and glycopeptides were retained fairly well on the HILIC column. The glycopeptides were fractionated from the non-glycosylated peptides by late elution and the five glycopeptides representing the individual RNase B glycoforms could be separated (see insert).

gradient, whereas the glycopeptides were better retained on the HILIC column and eluted separately. In addition, the five glycopeptides (Man 5–9) bearing the five high mannose glycoforms were resolved enabling relative quantitation of the glycoforms.

This approach has been used to quantitatively monitor variations in the microheterogeneity of glycoproteins i.e., antennary and sialylation profile of recombinant human interferon-γ from Chinese hamster ovary cell cultures.[44] Here, an initial separation of glycopeptides and non-glycosylated peptides was performed using RP-HPLC. Off-line HILIC-MS can also be used for prefractionation of glycopeptides from non-glycosylated peptides originating from crude mixtures, since glycopeptides in general will be more hydrophilic and therefore better retained on HILIC. However, due to the likely presence of hydrophilic non-glycosylated peptides in the glycopeptide fractions, a secondary separation step is required (e.g., online RPLC-MS).

23.3.2.3 Online HILIC-MS of Glycans

The use of high-polar organic/low aqueous solvent content in the HILIC mobile phase is ideal for direct ESI-MS detection, and this online approach has been used increasingly in recent years (a review by Nguyen and Schug in 2008 summarized fundamentals and applications of the technique[45]). The analysis of released glycans has been achieved with great success using online HILIC-ESI-MS with various stationary phases (e.g., ZIC-HILIC, TSK-Gel amide 80, and PolyHydroxyEthyl A), glycan types (e.g., *N*- and *O*-linked), and glycan derivatization levels (e.g., free, 2-AB or ANTS labeled)[6–48] as recently reviewed.[49] Regular HILIC conditions are used for online approaches: high concentration of organic solvent and low concentration of volatile salts and weak acids (i.e., formic acid or acetic acid). It is important, that the amount of ion pairing agents like TFA should be minimized, because it reduces the MS detection sensitivity significantly.[50] The elution of glycans is normally observed at above 50% ACN. The detection is typically performed in positive ionization mode; however, negative ionization mode has also proved efficient in particular for sialylated and sulfated glycans.[36,51] Most studies using online HILIC-ESI-MS use flow rates of 40–200 μL/min, but nanoflows (300 nL/min) have also been reported, where detection limits in the low femtomol range have been achieved.[46,48] Lately, a chip-based HILIC-ESI-MS platform has been introduced, where glycosaminoglycans are separated and profiled in negative ion mode.[52]

23.3.2.4 Online HILIC-MS of Glycopeptides

In the last few years in the field of functional glycomics and glycoproteomics, reports of glycopeptide separation using HILIC with online MS detection are moving toward site-specific approaches. Glycopeptide profiling is expected to be used increasingly in the future. Takegawa et al. showed that *N*-glycopeptides of human serum IgG could be profiled from a peptide mixture without previous enrichment using ZIC-HILIC on an analytical-scale column.[51] This was possible due to the strong retention of glycopeptides compared to the non-glycosylated peptides, which eluted early in the gradient. The same investigators tested the approach by applying a tryptic peptide mixture of α-1-acid glycoprotein to a similar HILIC-ESI-MS setup.[36] The sialylated and neutral glycopeptides were well-separated at the end of the gradient, and the glycopeptides were efficiently characterized using ESI-ion trap-MS in a negative ionization mode. Recently, capillary HILIC-ESI-MS has been used to profile 8 *O*- and 105 *N*-linked Glu-C digested glycopeptides from recombinant human erythropoietin starting from as low as 150 ng protein.[53] In a different strategy, pronase, which is a mixture of bacterial proteases, has been used to generate very small glycopeptides. Pronase is known to cleave the proteins to single amino acids and small peptides, but due to the steric hindrance of the glycan, the peptide part around the glycosylation site usually contains 3–5 amino acid residues. To demonstrate the potential of this approach, Wuhrer and coworkers showed that pronase digested RNase B glycopeptides could efficiently be analyzed by nano-HILIC-ESI-MS to identify both the peptide and the glycan.[54] In order to obtain peptide sequence information of glycopeptides using collision induced dissociation (CID) fragmentation, multistage MS/MS, is required since the first fragmentation (MS^2) often gives information only about the glycan structure. A subsequent fragmentation of the deglycosylated fragment in MS^3 or MS^4 results in peptide fragmentation information. To avoid multiple ion isolation/fragmentation cycles, ETD has been demonstrated to provide valuable peptide sequence information directly at the MS^2 stage and can consequently be used in combination with CID to yield complimentary ions in MS/MS.[17] *O*-linked glycopeptides generated from pronase digestion of the β-chain of human choriogonadotropin have similarly been applied to nano-HILIC-ESI-MS, which allowed identification of the four *O*-glycosylasion sites of the protein as well as the structure of the attached glycans.[49] Finally, an automated setup using the pronase treatment approach has recently been presented, where glycoproteins are digested on-column with pronase and the resulting glycopeptides are subsequently collected on a graphitized carbon column followed by separation and detection by HILIC-ESI-MS.[55]

23.4 ANALYSIS OF PHOSPHOPEPTIDES USING HILIC

Despite HILIC being utilized in glycopeptide analysis for a number of years, this chromatographic technique has only quite recently been extensively applied in the field of phosphopeptide analysis and fractionation. Like glycosylation, phosphorylation increases the hydrophilicity of a given peptide, making peptides containing this PTM suitable for HILIC. Due to their sub-stoichiometric abundance and generally lower ionization efficiencies compared to their non-phosphorylated counterparts,[56] phosphopeptides are rarely detected in complex mixtures, and specific enrichment strategies have therefore been developed to circumvent these problems.

23.4.1 Phosphopeptide Enrichment Methods

23.4.1.1 Immobilized Metal Affinity Chromatography

Immobilized metal affinity chromatography (IMAC) is one of the most extensively used enrichment methods prior to mass spectrometric analysis of protein phosphorylation. It was first introduced for separation of phosphoproteins,[57] but was later adapted in phosphopeptide enrichment.[58] IMAC is based on the affinity of the anionic phosphoryl group for metal ions, e.g., Fe^{3+} or Ga^{3+}. To facilitate an easy enrichment procedure, these metal ions are chelated to a solid resin such as nitrilotriacetate coated beads. Although effectively capturing phosphopeptides, IMAC suffers from co-enrichment of acidic peptides, but have nevertheless successfully been applied to phosphopeptide enrichment in several large-scale phosphoproteomics studies.[59,60]

23.4.1.2 Strong Cation Exchange

Traditionally, strong cation exchange (SCX) has been a popular first dimension in 2D LC-MS setups, whether off- or online, as in the so-called MudPIT strategy.[61] Most typical tryptic peptides only contain a single basic residue (lysine or arginine). In the acidic buffers used for SCX fractionation, the phosphorylated forms of these peptides will therefore overall be neutral or negatively charged due to the negatively charged phosphoryl group. These phosphopeptides thus elute in the flow-through or very early in the SCX gradient. This effect has successfully been exploited for phosphopeptide enrichment by Beausoleil et al.[62] However, all peptides containing histidines or missed cleavages will be positively charged and will elute later in the gradient, reducing the specificity of the method.

23.4.1.3 Titanium Dioxide

Recently, TiO_2 was introduced as a chromatographic resin with phosphopeptide-specific affinity in an online 2D-LC-MS setup.[63] As for IMAC, the resin binds acidic peptides as well as phosphopeptides, but this unspecific binding was shown to be abrogated by the use of very acidic buffers (<pH 1) and addition of specific organic acids (e.g., 2,5-dihydroxybenzoic acid).[64] Therefore, TiO_2 has gained widespread use in large-scale proteomic experiments.[65]

23.4.1.4 Other Phosphopeptide Enrichment Methods

Antibodies targeting phosphorylated amino acid residues have been successful in specific enrichment of tyrosine-phosphorylated peptides.[66,67] A phosphopeptide precipitation method using calcium phosphate prior to IMAC has also been devised.[68] Finally, phosphoproteomic researchers have developed various chemical derivatization strategies based on phosphoamidate chemistry[69] or beta elimination of the phosphoryl group on serines and threonines followed by coupling to a stationary phase.[70,71] The chemical reactions performed in these strategies, however, usually introduce further complexity into the already very complex phosphopeptide samples, and enrich only for primarily monophosphorylated peptides, and have thus had limited use in large-scale phosphoproteomics.

23.4.2 HILIC in Multidimentional LC-MS

As previously mentioned, up to 30% of all eukaryotic proteins are prone to phosphorylation at one or multiple sites. Consequently, the phosphoproteome may be as complex as the "standard" proteome. Efficient fractionation strategies, which can sort the proteomic samples according to specific physicochemical properties, are therefore required to reduce the overall complexity of the sample. Multidimensional LC peptide fractionation coupled to MS has thus become the method of choice in phosphoproteomics. In 2005, Gilar et al. investigated the orthogonality of different 2D LC-MS setups by comparing the elution profiles of a peptide mixture containing ~200 different species using size exclusion chromatrography, RPLC under acidic and basic conditions, SCX, and HILIC on a bare silica column.[72] The combination of HILIC and RPLC showed a higher peak capacity and orthogonality compared to SCX-RPLC, in which peptides eluted in distinct clusters according to their charge. Despite having not actually coupled the two chromatographic dimensions experimentally, this study showed the potential of HILIC as the first dimension in a 2D-LC setup in proteomics.

23.4.3 HILIC in Phosphorylation Analysis

23.4.3.1 Early Applications of HILIC in Phosphoprotein/Peptide Analysis

One of the first applications of HILIC for the separation of phosphorylated peptides was published in 1992. Here, HILIC was used to separate a ^{32}P-labeled phosphorylated peptide from its unreacted, non-phosphorylated counterpart, γ-^{32}P-ATP, and ^{32}P-orthophosphate after an in vitro tyrosine kinase reaction.[73] The different species in the reaction mixture were separated on a PolyHydroxyEthyl A column in triethylammonium phosphate (TEAP) buffers at pH 2.8. A gradient from 90% ACN, 4 mM TEAP to 0% ACN, 10 mM TEAP was first applied. During this gradient, the phosphorylated product could be resolved from the non-phosphorylated substrate and the ^{32}P-orthophosphate. The elution of the γ-^{32}P-ATP required an increase in TEAP concentration up to 800 mM.

Five years later, in 1997, histone H1.1 variants were fractionated on a PolyCat A, a weak cation exchange column under HILIC conditions (a methanephosphonic acid/triethylamine buffer with 70% ACN at pH 3.0).[74] Elution with an increasing sodium perchlorate gradient facilitated the separation of the histones into their distinct non-, mono-, doubly, and triply phosphorylated forms. It was shown that this separation could not be achieved running the column under regular cation exchange conditions (sodium phosphate buffer with increasing sodium perchlorate gradient). No mass spectrometric detection was used for the identification of the phosphorylation state of the histones—instead the histones were treated with alkaline phosphatase to remove the phosphate groups. By comparison of the chromatographic behavior before and after the treatment, the phosphorylation states could be deduced.

23.4.3.2 Application in Phosphoproteomics

The high potential of HILIC in fractionation of phosphopeptides was first demonstrated by McNulty and Annan.[75] Phosphopeptides were separated on a TSK-Gel amide 80 carbamoyl-derivatized silica column and analyzed by RPLC-MS on a linear ion trap MS. This fractionation resulted in a similar number of phosphopeptide identifications as obtained using SCX-RPLC-MS, but had the advantage of using volatile salt-free solvents and buffers (e.g., ACN and TFA), which were directly compatible with the next IMAC-based phosphopeptide enrichment. A reduction in sample complexity was achieved by eluting a significant fraction of the non-phosphorylated peptides in the flow through from the HILIC column, and furthermore fractionating the later eluting phosphopeptides using a decreasing organic gradient. IMAC is generally known to co-purify acidic peptides, but the reduction in sample complexity resulted in very pure phosphopeptide fractions (>99%). However, less than 10% of the phosphopeptides with multiple phosphorylation sites were identified, which is significantly less than generally observed in large-scale experiments using SCX.[60,65] In total, more than 1000 unique phosphorylation sites were identified from 300 μg of calyculin A-treated HeLa cells.

The combination of HILIC and IMAC prior to RPLC-MS was also employed in a very similar study, to investigate the phosphoproteome in the yeast *Saccharomyces cerevisiae* after DNA damage.[76] Here, the analysis resulted in the identification of more than 8000 unique phosphopeptides. However, these numbers cannot be compared directly with McNulty and Annan's, as the type of sample, sample amounts, database search algorithms, and validation criteria were not the same. The potential of the strategy was further emphasized by the analysis of two low-abundant proteins (Rad9 and Mrc1). Whereas immunoprecipitation of these specific proteins from 500 mg of yeast sample followed by phosphopeptide analysis by LC-MS (without HILIC fractionation) resulted in 40 phosphopeptide identifications from Rad9 and

31 from Mrc1, respectively. LC-MS analysis of HILIC fractionation of 6 mg sample (80 times less than that without HILIC fractionation) allowed for the identification of 21 phosphopeptides from Rad9 and 14 from Mrc1, respectively. For comparison, SCX fractionation of 50 mg of sample only resulted in five and two phosphopeptide identifications, respectively.

The buffers for HILIC and RPLC are not directly compatible. Therefore, automated online coupling of these two chromatographic dimensions cannot be achieved. However, Boersema et al. recently devised a semi-automated HILIC-RPLC-MS setup.[77] A ZIC-HILIC microliter flow system was coupled to a microliter flow-scale fraction collector to facilitate mixing of the HILIC output with formic acid buffer. Hereby, the organic concentration of HILIC fractions is decreased to a level compatible with RPLC. The distribution of peptides containing different PTMs was evaluated in this study, and while selectively enriching for *N*-acetylated peptides, no enrichment of phosphorylated peptides was achieved. As the authors speculated, the zwitter-ionic character of the functional groups on the chromatographic material might cause this effect: The negative sulfate group could potentially repel the phosphate group on the peptide, making the phosphopeptides elute along with the unmodified peptides.

23.4.4 Electrostatic Repulsion Hydrophilic Interaction Chromatography

Unless the analytes and/or the chromatographic material are completely uncharged at the pH used for the LC separation, the separation of the analytes in HILIC will be influenced by electrostatic interactions. The use of ion exchange material in HILIC mode for the retention of analytes with the same charge as the stationary phase through hydrophilic interactions has been termed electrostatic repulsion-hydrophilic interaction chromatography (ERLIC) by Alpert in 2008.[78] In this publication, ERLIC was performed on a PolyWAX LP column separating a peptide standard mixture. On this weak anion exchange column, the type of salt in the buffers strongly affected the retention of the different types of peptides in the standard mixture. Basic peptides were retained strongly in TEAP-containing buffers, while eluting very early in sodium methylphosphonate (Na-$MePO_4$) buffers. This was explained by the phosphate ion in the TEAP buffer acting as a counter-ion to the positive stationary phase. The second negative charge on the phosphate ion is free to attract positively charged basic amino acids, retaining the peptides analytes. Exchanging TEAP with Na-$MePO_4$ as the buffer salt enables negative amino acids to interact directly with the stationary phase, because methylphosphonate only has a single negative charge. Single phosphorylated peptides could be isocratically separated from non-phosphorylated peptides and eluted using 70% ACN, 20 mM Na-$MePO_4$ at pH 2.0. Multiply phosphorylated peptides, however, required a gradient of increasing salt (TEAP) and decreasing ACN concentration to elute. Hereby, peptides varying in phosphorylation number could be separated in discrete fractions.

Ytterberg et al. demonstrated the utility of HILIC and ERLIC in a micro-column setup by packing PolyHydroxyEthyl A and PolyWAX LP material in pipette tips.[79] In this work, samples from standard proteins as well as a digest of saliva proteins were selectively enriched for phosphopeptides. These were retained efficiently on the micro-columns using standard HILIC buffers. The PolyWAX LP material also used in Alpert's work showed the highest potential for phosphopeptide fractionation. The interaction between phosphopeptides and the resin could be gradually abrogated using decreasing pH, followed by decreasing ACN concentration, and finally increasing salt concentration. In this micro-column setup, Ytterberg et al. used stepwise elution to elute peptides containing different numbers of phosphorylations in separate fractions.

23.4.5 Multiphosphopeptide Enrichment Prior to HILIC Fractionation

The work of McNulty and Annan showed the high potential of HILIC as the first dimension in an off-line 2D-LC-MS setup, but also resulted in the identification of a low number of multiply phosphorylated peptides. Therefore, the evaluation of combining HILIC with a preceding multiphosphopeptide enrichment step was initiated. Such an enrichment technique, sequential elution from IMAC (SIMAC) has previously been published.[80] This technique is based on the observation that mainly monophosphorylated peptides are eluted from the IMAC resin under acidic conditions, while multiply phosphorylated peptides are retained and first eluted when using highly basic buffers. Being able to selectively separate multiphosphorylated from monophosphorylated peptides allows for separate sample preparation and mass spectrometric analysis for the two types of samples. This is advantageous because the multiphosphorylated species are particularly difficult to analyze. Adsorption to surfaces during sample handling,[81] poor ionization efficiencies, or limited peptide fragmentation in CID due to multiple neutral losses of H_3PO_4 decreases the number of identifications of these peptides. By reducing the number of handling steps and analyzing the sample with LC-MS setups specifically optimized for multiply phosphorylated peptides (e.g., multistage activation[82]), more identifications can be obtained.[80]

HILIC combined with an initial multiphosphopeptide enrichment step generating two fractions—one containing mainly multiply phosphorylated peptides and one containing non- and monophosphorylated peptides—was therefore very recently investigated.[83] The HILIC method was very similar to McNulty and Annan's except that TiO_2 was used for enrichment of phosphopeptides from the HILIC fractions instead of IMAC. Furthermore, an LTQ-Orbitrap XL capable of performing multistage activation was used for the LC-MS analysis. This method was applied to a sample from epidermal growth factor stimulated HeLa cells, which had been subjected to stable isotope labeling with amino acids in cell culture (SILAC).[84] The cells were lysed and the proteins were fractionated into membrane associated and soluble prior to proteolysis and subsequent peptide fractionation by SIMAC and HILIC. From less than 500 μg of total protein, more than 4600 unique phosphopeptides were identified. With 21% of the peptides having two or more phosphorylations, the recovery of multiple phosphorylated peptides was significantly improved. These results, along with the above-mentioned studies, emphasize the considerable potential for HILIC as an alternative to SCX as a prefractionation step in large-scale phosphoproteomics.

23.5 FUTURE PROSPECTS

The proteomic field has a strong focus on PTM characterization, because protein modification involves modulation of many protein functions. Therefore, a considerable effort has been put into expanding the classical proteomics strategies to enable PTM mapping, the so-called modification-specific proteomics. Taking the complementary nature of HILIC compared to standard chromatographic techniques into account, as well as its MS compatibility, it is not a surprise that HILIC has gained popularity in proteomics. HILIC is especially suitable for separation and enrichment of hydrophilic PTMs such as glycosylation and phosphorylation. The ability to retain peptides containing these hydrophilic modifications is a very attractive feature of HILIC and it has greatly expanded the list of biomolecules that can be analyzed using MS. However, it seems that the full potential of HILIC is not yet exploited and significant improvements must still be obtained before the glycopeptides and phosphopeptides can be routinely analyzed by HILIC-MS. In particular, it is important to establish the retention mechanisms of different types of PTMs (i.e., glycopeptides and phosphopeptides) on different HILIC phases and the influence of various solvents and additives. Understanding the binding mechanisms, the selectivity of HILIC can be optimized by designing new stationary phases and/or by using altered mobile phase conditions. For enrichment purposes, this would result in fractions with higher concentrations of the desired PTM proteins or modified peptides, and for chromatographic purposes, this knowledge would allow prediction of the conditions for optimal separation. In addition, full implementation of HILIC into the proteomic field will require development of methods with the potential for automatization/high throughput as well as high sensitivity, reproducibility, and robustness. It is evident from the data already available that HILIC is a worthy complement to RPLC. Thus, it is expected that HILIC gradually will move into the proteomic field as a new and complementary technique in the separation tool-box that will be used primarily for the analysis of hydrophilic PTMs like glycosylation and phosphorylation, but also for unmodified hydrophilic peptides that are difficult to analyze with the current existing techniques.

ACKNOWLEDGMENT

We would like to thank Tina Nielsen and Jimmy Ytterberg for contribution of data.

REFERENCES

1. Apweiler, R.; Hermjakob, H.; Sharon, N. *Biochim. Biophys. Acta* 1999, *1473*, 4–8.
2. Varki, A. *Glycobiology* 1993, *3*, 97–130.
3. Manning, G.; Plowman, G. D.; Hunter, T.; Sudarsanam, S. *Trends Biochem. Sci.* 2002, *27*, 514–520.
4. Manning, G.; Whyte, D. B.; Martinez, R.; Hunter, T.; Sudarsanam, S. *Science* 2002, *298*, 1912–1934.
5. Hubbard, M. J.; Cohen, P. *Trends Biochem. Sci.* 1993, *18*, 172–177.
6. Hunter, T.; Sefton, B. M. *Proc. Natl. Acad. Sci. U.S.A.* 1980, *77*, 1311–1315.

7. Hunter, T. *Cell* 2000, *100*, 113–127.
8. Alpert, A. J. *J. Chromatogr.* 1990, *499*, 177–196.
9. Hao, Z.; Xiao, B.; Weng, N. *J. Sep. Sci.* 2008, *31*, 1449–1464.
10. Hirabayashi, J. *Glycoconj. J.* 2004, *21*, 35–40.
11. Bunkenborg, J.; Pilch, B. J.; Podtelejnikov, A. V.; Wisniewski, J. R. *Proteomics* 2004, *4*, 454–465.
12. Packer, N. H.; Lawson, M. A.; Jardine, D. R.; Redmond, J. W. *Glycoconj. J.* 1998, *15*, 737–747.
13. Larsen, M. R.; Hojrup, P.; Roepstorff, P. *Mol. Cell. Proteomics* 2005, *4*, 107–119.
14. Larsen, M. R.; Jensen, S. S.; Jakobsen, L. A.; Heegaard, N. H. H. *Mol. Cell. Proteomics* 2007, *6*, 1778–1787.
15. Zhang, H.; Li, X. J.; Martin, D. B.; Aebersold, R. *Nat. Biotechnol.* 2003, *21*, 660–666.
16. Sparbier, K.; Koch, S.; Kessler, I.; Wenzel, T.; Kostrzewa, M. *J. Biomol. Tech.* 2005, *16*, 407–413.
17. Wuhrer, M.; Catalina, M. I.; Deelder, A. M.; Hokke, C. H. *J. Chromatogr. B Anal. Technol. Biomed. Life Sci.* 2007, *849*, 115–128.
18. Liu, X.; Ma, L.; Li, J. *J. Anal. Lett.* 2008, *41*, 268–277.
19. Guile, G. R.; Harvey, D. J.; O'Donnell, N.; Powell, A. K.; Hunter, A. P.; Zamze, S.; Fernandes, D. L.; Dwek, R. A.; Wing, D. R. *Eur. J. Biochem.* 1998, *258*, 623–656.
20. Thaysen-Andersen, M.; Thogersen, I. B.; Nielsen, H. J.; Lademann, U.; Brunner, N.; Enghild, J. J.; Hojrup, P. *Mol. Cell. Proteomics* 2007, *6*, 638–647.
21. Thaysen-Andersen, M.; Thogersen, I. B.; Lademann, U.; Offenberg, H.; Giessing, A. M.; Enghild, J. J.; Nielsen, H. J.; Brunner, N.; Hojrup, P. *Biochim. Biophys. Acta* 2008, *1784*, 455–463.
22. Omaetxebarria, M. J.; Hagglund, P.; Elortza, F.; Hooper, N. M.; Arizmendi, J. M.; Jensen, O. N. *Anal. Chem.* 2006, *78*, 3335–3341.
23. Yu, Y. Q.; Gilar, M.; Kaska, J.; Gebler, J. C. *Rapid Commun. Mass Spectrom.* 2005, *19*, 2331–2336.
24. Yu, Y. Q.; Fournier, J.; Gilar, M.; Gebler, J. C. *Anal. Chem.* 2007, *79*, 1731–1738.
25. Wada, Y.; Tajiri, M.; Yoshida, S. *Anal. Chem.* 2004, *76*, 6560–6565.
26. Tajiri, M.; Yoshida, S.; Wada, Y. *Glycobiology* 2005, *15*, 1332–1340.
27. Anumula, K. R.; Dhume, S. T. *Glycobiology* 1998, *8*, 685–694.
28. Ciucanu, I.; Costello, C. E. *J. Am. Chem. Soc.* 2003, *125*, 16213–16219.
29. Thaysen-Andersen, M.; Mysling, S.; Hojrup, P. *Anal. Chem.* 2009, *81*, 3933–3943.
30. Hagglund, P.; Bunkenborg, J.; Elortza, F.; Jensen, O. N.; Roepstorff, P. *J. Proteome Res.* 2004, *3*, 556–566.
31. Hagglund, P.; Matthiesen, R.; Elortza, F.; Hojrup, P.; Roepstorff, P.; Jensen, O. N.; Bunkenborg, J. *J. Proteome Res.* 2007, *6*, 3021–3031.
32. Picariello, G.; Ferranti, P.; Mamone, G.; Roepstorff, P.; Addeo, F. *Proteomics* 2008, *8*, 3833–3847.
33. Gonzalez, J.; Takao, T.; Hori, H.; Besada, V.; Rodriguez, R.; Padron, G.; Shimonishi, Y. *Anal. Biochem.* 1992, *205*, 151–158.
34. Calvano, C. D.; Zambonin, C. G.; Jensen, O. N. *J. Proteomics* 2008, *71*, 304–317.
35. Hemstrom, P.; Irgum, K. *J. Sep. Sci.* 2006, *29*, 1784–1821.
36. Takegawa, Y.; Deguchi, K.; Ito, H.; Keira, T.; Nakagawa, H.; Nishimura, S. I. *J. Sep. Sci.* 2006, *29*, 2533–2540.
37. Anumula, K. R. *Anal. Biochem.* 2006, *350*, 1–23.
38. Karlsson, N. G.; Hansson, G. C. *Anal. Biochem.* 1995, *224*, 538–541.
39. Wang, W. T.; Erlansson, K.; Lindh, F.; Lundgren, T.; Zopf, D. *Anal. Biochem.* 1990, *190*, 182–187.

40. Rudd, P. M.; Colominas, C.; Royle, L.; Murphy, N.; Hart, E.; Merry, A. H.; Hebestreit, H. F.; Dwek, R. A. *Proteomics* 2001, *1*, 285–294.
41. Wuhrer, M.; Koeleman, C. A.; Deelder, A. M.; Hokke, C. H. *FEBS J.* 2006, *273*, 347–361.
42. Maslen, S.; Sadowski, P.; Adam, A.; Lilley, K.; Stephens, E. *Anal. Chem.* 2006, *78*, 8491–8498.
43. Maslen, S. L.; Goubet, F.; Adam, A.; Dupree, P.; Stephens, E. *Carbohydr. Res.* 2007, *342*, 724–735.
44. Zhang, J.; Wang, D. I. *J. Chromatogr. B Biomed. Sci. Appl.* 1998, *712*, 73–82.
45. Nguyen, H. P.; Schug, K. A. *J. Sep. Sci.* 2008, *31*, 1465–1480.
46. Wuhrer, M.; Koeleman, C. A.; Deelder, A. M.; Hokke, C. H. *Anal. Chem.* 2004, *76*, 833–838.
47. Charlwood, J.; Birrell, H.; Bouvier, E. S.; Langridge, J.; Camilleri, P. *Anal. Chem.* 2000, *72*, 1469–1474.
48. Wuhrer, M.; Koeleman, C. A. M.; Hokke, C. H.; Deelder, A. M. *Int. J. Mass Spectrom.* 2004, *232*, 51–57.
49. Wuhrer, M.; de Boer, A. R.; Deelder, A. M. *Mass Spectrom. Rev.* 2009, *28*, 192–206.
50. Naidong, W. *J. Chromatogr. B Anal. Technol. Biomed. Life Sci.* 2003, *796*, 209–224.
51. Takegawa, Y.; Deguchi, K.; Keira, T.; Ito, H.; Nakagawa, H.; Nishimura, S. *J. Chromatogr. A* 2006, *1113*, 177–181.
52. Staples, G. O.; Bowman, M. J.; Costello, C. E.; Hitchcock, A. M.; Lau, J. M.; Leymarie, N.; Miller, C.; Naimy, H.; Shi, X.; Zaia, J. *Proteomics* 2009, *9*, 686–695.
53. Takegawa, Y.; Ito, H.; Keira, T.; Deguchi, K.; Nakagawa, H.; Nishimura, S. *J. Sep. Sci.* 2008, *31*, 1585–1593.
54. Wuhrer, M.; Koeleman, C. A.; Hokke, C. H.; Deelder, A. M. *Anal. Chem.* 2005, *77*, 886–894.
55. Temporini, C.; Perani, E.; Calleri, E.; Dolcini, L.; Lubda, D.; Caccialanza, G.; Massolini, G. *Anal. Chem.* 2007, *79*, 355–363.
56. Gropengiesser, J.; Varadarajan, B. T.; Stephanowitz, H.; Krause, E. *J. Mass Spectrom.* 2009, *44*, 821–831.
57. Andersson, L.; Porath, J. *Anal. Biochem.* 1986, *154*, 250–254.
58. Neville, D. C.; Rozanas, C. R.; Price, E. M.; Gruis, D. B.; Verkman, A. S.; Townsend, R. R. *Protein Sci.* 1997, *6*, 2436–2445.
59. Gruhler, A.; Olsen, J. V.; Mohammed, S.; Mortensen, P.; Faergeman, N. J.; Mann, M.; Jensen, O. N. *Mol. Cell. Proteomics* 2005, *4*, 310–327.
60. Villen, J.; Beausoleil, S. A.; Gerber, S. A.; Gygi, S. P. *Proc. Natl. Acad. Sci. U. S. A.* 2007, *104*, 1488–1493.
61. Washburn, M. P.; Wolters, D.; Yates, J. R., 3rd. *Nat. Biotechnol.* 2001, *19*, 242–247.
62. Beausoleil, S. A.; Jedrychowski, M.; Schwartz, D.; Elias, J. E.; Villen, J.; Li, J.; Cohn, M. A.; Cantley, L. C.; Gygi, S. P. *Proc. Natl. Acad. Sci. U.S.A.* 2004, *101*, 12130–12135.
63. Pinkse, M. W.; Uitto, P. M.; Hilhorst, M. J.; Ooms, B.; Heck, A. J. *Anal. Chem.* 2004, *76*, 3935–3943.
64. Larsen, M. R.; Thingholm, T. E.; Jensen, O. N.; Roepstorff, P.; Jorgensen, T. J. *Mol. Cell. Proteomics* 2005, *4*, 873–886.
65. Olsen, J. V.; Blagoev, B.; Gnad, F.; Macek, B.; Kumar, C.; Mortensen, P.; Mann, M. *Cell* 2006, *127*, 635–648.
66. Rush, J.; Moritz, A.; Lee, K. A.; Guo, A.; Goss, V. L.; Spek, E. J.; Zhang, H.; Zha, X. M.; Polakiewicz, R. D.; Comb, M. J. *Nat. Biotechnol.* 2005, *23*, 94–101.
67. Schmelzle, K.; Kane, S.; Gridley, S.; Lienhard, G. E.; White, F. M. *Diabetes* 2006, *55*, 2171–2179.
68. Zhang, X.; Ye, J.; Jensen, O. N.; Roepstorff, P. *Mol. Cell. Proteomics* 2007, *6*, 2032–2042.
69. Zhou, H.; Watts, J. D.; Aebersold, R. *Nat. Biotechnol.* 2001, *19*, 375–378.

70. Oda, Y.; Nagasu, T.; Chait, B. T. *Nat. Biotechnol.* 2001, *19*, 379–382.
71. Adamczyk, M.; Gebler, J. C.; Wu, J. *Rapid Commun. Mass Spectrom.* 2001, *15*, 1481–1488.
72. Gilar, M.; Olivova, P.; Daly, A. E.; Gebler, J. C. *Anal. Chem.* 2005, *77*, 6426–6434.
73. Boutin, J. A.; Ernould, A. P.; Ferry, G.; Genton, A.; Alpert, A. J. *J. Chromatogr.* 1992, *583*, 137–143.
74. Lindner, H.; Sarg, B.; Helliger, W. *J. Chromatogr. A* 1997, *782*, 55–62.
75. McNulty, D. E.; Annan, R. S. *Mol. Cell. Proteomics* 2008, *7*, 971–980.
76. Albuquerque, C. P.; Smolka, M. B.; Payne, S. H.; Bafna, V.; Eng, J.; Zhou, H. *Mol. Cell. Proteomics* 2008, *7*, 1389–1396.
77. Boersema, P. J.; Divecha, N.; Heck, A. J.; Mohammed, S. *J. Proteome Res.* 2007, *6*, 937–946.
78. Alpert, A. J. *Anal. Chem.* 2008, *80*, 62–76.
79. Ytterberg, J. A.; Ogorzalek-Loo, R. R.; Boontheung, P.; Loo, J. A. *Poster Presentation at the 55th ASMS Conference on Mass Spectrometry and Allied Topics* June 3–7, 2007, Indianapolis, IN.
80. Thingholm, T. E.; Jensen, O. N.; Robinson, P. J.; Larsen, M. R. *Mol. Cell. Proteomics* 2008, *7*, 661–671.
81. Jensen, S. S.; Larsen, M. R. *Rapid Commun. Mass Spectrom.* 2007, *21*, 3635–3645.
82. Schroeder, M. J.; Shabanowitz, J.; Schwartz, J. C.; Hunt, D. F.; Coon, J. J. *Anal. Chem.* 2004, *76*, 3590–3598.
83. Engholm-Keller, K.; Jensen, S. S.; Larsen, M. R. *Poster Presentation at the 57th ASMS Conference on Mass Spectrometry and Allied Topics* May 31–June 4, 2009, Philadelphia, PA.
84. Ong, S. E.; Blagoev, B.; Kratchmarova, I.; Kristensen, D. B.; Steen, H.; Pandey, A.; Mann, M. *Mol. Cell. Proteomics* 2002, *1*, 376–386.

Index

B

C

D

E

F

G

H

L

M

R

V

W

Z